CONVERSIONS BETWEEN U.S. CUSTOMARY UNITS AND SI UNITS (Continued)

U.S. Customary unit		Times conversion factor		Equals SI unit	
		Accurate	Practical		
Moment of inertia (area)					
inch to fourth power	in.^{4}	416,231	416,000	millimeter to fourth power	mm^4
inch to fourth power	in.^{4}	0.416231×10^{-6}	0.416×10^{-6}	meter to fourth power	m^4
Moment of inertia (mass)					
slug foot squared	slug-ft^2	1.35582	1.36	kilogram meter squared	kg·m^2
Power					
foot-pound per second	ft-lb/s	1.35582	1.36	watt (J/s or N·m/s)	W
foot-pound per minute	ft-lb/min	0.0225970	0.0226	watt	W
horsepower (550 ft-lb/s)	hp	745.701	746	watt	W
Pressure; stress					
pound per square foot	psf	47.8803	47.9	pascal (N/m^2)	Pa
pound per square inch	psi	6894.76	6890	pascal	Pa
kip per square foot	ksf	47.8803	47.9	kilopascal	kPa
kip per square inch	ksi	6.89476	6.89	megapascal	MPa
Section modulus					
inch to third power	in.^{3}	16,387.1	16,400	millimeter to third power	mm^3
inch to third power	in.^{3}	16.3871×10^{-6}	16.4×10^{-6}	meter to third power	m^3
Velocity (linear)					
foot per second	ft/s	0.3048*	0.305	meter per second	m/s
inch per second	in./s	0.0254*	0.0254	meter per second	m/s
mile per hour	mph	0.44704*	0.447	meter per second	m/s
mile per hour	mph	1.609344*	1.61	kilometer per hour	km/h
Volume					
cubic foot	ft^3	0.0283168	0.0283	cubic meter	m^3
cubic inch	in.^{3}	16.3871×10^{-6}	16.4×10^{-6}	cubic meter	m^3
cubic inch	in.^{3}	16.3871	16.4	cubic centimeter (cc)	cm^3
gallon (231 in.^{3})	gal.	3.78541	3.79	liter	L
gallon (231 in.^{3})	gal.	0.00378541	0.00379	cubic meter	m^3

*An asterisk denotes an *exact* conversion factor

Note: To convert from SI units to USCS units, *divide* by the conversion factor

D0224663

Temperature Conversion Formulas

$$T(°C) = \frac{5}{9}[T(°F) - 32] = T(K) - 273.15$$

$$T(K) = \frac{5}{9}[T(°F) - 32] + 273.15 = T(°C) + 273.15$$

$$T(°F) = \frac{9}{5}T(°C) + 32 = \frac{9}{5}T(K) - 459.67$$

ENGINEERING APPLICATIONS IN
SUSTAINABLE DESIGN
AND DEVELOPMENT
SI Edition

Bradley A. Striebig

Professor of Engineering

James Madison Univerisity

Adebayo A. Ogundipe

Assistant Professor of Engineering

James Madison University

Maria Papadakis

Professor of Integrated Science and Technology and Geographic Science

James Madison University

Australia • Brazil • Japan • Korea • Mexico • Singapore • Spain • United Kingdom • United States

CENGAGE
Learning®

Engineering Applications in Sustainable Design and Development, SI Edition
Bradley A. Striebig, Adebayo A. Ogundipe, and Maria Papadakis

Publisher, Global Engineering: Timothy L. Anderson

Senior Developmental Editor: Hilda Gowans

Development Editor: Eavan Cully

Media Assistant: Ashley Kaupert

Team Assistant: Sam Roth

Marketing Manager: Kristin Stine

Director, Content and Media Production: Sharon L. Smith

Content Project Manager: D. Jean Buttrom

Production Service: RPK Editorial Services, Inc.

Copyeditor: Betty Pessagno

Proofreader: Shelly Gerger-Knechtl

Indexer: Shelly Gerger-Knechtl

Compositor: MPS Limited

Senior Art Director: Michelle Kunkler

Internal Designer: Harasymczuk Design

Cover Designer: Harasymczuk Design

Cover Image: Getty Images

Intellectual Property

   Analyst: Christine Myaskovsky

   Project Manager: Sarah Shainwald

Text and Image Permissions Researcher: Kristiina Paul

Senior Manufacturing Planner: Doug Wilke

Library of Congress Control Number: 2014954398

ISBN-13: 9781133629788

Cengage Learning
20 Channel Center Street
Boston, MA 02210
USA

Cengage Learning is a leading provider of customized learning solutions with office locations around the globe, including Singapore, the United Kingdom, Australia, Mexico, Brazil, and Japan. Locate your local office at: **www.cengage.com/global**.

Cengage Learning products are represented in Canada by Nelson Education Ltd.

To learn more about Cengage Learning Solutions, visit **www.cengage.com/engineering**.

Purchase any of our products at your local college store or at our preferred online store **www.cengagebrain.com**.

Unless otherwise noted, all items © Cengage Learning.

Printed in Canada
Print Number: 01   Print Year: 2014

For Preston and Zachary

—Bradley A. Striebig

To all my Teachers

—Adebayo A. Ogundipe

In memory of Pete Papadakis

—Maria Papadakis

Contents

CHAPTER 9   Industrial Ecology   494

CHAPTER 10   Life Cycle Analysis   548

The purpose of *Engineering Applications in Sustainable Design and Development* is to provide a problem-based quantitative approach for a textbook on sustainable design and development. The book is comprehensive in nature and can be used for an introductory text in sustainable development that is suitable for any field of engineering.

Sustainability is important in manufacturing, construction, planning, and design. Allenby *et al.* state that: "Sustainable engineering is a conceptual and practical challenge to all engineering disciplines." The teaching of sustainability has sometimes been pigeonholed into graduate level courses in Industrial Ecology or Green Engineering. Environmental engineering and chemical engineering textbooks may cover some basic concepts of sustainability, but the extent and breadth of knowledge is insufficient to meet the multifaceted demand required by engineering sustainable processes and products.

Crittenden suggests that sustainable solutions include the following important elements/steps: (a) translating and understanding societal needs into engineering solutions such as infrastructures, products, practices, and processes; (b) explaining to society the long-term consequences of these engineering solutions; and (c) educating the next generation of scientists and engineers to acquire both the depth and breadth of skills necessary to address the important physical and behavioral science elements of environmental problems and to develop and use integrative analysis methods to identify and design sustainable products and systems.

Unfolding environmental events, like the flooding in New York after Hurricane Sandy, the BP oil spill in the Gulf of Mexico, and our changing climate, mean that sustainability courses will increasingly be needed in engineering studies. Sustainability courses are already a requirement for most graduates of European, Chinese, and Indian engineering programs. Chemical engineering, construction engineering, energy engineering, industrial engineering, and mechanical engineering, which have not historically included a required environmental engineering course, now require a fundamental understanding of sustainability indicators and metrics in order for students to be prepared for in international opportunities. Sustainability is most often covered in existing environmental engineering courses, however these courses are typically limited to civil and environmental engineering majors. Introductory environmental engineering courses often have objectives focused more upon historical perspectives in remediation and large-scale treatment systems than upon forward-looking sustainability concepts.

The Association for the Advancement of Sustainability in Higher Education (AASHE) identified over 500 programs in 2014 that address sustainability, including over 60 B.S. engineering programs, over 80 M.S. engineering programs, and nearly 60 engineering Ph.D. programs. *Engineering Applications in Sustainable Design and Development* introduces applications of fundamental environmental science and sustainability indicators that are being broadly adopted by industry and organizations to make informed resource management and design decisions.

How to Use *Engineering Applications in Sustainable Design and Development*

Engineering Applications in Sustainable Design and Development is designed to be the first introduction of these fields to students with the appropriate mathematical and science foundation for study in undergraduate or master's level engineering programs. Prerequisites for such a course are the foundational courses in calculus, chemistry, and physics. *Engineering Applications in Sustainable Design and Development* is most appropriate for second year undergraduate to first year graduate students in engineering and technology programs. The textbook is suitable for junior/senior level courses in chemical, civil, environmental, general, industrial, or mechanical engineering programs.

The authors of this book have drawn on their various backgrounds in engineering, policy, and technology to assemble a comprehensive textbook on sustainable design and development. *Engineering Applications in Sustainable Design and Development* may be used for a two-semester sequence in sustainable aspects of design and engineering, or individual chapters may be taught in a wide variety of one-semester courses.

Understanding sustainable design and related fields like basic climate science, green building, and life cycle analysis is becoming increasingly important for practicing engineers and scientists. This curriculum introduces a new approach to sustainability that includes foundational knowledge of environmental impact assessment methods, life cycle analysis, and energy considerations that are being adopted in many accredited engineering and technology programs. This curriculum is focused upon applying engineering principles to real-world design and problem analysis. It includes specific step-by-step examples and case studies for solving complex conceptual and design problems courses on sustainable design and engineering. *Engineering Applications in Sustainable Design and Development* applies the principles of sustainable design to issues in both developed and developing countries.

This course is also designed to meet ABET accreditation standards and prepare graduates for the FE examination. ABET related course outcomes might include:

TABLE 1 Expected outcomes and related ABET criteria for sustainability-focused courses

CHAPTER	EXPECTED COURSE OUTCOMES	RELATED ABET CRITERIA
1	Define sustainability in relationship to development indexes, consumption and population trends.	a, f, h, j
2	Perform calculations involving conventional units utilized in engineering	a, e
2	Solve basic equilibrium problems in environmental chemistry related to pH and solubility	a, c, e
3	Prepare mass balance equations to determine the impacts of pollutants upon the environment	c, e
3	Solve mass balance problems related to determine the impacts of pollutants upon the environment	a, c, e
4	Calculate and describe the impact of anthropogenic emissions on the oxygen content in natural aqueous environments	a, c, e
4	Describe the impact of anthropogenic sources on water quality	a, c, f, h, j, k
5	Describe the impact of anthropogenic sources on air quality and human health	a, c, f, h, j, k
6	Describe the relationship between community sustainability, global climate change, environmental impacts, economic projects, and fossil fuel emissions	f, h, i, j
7	Develop frameworks for conceptualizing complex, open system problems, and the inter-relationship of environmental, energy, economic, health, technological, and cultural factors	c, f, h, i, j
7	Describe the relationship between global, regional and local environmental impacts, and economic factors	f, h
8	Predict and feel concern for the biological and environmental effects of the design of man-made devices	f, h
8	Develop frameworks for conceptualizing complex, open system problems, and the inter-relationship of environmental, energy, economic, health, technological, and cultural factors	c, e, h
9	Prepare mass balance equations to track materials flows in manufactured products and emissions	a, e, h
9	Solve mass balance problems related to impacts of industrial processes	a, e, h
10	Implement sustainability tools, such as life cycle assessment when conducting systems analysis	a, c, h
10	Model total material cycles (i.e. product cradle-to-grave life including design, manufacturing, and disposal phases) when developing products and processes	a, h
11	Describe green building criteria and evaluate long term sustainability issues associated with built infrastructure	c, h, j, k
12	Identify criteria for evaluating social considerations in sustainable development	F, h, j

Features of the Book

This textbook includes learning objectives, case studies of developed and developing nations, and detailed examples. We use a developmental approach to understanding sustainable design. Some topics are briefly introduced and defined in early chapters of the text, then are further defined and developed in subsequent chapters. The book focuses extensively on topics related to climate change. The story of how climate change occurs, and how technologies are being developed to mitigate climate change unfold throughout the text. Starting with a basic understanding of the composition of the atmosphere, the text moves into mass and energy balances, the fate and transport of CO_2, mitigation options, and how life cycle analysis can be used to evaluate how design and development affect green house gas emissions.

Case studies from both developed and developing nations enhance student interest and understanding of many complex issues. International case studies and analysis of global issues, such as access to clean water and affects of climate change, bring current events and issues into the classroom.

Each chapter of *Engineering Applications in Sustainable Design and Development* includes both qualitative and quantitative problems that cover a range of difficulty and complexity.

An online solution manual and instructor website to download figures and spreadsheets is also available for instructors.

MindTap

This textbook is also available online through MindTap, a personalized learning program. Students who purchase the MindTap version have access to the book's MindTap Reader and are able to complete homework and assessment material online, through their desktop, laptop, or iPad. If your class is using a Learning Management System (such as Blackboard or Moodle) for tracking course content, assignments, and grading, you can seamlessly access the MindTap suite of content and assessments for this course.

In MindTap, instructors can:

- Personalize the Learning Path to match their course syllabus by rearranging content or appending original material to the textbook content
- Connect a Learning Management System portal to the online course and Reader
- Customize online assessments and assignments
- Track student progress and comprehension
- Promote student engagement through interactivity and exercises

Additionally, students can listen to the text through ReadSpeaker, take notes and highlight content for easy reference, and check their understanding of the material.

ACKNOWLEDGMENTS

The authors are grateful to their colleagues and students for the contributions they have made to the development of this textbook. Although there are too many contributors to name, a few deserve special mention. Dr. Samuel Morton III, Dr. Justin Henriques, Dr. Elise Barrella, and Dr. Zachary Bortolot deserve special mention for their material contributions to the textbook. Thomas Kaisen, Matt Wisniewski, Emily Fuller, Hillary Benedict Winchester, and Marianna Ford deserve special mention for their review and editing of the manuscript. We would like to thank our Cengage reviewers, especially Michael Robinson, Kenneth Reid, Brett Tempest, and Moses Karakouzian, who provided careful insights and many constructive suggestions for strengthening the text. We hope that our revisions achieve your desired results. Hilda Gowans was a model of patience as she guided us through the many stages of producing a textbook; we simply could not have done it without her wise counsel. Eavan Cully and Rose Kernan from Cengage cheerfully managed the final production of this book. A very special thank you to all those that have been involved with this text's development and production.

Additionally, Dr. Striebig would like to thank two mentors, Dr. Raymond Regan and Dr. Robert J. Heinsohn, for encouragement throughout his career and with the development of this curriculum and textbook. He would like to acknowledge the support and endurance of his wife, Abigail, and family, especially his children Preston and Zachary who are a constant inspiration for the hope and potential of future generations.

Dr. Ogundipe will be forever indebted to Dr. Washington Braida; a terrific teacher and friend who inspired him and pointed him in the right direction. He would also like to thank Tola, Safiyyah, Haneef, and Sumayyah for being the reason.

Dr. Papadakis owes her love of technology to her father, Pete Papadakis, a career-long experimental stress and design engineer. Tom Walls, Robert Zetzl, Gladys Good, and Donald Clodfelter had a bigger impact than they could possibly know, and innately understood the needs of K-12 girls in STEM decades before this issue came to national prominence. Her husband Fred Copithorn is a man of exquisite patience, and to her entire family she says "με αγάπη."

Bradley A. Striebig
Adebayo A. Ogundipe
Maria Papadakis

Preface to the SI Edition

This edition of *Engineering Applications in Sustainable Design and Development* has been adapted to incorporate the International System of Units (*Le Système International d'Unités* or SI) throughout the book.

LE SYSTÈME INTERNATIONAL D'UNITÉS

The United States Customary System (USCS) of units uses FPS (foot–pound–second) units (also called English or Imperial units). SI units are primarily the units of the MKS (meter–kilogram–second) system. However, CGS (centimeter–gram–second) units are often accepted as SI units, especially in textbooks.

USING SI UNITS IN THIS BOOK

In this book, we have used both MKS and CGS units. USCS units or FPS units used in the US Edition of the book have been converted to SI units throughout the text and problems. However, in case of data sourced from handbooks, government standards, and product manuals, it is not only extremely difficult to convert all values to SI, it also encroaches upon the intellectual property of the source. Some data in figures, tables, and references, therefore, remains in FPS units. For readers unfamiliar with the relationship between the FPS and the SI systems, a conversion table has been provided inside the front cover.

To solve problems that require the use of sourced data, the sourced values can be converted from FPS units to SI units just before they are to be used in a calculation. To obtain standardized quantities and manufacturers' data in SI units, the readers may contact the appropriate government agencies or authorities in their countries/regions.

INSTRUCTOR RESOURCES

The Instructors' Solution Manual in SI units is available through your Sales Representative or online through the book website at www.login.cengage.com. A digital version of the ISM and PowerPoint slides of figures, tables, and examples and equations from the SI text are available for instructors registering on the book website.

Feedback from users of this SI Edition will be greatly appreciated and will help us improve subsequent editions.

Cengage Learning

About the Authors

BRADLEY A. STRIEBIG

Professor of Engineering, James Madison University, Harrisonburg, Virginia
Professor Striebig earned his Ph.D. from the Pennsylvania State University, a top-50 engineering school. He has served as editor on major journals in his subject area. He has led major, funded, award-winning research activities. He has written several book chapters, numerous peer-reviewed journal articles, and presented at many peer-reviewed conferences.

ADEBAYO A. OGUNDIPE

Assistant Professor of Engineering, James Madison University, Harrisonburg, Virginia
Professor Ogundipe has previously held academic positions at Stevens Institute of Technology and the Polytechnic Institute of New York University. He has a Master's degree from Stevens Institute of Technology. His current areas of specialization and scholarship include Life Cycle Analysis, Industrial Ecology, and developing methods for assessing sustainability.

MARIA PAPADAKIS

Professor of Integrated Science and Technology and Geographic Science
Professor Papadakis is a political economist with expertise in energy management and the role of energy in sustainable development. Her research has been published in specialized reports of the National Science Foundation and in such journals as *Evaluation and Program Planning, Journal of Technology Transfer, The Scientist*, and the *International Journal of Technology Management*.

ENGINEERING APPLICATIONS IN

SUSTAINABLE DESIGN

AND DEVELOPMENT

SI Edition

CHAPTER 1

Sustainability, Engineering, and Design

FIGURE 1.1 A high-resolution photo of our planet showing various ecosystems and weather patterns. Many believe the first images of Earth taken from space had a profound effect on how people in general perceived the interconnectedness between people, the planet, and future prosperity.

Source: NASA Goddard Space Flight Center Image by Reto Stöckli (land surface, shallow water, clouds). Enhancements by Robert Simmon (ocean color, compositing, 3D globes, animation). Data and technical support: MODIS Land Group; MODIS Science Data Support Team; MODIS Atmosphere Group; MODIS Ocean Group Additional data: USGS EROS Data Center (topography); USGS Terrestrial Remote Sensing Flagstaff Field Center (Antarctica); Defense Meteorological Satellite Program (city lights).

It is known that there are an infinite number of worlds, simply because there is an infinite amount of space for them to be in. However, not every one of them is inhabited. Any finite number divided by infinity is as near nothing as makes no odds, so the average population of all the planets in the Universe can be said to be zero. From this it follows that the population of the whole Universe is also zero, and that any people you may meet from time to time are merely products of a deranged imagination.

—Douglas Adams, from *The Restaurant at the End of the Universe* (1980, p. 142)

GOALS

THE EDUCATIONAL GOALS OF THIS CHAPTER are to define sustainability and understand how social norms influence discussions about sustainability. We also examine how population changes and resource consumption have created the need for engineers, economists, scientists, and policymakers to consider sustainability in the design of products, infrastructure, and systems. The key concepts that are used to quantitatively consider sustainable design include the human development index, population growth models, and the ecological footprints analysis. This chapter also provides a greater context for the social and economic factors that shape successful engineering design. In this chapter we explore the ethical basis of human-centered design as a way of meeting the essential needs of the poor, which is an explicit element of sustainable development. In addition, we explain the dynamics of the adoption and diffusion of innovations, which is a critical prerequisite to the widespread social impact of more sustainable practices, products, and processes. Finally, we address the economic concepts that help us understand why achieving greater environmental sustainability can be a challenge and the role of governmental policymaking in surmounting those obstacles.

OBJECTIVES

At the conclusion of this chapter, you should be able to:

- Define and discuss different definitions of sustainability, sustainable design, and sustainable development.

- Calculate and relate the Human Development Index to indexes for lifespan, education, and income.

- Discuss ethical frameworks and engineering ethics in relationship to sustainability.

- Evaluate global trends in population and describe how those trends challenge engineers to develop sustainable products, infrastructure, and systems.

- Define and evaluate the carrying capacity of systems of various scales.

- Define and discuss quantitatively the indicators of sustainable design, including the ecological footprint and the Impact, Population, Affluence, and Technology [IPAT]equation.

- Explain the different ethical principles that inform sustainable development, and discuss how these affect engineering design.

- Give examples of successful and unsuccessful technologies appropriate for meeting the essential needs of the poor, and explain the reasons for their success or failure.

- Apply the principles of people-centered design to an engineering design problem.

- For a given innovation, summarize and analyze the social, cultural, technical, and economic factors that affect its potential adoption and diffusion.

1.1 Introduction

Genetically modern humans appeared on Earth about 200,000 years ago, and biologically and behaviorally modern humans appeared about 70,000 years ago. The number of people and their effects on the planet were negligible for most of the history of the planet (Figure 1.2).

The number of humans on the planet remained very small until a few hundred years ago when advances in farming, energy, and mechanization took place, allowing the human population to increase exponentially (see Figure 1.3). Rapid changes in technology allowed humans to live longer; the decreasing death rates contributed to the high rate of human population growth over the past thousand years. Some time shortly after the year 1800, the world population reached 1 billion people for the first time (UN, 1999).

Demographers, people who study trends in population, say we are likely heading toward a world population of 7.9 to 10.9 billion over the next century (UN, 2012a). While the human population on the planet is growing, natural resources that we have relied on for food, energy, and water are shrinking owing to the increasing human consumption of those resources. The human species has had a profound environmental impact on the planet, threatening the Earth's biodiversity, climate, energy resources, and water supply.

In the industrialized world, many people move faster, eat more, know more, and live in larger homes than even royalty could have dreamed of only a few centuries

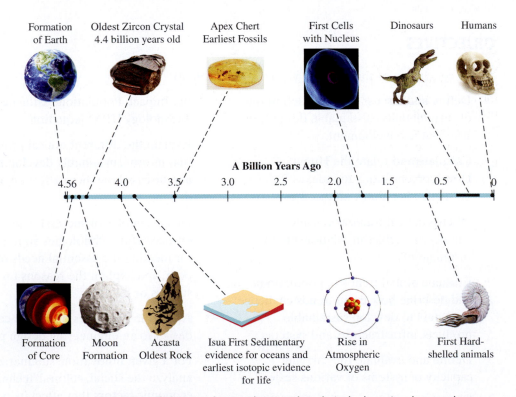

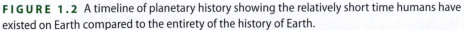

FIGURE 1.2 A timeline of planetary history showing the relatively short time humans have existed on Earth compared to the entirety of the history of Earth.

Source: Based on http://www.geology.wisc.edu/zircon/Earliest%20Piece/Images/28.jpg; leonello calvetti/Shutterstock.com; Johan Swanepoel/Shutterstock.com; Ortodox/Shutterstock.com; Imfoto/Shutterstock.com; falk/Shutterstock.com; oorka/Shutterstock.com; Sebastian Kaulitzki/Shutterstock.com; Number001/Shutterstock.com; DM7/Shutterstock.com; Empiric7/Shutterstock.com.

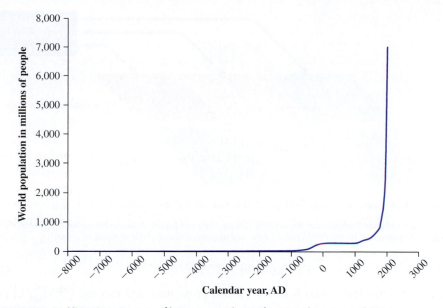

FIGURE 1.3 Historic estimates of human population from pre-history until 2011.

Source: Based on BC: *Kremer, Michael (1993). "Population Growth and Technological Change: One Million B.C. to 1990" in The Quarterly Journal of Economics 108(3): 681–716. AD 0-1990: United Nations Population Division Report, The World at Six Billion. AD 1995-2012: US Census Bureau Data: The World Population Clock.

ago. Yet despite the great advances in science, technology, government, economics, education and medicine over the past hundred years, these resources are not distributed equally on the planet. Economic, scientific, and technological advances have increased the lifespan and improved access to many marvelous things in the industrialized world, but this overall increase in the standard of living has failed to raise many people out of poverty. The standard of living relates income, comfort, and material goods to the socioeconomic classification of people. As we will see later in this chapter, those who have not benefited from modern science, technology, and industrialization may not be able to meet their basic needs for food, clothing, shelter, water, and sanitation.

1.2 Human Development Index

The United Nations Development Programme (UNDP) devised a **Human Development Index (HDI)** that is based on three dimensions: life expectancy, education, and income. These dimensions are combined into a single comparable value, as illustrated in Figure 1.4 (UN, 2011a). The HDI is calculated using the data reported each year by the United Nations and the following equations (1.1) to (1.6.)

Life Expectancy (LE) at birth uses the 2011 Life Expectancy Index:

$$\text{Life Expectancy Index (LEI)} = (\text{LE} - 20)/(83.2 - 20) \tag{1.1}$$

The Education Index (EI) is based on the Mean Years of Schooling Index (MYSI) and Expected Years of Schooling Index (EYSI), where

$$\text{MYSI} = \text{Mean Years of Schooling}/13.2 \tag{1.2}$$

$$\text{EYSI} = \text{Expected Years of Schooling}/20.6 \tag{1.3}$$

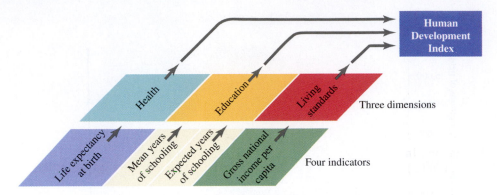

FIGURE 1.4 Components of the Human Development Index.

Source: Based on Human Development Report 2011 Sustainability and Equity: A Better Future for All. United Nations Development Programme. ISBN: 9780230363311. http://hdr.undp.org/en/media/HDI_EN.png.

$$EI = \{(MYSI \times EYSI)^{0.5}\}/0.951 \qquad (1.4)$$

And the Income Index (II) is based on the gross national income (GNI_{pc}) at purchasing power parity per capita, which is an estimate and standardization of each individual's income in a country:

$$II = \{\ln(GNI_{pc}) - \ln(100)\}/\{\ln(107{,}721) - \ln(100)\} \qquad (1.5)$$

The Human Development Index is determined from the geometric mean of the Life Expectancy Index, the Education Index, and the Income Index:

$$HDI = (LEI \times EI \times II)^{1/3} \qquad (1.6)$$

Based on this index, the United Nations categorizes countries as "Very High" Human Development, High Human Development, Medium Human Development, and Low Human Development.

EXAMPLE 1.1

Calculate the Human Development Index for Benin and compare it to the HDI of the five other countries listed in Table 1.1.

TABLE 1.1 Component values of the Human Development Index for selected countries

COUNTRY AND 2011 DATA	LIFE EXPECTANCY AT BIRTH (YEARS)	EXPECTED YEARS OF SCHOOLING	MEAN YEARS OF SCHOOLING	GNI PER CAPITA IN PPP TERMS (CONSTANT 2005 INTERNATIONAL $)
Benin	56.1	4.6	0.6	1,137
Costa Rica	79.3	11.7	8.3	10,497
India	65.4	10.3	4.4	3,468
Jordan	73.4	13.1	8.6	5,300
Norway	81.1	17.3	12.6	46,582
United States	78.5	16	12.4	43,017

Source: Based on Human Development Report 2011 Sustainability and Equity: A Better Future for All. United Nations Development Programme. ISBN: 9780230363311.

For Benin, the Life Expectancy Index can be calculated using Equation (1.1) from the Life Expectancy at birth:

$$(LEI) = (LE - 20)/(83.2 - 20)$$

$$(LEI) = (56.1 - 20)/63.2$$

$$(LEI) = 0.571$$

In order to calculate the Education Index, we first need to calculate the Mean Years of Schooling Index (MYSI) and the Expected Years of Schooling Index (EYSI) from Equations (1.2) and (1.3), respectively:

$$MYSI = \text{Mean Years of Schooling}/13.2$$

$$MYSI = 0.6/13.2 = 0.045$$

$$EYSI = \text{Expected Years of Schooling}/20.6$$

$$EYSI = 4.6/20.6 = 0.223$$

Substituting into the equation for the education index yields

$$EI = \{(MYSI \times EYSI)^{0.5}\}/0.951$$

$$EI = \{(0.045 \times 0.223)^{0.5}\}/0.951 = 0.106$$

We can calculate the Income Index (II) from the gross national income (GNI_{pc}) at purchasing power parity per capita, using Equation (2.5):

$$II = \{\ln(GNI_{pc}) - \ln(100)\}/\{\ln(107{,}721) - \ln(100)\}$$

$$II = \{\ln(1137) - \ln(100)\}/\{\ln(107{,}721) - \ln(100)\} = 0.348$$

We can then use the Life Expectancy Index, Education Index, and Income Index to calculate the HDI using Equation (1.6):

$$HDI = (LEI \times EI \times II)^{1/3}$$

$$HDI = (0.571 \times 0.106 \times 0.348)^{1/3} = 0.276$$

The HDI of 0.276 is much less than 1. Using the 2011 data from the United Nations for the countries given yields the HDI for selected countries shown in Table 1.2.

TABLE 1.2 Calculated values for the subparts of the Human Development Index

COUNTRY AND 2011 DATA	LEI	EI	II	HDI	2011 HDI RANKING OF 187 COUNTRIES
Benin	0.571	0.106	0.348	0.276	167
Costa Rica	0.938	0.629	0.667	0.732	69
India	0.718	0.333	0.508	0.539	134
Jordan	0.845	0.652	0.569	0.688	95
Norway	0.967	0.955	0.880	0.929	1
United States	0.926	0.939	0.869	0.897	4

Source: Based on Human Development Report 2011 Sustainability and Equity: A Better Future for All. United Nations Development Programme. ISBN: 9780230363311.

From Table 1.2 we can see significant gaps in resources associated with life expectancy, education, and income. Norway had the highest HDI score in 2011 followed by the United States. Both Norway and the United States are listed in the United Nations' Very High Human Development category. A person living in Costa Rica has a similar life expectancy as the United States and Norway, but would have lower education and income expectations. The United Nations classifies Costa Rica as a High Human Development country. Jordan and India have lower life expectancy and educational indexes and significantly lower income indexes; both of these countries are listed in the United Nations' Medium Human Development category. Benin has a much lower index score in each category than all the previous countries we mentioned, as is typical of many sub-Saharan African nations. This discrepancy in development is illustrated in Figure 1.5. Benin and other countries with little infrastructure, challenged educational systems, low life expectancy, and low expected income values are categorized as Low Human Development countries by the United Nations.

Figure 1.5 illustrates the uneven distribution and ranking of HDIs. By most definitions, in the year 2012, a total of 2.8 billion people lived in poverty or had income levels of less than 2 U.S. dollars per day. Nearly 1.4 billion lived in extreme poverty, earning less than 1.25 U.S. dollars per day (UN, 2012b). Over 850 million people were undernourished and lacked access to food. Approximately 2.5 billion people lacked access to either clean water or sanitation (UN, 2012b). These numbers illustrate the need for a large percentage of the world's population to improve their standard of living. Population numbers alone do not tell the whole story of resource consumption, the uneven distribution of scarce resources, and

HDI: Human Development Index (HDI) value
(2012)

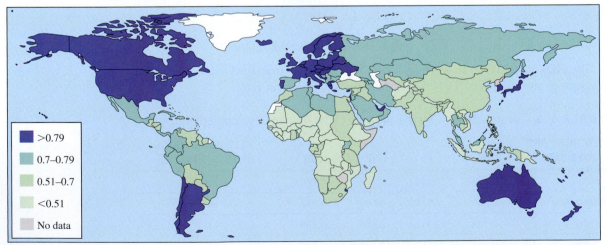

Legend:
>0.79
0.7–0.79
0.51–0.7
<0.51
No data

FIGURE 1.5 A map of country rankings based on the United Nations' Human Development Index where dark blue represents Very High Human Development countries, light blue High Human Development countries, dark green Medium Development countries, and light green Low Human Development countries.

Source: Based on the Human Development Report 2011 Sustainability and Equity: A Better Future for All. United Nations Development Programme. ISBN: 9780230363311.

the desire of many people living in poverty to improve their access to food, water, energy, education, and economic development.

Mahbub ul Haq (1934–1998), the founder of the *Human Development Report*, said that the purpose of development is:

> *to enlarge people's choices. In principle, these choices can be infinite and can change over time. People often value achievements that do not show up at all, or not immediately, in income or growth figures: greater access to knowledge, better nutrition and health services, more secure livelihoods, security against crime and physical violence, satisfying leisure hours, political and cultural freedoms and sense of participation in community activities. The objective of development is to create an enabling environment for people to enjoy long, healthy and creative lives.*

Sustainable development in one sense is the desire to improve the worldwide standard of living while considering the effects of economic development on natural resources. Since 1990, significant strides have been taken to decrease the percentage of the world's population living in poverty (Figure 1.6). The most significant gains have come from the industrialization of large population centers in Asia.

As economic centers and industrial centers continue to develop and transform our landscape, more and more people are looking to these centers as a means to improve their standard of living. As a result, current trends show that populations are migrating toward more centralized cities and urban areas (Figure 1.7). This rural-to-urban migration is putting a significant strain on the regions surrounding these cities and mega-cities (cities with more than 10 million people) (UN, 2008). Many countries that are becoming more industrialized are struggling to develop the infrastructure required to provide food, water, sanitation, and shelter for the rural migrants. Peri-urban areas are substantially increasing. **Peri-urban areas**, characterized by very high population densities, lack the infrastructure to distribute energy, water, and sanitation services. Peri-urban areas severely strain natural resources, especially water and energy.

1.3   Sustainable Development and Social Ethics

In many ways, the narrative that informs our contemporary understanding of sustainable development began over 50 years ago. Scientists, environmentalists, and economists identified a number of environmental and economic challenges associated with the unprecedented increase in global population growth and overconsumption described in the previous section. Connecting these threads, the United Nations requested that the World Commission on Environment and Development formulate "a global agenda for change" (discussed in more detail in Chapter 7). The commission articulated the concept of sustainable development in its holistic report *Our Common Future* (WCED, 1987 p. 41):

> *Sustainable development is development that meets the needs of the present without compromising the ability of future generations to meet their own needs. It contains within it two key concepts: (1) the concept of "needs", in particular the essential needs of the world's poor, to which overriding priority should be given; (2) the idea of limitations imposed by the state of technology and social organization on the environment's ability to meet present and future needs.*

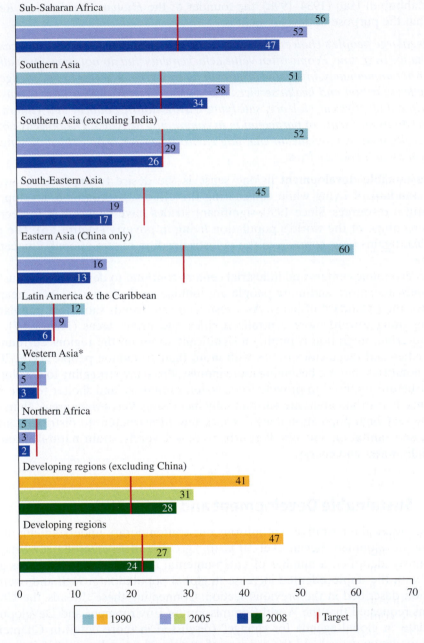

Proportion of people living on less than $1.25 a day in 1990, 2005, and 2008 (Percentage)

Sub-Saharan Africa
- 56
- 52
- 47

Southern Asia
- 51
- 38
- 34

Southern Asia (excluding India)
- 52
- 29
- 26

South-Eastern Asia
- 45
- 19
- 17

Eastern Asia (China only)
- 60
- 16
- 13

Latin America & the Caribbean
- 12
- 9
- 6

Western Asia*
- 5
- 5
- 3

Northern Africa
- 5
- 3
- 2

Developing regions (excluding China)
- 41
- 31
- 28

Developing regions
- 47
- 27
- 24

Legend: ▮ 1990   ▮ 2005   ▮ 2008 | Target

* The aggregate value is based on 5 of 13 countries in the region.

Note: No sufficient country data are available to calculate the aggregate values for Oceania.

FIGURE 1.6 Since 1990, there has been a significant decrease in the number of people living in economic poverty, defined as subsistence on less than one dollar per day.

Source: Based on The Millennium Development Goals Report 2012. United Nations.

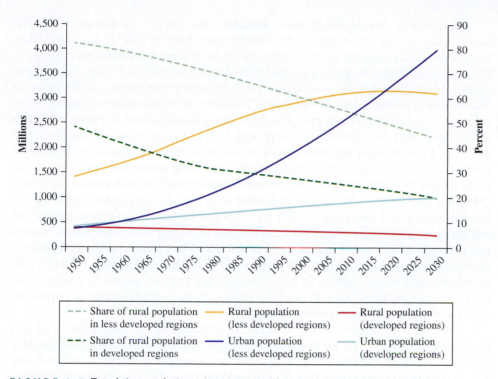

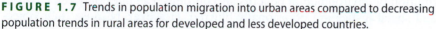

FIGURE 1.7 Trends in population migration into urban areas compared to decreasing population trends in rural areas for developed and less developed countries.

Source: Based on UN. (2008). Trends in Sustainable Development: Agriculture, rural development, land desertification and drought. United Nations Department of Economic and Social Affairs Division for Sustainable Development.

This definition of sustainable development is often referred to as the **Brundtland definition**, named after the chairman of the UN commission that produced the report. Importantly, the report and the definition of sustainable development linked the three key tenets (known as the three pillars) of sustainability: the environment, society, and the economy. Although the report was published a quarter of a century ago, it still serves as an important reference point for discussion of the definition, motivation, and challenges associated with sustainable development.

There are implicit ethical dimensions of the definition of sustainable development that generally build on the concepts of equity, fairness, and justice. It is important for the engineer or designer to be able to identify these ethical dimensions of sustainability, as they often directly inform the policies and regulations that influence engineering design decisions.

In the context of the Brundtland definition, intergenerational equity refers to the use of natural resources in such a way as to take into consideration the needs of both the present and future generations. The UN Rio Declaration in 1992 defined **intergenerational equity** more broadly, stating that "the right to development must be fulfilled so as to equitably meet developmental and environmental needs for both present and future generations." Intergenerational equity, then, is the ethical obligation of current societies to consider the welfare of future societies in the context of natural resource use and degradation (Makuch and Pereira, 2012). Engineers play a critical role in facilitating intergenerational equity, as we design systems, processes, and products that can ensure that natural resources will be conserved in such a way that future generations can use them.

Sustainable development also includes the notion of intragenerational equity, that is, the need for equity within members of the same generation. Intragenerational equity is often referenced in instances of economic inequity, for example, between developed and developing countries. In the Brundtland definition, it provides the motivation for prioritizing the needs of the poor. In international treaties such as the Kyoto Protocol, intragenerational equity is the principle used to define common but differing responsibilities between countries at various stages of economic development. In the context of the Kyoto Protocol, developed and developing countries do not have identical obligations to mitigate climate change. Rather, their requirements for action under the treaty are equitably scaled to reflect differences in their economic and infrastructure capacities (Makuch and Pereira, 2012). **Social justice** is similar to intragenerational equity, but it is concerned more specifically with the fair distribution (or sharing) of the advantages and disadvantages (or benefits and burdens) that exist within society. An example of the social justice issue is discussed in Chapter 8. Energy poverty, discussed in detail in Chapter 8, is one example of the relationship between sustainability and social justice. In this example, the lack of access to energy is an unfair burden on those who do not have it, as it denies people important opportunities for improving their quality of life.

Given the evolution of the concept of sustainable development, it is not surprising that arguments for sustainability hinge upon the concept of social justice. In the context of sustainable development, environmental justice deals with the fair distribution of environmental benefits and burdens. The lines between environmental and social justice are occasionally blurred. However, because their objectives may at times differ, it is useful to keep them distinct.

BOX 1.1 Formative Works Related to Sustainable Development

In 1956 a geoscientist who worked at Shell research lab named Dr. M. King Hubbert suggested that oil production in the United States would peak and followed a bell-shaped curve (Hubbert and American Petroleum, 1956). His work became a cornerstone of the notion of peak oil (i.e., the point in time where the rate of total petroleum extraction begins to decline permanently) and underlined the point that oil was a finite and depleting resource used as a fuel for economic development.

A few years later (Figure 1.8), the marine biologist Rachel Carson published *Silent Spring* (1962), which chronicled the negative impacts associated with pesticides and facilitated the ban of the pesticide DDT a decade later. This book is often cited as helping to begin the environmental movement in America.

In *The Tragedy of the Commons* (1968), ecologist Dr. Garret Hardin explored some of the moral and social challenges related to population growth and the management and use of natural resources.

The Limits to Growth (1972), by Donella Meadows, Dennis Meadows, Jørgen Randers, and William W. Behrens III, provided the first effort to holistically model the global system with respect to five key indicators of sustainability: world population, industrialization, pollution, food production, and resource depletion.

In *Small Is Beautiful* (1973), E. F. Schumacher critiqued modern economics, for example, for their treatment of natural resources such as oil as nondepleting. His work was has since been applied to engineering design through appropriate technology.

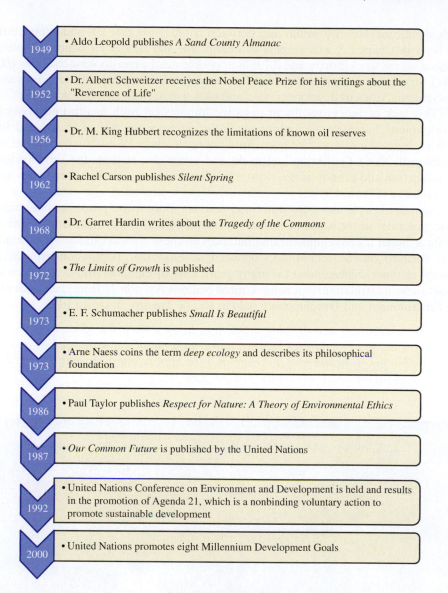

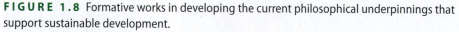

FIGURE 1.8 Formative works in developing the current philosophical underpinnings that support sustainable development.

Source: Bradley Striebig.

1.4 Sustainable International Development and the Essential Needs of the Poor

The Brundtland definition of sustainable development includes a focus on meeting the essential needs of the poor, such as basic human needs for food, water, and shelter. However, essential needs also include opportunities for the educational advantages and economic productivity that improve quality of life.

Developing communities often face a systemic lack of access to services that would meet their basic needs. For example, over 780 million people worldwide lack access to improved sources of drinking water, and 2.5 billion lack access to improved sanitation, with significant disparity within and between countries (UNICEF and WHO, 2012). As seen in Chapter 8 with respect to energy use and access, 1.3 billion people are without access to electricity, and 2.7 billion people need clean cooking facilities (IEA, 2011). Significant disparities also exist. For example, 95% of those without access are in sub-Saharan Africa or developing Asia, and 84% live in rural areas (IEA, 2011).

This lack of access results in negative impacts on both health and economic development. For example, indoor air pollution from the use of traditional fuels such as charcoal and wood for household energy leads to 1.5 million deaths per year (Reinhardt, 2006). Gathering fuel wood and water also reduces the time available for education and economic activity—a burden disproportionally shared by women and children (Figure 1.9).

As more people use more water and produce more waste products that contaminate potential water supplies, engineers, scientists, and policymakers have tried to create a model for development that balances all these considerations. World leaders have focused on the relationship between development, population growth, and natural resource management for many years. One of the most profound statements about these interrelationships is the United Nations Agenda 21 from the **Conference on Environment and Development**, held in Rio de Janeiro, Brazil, in 1992:

Humanity stands at a defining moment in history. We are confronted with a perpetuation of disparities between and within nations, a worsening of poverty, hunger, ill health and illiteracy, and the continuing deterioration of the ecosystems on which we depend for our well-being.

FIGURE 1.9 Gathering fuel wood and water in Benin, West Africa. The physical labor of gathering fuel wood for cooking and collecting water is often women and children's work in low-income countries. This represents a high social cost, as it takes them away from opportunities for schooling or engaging in business endeavors.

Source: Bradley Striebig.

The statement continues:

Human beings are at the center of concerns for sustainable development. They are entitled to a healthy and productive life in harmony with nature. The right to development must . . . meet developmental and environmental needs of present and future generations. All States and all people shall cooperate in the essential task of eradicating poverty as an indispensable requirement for sustainable development. . . . To achieve sustainable development and a higher quality of life for all people, States should reduce and eliminate unsustainable patterns of production and consumption and promote appropriate demographic policies.

In the year 2000, the United Nations specified eight **Millennium Development Goals** to address prior to the year 2015:

1. Eradicate extreme poverty and hunger.
2. Achieve universal primary education.
3. Promote gender equality and empower women.
4. Reduce childhood mortality.
5. Improve maternal health.
6. Combat HIV/AIDS, malaria, and other diseases.
7. Ensure environmental sustainability.
8. Develop a global partnership for development.

The Millennium Development Goals are focused on increasing the standard of living for the world's existing population and also for future generations as the world population continues to grow. The United Nations' overarching objective is to achieve sustainable development and reduce poverty across the planet. But how can sustainable development be achieved if there are more people, using more resources? In order to answer this complex question, we must first examine the definition of sustainability and understand the context associated with the term.

The term *sustainable* or *sustainability* is widely used in a variety of applications, contexts, and marketing materials. Most of the uses of the word "sustainable" infer a qualitative comparison to something "other"; for example, our new green Excellon automobile is the sustainable solution to yesterday's Sports Utility Vehicle (SUV). There is little or no quantifiable way to compare and contrast the advantages and disadvantages of one product compared to another. Nor is there an attempt to explain how a product can be sustainably produced, sold, and disposed of for any significant period of time. The marketing of "sustainable" goods and products also typically infers that the "sustainable" product is morally superior to the nonsustainable product.

Do you believe that sustainable products are morally superior to nonsustainable products? If so, what does this belief imply about the developed world's largely consumer-based economic system of retail merchandise? How are technology and moral convictions woven into the fabric of our definitions of sustainable design?

The importance of defining quantitative measures of sustainability for design will be addressed in this chapter and throughout subsequent chapters. In order to determine if a product or process is sustainable, we must examine the unit processes involved in manufacturing, including the consumption of raw materials, design life, cost, and ultimate disposal of a product. The context in which the product and process are developed should also be considered together with the societal implications of worldwide population growth and consumption patterns as well as with an evaluation of how an individual's environmental footprint may be affected by the product or process.

1.5   Engineering and Developing Communities

Engineers and designers play a key role in developing solutions for meeting the essential needs of the poor. However, to do this well, we must pay attention to the technical, cultural, economic, and environmental contexts that can affect project outcomes. Countless engineering and design efforts in developing communities have failed to meet expectations for sustained (and positive) societal impact. For example, in sub-Saharan Africa, 35% of rural water systems are nonfunctioning, with some countries experiencing an operational failure rate of 30 to 60% (Harvey and Reed, 2007). These high failure rates generally do not occur because of technical design flaws, but because of the failure to incorporate other salient factors through the design process, including social, economic, and environmental influences. Point-of-use (POU) water treatment technologies, like those described in Table 1.3, may significantly reduce the risk of exposure to pathogenic organisms in drinking water.

Engineers and designers are now better equipped to incorporate these key social, economic, and environmental factors into their design and implementation decisions than they were several decades ago. Generally, two dimensions of design require a shift in thinking for sustainable engineering in developing countries. The first dimension involves a shift in thinking with regard to *product*, and the second, with regard to *process*.

TABLE 1.3 Point-of-use (POU) appropriate water treatment technologies considered for implementation in the model home near Kigali, Rwanda

TECHNOLOGY	DESCRIPTION	ADVANTAGE	DISADVANTAGE	REFERENCES
Biosand™	Sand filtration	• High removal efficiency for microorganisms	• Needs continual use and regular maintenance • Cost	Duke et al. (2006); Stauber et al. (2006)
Filtron™	Ceramic filter	• High removal efficiency for microorganisms • Sized for individual homes • Relatively inexpensive	• Requires fuel for construction • Limited lifetime • Requires regular cleaning	Bielefeldt et al. (2010); Brown and Sobsey (2010); Clausen et al. (2004); Striebig et al. (2008); van Halem et al. (2009)
SODIS™	Solar water disinfection	• Highly effective • Inexpensive • Can reuse a waste product (PET bottles)	• Long treatment time (6 to 48 hours) • Does not remove other potential pollutants • Requires warm climate and sunlight	Conroy et al. (2001); Kehoe et al. (2001); Mania et al. (2006); Meierhofer and Wegelin (2002); Sommer et al. (1997)

Source: Based on Striebig, B., Atwood, S., Johnson, B., Lemkau, B., Shamrell, J., Spuler, P., Stanek, K., Vernon, A., Young., J. 2007. "Activated carbon amended ceramic drinking water filters for Benin." Journal of Engineering for Sustainable Development. 2(1):1–12.

In terms of products, engineers and designers have transitioned to the concept of **appropriate technology**. While this term has many differing meanings, it is used here to refer generally to engineering design that takes into consideration the key local social, economic, environmental, and technical factors that influence the success or failure of a design solution. That is, a technology (or design) is appropriate "when it is compatible with local, cultural, and economic conditions (i.e., the human, material and cultural resources of the economy), and utilizes locally available materials and energy resources, with tools and processes maintained and operationally controlled by the local population" (Conteh, 2003, p. 3). Appropriate technologies commonly considered when building homes or community buildings are briefly described in Table 1.4, and many of these technologies are described in more detail in the green building section in Chapter 11.

What does appropriate technology look like in practice? Consider, for example, the design of a point-of-use water supply and purification system for a rural household. In a high-income country, engineers would likely base their design decision on the amount of water needed by the household, size the system components for filtration, disinfection, and pumping accordingly, and then balance component quality and selection against the budgetary constraints of the household. In the case of a low-income country, engineers must still make these technical design decisions about system size and affordability, but they must also consider:

- Are parts readily available, either locally or nationally, if a component were to fail?
- Are there individuals who have the necessary skill or technical training to repair the component or system if it were to fail?

TABLE 1.4 Applications and uses of appropriate technology for medium- and low-income indexed countries

APPLICATION	TECHNOLOGY
Building design	Right-sized homes that maximize storage, comfort, social interactions, and use while minimizing the use of materials and energy.
	Natural ventilation can be integrated into a design by incorporating porches, central courtyards, other outside features, and strategically placed windows.
	Passive solar design maximizes exposure to the sun and takes advantage of the natural energy characteristics of building materials and air that are exposed to the energy of the sun.
	Overhangs take advantage of the thermal properties of the sun during the winter months while minimizing the sun's impact during the warmer summer months.
Power generation	Biogas power generation or microbial fuel cell technology can be integrated with waste management.
	Simple wind turbines may be made out of containers that would otherwise be disposed of as solid waste.
Material use in building construction	Appropriate, local, nontoxic, and reusable materials such as lime-stabilized rammed earth blocks, adobe, and straw bales can be promoted for building construction.
Stormwater management	Green roofs built on top of residential, commercial, and industrial structures not only effectively manage stormwater but also have benefits of reducing a building's energy consumption and regional urban heat island effect.
Water supply	Rainwater harvesting can assist groundwater recharge and provide all or a portion of domestic, commercial, and agricultural needs. It can also be incorporated into a building's cooling system.
Water treatment	*Moringa oleifera* tree seeds can be used to reduce turbidity.
	Other point-of-use treatment technologies described in Table 1.3 may be useful, especially in areas where power interruptions are frequent and those power interruptions frequently result in contamination of centralized piped drinking water supplies.

Source: Based on Hazeltine and Bull, Field Guide to Appropriate Technology, 2003.

- Would members of the household readily understand how to use this system?
- What is the local availability of required infrastructures, such as electric power?

There is a significant body of literature on how to develop and design appropriate technology solutions for developing communities. These solutions tend to be low-cost, culturally sensitive, and community-focused. They are also usually made and sourced from local supply chains. An added benefit of appropriate technology is that it can often be used as a tool for building human resource capacities in key areas, such as in electrical and mechanical skills and training. A wealth of information is now available on emerging appropriate technologies and addresses a wide range of essential human needs—for water, sanitation, energy, shelter, and so on—in developing communities. Global knowledge sharing about appropriate technology has greatly increased with the development of several online communities, such as *Engineering for Change* and *Appropedia* (Box 1.2).

The second shift in thinking about sustainable engineering is about the design process itself and the implementation plan, which are equally important to the technical design. The sustainable design process involves working with local community members to co-define the problem and its possible solutions, as well as to develop sound implementation plans. Community involvement is a hallmark of successful engineering design projects in low-income communities (Figure 1.10). It is through this process that appropriate technology can be used as a means of capacity building. In Bangladesh, an effort to provide solar energy and household lighting resulted in microlending opportunities that built local banking and financing capabilities (Box 1.3).

FIGURE 1.10 A Kenyan *Jiko*. This Kenyan cookstove is made of ceramic and tin and is an improved design that conserves scarce and costly charcoal and wood fuel. The final design resulted from collaboration between the women who use them and the Massachusetts Institute of Technology engineers working on the project.

Source: Bradley Striebig.

BOX 1.2 Online Resources for Appropriate Technology Solutions

Working on engineering solutions to meet the challenges in a developing community? Here are great resources to begin your exploration of possible strategies and ideas.

Engineering for Change (E4C): Founded by the American Society of Mechanical Engineers (ASME), Institute of Electrical and Electronics Engineers (IEEE), and Engineers Without Borders—USA (EWB-USA), E4C is a community of engineers, technologists, social scientists, non-governmental organizations (NGOs), local governments, and community advocates who work to develop locally appropriate and sustainable solutions for pressing humanitarian challenges. They maintain a "Solutions Library," which is a catalogue of appropriate technology solutions and case studies.

Website: www.engineeringforchange.org

Appropedia: This is a wiki that enables users to catalogue and collaborate on sustainability, appropriate technology, and poverty-reduction solutions. As with all crowd-sourced and community-managed wikis, the information on the site is best used for idea generation, as technical details are generally not independently validated.

Website: www.appropedia.org

Solar Cookers World Network: Another wiki-based site, the Solar Cookers World Network enables knowledge sharing on the design and construction of solar cookers.

Website: www.solarcooking.wikia.com

BOX 1.3 Grameen Shakti (*Grameen Energy*)

Founded in 1996, Grameen Shakti is one of the world's largest suppliers of solar technologies, having installed nearly 100,000 solar photovoltaic systems in homes at a current rate of approximately 3500 per month. Grameen Shakti is part of the Grameen Bank family, a microfinance enterprise for which its founder, Muhammad Yunus, won the Nobel Peace Prize.

Microfinance goes by several names, including microcredit and microlending. It operates by making extremely small loans to the poor without collateral or contract, and it relies on community peer pressure to assure loan repayment. Loans are given to individuals as investments that will allow borrowers an opportunity to surmount their poverty. An example is lending a woman money to purchase a sewing machine with which she can become a seamstress and earn an income. Indeed, the vast majority of Grameen microcredit is given to women.

With respect to Grameen Shakti, customers purchase their energy systems from the organization on a payment plan, usually a monthly installment over two or three years. Staff visit monthly to collect fees and perform maintenance on the systems. A typical photovoltaic system is used to power four light bulbs for four hours at night, enabling children to study and do their homework. Grameen Shakti also works with other renewable energy technologies, such as biogas applications and cook stoves.

Grameen Shakti is a unique social business in that it offers a complete package of technology integration: financing options to purchase the system, operational support of the technology, maintenance if the system breaks down, and technical advice on how to convert the product into a money-making tool. It is also unique in that it is location specific, providing remote, rural areas of Bangladesh with access to renewable energy technologies. This allows Grameen Shakti to make the design of the social aspects of its business sensitive to sociocultural factors that contribute to its success.

EXAMPLE 1.2 EWB House in Rwanda

Rwanda is a small country that lies in the heart of Central Africa. The capital of Rwanda is Kigali, a city that has grown very rapidly after the civil unrest that racked the country in the mid-1990s. As of 2002, Kigali City had 131,106 households, a total population of 604,966 (approximately 56% of which is age 20 years or under), and an annual growth rate of 10% (i.e., population will double every seven years) (Rwanda Ministry of Infrastructure, 2006).

The average Rwandan citizen consumed only 15.2 liters of water per day in 2002. The water cost was about a nickel (U.S.) each day. This may not sound like much, but that cost was more than 10% of the average daily wage! Water supply is still limited in Rwanda. Children are usually responsible for collecting the water. In Kimisange, a peri-urban area surrounding Kigali, the children carry the water over 0.25 kilometers from the nearest public water source, called a tapstand. However, this tapstand often runs dry, forcing the children to collect standing surface water in locations such as the one shown in Figure 1.11. The children may carry the water nearly 2 kilometers uphill back to their home. Cooking fuel is also in short supply in Rwanda due to deforestation. Cooking fires with poor fuel continue to result in 10 to 11 deaths each day in many low-development countries such as Rwanda. Sewage in Rwanda is often discharged directly into the streets. The raw sewage contaminates the local water supply. The lack of access to clean water, energy, and sanitation has had a profound negative effect on human health. The average life expectancy in Rwanda in 2002 was only 47.3 years. The infant mortality rate was 89.61 infant deaths per 1,000 live births (WHO/UNICEF, 2005). Waterborne disease is suspected in killing one out of every five children born before the age of 5.

In 2004, an international group consisting of architects, engineers, and scientists led a project to demonstrate a low-income model home with sustainable on-site water, sanitation, and renewable sources of cooking fuel and fertilizer.

FIGURE 1.11 Photo showing typical housing, agriculture, and water sources in the peri-urban area near Kigali, Rwanda.

Source: Bradley Striebig.

The local population of Kigali does not have the financial resources to invest in sanitation systems because they suffer from extreme poverty. Traditional centralized collection and treatment systems in the United States cost the average consumer 0.62 U.S. dollars per day (CIA, 2007). This would be nearly an entire day's wages or more for most people living in Kigali, so it is not a practical solution for Rwanda or most of the developing world. Existing centralized treatment facilities are expensive and require large amounts of energy for treatment and pumping. Ultimately, the traditional approach would not yield economically or environmentally sustainable solutions for sanitation or resource recovery in Kigali. Decentralized technologies are subject to failure due to poor maintenance and the costly disposal of residual wastes.

A group from Engineers Without Borders—USA and the Kigali Institute of Science and Technology worked with a community in Kigali to develop and optimize a scalable low-cost sustainable home. The home is made from low-cost bamboo reinforced and lime-stabilized earthen blocks, as shown in Figure 1.12. These blocks were produced at the site of the model home and will provide good insulation, keeping the interior warm during cool evenings and cool during warm days in Rwanda's very mild climate. Recovery of biogas will be used for cooking fuel and will help reduce deforestation due to the need to harvest wood for cooking. Rainwater will be collected from rooftops. This water can be used for drinking water, cooking, and sanitation. Drinking water is treated in model homes using a point-of-use treatment technology. Three treatment technologies (shown in Table 1.3) were considered for drinking water purification.

Construction and implementation of the model home occurred in 2007. Community-focused projects, such as this one, directly address the eight Millennium Development Goals (MDG) set by the United Nations (UN, 2000).

Millennium Development Goal 1: Eradicate extreme poverty and hunger

Problems of poverty are inextricably linked to the availability and quality of water. Improving access to sanitation and the quality of water in the watershed can make a major contribution to eradicating poverty. Reliable shelter and biogas cooking fuel has the potential to reduce poverty and hunger.

FIGURE 1.12 Interlocked earth-block construction methods use inexpensive soil-based construction blocks that are made primarily from materials available near the construction site and use locally harvested bamboo reinforcement for construction.

Source: Photo used by permission of Chris Rollins.

Millennium Development Goal 2: Achieve universal primary education

Implementing safe sanitation practices and improving the water quality may reduce absenteeism and help students concentrate more fully on their education. Students and staff may benefit from the significant reduction of exposure to pathogenic organisms. It is expected that there will be a significant decrease in illness and absence from school due to the reduced exposure to pathogenic organisms. Furthermore, less time may be spent finding fuel for cooking since wood can be replaced by biogas.

Millennium Development Goal 3: Promote gender equality and empower women

Millennium Development Goal 4: Reduce childhood mortality

Millennium Development Goal 5: Improve maternal health

Millennium Development Goal 6: Combat HIV/AIDS, malaria, and other diseases

Water development projects can address Millennium Development Goals 3, 4, 5 and 6. Increasing access to water can decrease the amount of time spent by young women retrieving water for use in the home, thereby increasing the likely hood those young women will do well in school and stay in school. Millions of children under the age of 5 die each year from preventable water-related diseases. The UN *Water World Development Report* states that, of all the people who died of diarrhea infections in 2001, 70% (or 1.4 million) were children. People weakened by HIV/AIDS are likely to suffer the most from the lack of a safe water supply and sanitation, especially from diarrhea and skin diseases. Access to pathogen-free drinking water may reduce childhood mortality and also reduce the suffering of people weakened by HIV/AIDS.

Waterborne diseases spread pathogens by the ingestion of urine- or feces-contaminated water. Typhoid fever, amoebic dysentery, schistosomiasis, and cholera are just a few of the diseases spread by contaminated water. Maternal mortality rates were estimated by the World Health Organization (WHO) to be 1 maternal death per 10 births. It is estimated that 203 of every 1,000 children die before the age of 5 in Rwanda. The model-building design increases access to water for drinking and hand washing. Promoting proper sanitation and providing the technology to implement point-source water treatment in the community will likely decrease childhood and maternal mortality rates in Kigali, Rwanda.

Millennium Development Goal 7: Ensure environmental sustainability

The earth-block model home addressed many of the indicators of sustainable development by meeting the current needs for wastewater treatment in the urban fringe of Kigali while conserving resources for future generations. The strategy reduced the spread of waterborne disease, while effectively treating wastewater so as not to exceed the assimilative capacity of native wetlands and also reduce the dependency on imported materials such as concrete and cement.

Millennium Development Goal 8: Develop a global partnership for development

This project involved partnerships with the following organizations: the German Development Corporation (DED), Gonzaga University, Engineers Without Borders—USA, Tetra-Tech, Inc., the city of Kigali, the Kigali Institute for Science and Technology, and the National University of Rwanda.

FIGURE 1.13 Completed model home in Rwanda made from earth-block construction with rainwater harvesting system shown in the foreground that is sitting on top of the visible portion of an underground biogas digester.

Source: Photo used by permission of Chris Rollins.

1.6 Definitions of Sustainability

The *Merriam-Webster Dictionary* defines **sustainable** as "capable of being sustained." This first dictionary definition of sustainable does not shed much light on our discussion. The second definition listed begins to illuminate our topic: "of, relating to, or being a method of harvesting or using a resource so that the resource is not depleted or permanently damaged." Within this definition, we begin to see some key topics, including resource depletion and the term *damage* associated with nonsustainable practices.

The United States Environmental Protection Agency (EPA) provides a more useful working definition of sustainability:

> **Sustainability** *is based on a simple principle: Everything that we need for our survival and well-being depends, either directly or indirectly, on our natural environment. Sustainability creates and maintains the conditions, under which humans and nature can exist in productive harmony, that permit fulfilling the social, economic and other requirements of present and future generations. Sustainability is important to making sure that we have and will continue to have, the water, materials, and resources to protect human health and our environment. (Federal Register, 2009)*

Within the EPA definition, we see the words "harmony" and "protect" being applied to a relationship between humans and nature, which infers a moral virtue associated with sustainability. When we begin to think about sustainability in moralistic terms, we venture into the world of ethics and conflicting or sometimes contradictory moral quandaries.

For our simple analysis, we can describe **morals** as the values people adopt to guide the way they ought to treat each other. When we have a conflict between morals, we can use ethics to guide us toward the best outcome based on our ethical reasoning. **Ethics**, therefore, provides a framework for making difficult choices when we face a problem involving moral conflict. These working definitions are much easier to understand when we evaluate a few of the following examples of the applications of an ethical code.

BOX 1.4 The National Society of Professional Engineers Code of Ethics for Engineers

Fundamental Canons: Engineers, in the fulfillment of their professional duties, shall:

1. Hold paramount the safety, health, and welfare of the public.
2. Perform services only in areas of their competence.
3. Issue public statements only in an objective and truthful manner.
4. Act for each employer or client as faithful agents or trustees.
5. Avoid deceptive acts.
6. Conduct themselves honorably, responsibly, ethically, and lawfully so as to enhance the honor, reputation, and usefulness of the profession.

One early ethical system is that of the **categorical imperative** formulated by Immanuel Kant (1724–1804). Kant suggested that an act is either ethical or unethical, if, when it is universalized, it makes for a better world. An example can be found in the Engineering Code of Ethics shown in Box 1.4 that states "the engineer shall hold paramount the health, safety, and welfare of the public" (NSPE, 2007). The concepts presented in the EPA and Merriam-Webster definitions of sustainability both suggest that conditions for the planet and humans inhabiting the Earth would be better if sustainable practices were adopted. However, the Engineering Code of Ethics takes a human-centered view of a system, holding the "public" good in the highest regard. This might be thought of as an anthropocentric ethical framework in which nature is considered and valued based solely on the goods that nature can provide to humans or the "public."

Long before Kant, Aristotle appears to support this anthropocentric view in his statement that "plants exist to give food to animals, and animals, to give food to men. . . . Since nature makes nothing purposeless or in vain, all animals must have been made by nature for the sake of man." (Vesilind et. al., p. 71) While the anthropocentric view is simple and concise, it does not seem to adequately reflect the feeling toward animals and nature held in most modern societies. For example, it is illegal in the United States and many other countries to encourage or promote animal fights for one's own pleasure.

This change in ethics was taking place in the Western world during the 1800s and was spearheaded by prominent writers and philosophers of the time. The growth of ethical arguments to include animals and nature has been termed *existentialist ethical thinking*, which extends the moral community to include creatures other than humans. Aldo Leopold articulates this viewpoint in the *Sand County Almanac*. Leopold's work led to the development of a new ethical framework called the **land ethic**, which encourages people to extend the thinking about communities to which we should behave ethically to include soil, water, plants and animals, or collectively, the land. Paul Taylor (1981, p. 207) describes a **biocentric outlook** that values all living things in Earth's community, so each organism is "a center of life pursing its own good in its own way", and all organisms are interconnected. Arne Naess (1989, p. 166) took the biocentric outlook one step further when he wrote: "The right of all the forms to live is a universal right which cannot be quantified.

No single species of living being has more of this particular right to live than any other species." The ethical framework in which humans have no greater importance than any other component of our world is sometimes referred to as deep ecology. The **deep ecology** ethic lies at the opposite side of the spectrum from an anthropocentric ethical code in evaluating the relationships between humankind and nature. However, most societies do not hold to the deep ecology worldview, or paradigm. If modern societies did apply the deep ecology ethic, then modern medicine would not try to kill pneumonia bacteria with antibiotics while saving a human life.

So are there any ethical systems that modern societies have adopted? Certainly there are, and in the engineering profession, we are expected to uphold to specific ethical canons. Aldo Leopold suggests that "a thing is right when it tends to preserve the integrity, stability, and beauty of the biotic community. It is wrong when it tends otherwise." Certainly Leopold considered the human species to be part of this biotic community. Vesilind and Gunn (1998, p. 466) suggest another approach, which they describe as the **environmental ethic** — "recognizing that we are, at least at the present time, unable to explain rationally our attitude towards the environment and that these attitudes are deeply felt, not unlike the feeling of spirituality." Furthermore, this approach embodies a sense of obligation to future generations of our species. When we consider the future conditions of our planet and species, we are using an intergenerational ethical model. If we consider future generations, we are perhaps making a moral choice to preserve and protect the things we value but have difficulty explaining.

Stewart Collis takes an agnostic approach to ethics:

> Both polytheism and monotheism have done their work. The images are broken; the idols are all overthrown. This is now regarded as a very irreligious age. But perhaps it only means that the mind is moving from one state to another. The next state is not belief in many gods. It is not a belief in one god. It is not a belief at all — not a conception of the intellect. It is an extension of consciousness so that we may feel God. (Collis, 1954, p. 72)

In his 2002 published letter, Pope John Paul II suggested an intergenerational and environmental ethical framework that blends advances in science and technology:

> We therefore invite all men and women of good will to ponder the importance of the following ethical goals: To think of the world's children when we reflect on and evaluate our options for action. To be open to study the true values based on the natural law that sustain every human culture. To use science and technology in a full and constructive way, while recognizing that the findings of science have always to be evaluated in the light of the centrality of the human person, of the common good and of the inner purpose of creation. Science may help us to correct the mistakes of the past, in order to enhance the spiritual and material well-being of the present and future generations. It is love for our children that will show us the path we must follow in the future.

Throughout the ages we have blended our ethical, moral, and spiritual beliefs in an effort to apply definitions to the actions we take both individually and as a species. The most commonly referenced definition of sustainability is derived from the Brundtland Commission's report on practices for sustainable development and approaches to reduce the number of people living in poverty. The report of this commission, called *Our Common Future*, defines sustainable development as "development that meets the needs of the present without compromising the ability of future generations to meet their own needs.

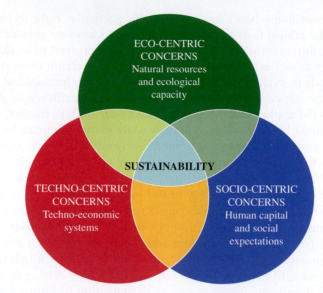

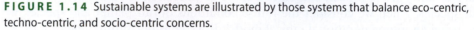

FIGURE 1.14 Sustainable systems are illustrated by those systems that balance eco-centric, techno-centric, and socio-centric concerns.

Source: Based on http://www.planetmattersandmore.com/tag/sustainable-design/.

Within this definition of sustainability, we see that human needs are placed somewhat above the needs of other animals and plants. However, this definition recognizes the inherent value of the natural world and the role of the natural world in meeting the basic needs of humanity. Furthermore, this definition relies heavily on an intergenerational ethical obligation to protect the natural world so that future generations will not live in an impoverished planet. Engineers, scientists, technicians, and policymakers are charged with the role of identifying technologies that can meet these needs and improve the standard of living on the planet, in spite of significant resource constraints.

Sustainable design is the design of products, processes, or systems that balance our beliefs in the sanctity of human life and promote an enabling environment for people to enjoy long, healthy, and creative lives, while protecting and preserving natural resources for both their intrinsic value and the natural world's value to humankind. The engineer who practices sustainable design must have a grasp of the social, economic, and environmental consequences of their design decision and a thorough understanding of the scientific principles of the technology available, as illustrated in Figure 1.14. The EPA recently defined sustainability as "the continued protection of human health and the environment while fostering economic prosperity and societal wellbeing" (Fiksel, 2012, p. 5). The only way we can actually achieve sustainable development is for scientists, engineers, technicians, and policymakers to develop and apply more efficient technologies that improve the standard of living and are adaptable to the global marketplace. Furthermore, we need to identify and relate the limitations of our natural resources to our desire for continual development.

1.7   Populations and Consumption

The relationship between consumption of a natural resource, waste production, and regeneration of that resource is described mathematically as the carrying capacity. Dr. Paul Bishop (1999, p. 574) defines the **carrying capacity** as "the maximum rate of resource consumption and waste discharge that can be sustained indefinitely in a given region without progressively impairing the functional integrity and productivity of the relevant ecosystem." A sustainable economic system operating within the Earth's carrying capacity demands the following.

The usage of renewable resources is not greater than the rates at which they are regenerated.

The rates of use of nonrenewable resources do not exceed the rates at which renewable substitutes are developed.

The rates of pollution or waste production do not exceed the capacity of the environment to assimilate these materials.

In order to calculate the carrying capacity of the Earth and its resources, we must be able to make some predictions about human population and consumption patterns. The world's steadily increasing population has put stress on available natural resources, including food, water, energy, phosphorus, fossil fuels, and precious metals. Both human population and consumption of resources are *increasing* at an *increasing rate*. Thus, linear mathematical models do not accurately estimate population trends; instead, we must apply an exponential model to estimate population growth and resource consumption patterns.

Exponential growth occurs when the rate of change, dA/dt, is proportional to the instantaneous value of A at some time t.

$$\frac{dA}{dt} = kA \tag{1.7}$$

We can integrate this expression with respect to time, which yields

$$A_{(t)} = A_o \exp(k(t - t_o)) \tag{1.8}$$

where A_o is the value of A at our initial time, t_o. The variable k is the exponential growth rate constant and typically has units of [1/time].

EXAMPLE 1.3   Historical World Population Growth

The world's population was approximately 370 million in AD 1350. By the year 1804, the world's population had reached 1 billion people. Find the exponential growth rate during that time period.

The term k in Equation (1.8) represents the population growth constant. We can rearrange Equation (1.8) to solve for the rate constant:

$$A_{(t)} = A_o \exp(k(t - t_o))$$

$$\ln\left(\frac{A_{(t)}}{A_o}\right) = k(t - t_o)$$

$$k = \left(\ln\left(\frac{A_{(t)}}{A_o}\right)\right)\left(\frac{1}{(t - t_o)}\right)$$

In this example, the variables are

- Initial population $= A_o = 370 \times 10^6$ people
- Initial time $= t_o = 1350$
- Final population $= A_{(t)} = 1 \times 10^9$ people
- Final time $= t = 1804$

Rearranging Equation (1.8) and substituting the appropriate values into our exponential equation yields

$$k = \left[\ln \left(\frac{1 \times 10^9}{370 \times 10^6} \right) \right] \left(\frac{1}{1804 - 1350} \right) = 0.0022 \; \frac{1}{\text{year}} \text{ or } 0.22\%$$

EXAMPLE 1.4 Current World Population Growth

The world's population was estimated to be 6 billion in 1999, but by 2012 the world's population had already grown to 7 billion people. What is the world's current population growth rate? When should we expect the population to reach 8 billion people if the growth rate between 1999 and 2012 remains constant?

In this example, the variables are

- Initial population $= A_o = 6 \times 10^9$ people
- Initial time $= t_o = 1999$
- Final population $= A_{(t)} = 7 \times 10^9$ people
- Final time $= t = 2012$

From example 1.3 the exponential growth rate constant, k, is found from:

$$k = \left(\ln \left(\frac{A_{(t)}}{A_o} \right) \right) \left(\frac{1}{(t - t_o)} \right)$$

$$k = \left[\ln \left(\frac{7 \times 10^9}{6 \times 10^9} \right) \right] \left(\frac{1}{2012 - 1999} \right) = 0.0119 \; \frac{1}{\text{year}} \text{ or } 1.19\%$$

Then, using the current rate constant and Equation (1.8), we can solve for the time when we expect the population to reach 8 billion as

$$\ln \left(\frac{A_{(t)}}{A_o} \right) = k(t - t_o)$$

$$t = \left[\ln \left(\frac{A_{(t)}}{A_o} \right) \right] \left(\frac{1}{k} \right) + t_o = \left[\ln \left(\frac{8 \times 10^9}{7 \times 10^9} \right) \right] \left(\frac{1}{0.0119} \right) + 2012 = 11.2 + 2012 = 2023$$

So, the population may reach 8 billion in 2023.

One of the most famous demographers of all time, Thomas Robert Malthus (1766–1834), wrote the following in his paper "An Essay on the Principle of Population" (1798):

The power of population is so superior to the power of the earth to produce subsistence for man, that premature death must in some shape or other visit the human race.

Thomas Malthus made this argument because he believed the human population would very soon outpace the production of food and the regeneration of natural resources. Malthus believed this would create a catastrophe that would surely reduce the population to a number that would be more sustainable. What Malthus did not foresee was a change in technology that would allow for the use of mechanical power and industrial fertilizer. These technological changes allowed for greater production of food than Malthus could conceive based upon the technology that was available during his lifetime.

The Malthusian Catastrophe, as Malthus's hypothesis was called, seemed to be reaching fulfillment during Malthus's life since the signs of his times were quite bleak. London was becoming a huge city—one of the largest on the planet at that time. The famed Thames River had become an open sewer and a source for miasma, a sickness associated with ill winds and rotten smells. As the population of London grew, the waste from this population accumulated in the Thames. In the summer of 1848, a deadly cholera outbreak in London killed at least 14,600 people, according to published records. Shortly after that outbreak, in 1854, yet another outbreak of cholera occurred. At this time, Dr. John Snow began to trace the cause of the disease by mapping the cases and neighborhoods where the cholera was occurring. Dr. Snow was able to trace the cases to a community well, known as the Broad Street Well. He was able to convince people to remove the handle to the well pump, and shortly thereafter, the cholera cases in that community subsided. Thus disease was linked for the first time with water contaminated by human waste. In spite of Dr. Snow's efforts, 10,675 deaths from cholera were reported in 1854. The link between the contaminated water and cholera soon led to acceptance of the germ theory and to important advances in public health, epidemiology, engineering, and water treatment, which eventually had a profound influence on the standard of living and lifespan throughout the rapidly industrializing nations.

Cholera still plagues those nations with inadequate water and sanitation systems. In Zimbabwe in December of 2008, for example, a massive cholera outbreak occurred that is believed to have infected more than 57,000 people and resulted in more than 3,000 deaths. The continuing closing of several local hospitals and the scarcity of basic medical commodities such as medicines and health personnel are believed to have been a major contributor to the spread. The state media reported that most of the capital city of Harare had been left without water after the city ran out of chemicals for its treatment plant. Chlorination has been shown to effectively prevent cholera, as shown in Figure 1.15, which illustrates how implementing chlorination in the United States starting in the 1920s greatly reduced the incidence of waterborne disease. Yet Zimbabwe, Haiti, and many other low- and medium-income countries still lack the infrastructure to adequately provide clean water. Despite readily available and effective methods of treating drinking water, major outbreaks of cholera still occur regularly in West Africa and Haiti (Figure 1.16).

Malthus published his dissertation in 1798 when the world's population was just approaching 1 billion people. In 2012, when the world's population

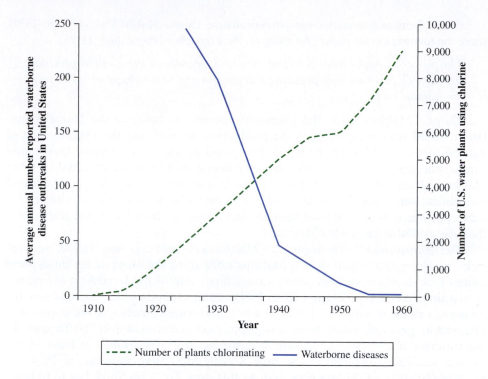

FIGURE 1.15 Chlorination of water supplies is an effective method of preventing cholera.

Source: Based on G.L. Culp and R.L. Culp, New Concepts in Water Purification, Litton Educational, New York, 1974.

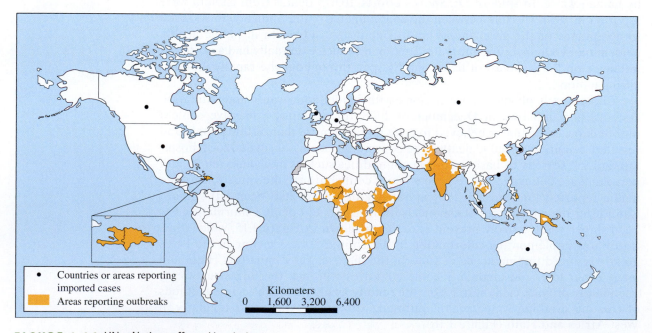

FIGURE 1.16 UN—Nations affected by cholera.

Source: Based on World Health Organization (2012) http://gamapserver.who.int/mapLibrary/Files/Maps/Global_Cholera_ITHRisk_20120118.png.

surpassed 7 billion people many economists, environmentalists, and scientists were once again gravely concerned that we would not be able to retain and improve the standard of living as the human population put greater strain on available natural resources.

In 1968, Garrett Hardin published an essay called "The Tragedy of the Commons" that discussed the problems with Thomas Malthus's original arguments and dates, but Hardin generally supported the notion that either population or the standard of living must be limited by the available natural resources. Hardin believed the first reason that there would be a great tragedy related to the environment is that it is not mathematically possible to maximize for two (or more) variables at the same time.

The second reason for Hardin's pessimism springs directly from biological facts. To live, any organism must have a source of energy (for example, food). This energy is utilized for two purposes: maintenance and work. For man, the maintenance of life requires about 1600 kilocalories a day "maintenance calories". Anything that he does over and above merely staying alive will be defined as work and is supported by "work calories" that he takes in. Work calories are used not only for what we call work in common speech; they are also required for all forms of enjoyment—from swimming and automobile racing to playing music and writing poetry. If our goal is to maximize population, it is obvious what we must do: We must make the work calories per person approach as close to zero as possible. No gourmet meals, no vacations, no sports, no music, no literature, no art. (Hardin, 1968, p. 1)

Both Malthus and Hardin present a formidable argument: increasing populations lead to more difficulty in accomplishing the United Nations goals associated with sustainable development. The National Academy of Engineers has recognized that the tasks associated with sustainable development represent the grandest challenges to engineers in this century. These challenges are illustrated in Figure 1.17.

 Make solar energy economical

 Provide energy from fusion

 Develop carbon sequestration methods

 Manage the nitrogen cycle

 Provide access to clean water

 Restore and improve urban infrastructure

 Advance the health informatics

 Engineer better medicines

 Reverse-engineer the brain

 Prevent nuclear terror

 Secure cyberspace

 Enhance virtual reality

 Advance personalized learning

 Engineer the tools of scientific discovery

FIGURE 1.17 Grand Challenges in engineering for the next century.

Source: Based on National Academy of Engineering http://www.engineeringchallenges.org/; c12/Shutterstock.com; Roman Sigaev/Shutterstock.com; Andriano/Shutterstock.com; Brian A Jackson/Shutterstock.com; saiva/Shutterstock .com; hxdbzxy/Shutterstock.com; wavebreakmedia/Shutterstock.com; Pressmaster/Shutterstock.com; ollyy/Shutterstock .com; hxdyl/Shutterstock.com; wongwean/Shutterstock.com; Syda Productions/Shutterstock.com; racorn/Shutterstock .com; Sergey Nivens/Shutterstock.com.

1.8 Technical Approaches to Quantifying Sustainability

It is vitally important that we develop quantitative methods of determining which technologies of the future truly represent progress toward more sustainable development. In this book we will explore several methods of developing and evaluating sustainably designed products and infrastructure. In order to effectively develop quantitative relationships, we must understand the fundamental units of measure that will help us compare and contrast the sustainable technologies of today and tomorrow.

Various terms are used to describe measures of sustainability and their interrelationships to one another. More specifically, a **measure** of sustainability is a value that is quantifiable against a standard at a point in time. A sustainability **metric** is a standardized set of measurements or data related to one or more sustainability indicators. A sustainability **indicator** is a measurement or metric based on verifiable data that can be used to communicate important information to decision makers and the public about processes related to sustainable design or development (Biodiversity Indicators Partnership, 2010). Sustainability indicators are those measurable aspects of economic, environmental, or societal systems that are useful for monitoring the continuation of human and environmental well-being (Siksel, 2012). The EPA has suggested four major categories of indicator outcomes:

> *Adverse Outcomes Indicator (AOI)—indicates destruction of value due to impacts on individuals, communities, business enterprises, or the natural environment.* The adverse outcomes are discussed in detail in Chapters 2 through 6 of this textbook.

> *Resource Flow Indicator (RFI)—indicates pressures associated with the rate of consumption of resources, including materials, energy, water, land, or biota.* The resource flows are discussed in detail in Chapters 2 and 7 through 10.

> *System Condition Indicator (SCI)—indicates the state of the system in question, that is, individuals, communities, business enterprises, or the natural environment.* The system condition is discussed in more detail in Chapters 7, 10, and 11.

> *Value Creation Indicator (VCI)—indicates the creation of value (both economic and well-being) through enhancement of individuals, communities, business enterprises, or the natural environment.* Value Creation typically lies outside the realm of the traditional engineering sciences and interfaces closely with the social and economic sciences. Value creation is discussed in detail in Chapters 7 and 12.

The relationships between these four indicator categories are illustrated in Figure 1.18. The resource flows have value, and the value is distributed by human actions and system relationships among natural capital, economic capital, and human capital. We may attempt to measure various indicators within the complex system. Examples of specific indicators of each indicator category are shown in Table 1.5. These individual indicators may be combined to form more complex tools that make up a sustainability index. A sustainability **index** is a numerical-based scale used to compare alternative designs or processes with one another. Examples of sustainability indexes include the IPAT equation, the environmental footprint, and other metrics and indexes described in more detail in Chapters 9 and 10.

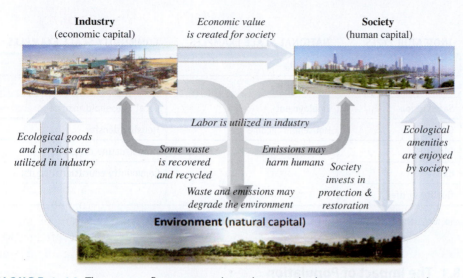

FIGURE 1.18 The resource flows, system dependence, and values among economic, human, and natural capital. Value, or capital, may be moved from one area of sustainability to another, maintaining a balance of capital resources for future generations in an intrinsic property of sustainability.

Source: US EPA (2012) A Framework for Sustainability Indicators at EPA. Figure 3, page 10.

TABLE 1.5 The major categories of sustainability indicators and examples of measureable properties.

INDICATOR CATEGORY	INDICATOR TYPES	NATIONAL SCALE EXAMPLES	COMMUNITY SCALE EXAMPLES
Resource Flow Indicators	Volume	Greenhouse gas emissions	Greenhouse gas emissions
	Intensity	Material flow volume	Material flow analysis
	Recovery	Resource depletion rate	Water treatment efficacy
	Impact		Recycling rate
	Quantity		Land use
Value Creation Indicators	Profitability	Cost (reduction)	Cost (reduction)
	Economic output	Fuel efficiency (gain)	Fuel efficiency (gain)
	Income	Energy efficiency (gain)	Energy efficiency (gain)
	Capital investment		Vehicle use (miles per capita)
	Human development		
Adverse Outcome Indicators	Exposure	Health impacts of air pollution	Health impacts of air pollution
	Risk	Public safety	Public safety
	Incidence	Life cycle footprint of energy use	Sewer overflow frequency
	Impact		
	Loss		
	Impairment		

(Continued)

TABLE 1.5 (*Continued*)

INDICATOR CATEGORY	INDICATOR TYPES	NATIONAL SCALE EXAMPLES	COMMUNITY SCALE EXAMPLES
System Condition Indicators	Health	Air quality	Air and water quality
	Wealth	Water quality	Local employment
	Satisfaction	Employment	Local household income
	Growth	Household income	Housing density
	Dignity		Infrastructure durability
	Capacity		Community educational equity
	Quality of life		

Source: Based on US EPA (2012) A Framework for Sustainability Indicators at EPA. Table 1, page 11.

1.8.1 The Impact of Population

An early attempt to develop a relationship among technology, society, and economic factors was developed during the 1970s by Barry Commoner, Paul Ehrlich, and John Holden (1971). They postulated that human impact on the environment, I, would be related to population, P, affluence, A, and technology, T:

$$I = P \times A \times T \tag{1.9}$$

Decades ago, scientists, philosophers, and economists were searching for a formula to create or define a scenario for sustainable development. In this chapter, we have demonstrated that population is increasing rapidly, and as a result, environmental impacts due to human causes are also increasing, as the IPAT equation (as it has come to be known) would predict.

There are large gaps in affluence between those living in highly economically developed countries and those living in low-development countries on the HDI scale. Countries that are low on the HDI scale have high levels of poverty, high rates of mortality, and limited opportunities for most people. As noted earlier, the United Nations has set goals, called the Millennium Development Goals, to decrease poverty or increase opportunity and affluence, particularly in Low Index Developing Countries. As these countries develop and achieve more affluence, the IPAT equation will again suggest that we are likely to see an increase in negative environmental impacts due to increasing affluence.

Scientists and engineers have an opportunity to significantly influence the last term in the IPAT equation. In a global society, if human population and affluence continue to increase, environmental impacts will also increase, unless breakthroughs are made in technology and understanding. Thus, sustainable development seeks to balance the demands placed on the planet by society, the environment, and economic development. The IPAT equation indicates that the only way to achieve sustainable growth without creating much greater environmental impact is to use current technologies more efficiently or to develop new technologies that have a lower overall environmental impact.

1.8.2 The Ecological Footprint

Prior to the Industrial Revolution, the number of humans and their impact on the planet were relatively small. As human population growth and resource consumption accelerated, the impact on natural resources also accelerated, to the point where many scientists believe we have already passed the sustainable carrying capacity of the planet. Scientists estimate that sometime in the mid-1970s, demand on natural resources was outpaced by what resources the planet could produce.

Mathis Wackernagel, president of the Global Footprint Network, and William Rees, a professor at the University of British Columbia, developed the concept of an **ecological footprint (EF)** to calculate and illustrate the relationship between the consumption and supply of natural resources (Ewing et al., 2010). The ecological footprint measures how much land and water area a human population requires to produce the resources it consumes and to absorb the carbon dioxide emissions using prevailing technology.

The ecological footprint for production of goods (EF_P) is expressed as

$$EF_P = P/Y_N \times YF \times EQF \tag{1.10}$$

where P is the amount of a product harvest or carbon dioxide emitted, Y_N is the national average yield for P (or its carbon uptake capacity), and YF and EQF are the yield factor and equivalence factor, respectively, for the land-use type in question. The yield factors are related to average regional land production compared to worldwide averages. At its essence, the ecological footprint concept creates an average of all different types of biomes, ecosystems, and human development and normalizes all these factors into a standard unit area, such as a hectare (ha). A global hectare (gha) accounts for the availability of all global resources and associated human production if it were to be distributed across one global hectare unit. In reality, global resources are not equally distributed throughout the Earth, so the global hectare unit is not an actual measure but a theoretical model that creates a common unit for counting resources and human impacts.

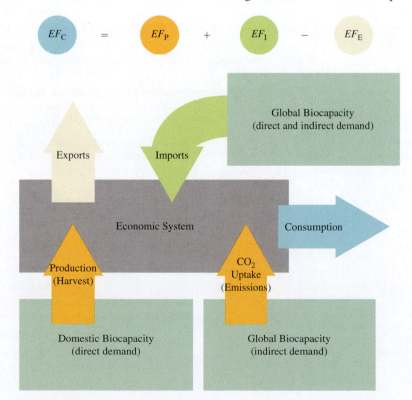

FIGURE 1.19 Schematic of the mathematical model for calculating the ecological footprint, including biocapacity parameters, production parameters, consumption parameters, and industrial emissions.

Source: Based on Ewing B., A. Reed, A. Galli, J. Kitzes, and M. Wackernagel. 2010. Calculation Methodology for the National Footprint Accounts, 2010 Edition. Oakland: Global Footprint Network.

The ecological footprint for consumption, EF_C, is a function of the factors illustrated in Figure 1.19 associated with production, and also the import and export of commodities, where

$$EF_C = EF_P + EF_I + EF_E \tag{1.11}$$

Details of the ecological footprint calculations and coefficients are available in the literature in the *Calculation Methodology for the National Footprint Accounts*. The ecological footprint is most often calculated on a nationwide basis. A nation's ecological footprint takes into account the nation's carbon dioxide emissions, forestry practices, fish catch, and amount of land used for crop production, grazing, and the built environment (or urban land use) as shown in Figure 1.20.

The ecological footprint calculations show that the land area required to sustain our current average lifestyle has more than doubled since 1961 (Figure 1.21). In 2008, it was estimated that humanity used 1.5 planet's worth of resources (Grooten, 2012)! Or, put another way, by September 2008, humanity would have exhausted the Earth's available resources, so from October through December 2008, humanity started depleting the Earth's resources. The rate of resource consumption continues to increase.

In 2012 more than 80% of the world's population lived in countries that exceed the supply of resources within their borders each year. The desire to achieve higher levels of economic development will likely lead to higher levels of resource consumption as illustrated in Figure 1.22, which shows that the ecological footprint is higher for countries with a higher HDI value. Generally, this implies

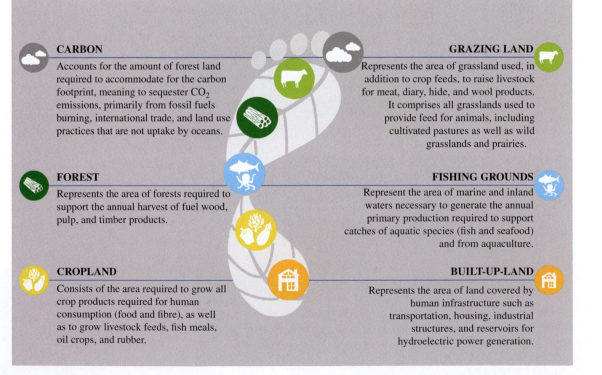

CARBON
Accounts for the amount of forest land required to accommodate for the carbon footprint, meaning to sequester CO_2 emissions, primarily from fossil fuels burning, international trade, and land use practices that are not uptake by oceans.

FOREST
Represents the area of forests required to support the annual harvest of fuel wood, pulp, and timber products.

CROPLAND
Consists of the area required to grow all crop products required for human consumption (food and fibre), as well as to grow livestock feeds, fish meals, oil crops, and rubber.

GRAZING LAND
Represents the area of grassland used, in addition to crop feeds, to raise livestock for meat, diary, hide, and wool products. It comprises all grasslands used to provide feed for animals, including cultivated pastures as well as wild grasslands and prairies.

FISHING GROUNDS
Represent the area of marine and inland waters necessary to generate the annual primary production required to support catches of aquatic species (fish and seafood) and from aquaculture.

BUILT-UP-LAND
Represents the area of land covered by human infrastructure such as transportation, housing, industrial structures, and reservoirs for hydroelectric power generation.

FIGURE 1.20 The ecological footprint consists of measurements of average utilization of resources consisting of carbon emissions, cropland, grazing land, forests , urbanization (or built-up land, and fish caught based on data for individual nations.

Source: Living Planet Report 2012: Biodiversity, biocapacity and better choices. Global Footprint Network, Oakland CA. USA. Figure 22. Page 22.

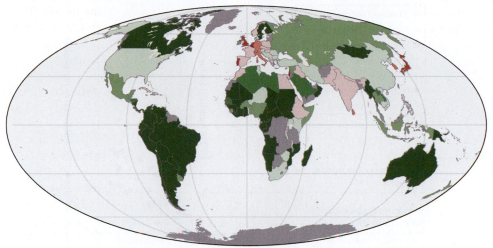

1961

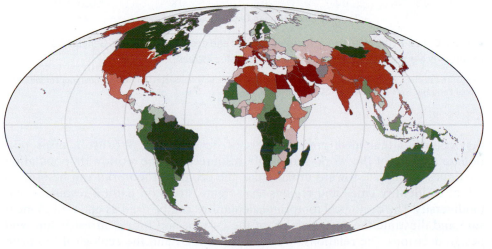

2006

- ■ Footprint more than 150% larger than biocapacity
- ■ Footprint 100–150% larger than biocapacity
- ■ Footprint 50–100% larger than biocapacity
- ■ Footprint 0–50% larger than biocapacity
- ■ Biocapacity 0–50% larger than Footprint
- ■ Biocapacity 50–100% larger than Footprint
- ■ Biocapacity 100–150% larger than Footprint
- ■ Biocapacity more than 150% larger than Footprint
- ■ Insufficient data

FIGURE 1.21 Areas that exceed and have excess capacity in the rate of consumption on resources.

Source: Based on The Ecological Wealth of Nations: Earth's biocapacity as a new framework for international cooperation. 2010. Global Footprint Network. Oakland CA. USA. Figure 9. Page 21.

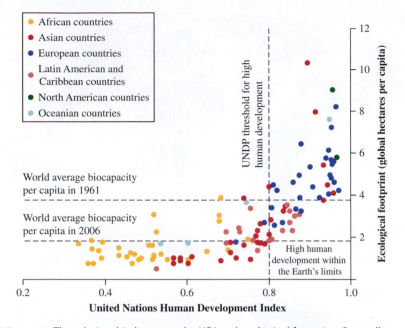

FIGURE 1.22 The relationship between the HDI and ecological footprint. Generally, a country's ecological footprint increases as the HDI value increases.

Source: The Ecological Wealth of Nations: Earth's biocapacity as a new framework for international cooperation. 2010. Global Footprint Network. Oakland CA. USA. Figure 7. Page 15.

that countries that have greater levels of development also use a disproportionate amount of the world's available resources. As the worldwide population continues to migrate toward urban areas, the ecological footprint can be used as a tool to identify cities and regions that create stress on natural resources and biodiversity. The ecological footprint calculations can be used as one tool to measure and illustrate the sustainability of the current and future infrastructure and design decisions. The relationship between the HDI and the ecological footprint also demonstrates the great challenge facing engineers and scientists: to reduce poverty and achieve higher levels of development while simultaneously trying to reduce the environmental footprint historically associated with greater levels of development.

You can calculate your own personal ecological footprint at websites such as the footprint network (www.footprintnetwork.org). This footprint is based on the National Footprint Accounts of each nation. The calculator is an interactive tool to help you learn about your ecological footprint and show how your choices influence your environmental impact. There are several environmental footprint calculators; those based on the Ecological Footprint Standards should yield comparable estimates of your land use, carbon dioxide emissions, and environmental impact.

1.8.3 People-Centered Design

As discussed in the previous sections, engineering design for sustainable development in low-income communities requires that we address the dynamics between social and technical systems. "People-centered design" adds layers of complexity to engineering; as a consequence, tools are available to help us consider both technical and nontechnical design factors. These tools address the selection, design, and implementation of engineering design solutions. For example, **capacity factor analysis** is a tool that assists with the development of sustainable water, sanitation, and household energy solutions by evaluating a community's capacity to manage its own technology (Henriques and Louis, 2011).

Because of the need to integrate social considerations into both the *product* and *process* of developing solutions, engineers and designers are increasingly turning to participatory and community-based design tools. **Participatory design** focuses on directly involving individuals who use the technology or information as co-designers, and on enabling those affected by a design to drive the collaborative design process (Jesper, Simonsen, and Robertson, 2012). **Community-based design** can be thought of as participatory design at the community scale (DiSalvo et al., 2012). In the context of international development, **participatory development** is a bottom-up, people-centered approach aimed at cultivating the full potential of people, especially the poor, at the grassroots level (Forsyth, 2005).

1.8.4 IDEO's Human-Centered Design Toolkit

IDEO, a design and innovation consulting firm, uses *design thinking* as a tool for problem solving and people-centered design. In their guide for educators, IDEO (2011a) defines design thinking as being human-centered and focused on understanding the needs and motivations of key stakeholders. It is a collaborative methodology that requires the participation of multiple individuals in order to gain multiple perspectives about a particular problem and the possible solution space. It is experimental because the process encourages exploring and trying out new ideas or approaches. As seen in Table 1.6, IDEO structures its *design process*—which operationalizes design thinking—into five key phases: discovery, interpretation, ideation, experimentation, and evolution.

TABLE 1.6 IDEO's Five Design Process Phases

IDEO PHASE	EXAMPLE METHOD OR APPROACH
1. Discovery	Defining and researching (including through fieldwork) the problem and gathering inspiration (e.g., contextual immersion, or learning from individuals, groups, experts, peers, etc.)
2. Interpretation	Telling stories that capture learning, finding themes, and framing opportunities
3. Ideation	Idea generation and refinement
4. Experimentation	Prototyping and testing
5. Evolution	Integrating feedback, defining success

Source: Based on IDEO (2011).

For IDEO, human-centered design is "a process and a set of techniques used to create new solutions for the world. Solutions include products, services, environments, organizations, and modes of interaction" (IDEO, 2011b). IDEO holds three

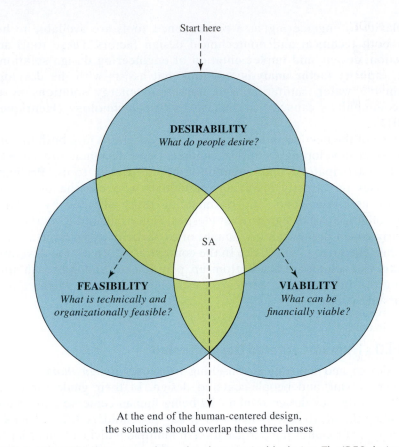

Start here

DESIRABILITY
What do people desire?

SA

FEASIBILITY
*What is technically and
organizationally feasible?*

VIABILITY
*What can be
financially viable?*

At the end of the human-centered design,
the solutions should overlap these three lenses

FIGURE 1.23 IDEO's three core values related to sustainable design. The IDEO design process requires that we create solutions that meet people's desires, that are technically and organizationally feasible, and that are economically viable.

Source: Based on IDEO (2011).

key values related to sustainable design: human desirability, technical feasibility, and technical viability (Figure 1.23).

While the IDEO process is similar to other engineering design and system analysis processes, it is absolutely distinctive for its focus on creating collaborative environments that facilitate creativity and innovation. Given the essential role that social factors play in engineering design for development, the approach adapts well to the contexts of low-income communities. IDEO partnered with the Bill and Melinda Gates Foundation to adapt its methodology (used successfully by large corporations) to develop a toolkit for nonprofit organizations, social enterprises, and individuals who create solutions for those living on less than 2 U.S. dollars per day.

The methodology of the toolkit begins by first defining a specific design challenge and then following through with an exploration of three themes known as Hear, Create, and Deliver (Table 1.7). For each theme, the toolkit describes a number of techniques that can be used to accomplish the goals of the theme. The toolkit is a powerful resource for engineers who want to incorporate significant social considerations in their design.

TABLE 1.7 IDEO Toolkit Design Method

THEME	DESCRIPTION	STEPS
Hear	During the Hear phase, your design team will collect stories and inspiration from people. You will prepare for and conduct field research.	1. Identify a design challenge 2. Recognize existing knowledge 3. Identify people to speak with 4. Choose research methods (e.g., expert interview, in context immersion, etc.) 5. Scenario-based questions (or Sacrificial Concepts) 6. Develop your mindset
Create	In the Create phase, you will work together in a workshop format to translate what you heard from people into frameworks, opportunities, solutions, and prototypes. During this phase, you will move together from concrete to more abstract thinking in identifying themes and opportunities, and then back to the concrete with solutions and prototypes.	1. Develop the approach (e.g., participatory co-design, emphatic design, etc.) 2. Share stories 3. Identify patterns (e.g., create frameworks) 4. Create opportunity areas 5. Brainstorm new solutions 6. Make ideas real (i.e., prototyping) 7. Gather feedback
Deliver	In the Deliver phase, you will begin to realize your solutions through rapid revenue and cost modeling, capability assessment, and implementation planning. This will help you launch new solutions into the world.	1. Develop a sustainable revenue model 2. Identify capabilities required for delivering solutions 3. Plan a pipeline of solutions 4. Create an implementation timeline 5. Plan mini-pilots and iterations 6. Create a learning plan (e.g., track indicators, evaluate outcomes, etc.)

Source: Based on IDEO (2011).

1.9 The Difficulty of Environmental Valuation

Environmental sustainability challenges traditional economic approaches in yet another way, which is how we monetarily value and account for the impacts of our actions on natural resources and a clean environment. We see this with the emergence of the ecosystem services concept (see Chapter 7) as well as sustainable development's "three pillars" of society–environment–economy.

Several important changes in economic culture are taking place throughout the world today in order to accommodate the growing emphasis on **social and environmental accounting** by business enterprises. Generally, this term refers to a formal assessment of how business and economic activities are benefiting people and the environment. In business, two terms are used somewhat synonymously to reflect this notion: **corporate social responsibility** and **triple bottom line**.

The British visionary John Elkington has been an advocate of corporate social responsibility for nearly three decades; he also coined the term *triple bottom line*. Both notions stress the idea that business accountability to investors and society involves more than just the traditional financial reporting of profits and losses. Instead, firms also build (or destroy) social and environmental capital, which they can and should account for as standard practice through their corporate reporting systems. In this manner, businesses can more realistically reflect how their operations holistically affect social, environmental, and economic conditions. Triple bottom line thus represents business outcomes in which "people, planet, and profit" benefit simultaneously.

Social and environmental accounting requires new methods, performance measures, and metrics. The ecological footprint is one such indicator, as is the United Nations' *System of Integrated Environmental and Economic Accounting* for a country's macroeconomy. Most initiatives to create social and environmental accounting techniques are global in nature. Businesses around the world—especially transnational corporations—should ideally use common terminology and methods for assessing their social and environmental business practices and outcomes. Examples include the Global Reporting Initiative's efforts to foster standardized sustainability reporting for businesses, and the ISO 14000 and ISO 26000 standards from the **International Organization for Standardization (ISO)**. As yet, however, there are no standard, universally adopted protocols.

As with social and environmental accounting, valuation of ecosystem services challenges traditional economic and financial concepts and methods. Because most ecosystem services are not directly bought and sold, no market and no price exist for them. Assigning economic value becomes a complex effort and is based on methods of **nonmarket valuation** (assigning monetary value that is not based on the price of actually buying and selling the good or service in question). With individual services, such as water purification, valuation is relatively straightforward and is estimated from the cost of operating water treatment and supply facilities and/or the supply of potable water. The process becomes more difficult once we allow for the complexity of ecological systems, such as the role of healthy forests in preventing soil erosion or water contamination. Monetary valuation of ecosystems is not a venal process of "putting a price tag on nature," but, rather, a way of illustrating the opportunity costs associated with exploiting nature in a manner that jeopardizes more easily measurable, conventional, and long-term profits.

1.10 Summary

Only in the recent history of the Earth have there been enough humans present to significantly affect natural ecosystems. With the advances in modern energy use and technology, both the population and the impacts associated with human population have increased. Based on the global environmental footprint, humans have exceeded the biocapacity of the Earth, and the Earth's natural resources may be significantly reduced for future generations. A focus on defining scientifically measurable sustainable product design, process design, and development will be necessary if future generations are to achieve the same standards of living as people currently living in countries with a High Human Development ranking.

The science of sustainability is complex and requires analysis of systems with several dimensions. Science and engineering use mathematical tools to measure various economic, environmental, and social indicators and metrics. Because sustainability is multidimensional, consisting of economic, environmental, and social components, various indexes have been created to help the public and decision makers compare and contrast the complex issues related to sustainable design and development. In order to properly understand how to develop, use, and interpret the sustainability indexes, it will be helpful to have a good understanding of the foundational principles and measures used to develop these indexes.

Chapters 2 through 5 cover the fundamental science and engineering principles used to measure impacts related to sustainability. Chapters 6 and 7 describe additional measures used to describe how natural resources, the climate, and anthropogenic energy development impact an individual's environmental footprint and the Earth's biocapacity. Chapters 8 and 9 discuss how engineers and scientists can use standardized sustainability indexing tools, such as Life Cycle Analysis (LCA) to compare and contrast alternative designs. Chapters 11 and 12 describe the current method used to improve sustainability through low-impact development and the use of economic tools to illustrate the costs and benefits of introducing sustainable design principles in process design and infrastructure development.

In the next chapter, we will begin by developing the scientific foundational tools to describe individual sustainability indicators and add layers of sustainability metrics so that proper sustainability indexes may be applied and interpreted with the help of the tools described throughout the subsequent chapters.

References

Bielefeldt, A. R., Kowalski, K., Schilling, C., Schreier, S., Kohler, A., and Summers, R. S. (2010). "Removal of virus to protozoan sized particles in point-of-use ceramic water filters." *Water Research* 44:1482–1488.

Biodiversity Indicators Partnership. (2010). *Guidance for national biodiversity indicator development and use.* UNEP World Conservation Monitoring Centre, Cambridge, UK.

Bishop, P. (1999). *Pollution Prevention: Fundamentals and Practice.* New York: McGraw-Hill.

Brown, J., and Sobsey, M. D. (2010). "Microbiological effectiveness of locally produced ceramic filters for drinking water treatment in Cambodia." *Journal of Water and Health* 8(1):1–10.

Carson, R. (1962). *Silent Spring.* Houghton Mifflin, New York.

CIA World FactBook (2006–2007). Retrieved October 29, 2006, March 24, 2007, at www .cia.gov/cia/publications/factbook/geos/rw.html.

Clasen, T., Beown, J., Suntura, O., and Collin, S. (2004). "Safe household water treatment and storage using ceramic drip filters: A randomized controlled trial in Bolivia." *Water Science and Technology* 50(1):111–115.

Collis, J. S. (1954). *The Triumph of the Tree.* New York: Viking.

Conroy R. M., Meegan M. E., Joyce T. M., McGuigan K. G., and Barnes, J. (2001) "Use of solar disinfection protects children under 6 years from cholera." *Archives of Diseases in Childhood* 85:293–295.

Conteh, A. (2003). *Culture and the Transfer of Technology: Field Guide to Appropriate Technology.* B. Hazeltine and C. Bull (Eds.). New York: Academic Press.

Culp, G. L., and Culp, R. L. (1974). *New Concepts in Water Purification.* New York: Litton Educational.

DiSalvo, Carl, Clement, A., et al. (2012). "Communities: Participatory design for, with, and by communities. *Routledge International Handbook of Participatory Design.*" Jesper Simonsen and T. Robertson. New York: Routledge.

Duke, W., Nordin, R., Baker, D. and Mazumder, A. (2006). "The use and performance of BioSand filters in the Artibonit Valley of Haiti: A field study of 107 households." *Rural Remote Health* 6(3):570.

Ehrlich, P. R., and Holdren, J. P. (2009, October). "Impact of population growth." *Science.* 171(3977):1212–1217.

EIA. (2009). *International Energy Outlook 2009.* Energy Information Administration, Office of Integrated Analysis and Forecasting, U.S. Department of Energy. Washington, DC.

Ewing B., A. Reed, A. Galli, J. Kitzes, and Wackernagel, M. (2010). *Calculation Methodology for the National Footprint Accounts*, 2010 ed. Oakland, CA: Global Footprint Network.

Federal Register. (2009). Executive Order 13514. Federal Leadership in Environmental, Energy, and Economic Performance onto the Agency's Green Purchasing Plan. Washington, D.C. 74(194):52117–52127.

Fiskel, J., Eason, T., and Frederickson, H. (2012). A Framework for Sustainability Indicators at EPA. National Risk Management Research Laboratory. Office of Research and Development, U.S. Environmental Protection Agency. EPA/600/R/12/687.

Forsyth, T. (2005). *Encyclopedia of International Development.* New York: Routledge.

Grooten, M. (2012). *Living Planet Report* 2012: "Biodiversity, biocapacity and better choices." Oakland, CA: Global Footprint Network.

Hardin, G. J. (1968). "The tragedy of the commons: The population problem has no technical solution; it requires a fundamental extension in morality." *Science* 162(859):8.

Harvey, P. A., and Reed, R. A. (2007). "Community-managed water supplies in Africa: Sustainable or dispensable?" *Community Development Journal* 42(3):365–378.

Hazeltine, B., and Bull, C. *Field Guide to Appropriate Technology,* 2003.

Henriques, J. J., and Louis, G. E. (2011). "A decision model for selecting sustainable drinking water supply and greywater reuse systems for developing communities with a case study in Cimahi, Indonesia." *Journal of Environmental Management* 92(1):222.

Hubbert, M. K., (1956). Nuclear energy and the fossil fuels. American Petroleum Institute, Shell Development Co., San Antonio, TX.

IDEO. (2011a). *Design Thinking for Educators.* v.1.

IDEO. (2011b). Human centered design: Toolkit. IDEO and Bill and Melinda Gates Foundation.

IEA. (2011). Energy for all: financing access for the poor. Special early excerpt of the World Energy Outlook 2011. International Energy Agency (IEA) and OECD.

Simonsen, J., and Robertson, T. (2012). *Routledge International Handbook of Participatory Design.* Preface. New York: Routledge.

Keenan, M. (2002). *From Stockholm to Johannesburg: An Historical Overview of the Concern of the Holy See for the Environment.* Vatican City.

Kehoe, S. C., Joyce T. M., Ibrahim P., Gillespie J. B., Shahar R. A., and McGuigan K. G. (2001). "Effect of agitation, turbidity, aluminum foil reflectors and volume on inactivation efficiency of batch-process solar disinfectors." *Water Research* 35(4): 1061–1065.

Koebel, C. Theodore, Papadakis, Maria, Hudson, Ed, and Cavell, Marilyn. (2003). "The Diffusion of Innovation in the Residential Building Industry." Prepared for U.S. Department of Housing and Urban Development, Office of Policy Development and Research; Center for Housing Research, Virginia Polytechnic Institute and State University, Blacksburg, VA.

Kremer, Michael. (1993). "Population growth and technological change: One million B.C. to 1990." *The Quarterly Journal of Economics* 108(3):681–716.

Leopold, A. (1949). *A Sand County Almanac and Sketches Here and There.* New York: Oxford University Press.

Makuch, Karen E., and Pereira, R. (2012). *Environmental and Energy Law.* Chichester, UK: John Wiley.& Sons, Inc.

Malthus, T. (1888), Essay on the Principle of Population. J. Johnson, London.

Mania S. K., Kanjura R., Singha I.S.B., and Reed R. H. (2006). "Comparative effectiveness of solar disinfection using small-scale batch reactors with reflective, absorptive and transmissive rear surfaces." *Water Research* 40(4):721–727.

Measows D. H. et al. (1972) *The Limits to Growth*, First Edition. Universe Books.

Meierhofer, R., and Wegelin, M. (2002, October). "Solar water disinfection—A guide for the application of SODIS." Swiss Federal Institute of Environmental Science and Technology (EAWAG) Department of Water and Sanitation in Developing Countries (SANDEC).

Naess, A. (1989). *Ecology, Community and Lifestyle: Outline of an Ecosphere.* Cambridge University Press, UK. p166.

National Research Council. (2010). *Advancing the Science of Climate Change, America's Climate Choices: Panel on Advancing the Science of Climate Change.* National Research Council. National Academies Press. Washington, D.C.

National Society of Professional Engineers (NSPE). (2007). *Code of Ethics for Engineers.* Alexandria , VA. Publication #1102.

Reinhardt, E. (2006). "Fuel for Life"—Household Energy and Health. *UN Chronicle* 43(2).

Rwanda Ministry of Infrastructure. (2006). Kigali Conceptual Master Plan. Existing Conditions Analysis by the Conceptual Master Plan Team: OZ Architecture, EDAW, Tetra Tech, Sypher, ERA, Geopmaps, Engineers Without Borders—USA and Water for People.

Schumacher, E. F. (1973). *Small Is Beautiful: Economics as if People Mattered.* Blond & Briggs, London.

Snow, J. (1855). *On the Mode of Communication of Cholera.* John Churchill, London.

Sommer, B., Mariño, A., Solarte, Y., Salas, M. L., Dierolf, C., Valiente, C., Mora, D., Rechsteiner, R., Setter, P., Wirojanagud, W., Ajarmeh, H., Al-Hassan, A., and Wegelin M. (1997). "SODIS— an emerging water treatment process." *Journal of Water Supply Research and Technology, Aqua* 46(3):127–137.

Stauber, C., Elliott, M., Koksal, F., Ortiz, G., DiGiano, F., and Sobsey, M. (2006). "Characterization of the biosand filter for E. coli reductions from household drinking water under controlled laboratory and field use conditions." *Water Science and Technology* 54(3):17.

Striebig, B., Atwood, S., Johnson, B., Lemkau, B., Shamrell, J., Spuler, P., Stanek, K., Vernon, A., and Young., J. 2007. "Activated carbon amended ceramic drinking water filters for Benin." *Journal of Engineering for Sustainable Development* 2(1):1–12.

Taylor, P. (1981). "The ethics of respect for nature." *Environmental Ethics.* 3(3):197–218.

UKCIP. (2012). United Kingdom Climate Impacts Program. Tools. Retrieved October 31, 2012, at www.ukcip.org.uk/tools.

UN. (1992). Agenda 21. United Nations Conference on Environment and Development. Rio de Janeiro, Brazil, June 3 to 14, 1992.

UN. (1999). *The World at Six Billion*. Population Division. Department of Economic and Social Affairs. United Nations Secretariat. ESA/WP.154.

UN. (2000). Millennium Declaration. UN A/Res/55/2.

UN. (2008). Trends in Sustainable Development: Agriculture, rural development, land desertification and drought. United Nations Department of Economic and Social Affairs Division for Sustainable Development.

UN. (2011a). *Human Development Report* 2011. "Sustainability and Equity: A Better Future for All." United Nations Development Programme.

UN. (2011b). Report of the Special Rapporteur on the promotion and protection of the right to freedom of opinion and expression, Frank La Rue. Human Rights Council, Seventeenth session, Agenda item 3, Promotion and protection of all human rights, civil, political, economic, social and cultural rights, including the right to development, United Nations General Assembly.

UN. (2012a) Population Matters for Sustainable Development. The United Nations Population Fund.

UN. (2012b) *The Millennium Development Goals Report* 2012.

UN. (2012c). Glossary of climate change acronyms. Retrieved October 31, 2012, at http://unfccc.int/essential_background/glossary/items/3666.php.

UNDP. (2009). Human Development Report 2009: Overcoming barriers—human mobility and development.

UNEP-WCMC (2011) Developing ecosystem service indicators: Experiences and lessons learned from sub-global assessments and other initiatives. Secretariat of the Convention on Biological Diversity, Montreal, Canada. Technical Series No. 58, 118p.

UNICEF and WHO. (2012). Progress on Drinking Water and Sanitation: 2012 Update, UNICEF and World Health Organization (WHO).

U.S. Census Bureau Data: The World Population Clock. www.census.gov/popclock/

U.S. Energy Information Administration. (2012). Annual Energy Review 2011. DOE/EIA-0384(2011), September 2012. www.eia.gov/aer.

U.S. EPA. (2013) United States Environmental Protection Agency: Sustainability. Retrieved April 8, 2013, at www.epa.gov/sustainability/index.htm.

van Halem, D., van der Laan, H., Heijman, S.G.J., van Dijk, J. C., and Amy, G. L. (2009). "Assessing the sustainability of the silver-impregnated ceramic pot filter for low-cost household drinking water treatment." *Physics and Chemistry of the Earth*, Parts A/B/C, 34(1–2):36–42.

Vesilind, P. A., and Gunn, A. S. (1998). *Engineering, Ethics, and the Environment*. New York: Cambridge University Press.

Vessilind, P. A. et. al. 2010. *Introduction to Environmental Engineering*, Third Edition. Cengage Learning, Stamford CT. page 71.

WCED. (1987). *Our Common Future*. World Commission on Environment and Development. New York: Oxford University Press.

World Health Organization/UNICEF Joint Monitoring Programme for Water Supply and Sanitation. (2005). *Water for Life: Making It Happen*.

Key Concepts

Demographers	Peri-urban areas
Human Development Index	Bruntland definition
Sustainable Development	Inter-generational equity

Social justice
Conference on Environment and
    Development
Millennium Development Goals
Appropriate technology
Sustainability
Morals
Ethics
Categorical imperative
Land ethic
Biocentric outlook
Deep ecology
Environmental ethic
Sustainable design

Carrying capacity
Exponential growth
Measure
Metric
Indicator
Sustainability index
Ecological footprint
Capacity factor analysis
Participatory design
Community-based design
Social and environmental accounting
International Organization for
    Standards (ISO)

Active Learning Exercises

ACTIVE LEARNING EXERCISE 1.1: Expectations

The next time you drink from a water fountain or buy a bottle of water, what are your expectations about the safety of the water? Who if anyone, makes a profit from the sale of tap water? Who, exactly, would be responsible for fulfilling these expectations?

ACTIVE LEARNING EXERCISE 1.2: Obligations

Pretend you have purchased bottled water from a vending machine, and answer the following questions:

- Where did the water originate?
- Where did the plastic materials originate?
- How much did the bottled water cost compared to tap water? Who makes a profit, if anyone, from the sale of bottled water?
- Does the bottled water present less risk than drinking tap water? Explain the factors you have considered.
- What are the limitations of providing bottled water to meet the need for drinking water in Rwanda? How does cost, waste production, and social justice factor into this equation?

ACTIVE LEARNING EXERCISE 1.3: Political perspective

Recently, a gubernatorial candidate in the state of New Hampshire ran on a single issue—to stop the disposal of wastewater sludge on land in New Hampshire. Wastewater sludge is composed primarily of the microorganisms that are used in wastewater treatment plants to remove pollutants and water. Suppose you had the opportunity to ask him three questions during a public panel discussion. What would they be, and how would they relate to sustainable design?

ACTIVE LEARNING EXERCISE 1.4: Ethics

In 1948 an air pollution event resulted in the deaths of 20 people and thousands of pets as people were order to evacuate their homes and leave their pets behind. The fact that pets suffered greatly in the Donora events has been almost completely ignored by all accounts. Why? Is this an ethically acceptable behavior? Why are we mostly concerned about only our own species?

Problems

1-1   Genetically, modern humans appeared on Earth about 200,000 years ago, and biologically and behaviorally, modern humans appeared about 70,000 years ago. The number of people and their effects on the planet were negligible, or as Douglas Adams says, "as near nothing as makes no odds," for most of the history of the planet. When did the planet's population reach 1 billion people? Assuming that the population has grown exponentially since that time, what was the time interval required to increase by 1 billion people—for up to 7 billion people, which was the approximate global population in 2012?

1-2   List the three dimensions and four categories used to calculate the Human Development Index (HDI) for a country.

1-3   For the countries listed in the accompanying table, calculate
a. The Life Expectancy Index
b. The Educational Index
c. The Income Index
d. The Human Development Index

COUNTRY AND 2011 DATA	LIFE EXPECTANCY AT BIRTH (YEARS)	EXPECTED YEARS OF SCHOOLING	MEAN YEARS OF SCHOOLING	GNI PER CAPITA IN PPP TERMS (CONSTANT 2005 INTERNATIONAL $)
Australia	81.9	18.0	12.0	34,431
China	73.5	11.6	7.5	7,476
Ireland	80.6	18.0	11.6	29,322
Kenya	57.1	11.0	7.0	1,492
South Africa	52.8	13.1	8.5	9,469

Source: Based on UN. (2011a) Human Development Report 2011 Sustainability and Equity: A Better Future for All. United Nations Development Programme.

1-4   For the countries listed in the accompanying table, calculate
a. The Life Expectancy Index
b. The Educational Index
c. The Income Index
d. The Human Development Index
e. Determine the United Nations' development category for the country

COUNTRY AND 2011 DATA	LIFE EXPECTANCY AT BIRTH (YEARS)	EXPECTED YEARS OF SCHOOLING	MEAN YEARS OF SCHOOLING	GNI PER CAPITA IN PPP TERMS (CONSTANT 2005 INTERNATIONAL $)
Canada	81.0	16.0	12.1	35,166
Japan	83.4	15.1	11.6	32,295
Mexico	77.0	13.9	8.5	13,245
Nigeria	51.9	8.9	5.0	2,069
United Kingdom	80.2	16.1	9.3	33,296

Source: Based on UN. (2011a) Human Development Report 2011 Sustainability and Equity: A Better Future for All. United Nations Development Programme.

1-5 For the country listed in the table, calculate
   a. The Life Expectancy Index
   b. The Educational Index
   c. The Income Index
   d. The Human Development Index

COUNTRY AND 2011 DATA	LIFE EXPECTANCY AT BIRTH (YEARS)	EXPECTED YEARS OF SCHOOLING	MEAN YEARS OF SCHOOLING	GNI PER CAPITA IN PPP TERMS (CONSTANT 2005 INTERNATIONAL $)
Benin	56.1	3.3	9.2	1,364
Costa Rica	79.3	11.7	8.3	10,497
India	65.4	10.3	4.4	3,468
Malta	79.6	14.4	9.9	21,460
New Zealand	80.7	18.0	12.5	23,737
Rwanda	55.4	11.1	3.3	1,133

Source: Based on UN. (2011a) Human Development Report 2011 Sustainability and Equity: A Better Future for All. United Nations Development Programme.

1-6 What are the HDI categories defined by the United Nations? For each of the four categories, describe what you think people may drink, eat, and wear, the type of homes they may live in, the types of school they are likely to attend, and the type of transportation they are most likely to use.

1-7 Describe, using your own words, the purpose of human development.

1-8 Define the following terms.
   a. Urban
   b. Suburban
   c. Peri-urban
   d. Rural

1-9   What is the "Brundtland definition" of sustainable development?

1-10   How would you describe sustainability to a 12-year-old student at your local school?

1-11   Look up and describe one of the formative written works related to sustainable development. Research this work more and summarize the main premise of the work in a short 500-word essay.

1-12   Create a sustainability indicator (similar to the HDI or the ecological footprint.) What would actions, processes or things would you measure for your indicator. How would you collect and find the data for your indicator? What are the advantages and disadvantages of your proposed indicator? Note: You might find helpful information on the Unite Nation Development Program (UNDP) web page, the UNCEF web page, the World Bank web page, and the U.S. Environmental Protection Agency web page.

1-13   What characteristics define unsustainable development? Make a table of characteristics that might negatively affect development. Mark which of these characteristics are important in the following.
   a. Very High Human Development countries
   b. Low Human Development countries
   c. Both Very High and Low Development countries

1-14   Create a schematic or cartoon that relates the following trends
   a. Human population
   b. Resource consumption
   c. Educational resources
   d. Economic resources

1-15   List and describe in your own words the eight United Nations Millennium Development Goals. Specifically describe the economic, environmental, social, and technical challenges associated with meeting each of the goals within the next five years.

1-16   Do you believe that sustainable products are morally superior to nonsustainable products? If so, what does this belief imply about the developed world's largely consumer-based economic system of retail merchandise? How are technology and moral convictions woven into the fabric of our definitions of sustainable design?

1-17   If you had to live on $2 per day, how would you meet your basic needs for food, shelter, water, sanitation, and other needs?
   a. Determine from recent utility bills how much you spend per day on
      i)   Water
      ii)   Sanitation (sewer or wastewater company bill)
      iii) Garbage collection services
      iv)   Energy
      v)   Heating/Cooling
      vi)   Communications (phone, cell phone, Internet, etc)
      vii) Food
      viii) Shelter (based on rent or mortgage payment)
      ix)   Entertainment
   b. Determine your total daily expenditure.
   c. If you were to pay 25% of your income on taxes, how much would your income need to be each year to pay for your daily expenses?
   d. With what level of the Human Development Index would this income be associated?

1-18   Imagine you are part of a company designing a school for a low-income country (based on the countries' HDI). Use online resources to help address the following questions for the design of the proposed school.

     a. Are parts readily available, either locally or nationally, if a component were to fail?

     b. Are there individuals who have the necessary skill or technical training to repair the component or system if it were to fail?

     c. Would members of the household readily understand how to use this system?

     d. What is the local availability of required infrastructures, such as electric power?

1-19   Compare and contrast the definition of sustainability as defined by the *Merriam-Webster dictionary*, the United States Environmental Protection Agency, and the Bruntland Commission's *Our Common Future*.

1-20   Describe how concepts of sustainability might be applied to the fundamental canons of the National Society of Professional Engineers.

1-21   It took about 12 years, between 2000 and 2012, for the world population to increase by 1 billion people. In contrast, the world's population was estimated to be 300 million people in the year AD 0. By the year 1500, the world's population was estimated to be 500 million.

     a. Assuming exponential growth, what was the percentage of world' population growth rate, between 2000 and 2012?

     b. Assuming exponential growth, what was the percentage of world' population growth rate, between AD 0 and 1500?

     c. How many times greater was the population growth rate in the twentieth century compared to that between AD 0 and 1500?

1-22   The world population in 1850 has been estimated at about 1 billion. The world population reached 4 billion in 1975. What was the percentage of the exponential growth rate during this time?

1-23   Tuition at a university rose from $1,500/year in 1962 to $25,000/year in 2010.

     a. What exponential growth rate characterized that period of time?

     b. If that rate of growth were to continue until 2050 (when your children might be paying tuition), what would the tuition be?

1-24   In 1999, RSU tuition was $1,963 per semester. In 2009, RSU tuition was $3,622 per semester. This increase is represented by an exponential growth rate of 6.1%; if tuition rates increase exponentially, what value is closest to the in-state semester tuition cost predicted in 2035?

1-25   In 2007, the world's population was estimated to be 6.7 billion. The UN forecasts the population will begin to level off at 9.2 billion in 2050. What will be the population growth rate (in percent) over this time period?

1-26   It has been estimated that 139.2×10^6 m^2 of rainforest is destroyed each day. Assume that the initial area of tropical rainforest is 20×10^{12} m^2.

     a. What is the exponential rate of rainforest destruction in units of 1/days?

     b. If there were 24.5×10^{12} m^2 of tropical rainforest on Earth in 1975, how much tropical rainforest would be left on Earth in 2015 if the exponential rate of destruction determined in part (a) stayed constant over this time interval?

     c. If tropical rainforests remove 0.83 kg(C)/m^2-year from the atmosphere, how much less carbon [kg(C)] would be removed in 2025 compared to that removed in 1975?

1-27 The world's population 10,000 years ago has been estimated at about 5 million. What exponential growth rate would have resulted in the population in 1800, which is estimated at 1 billion? Had that rate continued, what would have been the world's population in 2010?

1-28 In 2007, the population of the world's 50 least-developed countries was estimated to be 0.8 billion. The UN expects the population in these countries to grow exponentially at 1.75% until 2050. What is the predicted population of the least-developed countries in 2050?

1-29 What must engineers hold paramount in their designs according to most professional ethics codes?

1-30 Describe quantitatively (use numeric values) the differences between access to improved drinking water supplies in the United States compared to countries in Africa or the Caribbean.

1-31 List the UN Millennium Development Goals and describe briefly how they might relate to access to drinking water.

1-32 Energy derived from nuclear power has grown since 1970 according to the data from the U.S. Energy Information Administration that is summarized below.
  a. Plot the energy production from nuclear power between 1970 and 1990.
  b. Find the best curve fit for the plot (use either a linear fit, polynomial fit, or power function.) What is the equation for this best-fit curve?
  c. Take the mathematical or graphical derivative of the function from the plot and graph the rate of change (first derivative) of energy derived from nuclear power between 1970 and 1990 in Excel or a similar spreadsheet program.

YEAR	1958	1960	1962	1964	1966	1968	1970	1972	1974	1976	1978	1980	1982	1984
Billions Kilowatt-hours	0.2	0.5	2.3	3.3	5.5	12.5	21.8	54.1	114.0	191.1	276.4	251.1	282.8	327.6

Year	1986	1988	1990	1992	1994	1996	1998	2000	2002	2004	2006	2008	2010
Billions Kilowatt-hours	414.0	527.0	576.9	618.8	640.4	674.7	673.7	753.9	780.1	788.5	787.2	806.2	807.0

Source: Based on U.S. Energy Information Administration. (2012) Annual Energy Review 2011. DOE/EIA-0384(2011), September 2012. www.eia.gov/aer.

1-33 Thomas Malthus described a situation in which population could overcome the available supply of natural resources near the year 1800. Over 200 years later, scientists, policymakers, and demographers fear the same situation may be occurring—that we may exceed the biocapacity of the planet. Malthus's original arguments have been reworked in modern writings such as *The Tragedy of the Commons* and *The Population Bomb*. What role do scientists and engineers play in the debate about the likelihood of humans, based on our current lifestyle, exceeding the planet's biocapacity? Base your essay on economic, environmental, social, and technical parameters.

1-34 Describe which of the grand challenges of engineering most interests you. Frame the problems that must be overcome associated with the challenge you've selected in terms of the variables in the IPAT equation.

1-35 Sustainable development is extremely difficult, since the environmental footprint of a nation generally increases with increasing development. Use the IPAT equation and determine whether each variable is likely to increase, decrease, or remain unchanged if the HDI of a country increases. What must the response of each variable in the IPAT equation be (increase, decrease, or no change) if development is to be truly sustainable?

1-36 Describe how social and environmental accounting might be a useful tool in developing policies to promote sustainable development.

CHAPTER 2

Analyzing Sustainability Using Engineering Science

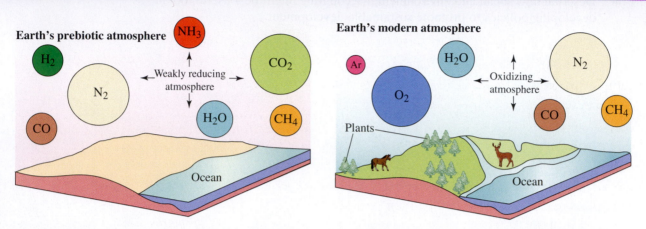

FIGURE 2.1 The Earth's surface undergoes constant change. Carbon dioxide concentrations were quite high until plant life developed and transformed the carbon dioxide in the atmosphere into oxygen as part of the photosynthetic process. Later on, animals would adapt to the increased oxygen concentration in the atmosphere and use the oxygen for energy and growth during respiration. The physical and chemical transformations in the Earth's surface are imperative for life as we know it.

Source: Re-draw.

I consider nature a vast chemical laboratory in which all kinds of composition and decomposition are formed.

—ANTOINE LAVOISIER

GOALS

The educational goals of this chapter are to develop an understanding of how the basic sciences, biology, chemistry, mathematics, and physics may be applied to develop quantitative indicators of sustainable design and development described in Chapter 1. Fundamental units of measure are defined and applied to solve engineering problems. This chapter examines the characteristics and units of measure used to define water quality. Principles of environmental chemistry are applied to understand the fate of acids and bases in the natural environment. Chemical principles that influence the fate and transport of chemicals in the environment are also examined. Inorganic and organic chemical compounds and their relationship to human health and societal uses are explored and considered in respect to their associated risk, along with the ethical responsibilities of engineers and scientists, to manage and minimize the risk factors associated with the use of modern chemicals in industry. These fundamental principles are used repeatedly in Chapters 3 through 6 to predict environmental impacts and also in Chapters 7 through 11 to compare and evaluate the sustainability of green design processes.

OBJECTIVES

At the conclusion of this chapter, you should be able to:

- Define, convert, and determine appropriate uses for conventional units of measure in environmental systems.

- Define the laws of chemical equilibrium.

- Calculate the ionic strength of a solution.

- Determine the hardness of water.

- Estimate activity coefficients and use activity coefficients to distinguish between ideal and nonideal solutions.

- Solve acid–base equilibrium problems both algebraically and graphically.

- Calculate chemical partitioning in the aquatic environment.

2.1 Introduction

Chemical spill on local highway—get the full story on this late-breaking news!

Have you heard this headline announced? If so, what is being inferred about "chemicals"? It is true that there have been many historical examples of chemicals that have had devastating environmental impacts. These include mercury pollution in Japan, arsenic contamination in Bangladesh, toxic waste contamination in Love Canal, New York, and the Chernobyl nuclear disaster in the former Soviet Union. Chemicals are also used to provide clean energy, disinfect and clean drinking water, purify food, manufacture medicines, and provide many other benefits associated with today's highly developed lifestyle. Chemicals support and comprise the molecules of life and the environment in which we live.

In order to get a quantitative understanding of what is happening in a system, we must also have knowledge of the units associated with the values provided. The topic of sustainability covers a very large breadth of information and spans multiple traditional fields, including chemistry, chemical engineering, energy engineering, environmental engineering and science, industrial engineering, and policy analysis. All of these fields have slightly different conventions in units and symbols, so if you are working in the field of sustainable engineering, you must become fluent in all these areas of study and also familiarize yourself with the units and terminology associated with the field. Both English standard units and the International System of Units (SI) are widely used in this field.

In analyzing various situations and responses, for example, we will discuss global energy scales and energy use, that deal with numbers on the order of terawatts of energy. We will also investigate the effects of substances in the environment at concentrations less than 1 microgram per liter of fluid. You will need to become conversant in the standard prefixes that describe the standard scales of measure in our subsequent analysis.

Chemicals in the environment affect biodiversity, human health, and the Earth's climate. The fundamental sciences of biology, chemistry, mathematics, and physics provide a basis for analyzing how we achieve sustainable designs and development. Mass and energy balances are helpful in tracking chemicals as they move from one repository to another. However, we must study the applications and principles of chemistry more closely to appreciate how these chemicals are transformed in the environment and also transform the Earth's environment.

For example, carbon dioxide levels in the atmosphere have increased by more than 30% since the start of the Industrial Revolution. This increase has changed our atmosphere and the energy balance that affects how our climate responds to the sun's radiation. The carbon dioxide levels have also had a profound influence on ocean chemistry. The oceans are becoming much more acidic due to the increased carbon dioxide in the atmosphere. The increasing acidity inhibits the production of corals and calcifying phytoplankton and zooplankton, which in turn negatively affects the marine food web. Quantitative evaluations of chemical reactions in the environment, or compositions and decompositions in nature as Monsieur Lavoisier stated, are required to understand how carbon dioxide added to the atmosphere will lead to changes in ocean chemistry and subsequent marine life.

2.2 Elemental Analysis

The periodic table shown in Figure 2.2 lists all the elements found on Earth. While this list is imposing, the periodic table groups the elements in a way that greatly helps us understand how those chemicals react and are stored in the

FIGURE 2.2 Periodic table of elements (also on front cover).

environment. Furthermore, of the 118 elements on the periodic table, there are a few elements that you'll get to know very well. All chemicals are important, but only a few of them are regularly encountered and significantly impact the environment.

Most life forms on the planet are composed primarily of five elements: carbon, hydrogen, oxygen, nitrogen, and phosphorus. Carbon is the most basic building block of life. It is the primary component of cell mass and also of fossil fuels. Carbon bonds strongly with hydrogen (and other elements) by sharing electrons to form relatively "small" molecules like methane (CH_4). Carbon is also a primary component of very large molecules that include oxygen, nitrogen, phosphorus, and chains of carbon molecules, like our DNA. Carbon also bonds strongly with oxygen, forming carbon dioxide (CO_2) in the gas phase. Carbon and oxygen compounds, called carbonates, are an important part of natural waters (which we will discuss in great detail in Chapters 4 and 6).

When we examine the elements in the Earth's atmosphere, we see in Table 2.1 that the atmosphere is composed of a mixture of gases. Nitrogen, in the form of N_2, and oxygen, in the form of O_2, are the two most prominent gases in our atmosphere. Argon, helium, neon, and krypton, with their place in the last column of the periodic chart, are "noble gases" and relatively inert. Although these noble gases are present in the atmosphere in relatively large amounts, they do not participate in environmentally or industrially common reactions.

TABLE 2.1 Composition of the dry standard atmosphere

GAS	FORMULA	VOLUME FRACTION (BY NUMBER OF MOLES)
Nitrogen	N_2	0.7808
Oxygen	O_2	0.2095
Argon	AR	0.0093
		VOLUME FRACTION (IN PARTS PER MILLION)
Carbon dioxide	CO_2	388 ppm
Neon	Ne	18 ppm
Helium	He	5.2 ppm
Methane	CH_4	1.5 ppm
Krypton	Kr	1.1 ppm
Hydrogen	H_2	0.5 ppm

Source: Based on Hart, J. (1988) Consider a Spherical Cow: A Course in Environmental Problem Solving. University Books.

While we are all aware of the importance of oxygen in the atmosphere, most of the mass of oxygen is present in the oceans, in water, in biomass, and in the Earth's crust, as shown in Table 2.2. Silica, aluminum, iron, calcium, magnesium, sodium, and potassium are all significant components of the Earth's crust and important to biological life, food production, and industry. Most of the Earth is blanketed in water, which contains more of the Earth's hydrogen, but we cannot overlook the importance of biomass as a chemical source of hydrogen as well.

TABLE 2.2 Major elements on the Earth's surface

ELEMENT	EARTH'S CRUST (PPM BY WEIGHT)	SEAWATER (PPM)	BIOMASS (PPM)
Oxygen, O	456,000	857,000	630,000
Silicon, Si	273,000	3	15,000
Aluminum, Al	83,600	0.01	500
Iron, Fe	62,200	0.01	1,000
Calcium, Ca	46,600	400	40,000
Magnesium, Mg	27,640	1,350	4,000
Sodium, Na	22,700	10,500	2,000
Potassium, K	18,400	380	20,000
Titanium, Ti	6,320	0.001	100
Hydrogen, H	1,520	108,000	80,000

Source: Based on Hart, J. (1988) Consider a Spherical Cow: A Course in Environmental Problem Solving. University Books.

Combinations of elements form molecules. The form that the molecules take in the natural environment usually requires the least energy for which the elements can exist. In the gas phase, less energy is required for two molecules of oxygen to share electrons than for an individual molecule of oxygen to exist alone. Most of the oxygen in the gas phase exists as O_2, although other forms, like O_3 or ozone, may also exist depending on surrounding environmental conditions. Chemical reactions, when they do occur, are balanced based on the number of atoms it takes to convert from one molecule to another. For example, when we add energy and ignite methane (CH_4) in the presence of oxygen (O_2), the resulting products are carbon dioxide, water and more energy:

$$\text{Energy for ignition} + CH_4 + O_2 \rightarrow CO_2 + H_2O + \text{more energy for heat or work} \quad (2.1)$$

Equation (2.1) is not balanced. The same number of carbon atoms occurs on each side of the equation, so the carbon atoms are "balanced" in this chemical equation. However, there are more hydrogen atoms on the right side of the equation than on the left, so hydrogen atoms must be added to the right-hand side of the equation. Similarly, the numbers of oxygen atoms also must be balanced. Two hydrogen atoms are added to the right-hand side of the equation, in the form of H_2O, so that there are four molecules of hydrogen on both sides of the equation:

$$\text{Energy for ignition} + CH_4 + O_2 \rightarrow CO_2 + 2\,H_2O + \text{more energy for heat or work} \quad (2.2)$$

The oxygen atoms must also be balanced. There are a total of four oxygen atoms on the right-hand side of the equation: two used to form the carbon dioxide molecule (CO_2) and two molecules of water (H_2O)—each of which contains one atom of oxygen. Thus, we need to add one more molecule of diatomic oxygen gas (O_2) to the left-hand side of the equation to balance all the elements:

$$CH_4 + 2\,O_2 \rightarrow CO_2 + 2\,H_2O \quad (2.3)$$

Atoms are incredibly tiny forms of matter. In practice as engineers, we don't really try to balance the exact number of atoms required to burn gasoline in an engine. We need to think on scales that are much larger than the atom. If we only needed an order of magnitude greater in scale than the atom for a meaningful mass of a chemical, we could use a dozen atoms. Even a dozen atoms is an unimaginably small amount of a substance. Instead of individual atoms or a dozen atoms, we talk in terms of moles, where one mole of atoms is equivalent to 6.02×10^{23} atoms. The mole then is just a chemical unit of measure that relates a significant mass of atoms for chemists, engineers, and scientists to use in balancing equations and performing calculations in chemistry.

$$1 \text{ mole} = 6.02 \times 10^{23} \text{ atoms} \tag{2.4}$$

We must use the molar mass of an element or molecule to find the mass of a given molecule associated with 1 mole of material.

$$\text{Mass [g]} = \text{number of moles} \times \text{molar mass [g/mole]} \tag{2.5}$$

EXAMPLE 2.1 | Atoms, Molecules, and Mass

Hydrogen (H_2) molecules and oxygen (O_2) molecules may combine to form water. If there were 24×10^{23} atoms of hydrogen (H) available to react with oxygen, how many moles of water could we produce, and how many moles of oxygen would we need to utilize all the hydrogen? How many grams of water would be produced?

The unbalanced equation looks like

$$H_2 + O_2 \rightarrow H_2O$$

Balancing the equation yields

$$2\,H_2 + O_2 \rightarrow 2H_2O$$

Four hydrogen atoms, or two hydrogen molecules (H_2), are needed to react with one molecule of oxygen gas (O_2). The reaction yields two molecules of H_2O.

The problem states that there are 24×10^{23} atoms of hydrogen as H.

$$\frac{24 \times 10^{23} \,\text{H atoms}}{2 \,\text{H atoms per molecule of } H_2} = 12 \times 10^{23} \,\text{molecules of } H_2$$

$$\frac{12 \times 10^{23} \,\text{molecules of } H_2}{6.02 \times 10^{23} \,\dfrac{\text{molecules}}{\text{mole}}} = 2.0 \,\text{moles of } H_2$$

from the balanced equation, 1.0 mole of O_2 is required to react with 2.0 moles of H_2, and this will yield 2.0 moles of H_2O.

$$2.0 \,\text{moles of } H_2 \times 18 \,\frac{\text{grams } H_2O}{\text{mole } H_2O} = 36 \,\text{grams } of\, H_2O$$

EXAMPLE 2.2 | Mass, Density, and Concentration

If you have 55.5 moles of distilled water in a 1-liter Nalgene® and bottle at standard temperature and pressure, what is the density of the fluid in g/liter and kg/m³?

$$55.5 \,\frac{\text{moles of } H_2O}{\text{liter}} \times 18 \,\frac{\text{grams}}{\text{mole}} = 1,000 \,\frac{\text{g}}{\text{L}}$$

$$1,000 \, \frac{g}{L} \times \frac{1 \, kg}{1,000 \, g} \times \frac{1,000 \, L}{m^3} = 1,000 \, \frac{kg}{m^3}$$

Most compounds found naturally in the environment are mixtures of elements and molecules. The air surrounding us is a mixture of nitrogen, oxygen, argon, carbon dioxide, and other gases, as shown in Table 2.2. Seawater contains high levels of dissolved salts, including sodium, chlorine, calcium, and magnesium. Even fresh water contains many of these same dissolved salts, but usually at lower concentrations than seawater. Both seawater and fresh water normally contain various concentrations of dissolved gases, including dissolved oxygen, carbon dioxide, and often several other dissolved gases (see Box 2.1). So, most of the water in water is

BOX 2.1 How Much Oxygen Do Fish Need to Breathe?

Mossy Creek, in Virginia, is the home to a trout fishing tourist industry. The trout species of fish (Figure 2.3) requires very clean, cool water as its habitat. Trout generally need about 8×10^{-3} *Kilograms* of dissolved oxygen gas (O_2) per cubic *meter* of water to remain healthy. How many milligrams of dissolved oxygen per liter of water are needed for the trout in Mossy Creek to stay healthy?

(NaCl) and sugar ($\sim C_6H_{12}O_6$) are very soluble in water.

$$8 \times 10^{-3} \, \frac{Kg \, of \, O_2}{1 \, m^3} \times \frac{10^6 \, mg}{kg} \times \frac{1 \, m^3}{1,000 \, L} = 8 \, \frac{mg}{L}$$

About 8 milligrams of dissolved oxygen are required in each liter of water for a healthy trout habitat. Trout populations become stressed at values below 8 mg/L.

While fish can breathe the small amount of oxygen in the water, we, as humans, cannot. The air consists of about 21% oxygen, while water only contains on the order of 10×10^{-6} parts mass of oxygen compared to the total mass of the solution. Since only a small amount of oxygen dissolves into the water, we say that oxygen, as O_2, is only slightly soluble in water. The term *solubility* is used to describe the amount of a substance that will dissolve in water. Substances like table salt

FIGURE 2.3 Brook trout require a high level of dissolved oxygen in the water for growth and reproduction. The dissolved oxygen concentration is sensitive to water temperature and the degree of mixing, so that cold, fast-moving water provides an ideal habitat for the brook trout species native to Virginia, and ponds and slower moving bodies of water do not have a high enough dissolved oxygen concentration in the water to allow this species to survive. Increasing surface temperatures and water pollution threaten to further decrease the trout habitat in temperate regions in North America and Europe.

Source: Bradley Striebig.

water, but not all of it! Or to be more precise, most of the liquid fluid we commonly call water is made up mostly of H_2O, but the fluid also contains other molecules. When we want to examine human-related (anthropogenic) impacts on the environment, we are concerned about very small quantities of mass that occur within the naturally occurring mixtures we call air, water, and soil. So, while most of the fluid in water is H_2O, it is the other chemical components that are of concern when we are trying to identify and evaluate the impacts of anthropogenic activities on a region's environment, society, and economy.

The mass of pollutants or compounds of concern in our environment may be very, very small—so small in fact, that these compounds may be much, much less than 1% of the total mass of the fluid in water. In order to avoid writing mass fractions of a pollutant in water that are very small decimal fractions, we define a unit fraction called a *part per million on a mass basis* (ppm_m) for compounds in water.

$$ppm_m = \frac{m_i}{m_{total}} \times 10^6 \qquad (2.6)$$

Notice that under conditions of standard pressure and temperature, water has a specific gravity of 1.00 or a density of 1.00 g/ml. When water comprises most of the fluid of concern, the specific gravity of the fluid is very nearly equal to 1.0, so

$$1 \; ppm_m = 1 \; mg/L \text{ for a fluid with a specific gravity} = 1.0 \; g/ml \qquad (2.7)$$

EXAMPLE 2.3   Calcium in Seawater

Calcium from dissolved limestone and other minerals is a common component of fresh water. If one liter of a water solution contains 0.100 grams of calcium, what is the concentration of calcium in the following units?

(a) Grams per liter:

$$\frac{0.100 \text{ grams}}{1 \text{ liter}} = 0.100 \; \frac{g}{L}$$

(b) Mass fraction:

From Example 2.2, we know that 1 liter of water weighs 1,000 grams; thus the mass fraction of calcium in water is

$$\frac{0.100 \text{ grams}}{1,000 \text{ grams}} = 0.000100 \; \frac{\text{grams of } Ca^{2+}}{\text{grams of water solution}}$$

(c) Mass percentage:

$$0.000100 \; \frac{\text{grams of } Ca^{2+}}{\text{grams of water solution}} \times 100\% = 0.01\% \text{ of the mass of water is } Ca^{2+}$$

(d) Part per million on a mass basis:

$$0.000100 \; \frac{\text{grams of } Ca^{2+}}{\text{grams of water solution}} \times 10^6 = 100 \; ppm_m$$

(e) Milligrams per liter:

$$\frac{0.100 \text{ grams}}{1 \text{ liter}} \times \frac{1000 \text{ milligrams}}{1 \text{gram}} = 100 \frac{\text{mg}}{\text{L}}$$

2.3   Solubility and Henry's Law Constant

Most gases are only slightly soluble in water. In dilute aqueous (or water) solutions, the concentration of a substance in the gas phase is linearly related to the concentration of that substance in the aqueous phase. The Henry's law constant (H_i) is the name given to the constant that describes the slope of this linear relationship:

$$C_i = H_i P_i \tag{2.8}$$

where

P_i = the partial pressure of the substance i in the gas. Typical units for the gas pressure are atmospheres [atm]

C_i = the molar fraction of the substance i in the aqueous solution [moles of i per liter of fluid]

H_i = Henry's law constant

TABLE 2.3 Henry's law constant, H_i, for oxygen and carbon dioxide in water in units of mol/L-atm.

TEMPERATURE (°C)	O_2	CO_2
0	0.0021812	0.076425
5	0.0019126	0.063532
10	0.0016963	0.053270
15	0.0015236	0.045463
20	0.0013840	0.039172
25	0.0012630	0.033363

Source: Based on Masters, G.M. and Ela, W. P. (2008) Introduction to Environmental Engineering and Science: Third Edition. Prentice Hall, NJ, USA.

EXAMPLE 2.4   Investing in a Trout Farm for Aquaculture

A local farmer wants to diversify his business by creating a pay-per-fish trout farm near Mossy Creek and raise trophy trout for tourists and local restaurants. The water temperature of the pond in July and August averages 25°C. Find the dissolved oxygen (DO) level in the pond water in order to determine if this is a good investment. (Assume that the total atmospheric pressure is 1 atm.)

The Henry's law values for oxygen and carbon dioxide are given in Table 2.3. The partial pressure of oxygen is given in Table 2.1:

$$P_{O_2} = 0.2095 \text{ atm}$$

Then the concentration of dissolved oxygen, C_{DO}, in the liquid phase can be determined form the Henry's law equation:

$$C_{DO} = P_{O_2} \times H_{O_2} = (0.2095 \text{ atm}) \left(0.0012630 \frac{\text{mol}}{\text{atm} - \text{L}} \right) = 0.0002646 \frac{\text{mol}}{\text{L}}$$

$$C_{DO} = 0.0002646 \frac{\text{mol}}{\text{L}} \times \frac{1,000 \text{ mmol}}{\text{mol}} \times \frac{32 \text{mg}}{\text{mmol}} = 8.467 \frac{\text{mg}}{\text{L}} \text{ of dissolved oxygen}$$

The Henry's law constant does not have standard units. In this example, H_i had units of [atm-L/mole$_i$]. However, the Henry's law constant units vary greatly from one academic field to another. Care must be taken to ensure that the units reported are correctly used in any application of this equation. The Henry's law constant varies with temperature, and these correlations can be estimated from data reported in the scientific literature. Further care must be taken, as the concentration of the substance in the fluid is nearly the same order of magnitude as the concentration of the fluid itself, the relationship becomes nonlinear, and the Henry's law value no longer can be used to model the relationship between highly concentrated gases and liquids.

2.4 The Ideal Gas Law

The units of concentration in the gas phase tend to vary slightly, but significantly, from those used in the liquid phase. It is essential to understand these differences. Fundamentally, the causes of these differences are (1) a gas is a compressible fluid and subject to deformation with small changes in pressure and (2) the density of air is much, much less than the density of water. The ideal gas law governs the relationship between the pressure, volume, mass, and temperature of a substance in the gas phase:

$$PV = nRT \tag{2.9}$$

where
P = absolute pressure (atm)
V = volume (L)
n = number of moles (moles)
T = absolute temperature (K)
R = universal gas constant (0.08206 l-atm/mole-K, or other units depending on the referenced source)

The numerical value of R is related to the units we use to define P, V, n, and T. R has a different numerical value if different units are used; for instance, another commonly value is

$$R = 8.314 \text{ (J/mol-K)} \tag{2.10}$$

As discussed previously, our atmosphere is composed of a mixture of various gases shown in Table 2.1. Each of these gases contributes to the total pressure the atmosphere exerts on the Earth. In the environmental sciences, we define standard conditions as a standard pressure of 1 atmosphere (atm), which is equivalent to 14.696 pounds per square inch (psi). The 1 atm (or 14.696 psi) is the approximate force exerted by the weight of the atmosphere on the Earth at sea level. The standard temperature is 25°C. Each component of the gas contributes to the total pressure or weight of the atmosphere. The pressure of each part of a gas that it exerts on its surrounding is called the **partial pressure** of that gas. When a system is at equilibrium, the partial pressure exerted by that gas is equal to the vapor pressure of the gas. When the partial pressure of each component of the gas is summed, the resultant pressure is the total pressure of the gas. The relationship between total pressure and the pressure of each component of the gas mixture is called Dalton's law of partial pressure:

$$P_t = \Sigma \, P_i \tag{2.11}$$

EXAMPLE 2.5 Atmospheric Pressure

The volume fractions of each of the major gases in the atmosphere are listed in Table 2.1. From this table, determine the partial pressure of nitrogen, oxygen, and argon at standard temperature and pressure (STP) in units of atm. What percentage of the total pressure of the dry atmosphere (neglecting water in the atmosphere) is made up of these three gases?

Nitrogen:	1 atm × 0.7808 = 0.7808 atm
Oxygen:	1 atm × 0.2095 = 0.2095 atm
Argon:	1 atm × 0.0093 = 0.0093 atm

Sum of the partial pressures of N, O, and Ar is 0.7808 atm + 0.2095 atm + 0.0093 atm = 0.9996 atm

$\dfrac{0.9996 \text{ atm}}{1.0000 \text{ atm}} \times 100\% = 99.96\%$ of the dry atmosphere is composed of these three gases.

For compounds in the gas phase, the pressure and volume that the substance occupies are related to the number of moles or mass of the compound. A volumetric basis is used for comparing the concentration of anthropogenic emissions and low-level, naturally occurring compounds in the atmosphere. The mole fraction of an individual substance, i, divided by the total moles of all compounds in the gas mixture is directly proportional to the volumetric fraction if pressure and temperature remain constant:

$$P_{total} V_{total} = n_{total} RT \tag{2.12}$$

$$P_i V_i = n_i RT \tag{2.13}$$

So

$$V_i / V_{total} = n_i / n_{total} = y_i \tag{2.14}$$

where y is the molar fraction of the individual component i in the gas phase. The molar gas fraction is commonly used in chemical engineering analysis where the chemical engineer is concerned with the change of a compound from one form to another, and those changes are several percentages of the total gas. However, as in aqueous solutions, the environmental sciences are usually concerned with far smaller concentrations. We will again use the concept of a part per million, but on a volume basis (ppm_v) to describe the atmospheric concentration of pollutants. It is extremely important to note that in the gas phase, parts per million is a volumetric fraction, not a mass fraction, and thus the conversions to a mass density is quite different.

EXAMPLE 2.6 | Historical Values of Carbon Dioxide

Prior to the Industrial Revolution, the atmosphere contained 0.00028 mole of carbon dioxide (CO_2) and 0.99972 mole of air (many from N_2, O_2, and Ar). What was the concentration of CO_2 expressed in ppm_v?

$$\frac{0.0028 \text{ mole } CO_2}{0.99972 \text{ mole of air} + 0.0028 \text{ mole } CO_2} = 0.0028 \text{ mole fraction of } CO_2 \text{ in air}$$

$$0.0028 \text{ mole fraction of } CO_2 \times 10^6 = 280 \text{ ppm}_v \text{ of } CO_2$$

In order to determine the mass concentration in a given volume of air, we must apply the ideal gas equation. In order to use the ideal gas law, we must make the following substitutions:

$$n_i = \frac{m_i}{MW_i} \tag{2.15}$$

where
m_i = mass of component i
MW_i = molar mass of component i

Also from the law of partial pressure:

$$P_i = y_i \times P_{total} \tag{2.16}$$

Substituting the above expressions into the ideal gas law yields

$$(y_i \times P_{total})V = \left(\frac{m_i}{MW_i}\right)RT \tag{2.17}$$

Rearranging yields

$$c_i = \frac{m_i}{V} = \left(\frac{P_i}{T}\right)\left(\frac{MW_i}{R}\right) = y_i\left(\frac{P_t}{T}\right) \tag{2.18}$$

Substituting the values (P_{total} = 1 atm, T = 289 K, R = 0.08206 l-atm/mole-K) for standard temperature and pressure and converting to units of mg/m³ yields

$$c_i\left[\frac{mg}{m^3}\right] = \frac{c_i[ppm_v]MW_i}{24.5} \tag{2.19}$$

And for nonstandard conditions

$$c_i \left[\frac{\mu g}{m^3} \right] = \text{ppm}_v \times MW_i \times \frac{1{,}000P}{RT} \tag{2.20}$$

Note how very different the conversion is between ppm_v and mg/m^3 in the gas phase and ppm_m and mg/L are in an aqueous solution.

The human species has adapted over history to be able to react to and detect extremely small quantities of pollutants associated with spoiled food and likely to cause disease. Historically, doctors practicing medicine would even make diagnoses based on their ability to smell conditions in a patient that indicated pathogenic bacteria were likely to be present. We are able to smell some compounds in air at concentrations on the order of a few parts per billion (ppb_v), or three orders of magnitude smaller than a ppm_v:

$$1 \ \text{ppm}_v = 1{,}000 \ \text{ppb}_v \tag{2.21}$$

EXAMPLE 2.7   Conversion of Gas Concentration between Volumetric Fractions and Mass Concentration

Table 2.4 shows the highest reported concentrations of carbon monoxide and nitrogen dioxide, criteria air pollutants, in 2008 in Fairfax County, a suburban area near Washington, D.C. Convert these concentrations from ppm_v and ppb_v to mg/m^3 and $\mu g/m^3$, respectively.

TABLE 2.4 Highest concentration carbon monoxide and nitrogen dioxide pollutant levels in the air in Fairfax, Virginia

CRITERIA POLLUTANT	VALUE	UNIT
CO	3.7	ppm_v
NO_2	13	ppb_v

Source: Based on Virginia Ambient Air Monitoring 2010 Data Report, 2011. Office of Air Quality Monitoring, Virginia Department of Environmental Quality.

From the periodic chart, find the molar mass, MW, of each element:

$MW_C = 12$ g/mole
$MW_N = 14$ g/mole
$MW_O = 16$ g/mole

(a) Find the concentration of carbon monoxide in the units specified.

Find the molar mass of one molecule of carbon monoxide:

$$12 \ \text{g/mole}(1) + 16 \ \text{g/mole}(1) = 28 \ \text{g/mole}$$

Substitute the molar mass for carbon monoxide and the concentration of carbon monoxide given in Table 2.4 into Equation (2.18):

$$c_{CO} \left[\frac{mg}{m^3} \right] = \frac{c_{CO}[\text{ppm}_v] \, MW_{CO}}{24.5} = \frac{(3.7 \ \text{ppm}_v)\left(28 \dfrac{g}{\text{mole}}\right)}{24.5} = 4.2 \ \frac{mg}{m^3}$$

(b) Find the concentration of nitrogen dioxide in the units specified.

Find the molar mass of nitrogen dioxide:

$$14 \text{ g/mole}(1) + 16 \text{ g/mole}(2) = 46$$

Convert from ppb$_v$ to ppm$_v$ of nitrogen dioxide:

$$13 \text{ ppb}_v \times \frac{1 \text{ ppm}_v}{1{,}000 \text{ ppb}_v} = 0.013 \text{ ppm}_v$$

Substitute the molar mass for carbon monoxide, and the concentration of carbon monoxide given in Table 2.4 into Equation (2.15):

$$C_{NO_2}\left[\frac{\mu g}{m^3}\right] = C_{NO_2}[\text{ppm}_v]\, MW_{NO_2}\left[\frac{g}{\text{mole}}\right]\frac{1{,}000(P)}{(RT)} = (0.013 \text{ ppm}_v)\left(46\frac{g}{\text{mole}}\right)\frac{1{,}000}{24.5} = 24\frac{\mu g}{m^3}$$

Notice that we could convert to mg/m³ and then convert from mg to μg and achieve the same results as using Equation (2.20).

2.5   Chemistry of Natural Systems

Carbon dioxide dissolves into water (see Figure 2.4) in accordance with Henry's law, as discussed earlier in the chapter. The dissolved carbon dioxide takes several forms in solution: dissolved carbon dioxide gas, carbonic acid (HCO_3^-), bicarbonate ion (HCO_3^-), and the carbon ion (CO_3^{2-}). Collectively, these make up the dissolved inorganic carbon (DIC) in water, which is sometimes designated with the symbol ($H_2CO_3^*$). The dissolved inorganic carbon compounds play a major role in regulating the acidity and basicity of natural and marine water.

$$CO_{2(g)} + H_2O \leftrightarrow H_2CO_3^* \tag{2.22}$$

$$[H_2CO_3^*] = [CO_{2(g)}] + [H_2CO_3] + [HCO_3^-] + [CO_3^{2-}] \tag{2.23}$$

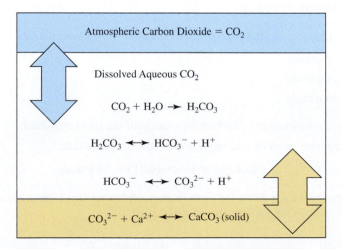

FIGURE 2.4 Predominant phases of inorganic carbon in natural systems.

Source: Bradley Striebig.

EXAMPLE 2.8 Carbon Dioxide Dissolution into Natural Waters

Use Henry's law to calculate the concentration of dissolved inorganic carbonates at standard temperature and pressure (STP) in solution if the atmospheric concentration of carbon dioxide is 390 ppm$_v$ CO_2 and the temperature is 15°C.

The Henry's law constant, from Table 1.7, for carbon dioxide at 15°C is

$$H_{CO_2} = 0.045463 \frac{mol}{L\text{-atm}}$$

The average partial pressure of carbon dioxide at STP is

$$P_{CO_2} = 390 \text{ ppm}_v = 390 \times 10^{-6} \text{ atm}$$

Using Henry's law yields

$$C_{H_2CO_3^*}\left[\frac{mol}{L}\right] = [H_2CO_3^*] = (H_{CO_2})(P_{CO_2}) = \left(0.045463 \frac{mol}{L\text{-atm}}\right)(390 \times 10^{-6} \text{ atm})$$

$$= 1.77 \times 10^{-5} \frac{mol}{L}$$

2.5.1 Law of Electroneutrality

The dissolved bicarbonate and carbonate ions are the major components of a natural water's alkalinity, or the abilities of water to neutralize acids. These ions also typically comprise a large percentage of the negative ions in a fresh-water solution. The amount of positive and negative dissolved ions in a solution must equal one another. The law of electroneutrality states that the sum of all positive ions (cations) in a solution must equal the sum of all the negative ions (anions) in a solution, so that the net charge of all natural waters is equal to zero:

$$\sum \text{cations} - \sum \text{anions} = 0 \tag{2.24}$$

Normality and equivalents represent a useful method of associating the charge of ions in solution. The number of charge equivalents (eq) associated with a compound is equal to the absolute value of the charge associated with the form of the dissolved ion. For example, the dissolved sodium ion, Na^+, has a charge of $+1$ or 1 equivalent. The calcium ion (Ca^{2+}) when dissolved in natural waters has a charge of $+2$ or 2 equivalents. The charge associated with some elements may depend on the pH or other dissolved ions in solution. Iron may be present in solution as the ferrous ion (Fe^{2+}) or ferric ion (Fe^{3+}). The normality of a solution is the number of equivalents per liter and can be determined by multiplying the concentration of a species, MW_i, by the number of equivalents, z_i:

$$N_i\left[\frac{eq}{L}\right] = \left(c_i\left[\frac{mol}{L}\right]\right)\left(z_i\left[\frac{eq}{mol}\right]\right) \tag{2.25}$$

The concept of an equivalent weight (*EW*) may also be useful in calculations involving aqueous solutions. The equivalent weight of any species, EW_i, is equal to the molar mass divided by the number of equivalents associated with the dissolved ion:

$$EW_i\left[\frac{g}{eq}\right] = \frac{MW_i\left[\frac{g}{mol}\right]}{z_i\left[\frac{eq}{mol}\right]} \tag{2.26}$$

2.5.2 Ionic Strength

The ionic strength, I, of a solution is the estimate of the overall concentration of dissolved ions in solution. The strength of the ionic interactions is strongly correlated with the square of the ionic charge of the individual ion.

$$I = \frac{1}{2} \sum_{\text{all ions}} c_i z_i^2 \qquad (2.27)$$

where
I = ionic strength [mol/L]

c_i = concentration of each individual ion [mol/L]

z_i = charge associated with each ion species, i

EXAMPLE 2.9   Estimating the Ionic Strength from Dissolved Ion Concentrations

Estimate the ionic strength of the Ganges River and the Dead Sea, given the data in Table 2.5.

TABLE 2.5 Comparison of major dissolved ion concentrations in the Dead Sea and Ganges River

WATER BODY	CONCENTRATION OF IONS [meq/L]						
	Cations				Anions		
	[Na$^+$]	[K$^+$]	[Ca^{2+}]	[Mg^{2+}]	[SO$_4^{2-}$]	[Cl$^-$]	[HCO$_3^-$]
Dead Sea	1,519	193	788	3,453	11	5,859	4
Ganges River	0.28	0.06	1.10	0.40	0.06	0.16	1.70

Source: Based on Meybeck, M., Chapman, D and Helmer, R. (Editors) (1989) Global Freshwater Quality: A First Assessment. Blackwell, London. 310pp.

Note that the concentrations are given in meq/L. To covert to mole/L, divide each ion by its associated charge and multiply by the appropriate unit conversion:

$$c_{\text{Na}^+} = 0.28 \, \frac{\text{meq}}{\text{L}} \times \frac{\text{eq}}{1{,}000 \, \text{meq}} \times \frac{1 \, \text{mol}}{1 \, \text{eq}} = 2.8 \times 10^{-4} \, \frac{\text{mol}}{\text{L}}$$

The same procedure can be used to determine the concentration of potassium ion K^+ in the solutions. For the diprotic ions, such as the calcium ion:

$$c_{\text{Ca}^{+2}} = 1.10 \, \frac{\text{meq}}{\text{L}} \times \frac{\text{eq}}{1{,}000 \, \text{meq}} \times \frac{1 \, \text{mol}}{2 \, \text{eq}} = 5.5 \times 10^{-4} \, \frac{\text{mol}}{\text{L}}$$

The same procedure can be used to determine the concentration of magnesium ion Mg^{2+} in the solutions. Notice that the absolute value of the charge is used for the negative ions in solution:

$$c_{\text{Cl}^-} = 0.16 \, \frac{\text{meq}}{\text{L}} \times \frac{\text{eq}}{1{,}000 \, \text{meq}} \times \frac{1 \, \text{mol}}{1 \, \text{eq}} = 1.6 \times 10^{-4} \, \frac{\text{mol}}{\text{L}}$$

The same procedure can be used to determine the concentration of other anions in the solutions. The charge associated with each ion and the sum of the terms, $c_i z_i^2$, are shown in Table 2.6.

TABLE 2.6 Tabular calculation of major dissolved ion concentrations and ionic strength components in the Dead Sea and Ganges River

WATER BODY	CATIONS				ANIONS		
	$[Na^+]$	$[K^+]$	$[Ca^{2+}]$	$[Mg^{2+}]$	$[SO_4^{2-}]$	$[Cl^-]$	$[HCO_3^-]$
z_i	1	1	2	2	2	1	1
	Concentration, C_i (mol/L)						
Dead Sea	1.519	0.193	0.394	1.726	5.5×10^{-3}	5.859	4×10^{-3}
Ganges River	2.8×10^{-4}	0.6×10^{-4}	5.5×10^{-4}	2.0×10^{-4}	0.3×10^{-4}	1.6×10^{-4}	1.7×10^{-4}
	$c_i z_i^2$						
Dead Sea	1.519	0.193	1.576	6.906	22×10^{-3}	5.859	4×10^{-3}
Ganges River	2.8×10^{-4}	0.6×10^{-4}	22×10^{-4}	8.0×10^{-4}	1.2×10^{-4}	1.6×10^{-4}	1.7×10^{-4}
	Summation						
Dead Sea	10.2				5.88		
Ganges River	3.34×10^{-3}				1.98×10^{-3}		
	Ionic Strength (mol/L)						
Dead Sea	8.04						
Ganges River	2.66×10^{-3}						

Source: Bradley Striebig.

The ionic strength for the solution can be determined using Equation (2.27):

$$I_{Dead\ Sea} = \frac{1}{2} \sum_{all\ ions} c_i z_i^2 = \frac{1}{2}(10.2 + 5.88) = 8.04 \frac{mol}{L}$$

$$I_{Ganges\ River} = \frac{1}{2} \sum_{all\ ions} c_i z_i^2 = \frac{1}{2}(3.34 + 1.98) \times 10^{-3} = 2.66 \times 10^{-3} \frac{mol}{L}$$

The Dead Sea is so called because the ionic strength of its waters is greater than that of ocean water due to its high concentration of salts; few life forms can survive in water with such large dissolved ion content. The Dead Sea's ionic strength is nearly four orders of magnitude greater than that of the Ganges River.

2.5.3 Solids and Turbidity

Because it is often difficult to account for and measure every dissolved ion in solution, several other methods for estimating ionic strength from simple laboratory or field measurements have been developed, as shown in Table 2.7 and Table 2.8. Several of these methods are based on laboratory procedures (see Figure 2.5) to measure the total dissolved solids (TDS). There are two types of materials in water that are described by the term *solids*: suspended solids and dissolved solids. **Suspended solids** are the materials that are floating or suspended in the water. These solids may consist of silt, sand, and soil or organic particles such as decaying leaves, algae, and bacteria. As you may expect, these

TABLE 2.7 Summary of procedures for calculating the solids content of a water sample

PARAMETER	ABBREVIATION	EQUATION
Total Solids	TS	$\dfrac{\text{(weight of dried unfiltered sample and flask} - \text{weight of empty sample flask)}}{\text{volume of water sample}}$ or TSS + TDS
Total Suspended Solids	TSS	$\dfrac{\text{(weight of dried sample on the filter} - \text{weight of filter only)}}{\text{volume of water sample}}$
Fixed Suspended Solids	FSS	$\dfrac{\text{(weight of dried sample and filter} - \text{weight of solids after heating to 550°C)}}{\text{volume of water sample}}$
Volatile Suspended Solids	VSS	TSS − FSS
Total Dissolved Solids	TDS	$\dfrac{\text{(weight of dried filtered sample and flask} - \text{weight of empty sample flask)}}{\text{volume of water sample}}$
Fixed Dissolved Solids	FDS	$\dfrac{\text{(weight of dried sample and flask} - \text{weight of flask after heating to 550°C)}}{\text{volume of water sample}}$
Volatile Dissolved Solids	VDS	TDS − FDS

Source: Bradley Striebig.

TABLE 2.8 Empirical relationships for estimating ionic strength from laboratory or field measurements

MEASURED PARAMETERS	EQUATION	RATIONALE	REFERENCE
Individual species concentration, C_i Individual species valence, z_i, and unaccounted for TDS, R	$I = \frac{1}{2}\Sigma C_i z_i^2 - 2.50 \times 10^{-5}\ R$ Where $R = \text{TDS}_{measured} -$ $\quad\quad \text{TDS}_{known\ species\ concentrations}$	Accounts for individual species concentrations	(Butler, 1982)
Total dissolved solids (TDS) [mg/l]	$I = 2.04 \times 10^{-5}(\text{TDS})$	Assumes the ionic strength to TDS ratio is equal to that of seawater	(Butler, 1982)
Total dissolved solids (TDS) in [mg/l]	$I = 2.50 \times 10^{-5}(\text{TDS})$	Assumes the ionic strength to TDS ratio is equal to that of a 40 g/mol monovalent salt	(Langelier, 1936)
Conductivity, κ in [μmho/cm = μS/cm; where S = siemens]	$I = 1.4$ to $1.6 \times 10^{-5}\ \kappa$	From data surveys	(Snoeyink and Jenkins, 1980)

Source: Based on Benjamin, M. M. (2001) Water Chemistry. McGraw-Hill. USA.

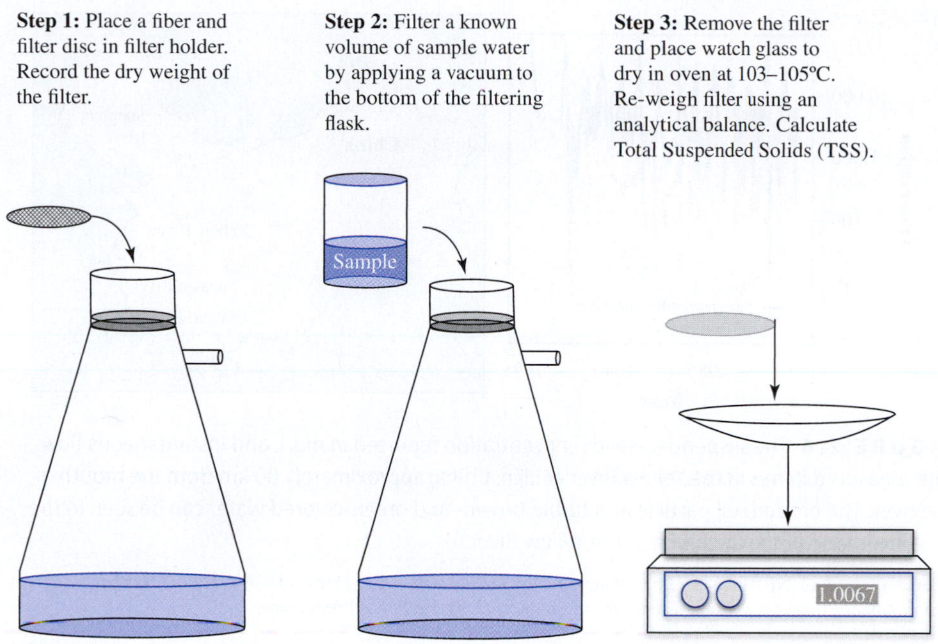

Step 1: Place a fiber and filter disc in filter holder. Record the dry weight of the filter.

Step 2: Filter a known volume of sample water by applying a vacuum to the bottom of the filtering flask.

Step 3: Remove the filter and place watch glass to dry in oven at 103–105°C. Re-weigh filter using an analytical balance. Calculate Total Suspended Solids (TSS).

Sample

1.0067

FIGURE 2.5 EPA-approved method for measuring total dissolved solids in water and wastewater.

Source: Bradley Striebig.

suspended solids make the water cloudy or turbid. The clarity or turbidity of the water can be measured optically as an indicator of the level of suspended solids in the water. Alternatively, these solid particles can be collected on a filter and by weighing the filter to obtain the mass of suspended solids per volume of water that passes through the filter (see Figure 2.5). The total suspended solids (TSS) are defined by the procedure as those particles greater than about 0.45 μm that are removed when the water is passed through a standardized paper filter and the filter is dried at 103°C. TSS are divided into the particles that are volatile, called **volatile suspended solids** (VSS) and fixed solids. The volatile solids are determined by the weight of any particles that evaporate after the filter is heated to 550°C. The TSS concentration of China's Yellow River is shown in Figure 2.6. The Yellow River contains a high amount of eroded soil particles that make the water very turbid compared to more pristine, unpolluted waters.

Unlike the suspended solids, dissolved solids cannot be seen by the naked eye. The **dissolved solids** consist of salts and minerals that have been dissolved through natural weathering of soils or through the anthropogenic process. Since these solids are dissolved, they are not removed or measured by filtering the water. The salts and minerals that were dissolved in the water sample will remain in the container, which can be weighed again to determine the mass of the dissolved solids in the water sample following the calculations outlined in Table 2.7. TDS is calculated by determining the mass of dissolved solids that remain in the flask after the water has passed through the filter, and the water from the flask has been evaporated by placing the flask in a furnace or oven at 103°C.

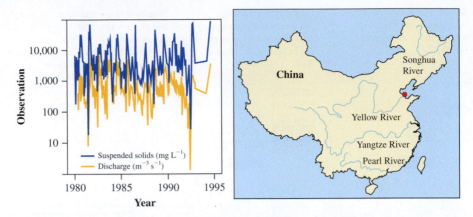

FIGURE 2.6 The suspended solids concentration reported in mg/L and instantaneous flow rate measured in m/s in the Yellow River at Lijin, China, approximately 80 km from the mouth of the river. The eroded soil particle and turbid brown- and green-colored water can be seen in the satellite image of the discharge of the Yellow River.

Source: Based on UNEP GEMS. (2006) Water Quality for Ecosystem and Human Health. United Nations Environment Programme Global Environment Monitoring System/Water Programme.

2.5.4 Water Hardness

Many concepts involving water quality, such as the definitions of TDS and TSS, have been derived from empirical laboratory procedures that have been used for several decades to indicate water quality and potential uses for water of that quality. Hardness is one such historically defined water quality parameter. As the name implies, it is related to the part of the water mixture, or more precisely, the aqueous solution, that will become a solid (and therefore "hard") when the water is heated. Typically, water hardness is a result of high levels of dissolved calcium and carbonate ions that are found in groundwater in limestone-rich geologic strata. Hardness may also be caused by magnesium, strontium, manganese, and iron. The dissolved calcium and carbonate ions have the unusual property of becoming less soluble in solution with increasing temperature when both ions are found in groundwater at high concentrations. As water is heated in home water heaters or industrial boilers, these ions combine and form a solid precipitate or "scale" that decreases the overall efficiency of these appliances and may eventually cause the water heaters and boilers to cease working. Hardness also reduces the effectiveness of soaps.

The specific definition of hardness in water is the sum of the concentration of the divalent cations (species with a charge of 2+) in water. For most waters, the most important species that contribute to hardness are calcium (Ca^{2+}) and magnesium (Mg^{2+}). Most public water in the United States ranges in hardness from about 25 to 150 mg/L of $CaCO_3$. Hardness is most often expressed in mg/L as $CaCO_3$, since it is typically assumed that most of the hardness in the water is associated with the formation of calcium carbonate solids. Water with a hardness greater than 150 mg/L of $CaCO_3$ is usually considered hard, as shown in Table 2.9, and may undergo a water "softening" process where calcium ions are replaced with single-valent cations such as sodium or potassium ions. Hardness values are shown for surface water in the United States and various other sampling locations in Figures 2.7 and 2.8.

TABLE 2.9 Water hardness classifications

HARDNESS RANGE [mg/L as CaCO₃]	DESCRIPTION
0–50	Extremely soft
50–100	Very soft
100–150	Soft to moderately hard
150–300	Hard
> 300	Very hard

Source: Based on Dufor and Becker, 1964.

Concentration of Hardness as Calcium Carbonate, in Milligrams per Liter

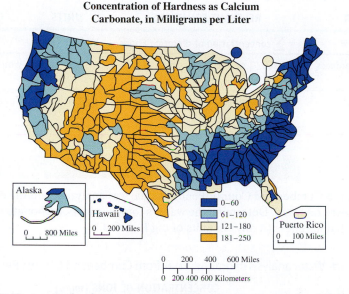

FIGURE 2.7 Mean water hardness values as calcium carbonate at water monitoring sites in 1975 in the United States (adapted from Briggs et al., 1977).

Source: Based on USGS, USGS Water-Quality Information, Water Hardness, http://water.usgs.gov/owq/hardness-alkalinity.html.

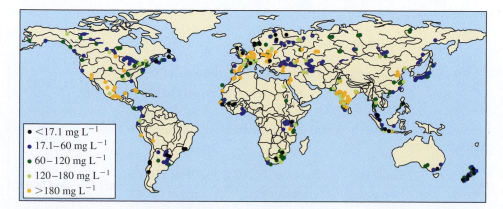

FIGURE 2.8 Water hardness values in mg/L reported at surface water monitoring stations on different continents.

Source: Based on UNEP GEMS. (2006) Water Quality for Ecosystem and Human Health. United Nations Environment Programme Global Environment Monitoring System/Water Programme.

Hardness is divided into three components: total hardness, carbonate hardness, and noncarbonated hardness. Total hardness is determined by the sum of the diprotic cations. Carbonate hardness represents the portion of the diprotic ions that can combine with carbonates to form scaling. It is equal to the smaller value between the total hardness and the total bicarbonate concentration expressed as mg/L $CaCO_3$. Noncarbonate hardness is present only if the concentration of the bicarbonate ion is less than the total hardness. Noncarbonated hardness is the difference between total hardness and carbonate hardness, as shown in Table 2.10.

TABLE 2.10 Summary of hardness calculations

PARAMETER	RELEVANT EQUATIONS	UNITS	COMMENT
Total hardness	$= \sum$ concentration of diprotic cations	mg/L as $CaCO_3$	
Carbonate hardness	$= [HCO_3^-]$ or = total hardness	mg/L as $CaCO_3$	Whichever value is smaller
Noncarbonated hardness	= total hardness − carbonate hardness	mg/L as $CaCO_3$	Or zero, if $[HCO_3^-]$ > total hardness

EXAMPLE 2.10   Determining Water Hardness

Spring water in Cumberland County, Pennsylvania, was analyzed, and the data shown in Table 2.11 were reported. Determine the water's total hardness, carbonate hardness, and noncarbonated hardness (if any) in units of mg/L as $CaCO_3$.

TABLE 2.11 Water analysis in groundwater from Cumberland County, Pennsylvania

WATER SOURCE	CONCENTRATION OF IONS [mg/L]						
	Cations			Anions			
	$[Na^+]$	$[Ca^{2+}]$	$[Mg^{2+}]$	$[NO_3^-]$	$[SO_4^{2-}]$	$[Cl^-]$	$[HCO_3^-]$
Spring Cu 29	13	95	13	24	34	12	300

Source: Based on Flippo, 1974.

Calculate the amount of calcium in solution as if it were to form calcium carbonate:

$$95\,\frac{\text{mg Ca}^{2+}}{\text{L}} \times \frac{1 \text{ mmol Ca}^{2+}}{40 \text{ mg}} \times \frac{1 \text{ mmol CaCO}_3}{1 \text{ mmol Ca}^{2+}} \times \frac{100 \text{ mg CaCO}_3}{1 \text{ mmol CaCO}_3} = 237.5\,\frac{\text{mg}}{\text{L}} \text{ as CaCO}_3$$

Similarly, calculate the amount of magnesium in solution as if it were to form calcium carbonate:

$$13\,\frac{\text{mg Mg}^{2+}}{\text{L}} \times \frac{1 \text{ mmol Mg}^{2+}}{24.4 \text{ mg Mg}^{2+}} \times \frac{1 \text{ mmol as if CaCO}_3}{1 \text{ mmol Mg}^{2+}} \times \frac{100 \text{ mg CaCO}_3}{1 \text{ mmol CaCO}_3} = 53.3\,\frac{\text{mg}}{\text{L}} \text{ as CaCO}_3$$

Total Hardness [mg/L as $CaCO_3$] = $[Ca^{2+}]$ + $[Mg^{2+}]$ as mg/L $CaCO_3$
= 237.5 + 53.3 = 326.8

Therefore, this particular groundwater is very hard.

Calculate the amount of bicarbonate in solution as if it were to form calcium carbonate:

$$300 \; \frac{\text{mg HCO}_3^-}{\text{L}} \times \frac{1 \text{ mmol HCO}_3^-}{61 \text{ mg}} \times \frac{1 \text{ mmol CaCO}_3}{1 \text{ mmol HCO}_3^-} \times \frac{100 \text{ mg CaCO}_3}{1 \text{ mmol CaCO}_3} = 491 \; \frac{\text{mg}}{\text{L}} \text{ as CaCO}_3$$

Since $[\text{HCO}_3^-]$ > total hardness, then the carbonate hardness = total hardness = 326.8 mg/L as $CaCO_3$.

Noncarbonated hardness = 0, since CH > TH.

2.5.5 Chemical Reactivity, Activity, and the Activity Coefficient

Carbon in the dissolved carbonates is an important negative dissolved ion in natural water chemistry. The other dissolved species in the solution may also influence the form and chemical behavior of the dissolved carbon in aqueous solutions. The ionic strength is used as an indicator of the influence of the dissolved ions in chemical reactions. This influence is determined by the calculation of the chemical activity of the dissolved ions in a solution.

The effect of dissolved ions on chemical reaction rates must be accounted for to accurately model and predict how chemicals will react in fresh water (with very low concentrations of dissolved ions) compared to saltwater solutions (with very high concentrations of dissolved ions). In order for one chemical to react with another in a solution, the two chemicals must first come in contact with one another. The chemical reactivity refers to the chemical's overall tendency to participate in a reaction.

In freshwater solutions, the likelihood of one atom, A, coming into contact with another, B, is directly related to the concentration of each compound $[A]$ and $[B]$ or to how many atoms of a given chemical are found within a given volume of water. The situation of fresh water is very similar to what chemists would define as an ideal solution or a solution where the dissolved ions behave independently of each other. The reaction rate of formation of AB, r_{AB}, is proportional to the concentration, as illustrated in Equations (2.28) and (2.29). The reaction rate constant, k_{AB}, relates the change in concentration to the rate of the reaction. The greater the number of each atoms, the more likely they are to contact one another and create a chemical reaction. The pure water case is analogous to two people trying to leave a football stadium through a single gate when the stadium is empty: The two people may meet at the gate (come into contact with one another) and pass through the door together (create a reaction) without anyone else interfering. In pure fresh water in the natural environment, the concentration of other chemical species is so low that other chemicals are unlikely to inhibit chemical reactions in most instances.

$$A + B \xrightarrow{\text{pure fresh water}} AB \qquad (2.28)$$

$$r_{AB} = k_{AB}[A][B] \qquad (2.29)$$

In solutions with significant concentrations of other dissolved ions in solution, or nonideal solutions, the dissolved ions may interfere or shield one reactant, C, from coming into contact with another reactant, D. The case with increased dissolved solids is analogous to two people on opposite sides of a stadium trying to leave a football stadium through a single gate when the football game is over and the stadium is full: The two people find it very difficult to meet at the gate (come into contact with one another) and pass through the door at the same time (create a reaction) because there are so many other people (dissolved ions) interfering with their progress. In salt water in the natural environment, the concentration of other dissolved ions is so high that other chemicals inhibit chemical reactions. Even if, in these two cases, the concentrations of all the species are identical:

$$[A] = [C] \text{ and } [B] = [D]$$

The rate of the reaction between compound C and D will be much slower than the reaction between A and B because of the interference of the other chemicals in the solution.

$$r_{AB} > r_{CD}$$

In order to account for the effects of other species in a solution, the concept of chemical activity has been developed. The chemical activity is a standardized measure of chemical reactivity within a defined system. The activity of a compound A is denoted by $\{A\}$. Standard state conditions are arbitrarily defined in environmental science as 25°C and the atmospheric pressure as equal to 1 bar (or 1 atm). (Other fields, such as oceanography, may define the standard state differently.) If the concentration of A in water is equal to 1 mol/L, then

$\{A\} = 1$ when under standard state conditions

If $\{A\} > 1$, then the system is not at standard state and has a greater chemical reactivity.

If $\{A\} < 1$, then the system is not at standard state and has a lower chemical reactivity.

At standard state, other conditions are also defined. The activity of the liquid solvent, in our case, water, is also defined as one, as indicated in Table 2.12.

$\{\text{pure liquid}\} \approx \{H_2O\} \approx 1$ at standard state for an ideal solution

TABLE 2.12 Chemical activity definitions of standard state conditions

STATE	CONCENTRATION	TEMPERATURE	PRESSURE
Solid	Pure		
Liquid	Pure	25°C	1 bar
Gas	Pure		
Solute	1.0 molar		

Since the standard conditions and activity have been defined, a relationship between activity and reactivity can be developed. The activity coefficient, γ_A, is used to relate the standard chemical activity and the conditional chemical reactivity. The activity coefficient is defined as the ratio of the reactivity per mole of A in a real system compared to the reactivity of A in the standard reference state.

$$\gamma_A = \frac{\text{real reactivity per mole of } A}{\text{standard activity per mole of } A} = \frac{\{A\}}{[A]} \quad (2.30)$$

Since both the activity of A, $\{A\}$ and the concentration of A, $[A]$ have units of mol/L, the activity coefficient is unitless.

Several empirical relationships have been developed to estimate the value of the activity coefficient based on the ionic strength of the aqueous solution, as shown in Table 2.13.

TABLE 2.13 Common approximations for individual ion activity coefficients at 25°C.

ACTIVITY CORRELATION EQUATION	NAME	COMMENTS
$\gamma_i = 10^{-0.51z_i^2 I^{0.5}}$	Debye–Hückel limiting law	$I \leq 5 \times 10^{-3}$ M
$\gamma_i = 10^{\frac{-(0.51z_i^2 I^{0.5})}{(1+0.16I^{0.5})}}$	Extended Debye–Hückel limiting law	$I \leq 0.1$ M
$\gamma_i = 10^{\frac{-(0.51z_i^2 I^{0.5})}{(1+I^{0.5})}}$	Guntelberg	Mixtures with $I \leq 0.1$ M
$\gamma_i = 10^{\frac{-(0.51z_i^2 I^{0.5})}{(1+I^{0.5})-0.2I}}$	Davies—for common ions such as Cl^- and OH^-	$I \leq 0.5$ M
$\gamma_i = 10^{\frac{-(0.51z_i^2 I^{0.5})}{(1+I^{0.5})-0.3I}}$	Davies—for common ions such as Cl^- and OH^-	$I \leq 0.5$ M
$\gamma_i = 10^{\frac{-(0.51z_i^2 I^{0.5})}{(1+1.5I^{0.5})}}$	Scatchard—for common ions such as Na^+, HCO_3^-, and CO_3^{2-}	$I \leq 0.1$ M

Source: Based on Benjamin, M.M. Water Chemistry: 1st Edition. McGraw-Hill.

EXAMPLE 2.11   Determination of the Activity Coefficient for the Ganges River

Estimate the activity coefficient associated with the sodium ion, calcium ion, and bicarbonate ion concentrations given in Table 2.5 in the Ganges River.

From Example 2.9, the ionic strength of the Ganges River water was determined to be $I = 2.66 \times 10^{-3}$ mol/L.

As shown in Table 2.13, the Debye–Huckel limiting law is valid for solutions with this approximate ionic strength. Substituting the values for the sodium ion into the Debye–Huckel limiting law yields

$$\log(\gamma_{Na^+}) = -0.5z_{Na^+}^2 I^{1/2} = -0.5(1)^2 (2.66 \times 10^{-3})^{1/2} = -0.0258$$

$$\gamma_{Na^+} = 0.942$$

Similarly, for the calcium ion:

$$\log(\gamma_{Ca^+}) = -0.5z_{Ca^+}^2 I^{1/2} = -0.5(2)^2 (2.66 \times 10^{-3})^{1/2} = -0.0103$$

$$\gamma_{Ca^{+2}} = 0.788$$

And for the bicarbonate ion:

$$\log(\gamma_{HCO_3^-}) = -0.5z_{HCO_3^-}^2 I^{1/2} = -0.5(-1)^2 (2.66 \times 10^{-3})^{1/2} = -0.0258$$

$$\gamma_{HCO_3^-} = 0.942$$

Notice that since the charge is squared, the activity coefficient for compounds with the same net charge, regardless of whether the charge is positive or negative, will have the same activity coefficient. The electrostatic forces that create the shielding are proportional to the square of the ionic strength, so the activity coefficient will change significantly for compounds with a different ionic charge. That is to say, the activity effects associated with the calcium ions are much greater than the activity effects we would expect to observe for the sodium or bicarbonate ions.

2.6 Equilibrium Models for Estimating Environmental Impacts

Many chemical processes reach equilibrium in the natural environment very quickly, particularly acid–base reactions. Equilibrium-based calculations and models are very useful for estimating the potential environmental impacts of many chemicals in the environment. Equilibrium models are also useful for designing systems to treat industrial and municipal gas, liquid, and solid waste streams. A basic understanding of equilibrium chemistry will help engineers design systems to reduce pollutant emissions, as well as understand the fate and transport of chemicals that are emitted into the environment. Engineers can make more informed choices about sustainable systems and sustainable design by applying the basic concepts of chemical equilibrium.

It is also important to recognize the limitations of equilibrium-based models. Environmental systems are rarely static; instead, these systems are dynamic, and equilibrium-based models cannot be expected to adequately represent each and every element of complex natural systems. The engineer or scientist must be aware of the limitations of the mathematical assumptions and simplifications. There are also those reactions that simply occur too slowly within the system of concern, particularly some oxidation–reduction reactions and many biogeochemical reactions. For these relatively slow reactions, the assumptions required to develop an equilibrium-based model are simply not valid. Finally, the dynamic environmental systems differ from the ideal conditions from which much of our basic data is derived, yielding a significant level of uncertainty in calculations. It is not unusual for the reported "standard" values of key parameters to differ by 0.5 to 10% under laboratory conditions, and extrapolating these uncertainties to field conditions is a difficult task. Nonetheless, the equilibrium-based models provide powerful tools to estimate the bounds of environmental conditions that might be expected from relatively fast reactions and the extent to which reactions may proceed. They may also indicate if a particular chemical transformation is possible.

The fundamental parameter for equilibrium-based models is the equilibrium constant, k. The equilibrium constant describes the proportionality between reactants and products that will occur when the reaction has reached its minimum energy state and is complete. For a given chemical reaction, the equilibrium constant is defined as

$$A + B \leftrightarrow C + D$$

$$k_{\text{equilibrium}} = \frac{\{\text{products}\}}{\{\text{reactants}\}} = \frac{\{C\}\{D\}}{\{A\}\{B\}} \tag{2.31}$$

2.6.1 Acid and Base Definitions

Acids and bases influence water quality by controlling the pH of the aqueous solution, which in turn affects the dissolution and precipitation of compounds, the solubility of gases, and even the interactions between chemicals and living organisms. Equilibrium conditions are used to model the effects of pH in the natural environment, including the effects of pH on the carbonate system, and vice versa. The pH of natural waters is important to biodiversity since most species are only tolerant of natural waters in the pH range from 6.5 to 8.5, as shown in Figure 2.9. However, the pH of rainfall is largely dependent on the concentration of pollutants in the atmosphere and the solubility of those pollutants, as discussed in the following examples. Airborne sulfates emitted from coal-burning power plants have the greatest impact on the pH of rainfall; consequently, the biodiversity of natural bodies of water may be threatened, as shown in Figure 2.10. Acidified rainfall is also discussed in more detail in Chapter 5.

An **acid** may be defined as any substance that can donate a hydrogen ion, H^+ (or proton). A **base** may be defined as any substance that can accept an H^+ ion (or proton). The acid that donates the hydrogen ion and the base that accepts the ion

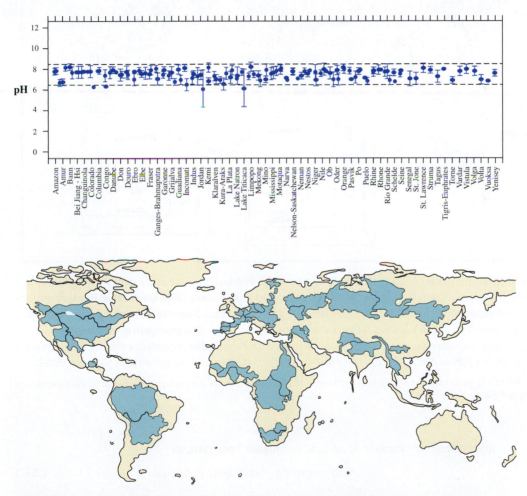

FIGURE 2.9 The mean pH (± 1 standard deviation) of major drainage basins in the world. The dashed lines indicate the approximate pH range suitable for protecting biodiversity in natural waterways. The shaded areas on the map show the areas that have been sampled as part of the Global Environment Monitoring System.

Source: Based on UNEP GEMS. (2006) Water Quality for Ecosystem and Human Health. United Nations Environment Programme Global Environment Monitoring System/Water Programme.

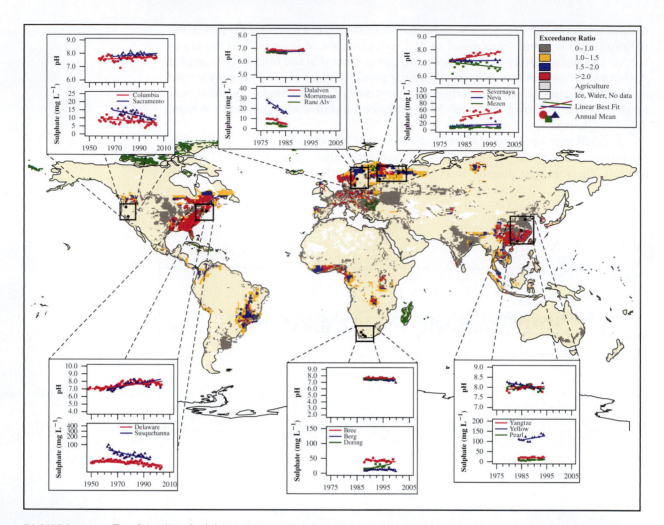

FIGURE 2.10 Trends in pH and sulphate concentrations in rivers around the world are overlaid onto a map showing the sensitivity of an area to acidification. Sensitivity is measured as an exceedance ratio. Exceedance ratios greater than one indicate areas that are sensitive to acid deposition (Bouwman et al., 2002). Note the general decreases in in-stream sulphate concentrations in the United States and Sweden, compared to increases in China and Russia. There was very little change in pH and sulphate concentrations in South Africa, which is considered at relatively low risk of acidification based on exceedance ratios. Changes in pH do not always parallel changes in sulphate concentrations, likely due to the effects of other acidifying chemicals such as nitrates.

Source: Based on UNEP GEMS. (2006) Water Quality for Ecosystem and Human Health. United Nations Environment Programme Global Environment Monitoring System/Water Programme.

are collectively known as an acid conjugate base pair, as

$$\text{Acid} \rightarrow \text{proton} + \text{conjugate base pair} \tag{2.32}$$

$$HA \rightarrow H^+ + A^- \tag{2.33}$$

or

$$\text{Base} + \text{proton} \rightarrow \text{conjugate acid pair} \tag{2.34}$$

$$A^- + H^+ \rightarrow HA \tag{2.35}$$

Either of the previous reactions may take place simultaneously, so that the net reaction becomes

$$HA \leftrightarrow H^+ + A^-$$
(2.36)

When written in this fashion, with the acid as a reactant and the base as a product, the acid equilibrium constant for the acid–base equilibrium is

$$k_a = \frac{\{H^+\}\{A^-\}}{\{HA\}} = \frac{\gamma_{H^+}[H^+]\gamma_{A^-}[A^-]}{\gamma_{HA}[HA]}$$
(2.37)

Acid equilibrium constants can be found online and in introductory chemisty textbooks. If the dissolved ion concentration in the solution is low, then we may assume that the solution is approximately ideal, $\gamma \cong 1$.

For an ideal solution, Equation (2.37) may be simplified to

$$k_a = \frac{\{H^+\}\{A^-\}}{\{HA\}} = \frac{[H^+][A^-]}{[HA]}$$
(2.38)

The water molecule may act as either an acid or a base (see Box 2.2). Substances that can either donate or receive a proton are called **ampholytes** (or they are described as amphoteric compounds). Equation (2.39) is directly analogous to Equation (2.38).

BOX 2.2 Acid and Base Properties of Pure Water

A review of the definitions of k_a and k_b will show many similarities. These equilibrium constants are related mathematically to the water equilibrium constant, k_w.

It will be helpful, in this proof of concept, to note that the actual form an acid takes in water is more complex than simply, H^+. The acid becomes linked to several water molecules, as illustrated in Figure 2.11, in what is appropriately called a chemical complex. The chemical water complex is a molecular group that is more accurately represented by the formula, $H_9O_4^+$. Rather than write this "complex" formula, acid in water may be abbreviated in the shorthand notation as H_3O^+ or H^+. It will be helpful to use the slightly more formal H_3O^+ and include water molecules in this example. If you proceed with the correct algebra, it will become apparent why we've neglected this nuance in nomenclature in earlier examples.

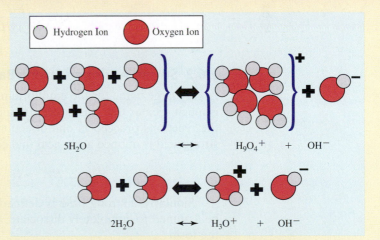

FIGURE 2.11 Visualization of how water partially dissociates to form very small amounts of acid molecules designated with the symbol H^+ and the hydroxide ion base OH^-, even in pure water solutions.

Source: Bradley Striebig.

However, because water is the solvent in natural aqueous solutions, its acid equilibrium constant is given a special subscript, k_w. Recall that the arbitrary standard state chosen for environmental science defines the activity of water, $\gamma_{HOH} = 1$. Values for k_w are well documented and vary slightly with temperature as shown in Table 2.14.

$$H_2O \leftrightarrow H^+ + OH^- \tag{2.39}$$

$$k_w = \frac{\{H^+\}\{OH^-\}}{\{HOH\}} = \{H^+\}\{OH^-\} = 10^{-14} \tag{2.40}$$

TABLE 2.14 Variation in k_w with variations in temperature

TEMPERATURE °C	LOG k_w
0	−14.93
10	−14.53
20	−14.17
30	−13.83
50	−13.26

Source: Based on Stumm and Morgan: Aquatic Chemistry: Chemical Equilibria and Rates in Natural waters, 3rd ed.

A base–equilibrium constant may also be written for the following general equilibrium equation:

$$B^- + H^+ \leftrightarrow BH \tag{2.41}$$

$$k_b = \frac{BH}{\{B^-\}\{H^+\}} = \frac{\gamma_{BH}[BH]}{\gamma_B[B^-]\gamma_{H^+}[H^+]} \tag{2.42}$$

2.6.2 Strong Acids and Strong Bases

A very large value for the equilibrium means that the concentration of the products is much greater than the concentration of the reactants. If $\{H^+\}\{A^-\} \gg \{HA\}$, then the reactions proceed in one direction only, and the acid completely dissociates. A strong acid is defined as an acid that completely dissociates when added to water:

$$HA \rightarrow H^+ + A^- \text{ and } \{HA\} \approx 0 \tag{2.43}$$

Similarly, a strong base is defined as one that has a large value of k_b and may be assumed to completely dissociate. A typical strong base equation may appear in the form

$$BOH \rightarrow B^+ + OH^- \text{ and } \{BOH\} \approx 0 \tag{2.44}$$

2.6.3 The Relationship Between pH and pOH

The components of the water equilibrium equation define the acidity and basicity of an aqueous solution. A neutral solution is defined as one in which the activity of the hydrogen ions is equal to the activity of the hydroxide ions: $\{H^+\} = \{OH^-\}$. The pH is a literal mathematical definition that is defined as the negative logarithm of the activity of the hydrogen ion in the solution. (Note that the symbol p in pX represents the mathematical function of the negative logarithm, p$X = -\log\{X\}$.)

$$pH = -\log\{H^+\} \qquad (2.45)$$

Similarly,

$$pOH = -\log\{OH^-\} \qquad (2.46)$$

The pH and pOH of a solution are related by the equilibrium expression for water, k_w, at standard conditions by Equation (2.40), so

$$k_w = \{H^+\}\{OH^-\} = 10^{-14} \qquad (2.47)$$

Applying the negative logarithm function to both sides of Equation (2.47) yields

$$-\log\{k_w\} = -\log\{H^+\} + -\log\{OH^-\} = -\log\{10^{-14}\} \qquad (2.48)$$

$$pk_w = pH + pOH = 14 \qquad (2.49)$$

There are several steps involved in modeling the effects of acids or bases in the natural environment.

Step 1: Define the system boundaries.

Step 2: Identify all the chemical species of interest.

Step 3: Write the constraining chemical equations for the system, including equilibrium equations, the electroneutrality equation, and mass balance equations.

Step 4: Make any simplifying assumptions that are possible.

Step 5: Algebraically solve the remaining independent equations and check the assumptions made to aid in solving the equations.

EXAMPLE 2.12   **What Is the pH of Pure Water?**

Determine the pH and pOH of pure water at 25°C and 1 bar.

Step 1: Define the system boundaries.

The container of the solution provides the system boundaries.

Step 2: Identify all the chemical species of interest.

$$H_2O, H^+, \text{ and } OH^-$$

Step 3: Write the constraining chemical equations for the system, including equilibrium equations, the electroneutrality equation, and mass balance equations.

The relevant equations are (2.40), (2.45), (2.46), and (2.49).

Step 4: Make any simplifying assumptions that are possible.

No simplifying assumptions are required to solve this problem.

Step 5: Algebraically solve the remaining independent equations and check the assumptions made to aid in solving the equations.

Substitute $\{H^+\} = \{OH^-\}$ into Equation (2.40):

$$k_w = \{H^+\}\{OH^-\} = \{H^+\}\{H^+\} = \{H^+\}^2 = 10^{-14}$$

Therefore, in a pure, neutral water solution:

$$\{H^+\} = \{OH^-\} = 10^{-7} \text{ mol/L}$$

The pH of the solution is determined by substituting the activity of the hydrogen ion into Equation (2.45):

$$pH = -\log\{H^+\} = -\log\{10^{-7}\} = 7$$

The pOH is found by using either Equation (2.46), since the hydroxide ion activity is known, or by using Equation (2.49):

$$pOH = -\log\{OH^-\} = -\log\{10^{-7}\} = 7$$

or

$$pk_w = pH + pOH = 14$$

$$pOH = 14 - pH = 14 - 7 = 7$$

EXAMPLE 2.13 **Determining the pH of a Solution to Which a Strong Acid Has Been Added**

Determine the pH for a 1-liter solution to which the strong acid, hydrochloric acid (HCl), has been added to produce a total acid concentration in the solution of 10^{-3} mol/L.

Step 1: Define the system boundaries.

The container of the solution provides the system boundaries.

Step 2: Identify all the chemical species of interest.

$$H_2O, H^+, Cl^-, \text{ and } OH^-$$

Step 3: Write the constraining chemical equations for the system, including equilibrium equations, the electroneutrality equation, and mass balance equations.

The relevant equations are (2.43) and (2.45).

The total concentration of the acid is given as: $H^+ = 10^{-3}$ mol/L.

Step 4: Make any simplifying assumptions that are possible.

It is assumed that HCl is a strong acid and completely dissociates. Since HCl is a strong acid:

$$HCl \rightarrow H^+ + Cl^-$$

Step 5: Algebraically solve the remaining independent equations and check the assumptions made to aid in solving the equations.

The pH of the solution is determined by substituting the activity of the hydrogen ion into Equation (2.45):

$$pH = -\log\{H^+\} = -\log\{10^{-3}\} = 3$$

EXAMPLE 2.14   Approximating the pH of Nitrogen Acidified Rain

Nitrous oxide (N_2O) is emitted to the atmosphere through naturally occurring biological transformations. The atmospheric concentration of nitrous oxide has increased from the anthropogenic use of nitrogen-based fertilizers, as shown in Figure 2.12. Although the concentration of nitrous oxides in the atmosphere is almost 1,000 times less than the concentration of carbon dioxide, it is an influential greenhouse gas. Nitrogen oxide has a very long residence time in the atmosphere, and it has a relatively large energy absorption capacity.

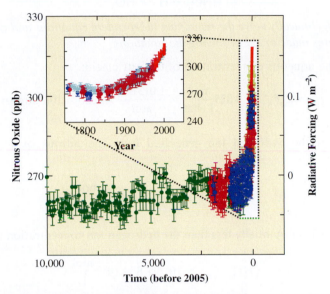

FIGURE 2.12 Atmospheric abundance of N_2O as determined from ice cores (various colors) and whole air (yellow line) samples. The inset contains changes since 1750.

Source: IPCC, 2007 Climate Change 2007: Working Group I: The Physical Science Basis, Fig SPM.1.

Nitrogen oxides can be removed by rainfall when the nitrogen oxides are absorbed by raindrops. The nitrogen oxides go through several reactions and form a strong acid, nitric acid (HNO_3), in the raindrop. Given today's average concentration of approximately 320 ppb N_2O in the atmosphere, what would be the pH of a raindrop, and the concentration of other dissolved ions in equilibrium with N_2O in the atmosphere?

Step 1: Define the system boundaries.

The system boundaries are defined by the raindrop in a standard atmosphere that contains 320 ppb N_2O.

Step 2: Identify all the chemical species of interest.

$$N_2O_{(g)}, H_2O, H^+, NO_3^-, HNO_3, \text{ and } OH^-$$

Step 3: Write the constraining chemical equations for the system, including equilibrium equations, the electroneutrality equation, and mass balance equations.

The relevant equations are (2.43) and (2.45).

The total concentration of the acid can be calculated from Henry's law:

$$k_H = 0.025 \frac{mol}{kg\text{-}bar}$$

$$c_{HNO_3(aq)} \cong c_{N_2O(aq)} = k_H P_{N_2O}$$

Step 4: Make any simplifying assumptions that are possible.

It is assumed that HNO_3 is a strong acid and completely dissociates. Since HNO_3 is a strong acid:

$$HNO_3 \rightarrow H^+ + NO_3^-$$

Step 5: Algebraically solve the remaining independent equations and check the assumptions made to aid in solving the equations.

Determine the aqueous concentration of the nitric acid:

$$c_{HNO_3(aq)} \cong c_{N_2O(aq)} = k_H P_{N_2O} = 0.025 \frac{mol}{L\text{-}bar} \times \frac{1.013\,bar}{atm} \times (320 \times 10^{-9}\,atm) = 8 \times 10^{-9} \frac{mol}{L}$$

The activity of the hydrogen ion generated from the nitrous oxide in the air is approximately

$$c_{HNO_3(aq)} \cong [H^+] = 8 \times 10^{-9} \frac{mol}{L}$$

Notice that this is substantially less than the hydrogen ion concentration in pure water:

$$[H^+]_{pure\ water} = 10^{-7} > 8 \times 10^{-9} \frac{mol}{L}$$

The pH of the solution is determined by substituting the activity of the greater of the hydrogen ion activities into Equation (2.45):

$$pH = -\log\{H^+\} \cong -\log\{10^{-7}\} = 7$$

In this example, even though nitric acid is a strong acid, not enough is added from average N_2O levels in the atmosphere to significantly change the pH of a pure water raindrop. Rainfall does acidify due to air pollutants in the atmosphere, but these occur at a more local or regional level, and the acidity largely comes from sulfur oxides emitted to the atmosphere. Atmospheric components that form a weaker acid may change the pH of rainfall, as will be illustrated in the next section.

2.6.4 Modeling Natural Waters That Contain a Weak Acid

An acid that only partially dissociates in an aqueous solution is called a weak acid. A weak acid has a small value for its equilibrium constant, k_a. The degree to which the acid dissociates depends on the amount of acid added to the solution and the overall pH of the solution if other dissolved ions are present. It should be noted that adding a weak acid to a solution may still produce a very acidic solution, with a very low pH, if enough of the acid is added to the solution.

A base that does not completely dissociate and has an associated small k_b value for its equilibrium constant is called a weak base and is analogous to a weak acid.

The pH of an aqueous solution controls the dissolved form of many weak acids and bases that are important to environmental systems, industrial systems, and human health. The pH can be thought of as a master variable that can be used in computer programs and graphical analyses to quickly identify the species of greatest concentration in aqueous systems. The same steps to solve acid–base equilibrium problems will be used, but more equations must be considered.

EXAMPLE 2.15   **An Algebraic Solution to Determining the pH of a Weak Acid**

Vinegar is formed biologically from the decomposition of sugars in water. Vinegar is a mixture of several organic compounds; the largest by concentration is acetic acid (CH_3COOH). The acid dissociates in solution to form hydrogen ions and acetate ions (CH_3COO^-). The acetate ion is commonly abbreviated as Ac^-. If acetic acid were added to a 1-liter flask of distilled water, so that the total concentration of all acetic acid species was 10^{-4} mol/L, what would the pH of the aqueous solution be? Acetic acid has a $pK_a = 4.7$.

Step 1: Define the system boundaries.

The system boundaries are walls of the 1-liter flask of water.

The total amount of the acetic acid species in the water = 10^{-4} mol/L.

Step 2: Identify all the chemical species of interest.

$$H_2O, HAc, H^+, Ac^-, \text{ and } OH^-$$

Step 3: Write the constraining chemical equations for the system, including equilibrium equations, the electroneutrality equation, and mass balance equations.

Relevant equilibrium equations are

$$k_w = \{H^+\}\{OH^-\} = 10^{-14}$$

$$k_a = \frac{\{H^+\}\{Ac^-\}}{\{HAc\}} = 10^{-4.7}$$

Relevant mass balance equation:

$$C_{HAc_{total}} = 10^{-4} = \{HAc\} + \{Ac^-\}$$

Electroneutrality equation:

$$\{H^+\} = \{AC^-\} + \{OH^-\}$$

Step 4: Make any simplifying assumptions that are possible.

If we add an acid to water, the hydrogen ion concentration increases, and the hydroxide ion concentration must decrease proportionally. Therefore, we will make the assumption that the hydroxide ion is small compared to the other possible variable, in order to simplify the algebraic analysis:

$$\{H^+\} \gg \{OH^-\}$$

The electroneutrality equation simplifies to:

$$\{H^+\} \cong \{AC^-\}$$

This example and others that occur in pure water are nearly ideal solutions; therefore, we may assume that the activity of the compound is equal to the concentration of the same compound, $\{A\} = [A]$.

Step 5: Algebraically solve the remaining independent equations and check the assumptions made to aid in solving the equations.

Substituting the above identity into the equilibrium expression yields:

$$k_a = \frac{\{H^+\}\{Ac^-\}}{\{HAc\}} = \frac{\{Ac^+\}\{Ac^-\}}{\{HAc\}} = 10^{-4.7}$$

Rearranging and solving for $\{HAc\}$ yield:

$$\{HAc\} = \frac{\{Ac^-\}^2}{10^{-4.7}}$$

Substituting the expression for $\{HAc\}$ into the mass balance equation yields a second-order polynomial equation that can be solved using the standard quadratic equation solution.

$$c_{HAc_{total}} = 10^{-4} = \{HAc\} + \{Ac^-\} = \frac{\{Ac^-\}^2}{10^{-4.7}} + \{Ac^-\}$$

$$10^{4.7}\{Ac^-\}^2 + \{Ac^-\} - 10^{-4} = 0$$

$$\{Ac^-\} = \{H^+\} = \frac{-1 \pm \sqrt{1^2 - 4(10^{4.7})(-10^{-4})}}{2(10^{4.7})} = 3.6 \times 10^{-5} \frac{mol}{L}$$

The activity of the undissociated acetic acid in solution can be found from the mass balance equation:

$$\{HAc\} = 10^{-4} - \{Ac^-\} = 6.4 \times 10^{-5} \frac{mol}{L}$$

Check the assumption made and the concentration of the hydroxide ion:

$$\{OH^-\} = \frac{k_w}{\{H^+\}} = \frac{10^{-14}}{3.6 \times 10^{-5}} = 2.8 \times 10^{-10} \frac{mol}{L} \ll 3.6 \times 10^{-5} \frac{mol}{L}$$

The pH of the solution is determined by substituting the activity of the hydrogen ion activities into Equation (2.45):

$$pH = -\log\{H^+\} \cong -\log\{3.6 \times 10^{-5}\} = 4.4$$

BOX 2.3 Acid Mine Drainage and Restoration

The Dents Run watershed is located in the northern part of Pennsylvania, as shown in Figure 2.13. This area is home to historic mining and forestry industries, as well as cold-water trout streams and one of the few elk herds east of the Mississippi River. Coal mining began in Dents Run in the late 1800s. These mining operations expanded to include underground "room-and-pillar" mines in the early 1900s. Larger mechanized "strip" mines began operating in the 1940s, and similar mines are still in operation in the region.

FIGURE 2.13 A map showing the Elk State Forest and the Dents Run watershed in northern Pennsylvania.

Source: Based on Cavazza, E.E., Malesky, T., and Beam, R. (2012) The Dents Run AML/AMD Ecosystem Restoration Project. Pennsylvania Department of Environmental Protection, Bureau of Abandoned Mine Reclamation.

The waste products from mining operations consist of a variety of waste rocks and minerals that become exposed to natural weathering processes (see Figure 2.14). Acid rock drainage (ARD) or acid mine drainage (AMD) refers to the acidic drainage that comes from mine waste rock, tailings, and remaining mining structures. The formation of ARD/AMD from existing and past mining

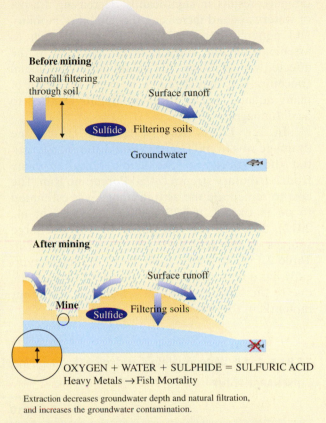

Extraction decreases groundwater depth and natural filtration, and increases the groundwater contamination.

FIGURE 2.14 Mining may expose sulfide-bearing minerals in the earth to natural weathering processes. The exposed sulfides react with oxygen and water to form sulfuric acid resulting in acid mine drainage (AMD).

Source: Based on Corcoran, E., C. Nellemann, E. Baker, R. Bos, D. Osborn, H. Savelli (eds). 2010. Sick Water? The central role of waste-water management in sustainable development. A Rapid Re-sponse Assessment. United Nations Environment Programme, UN-HABITAT, GRID-Arendal. www.grida.no.

operations represents a significant environmental and health threat in many areas throughout the world. For example, by some estimates, more than 7,000 kilometers of waterways east of the Mississippi River and 8,000 to 16,000 kilometers of waterways west of the Mississippi have been negatively impacted by acid drainage in the United States (Kim et al., 1982; U.S. Forest Service, 1993). The acid

(Continued)

BOX 2.3 **Acid Mine Drainage and Restoration** (*Continued*)

drainage results in significantly decreasing the pH of waterways and increasing the solubility, mobility, and aqueous concentrations of metals, such as arsenic, cadmium, copper, silver, and zinc in the affected waters. Lower pH values and higher metal concentrations may threaten or cause the death of many aquatic species in environments exposed to acid drainage.

In the Dents Run watershed, acid mine drainage was found from several different sources, but the majority of the acidic pollution came from a tributary stream called Porcupine Run. The cost to remediate the acid drainage was estimated to be in excess of $7.2 million (Cavazza et al., 2012). The pH and alkalinity of the discharge sources of the acid mine drainage are shown in Table 2.15. The water from these drainages typically has a very red color, as shown in Figure 2.15, due to the high levels of dissolved iron and other minerals in the water.

FIGURE 2.15 Discharge water from an abandoned underground mine flowing into Porcupine Run. The reddish/orange hue of the water is due in large part to the high concentration of iron and other dissolved minerals in the water.

Source: Cavazza, E.E., Malesky, T., and Beam, R. (2012) The Dents Run AML/AMD Ecosystem Restoration Project. Pennsylvania Department of Environmental Protection, Bureau of Abandoned Mine Reclamation. P.O. Box 8476 Harrisburg, PA. Figure 6, page 6.

TABLE 2.15 Water quality of acid mine discharge into Dents Run prior to and after remedial efforts

DISCHARGE POINT	PH		NEUTRALIZING CAPACITY (mg/L)	
	Pre-2010	Post-2012	Pre-2010	Post-2012
Dents Run upstream of Porcupine Run	6.3	6.6	9.6	12.4
Dents Run downstream of Porcupine Run	3.6	5.6	0	10.4
Porcupine Run at mouth	3.4	4.8	0	12.2
Tributary to Porcupine Run	3.9	5.85	0	16.3
Porcupine Run midstream	3.3	4.75	0	12.7
Porcupine Run headwaters	2.9	6.55	0	37.4
Dents Run at mouth	4.7	6.19	7.2	9.3

Source: Based on Cavazza, E.E., Malesky, T., and Beam, R. (2012) The Dents Run AML/AMD Ecosystem Restoration Project. Pennsylvania Department of Environmental Protection, Bureau of Abandoned Mine Reclamation.

The amount of acid from mine waste drainage was determined by analyzing Porcupine Run. Limestone ($CaCO_3$) forms the basic carbonate ion, CO_3^{2-} when dissolved in water, which may react with the H^+ acid to neutralize the acidic drainage. However, because limestone is only partially soluble, an excess amount of limestone must be added to the acid waste in order to bring the water to a pH that is acceptable to native freshwater species. Approximately 500,000 metric tons of limestone from nearby mines were added to neutralize 1.8 million cubic meters of mine waste from 14 AMD sites. The pH of the resulting drainage was increased to levels that are acceptable for aquatic life in Dents Run as a result of the chemical neutralization.

The total cost of the stream restoration process, the results of which are shown in Figures 2.16 and 2.17, was in excess of $14 million. The restoration effort and cost were shared among the project partners that included federal, state, and local government agencies, private foundations, the coal industry, and local grassroots organizations (Cavazza et al., 2012). The restored watershed has resulted in improved habitat for trout, elk (see Figure 2.17), and other species in the region. It also has created jobs in environmental restoration and maintained tourism in the area, where each year 75,000 people visit the region to see the elk herd.

FIGURE 2.16 Unvegetated mine waste shown in the left-hand-side photo, compared to the same area after undergoing the neutralization and restoration process near Porcupine Run.

Source: Cavazza, E.E., Malesky, T., and Beam, R. (2012) The Dents Run AML/AMD Ecosystem Restoration Project. Pennsylvania Department of Environmental Protection, Bureau of Abandoned Mine Reclamation. P.O. Box 8476 Harrisburg, PA. Figure 11, page 16.

FIGURE 2.17 Elk cooling off during the summer in a pond used to neutralize acid mine drainage in the Dents Run watershed.

Source: Cavazza, E.E., Malesky, T., and Beam, R. (2012) The Dents Run AML/AMD Ecosystem Restoration Project. Pennsylvania Department of Environmental Protection, Bureau of Abandoned Mine Reclamation. P.O. Box 8476 Harrisburg, PA. Figure 15, page 20.

2.7 Environmental Fate and Partitioning of Chemicals

Chemicals may move between three phases: gas, liquid, and solid (Figure 2.18). In the natural environment, chemicals usually exist in a mixture. For example, nitrogen, oxygen, argon, water vapor, and other gases make up the mixture we know as air. When water in the air moves from the gaseous state to the liquid state, through the process called condensation, the water falls out of the air in the form of precipitation. Soils consist of a complex mixture of silicon, calcium, carbon, nitrogen, phosphorus, and many other elements that are distributed unevenly in soils and sediments. These elements may dissolve and become a part of an aqueous mixture in runoff and groundwater. Dissolved minerals may also precipitate and fall out of an aqueous solution if the chemistry, pH, or temperature of the solution changes.

The movement of chemicals from one phase or mixture to another is governed by a variety of environmental chemistry principles. Two general types of reactions may occur within a system: Reduction–oxidation processes occur when the oxidation state of participating atoms change. Ionic reactions are those reactions, like acid–base reactions, where there is a change in ion–ion interactions and relationships. Ionic reactions also frequently occur when a metal ion reacts with a base in precipitation and dissolution reactions, as in

$$Fe^{2+}_{aq} + 2OH^- \leftrightarrow Fe(OH)_{2(s)} \qquad (2.50)$$

The dissolution of minerals is a large factor in determining the chemical composition of much natural water. Typically, precipitation and dissolution reactions occur much more slowly than acid–base reactions, so that both equilibrium considerations and the rate of the reaction is important in determining the extent and amount of chemical change that may occur within a system.

The extent of a compound's solubility is strongly influenced by ionic factors. Compounds that have a positive or negative charge tend to bond more strongly to water and may be more soluble (Figure 2.19). Compounds that bond strongly to water and tend to dissolve or stay in solution are called hydrophilic (water loving) compounds. Hydrophobic compounds are more strongly bonded to compounds with chemistries similar to those of their own and tend not to dissolve as much in water.

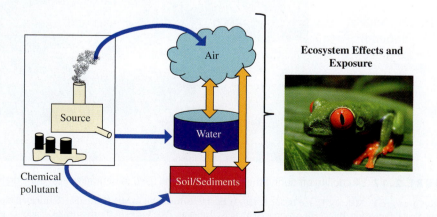

FIGURE 2.18 Pollutants may migrate from an anthropogenic or natural source through the air, water, or soil where they may come into contact with plants or animals.

Source: Ryan M. Bolton/Shutterstock.com.

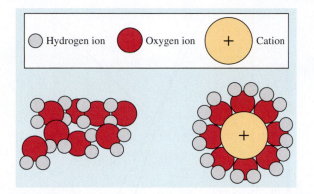

FIGURE 2.19 Compounds with an ionic charge as shown on the left tend to align and interact with water molecules (shown in red and gray), making ionic compounds more likely to dissolve into a solution.

Source: Bradley Striebig.

Salts are ionic compounds that dissociate completely in solution; examples are sodium chloride or table salt:

$$NaCl \rightarrow Na^+ + Cl^-$$

Salts therefore are very soluble compounds. The term *solubility* only has meaning in comparisons of one compound to another. In environmental chemistry, the solubility of carbonate compounds in water is often used as a reference point in comparing the solubility of different species and dissolved compounds. Some general guidelines for estimating chemical solubility are shown in Table 2.16.

Some metals may dissolve in water to form compounds that behave like acids or bases in solutions by abstracting a hydrogen or hydroxide ion from water, as in the case of the reaction of aluminum in solution shown in the following equations

In an acidic solution, aluminum is present in the form Al^{+3}:

$$Al(OH)_{3(aq)} + 6H_2O^+_{(aq)} \leftrightarrow Al^{+3}_{(aq)} + 6H_2O \qquad (2.51)$$

In a basic solution, aluminum reacts with water to form a complex aluminum hydroxide compound:

$$Al(OH)_{3(aq)} + OH^-_{(aq)} \leftrightarrow Al(OH)^-_{4(aq)} \qquad (2.52)$$

The solubility of a dissolved compound is influenced by the concentration of the given compound, the pH of the solution, and other dissolved components in solution. Phase diagrams for pure solutions illustrate the relationship between pH and

TABLE 2.16 Solubility guidelines for common inorganic compounds

ION	CHARACTERISTIC SOLUBILITY
Nitrate, NO^{3-}	Soluble
Chloride, Cl^-	Soluble, except AgCl, PbCl, and HgCl
Sulfate, SO_4^{2-}	Soluble, except $BaSO_4$ and $PbSO_4$; $AgSO_4$, $CaSO_4$, and $HgSO_4$ are only slightly soluble
Carbonate, CO^{3-}; Phosphate, PO_4^{3-}; Silicate, SiO_4^{4-}	Insoluble, except those of Na, K, and NH_4^+
Hydroxide, OH^-	Insoluble, exceptions include LiOH, NaOH, KOH, NH_4OH, $Ba(OH)_2$, $Ca(OH)_2$, and $Sr(OH)_2$
Sulfide, S^{2-}	Insoluble, except for alkali metal sulfides, $(NH_4)_2S$, MgS, CaS, and BaS
Sodium, Na^+; Potassium, K^+; Ammonium, NH_4^+	Soluble, except for iron (Fe) and compounds that contain these ions with a heavy metal

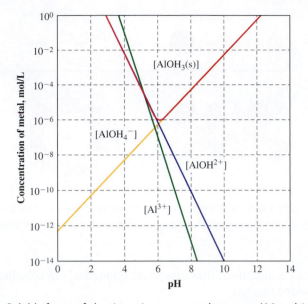

FIGURE 2.20 Soluble forms of aluminum in pure water between pH 0 and 14, and range in which aluminum would be expected to precipitate.

Source: Based on Tchobanoglous, G., Burton, F. L., and Stensel, H.D. (2003) Wastewater Engineering: Treatment and Reuse, Fourth Edition. The McGraw-Hill Companies, NY, USA.

solubility in Figure 2.20, where the solid lines approximate the total concentration of the stable residual aluminum concentration after precipitation of any insoluble components. Hydroxide ions or sulfur ions may be added to water to create less soluble forms of many metals in solution. Similar hydroxide and sulfur species solubility curves are shown for other common metals in Figures 2.21 to 2.23.

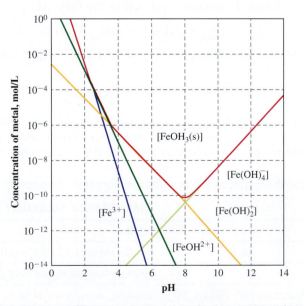

FIGURE 2.21 Soluble forms of ferric iron in pure water between pH 0 and 14, and range in which aluminum would be expected to precipitate.

Source: Based on Tchobanoglous, G., Burton, F. L., and Stensel, H.D. (2003) Wastewater Engineering: Treatment and Reuse, Fourth Edition. The McGraw-Hill Companies, NY, USA.

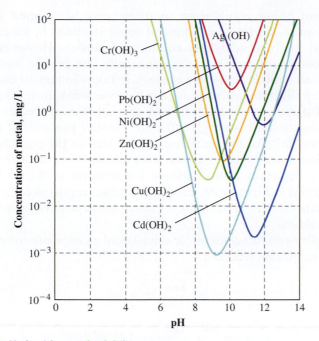

FIGURE 2.22 Hydroxide metal solubility curves.

Source: Based on Tchobanoglous, G., Burton, F. L., and Stensel, H.D. (2003) Wastewater Engineering: Treatment and Reuse, Fourth Edition. The McGraw-Hill Companies, NY, USA.

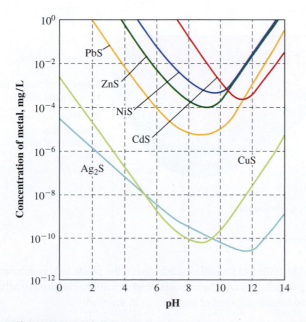

FIGURE 2.23 Sulfide metal solubility curves.

Source: Based on Tchobanoglous, G., Burton, F. L., and Stensel, H.D. (2003) Wastewater Engineering: Treatment and Reuse, Fourth Edition. The McGraw-Hill Companies, NY, USA.

Solid–liquid phase equilibrium solutions may be applied to calculating the dissolved and solid fractions of metal concentrations in water in a method similar to that used for acid–base equilibrium models. However, equilibrium reactions involving solids in water are generally less precise and have greater uncertainty than similar acid–base models. It takes longer to complete solid–liquid phase equilibrium processes in part because solids form a heterogeneous, or uneven, distribution in solution and the substances become less evenly mixed. Solid formation is strongly affected by other particles in the water, the degree of crystallinity of the particles, and the size of the particles in solution. Supersaturation, which creates faster removal reactions in what is called the labile phase, is generally a precondition for removal of the solid phase from solution, as shown in Figure 2.24. Finally, the values reported for solubility equilibrium constants in the literature vary much more widely than the acid–base equilibrium values.

The solubility–product constant, or solid–liquid equilibrium constant, is defined by the following general reaction:

$$A_aB_{b(s)} \overset{water}{\longleftrightarrow} aA^{b+} + bB^{a-} \tag{2.53}$$

$$k_{sp} = \frac{\{A^{b+}\}^a\{B^{a-}\}^b}{\{AB_{(s)}\}} = \{A^{b+}\}^a\{B^{a-}\}^b \tag{2.54}$$

Recall that the solid phase activity $\{AB_{(s)}\}$ is equal to one as defined by our standard state conditions. The equilibrium expressions and mass balance equations may be used to predict the solubility of compounds. Generally, the solubility of a given compound is reduced if the same ions are already present in solution; this effect is referred to as the common ion effect.

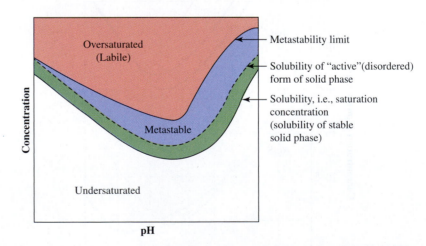

FIGURE 2.24 Illustration of the difference between the predicted equilibrium conditions to remove metals by precipitation and the oversaturated conditions required to achieve conditions that in practice produce precipitation of metal solids in solution.

Source: Based on Stumm and Morgan: Aquatic Chemistry: Chemical Equilibria and Rates in Natural waters, 3rd ed. John Wiley & Sons International.

EXAMPLE 2.16   Estimating the Solubility of Calcium Hydroxide

Calcium hydroxide, $Ca(OH)_2$, is widely used in industry as well as in some water treatment applications to modify the pH of solutions. Estimate the solubility of $Ca(OH)_2$ in distilled water at 25°C. Assume the solution acts as an ideal fluid. The solubility product constant is $k_{sp} = 10^{-5.19}$.

$$Ca(OH)_2 \leftrightarrow Ca^{2+} + 2(OH^-)$$

$$k_{sp} = [Ca^{2+}][OH^-]^2 = 10^{-5.19}$$

Inspecting the balanced chemical equation shows that two moles of hydroxide ion are formed for each mole of calcium ion formed. Using this relationships, let

$$[Ca^{2+}] = S \text{ and then } [OH^-] = 2S$$

Substituting the above into the equation for the solubility product constant yields

$$k_{sp} = [Ca^{2+}][OH^-]^2 = S(2S)^2 = 4S^3 = 10^{-5.19}$$

Solving for S yields

$$S = [Ca^{2+}] = 0.012 \text{ mol/L}$$

$$[OH^-] = 2S = 0.024 \text{ mol/L}$$

EXAMPLE 2.17   The Common Ion Effect for Fluoride in Water

Fluoride is commonly added to drinking water to help prevent tooth decay. Calculate the solubility of calcium fluoride, CaF_2, at 25°C if the water already contains 16 mg/L of the fluoride ion. Assume that the solution acts as an ideal-fluid. The solubility product constant is $k_{sp} = 10^{-10.50}$.

$$CaF_2 \leftrightarrow Ca^{2+} + 2F^-$$

$$k_{sp} = [Ca^{2+}][F^-]^2 = 10^{-10.50}$$

The molecular mass of fluoride is 19 g/mol. The pre-existing concentration of fluoride in the water is

$$[F^-_{(initial)}] = 16 \frac{mg}{L} \times \frac{mmol}{19 \text{ mg}} \times \frac{mol}{1000 \text{ mmol}} = 10^{-3.1}$$

Inspecting the balanced chemical equation shows that two moles of fluoride ion are formed for each mole of calcium ion formed. Using this relationships, let

$$[Ca^{2+}] = S \text{ and}$$

$$[F^-] = 2S + 10^{-3.1}$$

Substituting the above into the equation for the solubility product constant yields

$$k_{sp} = [Ca^{2+}][F^-]^2 = S(2S + 10^{-3.1})^2 = 10^{-10.50}$$

To simplify the algebraic solution, assume $2S \ll 10^{-3.1}$, then

$$k_{sp} = S(10^{-3.1})^2 = 10^{-10.50}$$

Solving for S yields

$$S = [Ca^{2+}] = 4.45 \times 10^{-5}$$

$2S = 8.90 \times 10^{-5} < 10^{-3.91}$ so the above assumption is close to this estimate, but $2S$ is not completely negligible. A more precise solution can be determined by iterating based on the previous solution. It will be assumed that:

$$S = [Ca^{2+}] = 3.75 \times 10^{-5} = 10^{-4.43}$$

Substituting into the equilibrium expression:

$$S(2S + 10^{-3.1})^2 = 3.75 \times 10^{-5}(2(3.75 \times 10^{-5}) + 10^{-3.1})^2 = 10^{-10.50}$$

then after adding CaF_2:

$$[F^-] = 2(3.75 \times 10^{-5}) + 10^{-3.1} = 0.00092 = 10^{-3.04} \text{ mol/L} = 17.4 \text{ mg/L}$$

Metals may have several forms in solution by reacting with other inorganic compounds, other organic compounds, and other particles in solution, as shown in Figure 2.25. In complex environmental systems, these complicating factors should be considered when evaluating the impacts of metals in the environment.

Other equilibrium-based coefficients may also be used to estimate the concentration, form, and fate of chemical compounds in the environment. In the most general form, the equilibrium concentrations of any compound, A, between two phases, x and y, can be compared by

$$A_{(x - \text{phase})} \leftrightarrow A_{(y - \text{phase})} \tag{2.55}$$

then

$$k_{A_{x-y}} = \frac{[A]_{x - \text{phase}}}{[A]_{y - \text{phase}}} \tag{2.56}$$

The fraction of the total amount of A at equilibrium in the generic x-phase is

$$f_{Ax} = \frac{\text{mass of } A \text{ in phase } x}{\text{total mass of } A} = \frac{(C_{Ax})(V_x)}{(C_{Ax})(V_x) + (C_{Ay})(V_y)} = \frac{1}{1 + \dfrac{C_{Ay} V_y}{C_{Ax} V_x}} \tag{2.57}$$

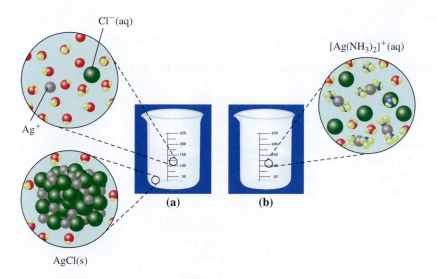

FIGURE 2.25 Silver chloride, AgCl(s), in distilled water is largely insoluble as shown in part (a). However, when the ammonium ion is added to water, the silver and ammonium ions form a complex molecule $[Ag(NH_3)_2]^+$ that dissolves into the aqueous solution in part (b).

Source: Based on Petrucci, Herring, Madura and Bissonnette. (2011) General Chemistry: Principles and Modern Applications 10th edition. Pearson Prentice Hall, Toronto, Canada.

The term V_y/V_x is simply the ratio of the two volumes, which can be defined as R_{xy}. Substituting the volume ratio and Equation (2.30) into Equation (2.31) yields

$$f_{Ax} = \cfrac{1}{1 + \cfrac{1}{(k_{A_{xy}})(R_{xy})}} = \frac{(k_{A_{xy}})(R_{xy})}{(k_{A_{xy}})(R_{xy}) + 1} \qquad (2.58)$$

Because our environment is not a homogeneous mixture of air and water, but a complex mixture that involves heterogeneous solid mixtures as well, various equilibrium approaches have been used to describe the gas–solid and liquid–solid phase equilibrium. Adsorption describes the process of a chemical adhering to a solid *surface*, whereas absorption describes the process of a chemical becoming incorporated into the bulk phase of a given volume of a liquid or solid. Because it is often difficult to determine precisely if a material is adsorbed on a surface or absorbs into a given material, the two processes are often described collectively as *sorption*. The *sorbate* is the substance that is transferred from one phase to another. The *sorbent* is the material into or onto which the sorbate is transferred. Sorption occurs due to the electrostatic or van der Waals force, chemical bonding or surface coordination reactions, or partitioning of the compound into the bulk phase of the sorbent.

Activated carbon is used in Brita drinking water filters, water treatment plants, and air pollution control facilities to remove chemical pollutants from either the gas or liquid phase. The Freundlich isotherm is used to estimate the portioning of pollutants from a more mobile phase in gas or water to the solid-activated carbon material. The Freundlich equation has the form

$$q = KC^{1/n} \qquad (2.59)$$

where

q = mass of sorbate sorbed per unit mass of sorbent (mg/g)

K = the Freundlich parameter that describes the partitioning $\left(\frac{\text{mg/g}}{(\text{mg/L})^{1/n}}\right)$

C = the equilibrium sorbate concentration in the aqueous solution (mg/L)

n = Freundlich isotherm intensity parameter (unitless)

The dimensionless octanol/water partition coefficient, K_{ow}, provides an indication of the solubility of organic compounds in water. A chemical that is insoluble in water is more likely to adhere to solid particles in soils and sediments. The octanol/water partition coefficient, K_{ow}, is defined by

$$K_{ow} = \frac{\dfrac{\text{mg chemical}}{\text{L octanol}}}{\dfrac{\text{mg chemical}}{\text{L water}}} \tag{2.60}$$

Octanol ($C_7H_{15}CH_2OH$) is an organic chemical with moderate polar and non-polar properties, so that a wide variety of organic compounds are soluble in octanol solutions. The octanol/water partition coefficient has been used in the pharmaceutical industry because the partitioning between octanol and water simulates the partitioning of drugs between water and fat; as such it is also a useful factor in estimating the bioaccumulation of organic compounds in animal fat. Smaller, more polar compounds tend to be more soluble in water, yielding a lower value for K_{ow}. Larger and fewer polar molecules tend to have a greater solubility in octanol and hence also tend to attach more readily to soil particles.

The organic carbon normalized partition coefficient, K_{oc}, is a useful parameter for predicting the partitioning of chemicals between soil and water and can be estimated from the octanol/water partition coefficient. The organic carbon normalized partition coefficient is defined as

$$K_{oc} = \frac{\dfrac{\text{mg chemical}}{\text{kg organic carbon}}}{\dfrac{\text{mg chemical}}{\text{L water}}} \tag{2.61}$$

Several empirical correlations have been developed for estimating the K_{oc} value based on a chemical's K_{ow} value

$$\text{Karickhoof et al. correlation: } \log(K_{oc}) = 1.00 \log(K_{ow}) - 0.21 \tag{2.62}$$

$$\text{Schwarzenbach and Westall correlation: } \log(K_{oc}) = 0.72 \log(K_{ow}) + 0.49 \tag{2.63}$$

If the decimal fraction of the organic carbon present in the soil f_{oc}, is known, then the sediment/water partition coefficient can be calculated from

$$K_p = f_{oc} K_{oc} = \frac{\dfrac{\text{mg chemical}}{\text{kg solids}}}{\dfrac{\text{mg chemical}}{\text{L water}}} \tag{2.64}$$

EXAMPLE 2.18   Chemical Partitioning in the Environment

Vinyl chloride is a colorless, odorless gas used in the production of polyvinyl chloride (PVC) plastics and in numerous industries. Vinyl chloride is also a known human carcinogen, and the EPA's maximum contaminant level (MCL) in drinking water is 0.002 mg/L, or 2 ppb. The vinyl chloride concentration in the soil outside of a plastic factory is 0.007 mg/kg, and the soil has a 15% organic content. Vinyl chloride (chloroethene) has a reported octanol–water coefficient equal to 4. The organic carbon normalized partition coefficient, K_{oc}, can be estimated using the procedures based on Equations (2.62) and (2.63) and the sediment/water partition coefficient. The sediment/water partition coefficient can be used to predict the possible equilibrium concentration of vinyl chloride in the groundwater.

$$\text{Karickhoof et al. correlation: } \log (K_{oc}) = 1.00 \log (K_{ow}) - 0.21$$

$$\log (K_{oc}) = 1.00 \log (4) - 0.21 = 0.4$$

$$K_{oc} = 2.5$$

$$\text{Schwarzenbach and Westall correlation: } \log (K_{oc}) = 0.72 \log (K_{ow}) + 0.49$$

$$\log (K_{oc}) = 0.72 \log (4) + 0.49 = 0.9$$

$$K_{oc} = 8.4$$

The sediment/water partition coefficient can be determined from the following empirical equations.

Using the Karickhoof approximation yields

$$K_p = f_{oc} K_{oc} = 0.15 * 2.5 = 0.37$$

Using the Schwarzenbach and Westall approximation yields

$$K_p = f_{oc} K_{oc} = 0.15 * 8.4 = 1.26$$

The partition coefficient in Equation 2.64 can be rearranged to yield

$$C_{water} \left[\frac{mg}{L} \right] = \frac{C_{soil} \left[\dfrac{mg_{pollutant}}{kg_{soil}} \right]}{K_p} = \frac{0.007 \left[\dfrac{mg_{pollutant}}{kg_{soil}} \right]}{0.37} = 0.019 \, \frac{mg}{L}$$

Using the values from the Schwarzenbach and Westall approximation yields $C_{water} = 0.006$ mg/L.

The expected range based in the groundwater would be 19 ppb $> C_{water} > 6$ ppb; in both cases this concentration is greater than the MCL of 2 ppb.

BOX 2.4 Metal Mobility in the Environment

For decades, lead has been used in making bullets and other ammunition because it has been readily available and is easy to form. However, exposure to lead in the environment has been linked to learning disabilities and other environmental and health concerns. The EPA estimates that roughly 7.26×10^7 kg of lead shot and bullets are released into the environment each year in the United States just from nonmilitary outdoor ranges. Concerns about the toxicity of lead in the environment have led to the development of a "green bullet" that replaces lead with tungsten munitions. Tungsten is a heavy metal with several unique physical and mechanical properties that make it appealing as a replacement for lead. In shooting ranges, where lead munitions have traditionally been fired, fragments of lead fall on the soil.

Once lead enters the soil, various factors will determine the extent of the actual hazard it might pose. The rate at which lead may move through the soil and groundwater depends on various factors, including soil type, soil pH, and rainfall volume. At a near-neutral pH, lead is relatively insoluble. However, if soils become more acidic (decreasing pH), lead solubility tends to increase, making it easier for the pollutant to spread over a larger area. The more easily the lead travels through the soil, the greater the impact it will have. However the solubility and transport of lead in the soil are generally limited by the reaction of lead with natural carbonate in the soil that forms lead carbonate (Cerussite), which is minimally soluble. In fact, the standard treatment recommended to limit lead transport in natural environments is to add pulverized limestone ($CaCO_3$) to raise the soil pH and increase the total soil carbonate concentration. The increased soil pH reduces the corrosion rate of Pb by lowering the concentration of available H^+ ions and reducing the solubility of the Cerussite that is formed.

Now that "green" bullets are being used in the same outdoor firing ranges, tungsten can be released into soils that already contain "immobilized lead." However, fate and transport research on tungsten and tungsten alloys has shown that tungsten corrodes through a process that leads to dissolved oxygen depletion and pH reduction in water and certain soils. The corrosion of tungsten can be represented by two reactions involving the oxidation of metallic tungsten and the reduction of dissolved molecular oxygen.

$$\frac{3}{2}O_2 + 5H_2O + 6e^- \rightarrow 8OH^- + 2H^+ \qquad (2.65)$$

$$W + 8OH^- \rightarrow WO_4^{2-} + 4H_2O + 6e^- \qquad (2.66)$$

The rate of dissolution, and by extension, the rate of acidification of the soil system, depend on several factors, including the chemical properties of the soil, level of oxygen in the soil, and the ionic capacity of the soil.

Results of experiments performed by some researchers suggest that when tungsten is released in some soils that contain immobilized legacy lead, the corrosion of tungsten can lead to a reduction in soil pH (acidification), leading to increased solubility and transport of lead through the soil and subsequent contamination and impact. This is an example of how trying to solve one environmental problem can lead to another one if proper analysis and modeling are not employed. In 2002, the National Center for Environmental Health of Centers for Disease Control and Prevention recommended toxicological and carcinogenesis studies. This recommendation was largely driven by the discovery of elevated tungsten concentrations in biological and drinking water samples in three different communities with leukemia clusters (CDC, 2003).

2.8 Summary

Chemicals in the environment move between the gas, liquid, and solid phases based on their chemical properties and other properties of the system. Chemicals in the gas phase move with local air movements and patterns and may be carried long distances and broadly dispersed. Chemicals that dissolve in water may move from the gas phase into the liquid phase and vice versa. Chemicals that have a preference for the liquid phase may move with local water movements through precipitation, runoff, infiltration, and groundwater movements. Chemicals that have an affinity for the solid phase or solid surfaces tend not to be very mobile, but may often persist in the environment with little change in concentration for extremely long time spans. The ability to assess and analyze risk associated with remediation or mitigation of potentially harmful chemicals in the environment must be considered, along with the transport and fate of those chemicals in the environment.

References

Benjamin, M. M. (2001). *Water Chemistry.* New York: McGraw-Hill.

Briggs, J. C., and Ficke, J. F. (1977). *Quality of Rivers of the United States, 1975 Water Year—Based on the National Stream Quality Accounting Network (NASQAN)*: U.S. Geological Survey Open-File Report 78–200.

Bouwman, A. F., Van Vuuren, D. P., Derwent, R. G., and Posch, M. (2002). "A global analysis of acidification and eutrophication of terrestrial ecosystems." *Water, Air and Soil Pollution* 141:349–382.

Butler, J. N. 1982. *Carbon Dioxide Equilibria and Their Applications.* Addison-Wesley, Reading, Mass.

Cavazza, E. E., Malesky, T., and Beam, R. (2012). The Dents Run AML/AMD Ecosystem Restoration Project. Pennsylvania Department of Environmental Protection, Bureau of Abandoned Mine Reclamation. P.O. Box 8476, Harrisburg, PA.

Centers for Disease Control and Prevention. National Center for Environmental Health. Cancer Clusters. Churchill County (Fallon). Nevada Exposure Assessment. [online] 2003.

Ceto, N., and Mahmud, S. (2000). *Abandoned Mine Site Characterization and Cleanup Handbook.* Washington, D.C.: U.S. Environmental Protection Agency. EPA 910-B-00-001

Clark, M. L., Sadler, W. J., and Ney, S. E. (2004). "Water-Quality Characteristics of the Snake River and Five Tributaries in the Upper Snake River Basin, Grand Teton National Park, Wyoming, 1998-2002." Scientific Investigations Report 2004-5017. U.S. Geological Survey, Reston, VA.

Dufor, C. N., and Becker, E. (1964). "Public Water Supplies of the 100 Largest Cities in the United States, 1962." U.S. Geological Survey, Water Supply Paper 1812.

Flippo, H. N., Jr. (1974). Springs of Pennsylvania. U.S. Department of the Interior, Geological Survey, Harrisburg, PA.

Hart, J. (1988). *Consider a Spherical Cow: A Course in Environmental Problem Solving.* University Science Books, Sausalito, CA.

IPCC. (2007). *Climate Change* 2007: "The Physical Science Basis." Contribution of Working Group I to the Fourth Assessment Report of the Intergovernmental

Panel on Climate Change. [Solomon, S., D. Qin, M. Manning, Z. Chen, M. Marquis, K. B. Averyt, M. Tignor, and H. L. Miller (Eds.)] Cambridge, UK: Cambridge University Press.

Jepson. J. N. (2003). *A Problem Solving Approach to Aquatic Chemistry*. New York: Wiley.

Langelier. W. F. (1936). "The analytical control of anti-corrosion water treatment." *Journal of American Water Works Association* 28(1936):1500–1521.

Masters, G .M., and Ela, W. P. (2008). *Introduction to Environmental Engineering and Science,* 3rd ed. Upper Saddle River, NJ: Prentice-Hall.

Meybeck, M., Chapman, D., and Helmer, R. (Eds.) (1989). *Global Freshwater Quality: A First Assessment*. London: Blackwell.

Nahle, Nasif. (2007). "Cycles of global climate change." Biology Cabinet Journal Online. Article no 295.

Snoeyink, V.L. and Jenkins, D.. (1980). *Water Chemistry.* New York: Wiley.

Stumm, W. and Morgan, J.J. (1996). *Aquatic Chemistry: Chemical Equilibria and Rates in Natural Waters,* 3rd ed. John Wiley & Sons Inc., New York, USA.

Tchobanoglous, G., Burton, F. L., and Stensel, H. D. (2003). *Wastewater Engineering: Treatment and Reuse*, 4th ed. New York: McGraw-Hill.

UNEP GEMS. (2006). *Water Quality for Ecosystem and Human Health*. United Nations Environment Programme, Global Environment Monitoring System/Water Programme.

U.S. EPA. (1994). *Acid Mine Drainage Prediction: Technical Document*. U.S. Environmental Protection Agency. Washington, D.C.: Office of Solid Waste. Special Waste Branch.

U.S. EPA. (2009). *National Primary Drinking Water Regulations*. EPA 816-F-09-004.

Virginia Department of Environmental Quality. (2009a). Virginia Ambient Monitoring Data Report.

Virginia Department of Environmental Quality. (2009b). Evaluation of Virginia's Air monitoring Network: A Report by the State Advisory Board on Air Pollution. November 20, 2009.

Virginia Ambient Air Monitoring 2010 Data Report. (2011). Office of Air Quality Monitoring, Virginia Department of Environmental Quality.

Visser, S. A., and Villeneuve, J. P. (1975). "Similarities and differences in the chemical composition of waters from West, Central, and East Africa." *Verh int. Verein. Theor. Angew. Limnology* 19:1416–1425.

Key Concepts

Henry's law constant
Partial pressure
Dissolved inorganic carbon
Law of electroneutrality
Ionic strength
Suspended solids
Volatile solids
Dissolved solids
Water hardness
Carbonate hardness
Noncarbonate hardness

Chemical reactivity
Chemical activity
Standard state activity
Activity coefficient
Solubility
Precipitate
Reduction-oxidation process
Ionic reactions
Dissolution of salts
Sorbent
Sorbate

Active Learning Exercises

ACTIVE LEARNING EXERCISE 2.1: Invasion of the toxic marbles

This exercise is designed to help visualize and reinforce the concept of ppm_m and ppm_v. All that is needed are a few marbles (preferably two different colors) and a drinking container. Any container will do, as long as the volume is known or can be assumed. If each of the marbles weighs 1 mg and six marbles of one color have $MW = 20$ g/mole while four marbles of another color have $MW = 40$ g/mole:

- What is the concentration of marbles in the container if it were filled with water in the given units?
  - mg/L
  - ppm_m
- What is the concentration of marbles in the 0.5-liter jar filled with air at STP in the given units?
  - mg/m^3
  - ppm_v

The conversions between units are an extremely important part of understanding quantitatively how we can assess measures of sustainability. Understanding these units and how to convert from one to another is a prerequisite to understanding environmental impacts, associating societal consequences of actions (or the lack of actions), and forecasting economic impacts. A deep understanding of unit analysis is also vital to performing Life Cycle Analysis based on material usage and transformation of those materials into different pathways. If that were not enough of a reason to understand and practice unit conversions, the bottom line is that, if you work in the field of environmental impact analysis, environmental policy, or sustainable engineering, converting from one unit basis to another is a vital skill you will need to be successfully employed. Furthermore, people outside this field of study will likely lack this skill and will need to pay you to perform these calculations!

ACTIVE LEARNING EXERCISE 2.2: Conversion of gas concentration between volumetric fractions and mass concentration, part II

The United States Environmental Protection Agency (U.S. EPA) reported the concentrations of criteria air pollutants in 2008 in Fairfax County, a suburban area near Washington, D.C. (see Table 2.17). Convert these concentrations from ppm_v and ppb_v to mg/m^3 and $\mu g/m^3$, respectively.

TABLE 2.17 Criteria pollutants observed in Fairfax, Virginia

CRITERIA POLLUTANT	VALUE	UNIT
O_3	107	ppb_v
SO_2	16	ppb_v

Source: Based on Virginia Ambient Air Monitoring 2010 Data Report, 2011. Office of Air Quality Monitoring, Virginia Department of Environmental Quality
http://www.deq.virginia.gov/Portals/0/DEQ/Air/AirMonitoring/Annual_Report_10.pdf

ACTIVE LEARNING EXERCISE 2.3: **Activity and concentration relationships**

The concepts of activity and reactivity are not difficult, but unfortunately, the language used to describe them can be intimidating. An example will help demonstrate the differences between activity and reactivity. Consider a slow reaction process, where compound X reacts slowly with compound Y to form XY:

$$X + Y \rightarrow XY$$

In the lab you are given a stop watch and a lab instrument that measures the mass of X and the mass of XY. You are asked to conduct an experiment to determine the activity coefficient for X in a sample of each of the following.
a) Pure fresh water
b) Estuary water, approximately a mixture of fresh water and salt water
c) Salt water

How would you set up your experiment?
Under what conditions would you conduct the experiment in the laboratory environment?
What information would each piece of equipment provide?

ACTIVE LEARNING EXERCISE 2.4: **Carbonate species in bottled soft drinks**

Bottled soft drinks, such as Coca-Cola® or Pepsi®, contain a high amount of dissolved carbon dioxide. The carbon dioxide escaping the liquid phase and going into the gas phase produces the bubbles or "fizziness" of these drinks.

- Using pH paper or a pH probe, measure the pH of your favorite soft drink.
- Which of the following assumptions do you hypothesize is true?

   - $[H^+] = [HCO_3^-]$

   or

   - $[H^+] = [H_2CO_3]$

- Draw a pC versus pH diagram for the soft drink that shows the concentrations of $[H^+]$, $[OH^-]$, $[H_2CO_3]$, $[HCO_3^-]$, and $[CO_3^{2-}]$.

Problems

2-1   Define the following terms or concepts.
a. Law of electroneutrality
b. Normality
c. Equivalents
d. Total dissolved solids (TDS)
e. Total suspended solids (TSS)
f. Volatile suspended solids (VSS)
g. Fixed solids
h. Total hardness
i. Carbonate hardness

j. Noncarbonated hardness

k. Chemical reactivity

l. Chemical activity

m. Standard state activity

n. Activity coefficient

o. Solubility

p. Precipitate

q. Reduction–oxidation process

r. Ionic reaction

s. Dissolution salts

t. Sorbent

u. Sorbate

2-2 Mass and energy balances are helpful to track chemicals as they move from one _____ to another; however, we must study the applications and principles of chemistry more closely to appreciate how these chemicals are transformed in the environment and also transform the Earth's environment.

2-3 How much have carbon dioxide levels in the atmosphere increased since the start of the Industrial Revolution?

2-4 What five elements are found in life on the planet?

2-5 Write the balanced equation for the combustion of methane (CH_4), in the presence of oxygen.

2-6 How many atoms are in a mole?

2-7 What is the numerical value of R in [J/mol-K] in the ideal gas law?

2-8 Define the partial pressure of a gas.

2-9 What is the relationship between the concentration of a gas in air in mg/m^3 and ppm_v at standard temperature and pressure?

2-10 What is the relationship between the concentration of a gas in air in mg/m^3 and ppm_v at nonstandard temperature and pressure?

2-11 Relate parts per billion (ppb_v) to ppm_v.

2-12 Use balanced chemical equations to describe the forms of dissolved carbon dioxide in aqueous solution. There should be a diprotic acid, a monovalent anion, and a divalent conjugate base.

2-13 The normality of a solution is the number of equivalents per liter and can be determined by multiplying the concentration of a species, MW_i, by the number of equivalents, z_i.

2-14 The concept of an equivalent weight, EW, may also be useful in calculations involving aqueous solutions. The equivalent weight of any species, EW_i, is equal to the molar mass divided by the number of equivalents associated with the dissolved ion.

2-15 The ionic strength, I, of a solution is the estimate of the overall concentration of dissolved ions in solution and is defined by what?

2-16 The activity coefficient is defined as the ratio of the reactivity per molecule of mol of A in a real system compared to the reactivity of A in the standard reference state.

2-17 The fundamental parameter for equilibrium base models is the equilibrium constant, k.

2-18 The pH of natural waters is important to biodiversity since most species are only tolerant of natural waters in the pH range from _____.

2-19 Phosphoric acid has the form H_3PO_4. List all the acids that may donate a hydrogen ion and the conjugate bases that may accept the proton for this acid compound,

2-20 What is the primary characteristic of an amphoteric compound?

2-21 Define the equilibrium constant for pure water.

2-22 What are the primary characteristics of a strong acid?

2-23 What are the primary characteristics of a strong base?

2-24 Write the equation that defines pH.

2-25 Write the equation that defines pOH.

2-26 Write the equation that relates k_w, pH, and pH.

2-27 What are the steps involved in modeling the effects of acids or bases in the natural environment?

2-28 Define the primary characteristic of a weak acid that partially dissociates in an aqueous solution. *Hint*: It may help to use the acid equilibrium constant in your description.

2-29 Define the primary characteristic of a weak base that partially dissociates in an aqueous solution. *Hint*: It may help to use the base equilibrium constant in your description.

2-30 The solubility-product constant, or solid-liquid equilibrium constant, is defined by what general reaction?

2-31 What is the fraction of the total amount of A at equilibrium in the generic x-phase?

2-32 The Freundlich isotherm is used to estimate the portioning of pollutants from a more mobile phase in gas or water to the solid activated carbon material. The Freundlich equation has what form?

2-33 The octanol/water partition coefficient, K_{ow}, is defined by _____.

2-34 The organic carbon normalized partition coefficient is defined as _____.

2-35 A 12-ounce can of soda contains about 40 grams of sugar. What is the concentration of sugar in a can of soda in mg/L?

2-36 A 2-ounce serving of espresso contains 100 mg of caffeine. Professor Coffy often has a 16-ounce iced latte with three shots of espresso latte before his 8 a.m. class.
  a. What is the concentration of caffeine in mg/L in a single shot of espresso?
  b. What is the concentration of caffeine in ppm in a single shot of espresso?
  c. What is the concentration of caffeine in his coffee drink?
   i) in mg/L
   ii) in ppm

2-37 There is about 5 mg of caffeine in each shot of decaf espresso. The barista is being paid under the table by Mrs. Coffy to change the espresso shots to decaffeinated espresso in Professor Coffy's 16-ounce latte drink cup from Problem 2-36.
  a. How much caffeine in mg does Dr. Coffy consume if he drinks decaffeinated lattes before class?
  b. How many 16-ounce decaf lattes must he drink before he consumes the amount of caffeine equivalent to his old three shots of caffeinated espresso (from 2-36)?

2-38 The average concentration of dissolved oxygen (DO) in the Shenandoah River was reported as 9.7 ppm_m in 2006. What is the concentration in mg/L in the river water?

2-39 Table 2.18 includes typical constituents in water in mg/L. Complete the table by converting the concentrations to units given in the table, and find the mass in 1,000 liters of water—the amount of water you would typically ingest over the course of a year.

TABLE 2.18 Typical constituents in water

CONSTITUENT	mg/L	ppm	MOLECULAR WEIGHT	mmol/L	Kg/YEAR
Bicarbonate (HCO_3)	75				
Carbonate (CO_3)	5				
Chloride (Cl)	35				
Sulfate (SO_3)	27				
Calcium (Ca)	11				
Magnesium (Mg)	7				
Potassium (K)	11				
Sodium (Na)	55				
Aluminum (Al)	0.2				
Fluoride (F)	0.3				

Source: Based on Tchobanoglous, G., Asano, T., Burton, F., Leverenz, H., Tsuchihashi, R. 2007. Water Reuse: Issues, Technologies, and Applications. McGraw-Hill.

2-40 The total volume of the oceans on Earth is 1.35×10^{18} m³. What are the masses of the following elements in the ocean in units of kg?
a. Oxygen (O) for which the concentration in the seawater is 857,000 ppm
b. Hydrogen (H) for which the concentration in the seawater is 108,000 ppm
c. Sodium (Na) for which the concentration in the seawater is 10,500 ppm

2-41 Air has a molecular weight of 28.967 g/mol. What is the density of air in units of g/m³ at 1 atm and 200°C?

2-42 The reported value of carbon dioxide in the atmosphere in 2010 was approximately 385 ppm_v. What is the concentration of carbon dioxide in the atmosphere in mg/m³?

2-43 The mass of the oceans is 1.4×10^{21} kg. (Assume the density of the ocean water on average is 1.03 g/ml.) The concentration of potassium (K) in the oceans is 380 ppm_m. What is the total mass of the potassium stored in the oceans?

2-44 The mass of the Earth's troposphere, the lower part of the atmosphere, is approximately 4.4×10^{18} kg. What would the total volume of the Earth's troposphere be in cubic meters (m³) if it were all under the constraints of standard temperature and pressure? Use the ideal gas law to calculate the average conditions for a standard state ($P = 1$ atm, $T = 298$ K, $R = 0.0821$ atm-L/mol-K). *Note:* The average molecular weight of air is 28.96.
a. The average mass of water in the troposphere is 1.3×10^{13} kg. Using your answer, calculate the concentration of water vapor in the troposphere in mg/m³?
b. Using your answer from part (b), what would be the concentration of water vapor in the troposphere in ppm_v?

2-45   Table 2.19 represents the National Ambient Air Quality Standards (NAAQS) air quality standard.
a. Express the standards in $\mu g/m^3$ at 1 atm of pressure and 25°C.
b. At the elevation of Denver, the pressure is about 0.82 atm. Express the standards in $\mu g/m^3$ at that pressure and a temperature of 5°C.

TABLE 2.19 National Ambient Air Quality Standards

POLLUTANT	AVERAGING TIME	LEVEL	STANDARD CONDITIONS ($\mu g/m^3$)	DENVER IN WINTER ($\mu g/m^3$)
Carbon monoxide (CO)	8-hour	9 ppm		
	1-hour	35 ppm		
Nitrogen dioxide (NO_2)	1-hour	100 ppb		
	Annual	53 ppb		
Ozone (O_3)	8-hour	0.075 ppm		
Sulfur dioxide (SO_2)	1-hour	75 ppb		
	3-hour	0.5 ppm		

Source: Based on U.S. EPA.

2-46   Use Henry's law to calculate the concentration of dissolved inorganic carbonates at standard temperature and pressure (STP) in a raindrop if the atmospheric concentration of carbon dioxide in the Ordovician Epoch was 2,240 ppm_v CO_2 as illustrated in Figure 2.26.

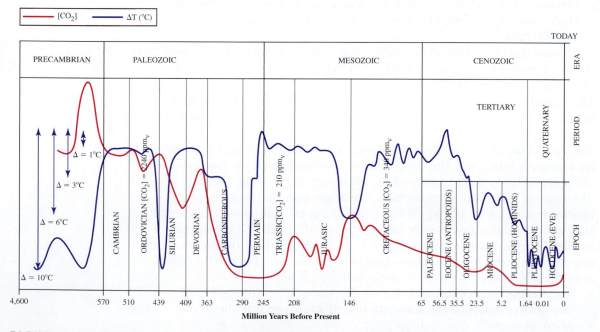

FIGURE 2.26 Fluctuation of carbon dioxide and temperature over geological timescales.

Source: Based on Nasif Nahle. 2007. Cycles of Global Climate Change. Biology Cabinet Journal Online. Article no 295.

2-47 The data shown in Table 2.20 were collected for ozone levels in the Los Angeles air basin. Plot the data recorded in Excel in ppm and $\mu g/m^3$. Identify the years Los Angeles was not in compliance with the NAAQS standards.

TABLE 2.20 Ozone concentrations reported in Los Angeles, CA (Historic Ozone Air Quality Trends, South Coast Air Quality Management District, http://www.aqmd.gov/smog/o3trend.html)

YEAR	BASIN MAXIMUM 1-HOUR AVERAGE (ppm)	BASIN MAXIMUM 8-HOUR AVERAGE (ppm)	YEAR	BASIN MAXIMUM 1-HOUR AVERAGE (ppm)	BASIN MAXIMUM 8-HOUR AVERAGE (ppm)
1976	0.38	0.268	1994	0.30	0.209
1977	0.39	0.284	1995	0.26	0.204
1978	0.43	0.321	1996	0.25	0.175
1979	0.45	0.312	1997	0.21	0.165
1980	0.41	0.288	1998	0.24	0.206
1981	0.37	0.266	1999	0.17	0.143
1982	0.40	0.266	2000	0.184	0.159
1983	0.39	0.245	2001	0.190	0.146
1984	0.34	0.249	2002	0.167	0.148
1985	0.39	0.288	2003	0.216	0.200
1986	0.35	0.251	2004	0.163	0.148
1987	0.33	0.210	2005	0.182	0.145
1988	0.35	0.258	2006	0.175	0.142
1989	0.34	0.253	2007	0.171	0.137
1990	0.33	0.194	2008	0.176	0.131
1991	0.32	0.204	2009	0.176	0.128
1992	0.30	0.219	2010	0.143	0.123
1993	0.28	0.195			

Source: Based on Historic Ozone Air Quality Trends, South Coast Air Quality Management District, http://www.aqmd.gov/smog/o3trend.html.

2-48 Butane and oxygen combine in the combustion process. Balance the following chemical equation that describes the combustion of butane:

$$C_4H_{10} + O_2 \rightarrow CO_2 + H_2O$$

a. How many moles of oxygen are required to burn 1 mol of butane?
b. How many grams of oxygen are required to burn 1 kg of butane?
c. At standard temperature and pressure, what volume of oxygen would be required to burn 100 g of butane?
d. What volume of air at STP is required to burn 100 g of butane?

2-49 What is the molarity of 25 g of glucose ($C_6H_{12}O_6$) dissolved in 1 L of water?

2-50 Wine contains about 15% ethyl alcohol (CH_3CH_2OH) by volume. If the density of ethyl alcohol is 0.79 kg/L, what is its molarity in wine? (Wine is an aqueous solution; that is, most of the rest of the volume of wine consists of water and the density of wine is 0.98 g/ml.)

2-51 A waste stream of 20,000 gal/min contains 270 mg/L of cyanide as NaCN.
a. What is the ionic strength of the solution in mmol/l?
b. What is the appropriate activity coefficient for dissolved cyanide (CN^-) in this solution if the sodium cyanide is the only dissolved species? (Use the Güntelberg approximation.)

2-52 Table 2.21 represents a "total analysis" of wastewater that has been reported. Note that the pH is not given.

TABLE 2.21 Data for analysis of wastewater in Problem 2-52.

CATION	CONCENTRATION (mg/L)	MOLAR MASS	CATION	CONCENTRATION (mg/L)	MOLAR MASS
NH_3	0.08	as N	F^-	21.2	as F
Na^+	18.3	as Na	Cl^-	24.1	as Cl
K^+	18.3	as K	HCO_3^-	15	as C
Ca^{+2}	1.7	as $CaCO_3$	SO_4^{-2}	20	as SO_4
			NO_3^-	2.0	as N
			NO_2^-	0.008	as N

a. Do a charge balance analysis on the data to see if such a solution would be electrically neutral.
b. If H^+ and OH^- are the only ions missing from the analysis, what must their concentration be? That is, what are the pH and pOH?

2-53 Calculate the activity coefficient and activity of each ion in a solution containing 300 mg/L $NaNO_3$ and 150 mg/L $CaSO_4$.

2-54 The major dissolved species in the Colorado River near Phoenix are given in mg/L in the accompanying Table 2.22.

TABLE 2.22 Approximate concentration of dissolved ions in the Colorado River near Phoenix, AZ.

CATION	CONCENTRATION (mg/L)	MOLAR MASS	CATION	CONCENTRATION (mg/L)	MOLAR MASS
Ca^{2+}	83	as Ca	SO_4^{-2}	250	as SO_4
K^+	5.1	as K	Cl^-	88	as Cl
H^{+-}	0.000065	as H	HCO_3^-	135	as $CaCO_3$

a. Express the concentration of Ca^{+2} in the Colorado River in moles of Ca^{+2} per liter and in milligrams per liter as $CaCO_3$.
b. The concentration of Na^+ is not given in the table. Assuming that Na^+ is the only significant species missing from the analysis, compute its

value based on the electroneutrality requirement. *Note*: The HCO_3^- concentration is expressed in terms of $CaCO_3$ hardness.

c. Calculate the total hardness, carbonate hardness, and noncarbonate hardness for the Colorado River using the data in the table.

2-55 For the water quality data shown in Table 2.23, find the following information

a. For Snake River
   i)   Calculate the ionic strength
   ii) Calculate the activity coefficient

b. For the Mississippi River
   i)   Calculate the ionic strength
   ii) Calculate the activity coefficient

c. For the Lower Congo River
   i)   Calculate the ionic strength
   ii) Calculate the activity coefficient

d. For the Dead Sea
   i)   Calculate the ionic strength
   ii) Calculate the activity coefficient

TABLE 2.23 Concentration of major dissolved ions in various water bodies

ION	SNAKE RIVER	MISSISSIPPI RIVER	LOWER CONGO, KINSHASA	GANGES RIVER	DEAD SEA
Cation	Concentration (mg/L)				
Ca^{2+}	16.1	162	10.8	88	63,000
Mg^{2+}	2.91	45	3.9	19	168,000
K^+	4.14	3		2.3	7,527
Na^+	28.4	21	14.2	6.4	35,000
Fe^{2+}	12				
Anions					
Cl^-	15.4	54	6.1	5.7	207,000
F^-	2.1				
HCO_3^-	73	124		104	244
SO_4^{2-}	25.2	100	7.8	371	2,110
Reference:	Clark et al., 2004	Maybeck et al., 1989	Visser and Villeneuve, 1975	Maybeck et al., 1989	Maybeck et al., 1989

2-56 Calculate the total hardness, carbonate hardness, and noncarbonate hardness for the Ganges River using the data in Table 2.23.

2-57 Calculate the total hardness, carbonate hardness, and noncarbonate hardness for the Mississippi River using the data in Table 2.23.

2-58 Calculate the total hardness, carbonate hardness, and noncarbonate hardness for the Dead Sea using the data in Table 2.23.

2-59 The following laboratory data, shown in Table 2.24, have been recorded for a 100-ml sample of water. Calculate the concentration of total solids, total dissolved solids, and total volatile solids in mg/L.

TABLE 2.24 Representative data for calculating solids concentration in amwater sample for Problem 2-59.

DESCRIPTION	UNIT	RECORDED VALUE
Sample size	ml	100
Mass of evaporating dish	g	34.9364
Mass of evaporating dish plus residue after evaporating at 105°C	g	34.9634
Mass of evaporating dish plus residue after evaporating at 550°C	g	34.9606
Mass of Whatman paper filter	g	1.6722
Mass of Whatman paper filter after evaporating at 105°C	g	1.6834
Mass of Whatman paper filter after evaporating at 550°C	g	1.6773

2-60 The following laboratory data, shown in Table 2.25, have been recorded for a 1-L sample of water. Calculate the concentration of total solids, total dissolved solids, and total volatile solids in mg/L.

TABLE 2.25 Representative data for calculating solids concentration in amwater sample for Problem 2-60.

DESCRIPTION	UNIT	RECORDED VALUE
Sample size	ml	1,000
Mass of evaporating dish	g	62.1740
Mass of evaporating dish plus residue after evaporating at 105°C	g	62.9712
Mass of evaporating dish plus residue after evaporating at 550 °C	g	62.8912
Mass of Whatman paper filter	g	1.4671
Mass of Whatman paper filter after evaporating at 105°C	g	1.7935
Mass of Whatman paper filter after evaporating at 550°C	g	1.6188

2-61 Find the hardness of the following groundwater for the well-water sample from a well in Pennsylvania.

WATER SOURCE	CONCENTRATION OF IONS (mg/L)						
	Cations			Anions			
	$[Na^+]$	$[Ca^{2+}]$	$[Mg^{2+}]$	$[NO_3^-]$	$[SO_4^{2-}]$	$[Cl^-]$	$[HCO_3^-]$
Spring Mf 1	16	51	15	25	24	12	197

2-62 Find the hardness of the following groundwater for the well water sample from a well in Pennsylvania:

WATER SOURCE	CONCENTRATION OF IONS (mg/L)						
	Cations			Anions			
	$[Na^+]$	$[Ca^{2+}]$	$[Mg^{2+}]$	$[NO_3^-]$	$[SO_4^{2-}]$	$[Cl^-]$	$[HCO_3^-]$
Spring Ln 12	4.7	48	7.3	25	26	9.2	114

2-63 Find the hydrogen concentration and the hydroxide concentration in tomato juice having a pH of 5.

2-64 What is $[H^+]$ of water (in mol/L) in an estuary at pH 8.2 with an ionic strength of $I = 0.05$ mol/L?

2-65 Calculate the normality of the following solutions:
a. 36.5 g/L hydrochloric acid [HCl]
b. 80 g/L sodium hydroxide [NaOH]
c. 9.8 g/L sulfuric acid [H_2SO_4]
d. 9.0 g/L acetic acid [CH_3COOH]

2-66 Find the pH of a solution containing 10^{-3} mol/L of hydrogen sulfide (H_2S), where $pK_a = 7.1$

2-67 The Henry's law constant for H_2S is 0.1 mol/L-atm, and

$$H_2S(aq) \leftrightarrow HS^- + H^+$$

The acid equilibrium constant for this reaction is $k_a = 10^{-7}$. If you bubble pure H_2S gas into a beaker of water, what is the concentration of HS^- if the pH is controlled and remains at 5?
a. mol/L
b. mg/L
c. ppm_m

2-68 What is the pH of a solution containing 1×10^{-6} mg/L of OH^- (25°C)?

2-69 Find the hydrogen ion concentration and the hydroxide ion concentration in baking soda solution with a pH of 8.5.

2-70 If 40 g of the strong base, NaOH, are added to 1 L of distilled water, what would the pH of the solution be, if
a. Activity effects are neglected
b. Activity effects are estimated using the Debye–Huckel approximation

2-71 Write a complete mathematical model for a closed system consisting of water only at equilibrium. Be sure to include starting materials, a species list, equilibria expressions, mass balances, a charge balance, and other constraints. Assuming activity effects are negligible, how do $[H^+]$ and $[OH^-]$ compare?

2-72 Estimate the pH of a solution containing 10^{-2} M of HCl.

2-73 Estimate the pH of a solution containing 10^{-4} M of HNO_3.

2-74 Use a graphical approach to estimate the pH of a solution containing 10^{-2} M of H_2CO_3.

2-75 Use a graphical approach to estimate the pH of pristine rainfall during the Ordovician Epoch when atmospheric CO_2 concentrations were approximately 2,240 ppm_v.

2-76 Acid rainfall due to sulfur emission led to the addition of 10^{-4} M of H_2SO_4 in rainfall over the northeastern United States during the 1970s and 1980s. Carbon dioxide levels at that time were approximately 375 ppm_v.

   a. Use the graphical approach to show the pH of rainfall over the northeastern United States during that time period.

   b. Estimate the concentration of carbonic acid and bicarbonate ion in the rainfall.

2-77 Calculate the solubility of magnesium carbonate ($MgCO_3$) at 25°C in distilled water. Assume the solution acts as an ideal-fluid. The solubility product constant is $k_{sp} = 4 \times 10^{-5}$.

2-78 Calculate the solubility of aluminum phosphate ($AlPO_4$) at 25°C in distilled water? Assume the solution acts as an ideal fluid. The solubility product constant is $k_{sp} = 3.2 \times 10^{-23}$.

2-79 Calculate the solubility of silver chloride (AgCl) at 25°C if the water already contains 10 mg/L of the chloride ion. Assume the solution acts as an ideal fluid. The solubility product constant is $k_{sp} = 3 \times 10^{-10}$.

2-80 Vinyl chloride is a colorless, odorless gas used in the production of polyvinyl chloride (PVC) plastics and in numerous industries. Vinyl chloride is also a known human carcinogen, and the U.S. EPA's maximum contaminant level (MCL) in drinking water is 0.002 mg/L or 2 ppb. The vinyl chloride concentration in the soil outside of a plastic factory is 0.007 mg/kg, and the soil has a 15% organic content. Vinyl chloride (chloroethene) has a reported octanol–water coefficient equal to 4. Estimate the organic carbon normalized partition coefficient, K_{oc}, using the estimation procedures based on Equations (2.73) and (2.74) and the sediment/water partition coefficient. Use the sediment/water partition coefficient to predict the possible equilibrium concentration of vinyl chloride in the groundwater.

2-81 The EPA's maximum contaminant level (MCL) in drinking water is 0.005 mg/L or 5 ppb for carbon tetrachloride. The carbon tetrachloride concentration in the soil outside of a factory is 0.01 mg/kg, and the soil has a 1% organic content. Carbon tetrachloride has a reported octanol–water coefficient equal to 400. Estimate the organic carbon normalized partition coefficient, K_{oc}, using the estimation procedures based on Equations (2.73) and (2.74) and the sediment/water partition coefficient. Use the sediment/water partition coefficient to predict the possible equilibrium concentration of carbon tetrachloride in the groundwater.

2-82 The EPA's maximum contaminant level (MCL) in drinking water is 0.1 mg/L, or 100 ppb for chlorobenzene. The chlorobenzene concentration in the soil outside of a factory is 0.01 mg/kg, and the soil has a 6% organic content. Chlorobenzene has a reported octanol–water coefficient equal to 700. Estimate the organic carbon normalized partition coefficient, K_{oc}, using the estimation procedures based on Equations (2.73) and (2.74) and the sediment/water partition coefficient. Use the sediment/water partition coefficient to predict the possible equilibrium concentration of chlorobenzene in the groundwater.

2-83 The EPA's MCL in drinking water is 0.002 mg/L or 2 ppb for aroclor. The Aroclor 1221 concentration in the soil outside of a factory is 0.028 mg/kg, and the soil has a 12% organic content. Aroclor 1221 has a reported octanol–water coefficient equal to 800. Estimate the organic carbon normalized partition coefficient, K_{oc}, using the estimation procedures based on Equations (2.73) and (2.74) and the sediment/water partition coefficient. Use the sediment/water partition coefficient to predict the possible equilibrium concentration of Aroclor 1221 in the groundwater.

CHAPTER 3

Biogeochemical Cycles

FIGURE 3.1 This is a famous view of the lower falls of the Yellowstone River in Yellowstone National Park in Wyoming. In this photo, we see the obvious liquid water cascading over the waterfall, along with the mist of fine water droplets and gas-phase water particles; close inspection also shows ice and snow, which is the solid phase of water along the canyon walls. The river has formed a canyon through the softer underlying stone. Not only does the water flow over the waterfall and continue on downstream, but you can also see the ongoing erosion along the side of the canyon that will carry soil and sediments downstream as well.

Source: Bradley Striebig

Everything flows and nothing stands still.

—PLATO

GOALS

Modern scientists recognize that there are more than the four elements—earth, fire, water, and wind—as defined by the ancient philosophers. Yet the words of these early philosophers of science still ring very true in many ways. Plato is credited with observing that "everything flows and nothing stands still." In this chapter, you will examine the flow of gas and liquids on the surface of the Earth. The chemical and physical properties of substances describe how those substances move from one part of the Earth to another. This chapter will examine the movement of water through biogeochemical cycles and how this movement transforms our everyday world.

The educational goals of this chapter are to introduce the concepts of biogeochemical cycles and their interrelationship with society. We begin with the water cycle to provide a tangible system with which students are somewhat familiar. The concept of a mass balance and how it may be used to identify sustainable and unsustainable uses of natural resources is also introduced via the water budget equation. Students will use basic principles of chemistry, physics, and math to define and solve engineering problems, in particular mass and energy balance problems.

OBJECTIVES

At the conclusion of this chapter, you should be able to:

- Define the laws of conservation for mass and energy.

- Determine system boundaries for solving mass and energy flow problems.

- Solve mass balance problems involving steady-state conservative substances.

- Define and discuss the importance of biogeochemical cycles.

- Describe how water may flow from one reservoir in the Earth to another.

- Describe the transformations and reservoirs that comprise the water cycle.

- Use the water budget equation to determine storage and withdrawal capacity.

- Describe the importance of biogeochemical cycles in determining the environmental impacts of anthropogenic activities.

- Describe how unsustainable water usage may lead to problems in communities and civilizations.

- Describe methods and the rationale for sustainable water consumption.

3.1   Introduction

Humans are an integral part of the Earth's environment. We are, however, the only species that can alter their surrounding in such a comprehensive fashion that the environment in today's cities does not remotely resemble the environment at the same location of a few hundred years ago. In the 4.5-billion-year history of the planet, the environmental changes made by humans since the Industrial Revolution in the mid-1800s have occurred in a geological blink of the eye. These radical changes to the environment are in part a consequence of many humans living a longer and more comfortable life.

The environment includes the world and all that surrounds us. However, for engineers and scientists, a more precise definition is required. Peavy, Rowe, and Tchobanglous (1985) define the environment as a system that "may take on global dimensions, may refer to a very localized area in which a specific problem must be addressed, or may, in the case of contained environments, refer to a small volume of liquid, gaseous, or solid materials within a . . . reactor." Engineering decisions and designs influence the environment in which we live. These design choices have significantly influenced our lifestyle, expectations, and the environment. We have altered the atmosphere (gas phase) and local watersheds (liquid phase), and we have created mountains of solid waste (solid phase). These lifestyle choices have led to more leisure time, longer lifespan, and less biodiversity. Engineers and scientists may use the science and math-based tools presented in this text to help forecast the impact of our engineering choices on the environment in which we live. By examining the relationship between engineering choices and their impacts on the environment, society, and the economy, engineers can make more informed decisions and create more sustainable infrastructure and manufacturing processes.

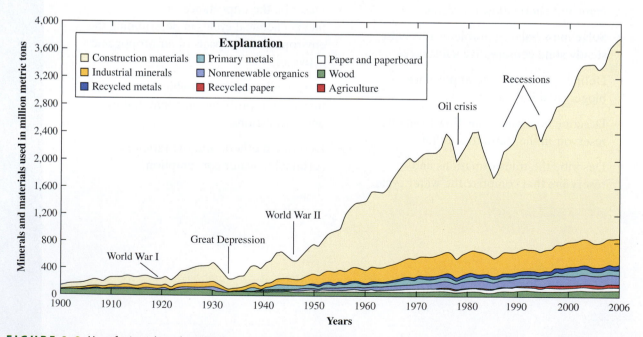

FIGURE 3.2 Use of minerals and materials in the United States.

Source: Use of Minerals and Materials in the United States From 1900 Through 2006, USGS Fact Sheet 2009-3008, April 2009

The anthropogenic (human-made) changes to the environment can be quantitatively described. The increase in the consumption of raw materials since 1900 is shown in Figure 3.2. This graph illustrates the increasing rate of consumption of raw materials as the world has industrialized. When the rate of consumption of natural resources exceeds the rate at which these resources can be regenerated, an unsustainable process is the result. The mass rate of material consumed can be compared to the mass rate of material being generated. If the mass rate of consumption is less than or equal to the mass rate of generation of that resource, then the process can be sustained indefinitely, or at least until the system changes. The mass of materials moving from one portion of the Earth to another can be measured, and the rates of production can be compared to the rate of consumption. Biogeochemical cycles describe the amount of material stored, produced, or consumed within a repository as well as the conversion of material from one repository to another.

3.2 Biogeochemical Cycles

Elements such as carbon, oxygen, hydrogen, nitrogen, and phosphorus are prerequisites for life. However, an overabundance of any of these elements in different parts of the ecosystem can cause an imbalance in the system and produce unwanted changes. Biogeochemical cycles help us understand and interpret how chemicals interact in the environment. Biogeochemical cycles are defined as "the biological and chemical reservoirs, agents of change and pathways of flow from one reservoir of a chemical on earth to another reservoir" (Mays, 2007, p. 57). These biogeochemical cycles help us understand what happens to important elements and compounds that influence our lifestyle. Perturbations, or changes to the water and carbon cycles, have long-range and important consequences for our environment and future generations.

Biogeochemical cycles describe the transport, storage, and conversion of compounds. The chemical form and process may determine how much of a compound is found in the atmosphere, oceans, or soil. Scientists and engineers use mathematical models, or descriptions, of these processes to predict how long compounds will reside in a certain reservoir and how they are converted to different reservoirs or different forms. The reservoir or repository is the place where these compounds may be found on Earth. For instance, most of the water on our planet is found in the oceans as shown in Table 3.1. Water is also found in the atmosphere but cycles

TABLE 3.1 Major repositories for carbon, nitrogen, and water

REPOSITORY	MAGNITUDE $\times 10^{12}$ kg		
	Carbon (C)	Water (H_2O)	Nitrogen (N)
Atmosphere	735	13,000	3.9×10^6
Geologic (soils, ice, groundwater, etc)	10^7	37,600	150
Ocean seawater	40,000	1.4×10^6	350
Dead organic mater	3,540	Negligible	400
Living organisms	562	3	7.8

Based on Harte, 1988. Pp. 238-262.

through the atmosphere rapidly, spending, on average, less than 10 days in the atmosphere before falling out as precipitation. Oceans are also an important chemical repository for water and carbon. Water and carbon cycle through the oceans very slowly.

The atmosphere is one of the Earth's major repositories for water. We can estimate the average length of time that water remains in the atmosphere from the flow to and from the atmosphere shown in Table 3.2. The length of time that a compound remains in a defined system, in this case the atmosphere, is referred to as the residence time, t_r. We can determine the residence time in any system by summing the mass of the water in the atmosphere, M, and dividing by the mass flow rate, F, of water from the system.

$$t_r = \frac{M}{F} \tag{3.1}$$

TABLE 3.2 Mean annual flows of water on Earth

FLOW	MAGNITUDE (10^{12} m^3/yr)
Precipitation on land	108
Precipitation on the sea	410
Evaporation from the land	62
Evaporation from the sea	456
Runoff	46

Based on Harte, 1988. Pp. 238-262.

EXAMPLE 3.1   How Long Does Water Remain in the Atmosphere?

In this case, the water that flows from the atmosphere is equal to the amount of precipitation on land and on the sea (see Table 3.2):

$$P_{total} = P_{precipitation\ land} + P_{precipitation\ in\ the\ sea} = 108 \times 10^{12}\ \frac{m^3}{yr} + 410 \times 10^{12}\ \frac{m^3}{yr} = 518 \times 10^{12}\ \frac{m^3}{yr} \tag{3.2}$$

The units, m^3/yr, represent the volumetric flow rate of water from the atmosphere, where m^3 is the volume per unit time—in this case one year. In order to convert the volumetric flow rate, we must multiply by the density of the water in the precipitation. We will assume that the mean or average density, ρ, of water in a raindrop is equal to 1,000 kg/m^3, which is the density of liquid water at 25°C and 1 atmosphere of pressure. Now we can calculate the mass flow rate:

$$F = P_{total} \times \rho = 518 \times 10^{12}\ \frac{m^3}{yr} \times 1,000\ \frac{kg}{m^3} = 518 \times 10^{15}\ \frac{kg}{yr} \tag{3.3}$$

As shown in Table 3.1, the mass of water in the atmosphere is $13,000 \times 10^{12}$ kg. Substituting the mass of water and mass flow rate into the equation for the residence time yields

$$t_r = \frac{M}{F} = \frac{13,000 \times 10^{15}}{518 \times 10^{15}} = 0.025\ \text{years} \tag{3.4}$$

Since it is difficult to relate 2.5×10^{-2} years to time spans discussed in everyday conversation, we will convert the time to more useful units. From our calculations, we can see that water spends, on average, about 9.1 days in the atmosphere.

$$t_r = 0.025 \text{ years} \times \frac{365 \text{ days}}{\text{year}} = 9.1 \text{ days} \qquad (3.5)$$

These cycles have changed little since the last ice age. Major shifts in biogeochemical cycles occurred over the Earth's history as discussed in Box 3.1. The change to a cooler climate influenced the mass extinction of the age of the large dinosaur reptiles and ushered in a new age dominated by a newer species—mammals. A significant climate change in the past illustrates how easily a dominant animal, like the Tyrannosaurus Rex, can become a victim of an environment that is changing. Species incapable of adapting to changes in their environment face extinction. The dinosaurs had relatively small brains and lacked the capacity and technology to radically change and adapt to their surrounding environment. Scientists are observing a climate now that is changing at a rate unequaled in human history. How species adapt to these changes and interact with the biogeochemical cycles will influence the extent of the extinction period that is currently happening. Biologists are documenting the greatest extinction since the age of the dinosaurs due to the changing climate and human interactions with the environment. Human activities will determine the fate of many species over the next few centuries and may even influence our own fate as a species.

BOX 3.1 Chaco Canyon Based upon *Collapse: How Societies Choose to Fail or Succeed* by Professor Jared Diamond (2005)

A beautiful high desert plateau runs along the border of New Mexico and Colorado. Amazing cliff dwellings from the Anasazi civilization such as those shown in Figure 3.3 can still be seen in this area at the Chaco Canyon National Historic Park and Mesa Verde National Park. The Anasazi civilization that ranged across the North American Southwest, as shown in Figure 3.4, constructed the tallest building in North America until skyscrapers were built in Chicago from steel girders in the 1880s.

Chaco Canyon's Anasazi society flourished from approximately AD 600 until sometime between AD 1150 and 1200. The narrow canyon at this time created environmental advantages for a fledgling civilization. In this arid region of the United States, precipitation is a relatively rare life-sustaining event. Runoff from surrounding areas was channeled through the narrow canyon, creating a reliable water supply in this otherwise

FIGURE 3.3 Anasazi ruins are still visible in parts of the southwestern United States. The complex ruins show a high level of technical sophistication and skill. The multistory buildings were some of the highest buildings constructed in North America.

Source: kravka/Shutterstock.com

(Continued)

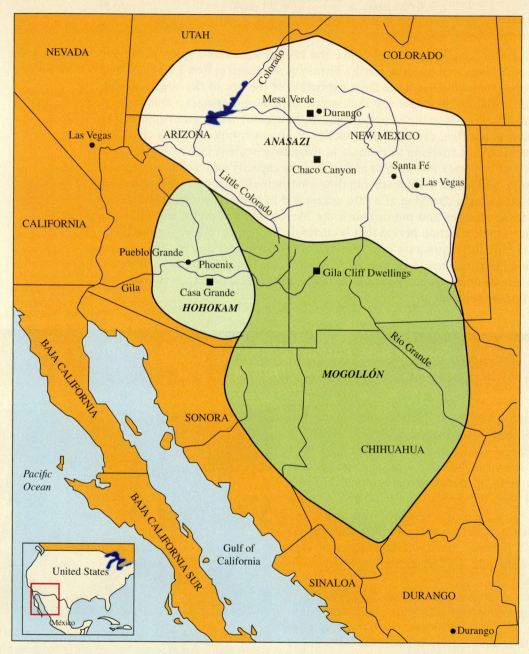

FIGURE 3.4 Map of the location and extent of the Anasazi civilization from AD 600 to approximately AD 1200 in the southwestern part of North America.

Source: Based on Angelakis, A.N., Mays, L.W., Koutsoyiannis, D., and Mamassis, N. (2012) Evolution of Water Supply Through the Millennia. IWA Publishing Alliance House, 12 Caxton Street, London, SW1H 0QS, UK.

arid region. The Anasazi relied on the runoff they collected to irrigate fields near their settlements. However, by about AD 900 deep channels called arroyos were formed that prohibited irrigation of crops unless the channels were completely filled with water. The nearby ecosystems were also degraded by deforestation. Deforestation occurred because the rate of logging was greater than the speed at which the slow-growing pinion and juniper pine trees grew.

Once the trees had been removed, the Anasazi could no longer use pine nuts as a source of protein in their diet or use pine timber for their buildings. The Anasazi population continued to grow in spite of the degraded environment. The population within Chaco Canyon received supplies in food and timber from satellite communities surrounding the canyon. Archaeologists determined that construction persisted in Chaco Canyon until approximately AD 1110. After this time, it has been theorized that the timber supplies were depleted in the surrounding area and the Chaco Canyon settlements began to decline and were eventually abandoned altogether.

Professor Jared Diamond, in his book *Collapse*, describes the conditions that led to the decline of the Anasazi civilization in and around Chaco Canyon after AD 1250. Diamond believes that population growth outstripped the rate of replenishment of the area's natural resources. Archaeological evidence shows that food, timber, and other natural resources were getting increasingly scarce as the Chaco Canyon population declined. Finally, water scarcity due to a drought that began in AD 1130 forced the residents of Chaco Canyon to abandon this region between AD 1150 and 1200.

According to Diamond, four significant factors led to the collapse of the Anasazi civilization in Chaco Canyon (and generally to other civilizations as well):

- Human-made environmental degradation occurred, caused by deforestation and arroyo cutting.
- Climate change associated with rainfall and temperature exacerbated the anthropogenic environmental factors.
- The Chaco Canyon civilization and population became unsustainable because of the fragile ecosystem in which the civilization developed.
- Societal and cultural factors apparently helped promote unsustainable practices within the community.

Diamond sees these characteristics as indicators of the potential collapse of a community. Sustainable societies must consider the needs of future generations and not just those of the current generation. The infrastructure and economic models of past and present societies that avoid the characteristics described by Professor Diamond are more sustainable.

Engineers and scientists continue to explore and research the biogeochemical cycles. Mathematical models that describe the chemical reactions, transformations, and pathways of water and carbon are being investigated to help populations adapt to our changing environments and avoid a fate similar to that of the residents of Chaco Canyon. Satellite technology and more accurate analyses of our atmosphere, oceans, and soil allow engineers and scientists to gain a better understanding of how biogeochemical cycles work during the present time. Analyzing biogeochemical cycles also reveals something about how these cycles worked thousands of years ago. Mathematical models of these

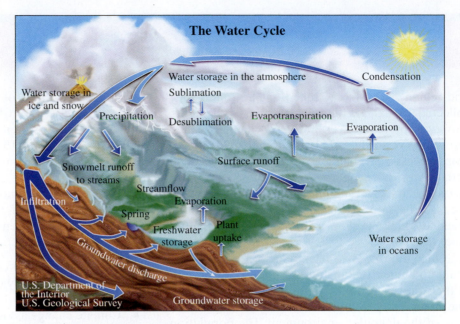

FIGURE 3.5 The hydrologic cycle.

Source: U.S. Dept. of the Interior, U.S. Geological Survey, John Evans, Howars Perlman, USGS, http://water.usgs.gov/edu/watercycleprint.html

systems based on data from the past and present allow scientists and engineers to predict how human behavior might impact these cycles and vice versa: specifically, how changes in biogeochemical cycles may impact human lifestyles in the future.

3.3 The Hydrologic Cycle

The atmosphere is a mixture of gases extending from the surface of the Earth toward space; the lithosphere is the soil crust that lies on the surface of the planet where we live; and the hydrosphere is the portion of the Earth that accounts for most of the water storage and consists of oceans, lakes, streams, and shallow groundwater bodies. Hydrology is "the science that treats the waters of the Earth, their occurrence, circulation, and distribution, their chemical and physical properties, and their reaction with the environment, including the relations to living things" (Mays, 2007, p. 191). The hydrologic cycle, or water cycle, describes the movement of water from one biogeochemical cycle to another (Figure 3.5). The hydrologic cycle is one example of a biogeochemical cycle.

3.4 Water Repositories

Oceans contain 97% of the water on the planet (Figures 3.6 and 3.7). They provide many definable separate ecosystems based on the chemistry of the water. Ocean water contains approximately 35 grams of dissolved minerals (or salts)

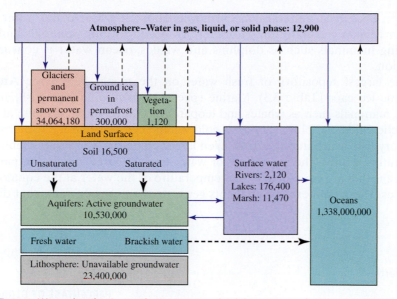

FIGURE 3.6 Water distribution with reserves in cubic kilometers and movement between biogeochemical repositories.

Source: Based on *Managing Water for Peace in the Middle East: Alternative Strategies.* Masahiro Murakami. United Nations University. 1995. Data source: Korzun, 1976.

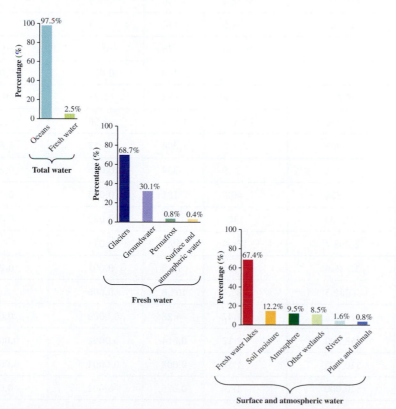

FIGURE 3.7 Water distribution in various repositories on Earth.

Source: Based on Arthurton, R., Barker, S., Rast, W., Huber, M., Alder. J., Chilton, J., Gaddis, E., Pietersen, K., and Zockler, C. (2007) Chapter 4: Water. State and Trends of the Environment: 1987-2007. Global Environment Outlook (GEO4) Environment for Development. United Nations Environment Programme. DEW/1001/NA.

per liter. Dissolved mineral concentrations above 2 g/liter make water unsuitable for human consumption. However, for ocean dwelling animals and plants, including mammals such as dolphins and whales, ocean water is consumed for hydration.

The largest repository of fresh water on the planet lies in the Arctic and Antarctic ice caps (Table 3.3). Marine species of microorganisms, fish, and large marine mammals such as whales and leopard seals have adapted to life at or near the Arctic. Microbial blooms in these cold waters feed fish, and the fish draw larger predatory fish, which in turn attract even larger predators. The water chemistry and temperature control the movement of water through large circulation patterns in the oceans that are governed by the temperature of the water at the equator and at the poles. These temperature gradients generate the ocean currents that distribute

TABLE 3.3 Water Reserves on Earth

RESERVE	DISTRIBUTION AREA (10^3km^2)	VOLUME (10^3km^3)	LAYER (m)	PERCENTAGE OF GLOBAL RESERVES Total water	Fresh water
World ocean	361,300	1,338,000	3,700	96.5	–
Ground water	134,800	23,400	174	1.7	–
fresh water	–	10,530	78	0.76	30.1
Soil moisture	–	16.5	0.2	0.001	0.05
Glaciers and permanent snow cover	16.227	24,064	1,463	1.74	68.7
Antarctic	13,980	21,600	1,546	1.56	61.7
Greenland	1,802	2,340	1,298	0.17	6.68
Arctic islands	226	83.5	369	0.006	0.24
Mountainous regions	224	40.6	181	0.003	0.12
Ground ice/ permafrost	21,000	300	14	0.022	0.86
Water reserves in lakes	2,058.7	176.4	85.7	0.013	–
Fresh	1,236.4	91	73.6	0.007	0.26
Saline	822.3	85.4	103.8	0.006	–
Swamp water	2682.6	11.47	4.28	0.0008	0.03
River flows	148,800	2.12	0.014	0.0002	0.006
Biological water	510,000	1.12	0.002	0.0001	0.003
Atmospheric water	510,000	12.9	0.025	0.001	0.04
Total water reserves	510,000	1,385,984	2,718	100	–
Total fresh water reserves	148,800	35,029	235	2.53	100

Source: Based on Shiklomanov 1993, Mays p. 4

the solar energy absorbed by the ocean waters. Yet again, few humans, excluding the Inuit and a few other tribes of the far north, have yet to adapt to the largest source of fresh water.

Groundwater is the largest source of fresh water actively used by the human species, accounting for 30.1% of fresh water on the planet and 99% of the fresh water available for human use. Groundwater reserves have accumulated for 4 million years without substantial impact from human activity. One ancient groundwater resource is the Ogallala Aquifer that humans have used to transform the midwestern part of North America, as shown in Figure 3.8. Water has

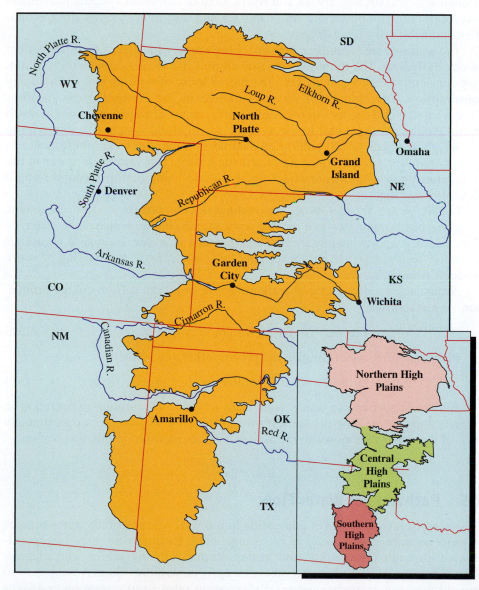

FIGURE 3.8 Extent of the Ogallala Aquifer.

Source: Based on USGS National Water-Quality Assessment (NAWQA) Program - High Plains Regional Groundwater (HPGW) Study, High Plains Aquifer System, http://co.water.usgs.gov/nawqa/hpgw/HPGW_home.html

been flowing into the Ogallala Aquifer near the base of the Rocky Mountains for 2 to 6 million years. It serves as a source of drinking water and irrigation water throughout the farms and cities of the midwest. Today, however, more water is being withdrawn from the aquifer than is entering. In 2005 the United States Geological Service (USGS) estimated that the total water stored in the aquifer was 3608 million cubic meters. While this is a vast amount of water, the USGS estimates that it is 9% (312 million cubic meters) less than predevelopment levels due to the rate of water use for irrigation of agricultural fields in the midwestern United States.

Only 0.3% of the world's total fresh water is available in surface waters. These surface waters range in size and availability from large sources like the Great Lakes to small seasonal streams, creeks, and rivers. Most of the world's population centers have been built around surface water sources, and they affect today's cities, economies, and health to a large degree.

The atmosphere is another repository for water, which has a profound effect on our daily lives. The humidity, or saturation of the local atmosphere by water, plays an important role in climate, the biodiversity of species, and even energy consumption. The amount of water contained in the atmosphere is startling. Even a small cloud consists of over 360 metric tonnes of water, or nearly as much as 100 elephants. The water in a very large storm cloud system like a hurricane can weigh as much as 100 million elephants! In total, the atmosphere contains approximately 1.3×10^{16} kg of water (Harte, 1988). The amount of water in the atmosphere has a significant effect on the world's population through rainfall patterns and climate.

Finally, water is also stored in biological components of our ecosystem, including plants, animals—and us! Our own brains contain nearly 70% water by mass, which is approximately the same percentage as a tree. Biological organisms use water to transport chemicals, regulate heat, and remove the buildup of waste products from cells.

Animals generate water and carbon dioxide as products from energy production, or metabolisms:

$$\text{Glucose + oxygen} \rightarrow \text{carbon dioxide + water} \qquad (3.6)$$

Plants consume water through the photosynthesis reaction:

$$\text{Carbon dioxide + water + photons} \rightarrow \text{plant tissue + oxygen`} \qquad (3.7)$$

Biological species use water to regulate temperature through perspiration in animals and transpiration in plants. Plant transpiration accounts for approximately 10% of the water that moves to the atmosphere, as discussed in the next section.

3.5   Pathways of Water Flow

Evaporation is the process of converting liquid water from surface water sources to gaseous water that resides in the atmosphere. Evaporation rates can be calculated through observation and varies dramatically by region and climate. Scientists use a 48-inch-diameter steel pan aptly named the "Class A Evaporation Pan" to measure evaporation rates. Evaporation rates from reservoirs and lakes

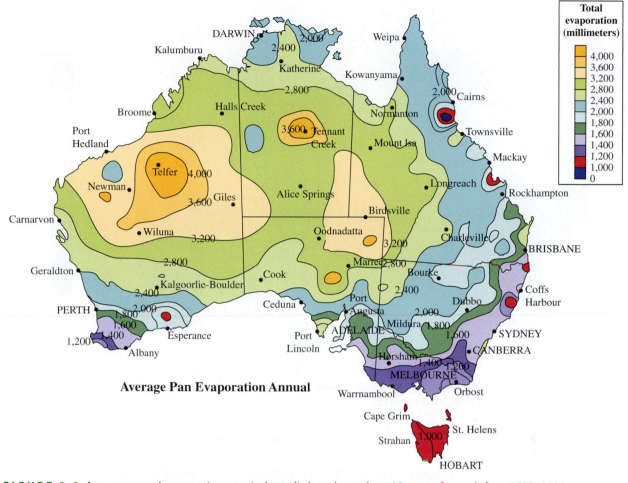

Average Pan Evaporation Annual

Total
evaporation
(millimeters)

	4,000
	3,600
	3,200
	2,800
	2,400
	2,000
	1,800
	1,600
	1,400
	1,200
	1,000
	0

FIGURE 3.9 Average annual evaporation rates in Australia based on at least 10 years of records from 1975–2005.

Source: Based on Australian Government Bureau of Meteorology: http://www.bom.gov.au/jsp/ncc/climate_averages/evaporation/index.jsp

can be significant sources of water loss in arid regions. Evaporation from surface waters accounts for 90% of the water that is transported into the atmosphere; most of that water evaporates from the surface of the oceans. The variations in evaporation rates and the change in the average soil moisture as measured by pan evaporation rates between 1970 and 2012 in Australia are shown in Figures 3.9 and 3.10.

Transpiration occurs when water is conveyed from living plant tissue, especially leaves, to the atmosphere. In wetter climates, transpiration remains fairly constant and is closely related to the growing season. In arid regions, transpiration is much more closely related to the depth of the roots for the plants growing in that region. Shallow-rooted plants may die from lack of moisture availability, while deep-rooted plants may continue to transpire water by utilizing deeper groundwater supplies for their growth and maintenance. During a growing season, a leaf will transpire many times more water than its own weight. An acre of corn gives

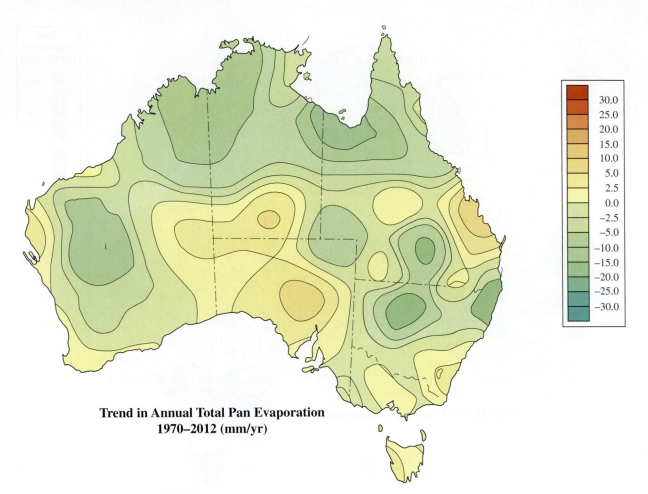

**Trend in Annual Total Pan Evaporation
1970–2012 (mm/yr)**

FIGURE 3.10 Trends in annual evaporation from soil, 1970 to 2012. Decreasing evaporation rates indicate that there is less moisture retained and available in the soils.

Source: Based on Australian Government Bureau of Meteorology: http://www.bom.gov.au/cgi-bin/climate/change/trendmaps.cgi?map=evap&area=aus&season=0112&period=1970

off about 11,400–15,100 L of water each day, and a large oak tree can transpire 151,000 L per year.

Evaporation and transpiration are often grouped into one term—evapotranspiration—that represents the overall pathways of water moving into the atmosphere. Evapotranspiration includes all evaporation and transpiration from plants. It varies with temperatures, wind, and ambient moisture conditions and can be calculated using a lysimeter (a buried soil-filled tank). Global evaporation rate estimates for Africa, Australia, India, and the United States are shown in Figures 3.11 to 3.15. The change in annual average evapotransporation rates across the United States is shown in Figure 3.16.

Condensation converts water in the gas phase to liquid water by cooling the water molecules. This process may result in the formation of clouds or fog, which is

(a) Annual *E* (1984–1998)

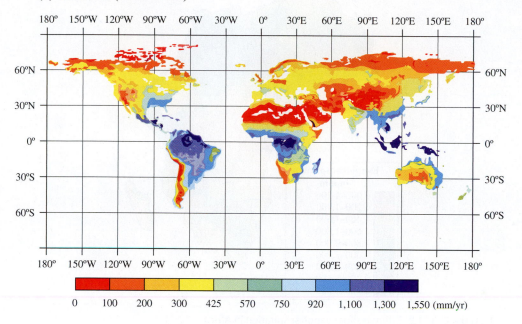

(b) Ration of annual *E* to annual precipitation (1984–1998)

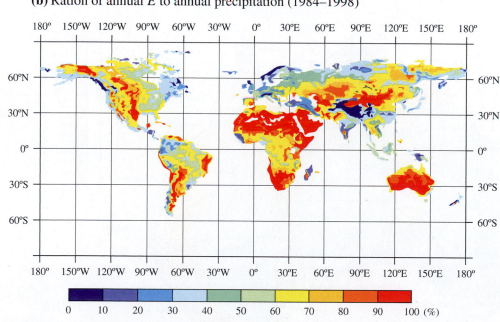

FIGURE 3.11 Global estimates of evapotranspiration in mm/yr compared to the ratio of annual precipitation. Red areas show regions where water loss is greater than precipitation, and blue areas show regions where excess water is available.

Source: Based on H. Yan, S.Q. Wang, D. Billesbach, W. Oechel, J.H. Zhang, T. Meyers, T.A. Martin, R. Matamala, D. Baldocchi, G. Bohrer, D. Dragoni, R. Scott. Global estimation of evapotranspiration using a leaf area index-based surface energy and water balance model. Remote Sensing of Environment, Volume 124, September 2012, Pages 581–595 http://dx.doi.org/10.1016/j.rse.2012.06.004

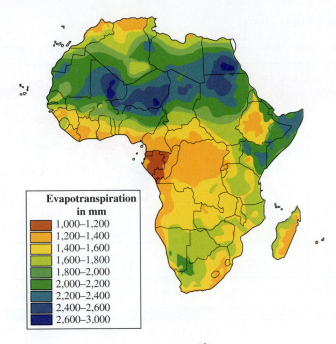

FIGURE 3.12 Estimates for evapotranspiration in Africa.

Source: Based on FAO: http://www.fao.org/nr/water/aquastat/watresafrica/index3.stm

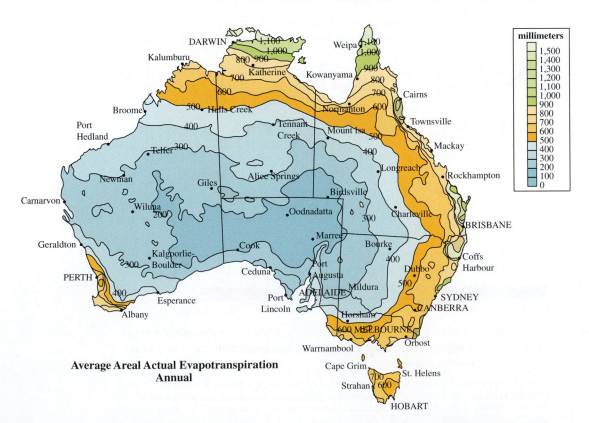

FIGURE 3.13 Estimates of evapotranspiration in Australia based on a standard 30-year climatology (1961–1990).

Source: Based on Australian Government Bureau of Meteorology: http://www.bom.gov.au/jsp/ncc/climate_averages/evapotranspiration/index.jsp

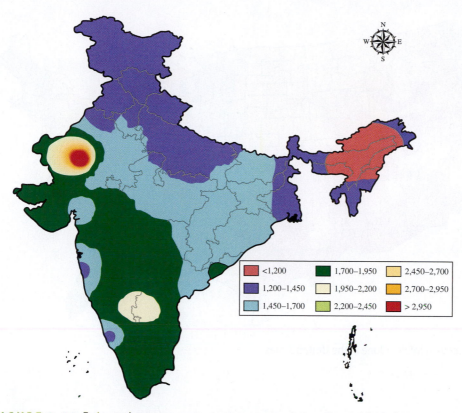

FIGURE 3.14 Estimated evapotranspiration rates and variations (in mm) for Indian conditions by the Penman–Montieth method.

Source: Based on Bapuji Rao, B., Sandeep, V.M., Rao, V.U.M. and Venkateswarlu, B. 2012. Potential Evapotranspiration estimation for Indian conditions : Improving accuracy through calibration coefficients. Tech. Bull. No 1/2012. All India Co-ordinated Research Project on Agrometeorology, Central Research Institute for Dryland Agriculture, Hyderabad.

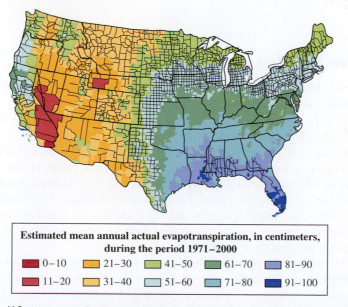

FIGURE 3.15 U.S. evapotranspiration rates in centimeters per year based on the years 1971–2000.

Source: Based on UC Santa Barbara Geography, http://www.geog.ucsb.edu/events/department-news /1169/new-maps-provide-crucial-information-for-water-managers/

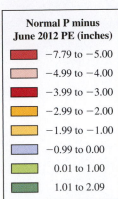

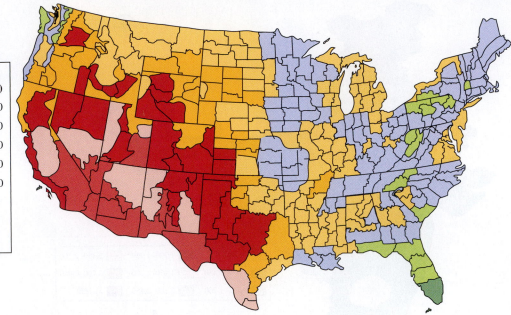

**Normal P minus
June 2012 PE (inches)**

■	−7.79 to −5.00
■	−4.99 to −4.00
■	−3.99 to −3.00
■	−2.99 to −2.00
■	−1.99 to −1.00
■	−0.99 to 0.00
■	0.01 to 1.00
■	1.01 to 2.09

FIGURE 3.16 Estimated effective rainfall in June in the United States.

Source: Based on NOAA

composed of very fine, lightweight water droplets. These droplets may collide and combine to form larger droplets that may eventually grow large enough to produce precipitation.

3.6 Precipitation

Water moves from the atmosphere to the surface of the planet through precipitation. Precipitation may occur when the atmosphere becomes completely saturated with water (100% humidity) and the droplets have enough mass to fall from the atmosphere. Evaporation from the oceans and eventual cooling of the water vapor from the oceans account for approximately 90% of the Earth's precipitation. Locations near the oceans generally receive greater rainfall than those in the interior of the continents as illustrated in Table 3.4. Variation in precipitation patterns is shown in Figures 3.17 to 3.21. However, some areas, such as the desert highland of Chile, are the most arid in the world, even though they are not far removed from the Pacific Ocean. Ocean currents, weather patterns, latitude, and geography are important influences on precipitation patterns.

Some fraction of precipitation seeps into the ground through a process called **infiltration**. Groundwater tables are replenished and sustainable when the rate of infiltration is equal to or greater than the rate of withdrawal from the groundwater table. Global estimates of groundwater recharge are shown in Figure 3.22. Infiltration estimates for various soil types are shown in Table 3.5.

TABLE 3.4 Average annual precipitation for selected U.S. cities from the National Climate Data Center

CITY	AMOUNT Millimeters
Phoenix, Arizona	193
El Paso, Texas	218
Los Angeles, California	302
Denver, Colorado	391
Salt Lake City, Utah	296
Fargo, North Dakota	498
Dallas, Texas	889
Chicago, Illinois	909
Portland, Oregon	922
Columbus, Ohio	960
Seattle, Washington	968
Kansas City, Missouri	970
Bangor, Maine	1,029
New York, New York	1,054
Atlanta, Georgia	1,265
Memphis, Tennessee	1,339
Miami, Florida	1,499

Source: Based on National Climate Data Center, National Oceanic and Atmospheric Administration, US Department of Commerce, Ashville, North Carolina, http://www.ncdc.moaa.gov/ol/climate/globalextremes.htm, September 2001.

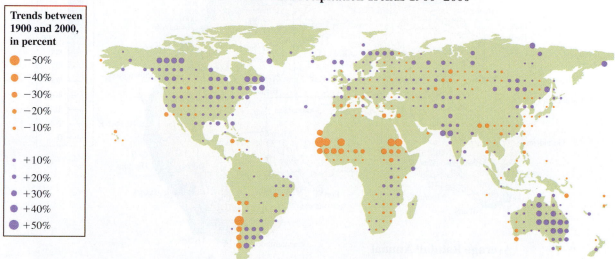

Annual Precipitation Trends 1900–2000

Trends between 1900 and 2000, in percent

- −50%
- −40%
- −30%
- −20%
- −10%
- +10%
- +20%
- +30%
- +40%
- +50%

FIGURE 3.17 Variations in global precipitation patterns from 1900 to 2000.

Source: Based on Arthurton, R., Barker, S., Rast, W., Huber, M., Alder. J., Chilton, J., Gaddis, E., Pietersen, K., and Zockler, C. (2007) Chapter 4: Water. State and Trends of the Environment: 1987-2007. Global Environment Outlook (GEO4) Environment for Development. United Nations Environment Programme. DEW/1001/NA.

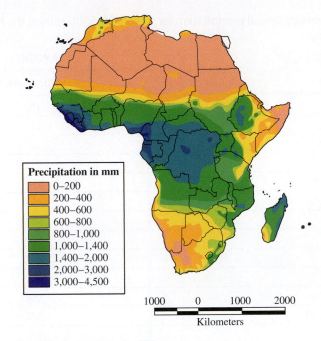

FIGURE 3.18 Precipitation patterns in Africa.

Source: Based on FAO: http://www.fao.org/nr/water/aquastat/watresafrica/index3.stm

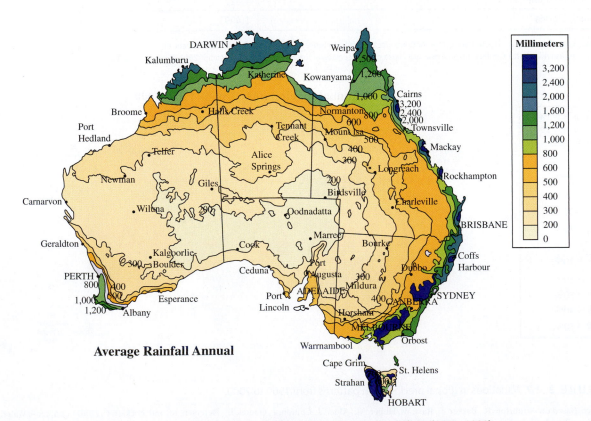

FIGURE 3.19 Precipitation patterns in Australia based on standard 30-year climatology (1961–1990).

Source: Based on Australian Government Bureau of Meteorology: http://www.bom.gov.au/jsp/ncc/climate_averages/rainfall/index.jsp

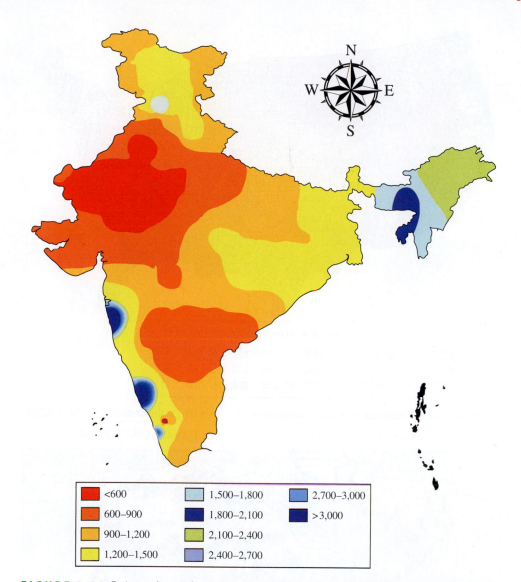

FIGURE 3.20 Estimated spatial variations in mean annual precipitation (mm) for India.

Source: Based on Bapuji Rao, B., Sandeep, V.M., Rao, V.U.M. and Venkateswarlu, B. 2012. Potential Evapotranspiration estimation for Indian conditions : Improving accuracy through calibration coefficients. Tech. Bull. No 1/2012. All India Co-ordinated Research Project on Agrometeorology, Central Research Institute for Dryland Agriculture, Hyderabad.

TABLE 3.5 Infiltration rate estimates for various soils

SOIL TYPE	BASIC INFILTRATION RATE (mm/hour)
Sand	less than 30
Sandy loam	20–30
Loam	10–20
Clay loam	5–10
Clay	1–5

Source: Based on FAO. 1988. Irrigation Water Management: Irrigation methods, http://www.fao.org/docrep/s8684e/s8684e0a.htm

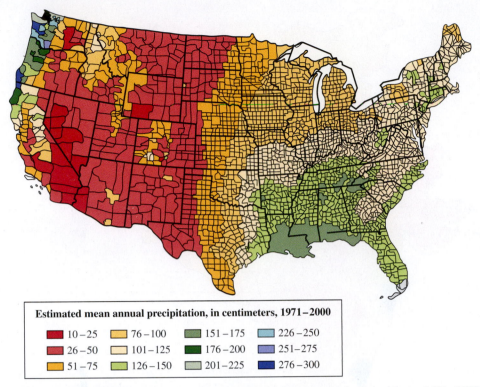

Estimated mean annual precipitation, in centimeters, 1971–2000

■ 10–25	■ 76–100	■ 151–175	■ 226–250
■ 26–50	■ 101–125	■ 176–200	■ 251–275
■ 51–75	■ 126–150	■ 201–225	■ 276–300

FIGURE 3.21 Estimated mean annual precipitation (cm) patterns in the United States, 1971–2000.

Source: Based on www.prism.oregonstate.edu: http://www.geog.ucsb.edu/img/news/2013/jawr12010_f2.gif

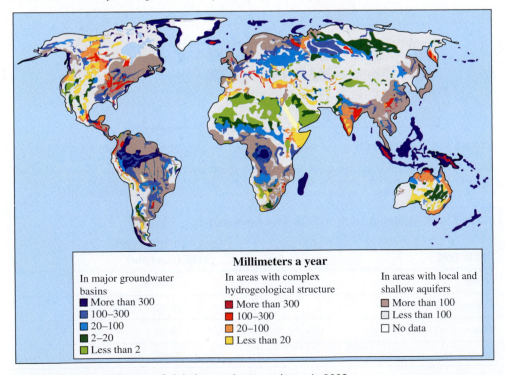

Millimeters a year

In major groundwater basins	In areas with complex hydrogeological structure	In areas with local and shallow aquifers
■ More than 300	■ More than 300	■ More than 100
■ 100–300	■ 100–300	☐ Less than 100
■ 20–100	■ 20–100	☐ No data
■ 2–20	■ Less than 20	
■ Less than 2		

FIGURE 3.22 Estimates of global groundwater recharge in 2008.

Source: Based on World Water Assessment Programme. 2009. The United Nations World Water Development Report 3: Water in a Changing World. Paris: UNESCO, and London: Earthscan.

3.7 Watersheds and Runoff

The **watershed** is the region that collects rainfall. When the height of the groundwater table is equal to the surface, a stream is formed. The precipitation may also flow over saturated land through **runoff** in small rivulets that may be collected into intermittent streams. The intermittent streams flow into groundwater-based streams or river systems. The river systems transport the water back to the oceans. Mays (2011) defines drainage basins, catchments, and watersheds as "three synonymous terms that refer to the topographic area that collects and discharges surface stream flow through one outlet or mouth." All the runoff within a watershed exits a single point downhill or downstream from all other points in the watershed. Figure 3.23 is a simple visualization of a watershed.

Watersheds can be very small when defined by a small parcel of land. The Amazon River collects water from the largest watershed in the world, which is 616 million hectares (Cech, 2010). Table 3.6 lists the 10 largest watersheds in the world and shows the area extent of the watershed and the **volumetric flow rate**, or volume of water exiting the watershed each day. Runoff formed from precipitation that is not absorbed by the soils is the source of much of the human freshwater drinking supply. Average long-term global runoff trends in a

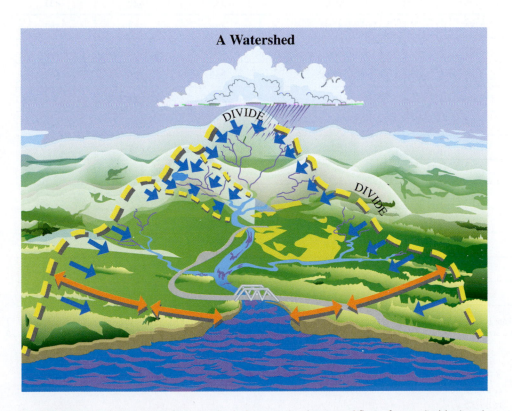

FIGURE 3.23 Illustration of a watershed, showing the downward flow of water (in blue) and the watershed boundaries (in yellow).

Source: Based on Recycle Works, http://www.recycleworks.org/images/watershed_800.jpg

TABLE 3.6 World's largest watersheds

RIVER	WATER RUNOFF VOLUME (km³/year)	WATERSHED POPULATION (MILLION PEOPLE)	WATERSHED LAND AREA (MILLION km²)		
			Minimum	Maximum	Average
Amazon	6.92	14.3	6,920	8,510	5,790
Ganges	1.75	439	1,389	1,690	1,220
Congo	3.50	48.3	1,300	1,775	1,050
Orinoco	1.00	22.4	1,010	1,380	706
Yangtze	1.81	346	1,003	1,410	700
La Plata	3.10	98.4	811	1,860	453
Yenisei	2.58	4.77	618	729	531
Mississippi	3.21	72.5	573	880	280
Lena	2.49	1.87	539	880	280
Ob	2.99	22.5	404	567	270
Mekong	0.79	75.0	505	610	376
Mackenzie	1.75	0.35	333	420	281
Amur	1.86	4.46	328	483	187
Niger	2.09	131	303	482	163
Volga	1.38	43.3	255	390	161
Danube	0.82	85.1	225	231	137
Indus	0.96	150	220	359	126
Nile	2.87	89.0	161	248	94.8
Amu Dana	0.31	15.5	77.1	118	56.7
Yellow	0.75	82.0	66.1	97	22.1
Dneiper	0.50	36.6	53.3	95	21.7
Syr Danya	0.22	13.4	38.3	75	26.2
Don	0.42	17.5	26.9	52	11.9
Murray	1.07	2.1	24	129	1.16

Source: Based on Revenga, C, Nackoney, J., oshino, E, Kura, Y., and Maidens, J. 2003. Watersheds of the World. IUCN, IWMI, Rasmar Convention Bureau and WRI¿World Resources Institute¿10 G Street NE¿Washington, DC 20002¿© 2003 World Resources Institute.

region along with human population in the region are illustrated in Figure 3.24. As shown in Figure 3.25, surface water supplies from runoff are the major source of water for all human uses including agriculture, drinking water, and energy production. Generally, regions of high human population have high water demands, but water demand is also related to affluence, agricultural practices, and industrial water consumption, as discussed in Box 3.2 and the following sections.

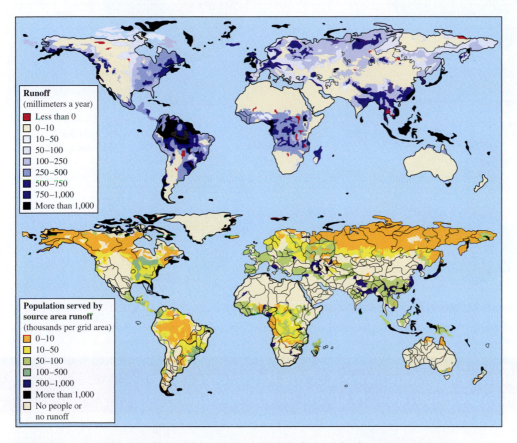

FIGURE 3.24 The runoff available in selected global watersheds compared to the population that uses the water. The consumptive use of water for agriculture, drinking water, industry, and other uses are called the water demand; areas of high population generally use a larger portion of the available runoff. The top map shows runoff-producing areas in absolute terms, with darker blue indicating areas that generate intense local runoff. This is the traditional view of the global distribution of the renewable water resource base. The bottom map shows the importance of all the world's runoff-producing areas as measured by the human population served. Thus runoff produced across a relatively unpopulated region, like Amazonia, while a globally significant source of water to the world's oceans, is much less critical to the global water resources base than runoff across a region like south Asia.

Source: Based on World Water Assessment Programme. 2009. The United Nations World Water Development Report 3: Water in a Changing World. Paris: UNESCO, and London: Earthscan.

Withdrawals by Supply Source

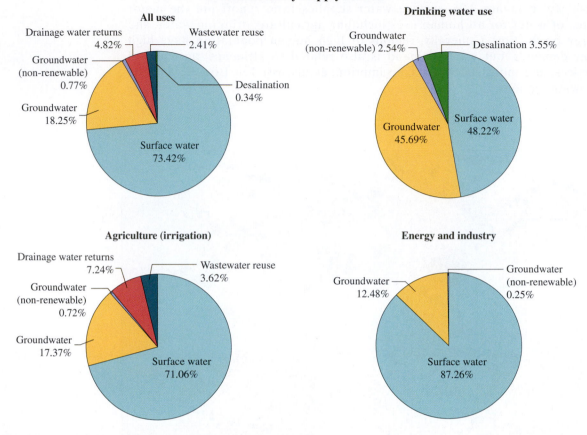

All uses

- Drainage water returns 4.82%
- Wastewater reuse 2.41%
- Groundwater (non-renewable) 0.77%
- Desalination 0.34%
- Groundwater 18.25%
- Surface water 73.42%

Drinking water use

- Groundwater (non-renewable) 2.54%
- Desalination 3.55%
- Groundwater 45.69%
- Surface water 48.22%

Agriculture (irrigation)

- Drainage water returns 7.24%
- Wastewater reuse 3.62%
- Groundwater (non-renewable) 0.72%
- Groundwater 17.37%
- Surface water 71.06%

Energy and industry

- Groundwater 12.48%
- Groundwater (non-renewable) 0.25%
- Surface water 87.26%

FIGURE 3.25 Global source and uses of drinking water for major water use sectors.

Source: Based on World Water Assessment Programme. 2009. The United Nations World Water Development Report 3: Water in a Changing World. Paris: UNESCO, and London: Earthscan.

BOX 3.2 Definitions of Key Components of the Hydrologic Cycle and Anthropogenic Impacts on the Hydrologic Cycle

NAME GIVEN TO A TYPE OF WATER	DEFINITION	SPACE AND TIME VARIABILITY	ROLE IN WATER RESOURCES SYSTEMS	MANAGEMENT CHALLENGES, VULNERABILITIES, AND OPPORTUNITIES
Green water	Soil moisture (nonproductive green water is evaporated from soil and open-water surfaces)	Very high over both dimensions	Direct support to rain-fed cropping systems	Highly sensitive to climate variability (both drought and flood); limited capacity to control Can be augmented by rainfall-harvesting techniques (many traditional and widely adopted) Weather and climate forecasts help in scheduling planting, harvest, supplemental irrigation, and other activities

(*Continued*)

BOX 3.2 Definitions of Key Components of the Hydrologic Cycle and Anthropogenic Impacts on the Hydrologic Cycle

NAME GIVEN TO A TYPE OF WATER	DEFINITION	SPACE AND TIME VARIABILITY	ROLE IN WATER RESOURCES SYSTEMS	MANAGEMENT CHALLENGES, VULNERABILITIES, AND OPPORTUNITIES
				Performance improved or compromised by land management
				Selection of improved crop strains for climate-proofing
Blue water (natural and altered)	Net or local groundwater recharge and surface runoff, streamflow	High over both dimensions	Farm ponds and check dams augment green water in rain-fed cropping systems	Highly sensitive to climate variability (both drought and flood) and ultimately climate change
			Source waters and entrained constituents delivered downstream within watershed	Some capacity to control
				Habitat management highly localized
				Many small engineering works can propagate strong cumulative downstream effects
				Poor land management heightens possibilities of flash flooding followed by dry streambeds
	Inland water systems (lakes, rivers, wetlands)	Decreased variability with increased size	Key resources over district, national, and multinational domains	Water losses through net evaporation occur naturally and through human use
			Important role in transport, waste management, and domestic, industrial, and agricultural sectors	Legacy of upstream management survives downstream (e.g., irrigation losses, pollution)
				Multiple-sector management objectives may be difficult to attain simultaneously
				Potential upstream–downstream conflicts (human to human; human to nature), including international
	Ground water (shallow)	Moderate over both dimensions; links to streams	Locally distributed shallow well systems serving drinking water and irrigation needs	Intimate connection to weather and climate means water yields subject to precipitation extremes
				Easily polluted
				Easily overused, resulting in temporary depletion; some loss of regional importance to oceans
	Fossil groundwater (deep)	Extremely stable	Critical (and often sole) source of water in arid and semi-arid regions	Large repositories of water but with limited recharge potential
				Use typically nonsustainable, leading to declining water levels and pressure, increasing extraction costs
				Low-replenishment rates mean pollution often effectively becomes permanent

(Continued)

BOX 3.2 Definitions of Key Components of the Hydrologic Cycle and Anthropogenic Impacts on the Hydrologic Cycle (*Continued*)

NAME GIVEN TO A TYPE OF WATER	DEFINITION	SPACE AND TIME VARIABILITY	ROLE IN WATER RESOURCES SYSTEMS	MANAGEMENT CHALLENGES, VULNERABILITIES, AND OPPORTUNITIES
Blue water (engineered)	Diversions, including reservoirs and inter-basin transfers Reuse waters	Stable to very stable	Critical (and often sole) source of water in arid and semi-arid regions Altered blue water balance as flows stabilized or redirected from water-rich times and places Multiple uses: hydropower, irrigation, domestic, industrial, recreational, flood control Secondary reuse as effluents in irrigation	Large quantities of water with high recharge potential Modified flow regime, with positive and negative impacts on humans and ecosystems Can destroy river fish habitat while creating lake fisheries by fragmenting habitat Natural ecosystem "cues" for breeding and migration removed Water supplies stabilized for use when needed most by society Sediment trapping, leading to downstream inland waterway, coastal zone problems Potential for introduction of exotic species Greenhouse gas emission from stagnant water Health problems (e.g., schistosomiasis) from stagnant water Social instability due to forced resettlement
Grey water	Recycled, reusable wastewater from residential, commercial, or industrial bathroom sinks, bathtub, shower drains, and clothes washing	Stable	May offset consumptive for nonpotable water use such as agriculture and landscaping or some industrial process waters	May be used for groundwater recharge May augment surface water reservoirs May be used to prevent saltwater intrusion in coastal areas Technical feasibility and risk still to be fully determined
Virtual water	Not an additional water system element	Stable, but linked to fluctuations in global economy	Water embodied in production of goods and services, typically with crops traded on the international market Not explicitly recognized as a water resources management tool until recently	Can implicitly offload water use requirements from more water-poor to more water-rich locations Particularly important where rain-fed agriculture is restricted and irrigation relies on rapidly depleting fossil groundwater sources
Desalination		Stable	Augmentation in water-scarce areas	Costly, special-use water supply, technologies rapidly developing for cost effectiveness

Source: Based on Adapted from World Water Assessment Programme. 2009. The United Nations World Water Development Report 3: Water in a Changing World. Paris: UNESCO, and London: Earthscan, and US EPA

3.8   Water Budget

The water budget balances the flows of water into and out of a watershed or system. The watershed boundaries determine the boundaries for the system of concern. In the case of the funnel system, the edges of the funnel, or the highest point, represent the boundaries of the watershed. Likewise, in natural watersheds, the highest points in those watersheds create the watershed boundaries. The boundaries can be determined by outlining the highest points on a topographic map. By connecting the dots between these highest points to the lowest point occurring in the stream that exits the watershed, the watershed boundaries can be determined.

We can sum all these terms to create a **water budget** for the watershed, where the sum of all the terms balance the inputs and outputs into a watershed:

$$\text{Inputs} = \text{outputs} + \text{storage} \tag{3.8}$$

$$P = R + I + E + T + S \tag{3.9}$$

Precipitation (P) is the input into the watershed on the left-hand side of the water budget equation. Water is removed from the watershed by evaporation (E), transpiration (T) [sometimes these terms are combined into the term evapotranspiration (ET)], runoff (R), and infiltration (I) into the soil. Water stored (S) in the system is also accounted for on the right-hand side of the equation. Civilization has tended to cluster and grow in or around watersheds where there are natural storage and plentiful water. Examples include the Egyptian civilization's growth around the Nile River, the location of present-day London around the Thames River, and almost any other city of significant size.

Humans have created aquifers to increase storage of water within a watershed in order to use this water for agriculture, industry, energy, or flood management as shown in Tables 3.7 to 3.10. Anthropogenic uses of water can significantly alter water-flow patterns within a watershed, even watersheds of significant size. Humans' water use may be consumptive, meaning that it leads to an extraction of water stored into the system, or also a water loss from the watershed.

Consumptive water use is defined as "water removed from available supplies without return to a water resources system (e.g., water used in manufacturing, agriculture, and food preparation that is not returned to a stream, river, or water treatment plant)" (Womach, 2005). Viessman and Hammer (1985) define two additional terms that help describe consumptive use:

Withdrawal use is "the use of water for any purpose which requires that it be physically removed from the source."

Nonwithdrawal use is "the use of water for any purpose which does not require that it be removed from the original source, such as water used for navigation."

Global water consumption rates and forecasted consumption rates are shown in Figure 3.26 see p. 152. Trends of water use for public supply, agriculture, electric production, and other industries in the United States between 1950 and 2000 are shown in Figure 3.27 see p. 153. Water demand

TABLE 3.7 Table water use, calorie return, and economic return on selected agricultural products

PRODUCT	KILOGRAM PER CUBIC METER	$ PER KILOGRAM	$ PER CUBIC METER	PROTEIN GRAMS PER CUBIC METER	KILOCALORIES PER CUBIC METER
Cereal					
Wheat	0.2–1.2	0.2	0.04–0.24	50–150	660–4,000
Rice	0.15–1.6	0.31	0.05–0.18	12–50	500–2,000
Maize	0.30–2.00	0.11	0.03–0.22	30–200	1,000–7,000
Legumes					
Lentils	0.3–1.0	0.3	0.09–0.30	90–150	1,060–3,500
Fava beans	0.3–0.8	0.3	0.09–0.24	100–150	1,260–3,360
Groundnut	0.1–0.4	0.8	0.08–0.32	30–120	800–3,200
Vegetables					
Potatoes	3.0–7.0	0.1	0.3–0.7	50–120	3,000–7,000
Tomatoes	5.0–20.0	0.15	0.75–3.0	50–200	1,000–4,000
Onions	3.0–10.0	0.1	0.3–1.0	20–67	1,200–4,000
Fruits					
Apples	1.0–5.0	0.8	0.8–4.0	Negligible	520–2,600
Olives	1.0–3.0	1.0	1.0–3.0	10–30	1,150–3,450
Dates	0.4–0.8	2.0	0.8–1.6	8–16	1,120–2,240
Other					
Beef	0.03–0.1	3.0	0.09–0.3	10–30	60–210
Fish (aquaculture)	0.05–1.0		0.07–1.35	17–340	85–1,750

Source: Based on Comprehensive Assessment of Water Management in Agriculture (2007) Water for Food, water for Life: A Comprehensive Assessment of Water Management in Agriculture. London: Earthscan, and Colombo: International Water Management Institute.

TABLE 3.8 Water consumption per metric ton of product produced

PRODUCT	WATER USE
Paper	80–2,000
Sugar	3–400
Steel	2–350
Petrol	0.1–40
Soap	1–35
Beer	8–25

Source: Based on Margat., J., and Andreassian, V. (2008) L'Eau, les Idees Receus. Paris, Editions le Cavalier Bleu.

TABLE 3.9 Typical household water use in the United States before and after installation of low flow apertenances and water conservation measures

USE	LITERS PER CAPITA: BEFORE	LITERS PER CAPITA: AFTER	SAVINGS: LITERS(S) DAILY WATER USE
Showers	43.9	33.3	10.6
Clothes washer	56.8	37.9	18.9
Toilets	70.0	31.0	39.0
Dishwasher	3.8	2.6	1.1
Baths	4.5	4.5	0.0
Leaks	36.0	15.1	20.8
Faucets	41.3	40.9	0.4
Other domestic uses	6.1	6.1	0.0
Total	262.3	171.5	90.8

Source: Based on Mays, p. 347

TABLE 3.10 Water use for energy production

POWER PROVIDER	GALLONS EVAPORATED PER kWh AT THERMOELECTRIC PLANTS	GALLONS EVAPORATED PER kWh AT HYDROELECTRIC PLANTS	WEIGHTED GALLONS EVAPORATED PER kWh OF SITE ENERGY
Western Interconnect	0.38 (1.4 L)	12.4 (47.0 L)	4.42 (16.7 L)
Eastern Interconnect	0.49 (1.9 L)	55.1 (208.5 L)	2.33 (8.8 L)
Texas Interconnect	0.44 (1.7 L)	0.0 (0.0 L)	0.43 (1.6 L)
U.S. Aggregate	0.47 (1.8 L)	18.0 (68.0 L)	2.00 (7.6 L)

Source: Based on Mays, pp. 348-349

in the United States peaked in the mid-1970s, at about the time water conservation, water recycling, and water regulations were established. In contrast, world water demand is expected to continue to climb into the foreseeable future. A portion of the worldwide demand is due to the related demand for more food to feed more people, for the globally increasing energy demand, and for other industrial needs. Water required for various energy generation scenarios is illustrated in Figure 3.28 see p. 154.

We can add the **consumption** term, **C**, to the output side of the water balance equation:

$$P = R + I + E + T + S + C \qquad (3.10)$$

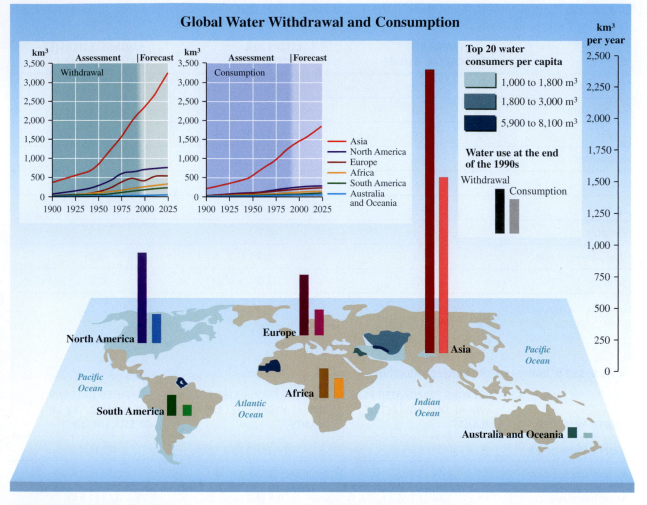

FIGURE 3.26 Global water use and consumption trends from 1900 to 2000 and expected forecast of water demand through 2025. The Asian continent has the highest rates of both water withdrawal and consumptive use.

Source: Based on UNEP/GRID-Arendal, Vital Water Graphics, Water withdrawal and consumption, http://www.grida.no/graphicslib/collection /vital-water-graphics. Data from Igor A. Shiklomanov, SHI (State Hydrological Institute, St. Petersburg), and UNESCO, Paris, 1999; World Resources 2000-2001: People and Ecosystems: the Fraying Web of Life, WRI, Washington DC, 2000; Paul Harrison and Fred Pearce, AAAS Atlas of Population 2001, AAAS, University of California Press, Berkeley.

In some cases, water can be extracted from other watersheds and rerouted to a different watershed via piping or aqueducts, as the Romans did, to increase the availability of water in their cities. In these cases, consumption from one watershed can lead to a net input of water into another watershed. Human interactions with the hydrologic cycle change the ecosystems within the affected watersheds. For instance, consumptive withdrawals from the Colorado River have been so great that in some years no flow from the Colorado River has reached the Pacific Ocean.

Engineers are currently examining water availability and have identified many water-scarce areas in both the developed and developing worlds where there are existing or expected water shortages. A sustainable level of consumption is one where the net storage term is greater than zero for a watershed. When the consumption terms increase from human demand to the point where the storage term becomes negative

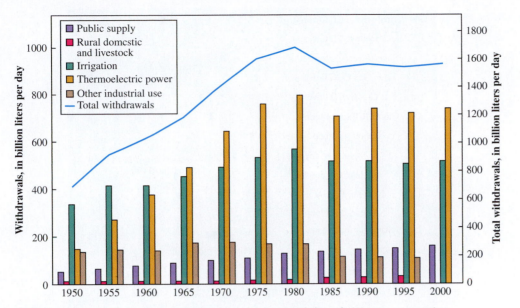

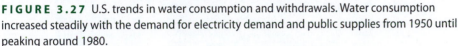

FIGURE 3.27 U.S. trends in water consumption and withdrawals. Water consumption increased steadily with the demand for electricity demand and public supplies from 1950 until peaking around 1980.

Source: From Hendrix/Thompson/Thompson/Graham, EARTH 2 (with CourseMate Printed Access Card), 2E. © 2015 Cengage Learning.

in value, then there is nonsustainable water use in the watershed. Archaeologists believe that water shortages have been a significant factor in the collapse of many civilized societies. Today in regions where consumptive withdrawals are greater than inputs into the watershed, water must be acquired from beyond the watershed boundaries, or water-rationing measures could be adopted to prevent the degradation and disruption of ecosystems and human systems within the watershed of concern.

The human uses of water may be part of a growing concern about global water scarcity, or the lack of available water supplies to meet human demands, as well as maintaining balanced ecosystem requirements. Human population growth, economic development, and advancing societal needs may outstrip the water available from current surface water and groundwater resources. "The rapid global rise in living standards combined with population growth presents the major threat to sustainability of water resources and environmental services." (World Water Assessment Programme, 2009). Technological advances in water development and energy development have the potential to help meet the growing water demand. At the same time, technological and industrial development may also increase the demand for water associated with agriculture, energy demand, and industrialization as illustrated in Tables 3.7 through 3.10. Engineers, scientists, and policymakers must work across regional and national boundaries to meet the future demand for water in many water-scarce areas, such as the American Southwest and all along the Mediterranean Sea, as shown in Figures 3.29 through 3.31 and discussed in Box 3.3: Malta's Water Worries.

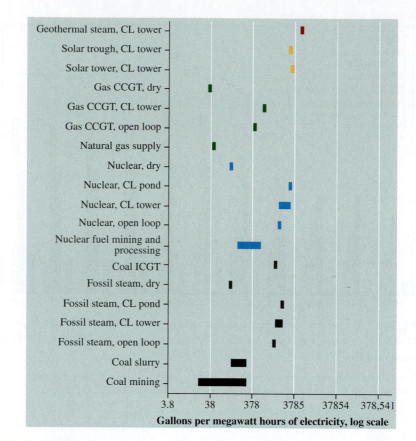

FIGURE 3.28 Consumptive water use for energy generation from selected generation scenarios in the United States in 2006. As shown in Figure 3.27, electivity supply is the largest use of water in the United States. This graphic shows the demand for closed-loop cooling (CL), combined cycle gas turbine (CCGT), and integrated gasification combined cycle (ICGT) generation of electricity.

Source: Based on World Water Assessment Programme. 2009. The United Nations World Water Development Report 3: Water in a Changing World. Paris: UNESCO, and London: Earthscan.

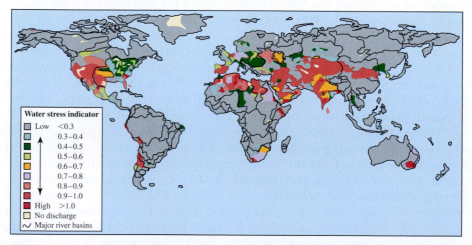

FIGURE 3.29 Water stress levels of major watersheds, worldwide in 2002.

Source: Based on World Water Assessment Programme. 2009. The United Nations World Water Development Report 3: Water in a Changing World. Paris: UNESCO, and London: Earthscan.

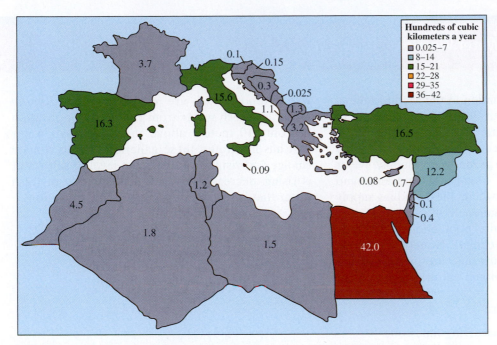

FIGURE 3.30 The values represent the overconsumption of available water resources in countries ordering the Mediterranean Sea for all uses of water from 2000 to 2005.

Source: Based on World Water Assessment Programme. 2009. The United Nations World Water Development Report 3: Water in a Changing World. Paris: UNESCO, and London: Earthscan.

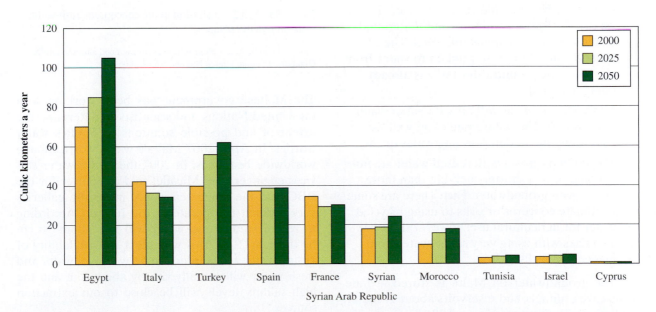

FIGURE 3.31 Water demand and consumption are likely to continue to increase for most Mediterranean countries in the foreseeable future.

Source: Based on World Water Assessment Programme. 2009. The United Nations World Water Development Report 3: Water in a Changing World. Paris: UNESCO, and London: Earthscan.

BOX 3.3 Malta's Water Worries

Malta is a small island nation in the Mediterranean Sea. It is home to about 400,000 people, most of whom receive fresh water from a limited groundwater supply. Every sector of Malta's economy depends on this limited groundwater supply. Water needs for agriculture, home use, industry, and tourism threaten to overwhelm the dwindling supply of water for this island nation. Malta's limited water resources make the entire population particularly susceptible to changing climate. Nonetheless, much of the population remains unaware of these issues and still clings to traditional cultural myths about their water supply.

Maltese Water Myths:

Myth: The natural occurring limestone removes salt from the groundwater, and thus as sea levels increase, there will be more fresh water. **Fact:** The limestone does not remove salt from the seawater. As sea level increases, the amount of available groundwater will decrease.

Myth: The groundwater table is replenished by water from the Nile River or other rivers on the European continent, and the groundwater supply is inexhaustible. **Fact:** The groundwater table is not linked to water from the Nile. The groundwater table is already being depleted.

Myth: Deeper wells will provide better water quality. **Fact:** The failure rate of all wells is increasing due to deteriorating water quality.

Myth: Crops grown with reused water are inferior and have a shorter shelf life than those grown with groundwater. **Fact:** There are some legitimate concerns or risks to using recycled water for agricultural use; however there are also risks with using very saline water from inadequate wells.

The groundwater on Malta is stored in limestone caves, tunnels, and reservoirs about 97 meters beneath the surface (see Figure 3.32). These ancient water repositories are called the *Ta' Kandja Galleries*. Fresh water from runoff and infiltration lies only 10 meters above the denser salt water that has infiltrated these galleries. Due to climate change and rising sea levels, the level of salt water is rising, decreasing the available supply of fresh water that is suitable for drinking and irrigation.

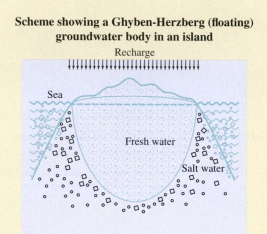

Scheme showing a Ghyben-Herzberg (floating) groundwater body in an island

FIGURE 3.32 Illustration of the conceptual fresh-water cone underlying Malta.

Source: Based on Malta water resources review. Food and Agriculture Organization of the United Nations, Rome, 2006.

The Maltese government has been working with the United Nations and scientists to determine the extent of and possible solutions to Malta's water worries. In April 2007 Malta's water worries made worldwide headlines. In 2007, the Malta water services engineer Paul Micallef told a reporter for the BBC that the rising sea would make the galleries very difficult to operate in the future. "According to recent studies, the water will rise about 96 cm by the year 2100. This will affect the availability of groundwater as the interface between sea water and fresh water will actually rise by about 1 m and the high salinity levels will be close to our extraction sources."

Groundwater Protected Areas in the Maltese Islands

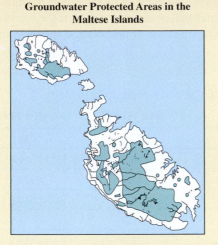

FIGURE 3.33 Area of Malta that is protected by the government for groundwater infiltration.

Source: Based on Malta water resources review. Food and Agriculture Organization of the United Nations, Rome, 2006.

Map Showing the Extent of Urbanization, 2000

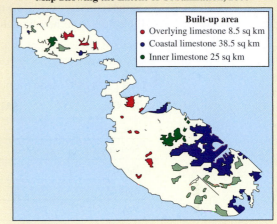

FIGURE 3.34 Areas of Malta that are already urbanized and lead to low rates of infiltration.

Source: Based on Malta water resources review. Food and Agriculture Organization of the United Nations, Rome, 2006.

Decreasing water supplies threaten Malta's food supply. Urbanization, which as increased runoff, has lead the government to create groundwater protection areas (Figure 3.33) to prevent development from encroaching upon groundwater recharge (Figure 3.34). Maltese farmers primarily use unregulated wells (called boreholes), which tap into the groundwater reservoirs, to irrigate their crops. Salty water prevents plants from obtaining the nutrients they need from the soil for growth. Plants and trees will die if they are irrigated with water that has too great a salt content. David and Mary Mallia, organic farmers interviewed for the 2007 BBC report, observed significant changes in their orchard productivity in recent years. "Since the rainfall has become less, the salinity is becoming higher and higher." In the past, normal salt concentration in the irrigation water would be 200 microsiemens. (A microseimen is a measure of electrical conductivity that is directly proportional to the amount

of salt in the water.) "This year (2007), with the lack of rain, it went up to 4000. It's not good for irrigation. If you water your trees with this water, it will kill them."

A panel of engineers and scientists with the Food and Agriculture Organization (FAO) of the United Nations (UN) reviewed the state of Malta's water resources. These engineers and scientists are working on a plan to develop a policy for water management that will be socially and politically acceptable and geared to tackling the complex water-related challenges. This review showed that, although the demand for groundwater is outstripping supply, water conservation measures could be effective. The UN report stated: "Groundwater quality can be protected and the mean sea-level aquifer stabilized. Policies and practices to do this need to be based on accurate information and acceptance that solutions must be applicable in the long-term. As such, political consensus and cross-party support is vital."

EXAMPLE 3.2 Malta Water Budget

We can apply the water budget equation to evaluate the urgency of water management measures for the island nation of Malta.

We first need to determine the average annual input, or precipitation, of water. We can estimate this by calculating the average rainfall that occurs over a year on the islands listed in Table 3.11. Note, however, that reported rainfall varies over the island, as shown in Table 3.12.

TABLE 3.11 Mean monthly values of rainfall and temperature in Malta

MONTH	RAINFALL (mm)	MAX TEMP °C	MIN TEMP °C
January	86.4	14.9	10.0
February	57.7	15.2	10.0
March	41.8	16.6	10.7
April	23.2	18.5	12.5
May	10.4	22.7	15.6
June	2.0	27.0	19.2
July	1.8	29.9	21.9
August	4.8	30.1	22.5
September	29.5	27.7	20.9
October	87.8	23.9	17.7
November	91.4	20.0	14.4
December	104.3	16.7	11.4

Source: Based on Malta water resources review. Food and Agriculture Organization of the United Nations, Rome, 2006.

TABLE 3.12 Reported rates of rainfall, evapotranspiration and effective rainfall (mm) for Malta, 1956–1991.

AUTHOR	RAINFALL	EVAPOTRANSPIRATION	RUNOFF	EFFECTIVE RAINFALL
Morris (1952) / Edelmann (1968)	522	392		130
ATIGA (Martin)	587	475		95
ATIGA (Verhoeven/ Gessel)	536	439		97
ATIGA (WWD Data)	551	431		120
FAO	587	437		150
Spiteri Staines (1987)	508	356	30	122
BRGM (1991)	551	348		203

Source: Based on Malta water resources review. Food and Agriculture Organization of the United Nations, Rome, 2006.

$$\text{Precipitation} = \frac{(86.4 + 57.7 + 41.8 + 23.2 + 10.4 + 2.0 + 1.8 + 4.8 + 29.5 + 87.8 + 91.4 + 104.3)}{12}$$

$$= \frac{45 \text{ mm}}{\text{month}} \tag{3.11}$$

We can convert the average monthly precipitation values expressed in mm/month to a yearly value expressed in m/year.

$$\frac{45 \text{ mm}}{\text{month}} \times \frac{12 \text{ months}}{\text{year}} = \frac{550 \text{ mm}}{\text{year}} \times \frac{1 \text{ m}}{1,000 \text{ mm}} = \frac{0.55 \text{ m}}{\text{year}} \tag{3.12}$$

We will assume this rainfall occurs equally over all the area of the islands, which is a reasonable assumption since the total land area is relatively small compared to much larger nations:

$$\text{Area} = 316 \text{ km}^2 = 316 \times \left(\frac{10^3 \text{ m}}{\text{km}}\right)^2 = 316 \times 10^6 \text{ m}^2 \tag{3.13}$$

$$\text{Average yearly rainfall over Malta} = \frac{0.55 \text{ m}}{\text{year}} \times 316 \times 10^6 \text{ m}^2$$

$$= 174 \times 10^6 \frac{\text{m}^3}{\text{year}} \tag{3.14}$$

We might first ask ourselves how much water is available per person if all the rainfall could be collected and used. Water consumption is normally expressed in liters per day, so we will convert our answer to liters per day:

$$\text{Population} = 398,000 + 2,400 \text{ per year} \tag{3.15}$$

Maximum consumption if all the rainfall were to be used is

$$\frac{\text{Average rainfall}}{\text{Population}} = \frac{\left(174 \times 10^6 \dfrac{\text{m}^3}{\text{year}}\right)}{398,000 \text{ people}} = \frac{430 \text{ m}^3}{\text{year}} \tag{3.16}$$

Converting to liters per day:

$$\frac{243 \text{ m}^3}{\text{year}} \times \frac{1,000 \text{ L}}{\text{m}^3} \times \frac{1 \text{ year}}{365 \text{ days}} = \frac{1,200 \text{ L}}{\text{day}} \tag{3.17}$$

This is more than enough water to meet demand, but it does not account for natural losses in the water budget equation. We will now examine those losses.

Losses include runoff and evapotranspiration listed in Table 3.13 in hm³/year. One cubic hectometer = 1×10^6 m.

$$\text{Yearly surface runoff to the sea} = 24 \text{ hm}^3 = 24 \times 10^6 \text{ m}^3 \tag{3.18}$$

Yearly actual evapotranspiration (assumed to be 68% of the total surface water)

$$= 105 \text{ hm}^3 = 105 \times 10^6 \text{ m}^3 \tag{3.19}$$

Not all the groundwater remains available for use. The groundwater mixes with sea water and also flows into the sea below the surface. The infiltration, I, and

TABLE 3.13 Balance of flow into and from the groundwater aquifers in Malta.

INFLOW	DESCRIPTION	hm₃/YEAR	COMMENTS
A	Precipitation	174	Based on an average annual rainfall of 550 mm
B	Surface runoff to the sea	24	Based on a variable catchment area runoff coefficient (excluding coastal built-up areas)
C	Actual evapotranspiration	105	Assumed as 68% of the total surface water
D	Natural aquifer recharge	45	B and C deducted from A
E	Artificial recharge from leaks	12	Estimated inflow from potable water and sewage network leakages
F	Total groundwater inflow	57	Sum of variables D and E
Outflow			
G	Water Services Corporation (WSC) groundwater abstraction	16	Official WSC extraction for hydrological year 2002/2003
H	Private groundwater outflow	15	Estimate based on water demand of various sectors (industry and agriculture)
I	Subsurface discharge to the sea	23	Estimate based on groundwater modeling
J	Total groundwater outflow	54	Sum of variables G, H, and I
Balance			
K	Total groundwater inflow	57	Equal to variable F
L	Total groundwater outflow	54	Equal to variable J
M	Balance	3	Inflow (K) less outflow (L)

Source: Based on Malta water resources review. Food and Agriculture Organization of the United Nations, Rome, 2006.

removal of water from the watershed by subsurface groundwater discharge to the sea is shown in Table 3.13. The water balance of each individual aquifer in Malta is shown in Table 3.14.

Rearranging and grouping evaporation and transpiration into one terms yields

$$S = P - R - ET - I = 174 \times 10^6 \frac{m^3}{year} - 24 \times 10^6 \frac{m^3}{year} - 105 \times 10^6 \frac{m^3}{year} - 23 \times 10^6 \frac{m^3}{year} \tag{3.20}$$

$$S = 22 \times 10^6 \frac{m^3}{year} \tag{3.21}$$

The remaining term, S, in the water budget equation accounts for all water available in BOTH groundwater and surface water storage.

TABLE 3.14 Water balance of flow into and from individual aquifers in Malta

GROUNDWATER CODE	GROUNDWATER BODY NAME	SIZE (km²)	INFLOW	OUTFLOW (hm²)	BALANCE	MAJOR EXTRACTION
MT001	Malta main mean sea level	216.6	34.27	36.65	−2.38	Abstraction for potable and agricultural purposes
MT002	Rabat-Dingli perched	22.6	4.64	4.62	0.02	Abstraction for agricultural purposes
MT003	Mgarr-Wardija perched	13.7	2.86	3.46	−0.59	Abstraction for potable and agricultural purposes
MT005	Pwales coastal	2.8	0.69	0.69	0.00	Abstraction for agricultural purposes
MT006	Mizieb mean seal level	5.2	1.11	0.96	0.15	Abstraction for potable and agricultural purposes
MT008	Mellieha perched	4.5	0.75	0.53	0.22	Abstraction for agricultural purposes
MT009	Mellieha coastal	2.9	0.69	0.38	0.31	Abstraction for agricultural purposes
MT010	Marfa coastal	5.5	0.89	0.62	0.27	Abstraction for agricultural purposes
MT011	Mqabba-Zurrieq perched	3.4	0.50	n/a	n/a	Abstraction for agricultural purposes
MT012	Comino mean sea level	2.7	0.52	0.30	0.22	Abstraction for agricultural purposes
MT013	Gozo mean sea level	65.8	8.66	9. 78	−1.12	Abstraction for potable and agricultural purposes
MT014	Ghajnielem perched	2.7	0.73	0.34	0.39	Abstraction for agricultural purposes
MT015	Nadur perched	5.0	1.15	0.58	0.57	Abstraction for agricultural purposes
MT016	Xaghra perched	3.0	0.71	0.33	0.38	Abstraction for agricultural purposes
MT017	Zebbug perched	0.4	0.10	0.03	0.07	Abstraction for domestic purposes
MT018	Victoria-Kercem perched	1.5	0.39	0.14	0.25	Abstraction for domestic purposes

Source: Based on Malta water resources review. Food and Agriculture Organization of the United Nations, Rome, 2006.

Once again we can estimate the available water supply if all the stored water were used by each person in Malta:

$$\text{Annual total storage/population} = \frac{\left(22 \times 10^6 \frac{m^3}{year}\right)}{398,000 \text{ people}} = \frac{55 \text{ m}^3}{year} \times \frac{1,000 \text{ L}}{m^3} \times \frac{1 \text{ year}}{365 \text{ days}}$$

$$= 151 \text{ L per day} \tag{3.22}$$

Therefore, 151 liters per day is the maximum consumptive use of water that is sustainable for Maltese society, without additional sources of water or water reuse.

In Table 3.15, we see that the total water use for Malta in 2003 was $58.641 \times 10^6 \frac{m^3}{year}$. This equates to a per capita consumption of = 400 L per capita day.

TABLE 3.15 Water use in Malta by sector and source (10^3 m^3)

USE	WATER SERVICES CORPORATION				PRIVATE		TOTAL
	BILLED	UNBILLED	GROUNDWATER	RO	TREATED EFFLUENT	RUNOFF HARVESTING	
	REVERSE OSMOSIS DESALINIZATION						
Domestic	12,620	3,686	1,000			2,000	19,306
Tourism	1,134	331	500	1,000			2,965
Farms	1,336	390	500				2,226
Agriculture			14,500		1,500	2,000	18,000
Commercial	1,247	364					1,611
Industrial	941	275	1,000		500		2,716
Government	818	239					1,057
Others	869	254					1,123
Total consumption	18,965	5,540	17,500	1,000	2,000	4,000	49,005
Real losses		9,636					9,636
Total + losses	18,965	15,176	17,500	1,000	2,000	4,000	58,641
WSC							
Total apparent losses	5,540	16%					
Total loss	15,176	44%					

Source: Based on Malta water resources review. Food and Agriculture Organization of the United Nations, Rome, 2006.

We can compute the percent over consumption of the available water resources in Malta.

Actual consumption/maximum sustainable consumption *100% is

$$58.641 \times 10^6 \, \frac{\text{m}^3}{\text{year}} \Big/ 22 \times 10^6 \, \frac{\text{m}^3}{\text{year}} \times 100\% = 266\% \qquad (3.23)$$

Therefore actual consumption is 266% beyond the sustainable withdrawals expected from rainfall.

We can compare the available sustainable water supply to actual water demand in Malta. In Example 3.3, it is shown that the water demand has outpaced the rate of replenishment of water to the groundwater supply on the island. Agricultural, industrial, and water demands in homes place a heavy burden on the limited water supply. Increasing water demands (see Figure 3.35) and a slowly changing climate that is becoming dryer have caused Malta to recognize the close connection between water supply, energy consumption, and resource use and the importance of the water–energy nexus. (A nexus is the point of interaction of one system with another.)

Malta has begun building and using desalination plants to supplement its natural water resources (see Figure 3.36). Desalinization is an energy-intensive

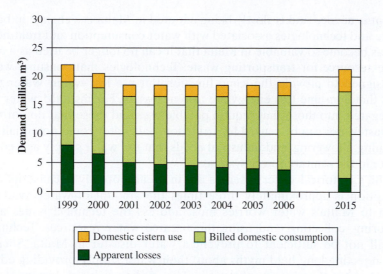

FIGURE 3.35 Maltese water consumption variations by year and expected demand.

Source: Based on Malta water resources review. Food and Agriculture Organization of the United Nations, Rome, 2006.

process that removes enough salt from seawater to make it suitable for drinking. Malta has had to increase the importation of oil from Middle Eastern countries; a significant portion of this energy must be used for desalinization. Increasing the use of energy has the potential to increase Malta's greenhouse gas emissions associated with fossil fuels. The greenhouse gases are regulated in the European Union and make the process of desalinization even more expensive. Furthermore, the emission of greenhouse gases due to desalinization may contribute to the changing climate that in part may increase Malta's water demand. This is known as a reinforcing spiral of cause and effect and will be explored in greater detail in subsequent chapters.

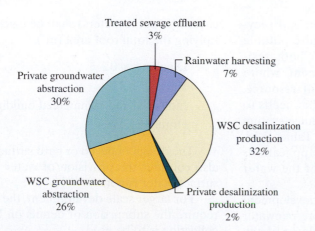

FIGURE 3.36 Maltese water production and sources.

Source: Based on Malta water resources review. Food and Agriculture Organization of the United Nations, Rome, 2006.

What is needed and is slowly being adopted in Malta is a change in basic infrastructure and technologies associated with water consumption and treatment. Fresh water has become so valuable in Malta that it can no longer be used as a cheap and effective resource for transporting waste. Technologies that consume water must be redesigned to provide the same function but use less water. Water treatment systems that consume less energy are also necessary. For instance, water uses may be segregated into those that require potable uses and those that do not require it. Some wastewater can be reused for irrigation. Salt water could potentially be used for cleaning, showering, and industrial needs that use water simply as a conveyance for waste in the sanitary sewer system.

Malta's historical water reserves are in great jeopardy because its estimated consumption is approximately three times the amount of water available. Any solution to Malta's water worries must address the technical issues associated with storing, collecting, and using their scarce water resources. Technical solutions will not be sufficient to prevent a water disaster in Malta. Social issues, such as rejecting long-held myths about their water supply, providing educational resources about the urgency of the problem, and ensuring government oversight and monitoring of the water resources must also be addressed. Private boreholes and wells are difficult to manage and regulate effectively. The FAO recommends significant investment in household rainwater collection systems, but private implementation is difficult to enforce (see Box 3.4). Neither technical nor social issues are likely to be addressed if the economic questions and concerns are not taken into account. Malta's economy relies on industrial productivity and a tourist industry, both of which create significant demands on Malta's limited water supply (Tables 3.16 and 3.17). However, if the environmental issues associated with the rising sea level due to climate change and anthropogenic depletion of the groundwater supplies are not addressed, the entire Maltese

BOX 3.4 **FAO Design Guidance and Recommendations for Malta**

Rainwater runoff should be collected and recycled (for those uses that do not require potable water). This requirement applies to both residential and nonresidential development, where the collected runoff may be a useful resource. Collection also reduces the amount that needs to be dealt with by the stormwater drainage system and so may have wider benefits (see Tables 3.18, 3.19, and 3.20). Plans submitted with applications should show the proposed location of the water cistern.

The FAO (2006) proposes new development be provided with a water cistern to store rainwater runoff from the built-up area. The volume of the cistern (in cubic meters) shall be calculated by multiplying the total roof area (m^2):

Dwelling $\times$ 0.3

Villas $\times$ 0.45

Industrial and commercial buildings $\times$ 0.45

Hotels $\times$ 0.6

The design of paved or hard surface areas should also consider the provision of water catchment for surface water runoff.

For larger scale development, the authority may require the submission of details on how the water collection is to be used.

TABLE 3.16 Agricultural demand of water for animal production

ANIMAL CATEGORY	NUMBER OF ANIMALS	DAILY WATER DEMAND (liters/day)	ANNUAL DEMAND (m³)
Pigs	73,067	Summer:20.25 Winter:13.5	4,50,000
Cattle			
Calves 1 year	4,909	Summer:139.5 Winter:94.5	2,09,500
Cattle 1–2 years	4,983	Summer:182.25 Winter:137.25	2,90,500
Cattle 2 years	8,093	Summer:195.75 Winter:150.75	5,11,800
Total	17,985		10,11,800
Sheep	14,861	Summer:18.0 Winter:13.5	85,500
Goats	5,374	Summer:18.0 Winter:13.5	31,000
Rabbits	55,254	Summer:3.0 Winter:1.5	45,000
Poultry			
Layers (others)	7,56,288	Summer:0.36 Winter:0.32	94,000
Broilers	11,84,157	Summer:0.22 Winter:0.28	1,08,000
Total			2,02,000
Equine	853	Summer:58.5 Winter:49.5	16,800
Total			18,42,100

Source: Based on Malta water resources review. Food and Agriculture Organization of the United Nations, Rome, 2006.

economy and society could be jeopardized in the near future. Sustainable use of Malta's water resources will require improved governance; improved awareness; staged and adaptive implementation; demand management; supply augmentation; the ensuring of equity and justice; and targeted interventions.

The FAO report suggests that significant progress must be made by 2015 to protect Malta's fragile water resources. The report describes four demand scenarios (based on current levels of demand and the subsequent projections that have emerged from discussions with various stakeholders). It also discusses different strategies for meeting future demand and achieving the vision shown in Tables 3.18 and 3.19.

TABLE 3.17 Water demand in Malta by the tourism industry

	1997/1998	1998/1999 (m³)	1999/2000
Total annual water production	40,772,926	37,963,808	36,604,128
Production to satisfy tourist demand	3,669,563	3,416,742	3,294,371
Per capita demand per day	0.324	0.293	0.321

Source: Based on Malta water resources review. Food and Agriculture Organization of the United Nations, Rome, 2006.

The water crisis on the small island of Malta is synonymous with issues other nations and communities must address in the near future. Large cities in arid regions of the United States, such as Los Angeles, Las Vegas, and Phoenix, are already facing several water shortages despite importing huge quantities of water from adjacent watersheds. Even Atlanta in the southeastern United States, a city located in a relatively wet climate, faces severe water shortages. Atlanta's water shortages have been created by an unusual geology. A large granite monolith beneath Atlanta

TABLE 3.18 Simplified plan for development of groundwater development in Malta

	STAGE 1 (YEARS PRIOR TO THE EARLY 1980s)	STAGE 2 (EARLY 1980s TO MID-2000s)	STAGE 3 (MID-2000s TO LATE 2000s)	STAGE 4 (LATE 2000s ONWARD)
Stages	Low-level irrigation meetings, local demand for vegetables, olives, grapes, wine, etc.	Agrarian boom initially by availability of drilling technology and subsequently sustained by EU-accession, land-based subsides. Large increase in area under olive cultivation and vineyards.	Symptoms of groundwater overexploitation increasingly apparent and starting to affect yields and crop quality.	Decline in overall groundwater use. Increased use of unconventional sources for irrigation.
Characteristics	Total irrigated area less than 500 ha. Use of spring water for irrigation and gradual development of the perched aquifer using shallow wells.	Total irrigated area increased to more than 250 ha. Large increase in use of supplemental irrigation. Deep boreholes used to exploit sea-level aquifers.	Peak reached in irrigated area and agricultural water use as a result of declining groundwater quality and introduction of regulations and tariffs.	Low-value crops no longer profitable. Area under irrigation declines. Agri-environment planning becomes the norm.
Impacts and sustainability	High-level sustainability. Some decline in water levels and spring flows from perched aquifer. Onset of a nitrate pollution problem.	Impacts of groundwater regulation, awareness campaigns, and improved planning starting to have an impact on sustainability.	Impacts of groundwater regulation awareness campaigns and improved planning starting to have an impact onsustainability.	A balance achieved between groundwater inflow and outflow. A strategic reverse created. Water quality improving steadily.
Interventions	Limited government support. No groundwater regulation.	Local market not protected. Groundwater regulation (including tariffs) and catchment planning introduced.	Local market not protected. Groundwater regulation (including tariffs) and catchment planning introduced.	Adaptive management of groundwater becomes the norm. Good environmental awareness established at all levels.

Source: Based on Malta water resources review. Food and Agriculture Organization of the United Nations, Rome, 2006.

TABLE 3.19 Expected demand and plans for Malta's water resources

WATER DEMAND SCENARIOS	POTENTIAL DEMAND SCENARIOS BASED ON CURRENT TRENDS	WATER SUPPLY STRATEGIES	STRATEGIES FOR MEETING CURRENT AND FUTURE DEMANDS (i.e., 2015)	ECONOMIC/SOCIAL IMPLICATIONS
I	Municipal demand remains fairly constant at current values or increases at about 1 to 2%/year, while agricultural demand increase reaches a maximum not exceeding 21 hm³ as projected.	I	Agriculture is given priority over the use of groundwater, and, consequently, the urban supply is increasingly sourced from RO plants.	The WSC has to source its supply increasingly from reverse osmosis desalination plants, with resulting price increases in the urban sectors. The quality of the domestic supply will increase. However, the country will be fully dependent on RO plants for the production of potable water, leaving the country vulnerable to fluctuations in the industrial and tourism sectors, which would be expected to reduce its economic competitiveness.
		II	No action is taken, and groundwater abstraction remains unregulated.	Over-abstraction will result in an increase in the salinity of groundwater abstracted from the sea-level aquifers and in the drying up of perched aquifers in the summer period. Degeneration in quality will make groundwater unsuitable for direct utilization, and extra treatment costs will be incurred by all sectors in the long term as groundwater becomes progressively unusable for all sectors.
		III	A reduction in groundwater abstraction is implemented in order to achieve a sustainable abstraction strategy allowing the setting up of a strategic groundwater reserve. The available abstractable groundwater quota is then allocated on a 50/50 basis between the WSC and all other users. Option involving artificial recharge of groundwater and improved rainwater harvesting will also have to be implemented in order to augment groundwater availability. Agri-environment schemes and smart irrigation techniques are used to encourage low-water-using farming systems.	The WSC will have to reduce the proportion of abstracted groundwater, while agriculture and industry will have to substantially increase the amount of recycled water used. Tariffs for the domestic/commercial sectors will be utilized to manage the sectoral demand. Agriculture will have to absorb a proportion of the cost of treating sewage effluent—possibly additional costs of desalination on the effluent treated to tertiary level. Groundwater will be viewed as a national strategic resource and will have a potential negative impact on the livelihood of some agricultural users. These drawbacks will be outweighed by the potential benefits to the economy as a whole. In the long term, better groundwater quality will result in decreasing treatment costs for all sectors.
II	Municipal demand remains fairly stable at current values or increases at about 1 to 2%/year, while agricultural demand decreases to pre-EU accession levels (of 10 hm³/year) driven mainly by market forces.	I	Agriculture is given priority over the use of groundwater, with the potable supply being increasingly sourced from RO plants. Effluent from wastewater treatment plants viewed primarily as an option to supplement water supply to the the agricultural/industrial sector, thus further reducing the pressures on groundwater.	The cost of the WSC supply will increase in proportion to the increased dependence on RO water that will be required in order to maintain potable quality standards.

Source: Based on Malta water resources review. Food and Agriculture Organization of the United Nations, Rome, 2006.

TABLE 3.20 Timeline for implementing Malta's water resources protection plan

YEAR	REQUIREMENT
2003	Directive transposed into national legislation.
	Identification of river-basin districts and of the competent authorities that will be empowered to implement the directive.
2004	Completion of the first characterization process and the first assessment of impacts on the river-basin districts.
	Completion of the first economic analysis of water use.
	Establishment of a register of protected areas for the river-basin districts.
2006	Environmental monitoring programs established and operational.
	Work program for the production of the first River Basin Management Plan established.
2007	Public consultation process on significant water management issues in the river-basin districts initiated.
2008	Publication of the first River Basin Management Plan for public consultation.
2009	First River Basin Management Plan finalized and published.
	Program of measures required in order to meet the environmental objectives of the directive finalized.
2012	Program of measures to be fully operational.
	Work program for the production of the second River Basin Management Plan published.
2013	Review of the characterization and impact assessment of the river-basin districts.
	Review of the economic analysis of water use.
	Interim overview of significant water management issues published.
2014	Publication of the second River Basin Management Plan for public consultation.
2015	Achievement of the environmental objectives specified in the first River Basin Management Plan
	second River Basin Management Plan finalized and published with revised program of measures.
2021	Achievement of the environmental objectives specified in the second River Basin Management Plan.
	third River Basin Management Plan to be published
2027	Achievement of the environmental objectives specified in the third River Basin Management Plan.
	fourth River Basin Management Plan to be published.

Source: Based on Malta water resources review. Food and Agriculture Organization of the United Nations, Rome, 2006.

prevents the city from easily tapping into groundwater supplies. Surrounding cities both upstream and downstream of Atlanta are concerned about the environmental health of their watersheds due to the amount of water the city of Atlanta demands from those surrounding watersheds.

Increasing populations and changing lifestyles that demand more water together with changing climates have profound effects on water supplies throughout the world.

3.9 Mass Balance and System Boundaries

Earlier in this chapter, we developed a water budget equation to model environmental data that could be used to predict the water resources available in a given area, or watershed. The boundaries of the system in the watershed were well defined. We can perform a similar analysis of any system, once we define the system boundaries. The law of conservation of mass states that mass cannot be created or destroyed. Balancing the mass flow into and out of a system allows engineers and scientists to quantitatively analyze the behavior of anthropogenic emissions in the environment. By accounting for the flow of water into and out of the watershed, we developed the water budget equation:

Mass of water flow into the watershed = mass of water flow out of the watershed

+ accumulation of water within the watershed (3.26)

We assumed that the climate was not changing and that the watershed was at steady state; that is to say, on average the values for the water budget equation did not change with time. However, in many instances, including even our own climate, conditions do change with time. In order to accommodate these changes in a mathematical model, we need to account for the change in the mass flow rate with respect to time, dM/dt. In its general form, in one dimension, the mass flow rate into and out of a system is defined by

$$\frac{dm}{dt} = M_{in} - M_{out} \pm M_{reaction}$$ (3.27)

The volume of flow into and out of a system is also defined by the boundaries of the system under examination. The volume within these boundaries is said to be the control volume (Figure 3.37). The control volume for our water budget equation is defined by the boundaries of the watershed. A control volume in the laboratory is defined by the boundaries of the reactor's walls.

For a given system, such as a wastewater treatment lagoon shown in Figure 3.37,

M_{in} = mass flow of any substance into the lagoon

M_{out} = mass flow of any substance out of the lagoon

$M_{reaction}$ = the production or degradation of the mass flow of any substance within the lagoon

dM/dt = accumulation or removal of a mass of any substance within the lagoon

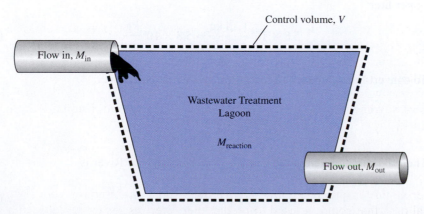

FIGURE 3.37 Schematic of a mass balance for a wastewater treatment process.

Source: Bradley Striebig.

A common example of the application of a material flow analysis occurs when two rivers join together at a point called the confluence and these rivers form one larger river that continues to flow toward an ocean.

EXAMPLE 3.3   Flow in the Blue and White Nile Rivers

The United Nations Food and Agriculture Organization (FAO) reported that the peak flow from the Blue Nile in August at Khartoum near the confluence of the two rivers is approximately 15 km³ per month. The FAO reported that the peak flow from the White Nile in August at Mogren near the confluence of the two rivers is approximately 2 km³ per month. The mass flow rates of these two rivers are additive, and the flow rate of the two combined rivers in August near Tamaniat, just downstream of the confluence of the Blue and White Nile Rivers, shown in Figure 3.38 is approximately:

$$M_{\text{in}} = M_{\text{out}} \tag{3.28}$$

$$M_{\text{White Nile}} + M_{\text{Blue Nile}} = M_{\text{Combined Nile Rivers}} \tag{3.29}$$

In units of km³/month:

15 km³ per month + 2 km³ per month = 17 km³ per month   (3.30)

In units of liters per second for the Blue Nile River:

$$15\,\frac{\text{km}^3}{\text{month}} \cdot \left(\frac{1000\,\text{m}}{\text{km}}\right)^3 \cdot \frac{1000\,\text{L}}{\text{m}^3} \cdot \frac{1\,\text{month}}{30\,\text{days}} \cdot \frac{1\,\text{day}}{24\,\text{hours}} \cdot \frac{1\,\text{hour}}{60\,\text{minutes}} \cdot \frac{1\,\text{minute}}{60\,\text{seconds}} = 5.8 \times 10^6\,\frac{\text{L}}{\text{s}}$$

(3.31)

In units of liters per second for the White Nile River:

$$2\,\frac{\text{km}^3}{\text{month}} \cdot \left(\frac{1000\,\text{m}}{\text{km}}\right)^3 \cdot \frac{1000\,\text{L}}{\text{m}^3} \cdot \frac{1\,\text{month}}{30\,\text{days}} \cdot \frac{1\,\text{day}}{24\,\text{hours}} \cdot \frac{1\,\text{hour}}{60\,\text{minutes}} \cdot \frac{1\,\text{minute}}{60\,\text{seconds}} = 0.8 \times 10^6\,\frac{\text{L}}{\text{s}}$$

(3.32)

To convert liters per second to kg/s for the Blue Nile River, we must multiply by the density of the fluid. For freshwater under standard conditions, the density of water is 1 kg per liter:

$$5.8 \times 10^6\,\frac{\text{L}}{\text{s}} \cdot \frac{1\,\text{kg}}{1\,\text{L}} = 5.8 \times 10^6\,\frac{\text{kg}}{\text{s}} \tag{3.33}$$

To convert liters per second to kg/s for the White Nile River:

$$0.8 \times 10^6\,\frac{\text{L}}{\text{s}} \cdot \frac{1\,\text{kg}}{1\,\text{L}} \cdot = 0.8 \times 10^6\,\frac{\text{kg}}{\text{s}} \tag{3.34}$$

Then the combined mass flow rate of water in the Nile River is

$$5.8 \times 10^6\,\frac{\text{kg}}{\text{s}} + 0.8\,10^6\,\frac{\text{kg}}{\text{s}} = 6.6\,10^6\,\frac{\text{kg}}{\text{s}} \tag{3.35}$$

Over six and half million kilograms of water flow each second through the Nile River channel during the peak flow season!

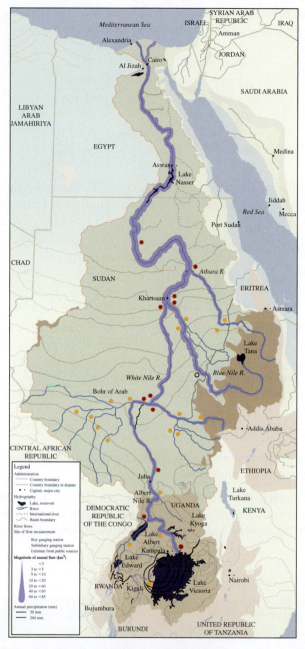

FIGURE 3.38 Map of the Blue and White Nile River basins.

Source: lapi/Shutterstock.com with data added by author.

The volumetric flow rate (km³ per month) must be multiplied by the density of the water in order to calculate the mass flow rate (kg/s). Since most of the water (H_2O) in water (the mixture of the aqueous solvent, H_2O, and the dissolved solids and gases in the liquid mixture) is water (H_2O), the density of the river water is approximately the same as the density of pure H_2O.

Chemicals or physical substances in water are often of interest in part because of their potential negative (or sometimes positive) environmental impact. A mass balance on the mass flow of any substances, $M_{substance}$, in a given flow of water, Q_{fluid}, can be related to the concentration of the substance, $c_{substance}$, where

$$M\left[\frac{mass}{time}\right] = Q\left[\frac{volume}{time}\right] c\left[\frac{mass}{volume}\right] \tag{3.36}$$

$$M_{subtance}\left[\frac{mg}{s}\right] = c_{substance}\left[\frac{mg}{L}\right] Q_{fluid}\left[\frac{1}{sec}\right] \tag{3.37}$$

Substituting the relationship between volume, V, and concentration, c, for the mass flow rate yields the general mass balance equation in terms of the substance concentration:

$$V\left(\frac{dc}{dt}\right) = cQ_{in} - cQ_{out} \pm \left(\frac{dc}{dt}\right)V \tag{3.38}$$

It is helpful to remember the following steps are when the general mass balance equation is being solved:

1. Draw a schematic or flow diagram of the system or process to be examined.
2. Define the system boundaries and control volume.
3. Define the variables and list the known data and assumptions needed to solve the equation.
4. Define and substitute the rate expressions for generation or decay of a substance.
5. Make any unit conversions necessary to ensure that the units within the mass balance equation are consistent.

In many instances, the substances of concern do not vary with time. Under this steady-state condition, $dc/dt = 0$. If there is no decay or generation of the substance within the control volume, then the mass balance equation simplifies to:

$$\sum_{inflow} c \cdot Q = \sum_{outflow} c \cdot Q \tag{3.39}$$

EXAMPLE 3.4   Total Suspended Solids in the Nile River at Steady State with No Reaction

The Nile River carries a tremendous amount of eroded soil particles in its waters during the wet season. Volumetric flow rates in the rivers and pollutant levels vary with the time of year and wet or dry season, as shown in Figure 3.39. The solid particles in the water are filtered and measured, and these particles are referred to as the total suspended solids (TSS). The Nile Basin Initiative Transboundary Environmental Action Project reported that the TSS concentration in the Blue Nile during the wet season just upstream of the confluence of the two rivers is approximately 7,000 mg/L. The TSS concentration in the White Nile during the wet season just upstream of the confluence of the two rivers is approximately 70 mg/L. Determine the concentration of TSS just downstream of the confluence of the Blue Nile and White Nile. For this example we will assume that there is no change in concentration with time (i.e., $dM/dt = 0$) and the

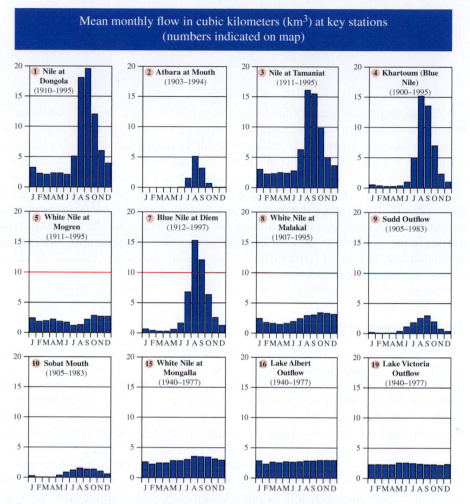

FIGURE 3.39 Monthly volumetric flow fluctuations in the Nile River basin.

Source: Based on FAO (2011) Synthesis Report: FAO – Nile Basin Project. GCP/INT/945/ITA 2004 to 2009. Food and Agriculture Organization of the United Nations, Rome, Italy.

total suspended solids do not react with anything in the river at the confluence of the two rivers (i.e., $M_{reaction} = 0$).

$$M_{TSSin} = M_{TSSout} \tag{3.40}$$

$$M_{TSS\ White\ Nile} + M_{TSS\ Blue\ Nile} = M_{TSS\ Nile\ River} \tag{3.41}$$

Substituting the volumetric flow rates and concentrations for each tributary of the Nile yields

$$Q_{White\ Nile}\,c_{TSS\ White\ Nile} + Q_{Blue\ Nile}\,c_{TSS\ Blue\ Nile} = Q_{Nile\ River}\,c_{TSS\ Nile\ River} \tag{3.42}$$

Rearranging:

$$c_{TSS\ Nile} = \frac{Q_{White\ Nile} \cdot c_{TSS\ White\ Nile} + Q_{Blue\ Nile} c_{TSSBlue\ Nile}}{Q_{Nile}} \tag{3.43}$$

$$c_{\text{TSS Nile}} = \frac{0.8 \times 10^6 \, \frac{L}{s} \cdot 70 \, \frac{mg}{L} + 5.8 \times 10^6 \, \frac{L}{s} \cdot 7{,}000 \, \frac{mg}{L}}{6.6 \times 10^6 \, \frac{L}{s}} \cong 6{,}000 \, \frac{mg}{L} \text{ of TSS in the Nile River}$$

(3.44)

The general mass balance equation can be used to examine the fate and transportation of chemicals from one biogeochemical repository to another. Many important biogeochemical cycles influence the sustainability of modern societal lifestyles. In the next section, we will look closely at the nitrogen and phosphorus cycles, in which natural and anthropogenic forces influence these elemental cycles. The fundamental mass balance equation is the basis of the climate models that account for the movement of oxygen and carbon dioxide from carbon stored in the ground in the form of fossil fuels to carbon and oxygen molecules stored in the air in the form of carbon dioxide. The amount of carbon dioxide in the atmosphere has a significant effect on the average surface temperature of the planet and, hence, the Earth's climate. The carbon and oxygen cycles in the Earth's air and water are discussed in more detail in Chapters 4, 5, and 6.

3.10 Summary

Our Common Future, a report of the Bruntland Commission for the United Nations (1987), defined sustainable development, as "development that meets the needs of the present without compromising the ability of future generations to meet their own needs." Water is a primary resource and a need for all generations. The World Health Organization (WHO) reports that 900 million people lack access to improved water supply, or a source of water protected from sewerage contamination. Nonimproved water sources are often contaminated, and 2.5 billion people lack access to proper sanitation. The lack of adequate sanitation results in pollution of the local and downstream watersheds and degradation of the local ecosystem, resulting in poor human health in those localities. WHO estimates that 25% of all preventable illnesses worldwide occur as a result of poor environmental quality.

Water will become scarcer as the planet's population continues to grow. Water scarcity occurs when there is insufficient water to meet the water demands for drinking water, washing, and cooking (see Figure 3.40). The WHO estimates that 20 liters per day of water located within 1 kilometer of the user's dwelling is the minimum amount of water necessary in most cultures to have reasonable access to water. Engineers are recognizing that the level of consumption of water is important for determining the requirements of the infrastructure we are developing for future generations. When consumption is greater than storage in a system, the level of water withdrawal is nonsustainable. Future generations will therefore be forced to ration water or find a technical solution to reuse water within the watershed. Otherwise they will face similar circumstances to those faced by Chaco Canyon's residents 1000 years ago.

Anthropogenic processes are changing rapidly, in turn changing the Earth's environment. These changes will make our current lifestyles more difficult to maintain with existing technology and projected population growth. Engineers have made choices in the past that did not consider the sustainability of systems as part of the design process. Today's engineers and scientists must utilize new technical changes that are currently underway, such as green building codes and incentives that value water conservation, to create a more sustainable infrastructure. Only then can they better meet the needs of today's generations and generations to come.

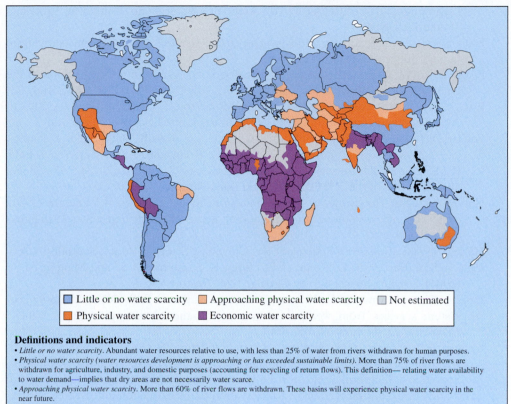

Definitions and indicators

- *Little or no water scarcity.* Abundant water resources relative to use, with less than 25% of water from rivers withdrawn for human purposes.
- *Physical water scarcity (water resources development is approaching or has exceeded sustainable limits).* More than 75% of river flows are withdrawn for agriculture, industry, and domestic purposes (accounting for recycling of return flows). This definition— relating water availability to water demand—implies that dry areas are not necessarily water scarce.
- *Approaching physical water scarcity.* More than 60% of river flows are withdrawn. These basins will experience physical water scarcity in the near future.
- *Economic water scarcity (human, institutional and financial capital from access to water even though water in natural is available locality to meet human demands)* water resources are abundant relative to water use, with less then 25% of water from rivers withdrawn for human purposes, but manupulation exises.

FIGURE 3.40 Areas and causes of existing water scarcity.

Source: Based on UNEP (2008), GRID-Arendal, Maps & Graphics Library, http://maps.grida.no/go/graphic/areas-of-physical-and-economic-water-scarcity

References

Arthurton, R., Barker, S., Rast, W., Huber, M., Alder. J., Chilton, J., Gaddis, E., Pietersen, K., and Zockler, C. (2007). *State and Trends of the Environment*: 1987–2007. Chapter 4: Water. Global Environment Outlook (GEO4) Environment for Development. United Nations Environment Programme. DEW/1001/NA. Australian Government Bureau of Meteorology: www.bom.gov.au/jsp/ncc/climate_averages/evaporation/index.jsp Product Code: IDCJCM006.

Azzopardi, E. (2001). The Development and Management of Water Resources in the Maltese Islands. Balearics 2015: Water and its management: Perspectives for the future. Palma de Mallorca, February 1–2, 2001.

Cech, T. V. (2010). *Principles of Water resources: History, Development, Management and Policy*. Third Ed. John Wiley & Sons, Inc. Hobokan, NJ. P. 74.

Chattopadhyay, N., and M. Hulme. (1997, November). "Evaporation and potential evapotranspiration in India under conditions of recent and future climate change." *Agricultural and Forest Meteorology*, 87(1):55–73.

Corcoran, E., Nellemann, C., Baker, E., Bos, R., Osborn, D., and Savelli, H. (Eds.). (2010). Sick Water? The central role of waste-water management in sustainable development. A Rapid Response Assessment. United Nations Environment

Programme, UN-HABITAT, GRID-Arendal. www.grida.no. Printed by Birkeland Trykkeri AS, Norway.

FAO. (1988). *Irrigation Water Management: Irrigation Methods*. Food and Agriculture Organization of the United Nations.

FAO. (2006). Malta water resources review. Food and Agriculture Organization of the United Nations, Rome, 2006. Viale delle Terme di Caracalla, 00100 Rome, Italy

FAO. (2011). Synthesis Report: FAO—Nile Basin Project. GCP/INT/945/ITA 2004 to 2009. Food and Agriculture Organization of the United Nations, Rome, Italy.

Fowler, D., Coyle, M., Skiba, U., Sutton, M. A., Cape, J. N., Reis, S., Sheppard, L. J., Jenkins, A., Galloway, J. N., Vitousek, P., Leech, A., Bouwman, A. F., Butterbach-Bahl, K., Dentener, F., Stevenson, D., Amann, M., and Voss, M. (2013). "The global nitrogen cycle in the 21st century." *Phil. Trans. Roy. Soc., Ser. B.*

Haber, F. (1920). The synthesis of ammonia from its elements. Nobel Lecture (1920), available at www.nobelprize.org/no-bel_prizes/chemistry/laureates/1918/haber-lecture.pdf.

Harte, J. (1988). *Consider a Spherical Cow*. University Sciences Books. Saulsalito, CA.

Hutson, S. S., Barber, N. L., Kenny, J. F., Linsey, K. S., Lumia, D. S., and Maupin, M. A. (2004). U.S. Geological Survey Circular 1268. Reston, VA.

Jenkins, W. A., Murray, B. C., Kramer, R. A., and Faulkner, S. P. (2010). "Valuing ecosystem services from wetlands restoration in the Mississippi Alluvial Valley." *Ecological Economics,* 69:1051–1061.

Kellogg, R. L., Lander, C.H., Moffitt, D. C., and Gollehon, N. (2000). "Manure Nutrients Relative to the Capacity of Cropland and Pastureland to Assimilate Nutrients: Spatial and Temporal Trends for the United States." United States Department of Agriculture, Natural Resources Conservation Service, Economic Research Service. Fort Worth, TX. Nps00-0579.

Leach, A. M., Galloway, J. N., Bleeker, A., Erisman, J. W., Kohn, R., and Kitzes, J. (2012). "A nitrogen footprint model to help consumers understand their role in nitrogen losses to the environment." *Environmental Development*, 1:40–66.

Linacre, E. (1994). "Estimating U.S. Class A Pan Evaporation from few climate data." *Water International,* 19(1):5–14.

Lins, H. F., and Slack, J. R. (1999). "Streamflow trends in the United States." *Geophysical Research Letters,* 26(2):227–230.

Matos, G.R. (2009). "Use of Minerals and Materials in the United States from 1900 through 2006." Fact Sheet. Minerals Information Team, U.S. Geological Survey, Reston, VA.

Mays, L. (2007). "Water sustainability of ancient civilizations in Mesoamerica and the American Southwest." *Water Science and Technology: Water Supply*, 7(1):229–236.

Mays. L., (editor). (2010). *Water Technology in Ancient American Societies*. Springer, Dordrecht Netherlands

Mays (2011) *Water Resources Engineering: Second Edition*. John Wiley & Sons, Inc. Hoboken, NJ. p. 4.

Molden, D. (Editor). Comprehensive Assessment of Water Management in Agriculture (2007). *Water for Food, Water for Life: A Comprehensive Assessment of Water Management in Agriculture*. London: Earthscan, and Colombo: International Water Management Institute.

Murakami, M. (1995). *Managing Water for Peace in the Middle East: Alternative Strategies*. United Nations University, Tokyo, Japan.

Peavy, H.S., Rowe, D. R., and Tchobanoglous, G. (1985). *Environmental Engineering*. McGraw Hill, Inc. New York, NY. p. 1.

Peterson, T. C., Golubev, V. S., and Groisman, P. Y. (1995). "Evaporation losing its strength." *Nature,* 10(26):687–688.

Rao, B., Sandeep, V.M., Rao, V.U.M., and Venkateswarlu, B. (2012). Potential evapotranspiration estimation for Indian conditions: Improving accuracy through calibration coefficients. *Tech. Bull.* No 1/2012. All India Co-ordinated Research Project on Agrometeorology, Central Research Institute for Dryland Agriculture, Hyderabad.

Revenga, C, Nackoney, J., Hoshino, E, Kura, Y., and Maidens, J. (2003). Watersheds of the World. IUCN, IWMI, Rasmar Convention Bureau and World Resources Institute, Washington, D.C. 2002 © 2003 World Resources Institute.

Smil, V. (2001). *Enriching the Earth: Fritz Haber, Carl Bosch, and the Transformation of World Food Production.* Cambridge, MA: MIT Press.

Sutton, M. A., et al. (2013). Our Nutrient World: The Challenge to Produce More Food and Energy with Less Pollution. Global Overview of Nutrient Management. Centre for Ecology and Hydrology, Edinburgh UK on behalf of the Global Partnership on Nutrient Management and the International Nitrogen Initiative. It is published by the Centre for Ecology and Hydrology - this is the suggested citation by the source

UNEP (United Nations Environment Programme). (2013). www.grida.no/graphicslib/collection/vital-water-graphics.

UNEP. (2008). http://maps.grida.no/go/graphic/areas-of-physical-and-economic-water-scarcity.

United Nations, Food and Agriculture Organization (UN FAO). (2013). www.fao.org/nr/water/aquastat/watresafrica/index3.stm.

U.S. Geological Survey. (2009, April). Use of Minerals and Materials in the United States from 1900 Through 2006. USGS Fact Sheet 2009-3008.

Viessman, W. Jr., Hammer, M. J. (1985). *Water Supply and Pollution Control.* Fourth Edition. Harper & Row, Publishers, New York, NY. p. 70.

Vitousek, P.M. Aber, J., Howarth, R.W., Likens, G.E., Matson, P.A., Schindler, D.W. Schlesinger, W. H., Tilman, G. D. (1997). "Human Alteration of the Global Nitrogen Cycle, Causes and Consequences." *Human Ecology.* Washington, D.C. Ecological Society of America.

Womach, J. (2005). Agriculture: A Glossary of Terms, Programs, and Laws, 2005 Edition. Congressional Research Service, The Library of Congress, Washington D.C. p. 62.

World Water Assessment Programme. (2009). *The United Nations World Water Development Report* 3: "Water in a Changing World." Paris: UNESCO, and London: Earthscan. ISBN: 978-9-23104-095-5. Map 10.5, p. 174.

Yan, H., S.Q. Wang, D. Billesbach, W. Oechel, J.H. Zhang, T. Meyers, T.A. Martin, R. Matamala, D. Baldocchi, G. Bohrer, D. Dragoni, R. Scott. (2012, September). Global estimation of evapotranspiration using a leaf area index-based surface energy and water balance model. *Remote Sensing of Environment*, 124:581–595. http://dx.doi.org/10.1016/j.rse.2012.06.004

Key Concepts

Biogeochemical cycle	Infiltration
Biogeochemical repository	Lithosphere
Atmosphere	Precipitation
Condensation	Residence time
Evaporation	Runoff
Hydrology	Transpiration
Hydrologic cycle	Watershed

Active Learning Exercises

ACTIVE LEARNING EXERCISE 3.1: Your water use

Many of our day-to-day activities depend on a constant and inexpensive supply of water. Just as residents of the Chaco Canyon community relied on water and imported resources to support their way of life, so do we today in the industrialized world.

In order to gain a better understanding of water consumption and its impacts, examine the consumptive use of water in your home or living space.

List the uses of water in your home. For each use, determine what happens to the water after it is used.

What part of the used water goes directly to the environment, for example, through watering the grass or washing the car?

What fraction of the used water goes to the sanitary sewer?

Is your wastewater treatment plant in the same region, or is the wastewater diverted to another location far away?

What effects might diverting water from your local area have on local streams, wetlands, or the groundwater?

ACTIVE LEARNING EXERCISE 3.2: Make your own watershed

Fold a piece of filter paper twice. Open up the folded paper into a cone shape and insert the paper into a funnel. Add sand or soil on top of the filter paper to simulate the effects of soil in a watershed. Wait for rain! Or if you are impatient you can use a garden sprinkling can to simulate rainfall. Take the funnel with the filter paper and sand into the rain. Any **precipitation (P)** that goes into your funnel falls into the watershed, the boundaries of which are the boundaries of the funnel. Before any **runoff (R)** is generated, first the water must saturate the sand and filter paper. The same is true for rainfall on land: The top level of soil must be saturated before runoff is generated. In this instance, rainfall that **infiltrates (I)** through the sand layer exits the funnel in a stream of water. In the same way, runoff water usually exits at the lowest point of the watershed through a stream or river. Some of the water may puddle in the funnel, as it may in a real watershed. The buildup of water in the funnel represents **storage (S)** of water in the watershed; the storage in an actual watershed may be in the form of a pond or lake.

Notice that when the precipitation event is complete, water remains stored in the funnel. First the water must drain through the sand filter, even after all the standing water has drained through the funnel. The sand is still wet. Water remains in the funnel due to the capillary force holding the water to the sand particles. Likewise in a real watershed, water drains slowly from lakes and storage areas. This is an important aspect in managing water runoff when developing infrastructure projects.

Water is stored in soil and in groundwater within an actual watershed. The availability of water in the soil is important for agricultural production, as plants use the water stored in the soil for growth. During photosynthesis and growth, plants release some of this water back into the atmosphere through transpiration. Plants play an important role in water management in a watershed by absorbing water from the soil; they store water and slowly release it back into the atmosphere. If the sand is left in a low-humidity environment, the miniature watershed will eventually dry out.

Where did the water go? The water has **evaporated (E)** into the air. Soil in a large watershed can become dry if precipitation does not occur regularly, and when the soil dries out plants cannot access water and cannot grow. During periods of insufficient precipitation, or a drought, water must sometimes be added to a watershed to provide **irrigation** to plants. Irrigation ditches similar to those used by the residents of Chaco Canyon are still widely used today and have been for thousands of years. In some sophisticated farming regions where a higher value is placed on water consumption, irrigation may be accomplished through high-tech computer-controlled systems providing water on demand. Water irrigation rates may be based on the measurements of moisture in the soil.

Precipitation is the only input into the headwaters in a watershed, as in our funnel, where no upstream rivers flow into the watershed. Water may exit the watershed through runoff, infiltration, evaporation, or transpiration. Water may also be stored in the watershed in soils, ponds, lakes, reservoirs (human-made lakes), and plants.

Problems

3-1 What percentage of materials used for physical goods in the United States came from recycled materials in the year 2000?

3-2 Define and describe a biogeochemical cycle.

3-3 List and describe important biogeochemical repositories.

3-4 Define the following terms:
   a. Atmosphere
   b. Condensation
   c. Evaporation
   d. Hydrology
   e. Hydrosphere
   f. Hydrologic cycle
   g. Infiltration
   h. Lithosphere
   i. Precipitation
   j. Residence time
   k. Runoff
   l. Transpiration
   m. Watershed

3-5 Describe the mathematical equation and variables used to define the residence time of a system.

3-6 Sketch the hydrologic cycle.

3-7 What is the difference between potable water and nonpotable water?

3-8 How are potable water and nonpotable water distributed?

3-9 What are the chemical equations that describe the biological transformation of water by plants and animals?

3-10 What is the annual average precipitation in your region in inches per year and millimeters per year?

3-11 Using topographic maps, Google Earth, or other geographic means, determine the boundaries of the watershed in which you live.

3-12 Write the water budget equation and define each variable.

3-13 Describe the differences between consumptive and nonconsumptive withdrawal and provide examples of each.

3-14 Estimate from the tables provided in the chapter how much water you consume each day through direct water use. How much water is this each year?

3-15 Estimate from your electric bill, or a family member's electric bill, how much water is used for production of the electricity you consume each day. How much water is this each year?

3-16 Estimate the residence time of water in the oceans.

3-17 Estimate the residence time of nitrogen in the atmosphere if the magnitude of flow of nitrogen from the atmosphere is 5×10^{12} kg(N)/year.

3-18 Estimate the residence time of carbon in the atmosphere assuming that the CO_2 flow rate to the atmosphere is approximately equal to the flow due to decomposition and combustion of terrestrial organic matter and that from animal respiration is 50×10^{12} kg(C)/year.

3-19 Define the term *biogeochemical cycle*.

3-20 What four factors does Jared Diamond conclude caused the collapse of the Chaco Canyon civilization?

3-21 Define the hydrologic cycle.

3-22 Sketch the hydrologic cycle for your watershed.

3-23 Sketch and label the repositories and transformation processes in the oxygen cycle.

3-24 Estimate evaporation rates in
a. Eastern North Carolina
b. Seattle, Washington
c. Southwest Arizona

3-25 Estimate the evapotranspiration rates in
a. Benin, West Africa
b. Central Tanzania, East Africa
c. Rwanda, East Africa
d. Sydney, Australia

3-26 Estimate precipitation in
a. Eastern North Carolina
b. Seattle, Washington
c. Southwest Arizona
d. Benin, West Africa
e. Central Tanzania, East Africa
f. Rwanda, East Africa
g. Sydney, Australia

3-27 Assuming runoff and infiltration are equal to 40% of the amount of precipitation, calculate the available water for withdrawal in
a. Eastern North Carolina
b. Seattle, Washington, assume runoff and infiltration are equal to 70%
c. Southwest Arizona, assume runoff and infiltration are equal to 20%
d. Benin, West Africa
e. Central Tanzania, East Africa
f. Rwanda, East Africa
g. Sydney, Australia

3-28 Describe the boundaries of your watershed and name the stream or river that provides the outflow from your immediate watershed. Use the World Resources Institute's website to find the average runoff and population served in your major drainage basin. See www.wri.org.

3-29 Describe the similarities and differences between the Chaco Canyon civilization at about AD 1110 and Maltese civilization today.

3-30 What remaining sources of water can be tapped into for meeting the needs of water-scarce regions of the planet?

3-31 Examine the map of physical and economic water scarcity. Does this map align closely with maps of precipitation? What areas are similar, and what areas are different? What, if anything, do you think is responsible for the differences between the water-scarcity maps and precipitation maps?

3-32 The average yearly precipitation in the Shenandoah Valley watershed is 36 inches. The average rate of evapotranspiration is 28 inches. Typical runoff through the Shenandoah River is 163 ft³/s (cfs). Reservoirs within the watershed store 28 million gallons of water. The area of the watershed is 7.5×10^{10} ft². Currently, consumptive use of water is 7 million gallons per day (MGD). There are 7.48 gallons per cubic foot.

    a. Determine the yearly infiltration rate to the groundwater table in MGD, based on the above information.

    b. If the 7 MGD accounted for all withdrawals from the groundwater table, what is the average net gain or loss from the groundwater table in MGD?

3-33 The Mitchell River catchment in Tropical North Queensland, Australia, spans 73,230 km² across the base of Cape York Peninsula. It incorporates five major river systems: the Mitchell, Alice, Palmer, Walsh, and Lynd. The Mitchell catchment is the largest watershed in Queensland in terms of average annual runoff, with a mean discharge of 11,300,000 megaliters (ML). The climate is tropical and monsoonal, with a mean annual rainfall of approximately 1,000 mm, with most precipitation occurring from November to April. Annual evapotranspiration rates in the region are approximately 800 mm. There are several dams within the watershed, with a total storage in all the dams of 425,779 ML. About 7,500 people live within the watershed. Springs and waterholes represent most of the permanent water in the far eastern part of the catchment. Permanent running water in the far east is associated only with the areas close to the rainforested wet tropics area. The majority of the catchment groundwater is within the Great Artesian Groundwater Province. A significant portion of this is overlain by the Karumba Basin, which lies on the west coast. This basin is the only feasible water supply in the area. The average water withdrawal from groundwater in the watershed is 55,229 ML. The quality of the groundwater in these provinces varies from excellent to saline. No sustainable yield studies are available for this area. A water management plan is currently being developed for the Mitchell catchment. To assist the watershed management association, you have been asked to do the following.

    a. Write the water budget equation.

    b. Show the volume of water available each year for each term in the water budget equation in units of m³.

    c. Compare the infiltration rate to the average withdraws of water in the watershed. Is this use sustainable?

3-34 Describe four Maltese water myths and the actual facts about water related to the myth.

3-35 What geologic feature contains most of Malta's groundwater supply?

3-36 Based on the provided figures, estimate approximately what percentage of Malta's surface area is protected for groundwater recharge. Also estimate what percentage of Malta was already urbanized in 2007.

3-37 Apply the water budget equation to evaluate the urgency of water management measures for the island nation of Malta.

    a. Determine the average annual input, or precipitation, of water. Estimate this total by calculating the average rainfall that occurs over a year on the islands from Table 3.2.

b. Convert the average monthly precipitation values expressed in mm/month to a yearly value expressed in m/year. Assume this rainfall occurs equally over all the area of the islands, which is a reasonable assumption since the total land area is relatively small compared to that of much larger nations.

c. How much water is available per person if all the rainfall could be collected and used? (Water consumption is normally expressed in liters per day, so convert your answer to liters per day.)

d. Not all the groundwater remains available for use. The groundwater mixes with seawater and also flows into the sea below the surface. The infiltration, I, and removal of water from the watershed by subsurface groundwater discharge to the sea are shown in Table 3.13. Calculate S in the water budget equation to account for all water available in BOTH groundwater and surface water storage.

e. Once again estimate the available water supply if all the stored water were used by the population in Malta.

   i Annual total storage/population

   ii We can compare the available sustainable water supply to actual water demand in Malta. In Table 3.15, we see that the total water use for Malta in 2003 was 58.641×10^6 m³/year.

f. Compute the percent over consumption of the available water resources in Malta.

g. The FAO report suggests that significant progress must be made by 2015 to protect Malta's fragile water resources. The report describes four demand scenarios (based on current levels of demand and the subsequent projections that have emerged from discussions with various stakeholders). It also discusses different strategies for meeting future demand and achieving the vision shown in Tables 3.16 and 3.17.

h. Does the proposed strategy seem like a good direction to ensure that future water resources are available in Malta?

i. Are the proposed strategies sufficient to ensure water availability for current and future development?

3-38 Find the topographic map section in your library and locate the USGS topographic quadrangle map for the region in which you live.

a. Determine your watershed boundaries.

b. Estimate the area of your watershed from the map.

3-39 Obtain a copy of the USGS National Resources Conservation Service soil map either online or in your library and determine the dominant soil type in your region. Estimate the infiltration rate based on the average soil type in your watershed.

3-40 Determine the annual average volumetric flow rate of runoff of the nearest stream to your watershed on the USGS webpage for streamflow gauges.

3-41 For your watershed:

a. Estimate the annual precipitation.

b. Estimate the annual evapotranspiration.

c. From your local water company, determine if there is any net storage in reservoirs within your watershed.

d. Solve the water budget equation.

e. What capacity of water can be sustainably withdrawn for your community?

f. Determine from your water company's website how much water is used within your community.

g. Does your community have a sustainable water supply? Why or why not?

3-42 A stormwater sewer empties snowmelt into a small stream channel. The storm sewer conveys 24 m³ of water per day of water containing 1,753 mg/L of nonreactive calcium chloride solids. The stream channel has a flow of 2,400 L per second and contains 84 mg/L of nonreactive calcium chloride.

a. What is the combined flow rate of the stormwater and streamwater in cubic meters per day just downstream of the two water sources?

b. What is the concentration of nonreactive calcium chloride in the stream in mg/L?

3-43 Wastewater from a natural gas extraction process that contains 4,350 mg/L of nonreactive dissolved solids is discharged to a receiving stream at a rate of 20 m³/day. The flow rate of the stream just upstream of the wastewater discharge is 700 m³/s and the upstream dissolved solids concentration is 14 mg/L.

a. What is the combined flow rate of the stream in cfs just downstream of wastewater discharge?

b. What is the concentration of dissolved solids in the stream in mg/L just downstream of wastewater discharge?

3-44 The volumetric flow rate of the north fork of a River was 241 m³/s as measured in Strasburg, Virginia. The south fork of the river had a flow rate of 732 m³/s near the confluence of the two forks at Front Royal. What was the combined flow rate of the Shenandoah River just downstream of the confluence?

3-45 The Columbia River is the tenth largest river in the United States. The Snake River is the largest tributary of the Columbia River. The Snake River discharges 1350 m³/s of water into the Columbia on average each year. The total dissolved solids, a nonreacting substance, in the Snake River averages approximately 200 mg/L. Just downstream of the discharge from the Snake River, the average flow rate of the combined rivers is 4800 m³/s. The total dissolved solids, a nonreacting substance, in the combined rivers averages 80 mg/L.

a. Determine the volumetric flow rate of the Columbia River in cfs just upstream of the confluence of the two rivers.

b. Determine the total dissolved solids concentration of the Columbia River in mg/L just upstream of the confluence of the two rivers.

3-46 A soft-drink bottling plant discharges a wastewater with a dissolved solids concentration of 10,000 mg/l NaCl at a rate of 100 Liters per minute to a stream. The stream above the discharge has a volumetric flow rate of 1×10^6 Liters per day and a dissolved solids concentration of 25 mg/l. Below the discharge point is a prime fishing spot, and the fish are intolerant of salt concentrations above 200 ppm.

a. Sketch the system.

b. Write the general equation for a mass balance.

c. What is the combined flow in MGD of the stream and wastewater downstream of the discharge?

d. What is the concentration of dissolved solids (as NaCl) in the combined stream downstream of the discharge

e. Are the fish in danger (show why or why not)?

CHAPTER 4

Water Quality Impacts

FIGURE 4.1 Water quantity, quality, and availability are closely related to sustainable development. At the time of writing this book, approximately one-third of global population lacked access to clean water. The burden of finding water often falls on children and women in society.

Source: Photo by Bradley Striebig

Water is life. A third of the nations on our shrinking planet, mostly in the developing world, suffer from water scarcity and the associated stress this scarcity places on societies. My home area of Kitale, Kenya shares in this suffering.

—GILBERT NALELIA

GOALS

THE EDUCATIONAL GOALS OF THIS CHAPTER are to describe basic water quality parameters and water quality improvement processes. This chapter will describe the parameters and methods used to determine if water is of sufficient quality for use as drinking water. Water quality parameters that influence human and ecosystem health in the environment will be examined. Mathematical models will be used to demonstrate how high levels of BOD or nutrients may adversely affect ecosystems. Difference in water quality between high-income nations and low- to medium-income nations will be demonstrated.

OBJECTIVES

At the conclusion of this chapter, you should be able to:

- Describe economic, environmental, and societal implications of water quality in high-income nations and low- to medium-income nations.

- Describe microbial pollutants in water.

- Calculate levels of dissolved oxygen in water and parameters that influence dissolved oxygen.

- Calculate biochemical oxygen demand (BOD) and nutrient levels in water.

- Describe and model the effects of BOD and nutrients on water quality and ecosystems.

- Describe basic centralized and decentralized water treatment technologies.

4.1 Introduction

Every year more people die from unsafe water than from all forms of violence, including war.

The most significant sources of water pollution are lack of inadequate treatment of human wastes and inadequately managed and treated industrial and agricultural waste.

Every day, 2 million tonnes of sewerage and other effluents drain into the world's waters.

The quality of water necessary for each human use varies, as do the criteria used to assess water quality.

Source: World Water Assessment Programme, 2009.

4.2 The Water Crisis

A silent humanitarian crisis kills an estimated 3,900 children every day (Bartram et al., 2005). The root of this unrelenting catastrophe lies in these plain, grim facts: The United Nations has estimated that nearly one-third of the people in the world do not have access to sanitation and more than one in ten have no source of safe drinking water (Figure 4.2). Far more people endure the largely preventable effects of poor sanitation and water supply than are affected by war, terrorism, and weapons of mass destruction combined. Yet those other issues capture the public and political imagination and resources in a way that water and sanitation issues do not. Most people find it hard to imagine defecating daily in plastic bags, buckets, open pits, and public areas because there is no alternative; nor can they relate to the everyday life of the 1.1 billion people without access to a protected well or spring within reasonable walking distance of their homes (Bartram et al., 2005).

Water is fundamental for life and health. The human right to water is a prerequisite to the realization of all other human rights. Many governments recognize the human right to access water and sanitation, but lack of or poor use of funding and the scarcity of a trained workforce in water treatment and delivery have prevented all people from gaining access to water and sanitation. The poverty of a large amount of the world's population is both a symptom and a cause of the water crisis that affects the poorest populations, ending in sickness, lost educational and employment opportunities, and for a staggeringly large number of people, early death (Gleick, 2002). Water-related diseases are among the most common causes of illness and death in developing countries, especially in sub-Saharan Africa (Figure 4.3) (Corcoran et al., 2010). The sad fact is that this disease burden is preventable. Furthermore, investing in water and sanitation services is expected to have positive economic benefits, with a 5 to 25 times return on each dollar invested (see Figure 4.4). In developing countries with poor water and sanitation systems, life expectancy is far lower than it is in industrialized countries. The causes of deaths are also quite different; infectious diseases account for more than 40% of deaths in developing countries, whereas in industrialized nations deaths are related to chronic disease and cancer (Kitawaki, 2002).

In addition to the Human Development Index, a **Water Poverty Index (WPI)** has also been developed; the WPI helps explain the relationship between poverty and five water-related criteria: resources, access, capacity, use, and environment (Lawrence et al., 2002; Sullivan, 2002). It is designed to allow a single measure to

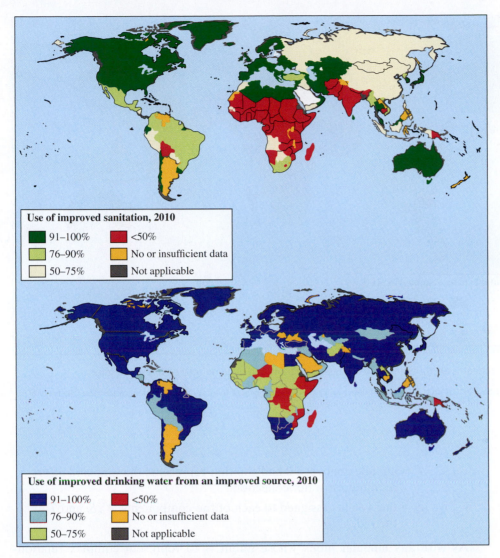

FIGURE 4.2 Percentage of population with access to improved water supply and sanitation.

Source: Based on WHO (2012) GLASS 2012 Report: UN-Water Global Analysis and Assessment of Sanitation and Drinking Water: The Challenge of Extending and Sustaining Services. World Health Organization. Geneva, Switzerland.

relate household welfare and access to water. The WPI value based on a ranking of 0 to 100 is given by the general equation:

$$WPI = \frac{1}{3}\left[w_a A_s + w_s S + w_t(100 - T)\right] \qquad (4.1)$$

where A_s = the percent of adjusted water availability (AWA) for the population. Calculated on the basis of ground and surface water availability related to ecological water requirements and a basic human requirement, plus all other domestic demands, as well as the demand for agriculture and industry. (The value of A should also recognize the seasonal variability of water supplies.)

S = the percent of the population with access to safe water and sanitation.

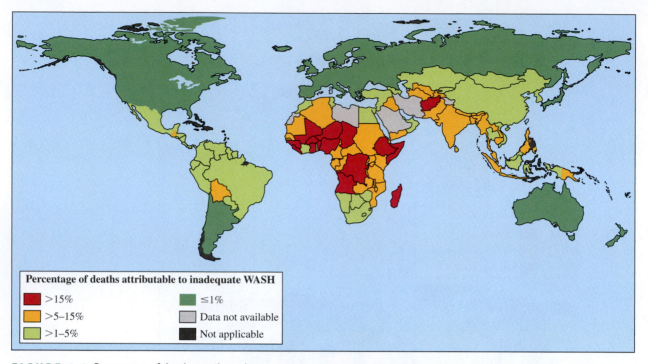

FIGURE 4.3 Percentage of deaths attributed to water access, sanitation, and hygiene-related disease or injury.

Source: Based on WHO (2012) GLASS 2012 Report: UN-Water Global Analysis and Assessment of Sanitation and Drinking Water: The Challenge of Extending and Sustaining Services. World Health Organization. Geneva, Switzerland.

T = an index between 0 and 100 that represents the time and effort required for water collection.

w_a, w_s, w_t = the weights assigned to each of the components, A_s, S, and T of the index.

As with any singular index whose intent is to represent complex relationships between multiple variables, the WPI does have limitations. The WPI may not adequately address all water-related issues, and issues such as "access" and "capacity" may not be clearly independent factors in some settings, but the WPI does provide a tool for comparing the economic, environmental, and social issues associated with access to water, as described in Box 4.1 (Heidecke, 2006; Komenic et al., 2009).

Policies and plans to increase access to improved drinking water and sanitation are a priority among many countries, including low- to medium-income countries (see Figure 4.8) (p. 193).

Infrastructure projects such as water and sanitation projects are expensive to implement, but also require revenue in order to operate and maintain (O&M) water and sanitation systems. Centralized water and sewer systems have proven to be very effective in improving and protecting water quality and reducing the spread of water-related disease. However, centralized systems require a large capital expense, sufficiently trained personnel to design, build, and operate the systems, and collection of revenue to pay for trained operators and O&M systems. Decentralized water and sanitation solutions exist and have also been shown to significantly improve

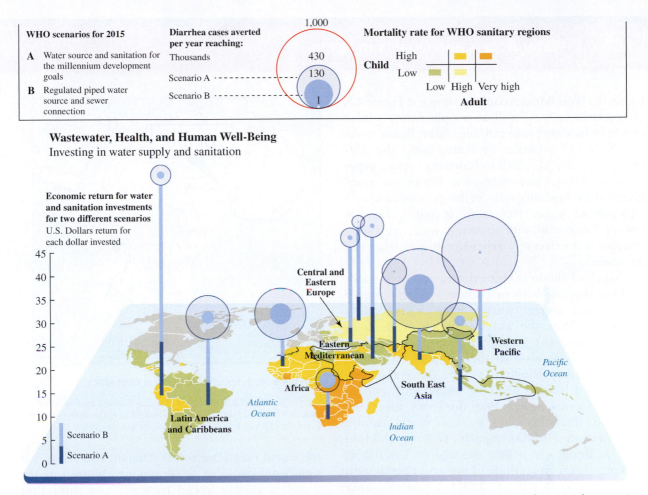

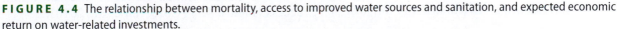

FIGURE 4.4 The relationship between mortality, access to improved water sources and sanitation, and expected economic return on water-related investments.

Source: Based on UNEP/GRID-Arendal, http://www.grida.no/graphicslib/detail/wastewater-health-and-human-well-being-investing-in-water-supply-and-sanitation_120c, data from Hutton, G., et al., Global cost-benefit analysis of water supply and sanitation interventions, Journal of Water and Health, 2007. http://www.who.int/water_sanitation_health/publications/2012/globalcosts.pdf

water quality and sanitation (see Box 4.2). Decentralized systems have a lower capital cost, are generally designed for lower O&M costs, and as such are often implemented to meet rural community needs.

Growing urban centers challenge engineers, scientists, and policymakers to implement an appropriate use of technology. Both the urban centers of Jakarta, Indonesia, and Sydney, Australia, generate approximately 1.2 to 1.3 million cubic meters of wastewater each day (Figure 4.9). Jakarta has a population of nearly 9 million people, and approximately 60% of them get their water from private decentralized wells. Most of the wastewater in Jakarta flows into decentralized septic tanks, which are improperly maintained. Existing sewage trenches, originally built to control flooding, have been partially clogged by silt and garbage. Increased runoff rates and lack of infrastructure create increasingly severe flooding events that often carry contaminants into the water supply. In addition, stagnant water remaining after the floods subside provides an ideal breeding ground for vectors,

BOX 4.1 Water Access and Quality in Benin

Benin is a West African country, shown in Figure 4.5, of approximately 8.5 million people, nearly a third of whom lack access to potable water. Benin ranks as 139 of 147 countries evaluated using the WPI (Lawrence et. al., 2002). Mortality rates, especially for infants and children in Benin, are much higher than mortality rates in the developed world. Centralized water treatment is not a feasible option for community drinking water in Benin because it is extremely expensive to construct and maintain.

Much of Benin is blessed with access to water through shallow wells or surface waters; however human and animal wastes have contaminated the water. Thus, the water in much of Benin is contaminated with bacteria and viruses as shown in Table 4.1. Because of their impact on human health, not surprisingly these are the major concerns about Benin's drinking water. Most people in Benin drink water that does not meet the standards recommended by the World Health Organization (WHO). Nearly 17% of children born in Benin die before the age of 5, which is evidence of the magnitude of the water and health quality issues (UNICEF, 2004). Child mortality rates in sub-Saharan Africa are the highest in

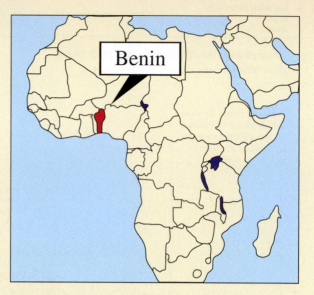

FIGURE 4.5 Location of Benin in West Africa

Source: Based on PEPPFAR (2006): Building Human Capacity in the Defense Force. The United States President's Emergency Plan for AIDS Relief (PEPPFAR) Benin (http://www.pepfar.gov/press/84892.htm)

the world (see Figure 4.6). Diarrheal disease is a significant cause of death among children under 5 and is closely linked to water and sanitation access (Figure 4.7) (Corcoran et al., 2010).

TABLE 4.1 Microbial and chemical contaminants measured in Benin water samples

CONTAMINANT	UNITS	CONCENTRATION IN BENIN WATER	WHO STANDARD	U.S. EPA STANDARD
Total coliforms	MPN/100 ml	>1,600	0	0
Fecal coliforms	MPN/100 ml	20	0	0
E. coli	MPN/100 ml	NA	0	0
Pathogens	MPN/100 ml	>8	0	0
Lead	μg/L Pb	4	10	15
Arsenic	μg/L As	ND	10	10
Nitrates	mg/L NO_3^--N	>30.0	50	10
Phosphate	mg/L PO_4^{3-}	0.19	NA	NA

Source: Based on Striebig, B., Atwood, S., Johnson, B., Lemkau, B., Shamrell, J., Spuler, P., Stanek, K., Vernon, A., Young., J. 2007. "Activated carbon amended ceramic drinking water filters for Benin." Journal of Engineering for Sustainable Development. 2(1):1-12.

In rural Benin, the people obtain clean drinking water primarily by boiling water or purchasing imported bottled water. Boiling water requires wood and native vegetation, depleting local resources and emitting smoke into households and the atmosphere. While treated water is available in a few locations, less than 10% of the population has treated water piped into their homes. Bagged and bottled water can be purchased in the marketplace, but this water is expensive, unsustainable, and sometimes can be tampered with, so that the water can unknowingly become contaminated. The preferred treatment for the area would be a cost-effective point-of-use technology. However, not all of Benin's population understands the links between contaminated drinking water, sanitation, and disease, thereby creating a need for an educational program.

Most early deaths due to water-related disease are the result of diarrheal illnesses caused by ingesting water contaminated by fecal matter, as well as by inadequate sanitation and hygiene. For public health purposes, water-related disease is typically grouped within the classifications defined in Table 4.2.

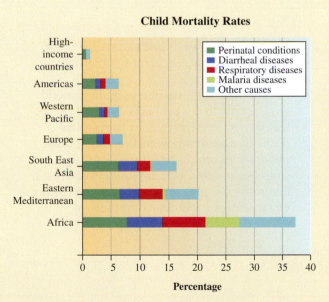

FIGURE 4.6 Childhood mortality rates among high-income and lower-income countries in different regions.

Source: Based on Corcoran, E., C. Nellemann, E. Baker, R. Bos, D. Osborn, H. Savelli (eds). 2010. Sick Water? The central role of wastewater management in sustainable development. A Rapid Response Assessment. United Nations Environment Programme, UN-HABITAT, GRID-Arendal. Data source: WHO, 2008

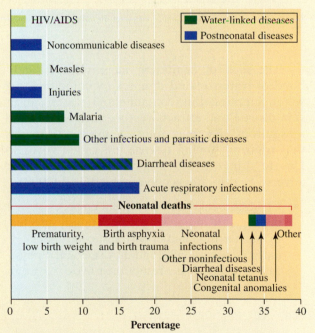

FIGURE 4.7 Water-related factors that contribute to childhood mortality rates.

Source: Based on Corcoran, E., C. Nellemann, E. Baker, R. Bos, D. Osborn, H. Savelli (eds). 2010. Sick Water? The central role of wastewater management in sustainable development. A Rapid Response Assessment. United Nations Environment Programme, UN-HABITAT, GRID-Arendal. Data source: WHO, 2008.

(Continued)

BOX 4.1 **Water Access and Quality in Benin** *(Continued)*

TABLE 4.2 Bradley classification of diseases related to water and sanitation issues

CATEGORY	EXAMPLE	INTERVENTION
Waterborne	Diarrheal disease, cholera, dysentery, typhoid, infectious hepatitis	Improve quality of drinking water, prevent casual use of unprotected sources
Water-washed	Diarrheal disease, cholera, dysentery, trachoma, scabies, skin and eye infections, acute respiratory infections	Increase water quantity used, improve hygiene
Water-based	Schistosomiasis, guinea worm	Reduce need for contact with contaminated water, reduce surface water contamination
Water-related (insect vector)	Malaria, onchocerciasis, dengue fever, Gambian sleeping sickness	Improve surface water management, destroy insect breeding sites, use mosquito netting

Source: Based on UNICEF (2008) UNICEF Handbook on Water Quality. United Nations Children's Fund (UNICEF) New York, New York, USA.

The lack of sanitation increases the likelihood of water-related diseases by contaminating water supplies; this contaminated water causes skin disease and attracts vectors, such as flies and mosquitoes, that may transfer disease. (Animals that transmit human disease are called vectors.) Unsafe drinking water may also lead to other adverse health effects cited by UNICEF (2008).

- Children weakened by frequent episodes of diarrhea are more likely to be seriously affected by malnutrition and opportunistic infections (such as pneumonia), and they can be left physically stunted for the rest of their lives.

- Chronic consumption of unsafe drinking water can lead to permanent cognitive damage.
- People with compromised immune systems (e.g., people living with HIV and AIDS) are less able to resist or recover from waterborne diseases. Pathogens that might cause minor symptoms in healthy people (e.g., *Cryptosporidium*, *Pseudomonas*, rotaviruses, heterotrophic plate count microorganisms) can be fatal for the immunocompromised.

which is likely to account for the increasing incidence of dengue fever and other water-related disease. As a result, only about 3% of the wastewater in Jakarta is treated prior to discharge (Figure 4.9). In contrast, Sydney has a population of about 4 million people and centralized water and sanitation systems that serve nearly the entire population. Nearly 100% of Sydney's wastewater is treated prior to discharge.

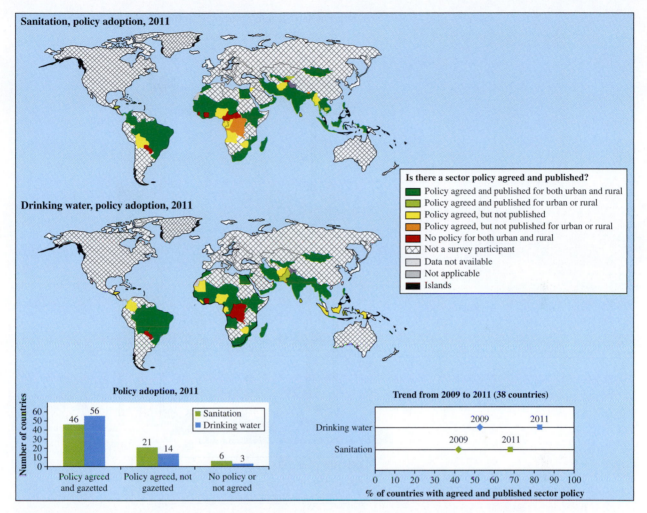

FIGURE 4.8 Water and sanitation policy development in low- to medium-income countries

Source: Based on WHO (2012) GLASS 2012 Report: UN-Water Global Analysis and Assessment of Sanitation and Drinking Water: The Challenge of Extending and Sustaining Services. World Health Organization. Geneva, Switzerland.

Only 40% of low- to medium-income countries report fiscal resources and responsibility for decentralized services. The low- to medium-income countries also lack adequately trained staff to operate and manage facilities in urban and rural areas, as illustrated by the data reported in Figures 4.10 and 4.11.

4.3 Water Quality Parameters

Water quality is influenced by natural and anthropogenic factors. Natural factors that influence water quality include weathering of bedrock and minerals, atmospheric deposition of dissolved gases and windblown particles, leaching of

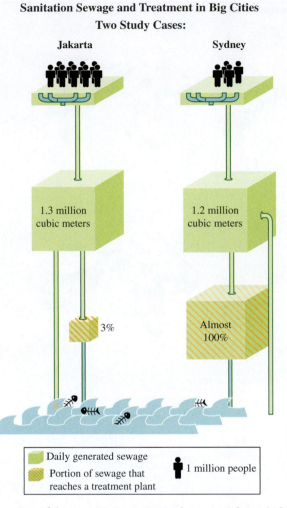

FIGURE 4.9 Comparison of the wastewater treatment between Jakarta, Indonesia, and Sydney, Australia.

Source: Based on Nellemann, E. Baker, R. Bos, D. Osborn, H. Savelli (eds). 2010. Sick Water? The central role of waste-water management in sustainable development. A Rapid Response Assessment. United Nations Environment Programme, UN-HABITAT, GRID-Arendal.

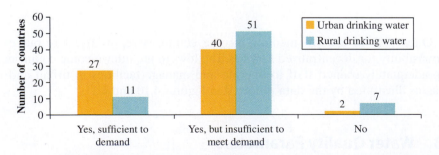

FIGURE 4.10 Availability of trained staff to meet water and sanitation needs reported by low- to medium-income countries.

Source: Based on WHO (2012) GLASS 2012 Report: UN-Water Global Analysis and Assessment of Sanitation and Drinking Water: The Challenge of Extending and Sustaining Services. World Health Organization. Geneva, Switzerland.

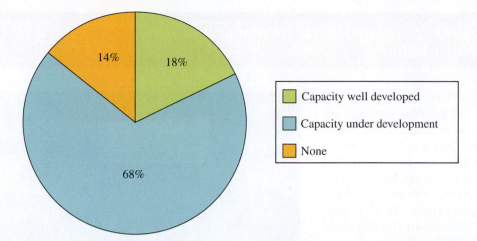

14%   18%

68%

- Capacity well developed
- Capacity under development
- None

FIGURE 4.11 Training capacity for artisans and technicians to meet rural sanitation needs.

Source: Based on WHO (2012) GLASS 2012 Report: UN-Water Global Analysis and Assessment of Sanitation and Drinking Water: The Challenge of Extending and Sustaining Services. World Health Organization. Geneva, Switzerland.

nutrients and organic matter from soils, biological processes, and the hydrologic factors discussed in Chapters 2 and 3. Declining water quality is a concern owing to the stresses imposed by increased human water consumption, agricultural and industrial discharges, and negative effects related to climate change. Poor water quality is a concern owing to the economic, environmental, and social impacts that occur if water supplies become too polluted for drinking, washing, fishing, or recreation.

Water quality is determined by measuring factors that are suitable to potential use. Drinking water should be pathogen free, but microbial content of the water may not be a concern for some industrial uses. Broad categories of water quality include microbial content, pH, solids content, ionic strength, dissolved oxygen, oxygen demand, nutrient levels, and toxic chemical concentrations. Water pollution may be any condition caused naturally or by human activity that adversely affects the quality of a stream, lake, ocean, or source of groundwater. Pollutants may also be defined as any harmful chemical or constituent present in the environment at concentrations above the naturally occurring background levels. Major water pollutants include microorganisms, nutrients, heavy metals, organic chemicals, oil, sediments, and heat. Water pollution issues affect every country in every economic category on Earth; however, the water issues vary with country and region, as illustrated in Figure 3.40. The types and origins of water pollutants likewise vary, but they all have negative economic, environmental, and societal impacts if the pollutants remain untreated, as illustrated in Figure 4.14 (p. 200). Pollutants discharged into the environment negatively affect water quality, as illustrated in Figure 4.15 (p. 201), but wastewater can be reused and treated to positively affect the economy, environment, and society with forethought and investment in basic infrastructure. Improving water quality directly impacts the UN Millennium Development Goal number 7—to ensure environmental sustainability—and indirectly affects many of the other goals.

BOX 4.2 Point-of-Use Decentralized Water Treatment Process for Porto-Novo, Benin

The Songhai Center (SC), a United Nations (UN) Center of Excellence for sustainable agricultural practices and technology, is headquartered in Porto-Novo, the capital of Benin. During a visit to Benin in 2004, Father Nzamujo Godfrey, director of the Songhai Center, identified the need for low-cost, sustainable point-of-use water treatment technology. Community-focused projects, such as this one, directly address the eight Millennium Development Goals set forth by the UN (2000).

The colloidal silver-impregnated ceramic water filter (CWF) technology was selected for potential implementation. Potters for Peace designed the chosen CWF technology to fit within a five-gallon plastic pail or clay container (see Figure 4.12). In addition to porous ceramic filtration, colloidal silver is used to inhibit bacterial growth (Fahlin, 2003; Clausen et al., 2004; and van Halem et al., 2009). The UN has cited the Filtrón in its *Appropriate Technology Handbook*, which is used by both the Red Cross and Doctors Without Borders (Lantagne, 2010). Currently, several countries around the globe have employed this low-cost, appropriate technology filter with good results. Most other water-treatment technologies require more energy (ultraviolet disinfection systems) or chemical additives (chlorination or other chemical disinfectants) as reported in Table 4.3. Energy and chemical-intensive disinfection systems may provide comparable or even better disinfection; however, the cost and availability of energy and chemical supplies are not sustainable within the community.

Ceramic water filters (CWFs) are manufactured using the process designed by Potters for Peace (Lantagne, 2010). This process uses a recipe for mixing clay and sawdust, forming the filters in a mold, and following a predetermined firing procedure in a high-temperature pottery kiln. The effectiveness of the colloidal silver-impregnated filters was evaluated using several methods. Potters for Peace recommends that the filters meet specific flow rate requirements in order to provide the required amount of water

FIGURE 4.12 Ceramic point-of-use water filter, called a Filtrón, produced at the Songhai Center for Sustainable Development in Porto-Novo, Benin.

Source: Photo by Bradley Striebig

needed each day and as an indicator of the pore integrity in the CWF. The CWFs were evaluated in the laboratory using several established procedures as indicators of drinking water quality, including the Hach LPT-MUG test for total coliform and *E. coli* and the Hach Pathoscreen test for hydrogen-sulfide producing organisms. The results of these tests are shown in Table 4.4. The Hach Pathoscreen test is widely accepted as a reasonable cost-effective indicator of potential contamination by mammalian wastes. If the water sample is contaminated, the water sample turns black indicating the presence of

TABLE 4.3 A comparison of point-of-use appropriate water treatment technologies

TECHNOLOGY	ADVANTAGE	DISADVANTAGE	FILTER TIME
Biosand™ sand filtration	High removal efficiency for microorganisms	Needs continual use and regular maintenance Cost	1–2 hours
Filtrón™ ceramic filter	High removal efficiency for microorganisms Sized for households Relatively inexpensive	Requires fuel for construction Limited lifetime Requires regular cleaning	2–8 hours
SODIS™ solar water disinfection	Highly effective Inexpensive Can reuse a waste product (PET bottles)	Long treatment time (12 to 48 hours) Does not remove other pollutants Requires warm climate and sunlight	12–48 hours

TABLE 4.4 Some laboratory results showing the potential reduction in microbial contaminants with the Filtrón ceramic water filters

CONTAMINANT	UNITS	FILTERED WATER	AVERAGE REMOVAL
Fecal coliforms	MPN/100 ml	< 2 ± 0	> 99.92%
Total coliforms	MPN/100 ml	< 2 ± 0	> 99.97%
E. coli	MPN/100 ml	< 2 ± 0	> 99.0%
Pathogens (H$_2$S-producing bacteria)	MPN/100 ml	< 2 ± 0	> 99.7%
Streptococci	MPN/100 ml	< 2 ± 0	NA
Amoeba	MPN/100 ml	37,000 ± 115,000	99.5

Source: Based on Striebig, B., Atwood, S., Johnson, B., Lemkau, B., Shamrell, J., Spuler, P., Stanek, K., Vernon, A., Young., J. 2007. "Activated carbon amended ceramic drinking water filters for Benin." Journal of Engineering for Sustainable Development. 2(1):1-12.

hydrogen-sulfide producing bacteria and the possible presence of pathogenic bacteria. Because this test is inexpensive, does not require special incubators (in the climate in Benin), and turns a jet black color, it is particularly useful as a tool to illustrate how debris-free, clear-colored water may still be contaminated. A negative LPT-MUG test for total coliforms indicates that there are likely to be few (if any) bacteria in the water. The LPT-MUG test procedure commonly used to provide reasonable indication water is safe for drinking and is likely free from bacteria in the United States.

As reported in the literature and discussed in personal communications, the Filtrón performed very well in removing biological constituents (Lantagne, 2010; Clausen et al., 2004). The micropore structure

(Continued)

of the ceramic filtration was able to prevent more than 99% of the total coliforms, fecal coliforms, *E. coli,* and pathogens from passing through the filter with the water in laboratory testing.

The Songhai Center has manufactured and sold approximately 400 ceramic water filters since the program was initiated. They sold approximately 60 filters in the first year. This number reflects start-up lessons, climatic considerations, and a conscious decision to build up an inventory for supply of the filters prior to developing a marketing plan. Also during this start-up phase, several manufacturing issues were identified and are still in the process of being addressed.

During follow-up visits in 2008 and 2009, U.S. university students and Songhai Center staff tested local water supplies and the effectiveness of the filters. Samples of local well water, the Songhai Center (unfiltered) tap water, Pur™ treated water from the local market, and well water were treated with the ceramic filters. Ten CWFs were randomly chosen from the Songhai Center's filter inventory for testing. Sample containers were prepared by cleaning with a dilute bleach solution, rinsing with filtered soapy water and rinsing one final time with bottled water. All water samples were prepared using sterilized sample vials. The water samples were evaluated using presence/absence methods with Hach test packets and prepared vials for total coliform bacteria and hydrogen sulfide-producing bacteria (Pathoscreen™). None of the field samples collected in Benin appeared dirty or contaminated to the naked eye (see Figure 4.13). In spite of being clear of dirt and turbidity, however, untreated water samples (well water and tap water) indicated the presence of total coliforms and hydrogen sulfide-producing bacteria, which are indicative of pathogen organisms and likely contamination by human or animal wastes. The treated water samples were all negative for all of the bacterial tests.

The data in Table 4.5, though limited in scope, was consistent with results from previous testing of water in Benin and of filtered water in both lab and field

FIGURE 4.13 The original water samples are both clear and free from debris. However, the analytical methods show that bacteria contaminate the untreated water, while no contamination was observed in the filtered water. These samples also provide a visual method for demonstrating contamination of water that otherwise appears "clean."

Source: Photo by Bradley Striebig.

tests of new filters. The ceramic filters performed very well as a point-of-use filter. The chemically treated Pur™ water was also free of bacteria. Although both methods of treating the water appeared successful, there were significant differences in the economic, cultural, and societal impacts of the two technologies.

Field tests of ceramic filters reported in the literature indicate that the average lifespan of a filter is approximately 2 years. If the filter is filled twice per day, it can easily supply about 20 liters per day. (Actual maximum capacity is approximately 45 to 75 liters per day.) The average household size in Benin is six family members who range widely in age. However, for comparing the costs of the methods, we have assumed only four members per household, which would provide a more conservative estimate of the daily water cost per household. The cost per liter of Pur-treated water in the local

TABLE 4.5 Presence of bacterial contamination in the untreated water samples compared to treated water samples

SOURCE	NO. OF SAMPLES	NO. INDICATING BACTERIAL PRESENCE (POSITIVE RESULT)		
		Total Coliform (LPT-MUG)	*E. coli*	Hydrogen-Sulfide Producing Bacteria (Pathoscreen™)
Well water	4	4	4	4
Tap water	10	6	1	6
Pur™ water	4	0	0	0
Filtered water	10	0	0	0

Source: Based on Striebig, Gieber, Cohen, Norwood, Godfrey, and Elliot. 2010. Implementation of a Ceramic Water Filter Manufacturing Process at the Songhai Center: A Case Study in Benin. International Water Association's World Water Congress in Montreal 2010.

marketplace was approximately 0.10 USD. The economic savings for treated water for those able to invest in the technology are self-evident. At the current sales price, using conservative estimates for water usage and household size, the filter would pay for itself in 37 days, as illustrated in Table 4.6 (Striebig et al., 2010).

Development and implementation of the CWF process was a complex issue. Considerations of transportation, scheduling, climatic variations, material availability, and educational resources should all be considered when planning a CWF implementa-

tion. Monitoring and evaluating the program have provided the initiative to continue process improvements and verified that the technology is effective. Addressing health education and maintenance procedures in the community was important if the filters were to have the maximum impact in protecting drinking water quality. A plan for long-term monitoring and evaluation should be considered prior to initial implementation, as this step has proven critical in making the CWF manufacturing process a sustainable solution to improving the quality of drinking water in Benin.

TABLE 4.6 Return on Investment (ROI) in a ceramic water filter compared to drinking bottled (or bagged) water available in the marketplace

TREATED WATER TYPE	LITERS PER DAY PER PERSON	PERSONS PER HOUSEHOLD	PURCHASE COST (USD)	COST PER HOUSEHOLD PER DAY (USD)	COST PER LITER (USD)	ROI (DAYS)
Pur	2	4	$0.11	$0.89	$0.11	
Ceramic filter	5	4	$31.11	$0.04	$0.01	37

Source: Based on Striebig, Gieber, Cohen, Norwood, Godfrey, and Elliot. 2010. Implementation of a Ceramic Water Filter Manufacturing Process at the Songhai Center: A Case Study in Benin. International Water Association's World Water Congress in Montreal 2010.

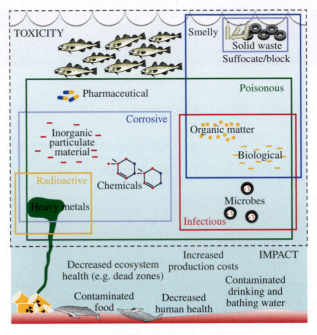

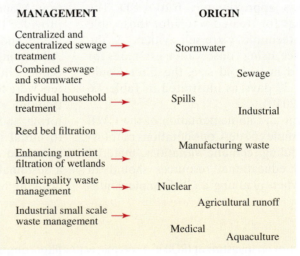

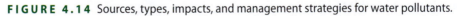

FIGURE 4.14 Sources, types, impacts, and management strategies for water pollutants.

Sources: Based on Corcoran, E., C. Nellemann, E. Baker, R. Bos, D. Osborn, H. Savelli (eds). 2010. Sick Water? The central role of waste-water management in sustainable development. A Rapid Response Assessment. United Nations Environment Programme, UN-HABITAT, GRID-Arendal.

4.3.1 Microorganisms in Water

Microorganisms are a natural part of the environment; in fact, they outnumber all other species combined on Earth. Microorganisms are the principal decomposers of natural and anthropogenic waste. Microorganisms convert organic waste in landfills to carbon dioxide, methane, and water. Microorganisms can also convert atmospheric nitrogen to ammonia through a process called nitrogen fixation, and they help fertilize soils. In their water treatment role, microorganisms are very helpful

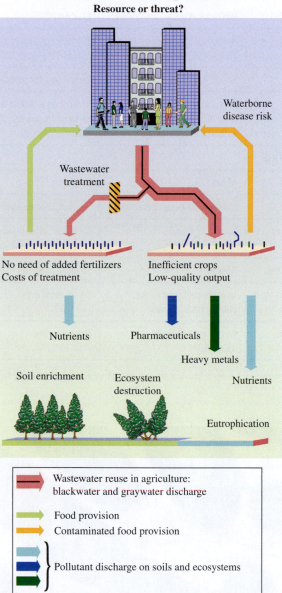

Wastewater in Urban Agriculture
Resource or threat?

Waterborne disease risk

Wastewater treatment

No need of added fertilizers
Costs of treatment

Inefficient crops
Low-quality output

Nutrients

Pharmaceuticals

Heavy metals

Nutrients

Soil enrichment

Ecosystem destruction

Eutrophication

Wastewater reuse in agriculture: blackwater and graywater discharge

Food provision

Contaminated food provision

} Pollutant discharge on soils and ecosystems

FIGURE 4.15 Potential pollutant effects, positive and negative, associated with urban wastewater.

Source: Based on Corcoran, E., C. Nellemann, E. Baker, R. Bos, D. Osborn, H. Savelli (eds). 2010. Sick Water? The central role of waste-water management in sustainable development. A Rapid Response Assessment. United Nations Environment Programme, UN-HABITAT, GRID-Arendal.

in remediating contaminated water, sediments, and soil in engineered systems. Microorganisms also consume oxygen in water sometimes to the exclusion of other species. A very small number, compared to all types of organisms, are pathogenic.

There are two broad classifications of microorganisms, prokaryotes and eukaryotes, the characteristics of which are summarized in Table 4.7. **Prokaryotes** have the simplest cell structure. Prokaryotes illustrated in Figure 4.16 are classified by their lack of a nucleus membrane that contains cellular DNA. Bacteria, blue-green algae, and archaea are classified as types of prokaryotes. **Eukaryotes**, illustrated in

TABLE 4.7 Characteristics of prokaryotic and eukaryotic microorganisms

CELL CHARACTERISTIC	PROKARYOTE	EUKARYOTE
Phylogenetic group	Bacteria, blue-green algae, archaea	Single cell: algae, fungi, protozoa Multicell: plants, animals
Size	Small, 0.2–3.0 μm	2–100 μm for single-cell organisms
Cell wall	Composed of peptidoglycan, other polysaccharides, protein, glycoprotein	Absent in animals and most protozoans; present in plants, algae, fungi; usually polysaccharide
NUCLEAR STRUCTURE		
Nuclear membrane	Absent	Present
DNA	Single molecular, plasmids	Several chromosomes
Internal membranes	Simple, limited	Complex, endoplasmic reticulum, golgi, mitochondria; several present
Membrane organelles	Absent	Several present
Photosynthetic pigments	In internal membranes; chloroplasts absent	In chloroplasts
Respiratory system	Part of cytoplasmic membrane	Mitochondria

Source: Based on Tchobanoglous, G., Burton, F.L., Stensel, H.D. (2003) Wastewater Engineering Treatment and Reuse: 4th Edition. McGraw-Hill, New York, NY, USA.

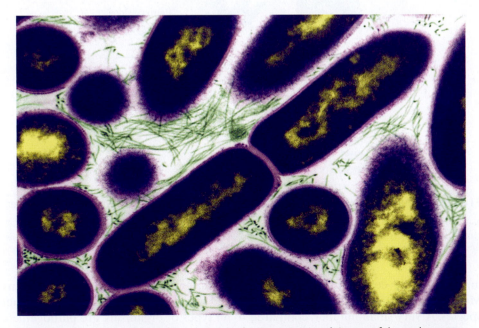

FIGURE 4.16 Color-enhanced transmission electron micrograph image of the prokaryote *Haemophilus ducreyi* bacteria associated with some types of food poisoning. The cellular DNA material is distributed and there is no distinct nucleus in the prokaryotic bacteria.

Source: David M. Phillips / Science Source

FIGURE 4.17 Stained light micrograph showing a small eukaryotic liver fluke (*Dicrocoelium*) that illustrates distinct nucleus, ribosomes, and mitochondria.

Source: M. I. Walker / Science Source

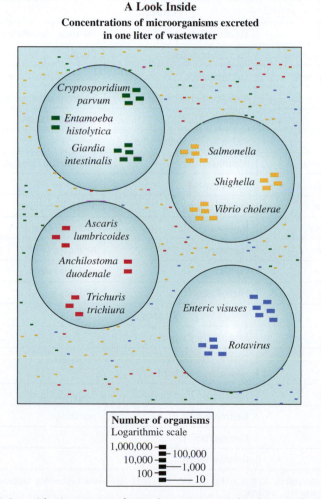

A Look Inside

Concentrations of microorganisms excreted in one liter of wastewater

FIGURE 4.18 Logarithmic estimate of typical microorganism levels in one liter of wastewater.

Source: Based on Corcoran, E., C. Nellemann, E. Baker, R. Bos, D. Osborn, H. Savelli (eds). 2010. Sick Water? The central role of waste-water management in sustainable development. A Rapid Response Assessment. United Nations Environment Programme, UN-HABITAT, GRID-Arendal.

TABLE 4.8 Potential infectious agents found in wastewater.

ORGANISM	DISEASE	SYMPTOMS
Bacteria		
Campylobacter jejuni	Gastroenteritis	Diarrhea
Escherichia coli	Gastroenteritis	Diarrhea
Legionella pneumophila	Legionnaires' disease	Fever, respiratory illness
Leptospira	Leptospirosis	Jaundice, fever
Salmonella	Salmonellosis	Food poisoning
Salmonella typhi	Typhoid fever	Fever, diarrhea, ulceration
Shigella	Shigellosis	Bacillary dysentery
Vibrio cholerae	Cholera	Diarrhea, dehydration
Yersinia enterocolitica	Yersiniosis	Diarrhea
Protozoa		
Balantidium coli	Balantidiasis	Diarrhea, dysentery
Cryptosporidium parvum	Cryptosporidosis	Diarrhea
Cyclospora cayetanesis	Cyclosporasis	Severe diarrhea and nausea
Entamoeba histolytica	Ambiasis (amoebic dysentery)	Severe diarrhea and abscesses
Giardia lamblia	Giadiasis	Diarrhea, nausea
Helminths		
Ascaris lumbrocoides	Ascariasis	Roundworm infestation
Enterobius vermicularis	Enterobiasis	Pinworm
Fasciola hepatica	Fascioliasis	Sheep liver fluke
Hymenolepis nana	Hymenolepiasis	Dwarf tapeworm
Tanenia saginata	Taeniasis	Beef tapeworm
T. solium	Taeniasis	Pork tapeworm
Trichuris triciura	Trichuriasis	Whipworm
Viruses		
Adenovirus	Respiratory disease	
Enteroviruses	Gastroenteritis, heart anomalies, meningitis	Polio, etc.
Hepatitis A virus	Infectious hepatitis	Jaundice, fever
Norwalk agent	Gastroenteritis	vomiting
Parvovirus	Gastroenteritis	
Rotavirus	Gastroenteritis	

Source: Based on Tchobanoglous, G., Burton, F.L., Stensel, H.D. (2003) Wastewater Engineering Treatment and Reuse: 4th Edition. McGraw-Hill, New York, NY, USA.

TABLE 4.9 Characteristics of selected microorganisms commonly found in wastewater

MICROORGANISM	SHAPE	SIZE, μm	RESISTANT FORM
Protozoa			
Cryptosporidium			
Oocysts	Spherical	3–6	Oocysts
Sporozoite	Teardrop	1–3 W × 6–8 L	Oocysts
Entamoeba histolytica			
Cysts	Spherical	10–155	Cysts
Trophozoite	Semispherical	10–20	Cysts
Giardia lamblia			
Cysts	Ovid	6–8 W × 8–14 L	Cysts
Trophozoite	Pear or Kite	6–8 W × 12–16 L	Cysts
Helminths			
Ancylostoma duodenale (hookworm) eggs	Elliptical or egg	36–40 W × 55–70 L	Filariform larva
Ascaris lummbicoides (roundworm) eggs	Lemon or egg	35–50 W × 45–70 L	Embryonated egg
Trichuris trichiura (whipworm) eggs	Elliptical or egg	20–24 W × 50–55 L	Embryonated egg
Viruses			
MS2	Spherical	0.022–0.026	Viron
Enterovirus	Spherical	0.020–0.030	Viron
Norwalk	Spherical	0.020–0.035	Viron
Polio	Spherical	0.025–0.030	Viron
Rotavirus	Spherical	0.070–0.080	Viron

Source: Based on Tchobanoglous, G., Burton, F.L., Stensel, H.D. (2003) Wastewater Engineering Treatment and Reuse: 4th Edition. McGraw-Hill, New York, NY, USA.

Figure 4.17, on page 203 are more complex organisms. Eukaryotes have a nucleus or nuclear envelope and endoplasmic reticulum, or interconnected organelles. Eukaryotes include protozoa, fungi, and green algae.

The most significant concern associated with drinking water quality is preventing pathogens from contaminating the water supply. Pathogenic organisms are typically excreted by human beings and other warm-blooded animals. Pathogens are usually classified as bacteria, protozoa, helminths, or viruses. Figure 4.18 on page 203 illustrates the approximate concentration of microorganisms found in a typical liter of wastewater. Tables 4.8 to 4.10 provide a range of expected concentrations of microorganisms in wastewater and the likely dose required for a human to develop symptoms associated with the pathogen.

Bacteria are unicellular organisms that do not have a nuclear membrane. Bacteria consist only of a cell wall, cell membrane, cytoplasm, and DNA. Membrane selectively allows passage of nutrients for growth and waste removal from the cell. Bacteria may be cylindrical to spherical in shape and range in size from about 0.1 to 5 μm. Some bacteria may form endospores in unfavorable environments that are resistant to the traditional water disinfection process. Both harmless and helpful

TABLE 4.10 Microorganism concentration and infectious dose in typical wastewater.

ORGANISM	CONCENTRATION	
	MPN/100 ml	**Infectious Dose**
Bacteria		
Bacterioides	10^7–10^{10}	
Coliform, total	10^7–10^9	
Coliform, fecal	10^6–10^8	10^6–10^{10}
Clostridium perfringens	10^3–10^5	1–10^{10}
Enterococci	10^4–10^5	
Fecal streptococci	10^4–10^7	
Pseudomonas aeruginosa	10^3–10^6	
Shigella	10–10^3	10–20
Salmonella	10^2–10^4	10–10^8
Protozoa		
Cryptosporidium parvum oocysts	10–10^3	1–10
Entamoeba histolytica cysts	10^{-1}–10	10–20
Giardia lamblia cysts	10^3–10^4	<20
Helminths		
Ova	10–10^3	
Ascaris lumbricoides	10^{-2}–10	1–10
Viruses		
Enteric virus	10^3–10^4	1–10
Coliphage	10^3–10^4	

Source: Based on Tchobanoglous, G., Burton, F.L., Stensel, H.D. (2003) Wastewater Engineering Treatment and Reuse: 4th Edition. McGraw-Hill, New York, NY, USA.

bacteria are typically found in the human intestinal tract. Pathogenic bacteria may be found in the intestines of infected individuals, and contact with their waste may spread the pathogenic bacteria. Domestic wastewater contains a wide variety of nonpathogenic and pathogenic bacteria (listed in Tables 4.8 to 4.10). Archaea are similar to bacteria in size and cellular components, but the cell wall, cell material, and RNA composition are different. Archaea are important in the anaerobic degradation of waste and are found in many environments not tolerated by bacteria.

Protozoa are motile, usually single-celled, microscopic eukaryotes, some examples of which are listed in Tables 4.8 to 4.10. The majority of protozoa are aerobic (use oxygen) heterotrophs. Some protozoa are aerotolerant anaerobes, and a few are anaerobic (i.e., they do not require oxygen for growth). Protozoa are often predatory as well and may be used to remove bacteria in wastewater treatment processes. Many protozoa are pathogenic, especially to individuals with compromised immune systems. *Cryptosporidium parvum*, cyclospora, and *giardia lamblia* are particularly problematic pathogens because they are resistant to traditional chlorine disinfection.

Helminths are parasitic worms that are transmitted by wastewater and wastewater solids; their characteristic size and shape are listed in Tables 4.8 to 4.10. Worldwide, they are one of the leading causes of disease. Helminths are transmitted in wastewater when an adult parasite living in an infected individual lays eggs that are excreted into untreated wastewater. The eggs range in size from 10 to 100 μm and can be removed by sedimentation or filtration. Some eggs, however, are resistant to the common water and wastewater disinfection processes.

Viruses are intracellular parasites composed of a nucleic acid core with an outer protein coating. Viruses cannot reproduce by themselves but require the host cell to reproduce. The extremely small size of viruses, 0.2–0.8 μm, makes them difficult to remove in traditional physical water treatment processes. Other characteristics of viruses are shown in Tables 4.8 to 4.10. Viruses may also be resistant to some chemical disinfection processes.

Algae are chlorophyll-containing eukaryotic organisms that produce oxygen through photosynthesis. Algae may be partially responsible for the negative taste and odor characteristics of some drinking water. When excessive levels of nutrients enter a water body, excessive algae reproduction rates (called **algae blooms**) may occur. The algae blooms result in a net reduction of dissolved oxygen available in the water for higher organisms as the algae decays. These algae blooms are frequently blamed for an increase in fish mortality rates during hot weather when oxygen levels are at their lowest, especially in temperate lakes.

Engineers and scientists often find it useful to characterize microorganisms by their metabolic classification, as the metabolic process may be the greatest concern or utility related to the microorganisms. Metabolism refers to the sum of all processes required for the organisms to convert other materials to energy for internal use for growth and reproduction of new cellular mass. All known life forms require carbon for growth, and microorganisms may be classified by the carbon source that they use. Adenosine triphosphate (ATP) is the primary chemical used to store energy within a cell. Nicotinamide adenine dinucleotide (NAD+) is the second most important chemical compounds used to store energy. Heterotrophs obtain carbon from other organic matter or proteins. In some instances, autotrophs obtain carbon from inorganic carbon sources like carbonates. Chemotrophs produce energy within the cells through chemical oxidation of electron donors in their environment. Phototrophs obtain energy for reproduction through photon capture and photooxidation of a food or substrate.

Microorganisms may also be classified by the source of oxygen they use. **Aerobic organisms** require molecular oxygen, (O_2), for respiration. Aerobic organisms typically utilize substrate more efficiently and more rapidly than nonaerobic organisms. Anaerobic organisms do not require molecular oxygen; they may obtain oxygen from inorganic ions, such as nitrates, sulfates, or proteins.

Microorganisms in water are difficult to enumerate owing to their small size and frequently uneven distribution in water sources. Pathogenic organisms, those we are most concerned about, are difficult to identify because they are so few in numbers, relative to all other species. A variety of indicator tests have been developed as a surrogate measure of pathogen measurement in water samples. The indicator tests focus on measuring a broad category of microorganisms that are numerous in water samples and much easier to measure than particular pathogenic organisms.

The ideal **indicator organisms** would be present to indicate fecal contamination by warm-blooded mammals and would be numerous in the sample. Ideally, the organisms would be more resistant to typical disinfection processes than comparable pathogenic organisms, but they should not be a health threat to laboratory personnel preparing samples and conducting the tests. The ideal organism would be easy to isolate and quantify. Unfortunately, the currently available tests do not meet all these ideals, so the engineer or scientist must recognize the potential shortcomings of the standardized test procedures, summarized in Table 4.11, that are available for determining water quality.

TABLE 4.11 Bacterial indicator organisms

INDICATOR ORGANISM	CHARACTERISTICS
Total coliform bacteria	Gram-negative rods that produce colonies in 24 to 48 h at 35°C
	Includes *Eschericha*, *Citobacter*, *Enterobacter*, and *Klebsiella*
Fecal coliform bacteria	Able to produce colonies at elevated incubation temp. (44.5°C for 24 ± 2 h)
Klebsiella	Included in fecal and total coliform tests. Cultured at 35°C for 24 ± 2 h
E. coli	Most representative of fecal sources
Bacteroides	Anaerobic organism proposed as a human-specific indicator
Fecal streptococci	Used with fecal coliforms to determine the source of fecal contamination
	Several strains are ubiquitous
Enterococci	Two strains, *S. faecalis* and *S. faecium*, are most human-specific organisms
Clostridium perfringens	Spore-forming anaerobe
	Desirable indicator for disinfection, past pollution or for "old" samples
P. aeruginosa	Present in large numbers in domestic wastewater
A. hydrophila	Can be recovered in the absence of immediate sources of fecal pollution

Source: Based on Tchobanoglous, G., Burton, F.L., Stensel, H.D. (2003) Wastewater Engineering Treatment and Reuse: 4th Edition. McGraw-Hill, New York, NY, USA.

The presence of indicator organisms in the existing standardized tests, listed in Table 4.12, do not necessarily mean enteric viruses or other pathogens are present in the water sample. The tests listed in the table are insufficient for some pathogens in water, especially those that are particularly resistant to disinfection, like *Giardia lamblia*. As a result, water-disease outbreaks have occurred sporadically in water systems that had passed the prescribed standard microbial test procedures. The science of microbial testing is quickly evolving, and there is greater interest, especially in bacteriophage tests, in indicating the possible presence of enteric viruses in water. Thus, it is likely that the tests listed in Table 4.12 may change as science advances.

Several procedures have been standardized that provide estimates of the type and number of indicator organisms listed in Table 4.12. Individual bacteria enumeration methods include direct microscopic counting, pour and spread plate counts, membrane filtration, and multiple tube fermentation. Colonies of bacteria may be quantified using the heterotrophic plate count. Specific bacteria species can be identified by staining or fluorescence methods.

The pour plate and spread plate methods may be used to culture, identify, and enumerate bacteria (see Figure 4.19). The process uses a series of sample dilutions that are then mixed with the warm agar medium placed into the Petri dish. The dishes are incubated under controlled conditions, and the bacterial colonies are counted based on the known sample volume added to each Petri dish. Values are typically reported in colony-forming units (cfu) per volume (ml), and since

TABLE 4.12 Standardized indicator-based criteria tests

WATER USE	INDICATOR ORGANISMS
Drinking water	Total coliform
Fresh water recreation	Fecal coliform
	E. coli and Enterococci
Saltwater recreation	Fecal and total coliform
	Enterococci
Shellfish-growing areas	Fecal and total coliform
Agricultural reclamation	Total coliform
Wastewater effluent	Total coliform
Disinfection	Fecal coliform
	MS2 coliphage

Source: Based on Tchobanoglous, G., Burton, F.L., Stensel, H.D. (2003) Wastewater Engineering Treatment and Reuse: 4th Edition. McGraw-Hill, New York, NY, USA.

the colonies must grow, they indicate only viable organisms and do not include nonviable organisms in the results.

The membrane filter technique, illustrated in Figure 4.20, is an effective technique for quickly estimating the number of indicator organisms in a sample. In this method, a known sample volume of water is passed through a filter, typically with a 0.45 μm pore size. Bacteria are retained on the filter, and the filter is placed into a Petri dish with an agar substrate and incubated for 24 hours. The viable colonies are counted after incubation. The membrane filter technique is commonly used for total coliform, fecal streptococcus, and E. coli.

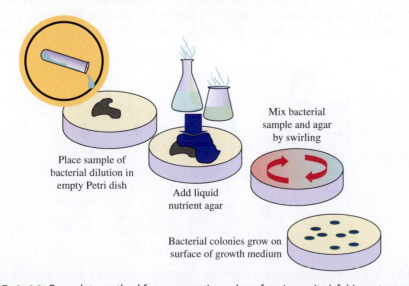

Place sample of bacterial dilution in empty Petri dish

Add liquid nutrient agar

Mix bacterial sample and agar by swirling

Bacterial colonies grow on surface of growth medium

FIGURE 4.19 Pour plate method for enumerating colony-forming units (cfu) in water samples.

Source: Bradley Striebig.

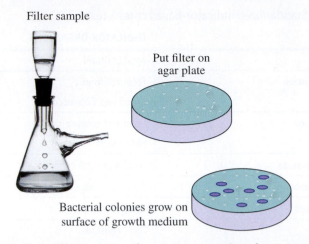

Filter sample

Put filter on
agar plate

Bacterial colonies grow on
surface of growth medium

FIGURE 4.20 Membrane filter technique for enumerating colony-forming units (cfu) in water samples.

Source: chromatos/iStock/Thinkstock

A rapid, statistical method for estimating the number of microorganisms in water is based on a process of dilution until extinction and is called either a multiple tube fermentation test or the most probable number (MPN) test. The MPN is based on statistical equations, such as the Poisson equation, or on MPN tables. The results are typically reported as MPN/100 ml. In this process, illustrated in Figure 4.21, a water sample is diluted by an order of magnitude in each of five sample vials. The resulting test thus covers three to five orders of magnitude. The numbers of tubes with a viable sample are counted, and the MPN is determined from a statistical equation or MPN tables approved for the method.

Quantitative information about the number of microorganisms can be gained from indicator tests; typically, these tests can be completed in 24 to 48 hours. Indicator tests are not yet ideal, and uncertainties in actual pathogen numbers exist, particularly those resistant to traditional water disinfection processes. Worldwide, pathogenic organisms in water are the leading cause of preventable mortality

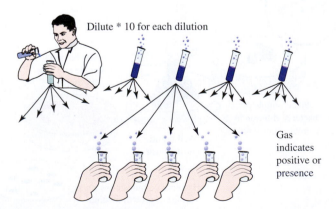

Dilute * 10 for each dilution

Gas
indicates
positive or
presence

FIGURE 4.21 Illustration of the multiple tube fermentation tests for determining the MPN per 100 ml of a water sample.

Source: Bradley Striebig.

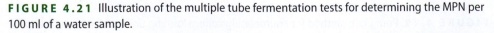

for children under 5 years of age. Water quality controls and treatment methods that decrease pathogen levels in water are extremely important. Most microorganisms in our environment are nonpathogenic, and many are considered helpful. Microorganisms help degrade a wide variety of "waste" products and pollutants in the environment. Engineers have harnessed this process to use microorganisms in water treatments, wastewater treatments, environmental remediation of hazardous wastes, and pharmaceutical products.

4.3.2 Dissolved Oxygen

The amount of oxygen in the standard atmosphere is stable at 20.95%. However, the amount of **dissolved oxygen (DO)** in water changes dramatically with the temperature of the water; the consumption rate or deoxygenation rate in the water due to uptake from microorganisms and higher organisms; and the re-aeration rate or how quickly oxygen is reabsorbed and mixed in the water body. The Henry's law constant value for oxygen was presented in Table 1.7 in Chapter 1. This information has been translated for standard conditions to a simple paper-based computer called a nomograph, shown in Figure 4.22. A straight line between any two points on the

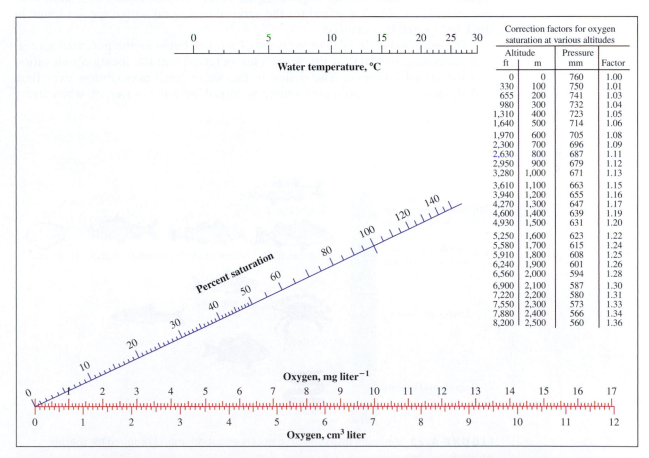

Correction factors for oxygen saturation at various altitudes			
Altitude		Pressure	
ft	m	mm	Factor
0	0	760	1.00
330	100	750	1.01
655	200	741	1.03
980	300	732	1.04
1,310	400	723	1.05
1,640	500	714	1.06
1,970	600	705	1.08
2,300	700	696	1.09
2,630	800	687	1.11
2,950	900	679	1.12
3,280	1,000	671	1.13
3,610	1,100	663	1.15
3,940	1,200	655	1.16
4,270	1,300	647	1.17
4,600	1,400	639	1.19
4,930	1,500	631	1.20
5,250	1,600	623	1.22
5,580	1,700	615	1.24
5,910	1,800	608	1.25
6,240	1,900	601	1.26
6,560	2,000	594	1.28
6,900	2,100	587	1.30
7,220	2,200	580	1.31
7,550	2,300	573	1.33
7,880	2,400	566	1.34
8,200	2,500	560	1.36

FIGURE 4.22 Nomograph that can be used to calculate the concentration of dissolved oxygen or percent saturation. Corrections are required to account for elevations above sea level.

Source: Based on Mortimer, C. H., (1981) The oxygen content of air-saturated fresh waters over ranges of temperature and atmospheric pressure of limnological interest. Mitt. int. Ver. Limnol. Pp. 22-23.

nomograph yields the value for the third point on the chart. The amount of dissolved oxygen in water is critical to animal life in aquatic systems. There is approximately 25 times more oxygen in air than in water, but due to the added weight of water, it requires about 800 times more energy to gain enough oxygen to support life in water than in air. Thus, the balance of dissolved oxygen in water is delicate, and small changes can and do result in large fish kills, especially in slow-moving rivers and lakes.

When oxygen concentrations are in equilibrium with the oxygen in the atmosphere, the aqueous solution is saturated with oxygen. The **saturated dissolved oxygen (DO$_s$)** concentration decreases with increasing water temperature. The DO$_s$ concentration increases with increasing atmospheric pressure. Since atmospheric pressure generally decreases with elevation, the DO$_s$ concentration in water tends to decrease as elevation increases. Thus, there are correction factors on the nomograph to account for variations in DO$_s$ in locations with different elevations.

Different species of organisms require different levels of DO in water. Generally, smaller organisms and those that feed near the bottom of waterways require less oxygen. Larger predatory species that feed on smaller organisms higher in the water column require higher levels of DO. Figure 4.23 shows the minimum level of DO required for the survival of several native species found in the Chesapeake Bay estuary.

The **oxygen deficit**, D, in water is the difference between the potential saturation concentration, DO$_s$, at the water's temperature and the location's elevation and the actual DO level. The deficit in the water level may change over time and distance when a pollutant source is mixed into the water or when living

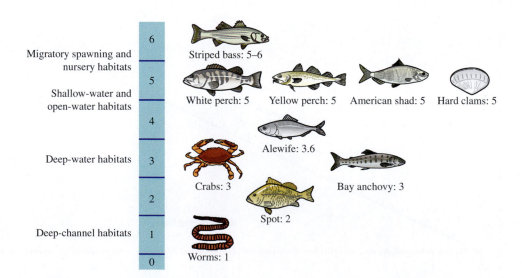

FIGURE 4.23 Minimum dissolved oxygen concentration (mg/L) required for species survival

Source: USEPA (2003) Ambient Water Quality Criteria for Dissolved Oxygen, Water Clarity and Chlorophyll a for the Chesapeake Bay and Its Tidal Tributaries. U.S. Environmental Protection Agency. Region III. Chesapeake Bay Program Office. Annapolis, Maryland, USA. EPA 903-R-03-002.

organisms consume oxygen and the rate of deoxygenation is greater than the re-aeration rate.

$$D = DO_s - DO \tag{4.2}$$

4.3.3 Biochemical Oxygen Demand

Microorganisms in the environment and in engineered systems remove oxygen in proportion to their initial number and rate of growth. The growth rate of microorganisms can be modeled, and from the growth rate, the oxygen uptake rate can be determined. In a closed system microorganisms go through four stages of growth: a lag phase, an exponential growth phase, a stationary phase, and a death phase. During the lag phase, the amount of nutrients and substrate is much greater than the number of microorganisms. Microbial growth is limited in the lag phase, as the microorganisms first manufacture enzymes necessary to metabolize the substrate and the number of microorganism present is small. Once the enzymes have been produced, growth becomes exponential as most of the microorganism's energy is used for growth and reproduction. In the exponential growth phase, the substrate and nutrients are still abundant and waste products from microbial growth do not yet inhibit growth. During exponential growth, the growth rate is proportional to the number of microorganisms. In the exponential phase, growth is governed by the exponential equations and is typically represented as

$$X = X_o e^{\mu t} \tag{4.3}$$

where

X = concentration of the microorganisms (g/m³)

μ = specific growth rate, with units of inverse time (time⁻¹)

The population tends to reach a maximum value and then plateaus, as shown in Figure 4.24. The population growth rate approaches zero as the substrate concentration diminishes or waste products build up to toxic levels in the closed system. As substrate levels continue to diminish and waste products increase, the number of viable organisms begins to decrease as the population enters a death phase in a closed system.

In many cases in the environment and engineered systems, growth is limited by the amount of substrate present in the system. A maximum specific growth

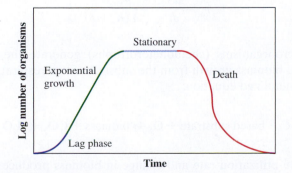

FIGURE 4.24 Stages of microbial growth in a closed system.

Source: Bradley Striebig.

rate can be estimated during substrate limited growth by the Michaelis–Menton equation:

$$r_{su} = -\frac{kXS}{K_s + S} \tag{4.4}$$

where

r_{su} = rate of substrate concentration change (g/m^3 · day)

k = the maximum specific substrate utilization rate (g of substrate/g of microorganism · day)

X = concentration of the microorganisms (g/m^3)

S = growth-limiting substrate concentration (g/m^3)

K_s = half-velocity degradation coefficient (g/g)

The half-velocity degradation coefficient, also called the Monod constant, corresponds to the substrate concentration at which the specific growth rate is one-half the maximum specific growth rate (i.e., ½k). The maximum specific **substrate** utilization rate is related to the maximum specific bacterial growth rate by

$$k = \frac{\mu_{max}}{Y} \tag{4.5}$$

where

μ_{max} = the maximum specific bacterial growth rate (g new cells/g cell · day)

Y = true substrate yield coefficient (g cell/ g substrate used)

The **substrate utilization rate** provides information about energy conversation from the substrate to cell maintenance and growth (see Figure 4.25). The fraction of energy from the substrate that is used for cell growth is equal to the substrate yield, Y. The substrate utilization rate may then also be written as

$$r_{su} = \frac{dS}{dt} = -\frac{\mu_{max}XS}{Y(K_s + S)} \tag{4.6}$$

Aerobic microorganisms (chemoheterotrophs) generate the energy for cell maintenance and biomass growth from the substrate from respiration according to the following generalized equation:

$$C - \text{based substrate} + O_2 \rightarrow \text{biomass} + CO_2 + H_2O \tag{4.7}$$

The substrate utilization rate and change in biomass produce a **carbonaceous oxygen demand (COD)** in water. Similarly, chemoautotrophs use carbon dioxide as a carbon source and inorganic matter as an energy source. Chemoautotrophs also

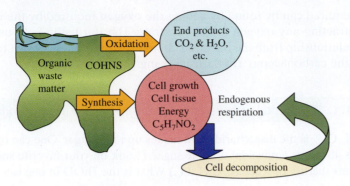

FIGURE 4.25 Conversion of organic waste matter to biomass and carbon dioxide through respiration.

Source: Bradley Striebig.

require energy in this energy conversion process according to the general equation for **nitrification**:

$$NH_3 + O_2 \rightarrow biomass + NO_3^- + H_2O + H^+ \tag{4.8}$$

The oxygen consumed for the nitrification of ammonia (NH_3) for the production of biomass is called the **nitrogenous oxygen demand (NOD)**.

Microorganisms convert a portion of the organic waste (represented by COHNS, where the amount of each element is unknown) into carbon dioxide to gain energy for cell maintenance through oxidation of the organic waste.

$$COHNS + O_2 + bacteria \rightarrow CO_2 + H_2O + NH_3 \tag{4.9}$$

A portion of the organic waste is converted into new biomass through synthesis. Biomass is represented by an average equation based on the elemental ratios of the components of cell tissue ($C_5H_7NO_2$).

$$COHNS + O_2 + bacteria + energy \rightarrow C_5H_7NO_2 \tag{4.10}$$

Microorganisms begin to convert their own cell tissue into energy for cell maintenance through **endogenous respiration** when substrate levels become very low.

$$C_5H_7NO_2 + 5O_2 \rightarrow 5CO_2 + NH_3 + 2H_2O \tag{4.11}$$

If the particular substrate in a system is known, a **theoretical oxygen demand (ThOD)** can be calculated in a series of three steps from the stoichiometric relationship. The first step is to write the equation describing the reaction for oxidation of the carbon based substrate. The second step is to balance the chemical equation. Usually, it is easiest to balance the number of carbon atoms on either side of the equation first and balance the hydrogen atoms next. The stoichiometric value for the oxygen required can be found by adding the oxygen required by the right-hand side and subtracting any oxygen in the substrate. The third step is to use the stoichiometric relationship from the balanced chemical equation to convert to the units required for the carbonaceous ThOD, usually mg/L.

EXAMPLE 4.1   Theoretical Carbonaceous Oxygen Demand of a Can of Soda

Sucrose ($C_{12}H_{22}O_{11}$), is the disaccharide that makes up table sugar. One can of soda typically contains 40 grams equivalent of table sugar. (Look on your favorite soda label to see if it contains sucrose or fructose, $C_6H_{12}O_6$.) What is the ThOD in one can of soda?

Step 1: Write the unbalanced equation for the reaction:

$$C_{12}H_{22}O_{11} + ?O_2 \rightarrow ?CO_2 + ?H_2O$$

Step 2: Balance the chemical equation

There are 12 atoms of carbon on the left-hand side of the equation, so 12 molecules of carbon dioxide will be needed to balance the carbon on the right-hand side of the equation.

There are 22 atoms of hydrogen on the left-hand side of the equation, so 11 molecules of water will be needed to balance the hydrogen on the right-hand side of the equation.

Adding the atoms of oxygen from carbon dioxide and water molecules on the right-hand side of the equation and subtracting the atoms of water contained in sucrose yields the atoms of oxygen needed:

$$12 \times 2 + 11 \times 1 - 11 = 24 \; \textit{additional atoms of oxygen are required}$$

The balanced chemical equation showing the molecular oxygen (O_2), required is

$$C_{12}H_{22}O_{11} + 12O_2 \rightarrow 12CO_2 + 11H_2O$$

Step 3: Use the balanced equation and unit conversions to find the carbonaceous ThOD in mg/L:

$$\frac{40 \text{ g sucrose}}{12 \text{ oz}} \times \frac{33.814 \text{ oz}}{L} \times \frac{\text{mole sucrose}}{342 \text{ g sucrose}} \times \frac{12 \text{ mole } O_2}{\text{mole sucrose}} \times \frac{32,000 \text{ mg } O_2}{\text{mole } O_2} = 1.27 \times 10^5 \frac{\text{mg}}{L}$$

Therefore, ThOD = 1.27×10^5 mg/L, so during the summer a truckload of soda that spills into a river or lake could have dramatic consequences due to the amount of oxygen that will need to be removed from the water as microorganisms in the water will have a carbonated soda party.

A variety of microorganisms (both chemoheterotrophs and chemoautotrophs) are present in the natural environment, and both use oxygen in water as an electron acceptor to convert a substrate to biomass. The amount of oxygen required by microorganisms to convert a substrate to biomass is called the **biochemical oxygen demand (BOD)**. The BOD is the most widely used water quality parameter to quantify water quality related to organic pollutants from public sewage, agricultural runoff, and industrial discharges. A measure of BOD provides a useful estimate of the overall strength of pollutants in a water sample. BOD measurements also provide estimates for the amount of oxygen demanded if a wastewater enters a river, lake, or engineered treatment system. A standardized procedure has been developed and widely used to measure the effectiveness of treatment processes and determine compliance with industrial and municipal discharge permits.

The standardized five-day BOD test is used to measure the oxygen consumed during the decomposition of organic waste in water. The test measures the total amount of oxygen required by microorganisms in five days of degradation in a sample containing an organic waste material. The sample bottles are either made of opaque glass or incubated in a way that prevents light from entering the bottle in order to prevent background algae from adding oxygen to the sample through photosynthesis. The standard BOD procedure consists of the following steps:

- Add a predetermined volume 1 to 300 ml of sample to a 300 ml bottle, according to Table 4.13.

TABLE 4.13 Approximate sample volumes required for standard five-day BOD test

SAMPLE TYPE	MINIMUM SAMPLE SIZE		MAXIMUM SAMPLE SIZE			
	Est. BOD (mn/L)	Sample size (ml)	Estimated BOD (mg/L)			At Sample Size (ml)
			Sea Level	305 m (1,000')	1524 m (5,000')	
Strong trade waste	600	1	2,460	2,380	2,032	1
	300	2	1,230	1,189	1,016	2
	200	3	820	793	677	3
Raw and settled sewage	150	4	615	595	508	4
	120	5	492	476	406	5
	100	6	410	397	339	6
	75	8	304	294	251	8
	60	10	246	238	203	10
	50	12	205	198	169	12
	40	15	164	158	135	15
Oxidized effluents	30	20	123	119	101	20
	20	30	82	79	68	30
	10	60	41	40	34	60
Polluted river waters	6	100	25	24	21	100
	4	200	12	12	10	200
	2	300	8	8	7	300

Source: Based on USEPA (2003) Ambient Water Quality Criteria for Dissolved Oxygen, Water Clarity and Chlorophyll a for the Chesapeake Bay and Its Tidal Tributaries. U.S. Environmental Protection Agency. Region III. Chesapeake Bay Program Office. Annapolis, Maryland, USA.

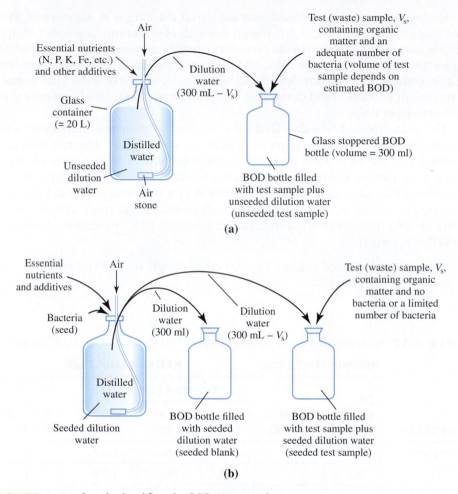

FIGURE 4.26 Standardized five-day BOD test procedure.

Source: Based on Tchobanoglous, G., Burton, F.L., Stensel, H.D. (2003) Wastewater Engineering Treatment and Reuse: 4th Edition. McGraw-Hill, New York, NY, USA.

- Fill the remaining volume of the 300 ml bottle with dilution water that has been saturated with oxygen and contains a phosphate buffer and nutrients as illustrated in Figure 4.26. The dilution water contains the phosphate buffer to keep changes in pH from the degradation of wastes from inhibiting substrate utilization. Nutrients are added in the dilution water to ensure that all the required elements for cell growth are present in the sample bottle. Microorganisms may also be added via the dilution water (called seeding the sample) in order to ensure that the number of organisms is sufficient in the sample to achieve logarithmic growth.
- Measure the initial DO of the bottle, DO_i.
- Incubate the bottle at 20°C in the dark.
- Measure DO after five days (± 4 hours) of incubation, DO_f.

Water samples that contain microorganisms, such as raw sewage, treated unchlorinated effluent from a wastewater treatment facility, or natural waters do not require the addition of extra microorganisms, called a sample seed, to perform the BOD test. However, water samples that do not contain microorganisms, such as chlorinated water from a treatment facility or industrial-strength wastewaters, require the addition of microorganisms through a seed. Tests that require a seed also require a sample blank that consists of a bottle containing only the sample seed and dilution water in addition to another sample that contains the water sample, the seed, and the dilution water. At least three blanks should be included with each set of BOD samples.

The five-day BOD of an *unseeded sample* is determined from:

$$BOD_5 = \frac{DO_i - DO_f}{P} \qquad (4.12)$$

where

BOD_5 = standard five-day BOD (mg/L)

DO = measured dissolved oxygen in the sample bottle (mg/L)

$$P = \frac{\text{volume of the sample (ml)}}{\text{volume (ml) of the sample + dillution water}} \overset{\text{def}}{=\!=} \frac{\text{volume of sample (ml)}}{300 \text{ (ml)}} \qquad (4.13)$$

The five-day BOD of a *seeded sample* is determined from

$$BOD_5 = \frac{(DO_i - DO_f) - (1 - P)(B_i - B_f)}{P} \qquad (4.14)$$

where B = measured dissolved oxygen in the blank bottle (mg/L).

EXAMPLE 4.2   Calculating the *BOD_5* of a Chlorinated Effluent from a Wastewater Treatment Plant

A wastewater treatment plant must maintain its effluent BOD_5 concentration at less than 20 mg/L to remain in compliance with its discharge permit. The laboratory provided the data shown in Table 4.14 to the facility engineer. Is the plant in compliance or not?

TABLE 4.14 Example BOD data

	SAMPLE	BLANK
Initial dissolved oxygen (mg/L)	8.06	8.12
Final dissolved oxygen (mg/L)	6.48	8.01
Volume of the sample (ml)	30	30
Volume of the BOD bottle (ml)	300	300

Using Equation (4.13) for the dilution factor and Equation (4.14) for the seeded sample:

$$P = \frac{30 \text{ ml}}{300 \text{ ml}} = 0.1$$

$$BOD_5 = \frac{(8.06 - 6.48) - (1 - 0.10)(8.12 - 8.01)}{0.01} = 14.8 \text{ mg/L}$$

There are several limitations associated with the BOD_5 test. The microorganisms must be readily able to degrade the waste. Special preparation of the sample is required

microorganisms in the water or a seed cannot degrade the chemicals in the water. Only wastes that are biodegradable can be measured; nonbiodegradable organic wastes and nonorganic wastes are not accounted for in the BOD test. Since specific chemical formulas for the waste mixture and microorganism composition are not typically known, the results cannot be used in stoichiometric equations. The test requires a relatively long period to complete, which may not represent the state of the systems at the completion of the test. Nonetheless, the BOD test is extremely helpful for predicting potential environmental impacts of organic wastes.

4.3.4 Nutrients in Water

Excessive levels of nutrients in water have caused major deterioration of many of the world's watersheds. Excessive nutrient levels have caused higher than normal levels of nitrogen-containing and phosphorus-containing compounds in water. All living organisms need a variety of organisms to support and maintain healthy ecosystems, including nitrogen and phosphorus. The ecosystem balance may be upset when high levels of nitrogen and phosphorus are discharged into waterways. Excessive nitrogen and phosphorus loading is usually due in large part to use of excessive fertilizer in agricultural practices and the discharge of municipal wastewater. The distribution of areas and sources that contribute to nitrogen and phosphorus levels into the Chesapeake Bay in the eastern United States is shown in Figures 4.27 and 4.28.

Excessive levels of nutrients may have adverse effects on the drinking water supply, recreational water use, aquatic life, and fisheries. Nitrate levels above 10 mg/L in drinking water have been linked to methemoglobinemia (Blue Baby Syndrome), which may lead to premature death in infants. **Eutrophication**, or the nutrient enrichment of aquatic ecosystems, is a natural process that can be

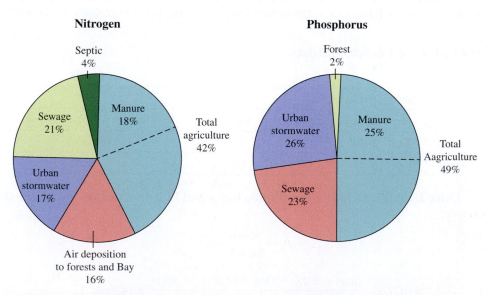

FIGURE 4.27 Sources of nitrogen and phosphorus in the Chesapeake Bay watershed and estuary.

Source: Based on Chesapeake Bay Foundation (2004) Manure's Impact on Rivers, Streams and the Chesapeake Bay: Keeping Manure Out of the Water.

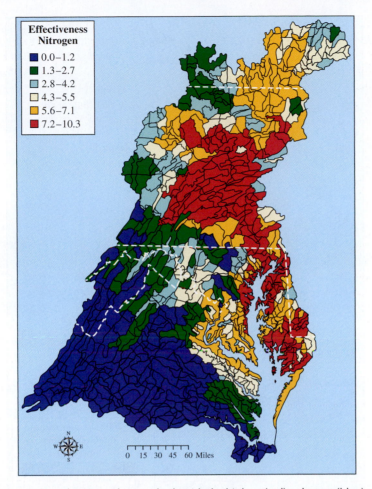

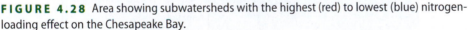

FIGURE 4.28 Area showing subwatersheds with the highest (red) to lowest (blue) nitrogen-loading effect on the Chesapeake Bay.

Source: USEPA. (2010) Chesapeake Bay Total Maximum Daily Load for Nitrogen, Phosphorus and Sediment. U.S. Environmental Protection Agency. Region 3. Water Protection Division. Philadelphia, PA, USA. Figure ES-2 Page ES-6.

accelerated by excessive anthropogenic emissions of nitrogen and phosphorus. Accelerated eutrophication increases the growth rate of aquatic plants, algae, and microorganisms.

Eutrophication often results in significantly increasing the oxygen deficit in bodies of water as the higher than normal levels of plant matter settle to the bottom of the water body. Microorganism concentrations increase as they use the dead plant matter as a substrate, and oxygen levels decrease as the plant matter decays. Eventually, oxygen levels may become completed depleted, which occurred in the summer of 2005 throughout large sections of the Bay (shown in Figure 4.29), changing the ecosystem from an aerobic-based ecosystem to an anaerobic system, or hypoxic, which may cause aquatic stress, change the pH of the water, and increase the release of ammonia, methane, and hydrogen sulfide. The increasing anoxic zones, levels of low DO, and hypoxic zones that have occurred since the 1950s owing to increased nutrient levels in the Chesapeake Bay are shown in Figure 4.30. Increased levels of diatoms and filamentous algae caused

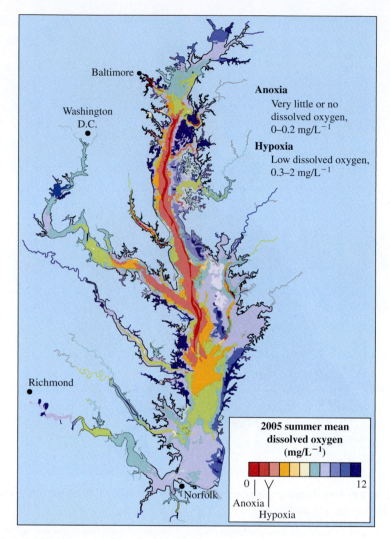

Anoxia
Very little or no
dissolved oxygen,
0–0.2 mg/L^{-1}

Hypoxia
Low dissolved oxygen,
0.3–2 mg/L^{-1}

FIGURE 4.29 Large areas of the Chesapeake Bay in the eastern United States suffer from low DO concentrations due in large part to excessive nutrient applications in the watershed.

Source: Based on Wicks, C., Jasin, D. and Longstaff. (2007) Breath of Life: Dissolved Oxygen in Chesapeake Bay. May 27, 2007 Newsletter. Chesapeake Bay Foundation. Annapolis, MD, USA.

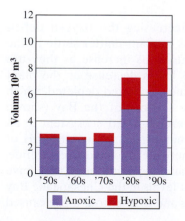

FIGURE 4.30 The increase in the volume of anoxic and hypoxic zones occurring in the Chesapeake Bay.

Source: Based on Chesapeake Bay Foundation (2004) Manure's Impact on Rivers, Streams and the Chesapeake Bay: Keeping Manure Out of the Water.

by eutrophication may clog water treatment plant filters, increase the risk associated with disinfection by-products from chlorination, and cause odor and taste problems with drinking water. The extensive growth of rooted macrophytes, phytoplankton, and mats of living and dead plant matter can interfere with swimming, boating, and fishing activities.

4.4 Modeling the Impacts of Water Pollutants

The effects of pollutant emissions can be modeled using the basic principles of mass and energy balances developed in early chapters. Water quality models are beneficial not only because they help engineers and scientists understand how water quality changes and may be changed by past events, but the models may be used to develop other models that can predict the level of treatment necessary to improve the environmental, economic, and societal benefits of a particular body of water. Water quality models can also be used to examine the effects of potential sources and develop appropriate policies and permits for potential wastewater emissions.

4.4.1 Modeling Dissolved Oxygen in a River or Stream

The biochemical oxygen demand and the impacts on the concentration of oxygen available for aquatic life forms can be modeled using a mass balance approach. The pollutant loading rate is determined from BOD measurements. However, the five-day BOD test only indicates the amount of organic matter that is degraded in five days (Figure 4.31), not the total organic content, or ultimate BOD, of the wastewater.

The oxygen consumption and change in organic content are proportional to the concentration of the organic content at any time, t, and is represented by a first-order rate constant, r_b:

$$-r_b = \frac{dL}{dt} = -kL \tag{4.15}$$

where

L = the oxygen demand associated with the organic content (mg/L)

k = reaction rate constant [days^{-1}]

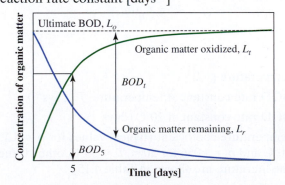

FIGURE 4.31 The total organic matter remaining in a water sample decreases with time and biological activity. The organic matter oxidized, which is measured by a change in oxygen concentration in the sample, is the reciprocal, or mirror image, of the organic matter remaining in the solution. The organic matter oxidized is equivalent to the BOD at the given time, and typically about two-thirds of the organic matter is oxidized by day five.

Source: Bradley Striebig.

Rearranging and integrating Equation (4.15) yield the exponential function that describes the change in organic content with time:

$$L_t = L_o e^{-kt} \tag{4.16}$$

where

L_o = the oxygen demand associated with the organic content at the initial time (mg/L)

L_t = the oxygen demand associated with the organic content at any time t (mg/L)

The BOD_t is equivalent to the difference between the initial organic content and the organic content at time t, as shown in Figure 4.31:

$$BOD_t = L_o - L_t \tag{4.17}$$

Substituting Equation (4.16) into Equation (4.17) yields

$$BOD_t = L_o - L_o e^{-kt} = L_o(1 - e^{-kt}) \tag{4.18}$$

The ultimate BOD is equivalent to the initial organic content of the wastewater, L_o, and is the maximum amount of oxygen that would be consumed by microbial degradation of the waste. The relationship between the measured BOD at five days, BOD_5, and the ultimate BOD, L_o, is

$$L_0 = \frac{BOD_5}{(1 - e^{-5k})} \tag{4.19}$$

The value for the BOD rate constant depends on how rapidly the waste is degraded, or how biodegradable the waste is, and the temperature. The change in the rate constant with temperature can be estimated from the following empirical correlations developed by Schroepfer, Robins, and Susag (1960):

$$k_T = k_{20}(\theta)^{T-20} \tag{4.20}$$

where

T = temperature (°C)

k_T = BOD rate constant at temperature, T (days^{-1})

k_{20} = BOD rate constant at 20°C (days^{-1})

θ = temperature coefficient (typical values for θ = 1.135 for $4 < T < 20$°C and θ = 1.056 for $20 < T < 30$°C, other values reported in the literature are shown in Table 4.15)

A conservative mass balance approach is useful for modeling a point-source pollutant discharge or tributary input into a river or stream, as illustrated in Figure 4.32. For simplicity, it is assumed that the river water and inputs are instantaneously well mixed at the junction point. The conservative mass balance equation may be used to find the ultimate BOD, dissolved

TABLE 4.15 Reported values for the temperature compensation coefficient used for carbonaceous BOD decay

θ, TEMPERATURE CORRECTION FACTOR	TEMPERATURE LIMITS (°C)	REPORTED REFERENCES IN U.S. EPA REPORT
1.047		Chen (1970)
		Harleman et al. (1977)
		Medina (1979)
		Genet et al. (1974)
		Bauer et al. (1979)
		JRD (1983)
		Bedford et al. (1983)
		Thomann and Fitzpatrick (1982)
		Velz (1984)
		Roesner et al. (1981)
1.05		Rich (1973)
1.03–1.06	(0–5)–(30–35)	Smith (1978)
1.075		Imhoff et al. (1981)
1.024		Metropolitan Washington Area Council of Governments (1982)
1.02–1.06		Baca and Arnett (1976)
		Baca et al. (1983)
1.04		Di Toro and Connolly (1980)
1.05–1.15	5–30	Fair et al. (1968)

Source: Based on Bowie, G. L, Mills, W. B., Porcella, D. B., Campbell, C. L., Pagenkopf, J. R., Rupp, G. L., Johnson, K. M., chan, P. W. H., Gherini, S. A. and Chamberlin, C. E. (1985) Rates, Constants, and Kinetics Formulations in Surface Water Quality Modeling: Second Edition. Environmental Research Laboratory. Office of Research and development, U.S. Environmental Protection Agency. Athens, Georgia, USA.

oxygen content, temperature, and combined flow rate of the combined river and input flows:

$$Q_{\text{river upstream}} + Q_{\text{input}} = Q_{\text{river downstream}} \quad (4.21)$$

and for any parameter C in the water:

$$Q_{\text{river upstream}}C_{\text{upstream}} + Q_{\text{input}}C_{\text{input}} = Q_{\text{river downstream}}C_{\text{downstream}} \quad (4.22)$$

As the water containing an organic waste moves downstream, the waste will be biologically degraded and the microorganisms utilizing the organic waste as a

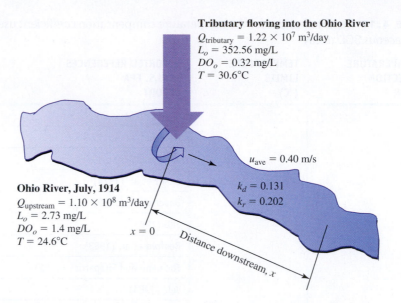

Tributary flowing into the Ohio River
$Q_{tributary} = 1.22 \times 10^7$ m³/day
$L_o = 352.56$ mg/L
$DO_o = 0.32$ mg/L
$T = 30.6°C$

$u_{ave} = 0.40$ m/s

Ohio River, July, 1914
$Q_{upstream} = 1.10 \times 10^8$ m³/day
$L_o = 2.73$ mg/L
$DO_o = 1.4$ mg/L
$T = 24.6°C$

$k_d = 0.131$
$k_r = 0.202$

$x = 0$

Distance downstream, x

FIGURE 4.32 A moving river system can be modeled with a mass balance approach. The input or tributary stream is considered to be instantaneously mixed with the upstream flow at the point of discharge, $x = 0$.

Source: Bradley Striebig.

substrate will consume oxygen. The rate of oxygen consumption, or deoxygenation, is proportional to the ultimate BOD at any time, t, multiplied by the deoxygenation rate constant for the moving water body:

$$r_d = k_d L_t \tag{4.23}$$

where

r_d = rate of deoxygenation

L_t = ultimate BOD remaining at time t some distance downstream [mg/L]

k_d = the deoxygenation rate constant (day⁻¹)

It is usually assumed that $kd = k$ in Equation (4.17). Then, substituting Equation (4.17) into Equation (4.23) yields the decay rate in terms of the ultimate BOD of the mixed water occurring at the point of mixing:

$$r_d = k_d L_o e^{-k_d t} \tag{4.24}$$

Note that the rate constant k_d must still be corrected for temperature if the water is not at 20°C. The assumption that $k_d = k$ is generally appropriate for slow-moving bodies of water, but it is a poor assumption for a shallow, rapidly moving stream. Additional values for k_d are reported in the literature for different bodies of water.

Re-aeration is the dissolution of molecular oxygen into water. The rate of reparation of a body of water is related to the difference between the saturated oxygen level and the actual amount of oxygen in solution, and the temperature, velocity, depth, and turbulence of water flow in the stream channel. The re-aeration rate, r_r, is directly proportional to the oxygen deficit, D. The gas diffusion rate into a liquid increases as the concentration of the gas in solution decreases. The oxygen deficit represents the difference between the saturation level and actual oxygen level in solution as stated in Equation (4.2); therefore:

$$r_r = k_r D \tag{4.25}$$

where k_r = re-aeration rate constant (day^{-1}). Several empirical correlations exist and are reported in Appendix G, which relates the re-aeration rate to the properties of the stream or river. O'Connor and Dobbins (1958) reported one correlation indicating that the re-aeration rate increases with increasing stream velocity, u, and decreases as the average depth, H, increases:

$$k_r = \frac{3.9u^{0.5}}{H^{1.5}} \tag{4.26}$$

where

k_r = re-aeration rate constant (day^{-1})

u = average stream velocity (m/s)

H = average stream depth (m)

The equation developed by O'Connor and Dobbins for streams and rivers was limited in depth and velocity. Churchill et al. (1962) and Owens et al. (1964) extended this range, and Covar (1976) compiled these relationships into Figure 4.33, which allows for an estimation of the re-aeration coefficient across a wide range of depths and velocities.

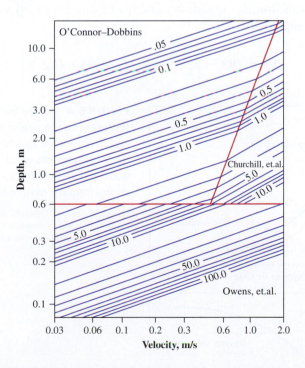

FIGURE 4.33 Re-aeration correlations compiled by Covar (1976) for estimating the re-aeration coefficient (day^{-1}) at a specific depth and velocity.

Source: EPA kinetic and constants report

The change in the oxygen deficit, dD/dt, as the water moves downstream from the point of discharge is equal to the rate of deoxygenation minus the rate of re-aeration:

$$\frac{dD}{dt} = k_d L_o e^{-k_d t} - k_r D \qquad (4.27)$$

The integration of the two curves yields the shape of the curves shown in Figure 4.34. If the organic waste concentration in the stream is relatively high, then the rate of deoxygenation is greater than the rate of re-aeration, and the dissolved oxygen concentration in the stream will decrease with time. However, eventually the organic waste concentration decreases, subsequently decreasing the rate of deoxygenation at the same time that the oxygen deficit has been increasing, which increases the rate of re-aeration. The point at which the rate of deoxygenation is equal to the rate of re-aeration is the critical point at which the lowest oxygen concentration is expected in the stream or river, as shown in Figure 4.35. After the critical point, the rate of re-aeration begins to decrease as the oxygen deficit decreases, and the rate of deoxygenation continues to decrease along with decreasing levels of organic waste in the water. The plot of the combined deoxygenation and re-aeration curves is called the oxygen sag curve. The curve is often also referred to as the Streeter–Phelps curve, named after the scientists who first described this process along the Ohio River in 1914. The oxygen sag curve, or Streeter–Phelps equation, represents the integral of Equation (4.27), which has the form

$$D = \frac{k_d L_o}{K_r - k_d}(e^{-k_d t} - e^{-k_r t}) + D_o e^{-k_r t} \qquad (4.28)$$

The actual level of dissolved oxygen (DO) in the moving water body can be determined by substituting the right-hand side of Equation (4.28), which is equal to the deficit into Equation (4.2):

$$DO = DO_s - \left[\frac{k_d L_O}{k_r - k_d}(e^{-k_d t} - e^{-k_r t}) + D_O e^{-k_r t}\right] \qquad (4.29)$$

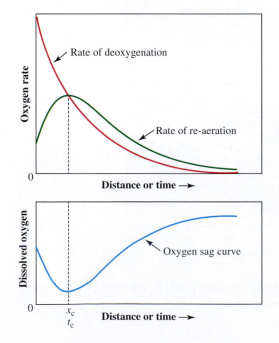

FIGURE 4.34 The individual curves representing the rate of deoxygenation and the rate of re-aeration in a moving body of water into which an organic waste has been introduced. The combined curves form the oxygen sag curve.

Source: Based on Masters,G. and Ela, W. P. (2007) Introduction to Environmental Engineering and Science: 3rd Edition. Prentice Hall.

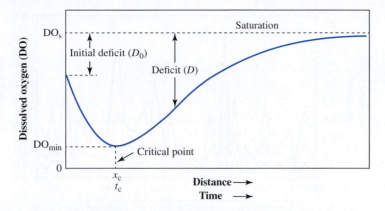

FIGURE 4.35 The oxygen sag curve and critical point, where the lowest oxygen concentration in the moving water body will occur.

Source: Based on Masters,G. and Ela, W. P. (2007) Introduction to Environmental Engineering and Science: 3rd Edition. Prentice Hall.

The greatest impact on the aquatic ecosystem occurs at the critical point, or maximum oxygen deficit. The **critical point** occurs at a specific time after the two streams have mixed together and can be determined by taking the derivative of Equation (4.29), which yields

$$t_c = \frac{1}{K_r - k_d} \ln \left\{ \frac{k_r}{k_r} \left[1 - \frac{D_o(k_r - k_d)}{k_d L_o} \right] \right\} \tag{4.30}$$

where t_c = critical time (days). The critical oxygen deficit can be found by substituting the resulting value for the critical time from Equation (4.30) into Equation (4.28).

The point, at some distance, x_c, downstream from the point of mixing is dependent on the average velocity of the stream or river:

$$x_c = u(t_c) \tag{4.31}$$

where u is the average stream velocity (m/day), and x_c [m] is the distance downstream from the mixing point where the critical deficit will occur.

EXAMPLE 4.3   A Mass Balance Approach to Modeling the Impact of Organic Waste on the Carbonaceous Oxygen Demand

In 1914, Streeter and Phelps (1925) studied the effects of organic waste loading on the oxygen content of the Ohio River. The researchers collected data over the course of a year over long stretches of the Ohio River. Their data have been modified slightly to suit this example, but are representative of the actual data they collected in 1914 (and sometimes their actual data are used). Streeter and Phelps published the completed report in Public Health Bulletin 146: A Study of the Pollution and Natural Purification of the Ohio River, by the United States Public Health Service in 1925. The long-term trends in the summer flow rate for the Ohio River are shown in Figure 4.36.

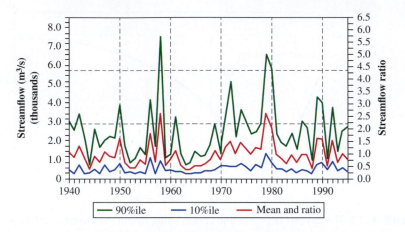

FIGURE 4.36 Long-term trends in mean, 10th, and 90th percentile statistics computed for summer (July–September) streamflow in the Ohio River at Louisville, Kentucky.

Source: US EPA (2000) Progress in Water Quality: An Evaluation of the National Investment in Municipal Wastewater Treatment. U. S. Environmental Protection Agency. Office of water. Washingotn DC, USA. EPA-832-R-00-008. Ch11

- Part I: Using the data set collected for July 1914 (illustrated in Figure 4.32), we will calculate the critical time and maximum oxygen deficit expected in the Ohio River. We will also plot the oxygen sag curve downstream from the point of mixing. Using the model, we will determine at what distance downstream the critical point occurs and where there might be a negative impact on the water quality in the Ohio River.
- Part II: Assuming that 90% of the BOD in the Ohio River and tributary have been removed as a result of passage of the Clean Water Act in 1972 and also assuming that present-day conditions in the Ohio River are as shown in Figure 4.32, we will calculate the critical time and maximum oxygen deficit expected in the Ohio River. The oxygen sag curve downstream from the point of mixing will also be shown graphically. The graphical analysis will show at what distance downstream the critical point occurs and where there might be a negative impact on the water quality in the Ohio River.

Solution for Part I:

The first step is to use conservative mass balance to determine the volumetric flow rate, organic waste loading rate, initial dissolved oxygen, and temperature that result from the two tributaries mixing together (see Figure 4.32).

From Equation (4.21),

$$Q_{downstream} = 1.10 \times 10^8 + 1.22 \times 10^7 = 1.22 \times 10^8 \; \frac{m^3}{day}$$

From Equation (4.22).

$$Q_{up}L_{o_{up}} + Q_{trib}L_{o_{trib}} = Q_{downstream}L_{o_{downstream}}$$

Rearranging yields

$$L_{o_{downstream}} = \frac{Q_{up}L_{o_{up}} + Q_{trib}L_{o_{trib}}}{Q_{downstream}} = \frac{(1.10 \times 10^8)(2.73) + (1.22 \times 10^7)(352.56)}{1.22 \times 10^8} = 41.6 \; \frac{mg}{L}$$

Similarly, for the initial dissolved oxygen:

$$DO_{o_{downstream}} = \frac{Q_{up}DO_{o_{up}} + Q_{trib}DO_{o_{trib}}}{Q_{downstream}} = \frac{(1.10 \times 10^8)(1.4) + (1.22 \times 10^7)(0.32)}{1.22 \times 10^8} = 1.28 \frac{mg}{L}$$

And the temperature of the Ohio River after mixing was

$$T_{o_{downstream}} = \frac{Q_{up}T_{o_{up}} + Q_{trib}T_{o_{trib}}}{Q_{downstream}} = \frac{(1.10 \times 10^8)(24.6) + (1.22 \times 10^7)(30.6)}{1.22 \times 10^8} = 25.8 \frac{mg}{L}$$

The saturated dissolved oxygen concentration for water at 25.8° may be determined from either Henry's law value corrected for temperature or the nomograph in Figure 4.22. A straight line drawn from the temperature through the 100% saturation level indicates that the DO_3 concentration would be 8.1 mg/L, as shown in Figure 4.37.

The initial deficit is determined from Equation (4.2):

$$D_o = 8.1 - 1.3 = 6.8 \frac{mg}{L}$$

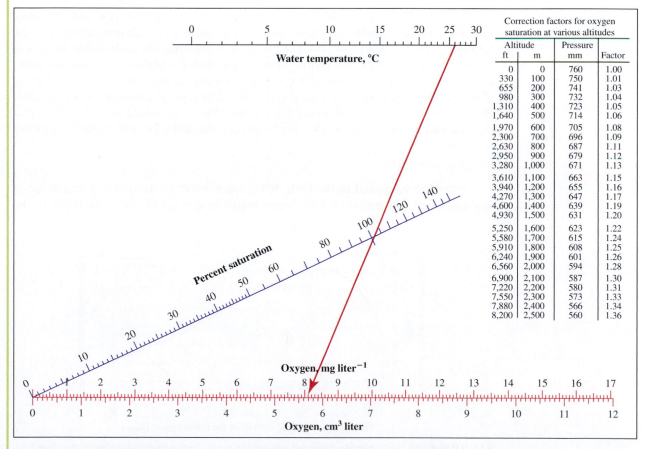

FIGURE 4.37 Determining the DO_3 value for conditions in the Ohio River in July 1914.

Source: Based on Adapted from Mortimer, C. H., (1981) The oxygen content of air-saturated fresh waters over ranges of temperature and atmospheric pressure of limnological interest. Mitt. int. Ver. Limnol. Pp. 22–23.

The deoxygenation rate constant, k_d, and the re-aeration rate constant, k_r, for this stretch of the Ohio River were reported by Streeter and Phelps and the values are provided in Figure 4.38. The known variables are now sufficient to solve the equation for the critical time:

$$t_c = \frac{1}{0.202 - 0.131} \ln\left[\frac{0.202}{0.131}\left(1 - (6.8)\frac{0.202 - 0.131}{0.131(41.6)}\right)\right] = 4.8 \text{ days}$$

The maximum deficit can be found at the critical point by substituting the above results into Equation (4.28):

$$D_c = \frac{0.131(41.6)}{0.202 - 0.131}(e^{-0.131(4.8)} - e^{-0.202(4.8)}) + 6.8(e^{-0.202(4.8)}) = 14.4 \frac{\text{mg}}{\text{L}}$$

An examination of this result shows that the Ohio River was bad news for fish in the summer of 1914. Closer inspection of the maximum deficit and the predicted dissolved oxygen concentration reveals one aspect that the basic mathematical equation does not account for as shown below:

$$DO_c = DO_s - D_c = 8.1 - 14.4 = -6.3 \frac{\text{mg}}{\text{L}}$$

The mathematical theory predicts a negative dissolved oxygen concentration. However, it is not possible in practice to have a negative DO concentration; instead the DO concentration would be zero. The initial DO concentration in the Ohio River was quite low to begin with, and shortly after mixing with the higher concentration waste containing tributary water, the water turned anaerobic, as illustrated in the oxygen sag curve for this system shown in Figure 4.38. If this data is truly representative of the Ohio River in 1914, and all indications are that it is, the Ohio River would have been devoid of any fish and the ecosystem would have been very unhealthy for both aquatic organisms and humans.

Solution to Part II:

What has happened in the Ohio River since 1914? Collection and treatment of sewage water and industrial wastewater began in earnest in the early 1900s in the

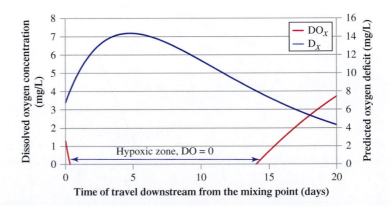

FIGURE 4.38 Approximate dissolved oxygen sag curve for the middle reaches of the Ohio River in July 1914 based on data collected by Streeter and Phelps.

Source: Bradley Striebig.

United States. As a result, significant changes in water quality in the Ohio River have occurred. It is estimated that the BOD levels in the water have decreased by approximately 90% as a result of treating municipal and industrial wastewater as part of the Clean Water Act. In July 2008, upstream dissolved oxygen levels were measured and were found to be significantly higher than in 1914. The water temperature in 2008 was similar to that in 1914. For the example below, we will assume that the same average flow rates occur in the river and tributary and that the deoxygenation and re-aeration constants also are the same as those Streeter and Phelps used in 1914 (as shown in Figure 4.32).

The ultimate BOD loading at the mixing point in 2008 is estimated by

$$L_{o_{downstream}} = \frac{Q_{up}L_{o_{up}} + Q_{trib}L_{o_{trib}}}{Q_{downstream}} = \frac{(1.10 \times 10^8)(0.3) + (1.22 \times 10^7)(35.2)}{1.22 \times 10^8} = 4.18 \frac{mg}{L}$$

Similarly, for the initial dissolved oxygen:

$$DO_{o_{downstream}} = \frac{Q_{up}DO_{o_{up}} + Q_{trib}DO_{o_{trib}}}{Q_{downstream}} = \frac{(1.10 \times 10^8)(6.44) + (1.22 \times 10^7)(5.76)}{1.22 \times 10^8} = 6.36 \frac{mg}{L}$$

And the temperature of the Ohio River after mixing in 2008 was

$$T_{o_{downstream}} = \frac{Q_{up}T_{o_{up}} + Q_{trib}T_{o_{trib}}}{Q_{downstream}} = \frac{(1.10 \times 10^8)(25.0) + (1.22 \times 10^7)(26.1)}{1.22 \times 10^8} = 25.1 \frac{mg}{L}$$

The saturated dissolved oxygen can be determined from the nomograph and is 8.2 for the 2008 conditions. The initial deficit is determined from Equation (4.2):

$$D_o = 8.2 - 6.36 = 1.8 \frac{mg}{L}$$

The known variables are now sufficient to solve the equation for the critical time:

$$t_c = \frac{1}{0.202 - 0.131} \ln\left[\frac{0.202}{0.131}\left(1 - (1.8)\frac{0.202 - 0.131}{0.131(4.16)}\right)\right] = 2.2 \text{ days}$$

The maximum deficit can be found from Equation (4.28):

$$D_c = \frac{0.131(4.16)}{0.202 - 0.131}(e^{-0.131(2.2)} - e^{-0.202(2.2)}) + 1.8(e^{-0.202(2.2)}) = 2.0 \frac{mg}{L}$$

The minimum expected dissolved oxygen content is

$$DO_c = DO_s - D_c = 8.2 - 2.0 = 6.2 \frac{mg}{L}$$

Figure 4.39 illustrates that the conditions in the Ohio River are much more conducive than in 1914 for both aquatic life and human use. Actual measurements of dissolved oxygen in 2008 show minimum oxygen levels in the range of 5.6 mg/L, which is very close to the levels predicted by the model Streeter and Phelps developed, since we have not accounted for additional BOD loading to the river downstream of this single tributary. The model also does not account for contributions to oxygen demand

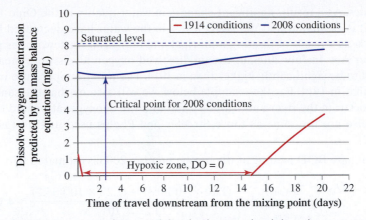

FIGURE 4.39 A comparison of expected dissolved oxygen levels based on reported conditions in 1914 and 2008 in the middle section of the Ohio River.

Source: Bradley Striebig.

from nitrogen-containing compounds or materials found in the sediments in the river. As Figure 4.40 clearly shows, it is quite apparent that the Clean Water Act regulations and subsequent wastewater treatment processes realized in the watershed profoundly improved the water quality in the Ohio River over the past century. The overall aquatic effects of the DO curve are summarized in Figure 4.41.

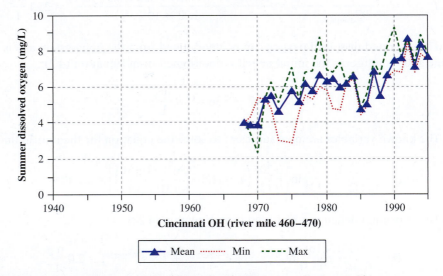

FIGURE 4.40 Long-term trends of DO in the Ohio River near Cincinnati, Ohio.

Source: US EPA (2000) Progress in Water Quality: An Evaluation of the National Investment in Municipal Wastewater Treatment. U. S. Environmental Protection Agency. Office of water. Washingotn DC, USA. EPA-832-R-00-008. Ch11

4.4.2 Modeling Oxygen Demand and Eutrophic Conditions in Temperate Lakes

The geometry, flow, and mixing conditions in lakes are substantially different from those in rivers and streams. Water quality models must be able to account for those differences. The structure, or **morphology**, of a typical lake is shown in

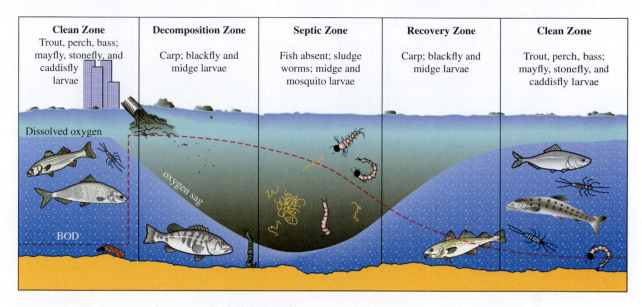

FIGURE 4.41 Aquatic and ecosystem impacts of oxygen demand, deoxygenation, and re-aeration.

Source: Based on Davis, M. L. and Masten, S. J. (2009) Principles of Environmental Engineering and Science: Second Edition. McGraw-Hill. New York, NY.

Figure 4.42. The important aspects of lake morphology related to water quality models are the regions where light penetrates the lake and where mixing with air occurs. Both light penetration and mixing are greater at the surface and decrease with depth. The light that penetrates the water also increases the water temperature and, as the temperature of the water changes, so does the density of the water, as illustrated in Figure 4.43. Water density and temperature differences create layers of water with different characteristics, or **thermal stratification**, in a lake.

The thermal stratification affects the biological, chemical, and physical processes that occur in the water body. Mass balance, energy balance, and equilibrium

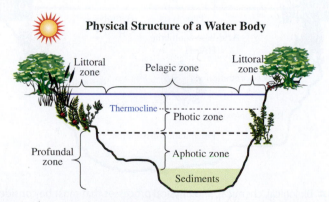

FIGURE 4.42 Physical structure of a typical lake.

Source: Bradley Striebig.

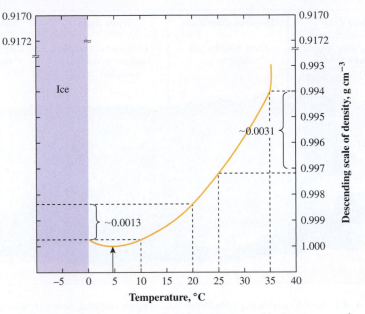

FIGURE 4.43 Change in freshwater density with temperature. Maximum water density occurs at about $+4°C$. Ice is much less dense than water and thus floats. Similarly, warmer water in the epilimnion may float over denser, cooler water in the hypolimnion. The rate of density change is also important to the thermal expansion of water due to global climate change—not the high rate of change in density of water between $+15$ and $+25°C$.

Source: Based on Goldman, C. R. and Horne, A. J. (1983) Limnology. McGraw-Hill.

models can be used to model the effects of the processes shown in Figure 4.44, as they occur in different strata in a lake or water body.

Oxygen distribution is correlated with light penetration and thermal differences in lakes. Oxygen is produced in the photic zone through photosynthesis. Oxygen at the surface of a water body is in equilibrium, or saturated, with oxygen

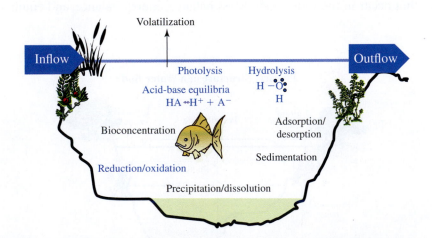

FIGURE 4.44 Biological, chemical, and physical processes that must be considered in modeling the water quality of a lake or water body.

Source: Bradley Striebig.

concentrations in the atmosphere. The depth to which the oxygen level is saturated is dependent on the amount of turbulence, or wind and velocity related mixing and the rate of deoxygenation due to organic wastes from both natural and anthropogenic sources in the water. Oxygen is consumed in the aphotic zone through respiration and degradation of the organic wastes.

Light penetrating the water leads to plant growth, photosynthesis, and heating of the water. The warmer water in a lake is less dense that the cooler, deeper water, and the warmer surface water floats above the deeper water. The uppermost water layer is called the **epilimnion**, and it is characterized by warmer, less dense water and usually has a higher DO concentration (see Figure 4.45). The lower layer of water is called the **hypolimnion**, and it is characterized by cooler, denser water and usually has a lower DO concentration. In many lakes, estuaries, and other water bodies, the hypolimnion may become anoxic, especially during summer months, often due in part to anthropogenic sources of organic wastes or nutrients. The epilimnion and hypolimnion are separated by the **metalimnion**, more commonly called the **thermocline**, which is the layer in which there is a rapid change in temperature and density over a very small change in depth.

In temperate regions in the midlatitudes, the stratification of lakes changes with seasonal changes in temperature (Figures 4.46 through 4.48). During the height of the summer, when the water temperature is at its peak, the dissolved oxygen concentration in the epilimnion may be quite low, in accordance with Henry's law. Dissolved oxygen in the hypolimnion will be even lower than oxygen levels in the epilimnion, due to decomposition of organic plant matter and other microbial processes. The very strong thermocline acts as a barrier, preventing water and oxygen

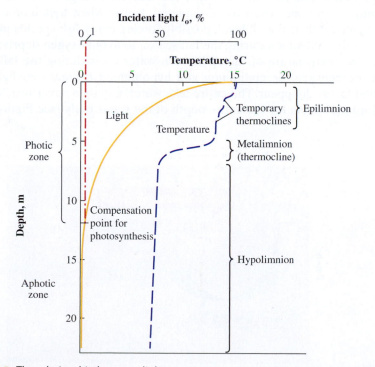

FIGURE 4.45 The relationship between light penetration, water temperature, and depth in a typical stratified lake or water body.

Source: Based on Davis, M. L. and Masten, S. J. (2009) Principles of Environmental Engineering and Science: Second Edition. McGraw-Hill. New York, NY.

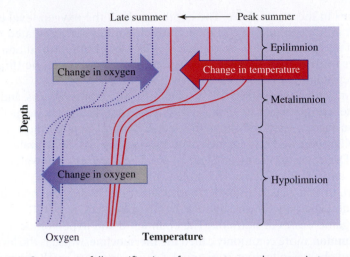

FIGURE 4.46 Summer-to-fall stratification of temperature and oxygen in temperate lakes.

Source: Bradley Striebig.

in the epilimnion from mixing with the water in the hypolimnion. As summer temperatures decrease and the season moves into fall, water temperatures in the epilimnion decrease owing to decreased exposure to solar energy as days grow shorter (Figure 4.47). Lower water temperatures coincide with increasing oxygen levels in the epilimnion. However, the strong thermocline still prevents mixing between the epilimnion and hypolimnion, and the dissolved oxygen concentration in the hypolimnion becomes even more depleted. Fall is when hypolimnion DO levels are typically at their lowest level, possibly forcing many fish species into shallower waters in the epilimnion during the fall season to avoid oxygen deprivation.

As the temperature of the epilimnion water cools during the fall, the density difference between the epilimnion and hypolimnion becomes negligible, and the stratified layers disappear. The entire lake water may then "overturn," representing a well-mixed system for the entire depth of the water body (see Figure 4.47). Thus,

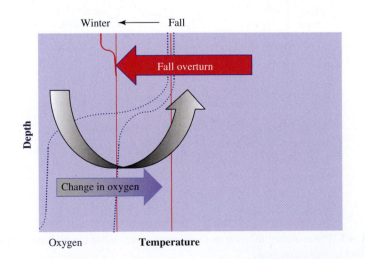

FIGURE 4.47 Fall overturns of temperature and oxygen in temperate lakes.

Source: Bradley Striebig.

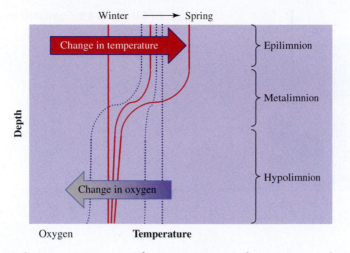

FIGURE 4.48 Spring-to-summer stratification progression of temperature and oxygen in temperate lakes.

Source: Bradley Striebig.

the oxygen concentration increases in the deeper water and is equal throughout the water column. Some of the nutrients that settled into the deeper, denser water during the summer months also gets re-suspended throughout the water column, where the nutrients become available and enhance aquatic plant and algae growth during the spring and summer months.

The cycle of temperature, density, oxygen, and nutrient stratification begins again as the lake's water temperature increases in spring. As the thermocline develops, plants begin photosynthesis, and oxygen levels remain quite high in the cool spring water in the epilimnion. As the thermocline is established and the epilimnion waters warm, dissolved oxygen levels begin to decrease in the hypolimnion (Figure 4.48). If excess nutrients have been added to the water body, plant growth accelerates; microorganisms in the hypolimnion and sediments consume oxygen at a higher rate if more organic matter is present from the previous year's growing season.

The thermal stratification of lakes is dependent on climate systems. Climate changes are likely to affect the dates associated with the fall and spring turnover cycle. If the lake water does not cool enough, the lakes may remain permanently stratified, thereby significantly affecting oxygen and nutrient profiles through the water column in the lake.

The thermal stratification process of a lake or water body should be understood in order to properly represent the system with chemical and physical mathematical models. For instance, if the water in the lake is not thermally stratified during the fall-to-spring overturn, then the entire lake may be represented as a constantly mixed flow reactor (CMFR), as illustrated in Figure 4.49. (The constantly stirred tank reactor (CSTR) is another common name used for the type of reactor shown in Figure 4.49.) During the period that the lake is thermally stratified, the volume of the lake representative of the epilimnion may still be modeled as if it were a CMFR. However, the hypolimnion should be modeled as a constant volume batch reactor without significant flow or mixing into the volume of the hypolimnion (Figure 4.50).

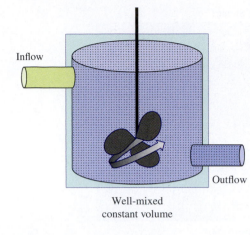

FIGURE 4.49 Constantly mixed flow reactor.

Source: Bradley Striebig.

Lakes may be classified based on their trophic status, as determined by the organic matter (usually represented by chlorophyll) or the nutrient concentration in the water column (Table 4.16). Oligotrophic lakes are characterized by low primary productivity, low levels of biomass, low N and P concentrations, and cool, clear water high in dissolved oxygen concentrations. Eutrophic and hypereutrophic lakes contain a large amount of plant matter that has high nutrient concentrations in the water column, low visibility, and a low dissolved oxygen concentration. The nutrients in lakes may be derived from internal recycling of organic matter or from external inputs of organic matter, nitrogen-containing wastes, or phosphorus-containing wastes. Depending on nitrogen and phosphorus levels in the water, either nutrient could limit plant growth and eutrophication. Generally, in water with the nitrogen-to-phosphorus ratio (N/P) greater than 7, phosphorus is the limiting nutrient; if the ratio is less than 7, nitrogen is the limiting nutrient. Accelerated eutrophication may significantly degrade the water quality in a lake and limit the water uses for drinking, fishing, or recreation.

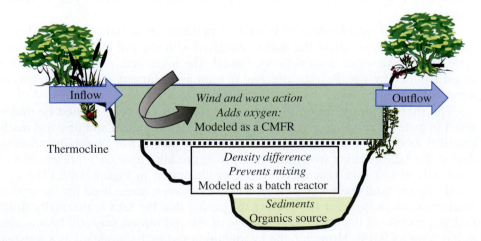

FIGURE 4.50 Idealized model of a stratified lake.

Source: Bradley Striebig.

TABLE 4.16 Typical nutrient levels, biomass, and productivity for different trophic classifications of lakes

TROPHIC CATEGORY	MEAN TOTAL PHOSPHORUS (mg/m³)	ANNUAL MEAN CHLOROPHYLL (mg/m³)	CHLOROPHYLL MAXIMA (mg/m³)	ANNUAL MEAN SECCHI DISC TRANSPARENCY (m)	SECCHI DISC TRANSPARENCY MINIMA (m)	MINIMUM OXYGEN (% sat)[a]	DOMINANT FISH
Ultra-oligotrophic	<4.0	1.0	2.5	12.0	6.0	<90	Trout, Whitefish
Oligotrophic	4.0–10.0	2.5	8.0	8.0	8.0	8.0	Trout, Whitefish
Mesotrophic	10–35	2.5–8	8–25	8–25	8–25	8–25	Whitefish, Perch
Eutrophic	35–100	8–25	25–75	25–75	25–75	25–75	Perch, Roach
Hypereutro-phic	>100.0	25.0	75.0	75.0	75.0	75.0	Roach, Bream

[a] % saturation in bottom waters depending on mean depth

Source: Based on Chapman, D. (1996) A Guide to Use of Biota, Sediments and Water in Environmental Monitoring - Second Edition Chapter 7 – Lakes. UNESCO/WHO/UNEP.

EXAMPLE 4.6   Phosphorus Loading and Classification of Greengots Lake

Greengots Lake in this example is based on a real 48,000-square-meter lake in a temperate region of the United States. The shallow lake is a human-made lake with an average depth of 1 meter, and it was built to operate a grist mill. The lake water is no longer used for the mill, but is now a source for drinking water and recreational fishing. The sources of water to the lake are a cold-water spring and the severely impaired Cooks Creek.

Cold water spring characteristics:

Flow rate = 10,600 m³/day
Dissolved oxygen = 4 mg/L
Phosphorus = 209 kg/year
Temperature = 15°C

Cooks Creek characteristics:

Flow rate = 61,000 m³/day
Dissolved oxygen = 3 mg/L
Phosphorus loading = 4,250 kg/year
Temperature = 23°C

Other phosphorus loads:

Row crops from surrounding farmland = 7,410 kg/yr
Pasture/hay = 700 kg/yr
Undeveloped areas = 227 kg/yr
Urban drainage = 1,020 kg/yr
Septic drainage = 200 kg/yr

Calculate:

- Outflow from the lake
- Hydraulic residence time in the lake
- Steady-state phosphorus concentration in the lake

Assume evaporation = precipitation and that the phosphorus settling rate can be estimated from an empirical correlation developed by Reckhow and Chapra (1983):

$$v_s\left[\frac{m}{yr}\right] = 11.6 + 0.2\left(\frac{Q\left[\frac{m^3}{yr}\right]}{A[m^2]}\right) \tag{4.32}$$

Solution:

$$Q_{out} = Q_{spring} + Q_{creek} + Q_{precipitation} - Q_{evaporation}$$

Since $Q_{precipitation} = Q_{evaporation}$, then

$$Q_{out} = 10{,}600\ \frac{m^3}{day} + 61{,}000\ \frac{m^3}{day} = 71{,}600\ \frac{m^3}{day}$$

The average hydraulic retention time in the lake is

$$t_r = \frac{V}{Q} = \frac{48{,}000\ m^2 \times 1\ m}{71{,}000\ \dfrac{m^3}{day}} = 0.67\ \text{days}$$

The removal rate of the phosphorus settling rate constant is determined from

$$v_s\left[\frac{m}{yr}\right] = 11.6 + 0.2\left(\frac{Q\left[\frac{m^3}{yr}\right]}{A\ [m^2]}\right) = 11.6 + 0.2\left(\frac{71{,}600\left[\frac{m^3}{yr}\right]}{48{,}000\ [m^2]}\right) = 11.9\left[\frac{m}{yr}\right]$$

$$k_s = 11.9\ \frac{m}{yr} * 1\ m = 11.9\ [\text{year}^{-1}]$$

Assuming the lake is at steady state:

$$\frac{dM}{dt} = 0 = (M_{in}) - (M_{out})$$

Rearranging and substituting the mass inputs on the right-hand side of the equation below, and the outputs on the left-hand side, yield

$$M_{spring} + M_{creek} + M_{crops} + M_{pasture} + M_{undeveloped} + M_{urban} + M_{septic} = M_{settling} + M_{outflow}$$

$$M_{spring} + M_{creek} + M_{crops} + M_{pasture} + M_{undeveloped} + M_{urban} + M_{septic} = k_s C_{lake} V + Q_{out} C_{lake}$$

$$209\ \frac{kg}{yr} + 4{,}250\ \frac{kg}{yr} + 7{,}410\ \frac{kg}{yr} + 700\ \frac{kg}{yr} + 227\ \frac{kg}{yr} + 1{,}020\ \frac{kg}{yr}$$

$$+ 200\ \frac{kg}{yr} = 11.9\ \frac{1}{yr} \cdot C_{lake} \cdot 48{,}000\ m^3 + 71{,}600\ \frac{m^3}{day}\ 365\ \frac{day}{yr}\ C_{lake}$$

$$C_{\text{lake}} = \frac{14{,}016 \ \dfrac{\text{kg}}{\text{yr}}}{26{,}705{,}000 \ \dfrac{\text{m}^3}{\text{yr}}} \times 10^6 \ \frac{\text{mg}}{\text{kg}} = 525 \ \frac{\text{mg}}{\text{m}^3}$$

As Table 4.16 shows, the concentration in the lake is hypereutrophic. Indeed, the actual lake on which the example is based is listed as an impaired water body, and while in winter it is capable of supporting trout, in the summer any trout in the lake die and the only native species found in the lake during the summer months are a type of bream.

4.5 Water Treatment Technologies

Few diseases are spread by water in highly economically developed countries. In contrast, in less economically developed countries, a safe, reliable supply of water may be unavailable and a sanitation infrastructure is often absent. Environmental engineers apply the basic principles of science and engineering to design water treatment systems for drinking water and treating human and industrially contaminated wastewater. The engineer's goal is to design water treatment systems to protect public health while balancing environmental, economic, social, and political constraints. A proper water treatment design requires knowledge of the constituents of concern, the impact of these constituents, the transformation and fate of the constituents, treatment methods to remove or reduce the toxicity of the constituents, and methods to dispose of or recycle treatment by-products.

Louis Pasteur explained the germ theory in the 1880s; however, sewer systems were being used for centuries before the theory was scientifically understood to remove the miasma of odorous sewage containing water away from urban centers. Cholera and typhoid were endemic problems in urban areas, including England and the United States, throughout the 1800s. Cholera and typhoid outbreaks did not begin to decrease until the discovery of chlorination for treating drinking water received widespread acceptance in 1908–1911. Mortality and water supply records kept by the Commonwealth of Massachusetts show how improved water filtration and chlorination led to a rapid decline in typhoid fever deaths (Figure 4.51). The United States Public Health Service adopted water quality-based standards in 1914 to improve water quality and reduce disease; this was the first of many policy measures the United States enacted over several decades to improve water quality (Figure 4.52).

Water movements have been altered to benefit communities and urban areas. A schematic of the modified "urban water cycle" is shown in Figure 4.53. The public and industry consume water through withdrawals from groundwater or surface water. The water is treated at a treatment plant or at the point-of-use. Wastewater is generated in homes and in industrial processes. This water may be collected in a community sewer system, and the wastewater should be treated prior to discharge in a receiving body of water. Stormwater in urban areas may also be collected to minimize flooding in urban areas. In some cases, stormwater runoff is emptied directly into receiving waters, or the stormwater may pass through treatment processes dedicated to treating only the runoff; in many urban areas, stormwater and sewage systems are mixed together into **combined sewer overflow (CSO)** that may or may not be treated prior to discharge.

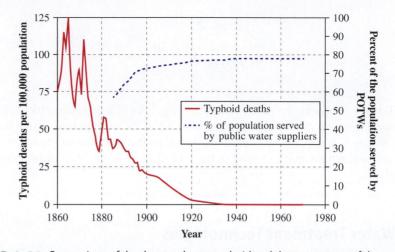

FIGURE 4.51 Comparison of death rates due to typhoid and the percentage of the population served by a treated public water supply in Massachusetts from 1860 to 1970.

Source: USEPA (2000) Chapter 2 - An examination of BOD Loadings before and after the CWA. Progress in Water Quality: An Evaluation of the National Investment in Municipal Wastewater Treatment. U. S. Environmental Protection Agency. Office of Water. Washington DC, USA. EPA-832-R-00-008.

4.5.1 Drinking Water Treatment Technologies

The highest quality water available should be used as the source of drinking water prior to treatment. Treatment technologies may be chemical or physical in nature and very simple or highly sophisticated, depending on the treatment needs and resources available. Groundwater from deep-well boreholes generally contains lower levels of microorganisms than surface waters (Table 4.17) and may require less sophisticated treatment than surface water.

TABLE 4.17 Fecal coliform levels in untreated domestic water sources in selected countries

COUNTRY	SOURCE	FECAL COLIFORMS/100 ml
Gambia	Open hand-dug wells, 15–18 m deep	Up to 100,000
Indonesia	Canals in central Jakarta	3,100–3,100,000
Lesotho	Streams	5,000
	Unprotected sprigs	900
	Water holes	860
	Protected springs	200
	Borehole	1
Uganda	Rivers	500–8,000
	Streams	2–1,000

Source: Based on UNICEF (2008) UNICEF Handbook on Water Quality. United Nations Children's Fund (UNICEF) New York, New York, USA.

	1948

Water Pollution Control Act of 1948

Water Pollution Control Act of 1948 authorized the U.S. Public Health Service to develop comprehensive basin plans for water pollution control and to encourage the adoption of uniform state laws. $100 million of loans were authorized to municipalities annually, but no appropriation for treatment facilities under this act was ever made. However, the act influenced the states to apply more control over the discharge of pollutants into their waters.

1949

1950

Water Pollution Control Act of 1956

Grants for assisting in the construction of municipal treatment works were authorized and, for the first time, funded with federal appropriations. The Surgeon General was directed to prepare comprehensive programs for pollution control in interstate waters in cooperation with states and municipalities, and the state was to prepare plans for prevention and control of water pollution. If there was no approved plan, no grant was to be made for constructing treatment facilities. $50 million annually in grants was authorized. Grants were limited to 30% of the cost of construction, or $250,000, whichever was smaller. Legislation in the states increasingly required secondary treatment for polluted waters.

1951

1952

1953

1954

1955

1956

Water Pollution Control Act of 1961

Comprehensive programs and plans for water pollution abatement and control were still required. Grants were limited to 30% of the cost of construction or $600,000, whichever was less, or $2.4 million for multiple municipal plants. At least half of the appropriation was to go to cities of 125,000 or less. The Congress advocated 85% removal of pollutants in the hearings.

1957

1958

1959

Water Quality Act of 1965

For the first time, each state was required to have water quality standards to receive grants, expressed as water quality criteria applicable to interstate waters. If the state did not develop standards, the Federal Water Pollution Control Administration (FWPCA) was required to do so. To comply with these standards and criteria, secondary treatment was increasingly necessary. Construction grants were raised to 30% of reasonable costs, and an additional 10% was allowed where the project conformed with a comprehensive plan for a metropolitan area. At least 50% of the first $100 million in appropriations had to go to municipalities of less than 125,000 population. Individual grants were limited to $1.2 million, with a limit of $4.8 million for multiple municipalities.

1960

1961

1962

Clean Water Restoration Act of 1966

The requirements for state water quality standards were continued. Each state planning agency receiving a grant was to develop an effective, comprehensive pollution control plan for a basin. The FWPCA, in a guideline, attempted to require states to conform to a national uniform standard of secondary treatment or its equivalent. This action was challenged and the guideline was not enforced. Secretary Udall stated at House hearings that the states had agreed to the requirement for secondary treatment. Grants for POTWs are set at 30% with an increase to 40% if the state paid 30%. The maximum could be increased to 50% if the state agreed to pay 25%. A grant could be increased by 10% if it conformed to a comprehensive plan for the metropolitan area. The limitation of $1.2 million and $4.8 million for grants was waived if the state matched equally all federal grants. At least 50% of the first $100 million in annual appropriations had to be directed to municipalities of <125,000 people.

1963

1964

1965

1966

1967

Water Quality Improvement Act of 1970

The Water Quality Improvement Act of 1970 did not contain any new provisions regarding required standards. The requirements for state water quality standards were continued. However, in hearings for the act, the authority of EPA to require uniform treatment limitations for discharges, such as secondary treatment, was questioned.

1968

1969

1970

1971

FIGURE 4.52 Timeline of major water quality initiatives in the United States from 1948 to 1996.

Source: USEPA (2000) Chapter 2 - An examination of BOD Loadings before and after the CWA. Progress in Water Quality: An Evaluation of the National Investment in Municipal Wastewater Treatment. U. S. Environmental Protection Agency. Office of Water. Washington DC, USA. EPA-832-R-00-008.

1972

1973

1974

Federal Water Pollution Control Act Amendments of 1972

The Federal Water Pollution Control Act of 1972 (later to be renamed the Clean Water Act) contained the first statutory requirement for a minimum of secondary treatment by all Publically owned treatment works (POTWs). The act also established the National Pollutant Discharge Elimination System (NPDES), under which every discharger of pollutants was required to obtain a permit. Under the permit each POTW is to discharge only effluent that had received secondary treatment. EPA defined secondary treatment in a regulation as attaining an effluent quality of at least 30 mg/L BOD_5, 30 mg/L TSS, and 85% removal of these pollutants in a period of 30 consecutive days.

1975

1976

1977

1978

1979

1980

Clean Water Act Amendments of 1977

The Clean Water Act Amendments of 1977 created the 301(h) program, which waived the secondary treatment requirement for POTWs discharging to a marine environment if they could show that the receiving waters would not be adversely affected. Extensive requirements had to be met before such a waiver could be issued.

1981

Clean Water Act Amendments of 1981, PL 97–117

The Clean Water Act Amendments of 1981 amended the Clean Water Act to the effect that "such biological treatment facilities as oxidation ponds, lagoons, ditches, and trickling filters shall be deemed the equivalent of secondary treatment." EPA is directed to provide guidance on design criteria for such facilities, taking into account pollutant removal efficiencies and assuring that water quality will not be adversely affected (Sec. 304(d)(4)). Regulations to this effect were published in final on September 20, 1984. Also, a notice was issued to solicit public comments on "problems related to meeting the percent removal requirements and on five options EPA was considering for amending the percent removal requirements."

1982

1983

1984

1985

1986

1987

1988

1989

National Municipal Policy, January 30, 1984

The EPA National Municipal Policy was published on January 30, 1984. It was designed to ensure that all POTWs met the compliance deadlines for secondary or greater treatment of discharges. The key to the policy is that it provides for POTWs that had not complied by the July 1, 1988 deadline to be put on enforceable schedules. The policy has been outstandingly successful and has resulted in significant increases in compliance.

Secondary Treatment Regulations, June 3, 1985

The secondary treatment regulation published in final on June 3, 1985 revised the previous regulations published in Title 40, Part 133, of the Code of Federal Regulations. Specifically, on a 30-day average, the achievement of not less than 85% removal of BOD_5, $CBOD_5$ and suspended solids for conventional secondary treatment processes was required. However, for those treatment processes designated by the Congress as being equivalent to secondary treatment (such biological treatment facilities as oxidation ponds, lagoons, ditches, and trickling filters), at least 65% pollution removal was required, provided that water quality was not adversely affected. Waste stabilization ponds were given separate suspended solids limits. Special consideration was provided for various influent conditions and concentration limits.

1990

1991

Secondary Treatment Regulations, January 27, 1989

This secondary treatment regulation allows adjustments for dry weather periods for POTWs serving combined sewers.

1992

1993

1994

1995

1996

FIGURE 4.52 (Continued)

FIGURE 4.53 Simple schematic of the modified urban water cycle.

Source: USEPA (2004) Primer for Municipal Treatment Systems.

Groundwater is usually the preferred source for drinking water (also called **potable water**), due to the typically low microbial constituent concentration. The water may be pumped from the groundwater aquifer into an elevated storage tank, so that gravity provides the force needed for fluid flow through remaining treatment processes. A gravity flow-based system ensures that some water will be available even in the event of a power outage. The groundwater may be aerated in the storage tank if the water contains elevated levels of dissolved gases, although this procedure is relatively uncommon. The water is typically passed through a granular filtration process. Filtration can improve both the physical and microbial quality of the water. Numerous types of granular filters are available, including slow sand filters and rapid sand filters.

Slow sand filters are characterized by a slow rate of filtration and the formation of an active layer called the **schmutzdecke**. Typically, slow sand filters are completely saturated with water. Large particles are removed in an upper course granular layer. In the lower finer granular layer, smaller particles are formed and predatory microorganisms may attack and remove pathogens in the schmutzdecke layer. Slow sand filters are usually operated in parallel, as the layers will become clogged over time, and they must be cleaned or drained, which temporarily upsets the active schmutzdecke layer, requiring some time for this layer to re-form and achieve high pathogen-removal efficiencies. Slow sand filters may be very effective and are relatively inexpensive and simple to operate and maintain.

Rapid sand filtration achieves higher velocities through the filter. The filter process is more evenly distributed throughout the depth of the granular media. These filters may remove 50% to 90% of larger pathogens, and the removal efficiency can be increased to 90% to 99% with the addition of a chemical coagulant. Rapid sand filters are periodically backwashed; that is, the flow through the filters is reversed in order to decrease clogging and pressure drop in the filters. The filters are operated in parallel to allow for frequent backwashing, and they may be used immediately after the backwashing procedure is completed.

The large-scale filtration methods are very successful at reducing pathogen concentrations in water. The filtration methods discussed above, however, may be disrupted by clogging and operational upset, and they do not remove viruses. Disinfection processes are the most effective and reliable method to reduce pathogens in drinking water. Disinfection processes are capable of destroying bacteria, viruses, and amoebic cysts that the three types of pathogenic organisms that may be problematic in water. Commonly used chemical disinfection agents used to treat water include free chlorine (OCl^-), combined chlorine (HOCl and OCl^-), ozone (O_3), chlorine dioxide (ClO_2), and ultraviolet irradiation. The effectiveness of the disinfection process is dependent on the pathogen's sensitivity to the particular disinfectant, the concentration of the disinfectant, the time of contact, and the possible presence of other substances in the water that could interfere with the disinfectant. In rural areas, chlorine is the most common and easily available disinfectant. It is recommended that the residual chlorine concentration in the water be 0.5 mg/L; a minimum contact time of 30 minutes is needed to ensure adequate disinfection.

Drinking water is usually stored at a high elevation, either on a rooftop, a drinking water tank, or an elevated structure, to ensure the availability of water in the event of a power outage. Fluoride may be added to drinking water to help prevent tooth decay if naturally occurring fluoride levels in the water are low. (In some cases, fluoride concentrations may be too high and may require fluorine to be removed from the water.) Finally, the water must enter a distribution system that remains under a constant positive pressure to prevent potentially contaminated water in the soil from seeping into the distribution lines.

4.5.2 Wastewater Treatment Technologies

Water contaminated by human waste, or sewage, increases the concentration of pathogens, organic waste, and nutrients in water. The negative effects of sewage were accepted long ago. Sewers were commonly used in ancient Roman cities. However, it was not until the outbreak of diseases in the 1800s and the scientific association of disease to germ theory that the collection of wastewater became a public policy concern. The treatment of wastewater is a relatively new practice that resulted from public policies such as those shown in Figure 4.54 for the United States.

A typical wastewater treatment process must be capable of operating 365 days a year, 24 hours a day. **Publicly owned treatment works (POTWs)** may be designed to remove or reduce the concentration of organic matter, suspended solids, nutrients, and bacteria in municipal wastewater generated from homes and businesses. Industrial wastewater treatment facilities may be designed to remove only one or more targeted contaminants from the wastewater produced on the industrial site. The industrial wastewater typically must meet legal permit requirements before the water can be discharged to the municipal sewer or a receiving water body.

Wastewater treatment processes are generally classified into primary, secondary, and tertiary treatment processes (see Table 4.18 and Figure 4.54) Large objects and

TABLE 4.18 Typical municipal wastewater treatment process classifications

PRELIMINARY	GROSS SOLIDS ARE REMOVED TO PREVENT EQUIPMENT DAMAGE
Primary	Removes a portion of the suspended solids and organic matter
Advanced primary	Includes chemical addition or filtration for enhanced solids and organic matter removal
Secondary	Uses biological and chemical processes for removal of biodegradable organic matter and suspended solids
Secondary with nutrient removal	Includes removal processes to reduce nitrogen and phosphorus
Tertiary	Removes residual suspended solids usually by filtration or microscreens. Usually includes disinfection and possible nutrient removal
Advanced	Removes dissolved and suspended materials when required for water reuse

large solid particles are removed in preliminary treatment to keep these objects from clogging pipes and damaging pumps in the treatment works. A majority of the suspended solids in the incoming wastewater and about a third of the organic BOD are removed in the primary treatment phase through settling. Biological degradation of the organic waste and removal of the microorganisms and most of the remaining solids occur in the secondary treatment stage. Secondary treatment often consists of a biological process followed by coagulation, flocculation, and settling of the biomass. Tertiary or advanced treatment processes may include nutrient removal, disinfection, or other specialized processes, depending on the constituents in the wastewater.

Primary wastewater treatment processes include screening, flow equalization, coagulation, flocculation, and clarification, which are similar to the operations used in drinking water treatment. Secondary treatment processes are traditionally designed for cost-effective organic waste removal by utilizing microorganisms to degrade the organic waste and gravity to remove the microorganisms.

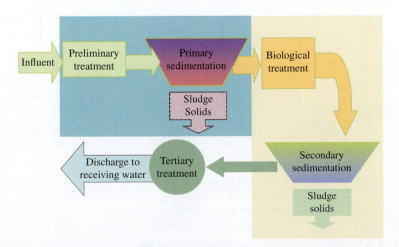

FIGURE 4.54 Typical treatment levels found in POTWs.

Source: Bradley Striebig.

Municipal wastewater treatment facilities receive varied amounts of wastewater with various concentrations of constituents. Wastewater treatment plants are designed to reduce total suspended solids, organic wastes, nutrients, and pathogens in the water. The collection and treatment of the wastewater may be most efficiently treated collectively at publicly owned plants. However, in rural areas and some developing nations, the infrastructure does not exist to provide proper treatment of domestic wastewaters. In this case, numerous decentralized processes may be applicable. One effective method is to create small collective activated sludge or suspended growth treatment systems for a small community. Septic systems, sand filters, constructed wetlands, and composting toilets are other effective processes used to treat domestic wastewater.

Many water supplies are facing increasing demands on a limited quantity of water. Consequently, the reuse of treated wastewater is increasingly being considered for urban community landscaping use, industrial purposes, agriculture, groundwater recharge, and augmentation of potable supplies. An example of how water might be reused to decrease demand on water sources is presented in Figure 4.55. Properly treated water may supplement existing water supplies in a wide variety of industrial, landscaping, and agricultural uses. Several countries whose water resources do not meet current demand have already adopted water reclamation and reuse strategies.

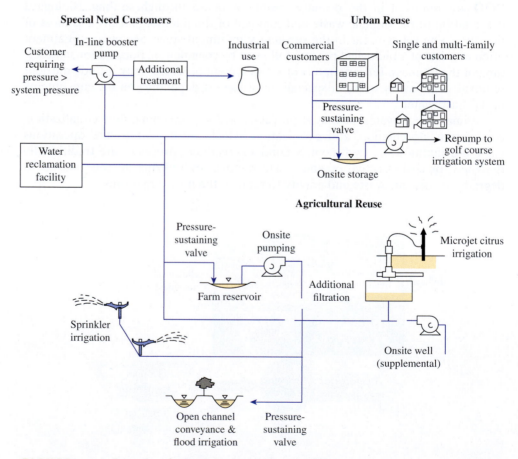

FIGURE 4.55 Examples of potential water reuse scenarios.

Source: US EPA (2004) Guidelines for Water Reuse. U. S. Environmental Protection Agency. Office of Wastewater Management, Office of Water, Washington, DC, USA.

4.6 Summary

Access to an appropriate quantity of water was discussed in detail in Chapter 3. In this chapter, we have investigated issues related to water quality. Water-related microorganisms in water mostly cause diseases, which are still endemic in large areas of the world. Microbial contamination of water is often caused by improper sanitation and lack of adequate wastewater treatment. Improper treatment of human and animal wastes also leads to environmental impacts from elevated levels of organic material and nutrients in water. Unnatural increases in organic material and nutrients decrease the available dissolved oxygen in water, negatively impacting many aquatic species in freshwater and marine ecosystems. Environmental engineers use mathematical models to estimate the causes, effects, and potential remediation strategies associated with water pollution. The degree of water quality required for drinking water differs substantially from the quality of water required for irrigation and some industrial uses. Water quality and water quantity are related in determining appropriate water reuse strategies that may be able to offset increasing water demand and greater global water scarcity. Methods to assess water reuse strategies are discussed in more detail in Chapters 7, 8, and 9.

References

American Public Health Association (APHA). (1995). *Standard Methods for the Examination of Water and Wastewater*, 19th ed. American Public Health Association/American Water Works Association/Water Environment Federation, Washington, D.C.

Baca, R.G., Waddel, W.W., Cole, C.R., Brandsletter, A., and Caerlock, D.B. (1973). EXPLORE-I: A river-basin water quality model. Battelle Laboratories, Richland, WA.

Bacca, R.G. and Arnett, R.C. (1976). A limnological model for eutrophic lakes and impoundments. Battelle, Inc., Pacific Northwest Laboratories, Richland, WA.

Bartram, J., Lewis, K., Lenton, R., and Wright, A. (2005). "Focusing on improved water and sanitation." *The Lancet* 365(9461):810–812.

Bauer, D.P., Rathbun, R.E. and Lowham, H.W. (1979). Travel time, unit-concentration, longitudinal dispersion, and reaeration characteristics of upstream reaches of the Yampa and Snake Rivers, Colorado and Wyoming. USGS Water Resources Investigation. 78–122.

Bedford, K.W., Sylees, R.M. and Libicki, C. (1983). "Dynamic advective water quality models for rivers." *ASCE Jour. Env. Engr.* 109:535–554.

Bowie, G.L, Mills, W.B., Porcella, D.B., Campbell, C.L., Pagenkopf, J.R., Rupp, G.L., et al. (1985). *Rates, Constants, and Kinetics Formulations in Surface Water Quality Modeling*, 2nd ed. Environmental Research Laboratory. Office of Research and Development, U.S. Environmental Protection Agency, Athens, GA.

Chapman, D. (1996). *A Guide to Use of Biota, Sediments and Water in Environmental Monitoring*, 2nd ed. Chapter 7—Lakes. UNESCO/WHO/UNEP.5.

Chen, C.W. (1970). "Concepts and utilities of ecological mode." *ASCE Journ. San. Eng. Div.* 96:SA5.

Chesapeake Bay Foundation. (2004). Manure's Impact on Rivers, Streams and the Chesapeake Bay: Keeping Manure Out of the Water. Annapolis, MD.

Churchill, M.A., Elmore, H.L. and Buckingham, R.A. (1962). "The prediction of stream reaeration rates." *ASCE, Journal Sanitary Engineering Division*. 88(SA4):1–46.

Clausen, T., Brown, J., Suntura, O., and Collin, S. (2004). "Safe household water treatment and storage using ceramic drip filters: A randomized controlled trial in Bolivia." *Water Science and Technology* **50**(1):111–115.

Corcoran, E., Nellemann, C., Baker, E., Bos, R., Osborn, D., and Savelli, H. (Eds.). (2010). *Sick water? The central role of wastewater management in sustainable development*. A Rapid Response Assessment. United Nations Environment Programme, UN-HABITAT, GRID-Arendal. *www.grida.no*. Printed by Birkeland Trykkeri AS, Norway.

Covar, A.P. (1976). "Selecting the proper reaeration coefficient for use in water quality models." Presented at the U.S. EPA Conference on Environmental Simulation and Modeling. April 19–22, Cincinnati, OH.

Di Toro, D.M. and Connolly, J.P. (1980). Mathematical models of water quality in large lakes. Part 2: Lake Erie. U.S. Environmental Protection Agency, Deluth, MN. EPA-600/3-3-80-065.

Fahlin, C., and Valdez, H. (2002). Belize Filtron Feasibility Report. University of Colorado—Engineers Without Borders.

Fair, G.M., Geyer, J.C. and Okun, D.A. (1968). *Water and Wastewater Engineering*. John Wiley & Sons, Inc. New York, NY.

Genet, L.A., Smith, D.J. and Sonnen, M.B. (1974). Computer program documentation for the dynamic estuary model. U.S. Environmental Protection Agency, Systems Development Branch, Washington, D.C.

Gleick, P. H. (2002). *Dirty Water: Estimated Deaths from Water-related Diseases 2000–2020*. Pacific Institute for Studies in Development, Environment and Security, Research Report. *www.pacinst.org*

Goldman, C. R., and Horne, A. J. (1983). *Limnology*. New York: McGraw-Hill.

Harleman, D.R.F., Dailey J.E., Thatcher, M.L., Najarian, T.O., Brocard, D.N. and Ferrara, R.A. (1977). User's manual for the M.I.T. transient water quality model network model – Including nitrogen-cycle dynamics for rivers and estuaries. R.M. Parsons Laboratory for Water Resources and Hydrodynamics, Massachusetts Institute of Technology, Cambridge, MA. For U.S. Environemntal Protection Agency, Corvallis, OR. EPA-600/3-77-010.

Heidecke, C. (2006). Development and Evaluation of a Regional Water Poverty Index for Benin. International Food Policy Research Institute. Environment and Production Division, Washington, D.C.

Imhoff, J.C., Kittle, J.L. Jr., Donigian, A.S. Jr. and Johanson, R.C. (1981). Users Manual for Hydrological Simulation Program – Fortran (HSPF). Contract 68-03-2895. U.S. Environmental Protection Agency, Athens, GA.

Kitawaki, H. (2002). "Common problems in water supply and sanitation in developing countries." *International Review for Environmental Strategies* 3:264.

Komenic, V., Ahlers, R., and van der Zaag, P. (2009). "Assessing the usefulness of the water poverty index by applying it to a special case: Can one be water poor with high levels of access?" *Physics and Chemistry of the Earth*, Parts A/B/C. 34(4–5):219–224.

Lantagne, D., Klarman, M., Mayer, A., Preston, K., Napotnik, J., and Jellison, K. (2010). "Effect of production variables on microbial removal in locally-produced ceramic filters for household water treatment." *International Journal of Environmental Health Research* 20(3):171–187.

Lawrence, P., Meigh, J., and Sullivan, C. (2002). The Water Poverty Index: An International Comparison. Keel Economics Research Papers. Department of Economics. Keele University, Keele, Staffordshire, UK.

Masters, G., and Ela, W.P. (2007). *Introduction to Environmental Engineering and Science*, 3rd ed. Prentice Hall. Upper Saddle River, NJ.

Medina, M.A., Jr. (1979). Level II – Receiving water quality modeling for urban stormwater management. U.S. Environmental Protection Agency. Municipal Environmental research Laboratory, Cincinnati, OH. EPA-600/2-79-100.

Metropolitan Washington Council of Governments. (1982). Application of HSPF to Seneca Creek watershed, Washington, D.C.

O'Conner, D.J. and Dobbins, W.E. (1958). "Mechanisms of reaeration in natural streams." *American Society of Civil Engineering Proceedings, Transactions*. 123(2934):641-684.

Owens, M., Edwards, R.W. and Gibbs, J.W. (1964). "Some reaeration studies in streams." *Int. J. Air Wat. Poll.* 8:469-486.

Reckhow, H. and Chapra, S. C. (1983). *Engineering Approaches for Lake Management. Vol 1. Data Analysis & Empirical Modeling*. Butterworth-Heinemann. London, UK.

Rich, L.G. (1973) *Environmental Systems Engineering*. McGraw Hill, New York, NY.

Roesner, L.A., Giguere, P.A. and Evenson, D.E. (1981). User's Manual for Stream Quality Model (QUAL-II). U.S. Environmental Protection Agency, Environmental Research Laboratory, Athens, GA. EPA-600/a-81-015.

Schroepfer, G.J., Robins, M.L., and Susag, R.H. (1960). "A reappraisal of deoxygenation rates of raw sewage, effluents, and receiving waters." *Water Pollution Control Federation Jour.*, 32(11):1212-1231.

Smith, D.J. (1978). WQRRS, Generalized Computer Program for River-Reservoir Systems. U.S. Army Corps of Engineers, Hydrologic Engineering Center (HEC), Davis, CA. User's Manual 401-100, 100A, 210pp.

Streeter, H.W. and Phelps, E. B. (1925). A study of the pollution and natural purification of the Ohio River, III. Factors concerned in the phenomena of oxidation and reaeration. United States Public Health Service. Public Health Bulletin No. 146.

Striebig, B., Atwood, S., Johnson, B., Lemkau, B., Shamrell, J., Spuler, P., et al. (2007). "Activated carbon amended ceramic drinking water filters for Benin." *Journal of Engineering for Sustainable Development* 2(1):1–12.

Striebig, B., Gieber, T., Cohen, B., Norwood, S., Godfrey, N., and Elliot, W. (2010). "Implementation of a Ceramic Water Filter Manufacturing Process at the Songhai Center: A Case Study in Benin." *International Water Association's World Water Congress*, Montreal.

Sullivan, C. (2002). "Calculating a Water Poverty Index." *World Development* 30(7): 1195–1210.

Tchobanoglous, G., Burton, F.L., and Stensel, H.D. (2003). *Wastewater Engineering Treatment and Reuse*, 4th ed. McGraw-Hill, New York:, NY.

Thomann, R.V. and Fitzpatrick, J.J. (1982). Calibration and verification of a mathematical model of eutrophication of the Potomac Estuary. Department of Environmental Services, Government of the District of Columbia.

UN. (2000). Millennium Declaration, UN A/Res/55/2.

UNEP. (2010). Clearing the Waters: A Focus on Water Quality Solutions. 2010. UNEP.

UNICEF. (2004). The Official Summary of the State of the World's Children. United Nations Children's Fund (UNICEF) New York.

UNICEF. (2007). The State of The World's Children. United Nations Children's Fund (UNICEF) New York.

UNICEF. (2008). UNICEF Handbook on Water Quality. United Nations Children's Fund (UNICEF) New York.

U.S. EPA. (2000). Progress in Water Quality: An Evaluation of the National Investment in Municipal Wastewater Treatment. U. S. Environmental Protection Agency, Office of Water, Washington, D.C.

U.S. EPA. (2004). Guidelines for Water Reuse. U. S. Environmental Protection Agency. Office of Wastewater Management, Office of Water, Washington, D.C.

U.S. EPA. (2004). Primer for Municipal Treatment Systems. U.S. Environmental Protection Agency, Office of Wastewater Management, Office of Water, Washington, D.C.

U.S. EPA. (2010). Chesapeake Bay Total Maximum Daily Load for Nitrogen, Phosphorus and Sediment. U.S. Environmental Protection Agency. Region 3. Water Protection Division. Philadelphia, PA.

U.S. EPA. (2012). *Climate Change Indicators in the United States,* 2012, 2nd ed. United States Environmental Protection Agency, Washington, D.C.

van Halem, D., van der Lann, H., Heijman, S. G. J., van Dijk, J. C., and Amy, G. L. (2009). "Assessing the sustainability of the silver-impregnated ceramic pot filter for low-cost household drinking water treatment." *Physics and Chemistry of the Earth* 34(2009):36–42.

Velz, C.J. (1984). *Applied Stream Sanitation.* John Wiley & Sons, Inc., New York, NY.

Wicks, C., Jasin, D., and Longstaff, B. (2007). "Breath of Life: Dissolved Oxygen in Chesapeake Bay." May 27, 2007 Newsletter. Chesapeake Bay Foundation, Annapolis, MD.

World Health Organization (WHO). (1993). *Guidelines for drinking-water quality,* 2nd ed. Volume 1, Recommendations. Geneva, Switzerland.

World Health Organization (WHO). (2012). GLASS 2012 Report: UN-Water Global Analysis and Assessment of Sanitation and Drinking Water: The Challenge of Extending and Sustaining Services. World Health Organization. Geneva, Switzerland.

World Water Assessment Programme. (2009). *The United Nations World Water Development Report 3: Water in a Changing World.* Paris: UNESCO, and London: Earthscan.

Key Concepts

Water-related disease
Drinking water
Sanitation
Centralized treatment
Decentralized treatment
Point-of-use treatment
Prokaryotes
Eukaryotes
Pathogen
Dissolved oxygen
Deoxygenation rate
Re-aeration rate
Saturated dissolved oxygen
Oxygen deficit
Aerobic organisms
Anaerobic organisms

Biochemical oxygen demand
Carbonaceous oxygen demand
Nitrogenous oxygen demand
Theoretical oxygen demand
Nutrients
Eutrophication
Hypoxia
Critical point
Aphotic zone
Photic zone
CMFR or CSTR
Epilimnion
Hypolimnion
Thermocline
Trophic level

Active Learning Exercises

ACTIVE LEARNING EXERCISE 4.1: **Analysis of ceramic water filter**

Describe what characteristics of the ceramic water filters that are being produced by the Songhai Centre in Benin are sustainable and why. What characteristics or aspects of the ceramic filter design are least sustainable? Why? Do this for each of the following aspects of sustainability:

a) Technology

b) Economic

c) Social

d) Environmental

ACTIVE LEARNING EXERCISE 4.2: **Sand filter in a bucket**

The biosand filter is a small point-of-use modification of a slow sand filter. The point-of-use filter is designed to treat water in the home or at the point of use. The general design of the biosand filter is illustrated in Figure 4.56.

1. An even smaller version of the biosand filter can be used to demonstrate the principles of how the filter is designed, fabricated, operated, and maintained. The materials list to build your own sand filter is shown in Table 4.19. Based on Figure 4.56 and the materials list, design a sand filter on paper. List each part and how they will fit together. Create engineering drawings of your design. Your drawing should include the following elements.
   a) A profile view that shows:
      i)   Dimensions and location of gravel
      ii) Dimensions and location of sand
      iii) Dimensions and location of PVC piping
      iv) Dimensions and location of the nozzle
      v)   Call out for any fittings required
   b) A top view
   c) A side view
2. Meet in your predetermined groups. Compare and discuss your individual drawings. Note the differences. Create as a group another sketch with a similar level of detail. This drawing should include some approximate dimensions, but it need not be exactly to scale.
3. Construct your filter from the available supplies. Submit in hard copy an AS BUILT drawing of your filter. The drawing should comply with professional drafting rules and standards:
   a) Professional appearance (either hand drawn or drawn using computer software)
   b) Border
   c) Title block

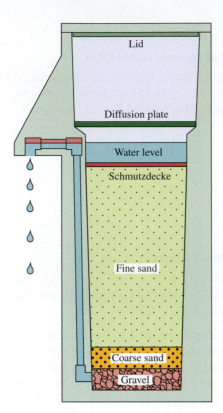

FIGURE 4.56 Schematic of a point-of-use biosand filter.

Source: Based on http://commons.wikimedia.org/wiki /File%3ABiosandFilter_Section.svg

d) Since this is a relatively simple machine, you need only show a cross section of your design (side view)

e) Use appropriate line weights and types (i.e., solid or hidden lines)

f) Use callouts to label off-the-shelf parts; these are components that can be purchased and used unmodified, such as spigots, 5-gallon paint buckets, and PVC fittings

g) Properly dimension all other components such as the height of the sand or gravel, height of the spigot, and PVC dimensions

4. Collect surface water from a local pond, puddle, or other available source. Test this water prior to filtering using the Hach MPN Pathoscreen methodology in Appendix H to estimate the number of organisms in 100 mml of the water.

a) Report the MPN for the unfiltered pond water.

b) Using the surface water and the Hach MPN Pathoscreen test procedure, evaluate how your filter works. You will need enough water to "rinse" your filter three times, before collecting your filtered water sample.

c) Compare your results to those of the rest of the class. Report the range of values reported in your section.

5. Using a spreadsheet and the materials list in Table 4.19, calculate how much your group's filter materials cost.

a) What parts are most expensive?

b) Where might you be able to achieve the greatest reduction in cost?

6. List the characteristics and qualities of the sand water filters that you believe are most sustainable. Do this for each of the following aspects of sustainability.

TABLE 4.19 Materials list for a bucket sand filter

MATERIAL	AMOUNT	COST ($ U.S.)	AMOUNT PER FILTER	COST PER FILTER
Sand	0.5 cu ft	3.38		
Gravel	0.5 cu ft	3.48		
Bucket	1	2.54		
White spigot	1	1.76		
Black spigot with gaskets	1	2.81		
PVC pipe	10 ft	1.68		
PVC cement	1 pack	7.51		
Epoxy	1 pack	5.98		
Male adapter	1	0.23		
Elbow	1	0.26		
Threaded elbow	1	0.57		
End cap	1	0.32		
O rings	10 pack	1.97		
Other				

   a) Technology
   b) Economic
   c) Social
   d) Environmental
7. List the characteristics and qualities of the sand water filters that you believe are least sustainable. Do this for each of the following aspects of sustainability.
   a) Technology
   b) Economic
   c) Social
   d) Environmental
8. Describe and show how you would improve the filter design. The drawing should comply with professional drafting rules and standards:
   a) Professional appearance (either hand drawn or computerized)
   b) Border
   c) Title block
   d) Since this is a relatively simple machine, you need only show a cross section of your design (or side view)
   e) Use appropriate line weights and types (i.e., solid or hidden lines)
   f) Use callouts to label off-the-shelf parts; these are components that can be purchased and used unmodified, such as spigots, 5 gallon paint buckets, and PVC fittings
   g) Properly dimension all other components such as the height of the sand or gravel, height of the spigot, and PVC dimensions

Problems

4-1 Describe the human use, human health, or environmental concerns associated with the following water quality parameters:
a. Total suspended solids
b. BOD
c. Nutrients
d. Pathogens

4-2 Describe the differences between prokaryotes and eukaryotes.

4-3 A poorly designed latrine is located upstream of a surface drinking water source. Two cubic meters of stormwater contaminated by the latrine and containing 10^{10} MPN/100 ml of salmonella flow into a shallow pool for collecting drinking water. The total volume of the drinking water pool after the contaminated water flows into it is 5 m^3. If the typical infectious dose for salmonella in this community is 100 organisms, how much water would someone have to drink to become ill due to the salmonella contamination?

4-4 A stream flowing through a rural agricultural region in Central America contains 10^{-1} MPN/100 ml of *Giardia lamblia* cysts. If a backpacking tourist drinks 5 L of water per day, how many days before the hiker is likely to become ill due to the salmonella contamination? Assume the typical infectious dose for *Giardia* in this community is 10 organisms.

4-5 What is the saturated dissolved oxygen concentration for water at sea level and 18°C?

4-6 What is the saturated dissolved oxygen concentration for water at 2,000 m and 15°C?

4-7 What is the percent saturation of water at sea level and 25°C containing 9.8 mg/L of dissolved oxygen? Would you expect the dissolved oxygen level to increase or decrease if the system were to move toward equilibrium?

4-8 What is the dissolved oxygen deficit for water containing 5 mg/L of dissolved oxygen at sea level and 18°C?

4-9 What is the dissolved oxygen deficit for water containing 5 mg/L of dissolved oxygen at 2,000 m and 15°C?

4-10 What is the theoretical carbonaceous oxygen demand for water containing 25 mg/L of propyl alcohol, C_3H_7OH?

4-11 What is the theoretical carbonaceous oxygen demand for water containing 100 mg/L of acetic acid, CH_3COOH?

4-12 Bacterial cells are often represented by the chemical formula $C_5H_7NO_2$. Compute the theoretical carbonaceous oxygen demand (in mg/L) in a 1 molar solution of cells in water.

4-13 A 5 ml sample of municipal sewage is used for a BOD_5 analysis, and 4.7 mg/L of oxygen is consumed after the sample is incubated for five days in the dark at 20°C. What is the BOD_5 of the sewage?

4-14 What is the BOD_5 of the domestic wastewater sample, given the following data collected from the laboratory.

WASTEWATER SAMPLE	VOLUME OF WASTEWATER (ml)	INITIAL DO OF THE WASTEWATER (mg/L)	VOLUME OF DILUTION WATER (ml)	INITIAL DO OF THE DILUTION WATER (mg/L)	FINAL DO (mg/L)
Sample A	2.0	3.0	298.0	10.5	6.2
Sample B	6.0	4.9	294.0	10.5	7.6
Sample C	10.0	3.8	290.0	10.5	8.2

4-15 What is the BOD_5 of the lake water sample, given the following data collected from the laboratory.

	VOLUME OF WASTEWATER (ml)	INITIAL DO OF THE WASTEWATER (mg/L)	VOLUME OF DILUTION WATER (ml)	INITIAL DO OF THE DILUTION WATER (mg/L)	FINAL DO (mg/L)
Sample A	100	6.0	200	9.2	5.4
Sample B	200	7.0	100	9.2	4.4
Sample C	300	8.1	0	9.2	6.1

4-16 Operators at the local wastewater treatment plant have reported the following data to determine how well a wastewater treatment plant is operating.
   a. What percentage of the BOD does the treatment plant remove?
   b. Is the plant in compliance with its permit if it is required to remove 85% of the BOD?

	VOLUME OF WASTEWATER (ml)	INITIAL DO OF THE WASTEWATER (mg/L)	VOLUME OF DILUTION WATER (ml)	INITIAL DO OF THE DILUTION WATER (mg/L)	FINAL DO (mg/L)
Influent to treatment facility	5	3.0	295	8.5	6.2
Effluent from treatment facility	25	8.2	275	8.5	5.9

4-17 A former scientist and opponent of dam removal in the Northwest has recently reversed his stance and now believes dam removal is the only way to help endangered salmon. He has stated that his position has changed because of global warming. If water temperatures increase from 20°C to 25°C in the Columbia River:
   a. What is the change in the saturated dissolved oxygen levels (in mg/L) that would be expected in the river?
   b. In what way does impounding (damming) a river reduce the dissolved oxygen? Please state the specific variable that might be affected.

4-18 A mixture of river water and effluent from a municipal wastewater treatment plant is completely mixed near the point of discharge. The dissolved oxygen concentration in the river at the mixing point is 5.6 mg/L, and the ultimate BOD of the mixed water is 24 mg/L. The temperature of the river is 21°C. The deoxygenation constant for the river is $k_d = 0.06$/day.
   a. Estimate the re-aeration coefficient assuming the river speed is 0.25 m/s and the average depth is 3 m.
   b. Find the critical time downstream at which the minimum DO occurs.
   c. Find the minimum DO downstream.

4-19 The flow rate of the Madison River is about 500 m³/s. The Madison River has an ultimate BOD value of 5.9 mg/L and DO of 6.2 mg/L. The Madison Wastewater Treatment Plant (WWTP) discharges 10 million gallons per day (MGD) of treated wastewater with an ultimate BOD of 20 mg/L and DO equal to 8.0 mg/L.
   a. What is the BOD (mg/L) immediately downstream of the discharge?
   b. What is the DO (mg/L) immediately downstream of the discharge?
   c. If the temperature for the mixed water at the point of discharge is 20°C, what is the initial oxygen deficit in mg/L?
   d. If the deoxygenation constant is 0.8 day⁻¹ and the re-aeration rate is 1.4 day⁻¹, what is the critical time [days]?
   e. What would the minimum dissolved oxygen concentration in the river be in mg/L?

4-20 The flow rate of the Spokane River is about m³/s. The Spokane River has an ultimate BOD value of 5 mg/L and DO of 6 mg/L. The Spokane Wastewater Treatment Plant (WWTP) discharges 50 million gallons per day (MGD) of treated wastewater, with an ultimate BOD of 30 mg/L and DO equal to 8 mg/L.
   a. What is the BOD [mg/L] immediately downstream of the discharge?
   b. What is the DO [mg/L] immediately downstream of the discharge?
   c. If the temperature for the mixed water at the point of discharge is 20°C, what is the initial oxygen deficit in mg/L?
   d. If the deoxygenation constant is 0.4 day⁻¹ and the re-aeration rate is 0.8 day⁻¹, what is the critical time [days]?
   e. What would the minimum dissolved oxygen concentration in the Spokane River be in mg/L?
   f. Plot the oxygen-sag curve for the river.

4-21 Sketch and briefly describe why thermal stratification occurs in a lake in Scotland over four seasons.

4-22 Bald Eagle Creek flows into Bald Eagle reservoir, which acts as a completely mixed reactor. Bald Eagle Creek has a flow rate of 0.20 m³/s and a dissolved oxygen concentration of 4.75 mg/L. Bald Eagle reservoir has a volume of 10⁴ m³. Oxygen is removed in the lake through biological processes, and the reaction rate, k_d, is 0.06 day⁻¹. Evaporation removes water from the surface of the lake at a rate of 0.002 m³/s. Determine the effluent flow rate and dissolved oxygen concentration for this completely mixed lake.

4-23 You have been assigned to conduct a site assessment for a lake in Montana. There is a new biofuels plant operating and discharging into the stream that feeds the lake and provides water for agricultural production of cattle and switchgrass. You observed high levels of algae floating on the surface of the lake, and the water was highly turbid. You also measured the DO at 1 m to be 5 mg/L, and at a depth of 20 m the DO was 1.0 mg/L.

    a. How would you describe the productivity level of the lake? Explain your reasoning.

    b. Similar lakes have had threatened Yellowstone Cutthroat Trout populations. But this lake does not. What changes would you suggest be made in managing the lake's water quality prior to reintroducing trout to the lake?

4-24 The Ganges River, shown in Figure 4.57, has an annual average flow rate of 12,105 m^3/s. The average temperature of the Ganges River was reported to be 24.6°C. The Ganges has become contaminated with raw sewage and industrial waste, so that the dissolved oxygen in the river upstream of sewage discharges is 4.6 mg/L. Suppose that the raw sewage that is untreated flows into the Ganges at a rate of 460 m^3/s, the temperature is 25°C, and the dissolved oxygen in the sewage stream is completely depleted (i.e., DO = 0). How would you develop a plan for improving water quality in the Ganges River?

4-25 Refer to Problem 4-25 for the following questions.

    a. Assuming the river is well mixed immediately at the point of discharge of the sewage stream, what is the DO in mg/L immediately downstream of the discharge?

    b. What is the initial oxygen deficit in the mixed sewage and river water in mg/L?

    c. Suppose the following data were collected to estimate the five-day BOD in the river and sewage water:

	VOLUME OF SAMPLE (ml)	INITIAL DO OF THE SAMPLE (mg/l)	VOLUME OF DILUTION WATER (ml)	INITIAL DO OF THE DILUTION WATER (mg/l)	FINAL DO (mg/l)
River water	25	4.6	275	7.8	4.7
Sewage	2	0	298	7.6	5.9

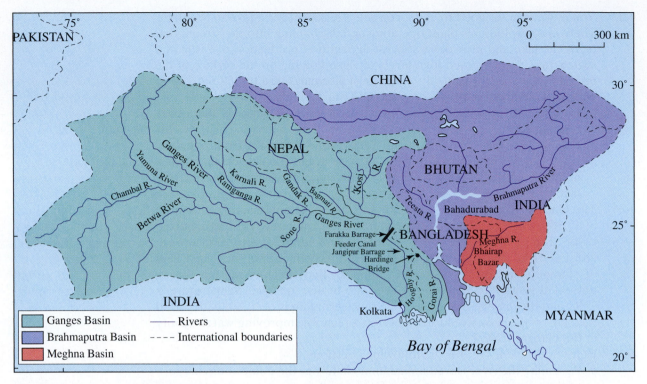

FIGURE 4.57 The Ganges River Basin.

Source: Based on Rahaman, M.M. (2006) The Ganges water conflict: A comparative analysis of the 1977 Agreement and the 1996 Treaty.

i)   Calculate the BOD of the river water.
ii) Calculate the BOD of the sewage stream.
iii) Calculate the ultimate BOD of the river, assuming $k_L = 0.20$ day^{-1}.
iv) Calculate the ultimate BOD of the sewage stream, assuming $k_L = 0.50$ day^{-1}.

d. Calculate the ultimate BOD of the mixed river and sewage stream.
e. Estimate the re-aeration constant assuming the mean depth of the river is 1.5 m and the mean velocity is 0.1 m/s.
f. Suppose the deoxygenation constant is 4.5*(k$_r$) day^{-1}. At what time downstream will the critical point occur?
g. What would the minimum dissolved oxygen concentration in the Ganges River be in mg/L?
h. Plot the oxygen sag curve for the river.
i. Will there be negative effects on the aquatic ecosystem?
j. What effects might there be on human health and recreation? Note that the Ganges is a source of drinking water and water for washing and also has a spiritual significance to the surrounding population.
k. If a dam were built to create a reservoir impounding part of the Ganges, what characteristic trophic state might you expect for the reservoir? Explain your answer.

l. Explain how you might design a series of water treatment facilities to treat the raw sewage. What processes would be required, and what positive impacts would treating the sewage have on water quality?

m. Suppose the upstream river water quality is improved so that the dissolved oxygen content of the river prior to receiving water from the sewage stream is equal to 90% of the saturated DO level. Suppose also that the sewage would be treated in a series of wastewater treatment facilities.

i) If the minimum allowable DO level were established to be 2.0 mg/L, what would the allowable ultimate BOD discharged from the waste-water treatment facilities be?

ii) What percent removal efficiency would be needed?

CHAPTER 5

Impacts on Air Quality

FIGURE 5.1 Air pollution over the city of Kathmandu, Nepal, from photochemical oxidation of nitrogen oxides and volatile organic compounds

Source: Jerome Lorieau Photography/Flickr Open/Getty Images

Our most basic common link is that we all inhabit this planet. We all breathe the same air. We all cherish our children's future. And we are all mortal.

—JOHN F. KENNEDY

You go into a community and they will vote 80 percent to 20 percent in favor of a tougher Clean Air Act, but if you ask them to devote 20 minutes a year to having their car emissions inspected, they will vote 80 to 20 against it. We are a long way in this country from taking individual responsibility for the environmental problem.

—WILLIAM D. RUCKELSHAUS, FORMER EPA ADMINISTRATOR, *NEW YORK TIMES*, NOVEMBER 30, 1988

GOALS

THE EDUCATIONAL GOALS OF THIS CHAPTER are to use basic principles of chemistry, physics, and math to define and solve engineering problems associated with air pollutant emissions, dispersion, and control. This chapter utilizes the fundamental physical and chemical models discussed in Chapter 2 to examine problems associated with air pollution emissions at local, regional, and global levels. Regulatory policies that limit air pollutants in the United States will be discussed as well as international regulations designed to protect the stratospheric ozone later. This chapter will describe the parameters and methods used to determine and predict air quality downwind from air pollution sources. Air quality constituents that influence human and ecosystem health in the environment will be examined. Mathematical models will be used to predict air pollutant emissions and transport and to identify potential health risks. Many of the fundamental principles discussed in this chapter are also applicable to carbon dioxide emissions discussed in more detail in Chapter 6. A basic understanding of air pollution pro-vides a foundation for an understanding of energy issues, industrial ecology, Life Cycle Analysis and the rationale for more efficient building practices discussed in Chapters 8 to 11 of the book.

OBJECTIVES

At the conclusion of this chapter, you should be able to:

- Describe the economic, environmental, and societal implications of air quality in high-income nations and low- to medium-income nations.

- Describe the chemical precursors and transformation that produce urban smog.

- Estimate emissions of air pollutants using emission factors.

- Estimate the dispersion of air pollutants downwind from a source.

- Describe the processes that produce air pollution from combustion processes.

- Describe the global effects of air pollutants, especially their environmental consequences.

5.1 Introduction

The average adult typically inhales 3 to 4 liters of air in each breath he or she takes. The air quality in a region is dependent on local and regional factors. The proximity to power plants, highways, industry, and mountains significantly influences local air quality. Global air circulation patterns, latitude, and air temperature profoundly influence regional air quality. Scientists and engineers use the principles of chemistry, math, and physics to calculate the amount of pollutants emitted from a process, how those pollutants may be transported to expose a population downwind from the source, what health effects may be incurred by the exposed population, and what technologies may be used to reduce emissions and subsequent exposure.

The composition of clean air is described in Table 2.1. Clean air is made up primarily of nitrogen (78.08%), oxygen (20.94%), and argon (0.93%). The remaining components of the air that are of concern as environmental pollutants are present in quantities less than 1% in the atmosphere. An air pollutant may be thought of as any chemical or substance that is present in the atmosphere in quantities that negatively impact human health, the environment, and the economy or unreasonably interfere with the enjoyment of life, property, or recreation.

Many air pollutants are emitted as a result of the process involved in producing energy for heat, work, or electricity. Urbanization patterns, climate change, population growth, and increasing energy have decreased air quality in many locations. Premature death has been attributed to poor air quality in high- to low-income countries in rural and in urban areas. Over 49,000 deaths per year are attributed to second-hand cigarette smoke indoors (CDC, 2008). The World Health Organization (WHO) estimates that 2 million people per year die prematurely from illness attributed to use of household solid fuels. WHO also estimates that 50% of pneumonia deaths among children under the age of 5 are related to particulate matter (PM) inhaled from indoor pollutants. However, improvements in air quality and decreased health effects have also been documented when regulations and investment in air treatment processes have been made to improve regional air quality.

5.1.1 History

One of the first laws against air pollution came in 1300 when King Edward I decreed the death penalty for burning of coal. At least one execution for that offense is recorded. But economics triumphed over health considerations, and air pollution became an appalling problem in England.

—Glenn T. Seaborg, Atomic Energy Commission Chairman, Speech,

Argonne National Laboratory, 1969

Air pollution was documented as a problem in preindustrialized England, when burning "sea coal" or peat resulted in toxic indoor air conditions. Subsequently, King Edward I passed the first laws to prevent air pollution from burning this "dirty" fuel. Later, in 1661, pamphlets were printed and distributed suggesting ways to reduce air pollution. The industrial age increased the demand for burning fuels for energy, work, and electricity, which has increased air pollutant emissions from solid and fossil fuel combustion.

Air pollution continues to be a very significant human health threat. In 1948, industrial air pollutants that included sulfur dioxide, nitrogen dioxide, and fluorine caused approximately 20 human deaths and thousands of reported cases of respiratory distress due to atmospheric conditions that led to very high concentrations of a toxic smog in the town of Donora, Pennsylvania in the United States. A few years later, in 1952, the "great smog" enveloped parts of London. The smog was formed when unusually still air conditions and cold weather combined to form a toxic mixture of gases. While acute effects were limited during the five-day event, more recent research estimates that this event caused more than 10,000 premature deaths.

Acid rainfall was first observed from plants damaged by the rain in the northeastern United States in the mid-nineteenth century. Decades of acid rainfall in the northeastern United States led to the acidification of many lakes and streams, damaging aquatic ecosystems and also causing architectural damage. These and other air pollution-related events resulted in the passage of numerous environmental laws designed to protect or improve air quality.

5.1.2 Regulations

Regulations are laws used to attempt to protect the local or global environment. Regulations are, by political necessity, a balance between social and commercial interests. Laws may be enacted at multiple levels of government, as illustrated in the regulatory pyramid shown in Figure 5.2.

Regulation of air pollutants in the United States has evolved since 1955, as illustrated in Figure 5.3, when Public Law 159 authorized funding for the U.S. Public Health Service (the forerunner to the EPA) to initiate research into air pollution. The 1963 **Clean Air Act** was the first federal regulation in the United States to allow regulations for the control of air pollutants. The 1970 Clean Air Act (CAA) was the first comprehensive step by the federal government to limit emissions of air pollutants from both stationary and mobile sources. Facilities that did not comply with the CAA would be subject to fines for violating emissions limits. The 1970 CAA also established **National Ambient Air Quality Standards (NAAQS)**. The NAAQS are goals set to achieve reasonable air quality in all regions of the United States. The 1977 amendments to the CAA were targeted at improving air quality in regions where concentrations of pollutants in the air were found to be higher than the target goals of the NAAQS. The additional amendments to the 1970 CAA passed in 1990 expanded programs to bring regions into compliance with the NAAQS, set more stringent standards for new sources, and included provisions for stratospheric ozone protection.

The CAA required all regions in the United States to develop **Prevention of Significant Deterioration (PSD)** standards to maintain air quality if in compliance

FIGURE 5.2 The regulatory pyramid illustrates that facilities must comply with all local, state, and federal regulations for a facility.

Source: Bradley Striebig.

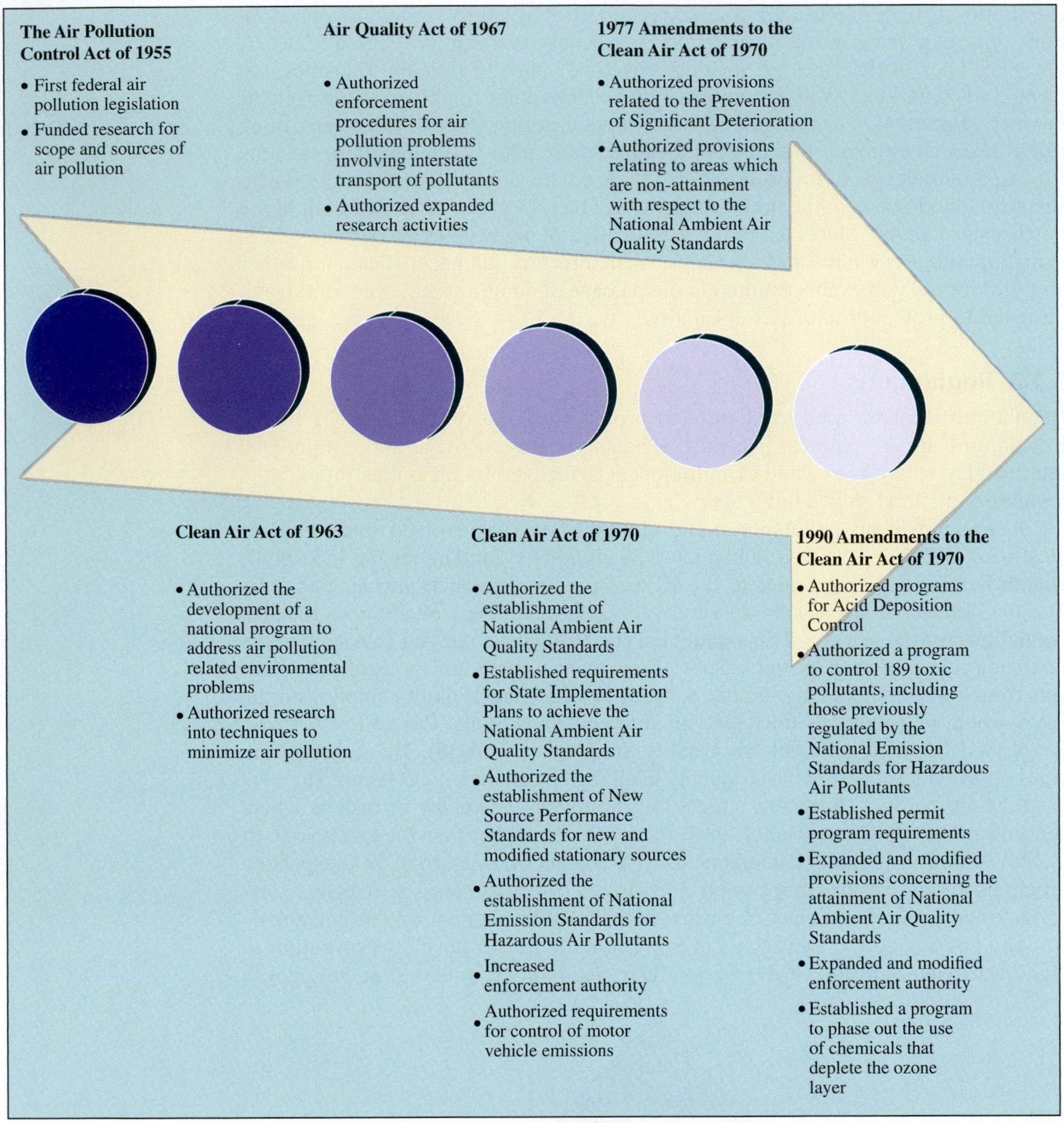

The Air Pollution Control Act of 1955

- First federal air pollution legislation
- Funded research for scope and sources of air pollution

Air Quality Act of 1967

- Authorized enforcement procedures for air pollution problems involving interstate transport of pollutants
- Authorized expanded research activities

1977 Amendments to the Clean Air Act of 1970

- Authorized provisions related to the Prevention of Significant Deterioration
- Authorized provisions relating to areas which are non-attainment with respect to the National Ambient Air Quality Standards

Clean Air Act of 1963

- Authorized the development of a national program to address air pollution related environmental problems
- Authorized research into techniques to minimize air pollution

Clean Air Act of 1970

- Authorized the establishment of National Ambient Air Quality Standards
- Established requirements for State Implementation Plans to achieve the National Ambient Air Quality Standards
- Authorized the establishment of New Source Performance Standards for new and modified stationary sources
- Authorized the establishment of National Emission Standards for Hazardous Air Pollutants
- Increased enforcement authority
- Authorized requirements for control of motor vehicle emissions

1990 Amendments to the Clean Air Act of 1970

- Authorized programs for Acid Deposition Control
- Authorized a program to control 189 toxic pollutants, including those previously regulated by the National Emission Standards for Hazardous Air Pollutants
- Established permit program requirements
- Expanded and modified provisions concerning the attainment of National Ambient Air Quality Standards
- Expanded and modified enforcement authority
- Established a program to phase out the use of chemicals that deplete the ozone layer

FIGURE 5.3 Timeline of major air quality legislation in the United States.

Source: Bradley Striebig.

with the NAAQS or to develop **State Implementation Plans (SIPs)** to improve air quality if the NAAQS goals were exceeded in the region. The NAAQS are defined for the six primary pollutants shown in Table 5.1 that had demonstrable negative human health impacts. Secondary standards are required for air pollutants that may have potentially negative environmental effects, including plant damage, decreased visibility, and architectural damage. The actions required by the state regulatory

TABLE 5.1 U.S. EPA primary and secondary NAAQS standards for primary pollutants

POLLUTANT [FINAL RULE CITE]		PRIMARY/ SECONDARY	AVERAGING TIME	LEVEL	FORM
Carbon monoxide [76 FR 54294, Aug. 31, 2011]		Primary	8-hour	9 ppm	Not to be exceeded more than once per year
			1-hour	35 ppm	
Lead [73 FR 66964, Nov. 12, 2008]		Primary and secondary	Rolling 3-month average	0.15 μg/m$^{3(a)}$	Not to be exceeded
Nitrogen dioxide [75 FR 6474, Feb. 9, 2010] [61 FR 52852, Oct. 8, 1996]		Primary	1-hour	100 ppb	98th percentile averaged over 3 years
		Primary and secondary	Annual	53 ppb$^{(b)}$	Annual mean
Ozone [73 FR 16436, Mar. 27, 2008]		Primary and secondary	8-hour	0.075 ppm$^{(c)}$	Annual fourth highest daily maximum 8-hr concentration, averaged over 3 years
Particle pollution [71 FR 61144, Oct. 17, 2006]	PM$_{2.5}$	Primary and secondary	Annual	15 μg/m^3	Annual mean, averaged over 3 years
			24-hour	35 μg/m^3	98th percentile, averaged over 3 years
	PM$_{10}$	Primary and secondary	24-hour	150 μg/m^3	99th percentile of 1-hour daily maximum concentrations, averaged over 3 years
Sulfur dioxide [75 FR 35520, June 22, 2010] [38 FR 25678, Sept. 14, 1973]		Primary	1-hour	75 ppb$^{(d)}$	99th percentile of 1-hour daily maximum concentrations, averaged over 3 years
		Secondary	3-hour	0.5 ppm	Not to be exceeded more than once per year

[a] Final rule signed October 1, 2008. The 1978 lead standard (1.5 μg/m^3 as a quarterly average) remains in effect until one year after an area is designated for the 2008 standard, except that in areas designated nonattainment for the 1978 standard remains in effect until implementation plans to attain or maintain the 2008 standard are approved.

[b] The official level of the annual NO$_2$ standard is 0.053 ppm, equal to 53 ppb, which is shown here for the purpose of clearer comparison to the 1-hour standard.

[c] Final rule signed March 1, 2008. The 1997 ozone standard (0.08 ppm, annual fourth highest daily maximum 8-hour concentration, averaged over 3 years) and related implementation rules remain in place. In 1997, EPA revoked the 1-hour ozone standard (012 ppm, not to be exceeded more than once per year) in all areas, although some areas have continued obligations under that standard ("anti-backsliding"). The 1-hour ozone standard is attained when the expected number of days per calendar year with maximum hourly average concentrations above 0.12 ppm is less than or equal to 1.

[d] Final rule signed June 2, 2010. The 1971 annual and 24-hour SO$_2$ standards were revoked in that same rulemaking. However, these standards remain in effect until one year after an area is designated for the 2010 standard, except in areas designated nonattainment for the 1971 standards remain in effect until implementation plans to attain or maintain the 2010 standard are approved.

Source: USEPA (2012) National Ambient Air Quality Standards (NAAQS). Air and Radiation, United State Environmental Protection Agency. http://www.epa.gov/air/criteria.html

agency for compliance with NAAQS goals are summarized in Table 5.2. The CAA resulted in significant reductions of the primary pollutants shown in Figure 5.4. Despite improvements in air quality in some regions and regulations of new sources, nearly 124 million people lived in counties that did not comply with one or more of the NAAQS criteria, as shown in Figure 5.5.

TABLE 5.2 Definition of attainment regions for NAAQS

ATTAINMENT REGION	UNCLASSIFIABLE REGION	NONATTAINMENT REGION
Air quality is in compliance with NAAQS	Insufficient data to determine compliance with NAAQS	Air quality violates NAAQS
PSD applies	PSD applies	New source review applies

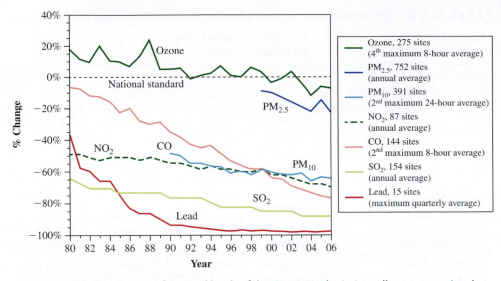

FIGURE 5.4 Comparison of national levels of the six principal priority pollutants to national ambient air quality standards, 1980–2006. National levels are averages across all sites with complete data for the time period.

Source: USEPA (2007) A Plain English Guide to the Clean Air Act. Office of Air Quality, Planning and Standards, United States Environmental Protection Agency. Research Triangle Park, NC, USA. EPA-456/K-07-001.

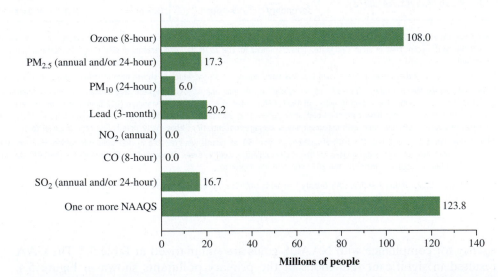

FIGURE 5.5 Number of people (in millions) living in counties with air quality concentrations above the level of the primary (health-based) National Ambient Air Quality Standards (NAAQS) in 2010.

Source: USEPA (2011) The Benefits and Costs of the Clean Air Act from 1990 to 2020: Summary Report. Office of Air and Radiation, United States Environmental Protection Agency. Research Triangle Park, NC, USA. EPA-454/R-12-001.

The Environmental Protection Agency has developed an **Air Quality Index (AQI)** to help communicate air quality trends to the public and to provide warnings to the public on days that a region is not in compliance with the NAAQS. The AQI relates daily air pollution concentrations for ozone, particle pollution, NO_2, CO, and SO_2 to health concerns for the general public. Each compound is given a numerical value, based on a scale, where 100 is equivalent to the NAAQS standard.

Air Quality Index (AQI) Values	Levels of Health Concern
0–50	Good
51–100	Moderate
101–150	Unhealthy for sensitive groups
151–200	Unhealthy
201–300	Very unhealthy
301–500	Hazardous

FIGURE 5.6 U.S. EPA Air Quality Index.

Source: USEPA (2009) Ozone and Your Health. Office of Air and Radiation, United States Environmental Protection Agency. Washington, DC.

An AQI score below 100 is considered acceptable, as illustrated in Figure 5.6, and a value greater than 100 is considered unhealthy, especially for sensitive populations, such as the elderly and people with asthma. Figure 5.7 shows the number of days on which the AQI exceeded 100 for 35 metropolitan cities in the United States.

$$AQI = \frac{C_{environment}}{C_{NAAQS}} \times 100 \tag{5.1}$$

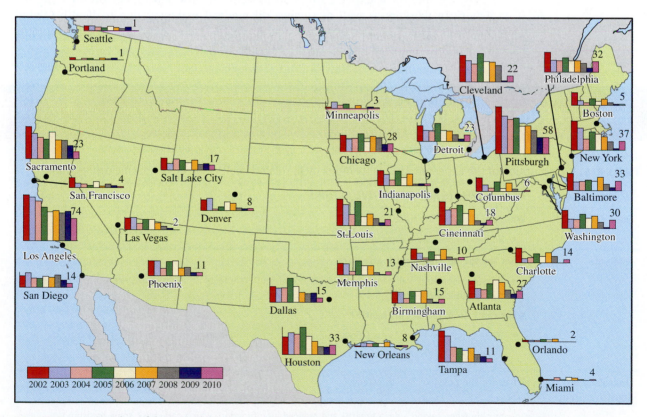

FIGURE 5.7 Number of days on which AQI values were greater than 100 during 2002–2012 in selected cities.

Source: USEPA (2011) The Benefits and Costs of the Clean Air Act from 1990 to 2020: Summary Report. Office of Air and Radiation, United States Environmental Protection Agency. Research Triangle Park, NC, USA.

EXAMPLE 5.1 | **Air Quality Index Values**

The ozone level in a city was measured at 86 ppb at 8:50 a.m. What AQI value should be reported on the 9:00 a.m. morning news broadcast for the city?

Solution:

$$AQI = \frac{C_{environment}}{C_{NAAQS}} \times 100 = \frac{86\ ppb_v\ \dfrac{1\ ppm_v}{1{,}000\ ppb_v}}{0.075\ ppm_v} \times 100 = 115$$

The AQI is level orange, between 101 and 150, and is unhealthy for sensitive groups.

The EPA performed a cost-benefit analysis of implementation of the 1990 CAA, evaluating the expected benefits of the CAA between 1990 and 2020. The findings show a significant reduction of over 230,000 early deaths associated with decreased exposure to air pollutants (Table 5.4). It is estimated that the direct benefits of the CAA will reach almost $2 trillion (U.S.) by the year 2020. Due to the high uncertainty associated with the benefits projection, a wide range in expected benefits is projected with a low estimate of benefits equal to three times the cost, the expected benefits equal to 30 times the cost, and the high estimate equal to 90 times the cost, and all benefits should exceed the $65 billion (U.S.) cost of implementing the CAA. Cleaner air is expected to lead to better health and higher productivity for American workers as well as savings on medical expenses due to health-related problems (Box 5.1). The expected health effects due to exposure to air pollutants are explained in more detail in the following section.

BOX 5.1 Health Effects and Actions for Ozone Actions Days

TABLE 5.3 Description of the expected Air Quality Index (AQI)-related health effects for specific ozone ranges

OZONE LEVEL	HEALTH EFFECTS AND PROTECTIVE ACTIONS
Good	**What are the possible health effects?** • No health effects are expected.
Moderate	**What are the possible health effects?** • Unusually sensitive individuals may experience respiratory effects from prolonged exposure to ozone during outdoor exertion. **What can I do to protect my health?** • When ozone levels are in the "moderate" range, consider limiting prolonged outdoor exertion if you are unusually sensitive to ozone.
Unhealthy for Sensitive Groups	**What are the possible health effects?** • If you are a member of a sensitive group,[a] you may experience respiratory symptoms (such as coughing or pain when taking a deep breath) and reduced lung function, which can cause some breathing discomfort.

What can I do to protect my health?

- If you are a member of a sensitive group,[a] limit prolonged outdoor exertion. In general, you can protect your health by reducing how long or how strenuously you exert yourself outdoors and by planning outdoor activities when ozone levels are lower (usually in the early morning or evening).

- You can check with your state air agency to find out about current or predicted ozone levels in your location. This information on ozone levels is available on the Internet at *www.epa.gov/airnow*.

Unhealthy

What are the possible health effects?

- If you are a member of a sensitive group,[a] you have a higher chance of experiencing respiratory symptoms (such as aggravated cough or pain when taking a deep breath) and reduced lung function, which can cause some breathing difficulty.

- At this level, anyone could experience respiratory effects.

What can I do to protect my health?

- If you are a member of a sensitive group,[a] avoid prolonged outdoor exertion. Everyone else—especially children—should limit prolonged outdoor exertion.

- Plan outdoor activities when ozone levels are lower (usually in the early morning or evening).

- You can check with your state air agency to find out about current or predicted ozone levels in your location. This information on ozone levels is available on the Internet at *www.epa.gov/airnow*.

Very Unhealthy

What are the possible health effects?

- Members of sensitive groups[a] will likely experience increasingly severe respiratory symptoms and impaired breathing.

- Many healthy people in the general population engaged in moderate exertion will experience some kind of effect. According to EPA estimates, approximately:

- Half will experience moderately reduced lung function.

- One-fifth will experience severely reduced lung function.

- 10 to 15% will experience moderate to severe respiratory symptoms (such as aggravated cough and pain when taking a deep breath).

- People with asthma or other respiratory conditions will be more severely affected, leading some to increase medication usage and to seek medical attention at an emergency room or clinic.

What can I do to protect my health?

- If you are a member of a sensitive group,[a] avoid outdoor activity altogether. Everyone else—especially children—should limit outdoor exertion and avoid heavy exertion altogether.

- Check with your state air agency to find out about current or predicted ozone levels in your location. This information on ozone levels is available on the Internet at *www.epa.gov/airnow*.

[a]Members of sensitive groups include children who are active outdoors; adults involved in moderate or strenuous outdoor activities; individuals with respiratory disease, such as asthma; and individuals with unusual susceptibility to ozone.

Source: USEPA (1999) Smog – Who Does It Hurt?: what You Need to Know About Ozone and Your Health. Office of Air and radiation, United States Environmental Protection Agency. Washington, DC, USA. EPA 452/K-99-001.

TABLE 5.4 Differences in key health effects outcomes associated with fine particles (PM$_{2.5}$) and ozone between the With-CAA90 and Without-CAA90 scenarios for the 2010 and 2020 study target years (in number of cases avoided, rounded to two significant digits). The table shows the reductions in risk of various air pollution-related health effects achieved by 1990 Clean Air Act Amendments programs with each risk change expressed as the equivalent number of incidences avoided across the exposed population.

HEALTH EFFECT REDUCTIONS (PM$_{2.5}$ & OZONE ONLY)	POLLUTANT(S)	YEAR	
		2010	2020
PM$_{2.5}$ adult mortality	PM	160,000	230,000
PM$_{2.5}$ infant mortality	PM	230	280
Ozone mortality	Ozone	4,300	7,100
Chronic bronchitis	PM	54,000	75,000
Acute bronchitis	PM	130,000	180,000
Acute myocardial infarction	PM	130,000	200,000
Asthma exacerbation	PM	1,700,000	2,400,000
Hospital admissions	PM, Ozone	86,000	135,000
Emergency room visits	PM, Ozone	86,000	120,000
Restricted activity days	PM, Ozone	84,000,000	110,000,000
School loss days	Ozone	3,200,000	5,400,000
Lost work days	PM	13,000,000	17,000,000

Source: USEPA (2011) The Benefits and Costs of the Clean Air Act from 1990 to 2020: Summary Report. Office of Air and Radiation, United States Environmental Protection Agency. Research Triangle Park, NC, USA.

5.2   Health Effects of Air Pollutants

The acute and chronic human health effects of air pollutants are often difficult to predict from direct exposure. Air pollutants enter the body through the respiratory system (Figure 5.8). The adverse health effects of most air pollutants are related to the cardiopulmonary system. **Acute**, short-term symptoms are often much more severe in individuals predisposed to adverse respiratory events. Common short-term effects include irritated mucous membranes, inflammation of the bronchial tubes, and possible airway restriction. **Chronic** or long-term effects of exposure to air pollutants may include chronic inflammation of the bronchial tubes, sustained airway restriction, or pulmonary emphysema. For example, pulmonary emphysema, which results in shortness of breath due to destruction of the alveoli membranes in the lungs, has been linked to exposure to airborne asbestos dust.

The human respiratory system consists of three distinct regions, each affected differently by different constituents in the air. Air enters the body through the nasopharyngeal region or exothoracic airway. The nasopharyngeal region includes the nose, mouth, and larynx. Water-soluble compounds in the air may be absorbed by the mucous membranes in the nasal cavity. Air flow restrictions in the larynx result in the deposit of large airborne particles in the mucous lining of the larynx, increasing the risk of malignant tumors in the larynx.

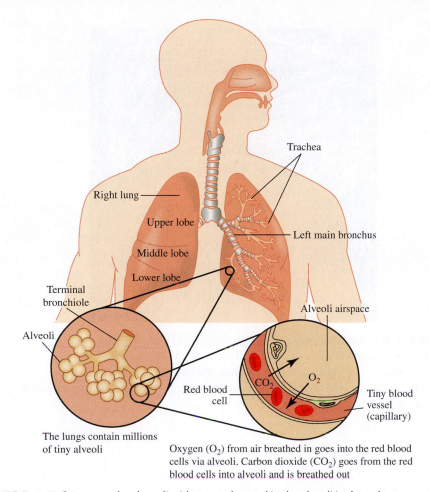

The lungs contain millions of tiny alveoli

Oxygen (O_2) from air breathed in goes into the red blood cells via alveoli. Carbon dioxide (CO_2) goes from the red blood cells into alveoli and is breathed out

FIGURE 5.8 Oxygen and carbon dioxide are exchanged in the alveoli in the pulmonary region of the respiratory system.

Source: Based on Patient.co.uk.

The tracheobronchial region of the respiratory system is characterized by the branching of the airway bronchi through the lungs, conveying the air to the pulmonary region. The tracheobronchial region includes the cilia, trachea, primary and secondary bronchi, and terminal bronchioles. The cilia are fine, hair-like particles (Figure 5.9) that line the entire region and help trap small to medium-size particles that make it past the larynx. The particles become covered in the mucous lining of the bronchi, and the cilia use a whip-like motion to propel the particles out of the respiratory system and into the throat. The bronchi are the conduits that move the air into and from the lungs. There are between 20 and 25 branching steps, or generations of branching, from the trachea (called the first-generation respiratory passage) to the alveoli (see Figure 5.10).

Carbon dioxide is released from the blood into the lungs, and oxygen is moved from the air to the bloodstream in the pulmonary region of the lungs (Figure 5.8). The pulmonary region consists of 300 to 500 million very small, interconnected groups of porous membranes called alveoli. The total air exchange area of the

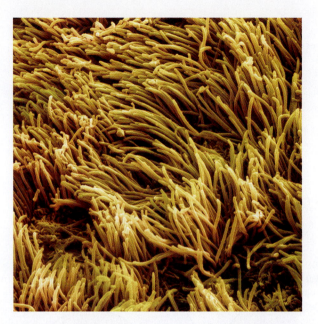

FIGURE 5.9 Colored scanning electron micrograph (SEM) of cilia (hairs) lining the inner surface of a bronchiole, one of the small airways in a lung. These cilia help to keep the lungs free of infection. Mucus in the lungs traps bacteria and foreign bodies, and the cilia waft it up to the throat, from where it can be swallowed or coughed up.

Source: Eye of Science / Science Source

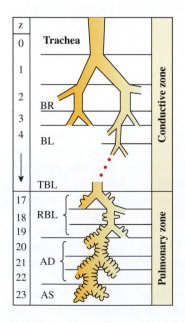

FIGURE 5.10 Idealization of the first 16 generations of branching (z) of the human airway.

Source: Based on Weibel E.R., 1962, Morphology of the human lung. Berlin: Springer-Verlag

TABLE 5.5 Oxygen concentrations in the air and symptoms associated with oxygen deficiency

% O_2 IN AIR	DESCRIPTION OF SYMPTOMS
20.9	Normal O_2 concentration in air
17	Hypoxia occurs with deteriorating night vision, increased heart beat, accelerated breathing
14–16	Very poor muscular coordination, rapid fatigue, intermittent respiration
6–10	Nausea, vomiting, inability to perform, unconsciousness
<6	Spasmodic breathing, convulsive movements, death within minutes

Source: Based on Heinsohn, 1991

alveoli region is about 50 square meters in the average adult. White blood cells, called macrophage, destroy most infectious bacteria and fungi, with some notable exceptions such as tuberculosis.

Respiratory distress symptoms are related to an upset in the process of air exchange in the lungs. Hypoxia is caused by an inadequate supply of oxygen to the body. The onset of hypoxia in a healthy adult can occur if oxygen levels in an enclosed space become too low as shown in Table 5.5. Cyanosis, a symptom of hypoxia, is caused when there is not enough oxygen in the blood and the blood turns blue. Hypercapnia is a condition that occurs when too much carbon dioxide builds up in the blood stream causing distress.

Other respiratory diseases may lead to irritation of the respirator system where air passages become irritated and are physiologically constricted. Constricted airways may lead to edema and secondary infections in the lungs. If cell tissue in the lungs is damaged, the airways may not be able to transfer oxygen and carbon dioxide from the blood leading to necrosis and edema. Excess proteins in the blood or infectious agents may cause fibrosis or stiffening of the lung tissue that may restrict airflow and lung function. Foreign particles or chemical exposure may cause oncogenesis, or tumor formation, in the lungs that could become malignant. Pneumonia is an inflammation of lung tissue condition usually caused by a bacterial or virus infection of the lungs. Pneumonia is not normally fatal in high-income countries, but it is still a leading cause of death, particularly for children and the elderly, in low-income countries.

5.2.1 Carbon Monoxide

Carbon monoxide (CO) is a colorless odorless gas emitted from combustion processes and subject to NAAQS standards, as shown in Table 5.6. Carbon monoxide is similar in size and molecular mass to diatomic oxygen (O_2) in the air. Carbon monoxide will replace oxygen in the bloodstream if it is present in air at high concentrations, as illustrated in Figure 5.11. The carbon monoxide replaces oxygen at the active hemoglobin site, thereby reducing oxygen transport to the brain, organs, and muscles. The degree of symptoms described in Table 5.6 associated with carbon monoxide depends on the extent of carbon monoxide that bonds with hemoglobin in the blood. As carbon monoxide levels increase in the bloodstream, exposed individuals experience headaches and reduced hand–eye coordination. If oxygen deprivation becomes severe due to elevated levels of carbon monoxide,

TABLE 5.6 Sources and effects of NAAQS criteria air pollutants

POLLUTANT	SOURCES	HEALTH EFFECTS
Ozone (O_3)	Secondary pollutant typically formed by chemical reaction of volatile organic compounds (VOCs) and NOx in the presence of sunlight.	Decreases lung function and causes respiratory symptoms, such as coughing and shortness of breath; aggravates asthma and other lung diseases leading to increased medication use, hospital admissions, emergency department (ED) visits, and premature mortality.
Particulate matter (PM)	Emitted or formed through chemical reactions; fuel combustion (e.g., burning coal, wood, diesel); industrial processes; agriculture (plowing, field burning); and unpaved roads.	Short-term exposures can aggravate heart or lung diseases leading to respiratory symptoms, increase medication use, hospital admissions, ED visits, and premature mortality; long-term exposures can lead to the development of heart or lung disease and premature mortality.
Lead	Smelters (metal refineries) and other metal industries; combustion of leaded gasoline in piston engine aircraft; waste incinerators; and battery manufacturing.	Damages the developing nervous system, resulting in IQ loss and impacts on learning, memory, and behavior in children. Cardiovascular and renal effects in adults and early effects related to anemia.
Oxides of nitrogen (NOx)	Fuel combustion (e.g., electric utilities, industrial boilers, and vehicles) and wood burning.	Aggravates lung diseases leading to respiratory symptoms, hospital admissions, and ED visits; increased susceptibility to respiratory infection.
Carbon monoxide (CO)	Fuel combustion (especially vehicles).	Reduces the amount of oxygen reaching the body's organs and tissues; aggravates heart disease, resulting in chest pain and other symptoms leading to hospital admissions and ED visits.
Sulfur dioxide (SO_2)	Fuel combustion (especially high-sulfur coal); electric utilities and industrial processes; and natural sources such as volcanoes.	Aggravates asthma and increases respiratory symptoms. Contributes to particle formation with associated health effects.

Source: USEPA (2012b) Our Nation's Air: Status and Trends through 2010. Office of Air Quality, Planning and Standards, United States Environmental Protection Agency. Research Triangle Park, NC, USA.

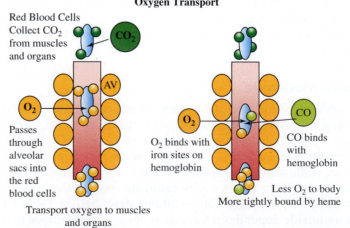

FIGURE 5.11 Oxygen transport of blood and carbon monoxide interference in oxygen transfer.

Source: Bradley Striebig.

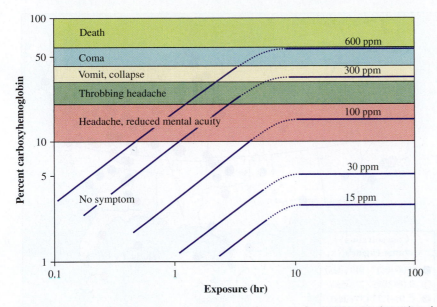

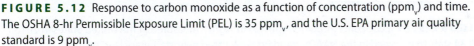

FIGURE 5.12 Response to carbon monoxide as a function of concentration (ppm$_v$) and time. The OSHA 8-hr Permissible Exposure Limit (PEL) is 35 ppm$_v$, and the U.S. EPA primary air quality standard is 9 ppm$_v$.

Source: Based on Seinfeld , J. H. (1986) Atmospheric chemistry and physics of air pollution. John Wiley & Sons, Inc. Somerset, NJ, USA.

symptoms may include fainting, coma, and death, as illustrated in Figure 5.12. Minor symptoms usually are alleviated within 2 to 4 hours if the exposure to carbon monoxide ceases. Individuals with cardiovascular disease may experience myocardial ischemia (reduced oxygen to the heart), often accompanied by chest pain (angina), when exercising or under increased stress. Even short-term CO exposure can lead to severe symptoms for people with heart disease. The majority of carbon monoxide emissions are due to mobile sources, especially in urban areas. Urban automobile exhaust is a major source of carbon monoxide in the atmosphere. Carbon monoxide concentrations of 5 to 50 ppm$_v$ have been measured in some urban areas, and concentrations above 100 ppm$_v$ have been measured along some heavily congested highways.

5.2.2 Lead

Lead is found in both manufactured products and the natural environment, and it has been used as an additive in gasoline to improve engine performance. Lead is subject to NAAQS standards (Table 5.1). Lead emissions in the United States decreased by 95% from 1980 to 1999, as lead was phased out as a fuel additive. Consequently, the concentration of lead found in the air in the United States decreased by 94% from 1980 to 1999.

Major emission sources of lead in the United States today include lead smelters, ore and metal processing facilities, and aviation fuel. Lead concentrations measured in ambient air in the United States in 2010 are shown in Figure 5.13. Individuals may be exposed to lead in drinking water, lead-contaminated food, or soil. Ingestion of dust from lead-based paint in older homes is one of the primary routes of lead exposure. Once ingested, lead accumulates in the bones. Medical symptoms are related to the amount of lead to which an individual is exposed.

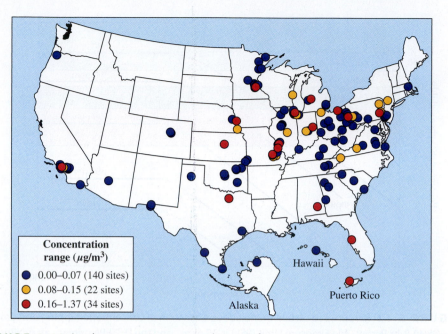

FIGURE 5.13 Lead concentrations reported in μg/m^3 measured in the United States in 2010. All 34 sites that exceeded the NAAQS standard are located near stationary lead sources.

Source: USEPA (2012b) Our Nation's Air: Status and Trends through 2010. Office of Air Quality, Planning and Standards, United States Environmental Protection Agency. Research Triangle Park, NC, USA.

Lead negatively affects the nervous system, kidney function, the immune system, the reproductive and development system, the cardiovascular system, and the oxygen-carrying capacity of the blood. Infants and children under the age of 5 are especially vulnerable to the neurological effects of lead exposure, which has been documented to cause behavioral problems, learning disorders, and lowered IQ.

5.2.3 Nitrogen Oxides

Nitrogen oxides (NOx) are a variety of reactive gases that include nitrous acid and nitrogen dioxide (NO$_2$). NO$_2$ is formed from the emissions of combustion processes from both mobile and stationary sources (see Figure 5.14). Nitrogen dioxide concentrations measured in ambient air in the United States in 2010 are shown in Figure 5.15. NOx's contribute to the formation of ground-level ozone and are also linked to direct adverse human health effects. NO$_2$ is a known irritant to the alveoli after exposures of 30 minutes to 24 hours and reportedly causes emphysema in animals. NOx in the air increases an individual's susceptibility to pulmonary infections. Studies show a connection between short-term NO$_2$ exposure and hospital visits for respiratory symptoms. Individuals are generally exposed to the highest level of NO$_2$ concentrations in automobiles and within 300 feet of a major highway, railroad, or airport. NOx reacts with ammonia, moisture, and other compounds to form small aerosols. These small particles may be transported deep into the respiratory system, causing or worsening respiratory symptoms. Regulations that require catalytic converters for motor vehicles have led to a 40% decrease in NO$_2$ concentrations in the United States since 1980. NOx also reacts with volatile organic compounds (VOCs) to form ozone in the presence of heat and sunlight, as illustrated by the chemical

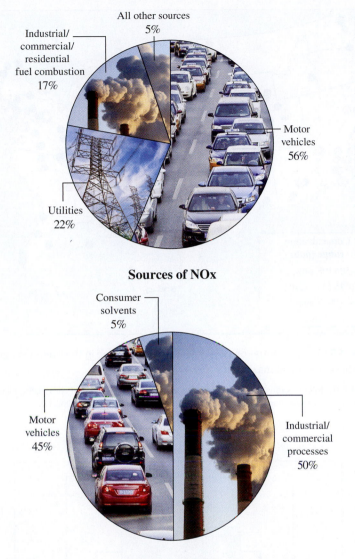

Sources of NOx

Sources of VOC

FIGURE 5.14 Sources of NOx and VOCs that are precursor compounds to ozone formation.

Source: USEPA (2003b) Ozone – Good Up High, Bad Nearby. Office of Air and radiation, United States Environmental Protection Agency. Washington, DC, USA. EPA 451/K-03-001; TonyV3112/Shutterstock.com; Shutter_M/Shutterstock.com; M. Shcherbyna/Shutterstock.com; M. Shcherbyna/Shutterstock.com

pathways shown in Figure 5.16. Ozone forms in the troposphere from reactions with NOx and organic pollutants. (It is discussed in more detail in the next section.)

5.2.4 Ozone

Ozone (O_3) is a permanent gas and beneficial compound in the Earth's stratosphere. The ozone in the stratosphere absorbs harmful ultraviolet irradiation from the sun. In the troposphere, the part of the atmosphere from the surface of the Earth to approximately 17 kilometers above the surface of the Earth, ozone is a highly reactive, short-lived gas. **Tropospheric ozone** is formed by the chemical reactions between NOx, VOCs, and light, which are the precursors for ozone formation.

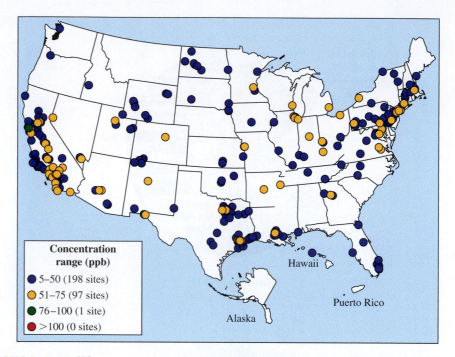

FIGURE 5.15 NO$_2$ concentrations reported in ppb$_v$ measured in the United States in 2010 (98th percentile of daily 1-hr maximum).

Source: USEPA (2012b) Our Nation's Air: Status and Trends through 2010. Office of Air Quality, Planning and Standards, United States Environmental Protection Agency. Research Triangle Park, NC, USA.

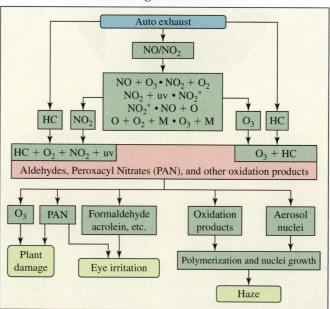

FIGURE 5.16 Atmospheric photochemical oxidation reactions of NOx and hydrocarbon-based VOCs.

Source: Based on Heinsohn, R.J. and Kabel, R.L. 1999. "Sources and Control of Air Pollution." Prentice Hall, NJ.

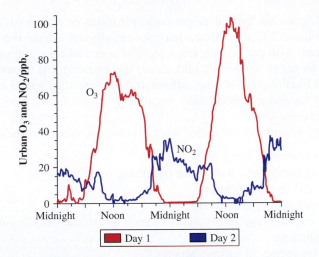

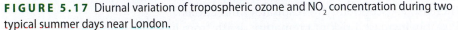

FIGURE 5.17 Diurnal variation of tropospheric ozone and NO_2 concentration during two typical summer days near London.

Source: Based on Palmer: Quantifying sources and sinks of trace gases using space-borne measurements: current and future science. Phil. Trans. R. Soc. A 2008 366, 4509-4528 http://rsta.royalsocietypublishing.org/content/366/1885/4509.full.pdf+html

Emissions from industrial facilities, combustion processes, motor vehicle exhaust, gasoline vapors, and chemical solvents are major sources of ozone precursors. Ozone is most likely to form on hot days in urban areas. Ozone concentrations increase during the daytime and decrease during evening hours (see Figure 5.17). Notice also that NO_2 reacts to form ozone during the daylight hours and NO_2 concentrations increase in the evening hours when ozone formation ceases, since there is no sunlight to catalyze the ozone formation reactions. Ozone may be transported long distances by wind, causing decreased air quality in rural areas as well.

Low levels of ozone may cause adverse health effects, and for this reason, it is regulated as a NAAQS priority pollutant (see Table 5.1). Ozone concentrations of 1 ppm_v constrict and irritate the airways in the lungs (Figure 5.18). Long-term

FIGURE 5.18 Photo of a healthy lung airway (left) and an inflamed lung airway (right). Ozone can inflame the lung's lining, and repeated episodes of inflammation may cause permanent lung damage.

Source: Stocktrek Images/Getty Images

effects are unknown for typical urban concentrations of ozone (0.05 to 0.2 ppm) illustrated in Figure 5.19, but ozone may potentially accelerate the aging of lung tissue. Individuals with preexisting respiratory illness, children, and older adults are particularly susceptible to ozone. Children are most susceptible to ozone because of their increased likelihood of having asthma and because they spend more time outdoors exposed to ozone. Ozone also negatively affects ecosystems by damaging sensitive vegetation.

Respiratory symptoms associated with exposure to ozone include:

- Difficulty breathing
- Shortness of breath and pain when breathing deeply
- Coughing and throat irritation
- Inflamed and damaged bronchi
- Worsened existing respiratory diseases such as asthma, emphysema, and chronic bronchitis
- Increased frequency of asthma attacks
- Increased susceptibility to respiratory infections
- Possibly increased risk of premature death from heart or lung disease
- Potentially damaged lungs without obvious symptoms

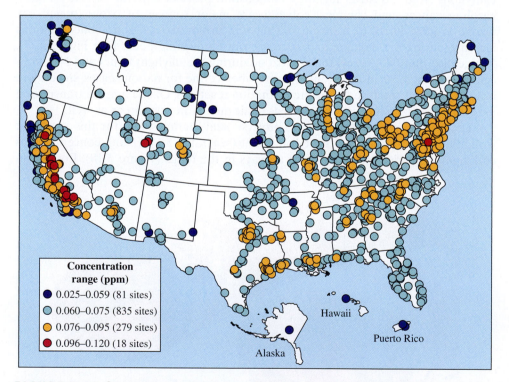

FIGURE 5.19 Ozone concentrations in ppm_v (fourth highest daily maximum 8-hour concentration) measured in the United States in 2010.

Source: USEPA (2012b) Our Nation's Air: Status and Trends through 2010. Office of Air Quality, Planning and Standards, United States Environmental Protection Agency. Research Triangle Park, NC, USA.

5.2.5 Particulate Matter (PM)

Particulate matter is any very small airborne solid or liquid mixture that may contain acids, organic chemicals, soil, or dust (see Table 5.7 and Figure 5.20). The size of the particle is directly related to the particle's potential to cause negative health effects. Particles larger than 10 micrometers (or microns) in size are typically removed in the nasopharyngeal region of the respiratory system and are not

TABLE 5.7 Constituents of atmospheric particles and their major sources

Aerosol species	PRIMARY (PM < 2.5 μm)		PRIMARY (PM > 2.5 μm)		SECONDARY PM PRECURSORS (PM < 2.5 μm)	
	Natural	Anthropogenic	Natural	Anthropogenic	Natural	Anthropogenic
Sulfate (SO_4^{2-})	Sea spray	Fossil fuel combustion	Sea spray	–	Oxidation of reduced sulfur gases emitted by the oceans and wetlands and SO_2 and H_2S emitted by volcanism and forest fires	Oxidation of SO_2 emitted from fossil fuel combustion
Nitrate (NO_3^-)	–	Mobile source exhaust	–	–	Oxidation of NOx produced by soils, forest fires, and lightning	Oxidation of NOx emitted from fossil fuel combustion and in motor vehicle exhaust
Minerals	Erosion and re-entrainment	Fugitive dust from paved and unpaved roads, agriculture, forestry, construction, and demolition	Erosion and re-entrainment	Fugitive dust from paved and unpaved roads, agriculture, forestry, construction, and demolition	–	–
Ammonium (NH_4^+)	–	Mobile source exhaust	–	–	Emissions of NH_3 from wild animals and undisturbed soil	Emissions of NH_3 from motor vehicles, animal husbandry, sewage, and fertilized land
Organic carbon (OC)	Wildfires	Prescribed burning, wood burning, mobile source exhaust, cooking, tire wear, and industrial processes	Soil humic matter	Tire and asphalt wear, paved and unpaved road dust	Oxidation of hydrocarbons emitted by vegetation (terpenes, waxes) and wildfires	Oxidation of hydrocarbons emitted by motor vehicles, prescribed burning, wood burning, solvent use, and industrial processes

(Continued)

TABLE 5.7 (*Continued*)

Aerosol species	PRIMARY (PM < 2.5 μm)		PRIMARY (PM > 2.5 μm)		SECONDARY PM PRECURSORS (PM < 2.5 μm)	
	Natural	Anthropogenic	Natural	Anthropogenic	Natural	Anthropogenic
Elemental carbon (EC)	Wildfires	Mobile source exhaust (mainly diesel), wood biomass burning, and cooking	–	Tire and asphalt wear, paved and unpaved road dust	–	–
Metals	Volcanic activity	Fossil fuel combustion, smelting and other metallurgical processes, and brake wear	Erosion, re-entrainment, and organic debris	–	–	–
Bioaerosols	Viruses and bacteria	–	Plant and insect fragments, pollen, fungal spores, and bacterial agglomerates	–	–	–

Dash (–) indicates either very minor source or no known source of component.

Source: USEPA (2009a) Policy Assessment for the review of the Particulate Matter national Ambient Air Quality Standards. Office of Air Quality Planning and Standards, United States Environmental Protection Agency. Research Triangle Park, NC, USA.

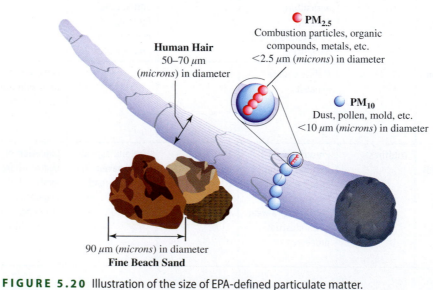

FIGURE 5.20 Illustration of the size of EPA-defined particulate matter.

Source: Based on USEPA (http://www.epa.gov/airquality/particlepollution/basic.html)

classified as particulate matter (PM) by the EPA. The EPA groups PM into two categories illustrated in Figure 5.21 and described as follows:

Inhalable coarse particles (PM_{10}), such as those found near roadways and dusty industries, are larger than 2.5 microns and smaller than 10 microns in diameter.

Fine particles ($PM_{2.5}$), such as those found in smoke and haze, are 2.5 micrometers in diameter and smaller. These particles can be directly emitted from sources such as forest fires, or they can form when gases emitted from power plants, industries, and automobiles react in the air (see Table 5.8).

Particulate matter is deposited in different regions of the respiratory system (see Figure 5.22). In the tracheobronchial region, particles are deposited by settling in the low-velocity regions of the bronchioles and alveolar spaces. Particles may also diffuse to the surface of the bronchial bifurcations and alveoli. Once in the alveoli, some particles will be attacked, encapsulated, and passed up the tracheo-bronchial tree through cilia action. If particles pass into the bloodstream, they may be dissolved or removed by the cellular defenses and transferred to the lymphatic drainage system. Some particles may be permanently retained where they may remain benign or in some cases cause lung damage though fibrosis or malignancy.

Exposure to particulate matter has been linked to a variety of adverse health effects, including:

- Increased respiratory symptoms such as airway irritation, coughing, and difficulty breathing
- Decreased lung function
- Aggravation of asthma
- Chronic bronchitis
- Negative cardiovascular symptoms such as irregular heartbeat and heart attacks
- Premature death in people with heart or lung disease

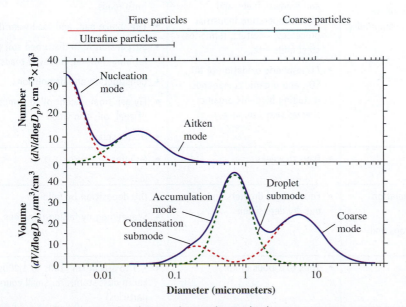

FIGURE 5.21 Particle-size distributions by number and volume.

Source: USEPA (2009a) Policy Assessment for the review of the Particulate Matter national Ambient Air Quality Standards. Office of Air Quality Planning and Standards, United States Environmental Protection Agency. Research Triangle Park, NC, USA.

TABLE 5.8 Characteristics of fine and coarse particulate matter

	FINE		COARSE
	Ultrafine	**Accumulation**	
Formation Processes	• Combustion, high-temperature processes, and atmospheric reactions		• Breakup of large solids/droplets
Formed by	• Nucleation of atmospheric gases including H_2SO_4, NH_3, and some organic compounds • Condensation of gases	• Condensation of gases • Coagulation of smaller particles • Reactions of gases in or on particles • Evaporation of fog and cloud droplets in which gases have dissolved and reacted	• Mechanical disruption (crushing, grinding, abrasion of surfaces) • Evaporation of sprays • Suspension of dusts • Reactions of gases in or on particles
Composed of	• Sulfate • Elemental carbon (EC) • Metal compounds • Organic compounds with very low saturation vapor pressure at ambient temperature	• Sulfate, nitrate, ammonium, and hydrogen ions • EC • Large variety of organic compounds • Metals: compounds of Pb, Cd, V, Ni, Cu, Zn, Mn, Fe, etc. • Particle-bound water • Bacteria, viruses	• Nitrates/chlorides/sulfates from HNO_3/HCl/SO_2 reactions with coarse particles • Oxides of crustal elements (Si, Al, Ti, Fe) $CaCO_3$, $CaSO_4$, NaCl, sea salt • Bacteria, pollen, mold, fungal spores, plant and animal debris
Solubility	• Not well characterized	• Largely soluble, hygroscopic, and deliquescent	• Largely insoluble and nonhygroscopic
Sources	• High-temperature combustion • Atmospheric reactions of primary, gaseous compounds.	• Combustion of fossil and biomass fuels, and high-temperature industrial processes, smelters, refineries, steel mills, etc. • Atmospheric oxidation of NO_2, SO_2, and organic compounds, including biogenic organic species (e.g., terpenes)	• Resuspension of particles deposited onto roads • Tire, brake pad, and road wear debris • Suspension from disturbed soil (e.g., farming, mining, unpaved roads) • Construction and demolition • Fly ash from uncontrolled combustion of coal, oil, and wood • Ocean spray
Atmospheric half-life	• Minutes to hours	• Days to weeks	• Minutes to hours
Removal processes	• Grows into accumulation mode • Diffuses to raindrops and other surfaces	• Forms cloud droplets and rains out dry deposition	• Dry deposition by fallout • Scavenging by falling raindrops
Travel distance	• <1 to 10s of km	• 100s to 1,000s of km	• <1 to 10s of km (100s to 1,000s of km in dust storms for small coarse particles)

Source: Based on Wilson, W.E. and Suh, H.H. (1997) Fine particles and coarse particles: Concentration relationships relevant to epidemiologic studies. Journal of the Air and Waste Management Association. Volume 47:1238-1249.

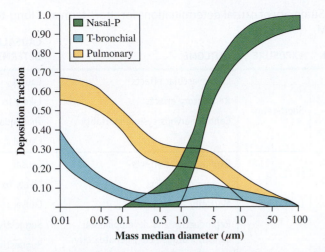

FIGURE 5.22 Predicted regional deposition of particles in the respiratory system for a tidal volumetric flow rate of 21 L/min. The shaded area indicates the variation resulting from two geometric deviations.

Source: Based on Heinsohn, R.J. and Kabel, R.L. 1999. "Sources and Control of Air Pollution". Prentice Hall, NJ. 4-33, adapted from Stahlhofen, W., Gebhart, J., Heyder, J., and Scheuch, G. (1983) New Regional Deposition Data of the Human Respiratory Tract. Journal of Aerosol Science. Volume 14:186-188.

Detailed information regarding the relationship between PM exposure and health effects is presented in the 2009 EPA report entitled "Integrated Science Assessment for Particulate Matter." The EPA found that PM less than 2.5 microns in diameter ($PM_{2.5}$) was likely to cause cardiovascular and respiratory effects and early mortality (Table 5.9).

Particulate matter also contributes to reduced visibility, or haze, as reported in Table 5.10. It may damage the environment by contributing to sedimentation in water bodies, depleting nutrients in soil, damaging sensitive plants, and consequently decreasing ecosystem diversity. PM may also be detrimental to architectural structures, potentially damaging or staining stone features, statues, and monuments.

5.2.6 Sulfur Oxides

Sulfur oxides (SOx) are a priority pollutant and are subject to NAAQS standards as shown in Table 5.1. Fossil fuel combustion accounts for 73% of SOx emissions. Other industrial sources account for 20% of SOx emissions. Sulfur oxides are highly soluble and are absorbed in the upper respiratory system. They irritate and constrict the airways. Studies also show a connection between short-term exposure and increased visits to emergency departments and hospital admissions for respiratory illnesses, particularly in at-risk populations, including children, the elderly, and asthmatics. Asthmatic patients may suffer brochioconstriction at concentrations of only 0.25 to 0.5 ppm_v. If particulate matter is present in the air along with sulfur oxides, the sulfur oxide sorbs onto the particulate matter, forming an aerosol that can be transported deep into the respiratory system and causes symptoms that are three to four times more severe than would be the case for SOx alone. The adverse human health effects associated with the exposure dose are illustrated in Figure 5.23.

TABLE 5.9 Summary of causal determinations for short-term and long-term exposure to $PM_{2.5}$ and PM_{10}

SIZE FRACTION	EXPOSURE	OUTCOME	CAUSALITY DETERMINATION
$PM_{2.5}$	Short-term	Cardiovascular effects	Causal
		Respiratory effects	Likely to be causal
		Central nervous system	Inadequate
		mortality	Causal
	Long-term	Cardiovascular effects	Causal
		Respiratory effects	Likely to be causal
		Mortality	Causal
		Reproductive and developmental	Suggestive
		cancer, mutagenicity, genotoxicity	Suggestive
$PM_{10-2.5}$	Short-term	Cardiovascular effects	Suggestive
		Respiratory effects	Suggestive
		Central nervous system	Inadequate
		mortality	Suggestive
	Long-term	Cardiovascular effects	Inadequate
		Respiratory effects	Inadequate
		Mortality	Inadequate
		Reproductive and developmental	Inadequate
		cancer, mutagenicity, genotoxicity	Inadequate

Source: USEPA (2009a) Policy Assessment for the review of the Particulate Matter national Ambient Air Quality Standards. Office of Air Quality Planning and Standards, United States Environmental Protection Agency. Research Triangle Park, NC, USA.

TABLE 5.10 Summary of causality determination for welfare effects for short-term and long-term exposure to $PM_{2.5}$ and PM_{10}

WELFARE EFFECTS	CAUSALITY DETERMINATION
Effects on visibility	Causal
Effects on climate	Causal
Ecological effects	Likely to be causal
Effects on materials	Causal

Source: USEPA (2009a) Policy Assessment for the review of the Particulate Matter national Ambient Air Quality Standards. Office of Air Quality Planning and Standards, United States Environmental Protection Agency. Research Triangle Park, NC, USA.

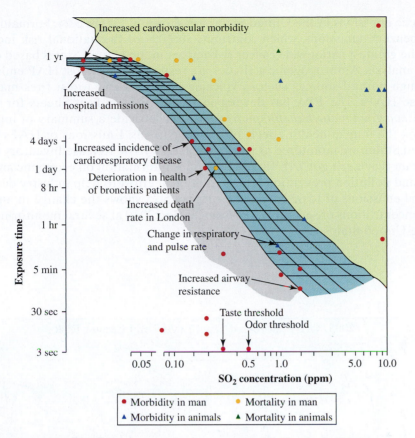

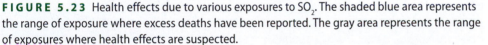

FIGURE 5.23 Health effects due to various exposures to SO_2. The shaded blue area represents the range of exposure where excess deaths have been reported. The gray area represents the range of exposures where health effects are suspected.

Source: Based on Seinfeld , J. H. (1986) Atmospheric chemistry and physics of air pollution. John Wiley & Sons, Inc. Somerset, NJ, USA.

5.2.7 Hazardous Air Pollutants

The EPA has developed a list of 188 compounds (see Appendix I) in addition to the six priority air pollutants that are potentially carcinogenic, mutagenic, or teratogenic. These 188 compounds are defined as **hazardous air pollutants (HAPS)** or, as they are sometimes called, toxic air pollutants. A partial list of suspected and confirmed carcinogenic compounds commonly used in industry is shown in Table 5.11. EPA's 2005 estimates of the increased cancer risk from

TABLE 5.11 List of commonly used industrial chemicals that are suspected or known carcinogens

COMMON SUSPECTED CARCINOGEN COMPOUND NAME			
Arsenic	Carbon Tetrachloride	1,2-Dibromoethane	Nitrosamine
Asbestos	Chloroform	1,2-Dichloroethane	Perchloroethylene
Benzene	Chromium	Inorganic lead	Polycyclic aromatic hydrocarbons
Cadmium	1,4–Dioxane	Nickel	Vinyl chloride

air toxic emissions in the United States are shown in Figure 5.24. Formaldehyde and benzene emissions, which contribute to 60% of the national risk increase, had the greatest nationwide cancer risk impact in the United States based on the 2005 analysis. Anthropogenic activities are responsible for most HAP emissions, although natural sources such as volcanic eruptions and forest fires may also release HAPs. The EPA has developed Health Effects Fact Sheets for HAPs (*www.epa.gov/ttn/atw/hlthef/hapintro.html*) that provide a summary of information on the human health effects of these air toxins. Emissions of HAPs in the United States are monitored as part of the EPA's Toxic Release Inventory (TRI) program (*www.epa.gov/tri*). The TRI provides information on economic analysis, risk, and pollution prevention information in an effort to help industry decrease annual emissions of toxic compounds. Figure 5.25 shows the change in ambient air concentrations of several HAPs from 2003 to 2010 at several monitoring sites in the United States.

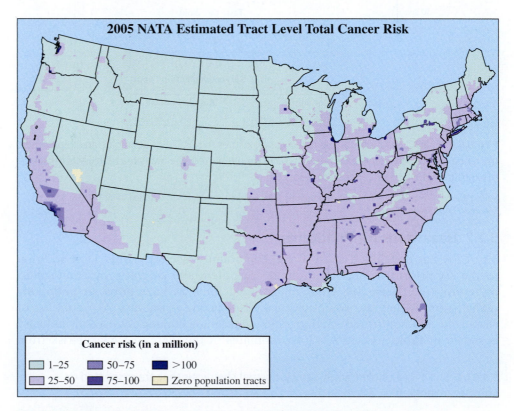

FIGURE 5.24 EPA 2005 assessment that characterizes the nationwide chronic lifetime cancer risk estimates and noncancerous hazards from inhaling air toxics.

Source: USEPA (2011b) An Overview of Methods for EPA's National-Scale Air Toxics Assessment. Office of Air Quality Planning and Standards, United States Environmental Protection Agency. Research Triangle Park, NC, USA. Cancer Map 1. http://www.epa.gov/ttn/atw/natamain/

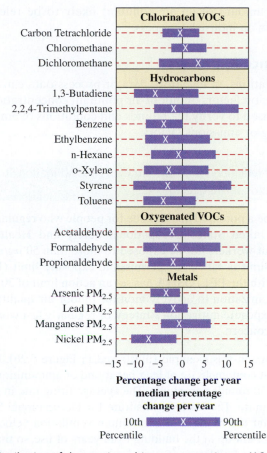

FIGURE 5.25 Distribution of changes in ambient concentrations at U.S. toxic air pollutant monitoring sites, 2003–2010 (percent change in annual average concentration).

Source: USEPA (2012b) Our Nation's Air: Status and Trends through 2010. Office of Air Quality, Planning and Standards, United States Environmental Protection Agency. Research Triangle Park, NC, USA. http://www.epa.gov /airtrends/2011/dl_graph.html

5.3 Estimating Emissions of Air Pollutants

Air pollutants may adversely affect human health, plant and animal health, architectural and historically significant structures, visibility, and ecosystem diversity. Some of these effects are well understood and have been the basis of environmental legislation and regulations, while other effects such as mixtures of compounds and endocrine disruptors are still being studied. Reducing the risk and negative impacts associated with air pollutants requires either reducing their emissions at the source or adding pollution control equipment to prevent emissions from escaping into the environment. Regardless, the ability to estimate and measure air pollutant emissions is a prerequisite to controlling emissions and reducing their negative impacts.

The emission rate of air pollutants must be determined for formulating emission control strategies, developing applicable permitand control programs, and identifying the possible effects of sources and mitigation strategies. Air pollutant emissions may be determined from direct measurement at the source, mass balance calculations based on chemical inventories, tabular emission factors for similar processes, or fundamental relationships based on similar industrial processes. **Emission**

factors express the amount of an air pollutant likely to be released based on a determinant factor in an industrial process.

5.3.1 Mass Balance Approach

Mass balance calculations on inventory data or process data can be useful to estimate pollutant emission factors or indoor air pollutant concentrations. Mass balance models can also be used to verify the bounds and conditions of emissions calculated from direct sampling or emissions factors.

EXAMPLE 5.2 **Predicting Lead Concentrations and Mitigation Strategies in an Indoor Firing Range**

Lead exposure may be a potential health issue for people who regularly work in indoor firing ranges. The United States Occupational Safety and Health Administration (OSHA) requires that workers not be exposed to more than 50 μg/m³ of lead over an 8-hour period. This limit is called the permissible exposure limit (PEL). In order to ensure compliance with the PEL, OSHA has set an action limit of 30 μg/m³. The action limit requires an organization to actively monitor indoor air quality, install an alarm system, develop an exposure mitigation strategy, or regularly test workers' blood levels for the pollutant of concern.

In the firing range in this example (illustrated in Figure 5.26), large-caliber rifles are typically used that contain 48 μg of lead per round of ammunition. The rifles fire at a maximum rate of 650 rounds per minute. The average firing rate in the range is closer to 150 rounds per minute. The National Institute for Occupational Safety and Health (NIOSH) requires that the room have a minimum ventilation velocity of 22.9 meters per minute. Lead has built up in the building from years of use, so that the background lead concentration in the building is 15 μg/m³, even when no one is firing a rifle in the range. Exhaust from the building passes through three stages of fabric high-efficiency particulate arrestor (HEPA) filters that remove 95% of overall lead emissions prior to being emitted to the atmosphere.

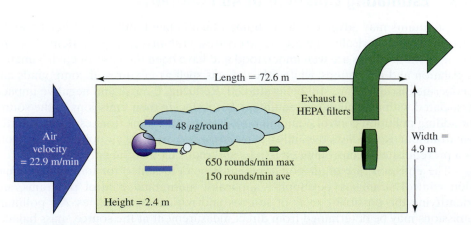

FIGURE 5.26 Schematic of lead emissions in an indoor firing range.

Source: Bradley Striebig.

Determine the following information about individual potential lead exposure in the firing range:

- Determine the average lead concentration from firing at the maximum firing rate, C_{max}, neglecting background lead levels.
- Determine the total lead concentration averaged over an 8-hour period, C_{tot}, when firing at the average firing rate, including background lead levels, C_b.
- What is the expected average emission rate from the range with the HEPA filters?

Solution:

Begin by calculating the volumetric flow rate of air in the range by multiplying the air flow velocity, v_{NIOSH} by the cross-section area, A_{cs}, perpindicular to the air flow:

$$v_{NIOSH} = 22.9$$

$$A_{cs} = 4.9\,\text{m} \times 2.4\,\text{m} = 11.9\,\text{m}^2$$

$$Q = v_{NIOSH} \times A_{cs} = 22.9\,\frac{\text{m}}{\text{min}} \times 11.9\,\text{m}^2 = 272\,\frac{\text{m}^3}{\text{min}}$$

Convert to m³/min:

The emission rate from firing the rifle at the maximum rate of 650 rounds per minute is

$$M = \left(48\,\mu g\,\frac{\text{Pb}}{\text{round}}\right)\left(650\,\frac{\text{rounds}}{\text{min}}\right) = 31{,}200\,\mu g\,\frac{\text{Pb}}{\text{min}}$$

Rearranging the equation for the relationship between mass emission rate, volumetric flow rate, and concentration yields

$$C_{max}\left(\frac{\mu g}{\text{m}^3}\right) = \frac{M}{Q} = \frac{\left(31{,}200\,\mu g\,\dfrac{\text{Pb}}{\text{min}}\right)}{\left(272\,\dfrac{\text{m}^3}{\text{min}}\right)} = 115\,\frac{\mu g}{\text{m}^3}$$

The lead concentrations at the maximum firing rate are expected to be well above the PEL, at least for short periods of time when firing at the maximum rate, even neglecting background concentrations.

Similarly, if we use an 8-hour average firing rate of 150 rounds per minute, then

$$M = \left(48\,\mu g\,\frac{\text{Pb}}{\text{round}}\right)\left(150\,\frac{\text{rounds}}{\text{min}}\right) = 7{,}200\,\mu g\,\frac{\text{Pb}}{\text{min}}$$

Rearranging the equation for the relationship between mass emission rate, volumetric flow rate and concentration yields

$$C_{ave}\left(\frac{\mu g}{\text{m}^3}\right) = \frac{M}{Q} = \frac{\left(7{,}200\,\mu g\,\dfrac{\text{Pb}}{\text{min}}\right)}{\left(272\,\dfrac{\text{m}^3}{\text{min}}\right)} = 26\,\frac{\mu g}{\text{m}^3}$$

The concentration from firing at the average flow rate is still near the action level, and the background concentration has not yet been included. The total expected concentration in the firing range, C_{tot}, averaged over an 8-hour period is

$$C_{tot} = C_{ave} + C_b = 26 \frac{\mu g}{m^3} + 15 \frac{\mu g}{m^3} = 41 \frac{\mu g}{m^3} > \text{action limit of } 30 \frac{\mu g}{m^3}$$

Therefore, further action must be taken within the firing range to reduce exposure to those working in the facility.

The average emissions to the environment from the facility may be determined from the average mass emission rate and subtracting the amount removed by the HEPA filters:

$$EM_{ave} = C_{tot}Q\left(1 - \frac{EF}{100}\right) = \left(41\frac{\mu g}{m^3}\right)\left(272 \frac{m^3}{min}\right)\left(1 - \frac{95\%}{100}\right)\left(\frac{60 \text{ min}}{hr}\right)\left(\frac{mg}{1,000\mu g}\right) = 33 \frac{mg}{hr}$$

5.3.2 Emission Factors

Direct analysis and mass balance approaches to determining emissions may not always be possible. Sampling can be very costly and may be highly variable due to short-term changes in process operations or expected major modifications to an industrial process. A mass balance approach may be limited by a lack of available data from similar processes and changes in inventory. Both direct measurement methods and mass balance approaches may not accurately reflect the variability of emissions over time. Incorporating safety factors when predicting emissions and modeling possible exposure and environmental impacts should account for this variability. Proposed new facilities may not have useful inventory data on which to base emissions and cannot be sampled directly. In these cases, the EPA has developed a comprehensive set of emission factors that attempt to relate the quantity of pollutants released to the atmosphere based on an activity associated with the release of that pollutant. These factors are called AP-42 emission factors, and they are averages of available data associated within a specific industrial category. A summary of the approaches to estimate emissions is provided in Table 5.12.

TABLE 5.12 Summary of methods for estimating source emissions

APPROACH TO DETERMINING EMISSIONS	COMMENTS
Direct sampling	Taken from actual or similar source
	May be expensive to determine change in emissions over time
	Continuous emission monitoring may be required for some processes
Mass balance	Must accurately define the feed stock and process conversion
	Useful for providing boundary conditions to check direct sampling and emission factor estimates
Emission factors	Average value from typical industry performance
	Found in EPA AP-42 tables
	Used for predicting emissions, impacts on ambient air quality, and as an input in dispersion models for new sources or major modifications to existing sources

The general equation for estimating emissions is

$$EM = AC \times EF \times \left(1 - \frac{ER}{100}\right) \tag{5.2}$$

where

EM = pollutant emission rate (mass/time)

AC = Activity or production rate (units are varied)

EF = AP-42 emission factor (units vary)

ER = percent overall emissions reduction efficiency, including removal and capture efficiency (%)

EXAMPLE 5.3   Estimating Emissions from Coffee Roasting Using U.S. EPA AP-42 Emission Factors

Professor Coffy buys his beans from the Coffee Mountain Roasting company that roasted 26 kg of coffee beans in 2010. Determine the following emissions if all 26 kg of beans were continuously roasted. Create a table that shows the emissions with and without a thermal oxidizer.

Solution

The AP-42 emission factors for coffee roasting are found in Chapter 9 of AP42, Volume I, 5th edition (*www.epa.gov/ttn/chief/ap42*) under Section 9.13, Miscellaneous Food & Kindred Products: Coffee Roasting. The emission factors for uncontrolled filterable particulate matter, VOCs, methane, CO, and CO_2 are provided for various types of coffee roasting, with and without thermal oxidation VOC control devices in the U.S. EPA tables reproduced in Tables 5.13 and 5.14. The emission factor for filterable particulate matter, EF_{PM}, is given in Table 5.13 as 0.66 pound of PM/ton of

---

TABLE 5.13 Emission factors for particulate matter for coffee-roasting operations[a]

SOURCE	SOURCE CLASSIFICATION CODE	FILTERABLE PM (mg/kg)	CONDENSIBLE PM (mg/kg)
Batch roaster with thermal oxidizer	SCC 3-02-002-20	27	ND
Continuous cooler with cyclone	SCC 3-02-002-28	6.35	ND
Continuous roaster	SCC 3-02-002-21	150	ND
Continuous roaster with thermal oxidizer	SCC 3-02-002-21	21	23
Green coffee bean screening, handling, and storage system with fabric filter[b]	SCC 3-02-002-08	13	ND

[a] Emission factors are based on green coffee bean feed. Factors represent uncontrolled emissions unless noted. SCC = Source Classification Code. ND = no data. D-rated and E-rated emission factors are based on limited test data; these factors may not be representative of the industry.

[b] Emission Factor Rating: E

Source: Based on USEPA (2011d) Compilation of Air Pollutant Emission Factors – Update 2011. United States Environmental Protection Agency. Washington, DC, USA. http://www.epa.gov/ttn/chief/ap42/index.html

coffee roasted. The activity of coffee roasting in tons can be determined from the given information:

$$AC_{coffee} = 26 \times 10^6 \, \frac{kg}{year}$$

Substituting the values into Equation (5.2) yields

$$EM_{PM} = AC \times EF \times \left(1 - \frac{ER}{100}\right) = \left(26 \times 10^6 \, \frac{kg}{year}\right)\left(150 \, \frac{mg \, of \, PM}{kg \, coffee}\right)$$

$$= 3.9 \times 10^9 \, \frac{mg}{yr} = 3900 \, \frac{kg}{yr}$$

Similar calculations may be performed for each major category of emissions, the results of which are shown in Table 5.14 and 5.15

Notice that for the coffee roasting company, filterable particulate matter, VOCs, methane, and carbon monoxide emissions are reduced by a thermal oxidizer air pollution control system. The thermal oxidizer converts the VOCs, and possibly some of the particulate matter, along with any supplemental fuel needed for the oxidation process into carbon dioxide, so that carbon dioxide emissions from the process with the thermal oxidizer would likely be greater than from the same process without a thermal oxidizer. Furthermore, the mass emission rate of carbon dioxide is two to three times greater than the other pollutants. While carbon dioxide is not a direct health threat and is currently unregulated in the United States, it is a prominent greenhouse gas.

TABLE 5.14 Emission factors for carbon containing compounds for coffee roasting operations.[a] Emission Factor Rating: D

SOURCE	SOURCE CLASSIFICATION CODE	VOC[b] (mg/kg)	METHANE (mg/kg)	CO (mg/kg)	CO$_2$ (mg/kg)
Batch roaster	SCC 3-02-002-20	195	ND	ND	40,823
Batch roaster with thermal oxidizer	SCC 3-02-002-20	11	ND	125	120,200
Continuous roaster	SCC 3-02-002-21	318	59[d]	340	27,215[c]
Continuous roaster with thermal oxidizer	SCC 3-02-002-21	36	34[d]	22	45,360

[a] Emission factors are based on green coffee bean feed. Factors represent uncontrolled emissions unless noted. SCC = Source Classification Code. ND = no data. D-rated and E-rated emission factors are based on limited test data; these factors may not be representative of the industry.

[b] Volatile organic compounds as methane. Measured using GC/FID.

[c] Emission Factor Rating: C

[d] Emission Factor Rating: E

Source: Based on USEPA (2011d) Compilation of Air Pollutant Emission Factors – Update 2011. United States Environmental Protection Agency. Washington, DC, USA. http://www.epa.gov/ttn/chief/ap42/index.html

TABLE 5.15 Emission estimates from a coffee roasting company with and without a thermal oxidizer for VOC control

EMISSION TYPE	EMISSION FACTORS (mg of pollutant/kg of coffee)		ESTIMATE EMISSION (kg/yr)	
	Without a Thermal Oxidizer	With a Thermal Catalytic Oxidizer	Without a Thermal Oxidizer	With a Thermal Catalytic Oxidizer
PM	150	21	3900	542
VOC	318	36	8255	943
Methane	60	34	1533	885
CO	340	22	8845	578
CO_2	27,215	45,360	707,000	1.2×10^6

Source: Based on U.S. EPA (2011d) Compilation of Air Pollutant Emission Factors - Update 2011. United States Environmental Protection Agency. Washington, DC.

The EPA has classified the reliability of the emission factors into the ratings shown in Table 5.16. Emission factors are given a rating of A to E, where A is based on the most reliable data and the data used for E-rated emissions factors are the least reliable. The ratings are subjective in nature and do not imply statistical error bounds or confidence intervals. The emission factors are appropriate for estimating source specific emissions impacts on area-wide pollutant inventories. Emission factors may also be used in dispersion models to help estimate the potential impacts of new sources or process modifications on individual exposure and ambient air quality. Regulatory authorities may use emission factors for screening sources that require compliance permits. Emissions factors may also be used to determine appropriate control and mitigation strategies.

TABLE 5.16 Description of ratings factors for U.S. EPA AP-42 emissions factors

RATING	DESCRIPTION
A	**Excellent.** Factor is developed from A- and B-rated source test data taken from many randomly chosen facilities in the industry population. The source population is sufficiently specific to minimize variability.
B	**Above average.** Factor is developed from A- or B-rated test data from a "reasonable number" of facilities. The facilities tested may not represent a random sample. The population is sufficiently specific to minimize variability.
C	**Average.** Factor is developed from A-, B-, and/or C-rated test data from a reasonable number of facilities. The facilities tested may not represent a random sample. The source population is sufficiently specific to minimize variability.
D	**Below average.** Factor is developed from A-, B-, and/or C-rated data from a small number of facilities. The facilities may not represent a random sample. There is evidence of variability.
E	**Poor.** Factor is developed from C- and D-rated test data. The facilities may not represent a random sample, and there is evidence of variability within the population.

Source: USEPA (1995) Compilation of Air Pollutant emission Factors – Volume I: Stationary Point and Area Sources. United States Environmental Protection Agency. Washington, DC, USA. AP-42 Fifth Edition.

The AP-42 emission factors are published and updated in the EPA's two-volume set entitled *Compilation of Air Pollutant Emission Factors* (*www.epa.gov/ttn/chief/ap42/index.html*). It should be noted that the absence of an emission factor for a process does not indicate that there are no emissions for the process. There may be more than one emission factor for a general category of processes, so table footnotes and details should be read carefully. The data are useful for estimating emissions, but are not accurate enough to indicate how process variations or changes might affect emissions. The emission factors do not imply or represent emissions limits or regulations for a facility or process.

Direct sampling should be conducted to confirm emission rates and concentrations if expected values are within an order of magnitude of regulatory thresholds for indoor air limits, ambient air quality compliance, or air toxic requirements. Emission factors are very useful for a first step in estimating the order of magnitude of expected emissions from a process. Emissions factors are also useful in developing airshed regional plans for ambient air quality control.

5.4   Dispersion of Air Pollutants

Air emissions may be transported, dispersed, or concentrated by meteorological conditions. Emissions may be modeled to determine the effects an emission source may have on NAAQS. Various types and complexities of models may be used to represent the movement of air pollutants.

Dispersion occurs owing to the mean air motion that moves pollutants downwind, turbulent fluctuations that disperse pollutants in multiple directions, and mass diffusion of the pollutants that causes pollutant migrations from areas of high to low concentration. The density, shape, and size of the pollutant also influence movement.

A simple three-dimensional box model, illustrated in Figure 5.27, may sometimes represent topographic and meteorological conditions. The box model is useful for estimating worst-case scenarios for exposure or for developing long-term average concentrations in a topographically bound air shed. The model is only a long-term average estimate of pollutants since it cannot account for short-term variations or turbulence.

The box model is most accurate when topographic boundaries are available, such as a city or town that lies in a valley between two mountain ranges. In this case, the width of the model box, W, is the distance between the two mountain ranges. The height of the box would be the height of the meteorological boundary layer. It is common for air to become stratified due to changes in density and temperature in a process similar to that discussed in Chapter 4 in

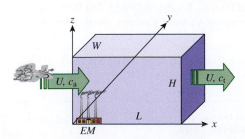

FIGURE 5.27 Box model for estimating air pollutant movement.

Source: Bradley Striebig.

regards to lake stratification. An inversion is a common meteorological process, especially in narrow valleys, that occurs and prevents mixing with air at higher elevations. The height of the box model, H, would be the height at which the atmospheric inversion occurs. A mass balance upon the emissions, EM (mass/time), can be developed to calculate the concentration, C_t, at some distance, L, downwind:

$$V \frac{dC}{dt} = EM + C_a Q_{in} - C Q_{out} \tag{5.3}$$

where

dC/dt = change in concentration with time

$V = H \times W \times L$ = volume of the box

C_a = ambient concentration of the pollutant in the air entering the box upwind of the emission source

C_t = concentration of the pollutant downwind of the emission source

Q = volumetric airflow rate

$$Q_{in} = Q_{out} = Q = U \times H \times W \tag{5.4}$$

where U = mean velocity of the airflow through the box.

Substituting the dimensions of the box and Equation (5.4) into Equation (5.3) yields

$$(H \cdot W \cdot L) \frac{dC}{dt} = EM + C_a(U)(H \cdot W) - C(U)(H \cdot W) \tag{5.5}$$

Under steady-state conditions, $dC/dt = 0$. Equation (5.5) can be solved for the steady-state concentration downwind, C_{ss}:

$$C_{ss} = C_a + \frac{EM}{U \cdot H \cdot W} \tag{5.6}$$

EXAMPLE 5.4   Ammonia Concentration in Perdu Valley

Perdu Valley is a 4-kilometer-wide agricultural valley where more than 5 million chickens are raised each year. The farms in the valley are at an elevation of 500 meters. The valley is bordered on two sides by mountains that are 1,200 meters high, which is the same height where inversions typically occur. The EPA has estimated that raising caged chickens in a flush house may result in 30 pounds of ammonia per year for 100 chickens. The average mean airflow velocity through the valley is 1.5 m/s, and the upwind ammonia concentration is 1.0 $\mu g/m^3$. What is the downwind steady-state concentration of ammonia in the valley in $\mu g/m^3$?

Solution

Determine the emission rate of ammonia in the valley.

$$EM = \frac{30 \text{ lb NH}_3}{yr - 100 \text{ chickens}} \times (5 \times 10^6) \text{ chickens} \frac{kg}{2.2 \text{ lb}} \frac{10^9 \, \mu g}{kg} = 6.82 \times 10^{14} \frac{\mu g}{yr}$$

Determine the steady-state concentration in the valley using Equation (5.6):

$$C_{ss} = C_a + \frac{EM}{U \cdot H \cdot W} = 1.0 \frac{\mu g}{m^3} + \frac{6.82 \times 10^{14} \frac{\mu g}{yr}}{\left(1.5 \frac{m}{s}\right)\left(31.536 \times 10^6 \frac{s}{yr}\right)(1,200[m])(4,000[m])}$$

$$= 6.1 \frac{\mu g}{m^3}$$

The box-model may also be used for variable emissions, where $dC/dt \neq 0$. Separating the variables and integrating with respect to time yields

$$\int_{C_o}^{C_t} \frac{dC}{(EM + U(H \cdot W)C_a) - U(H \cdot W)C} = \frac{1}{H \cdot W \cdot S} \int_0^t dt \tag{5.7}$$

Rearranging Equation (5.5) yields

$$U(H \cdot W)(C_{ss}) = U(H \cdot W)(C_a) + EM \tag{5.8}$$

Substituting the steady-state concentration for the ambient concentration in Equation (5.8) and multiplying both sides of the integrated equation by $U(HW)$ yields

$$\frac{C_t - C_o}{C_{ss} - C_o} = 1 - \exp\left\{\frac{-Ut}{L}\right\} \tag{5.9}$$

If the initial concentration is very small compared to the final concentration, then Equation (5.9) simplifies to

$$\frac{C_t}{C_{ss}} = 1 - \exp\left\{\frac{-Ut}{L}\right\} \tag{5.10}$$

The simplified box model can be used to model emissions in a room or to cross the defined boundaries of a facility. However, the simplified box model does not account for dispersion and variations with time. The box model also does not account for various meteorological conditions and other atmospheric phenomena.

Air motions that impact pollutant motion are strongly affected by the height at which the air pollutants are released or are transported in the atmosphere (see Figure 5.28). There are three general classifications of atmospheric transport.

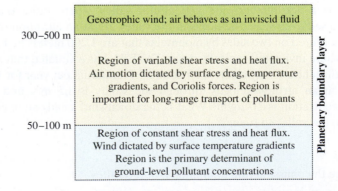

FIGURE 5.28 Earth surface effects on wind speed retardation in the planetary boundary layer.

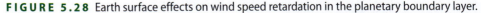

Source: Based on Seinfeld, J. H. (1986) Atmospheric chemistry and physics of air pollution. John Wiley & Sons, Inc. Somerset, NJ, USA.

Macroscopic transport occurs over scales of thousands of kilometers; these movements are closely related to global air circulation patterns. *Microscopic air circulation patterns* are most commonly modeled to determine air pollutant concentrations defined by local phenomena less than 10 kilometers of the source. *Mesoscopic transport* may be affected by both local and large-scale air circulation patterns for pollutants on the order of 100 kilometers from the source.

The **troposphere** is the region of the atmosphere closest to the Earth. The surface conditions do not typically affect air movements in the geostrophic region, which is about 300 to 500 meters above the Earth's surface. Weather, climate, and long-distance pollutant transport are strongly determined by geostrophic wind patterns. The planetary boundary layer is the layer between the Earth's surface and the upper atmosphere. Terrain and topography variations and temperature gradients strongly affect the friction losses and retardation of wind speed within the boundary layer. General atmospheric conditions must be understood before pollutant movements can be predicted.

Atmospheric conditions may be characterized by a few key parameters. The lapse rate is the negative temperature gradient of the atmosphere, and it is dependent on the weather, pressure, and humidity. The equation that represents the change in air pressure with elevation under isothermal (constant temperature) conditions can be written as

$$\frac{dP}{P} = -\frac{g}{RT_z}\,dz \tag{5.11}$$

where

P = atmospheric pressure of a parcel of air

g = gravitation constant

R = ideal gas law constant

dz = change in elevation

$T_z = T_0$ = constant temperature at the surface, $z = 0$ and at elevation $= z$

If the atmosphere is at a constant temperature, a relatively rare occurrence, Equation (5.11) may be integrated:

$$\int_{P_0}^{P_z} \frac{dP}{P} = \int_0^z \frac{-g}{RT}\,dz \tag{5.12}$$

The integrated form is

$$\frac{P_z}{P_0} = \exp\left(\frac{-g \cdot z}{R \cdot T}\right) \tag{5.13}$$

If you look at a local weather forecast or if you go from a lower elevation to a mountain for hiking or skiing, you may realize that the air temperature usually decreases with height in the planetary boundary layer. The pressure and density also vary with altitude according to

$$\frac{T_z}{T_0} = \left(\frac{P_z}{P_0}\right)^{\frac{(n-1)}{n}} = \left(\frac{\rho_z}{\rho_0}\right)^{\frac{(n-1)}{n}} \tag{5.14}$$

The change in temperature with respect to the change in elevation can be found by differentiating Equation (5.14):

$$\frac{dT}{dz} = -\frac{g}{R}\left(\frac{n-1}{n}\right) \tag{5.15}$$

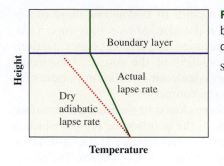

FIGURE 5.29 The actual lapse rate can be compared to the dry adiabatic lapse rate to determine the degree of stability.

Source: Bradley Striebig.

where

$$n = c_p/c_v$$

c_p = constant pressure specific heat

c_v = constant volume specific heat

Under the specific conditions of dry adiabatic air, then the **dry adiabatic lapse rate**, Γ, is

$$\Gamma = \left(-\frac{dT}{dz}\right)_{\text{dry adiabatic}} = \frac{0.0098°C}{m} = \frac{0.0054°F}{ft} \tag{5.16}$$

The amount of turbulence in the atmosphere is related to the actual lapse rate compared to the dry adiabatic lapse rate (Figure 5.29). If there is a high degree of turbulence in the atmosphere, the atmospheric conditions are classified as unstable. In an unstable atmosphere, if a pocket or parcel of air is disturbed, it is unlikely to approach a new equilibrium and will increase turbulence. A pocket of air that is disturbed in a stable atmosphere will be forced into a position of equilibrium by the surrounding air, and turbulence will be decreased. A stable atmosphere resists vertical mixing. The potential temperature gradient may be used to compare the dry adiabatic lapse rate and the actual or environmental lapse rate and to determine the stability classification shown in Table 5.17. The **potential temperature gradient** is defined as the actual lapse rate plus the dry adiabatic lapse rate:

$$\frac{\Delta\theta}{\Delta z} \cong \left(\frac{\Delta T}{\Delta z}\right)_{\text{actual}} + \Gamma \tag{5.17}$$

Unstable atmospheric conditions and strong vertical mixing characterize superadiabatic conditions. The environmental lapse rate is greater than the dry adiabatic lapse rate for stable atmospheres (Figure 5.30). A neutral atmosphere is typically a transitional condition where the environmental lapse rate is equal to the dry adiabatic

TABLE 5.17 Stability Classification of atmospheric conditions

$d\theta/dz$	STABILITY CLASSIFICATION	CATEGORY
< 0	Unstable	A, B, C
$= 0$	Neutral	D
> 0	Stable	E, F, isothermal or inversion

Source: Based on Heinsohn, R.J. and Kabel, R.L. 1999. "Sources and Control of Air Pollution." Prentice Hall, NJ..

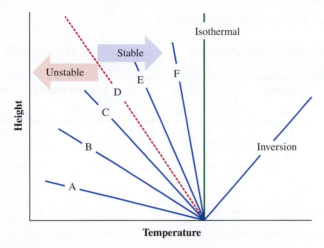

FIGURE 5.30 Stability classifications for the atmosphere.

Source: Based on Heinsohn, R.J. and Kabel, R.L. 1999. "Sources and Control of Air Pollution". Prentice Hall, NJ.

lapse rate. Stable atmospheric conditions tend to reduce vertical mixing and the environmental lapse rate is less than the dry adiabatic lapse rate. Isothermal conditions, when there is no temperature change with elevation, create very strongly stable atmospheric conditions. Inversions, which occur when the temperature increases with increasing altitude, are the most stable, and little mixing occurs during an inversion. The stability classification influences wind characteristics and wind speed (Table 5.18). Observations of atmospheric conditions may be used with the Pasquill stability classification system shown in Table 5.19 to estimate stability classifications.

TABLE 5.18 Beaufort scale of wind speed equivalents at 10-meter elevation

DESCRIPTION	SPECIFICATIONS	U_{10} (m/s)
Calm	Smoke rises vertically	< 0.5
	Smoke drifts but does not move wind vane	0.5–1.5
Light	Wind felt, leaves rustle, ordinary vanes moves	1.5–3
Gentle	Leaves and twigs in constant motion, extends light flag	4–5
Moderate	Raises dust and loose paper, small branches are moved	6–8
Fresh	Small trees may sway, crested wavelets form on lakes	8–10
	Large branches move, whistling heard in wires, umbrellas used with difficulty	11–14
Strong	Whole trees in motion, difficulty walking into wind	15–17
	Breaks twigs off tress, generally impedes progress	18–20
Gale	Slight structural damage occurs (chimneys and roof)	21–24
	Trees uprooted, considerable structural damage occurs	25–28
Whole gale	Rarely experienced, widespread damage	29–33
Hurricane	> 75 miles per hour	> 34

Source: Based on Heinsohn, R.J. and Kabel, R.L. 1999. "Sources and Control of Air Pollution." Prentice Hall, NJ.

TABLE 5.19 Meteorological conditions for Pasquill stability categories[a]

WIND SPEED @ 10 m	DAYTIME INCOMING SOLAR RADIATION			NIGHT, CLOUD COVER, OR THICKLY OVERCAST	
Class[b]:	Strong	Moderate	Slight	> 1/2 Low Clouds	< 3/8 Clouds
	(1)	(2)	(3)	(4)	(5)
< 2 m/s	A	A–B	B	Strongly stable	
2–3 m/s	A–B	B	C	E	F
3–5 m/s	B	B–C	B–C	D	E
5–6 m/s	C	C–D	C–D	D	D
> 6 m/s	C	D	D	D	D

[a] The neutral category, D, should be assumed for overcast conditions during the day or night. Category A is the most unstable and category F is the most stable, with category B moderately unstable and category E slightly stable.

[b] Class 1: clear skies, solar altitude greater than 60 degrees above the horizontal, typical of a sunny afternoon, very convective atmosphere; class 2: summer day with a few broken clouds; class 3: typical of a fall afternoon, summer day with broken low clouds, or summer day with clear skies and solar altitude from only 15 to 35 degrees above the horizon, class 4: can be used for a winter day.

Source: Based on Hanna, S.R., Briggs, G.A., and Hosker, R.P. Jr. (1982) handbook on Atmospheric Diffusion. Technical Information Center, U.S. Department of Energy.

EXAMPLE 5.5 Stability Classification Based on Temperature Measurements

Determine the environmental lapse rate and potential temperature for the early morning, at 6:00 a.m., 8:00 a.m., and 2:00 p.m. for the data presented in Table 5.20.

$$\frac{-dT}{dz} \cong \frac{-\Delta T}{\Delta z} = \frac{-(12 - 14)}{100} = 0.02°C/m$$

TABLE 5.20 Temperature measurements at a meteorological field station for Example 5.5

TIME	SURFACE TEMPERATURE (°C)	TEMPERATURE AT 100 m (°C)
6:00	12	14
8:00	15	14
14:00	22	18

The potential temperature from Equation (5.17) is

$$\frac{\Delta \theta}{\Delta z} \cong \left(\frac{\Delta T}{\Delta z}\right)_{actual} + \Gamma = 0.02°C/m + 0.0098°C/m = 0.03°C/m$$

From Table 5.17 and Figure 5.30 we see that a positive lapse rate is a stable atmosphere since the positive environmental lapse rate indicates that an inversion condition exists early in the morning. This is a very stable condition with very little vertical mixing.

EXAMPLE 5.6 Stability Classification Based on Meteorological Observations

The following conditions were observed near a location being considered for an industrial site (stability conditions are needed to model predicted emissions). Records show typical daytime conditions are moderate sunlight with broken clouds and wind can be felt that rustles leaves. Determine the likely average Pasquill stability category.

From Table 5.18, the wind description is indicative of light winds of 1.5 to 3 m/s. From Table 5.17, with moderate incoming solar irradiation, the stability category is most likely B, or possibly A, both of which are unstable conditions.

The stability conditions affect the shape and movement of air released from a smokestack. The amount of dispersion and downwind concentration of pollutants released from a stack is related to the exit velocity and temperature of air emitted from a smokestack, the mean velocity of the air moving past the stack, and the stability classification of the atmosphere. There are six characteristic plume shapes: fanning, looping, coning, fumigation, lofting, and trapping plumes (as illustrated in Figure 5.31).

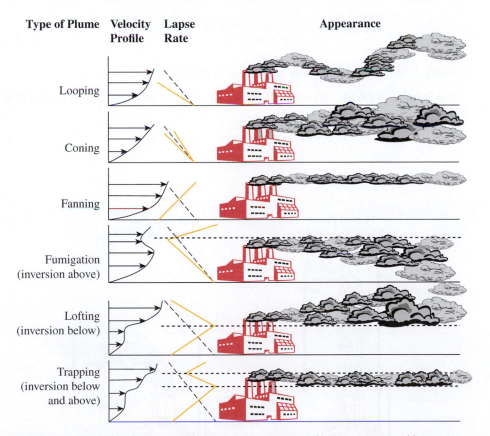

FIGURE 5.31 Types of plumes, typical mean air velocity profiles, environmental lapse rate (in yellow) compared to the dry adiabatic lapse rate (dashed line in gray), and the two-dimensional appearance of the plume type.

Source: Based on Slade, D.H. (ed.) 1968. Meteorology and atomic energy. US Atomic Energy Commission, Air Resources Laboratories, Environmental Sciences Services Administration, U.S. Department of Commerce, Washington, D.C.

The movement of air pollutants released from a smokestack is dependent on how high the smokestack is, particularly in relationship to the altitude that an inversion may occur. Pollutants being trapped in a fumigation plume below an inversion layer have caused the most devastating air pollution events, such as the 1948 event in Donora, Pennsylvania, described earlier in this chapter. Modern smokestacks are designed so that the combined height of the stack and the upward momentum of the air jet leaving the smokestack propel pollutants above the inversion layer.

Gases exiting a smokestack with a **specific height**, H_s, have a momentum and buoyancy that are characteristic of the gas emitted and the atmosphere they are being emitted into (Figure 5.32). As the plume rises, it also travels downwind and mixes vertically and laterally. The height of the **plume rise**, δh, is important for modeling the movement of pollutants and downwind exposure estimates. The **effective stack height**, H, is the sum of the actual stack height and the plume rise:

$$H = H_s + \delta h \tag{5.18}$$

There are numerous correlations for estimating the plume rise. The Briggs plume rise formula is a simplified model for use when the temperature of the exit gas plume is greater than the ambient temperature.:

$$\delta h \text{ m} = \frac{1.6(x)^{\frac{2}{3}}(F)^{\frac{1}{3}}}{U} \tag{5.19}$$

where U = wind speed at the stack height, H_s, in m/s.

The buoyancy of the gas emitted from the stack is accounted for in Equation (5.19) by the buoyancy flux term, F:

$$F \; \frac{\text{m}^4}{\text{s}^3} = \frac{g v_s D_s^2 (T_s - T_a)}{4 T_a} \tag{5.20}$$

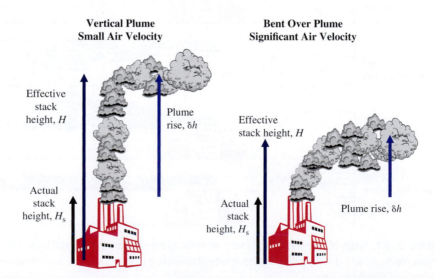

FIGURE 5.32 Plume rise from a smokestack into an atmosphere with a small air velocity and a large air velocity that cause the plume to bend.

where

v_s = exit velocity of gas from the smokestack (m/s)

D_s = inside diameter of the smokestack (m)

g = gravitational acceleration constant (m/s²)

T_s = temperature of the gas exiting the smokestack (K)

T_a = ambient air temperature at the height of the smokestack, H_s (K)

x = downwind distance from the stack source in (m) for a stable atmosphere or for a neutral or unstable atmosphere if $x < 3.5\ x*$

$$\delta h \text{ m} = \frac{1.6(3.5x*)^{\frac{2}{3}}(F)^{\frac{1}{3}}}{U} \tag{5.21}$$

$x*$ = the distance at which atmospheric turbulence begins to dominate parcel and pollutant movement

$$x* = 14F^{\frac{5}{8}} \text{ if } F < 55 \text{ and } x* = 34F^{\frac{2}{5}} \text{ if } F > 55 \tag{5.22}$$

The plume rise can be affected by the geometry of the building and the ratio of the exit gas to the mean air velocity. If the exit velocity is less than 1.5 times the mean air velocity, low-pressure wake regions can be created behind the smokestack or building. The low-pressure wake region may cause the pollutants to become trapped in the low-pressure eddy created by air passing by the smokestack or building. Wake effects may increase exposure near the sources and affect building ventilation. In this case, more detailed calculations are required. Proper stack placement and building ventilation should take into consideration topography, meteorology, and potential building wake effects.

Downwind air pollutant concentrations and subsequent exposure are affected by the stack characteristics and turbulent conditions in the atmosphere. Temperature gradients, convective air currents, and wind shear over surface features naturally cause turbulence. However, turbulence is very difficult to accurately model mathematically. Instead, statistical methods are used to estimate vertical and horizontal dispersion parameters based on the likely magnitude of turbulence associated with the aforementioned atmospheric stability classes. The stability class of the atmosphere governs the degree of vertical mixing. Eddy diffusion governs the degree of horizontal dispersion. There are four types of air pollutant dispersion models: physical, numerical, statistical, and Gaussian plume models. Only the Gaussian plume model will be described in this text, but the many resources listed in the References section of this chapter may be consulted for exploration of other air dispersion models.

A mass balance can be written that includes dispersion of air pollutants with respect to time and location.

$$\frac{\delta c}{\delta t} = -\frac{\delta(cU)}{\delta x} + EM + \frac{\delta\left[\frac{\delta(cD_x)}{\delta x}\right]}{\delta x} + \frac{\delta\left[\frac{\delta(cD_y)}{\delta y}\right]}{\delta y} + \frac{\delta\left[\frac{\delta(cD_z)}{\delta z}\right]}{\delta z} \tag{5.23}$$

where

c = concentration of the pollutant emitted from a smokestack

U = wind velocity in the x-direction

EM = source generation (or decay) term

D_x, D_y, and D_z are the effective dispersion coefficients along each axis that incorporate both eddy and molecular diffusion

Several simplifying assumptions may be applicable to many air pollution sources. If the source is constant and the gases are nonreactive, a steady-state assumption can be made, where $dC/\mathbf{dt} = 0$. We will also assume that bulk motion due to the mean air velocity governs air pollutant transport in the x-direction, so that dispersion in the x-direction is negligible. The mean air velocity, U, and the dispersion coefficients in the y- and z-directions will be assumed to be constant. Equation (5.23) can then be simplified to

$$U \frac{\delta c}{\delta x} = D_y \frac{\delta^2 c}{\delta y^2} + D_z \frac{\delta^2 c}{\delta z^2} \tag{5.24}$$

The following mathematical boundary conditions are required to solve the mass balance equation: The concentration at the initial point approaches infinity as the size of the release point approaches zero. The concentration approaches zero as the distance from the source approaches infinity. There is no diffusion at ground level. The simplified mass balance using the Gaussian solution to solve Equation (5.24) is visualized in Figure 5.33. The solution to Equation (5.24), based on the Gaussian function used to represent dispersion, is

$$c_i(x, y, z) = \frac{EM_{i,s}}{2\pi x (D_y D_z)^{\frac{1}{2}}} exp\left\{-\frac{U}{4x}\left(\frac{y^2}{D_y} + \frac{z^2}{D_z}\right)\right\} \tag{5.25}$$

The theoretical dispersion parameters, D_x and D_y, are in practice replaced with empirically based dispersion coefficients in the horizontal direction, σ_y, and the vertical direction, σ_z, where

$$\sigma_y^2 = \frac{2xD_y}{U} \tag{5.26}$$

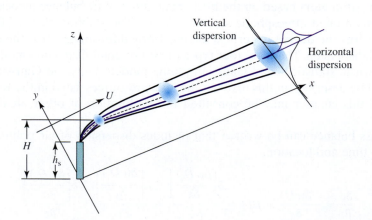

FIGURE 5.33 Visualization of the Gaussian solution to the pollutant dispersion mass balance equation for a constant mean air velocity, U, oriented along the x-direction. The blue lines represent horizontal dispersion in the y-direction. The gray lines represent vertical dispersion in the z-direction.

Source: Bradley Striebig.

TABLE 5.21 Approximate curve fit constants for calculating dispersion coefficients as a function of atmospheric stability classification and downwind distance from the air pollution source in a rural environment

STABILITY CLASS	A	B	x < 1 km C	D	F	x > 1 km C	D	F
A	213	0.894	440.8	1.941	9.27	459.7	2.094	9.6
B	156	0.894	106.6	1.149	3.3	108.2	1.098	2.0
C	104	0.894	61.0	0.911	0	61.0	0.911	0
D	68	0.894	33.2	0.725	−1.7	44.5	0.516	−13.0
E	50.5	0.894	22.8	0.678	−1.3	55.4	0.305	−34.0
F	34	0.894	14.35	0.740	−0.35	62.6	0.180	−48.6

Source: Based on Martin, D. O. 1976. The Change of Concentration Standard Deviation with Distance" Journal Air Pollution control Association. 26(2):145-147.

$$\sigma_z^2 = \frac{2xD_z}{U} \tag{5.27}$$

Then Equation (5.25) can be rearranged so

$$c_i(x, y, z) = \frac{EM_{i,s}}{2\pi U \sigma_y \sigma_z} \exp\left\{ -\frac{1}{2}\left(\frac{y}{\sigma_y}\right)^2 - \frac{1}{2}\left(\frac{z}{\sigma_z}\right)^2 \right\} \tag{5.28}$$

The dispersion coefficients are based on the previously discussed stability categories and are referred to as the Pasquill–Gifford coefficients. Different empirical correlations have been developed. General equations for the Pasquill–Gifford dispersion coefficients for rural areas are provided in Equations (5.29) and (5.30), and the coefficients for the empirical equations for rural dispersion in flat terrain are provided in Table 5.21. The equations suitable for urban areas are presented in Table 5.22, where the distance x has the units of kilometers.

$$\sigma_y = a(x)^b \tag{5.29}$$

TABLE 5.22 Pasquill–Gifford dispersion coefficients for urban sites, with downwind distance, x, measured in meters (m);

STABILITY	σ_y (m)	σ_z (m)
A–B	$0.32x(1 + 0.0004x)^{-0.5}$	$0.24x(1 + 0.0001x)^{0.5}$
C	$0.22x(1 + 0.0004x)^{-0.5}$	$0.20x$
D	$0.16x(1 + 0.0004x)^{-0.5}$	$0.14x(1 + 0.0003x)^{-0.5}$
E–F	$0.11x(1 + 0.0004x)^{-0.5}$	$0.08x(1 + 0.0015x)^{-0.5}$

Source: Based on Griffiths, R.F. (1994) Errors in the use of the Briggs parameterization for atmospheric dispersion coefficients. Atmospheric Environment. 28(17)2861-2865.

$$\sigma_z = c(x)^d + f \tag{5.30}$$

For air pollutants emitted from a smokestack, the x-axis is aligned with the prevailing wind direction. and the effective stack height must be considered, Thus, the general equation for Gaussian dispersion from an elevated point source becomes

$$c_i(x, y, z) = \frac{EM_{i,s}}{2\pi U \sigma_y \sigma_z} \exp\left\{ -\frac{1}{2}\left(\frac{y}{\sigma_y}\right)^2 - \frac{1}{2}\left(\frac{z - H}{\sigma_z}\right)^2 \right\} \tag{5.31}$$

The maximum ground-level concentration can be determined for estimating compliance with NAAQS and for estimating the maximum exposure to an individual downwind. The most conservative estimate assumes that pollutants cannot be dispersed into the ground, but are instead reflected by the ground (see Figure 5.34). For gaseous pollutants that are reflected by the ground, Equation (5.31) can be modified to account for the reflected particles:

$$c_i(x, 0, 0) = \frac{EM_{i,s}}{\pi U \sigma_y \sigma_z} \exp\left\{ -\frac{1}{2}\left(\frac{H}{\sigma_z}\right)^2 \right\} \tag{5.32}$$

Particles with a particle diameter less than 20 μm typically have a very low settling velocity and move with the mean air stream. Particles less than 20 μm in diameter may be modeled with Equation (5.31), but the particles may adhere to the ground and are reflected to a lesser degree than gaseous particles. Therefore, Equation (5.31) is unreliable for estimating the maximum concentration of particulate matter. Particles greater than about 20 μm in diameter may have a significant settling velocity that should be considered and require modifications to the general Gaussian equation for accurate estimates of downwind concentrations.

Atmospheric dispersion models may be used for developing state implementation plans, new source review, environmental audits, environmental assessments, liability cases, and risk assessments. The information required for the application of most models is similar and is listed in Table 5.23. Various air dispersion models have been developed and adapted to a particular use, depending on topographic conditions, the type of pollutants, and the expected application of model results. A few of the common models are listed in Table 5.24.

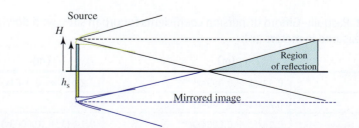

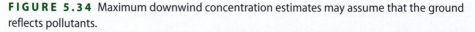

FIGURE 5.34 Maximum downwind concentration estimates may assume that the ground reflects pollutants.

Source: Bradley Striebig.

TABLE 5.23 Information required and available from air dispersion modeling

SOURCE INFORMATION	METEOROLOGICAL INFORMATION	RECEPTOR INFORMATION
Coordinates of source	Pasquill stability class	Coordinates
Stack height	Wind direction	Elevation of ground level at receptor
Stack top ID	Wind speed	Height of receptor above ground level
Stack gas exit velocity	Temperature	
Stack gas exit temp	Mixing height (urban)	
Directionally dependent building width	Mixing height (rural)	
Directionally dependent building height		
Elevation of stack base		

TABLE 5.24 Recommended air dispersion models for specific application

APPLICATION		SCREENING	REFINED SHORT TERM	REFINED LONG TERM
Flat or simple terrain	Rural	SCREEN	ISCST, OCD, BLP, FDM	ISCST or ISCLT
	Urban	SCREEN	ISCST, RAM, FDM	ISCST or ISCLT
Complex terrain	Rural	CTSCREEN VALLEY COMPLEX-I RTDM	CTDM PLUS	
	Urban	CTSCREEN VALLEY SHORTZ LONGZ		

EXAMPLE 5.7   Modeling Emissions from an Asphalt Processing Facility

A drum mix/hot mix asphalt plant uses a No. 2 fuel oil-fired dryer for asphalt production. The asphalt plant operates in a rural area, and you have been asked to quantify its impacts on the air quality at a mobile home park located 1,000 m directly downwind from the plant.

The plant produces 1.1×10^6 Mg of asphalt per year. The plant operates around the clock, 365 days a year, 24 hours per day. The production plant is 6 meters high, 30 meters wide, and 40 meters in length (aligned with the prevailing winds). Assume the pollutants are not reflected by the ground.

The 4 m³/s (actual) of exhaust from the plant exits the 1-meter-diameter stack at 1,000 K. The height of the stack is 10 meters high. The air pressure within the stack is essentially equal to the ambient air pressure of 1 atm.

The average ambient temperature measured at the stack height is 300 K. The average temperature measured at 110 meters on a nearby radio tower is 299.02 K. The prevailing wind speed is 5 m/s directly towards the mobile home park. The upwind maximum reported ambient air NO_2 concentration was 2 ppb.

Determine:

1. The mass emission rate of NOx in units of mg/s.

2. The plume rise from the stack in m.

3. The concentration of NOx directly downwind from the asphalt plant at the mobile home park. Calculate the concentration at ground level for gaseous pollutants that are not absorbed by the ground.

Solution

1. The EPA in Chapter 11 of the U.S. EPA AP-42 lists the emission factor for nitrogen oxide emissions from an asphalt plant. According to the AP-42 list, nitrogen oxide emissions will be 0.055 lb per ton. The emission rate is calculated from

$$EM = \left(1.1 \times 10^6 \, \frac{Mg}{year}\right)\left(0.055 \, \frac{lb}{ton}\right)\left(0.5 \, \frac{\frac{kg}{Mg}}{\frac{lb}{ton}}\right)\left(\frac{10^6 mg}{kg}\right)\left(\frac{yr}{31,536,000 \, s}\right)$$

$$= 960 \, \frac{mg}{s} \text{ of NOx}$$

2. The plume rise from the smokestack can be determined from Equation (5.19), but first the buoyancy flux and stability coefficient must be calculated.

The gas velocity exiting the smokestack is

$$v_s = \frac{Q}{\left(\frac{\pi D^2}{4}\right)} = \frac{4}{\left(\frac{\pi (1)^2}{4}\right)} = 5.1 \, \frac{m}{s}$$

The buoyancy flux may be determined from Equation (5.21):

$$F \frac{m^4}{s^3} = \frac{g v_s D_s^2 (T_s - T_a)}{4 T_a} = \frac{(9.81)(63.7)(1)^2 (1,000 - 300)}{4(300)} = 29$$

The potential temperature gradient is determined from Equation (5.17):

$$\frac{\Delta \theta}{\Delta z} \cong \left(\frac{\Delta T}{\Delta z}\right)_{actual} + \Gamma = \left(\frac{299.02 - 300}{110 - 10}\right) + 0.0098°C/m = 0°C/m$$

The atmospheric stability is neutral, so we must use Equation (5.22) to determine x^* in a neutral atmosphere for $F < 55$:

$$x^* = 14 F^{\frac{5}{8}} \quad \text{if} \quad F < 55 \quad \text{so} \quad x^* = 14(29)^{\frac{5}{8}} = 115 \, m$$

Then the plume rise is estimated from the Briggs equation (5.21):

$$\delta h \, m = \frac{1.6(3.5 x^*)^{\frac{2}{3}}(F)^{\frac{1}{3}}}{U} = \frac{1.6(3.5 \times 115)^{\frac{2}{3}}(29)^{\frac{1}{3}}}{5} = 54 \, m.$$

3. The general dispersion equation is used to estimate the downwind dispersion. From the potential temperature gradient, the atmosphere is characterized by the

neutral, D, stability category. The dispersion coefficients for a rural atmosphere are estimated from Table 5.21:

$$\sigma_y = a(x)^b = 68(1)^{0894} = 68 \text{ m}$$

$$\sigma_z = c(x)^d + f = 44.5(1)^{0.516} - 13 = 31.5 \text{ m}$$

Substituting the variables into the general dispersion equation (5.31) yields

$$C_{\text{NOx},(1000,0,0)} = \frac{EM}{2\pi U \sigma_y \sigma_z} \exp\left\{ -\frac{1}{2}\left(\frac{y}{\sigma_y}\right)^2 - \frac{1}{2}\left(\frac{z-H}{\sigma_z}\right)^2 \right\}$$

$$= \frac{960}{2\pi(5)(68)(31.5)} \exp\left\{ -\frac{1}{2}\left(\frac{0}{68}\right)^2 - \frac{1}{2}\left(\frac{0-(10+54)}{31.5}\right)^2 \right\}$$

$$= 1.8 \times 10^{-3} \frac{\text{mg}}{\text{m}^3} \text{ of NOx}$$

Converting the concentration one kilometer downwind at the trailer park to ppb as NO_2:

$$C_{\text{NOx},(1000,0,0)}[\text{ppb}] = \frac{24.5\left(2.8 \times 10^{-3} \frac{\text{mg}}{\text{m}^3}\right)}{46} \times 1{,}000 = 0.98 \text{ ppb NOx increase}$$

Since the maximum reported upwind NO_2 concentration was 2 ppb, the downwind concentration increases to 3 ppb. The asphalt plant in this location and atmospheric conditions will increase exposure to NO_2 by approximately 50% but will still be well below the NAAQS standards in Table 5.1.

5.5 Global Impacts of Air Pollutants

Scientific evidence has developed an absolute consensus that anthropogenic emissions have significantly altered our atmosphere at regional and global scales. There has been unprecedented international scientific cooperation to study global air pollutant effects and potential mitigation strategies.

Imagine, looking up at the sky one day and seeing a gaping hole in the atmosphere and looking directly into space, without the blue sky and clouds. Joe Farman, Brian Gardiner, and John Shankin were members of a 1985 British Antarctic Survey (BAS) that was studying the atmosphere in the lower part of the Southern Hemisphere. These scientists were looking for what was thought to be a relatively small effect of chlorinated chemicals on ozone concentration in the atmosphere released from air transportation proposed in 1974 by Drs. Frank Sherwood Roland and Maria J. Molina. Some satellite data were producing what was thought to be unreliable data representing ozone concentrations. The British Antarctic Survey sought to directly measure the ozone concentrations in the atmosphere over Antarctica. The British research team "looked" into the sky with their equipment to measure ozone, and they could not believe what they "saw." Their first presumption was that their instrumentation was not working correctly. Only after rechecking their instruments and data did they trust their own observations. The direct ozone measurements and satellite data showed a 40% decrease in ozone concentrations over Antarctica by 1984. By 1999, ozone concentrations in the Antarctic had

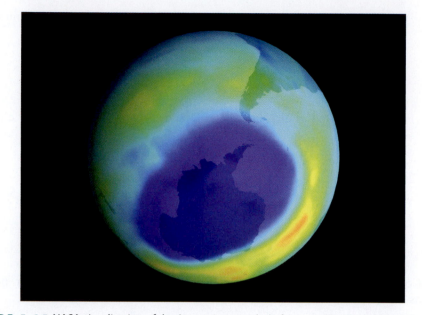

FIGURE 5.35 NASA visualization of the Antarctic ozone hole from satellite data collected on September 10, 2000.

Source: NASA/Goddard Space Flight Center, Scientific Visualization Studio

decreased by 57% in what had become known as the "Antarctic ozone hole," illustrated by the National Aeronautics and Space Agency (NASA) for the year 2000 in Figure 5.35. Drs. Rowland and Molina received a Nobel Prize for their work in atmospheric chemistry that explained how parts-per-billion concentrations of chlorine in atmosphere could react to destroy ozone.

Approximately 90% of ozone (O_3) in the Earth's atmosphere is found in the stratosphere (see Figure 5.36). The ozone in the atmosphere absorbs nearly all of the UV-c ultraviolet radiation from the sun, most of the UV-b irradiation, and about 50% of the UV-a irradiation (Figure 5.37).

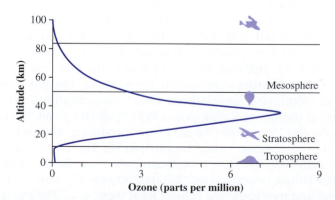

FIGURE 5.36 The concentration of ozone varies with altitude. Peak concentrations, an average of 8 molecules of ozone per million molecules in the atmosphere, occur between an altitude of 30 and 35 kilometers.

Source: NASA, http://ozonewatch.gsfc.nasa.gov/facts/SH.html

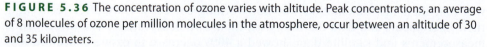

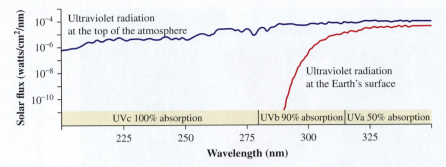

FIGURE 5.37 Solar ultraviolet radiation is largely absorbed by the ozone in the atmosphere—especially the harmful, high-energy UV-a and UV-b. The graph shows the flux (amount of energy flowing through an area) of solar ultraviolet radiation at the top of the atmosphere (top line) and at the Earth's surface (lower line). The flux is shown on a logarithmic scale, so each tick mark on the y-axis indicates 10 times more energy.

Source: NASA, http://ozonewatch.gsfc.nasa.gov/facts/SH.html

The "hole" in the stratospheric ozone layer has been mainly attributed to the widespread use and emissions of chlorofluorocarbons (some of which are illustrated in Figure 5.38). Chlorofluorocarbons (CFCs) were invented in the early 1930s and were used for a wide variety of commercial and industrial applications, including aerosol spray cans, coolants, refrigerants, and fire retardants (Figure 5.38). The simplest CFCs contain one or two carbon atoms and chlorine or fluorine, such as CFC_{-12} (CF_2Cl_2). The CFCs are relatively inert, insoluble, transparent, and do not degrade with time. They persist in the atmosphere until they migrate to the upper atmosphere where cold Antarctic stratospheric clouds convert inert chlorine (Cl) molecules to reactive chlorine forms that react with ozone to form diatomic oxygen (O_2). The production of reactive chlorine requires sunlight (Figure 5.39), which initiates the following reactions:

$$2Cl + 2O_3 \rightarrow 2ClO + 2O_2 \tag{5.33}$$

$$ClO + ClO + M \rightarrow Cl_2O_2 + M \tag{5.34}$$

$$Cl_2O_2 \xrightarrow{uv} Cl + ClO_2 \tag{5.35}$$

$$ClO_2 + M \rightarrow Cl + O_2 + M \tag{5.36}$$

The net reaction for chlorine in the stratosphere is

$$2O_3 \xrightarrow{Cl,uv,\&M} 3O_2 \tag{5.37}$$

Bromine, another active stratospheric halogen, goes through a similar process:

$$ClO + CBrO \rightarrow Br + Cl + O_2 \tag{5.38}$$

$$Cl + O_3 \rightarrow ClO + O_2 \tag{5.39}$$

$$Br + O_3 \rightarrow BrO + O_2 \tag{5.40}$$

Chlorine monoxide (ClO) forms when the South Pole begins to receive solar radiation in the spring. Chlorine monoxide (ClO) atoms combine to form a dimer (Cl_2O_2) that reacts with the sunlight to release diatomic oxygen and form more

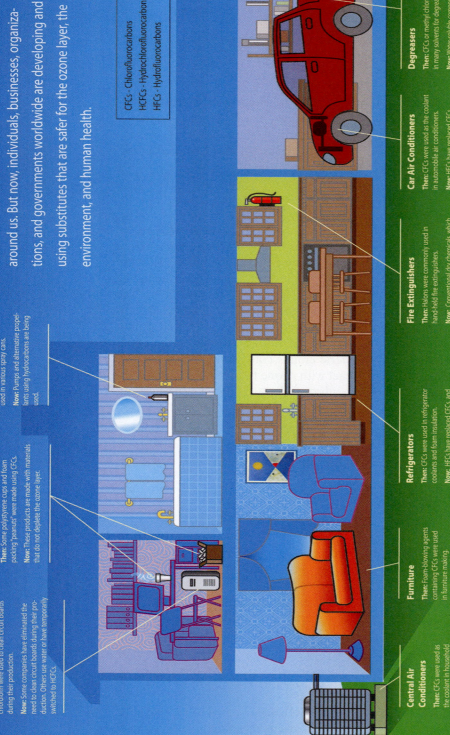

Prior to the 1980s, ozone-depleting substances were all around us. But now, individuals, businesses, organizations, and governments worldwide are developing and using substitutes that are safer for the ozone layer, the environment, and human health.

CFCs – Chlorofluorocarbons
HCFCs – Hydrochlorofluorocarbons
HFCs – Hydrofluorocarbons

Computers
Then: Solvents containing CFCs and methyl chloroform were used to clean circuit boards during their production.
Now: Some companies have eliminated the need to clean circuit boards during their production. Others use water or have temporarily switched to HCFCs.

Polystyrene Cups and Packing Peanuts
Then: Some polystyrene cups and foam packing "peanuts" were made using CFCs.
Now: These products are made with materials that do not deplete the ozone layer.

Aerosol Cans
Then: CFCs were the propellant used in various spray cans.
Now: Pumps and alternative propellants using hydrocarbons are being used.

Central Air Conditioners
Then: CFCs were used as the coolant in household air conditioners.
Now: HCFCs and HFCs have replaced CFCs.

Furniture
Then: Foam-blowing agents containing CFCs were used in furniture making.
Now: Water-blown foam is being used.

Refrigerators
Then: CFCs were used in refrigerator coolants and foam insulation.
Now: HFCs have replaced CFCs, and substitutes are on the horizon that will not deplete the ozone layer.

Fire Extinguishers
Then: Halons were commonly used in hand-held fire extinguishers.
Now: Conventional dry chemicals, which don't deplete the ozone layer, and water have replaced halons. HFCs are also used.

Car Air Conditioners
Then: CFCs were used as the coolant in automobile air conditioners.
Now: HFCs have replaced CFCs.

Degreasers
Then: CFCs or methyl chloroform were used in many solvents for degreasing.
Now: Water-soluble compounds and hydrocarbon degreasers that do not deplete the ozone layer are available for many applications.

FIGURE 5.38 CFC and other ozone-depleting substances used in the home prior to ratification of the 1987 Montreal Protocol.

Source: USEPA (2007b) Achievements in Stratospheric Ozone Protection: Progress Report. Office of Air and Radiation. U.S. Environmental Protection Agency. Washington, D.C.

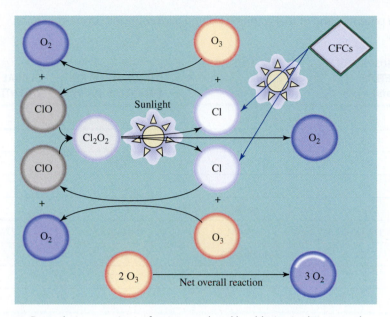

FIGURE 5.39 Degradation reactions of ozone, catalyzed by chlorine in the atmosphere.

Source: Based on R.S. Stolarski, (1985) A Century of Nature: Twenty-one discoveries that changed Science and the World. Chapter 18: A hole in the Earth's shield. University of Chicago Press.

reactive chlorine compounds that begin the chain reaction again. It has been estimated that one chlorine molecule may destroy over 100,000 ozone molecules before the chlorine is removed from the atmosphere. The chlorine atom acts as a catalyst in the chain reaction proposed by Rowland and Molina that ultimately converts ozone into the diatomic oxygen (O_2) which absorbs much less UV irradiation.

As scientific understanding of the process progressed, scientists observed that the ozone concentrations were reduced by 5% over the midlatitudes, not just in the Antarctic. Scientists have also observed the existence of a similar ozone depression in the Arctic as well as in the Antarctic. Anthropogenic chlorine sources are responsible for approximately 84% of chlorine in the stratosphere, with natural sources contributing only about 16%.

The relative contribution of an individual chemical to participate in stratospheric ozone depletion is given by that compound's ozone-depletion potential. A single CFC molecule is 4,680 to 10,720 times more effective at trapping heat than a single molecule of CO_2 (see Table 5.25). A decrease in ozone concentrations in the upper atmosphere from reactions with stratospheric chlorine results in more UV radiation reaching the Earth's surface. Increased exposure to UV radiation may increase the risk of skin cancer, eye damage, and suppression of the immune system. Increased UV radiation can also damage marine phytoplankton and crops, such as soybeans, and reduce the crop yield. Furthermore, CFCs and other ozone-depleting substances (ODS) have a disproportionate effect on the greenhouse effect by absorbing heat re-radiated from the Earth's surface. The global warming potential (GWP), which will be discussed in more detail in subsequent chapters, compares the CFC's effectiveness as a greenhouse gas to carbon dioxide.

As a result of these unprecedented observations of anthropogenic impacts on the Earth's atmosphere, the Montreal Protocol was adopted in September of

TABLE 5.25 Ozone-depletion potential and global warming potential for common ozone-depleting substances and their alternatives

HALOGEN SOURCE GASES	ATMOSPHERIC LIFETIME (years)	GLOBAL EMISSIONS IN 2008 (Kt/yr)[a]	OZONE-DEPLETION POTENTIAL (ODP)[c]	GLOBAL WARMING POTENTIAL (GWP)[c]
Chlorine gases				
CFC-11	45	52–91	1	4,750
CFC-12	100	41–99	0.82	10,900
CFC-13	85	3–8	0.85	6,130
Carbon tetrachloride (CCl_4)	26	40–80	0.82	1,400
HCFCs	1–17	385–481	0.01–0.12	77–2,220
Methyl chloroform (CH_3CCl_3)	5	Less than 10	0.16	146
Methyl chloride (CH_3Cl)	1	3600–460	0.02	13
Bromine gases				
Halon-1301	65	1–3	15.9	7,140
Halon-1211	16	4–7	7.9	1,890
Methyl bromide (CH_3Br)	0.8	110–150	0.66	5
Very short-lived gases (e.g., $CHBr_3$)	Less than 0.5	[b]	[b]very low	[b]very low
Hydrofluorocarbons (HFCs)				
HFC-134a	13.4	149 ± 27	0	1,370
HFC-23	222	12	0	14,200
HFC-143a	47.1	17	0	4,180
HFC-125	28.2	22	0	3,420
HFC-152a	1.5	50	0	133
HFC-32	5.2	8.9	0	716

[a] Includes both human activities (production and banks) and natural sources. Emissions are in units of kilotons per year (1 kiloton = 1,000 metric tons = 1 gigagram = 10^9 grams).

[b] Estimates are very uncertain for most species.

[c] 100-year GWPs. Values are calculated for emission of an equal mass of each gas.

Source: Based on Fahey, D. W. and Hegglin, M. I. (2010) Twenty Questions and Answers About the Ozone Layer: 2010 Update. United Nations Environment Programme.

1987. The **Montreal Protocol** was an international treaty, ultimately signed by 191 countries. The treaty phased out and banned the use and production of CFCs and ODSs. The protocol was amended in 1990 to end CFC and other ODS use by 2000 in economically developed countries, and funds were allocated to help developing countries take similar steps. Emissions of ODSs have been declining for some time, as shown in Figure 5.40, and the total inorganic chlorine concentration in the stratosphere appears to have peaked in 1997 and 1998. As a result of the international cooperation, the ozone layer has not significantly thinned since 1998

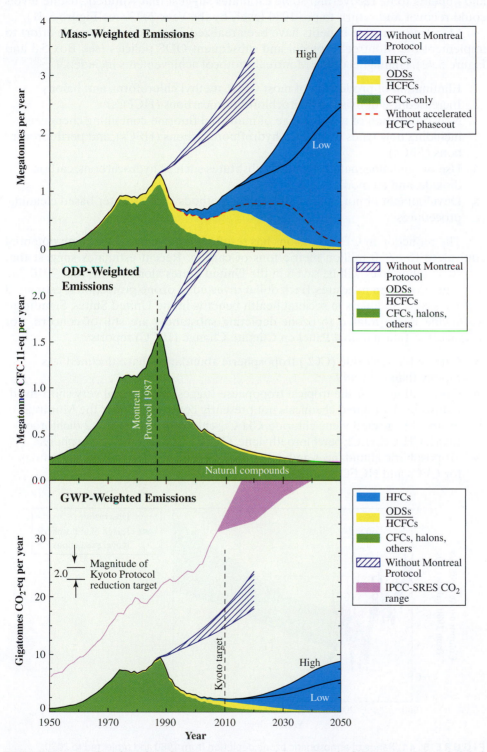

FIGURE 5.40 Trends and predictions in mass emissions, ozone-depletion potential, and global warming potential of ozone-depleting substances.

Source Based on Fahey, D. W. and Hegglin, M. I. (2010) Twenty Questions and Answers About the Ozone Layer: 2010 Update. United Nations Environment Programme.

and appears to be recovering. Some estimates suggest that Antarctic ozone levels could recover and return to pre-1980 levels by the year 2075 (see Figure 5.41).

Many measureable benefits have been realized from the collaborative effort to implement the Montreal Protocol and subsequent ODS policies (see Box 5.3 and Figure 5.42). Highlights of the Montreal Protocol achievements include:

1. Elimination of production of most CFCs, methyl chloroform, and halons
2. Increased use of existing hydrochlorofluorocarbons (HCFCs)
3. New production of a wide range of industrial fluorine-containing chemicals, including new types of HCFCs, hydrofluorocarbons (HFCs), and perfluorocarbons (PFCs)
4. Use of nonhalogenated chemical substitutes such as hydrocarbons, carbon dioxide, and ammonia
5. Development of not-in-kind alternative methods such as water-based cleaning procedures

The reduction in ODS emissions has resulted in a decreased warming potential equivalent to 8,900 million metric tons of carbon. Recent estimates suggest that over 6.3 million lives will be saved in the United States alone by the year 2016 due to ozone protection measures. Each dollar invested in ozone protection is estimated to have saved 20 dollars in societal health benefits in the United States. Significant worldwide issues related to ozone-depleting substances are still of concern. For instance, the International Panel on Climate Change (IPCC) reports:

- Carbon tetrachloride (CCl_4) tropospheric abundances have declined less rapidly than expected.
- Observations near the tropical tropopause suggest that several very short-lived industrial chlorinated chemicals not presently controlled under the Montreal Protocol (e.g., methylene chloride, CH_2Cl_2; chloroform, $CHCl_3$; 1,2 dichloroethane, CH_2ClCH_2Cl; perchloroethylene, CCl_2CCl_2) reach the stratosphere.
- Tropospheric abundances and emissions of HFCs, used mainly as substitutes for CFCs and HCFCs, continue to increase.

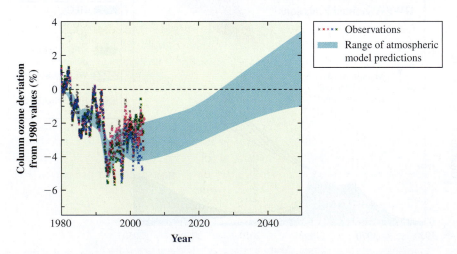

FIGURE 5.41 Range of atmospheric ozone depletion from 1980 and projected to 2050.

Source: Based on IPCC/TEAP, Bert Metz, Lambert Kuijpers, Susan Solomon, Stephen O. Andersen, Ogunlade Davidson, José Pons, David de Jager, Tahl Kestin, Martin Manning, and Leo Meyer (Eds) (2005) Safeguarding the Ozone Layer and the Global Climate System: Issues Related to Hydrofluorocarbons and Perfluorocarbons Summary for Policymakers and Technical Summary. IPCC, Geneva, Switzerland.

BOX 5.3 Changes in Mass Emissions, Ozone-Depletion Potential (ODP), and Global Warming Potential (GWP) as a Result of Implementing the Montreal Protocol and Other International Treaties

CFCs have been inexpensive and effective coolants and industrial chemicals that have been widely used and manufactured. As a result of increasing industrialization, CFC and chlorinated organic compound manufacturing was expected to increase in an exponential fashion had the effects of the CFCs on the ozone layer not been identified and quick action taken to change those trends (Figure 5.40). The Montreal Protocol was only the first of many sequential actions taken to reduce the levels of reactive chlorine in the stratosphere. Substitute chemicals have been developed to replace CFCs and other ODSs, but environmental impacts have been found to be associated with the replacement compounds (see Table 5.25 and Figure 5.40).

It is important to note that while the mass of total emissions from industrial use of hydrofluorocarbon (HFC) coolants and ODS replacements is expected to increase in the future, the environmental impacts of the replacement chemicals are likely to be much smaller. HFCs have no appreciable reaction in the stratosphere with ozone, so that the emissions of ozone-depleting substances are expected to continue to decrease even as overall industrial emissions increase. Unfortunately, the global warming potential of HFCs ranges from 133 to 14,200 times the GWP of CO_2, so the continued increase in these chemicals as illustrated in the bottom graph in Figure 5.40 is expected to contribute to the anthropogenically enhanced greenhouse gas effect and resultant climate change (which is discussed in Chapter 6).

5.6 Summary

The control and mitigation of pollutants in the air protect ecosystem diversity and human health, which in turn lead to economic and social impacts from the local to the global scale. National regulations and international treaties are attributed with saving millions of lives and providing significant economic and environmental benefits. Airborne pollutants in the atmosphere may undergo a variety of chemical transformations. Some air pollutants have direct environmental and health impacts, while others, like the precursors to photochemical smog and ozone-depleting substances, undergo reactions in the environment that yield adverse environmental or health effects. The mass balance principles and emission rates that have been discussed in prior chapters of this book can also be applied to calculate or estimate air pollutant emission rates. Physical and chemical transport models are useful tools that allow engineers and scientists to estimate the potential effects of air pollution and air pollution control technologies on local, regional, and global scales. Depending on the chemical and industrial process, air pollution control technologies may involve treating emissions from an industrial process or reducing or replacing the chemicals used in a process or product. Regulatory air pollution control policies have maintained or improved air quality in many regions of the world. For example, international regulations and the ban on CFCs have helped reduce the magnitude of the ozone hole. However, international agreements on the future control of greenhouse gases, discussed in the next chapter, have been much more difficult to agree upon and implement.

References

ATS. (2000). "What constitutes an adverse health effect of air pollution?" Official statement of the American Thoracic Society. *American Journal of Respiratory and Critical Care Medicine* 161:665–673.

Buonicore, A. J., and Davis, W. T. (1992). *Air Pollution Engineering Manual*. New York: Van Nostrand Reinhold.

CDC. (2004). The health consequences of smoking: A report of the Surgeon General. Centers for Disease Control and Prevention.

CDC. (2008). "Smoking-attributable mortality, years of potential life lost, and productivity losses—United States, 2000–2004." *Morbidity and Mortality Weekly Report* 57(45):1226–1228. Centers for Disease Control and Prevention.

Chou, M. S., and Lu, S. L. (1998). "Treatment for 1,3-butadiene in an air stream by biotrickling filter and a biofilter." *Journal of Air and Waste Management Association* 48(8):711–720.

Cooper, C. D., and Alley, F. C. (1986). *Air pollution Control: A design approach*. Prospect Heights, IL: Waveland Press.

Devinny, J. S., Deshusses, M. A., and Webster, T. S. (1999). *Biofiltration for air pollution control*. Boca Raton, FL: Lewis Publishers.

Fahey, D. W., and Hegglin, M. I. (2010). Twenty questions and answers about the ozone layer: 2010 Update. United Nations Environment Programme.

Griffiths, R. F. (1994). "Errors in the use of the Briggs parameterization for atmospheric dispersion coefficients." *Atmospheric Environment* 28(17):2861–2865.

Hanna, S. R., Briggs, G. A., and Hosker, R. P., Jr. (1982). *Handbook on atmospheric diffusion*. Technical Information Center, U.S. Department of Energy.

Heinsohn, R. J. (1991). *Industrial ventilation: Engineering principles*. New York: John Wiley & Sons.

Heinsohn, R. J., and Kabel, R. L. (1999). *Sources and control of air pollution*. Upper Saddle River, NJ: Prentice-Hall.

Hesketh, H. E. (1996). *Air pollution control: Traditional and hazardous pollutants*. Lancaster, PA: Technomic Publishing Co.

IPCC/TEAP, Bert Metz, Lambert Kuijpers, Susan Solomon, Stephen O. Andersen, Ogunlade Davidson, José Pons, et al. (Eds.) (2005). Safeguarding the ozone layer and the global climate system: Issues related to hydrofluorocarbons and perfluorocarbons. Summary for policymakers and technical summary. IPCC, Geneva, Switzerland.

Noll, K. E. (1999). *Fundamentals of air quality systems: Design of air pollution control devices*. American Academy of Environmental Engineers. Annapolis, MD.

Palmer, P. (2008). "Quantifying sources and sinks of trace gases using space-borne measurements: Current and future science." *Philosophical Transactions of the Royal Society* A 366:4509–4528.

Pasquill, F (1961). "The estimation of the dispersion of windborne material." *The Meteorological Magazine* 99(1063):33–49.

Richards, J. (1995). *Control of particulate emissions*. North Carolina State University, Raleigh, NC.

Seinfeld, J. H. (1986). *Atmospheric chemistry and physics of air pollution*. New York: John Wiley & Sons.

Slade, D. H. (Ed.) (1968). Meteorology and atomic energy. U.S. Atomic Energy Commission, Air Resources Laboratories, Environmental Sciences Services Administration, U.S. Department of Commerce, Washington, D.C.

Steigerwald, J. E. (1999). RACT/BACT/LAER Clearinghouse Clean Air Technology Center annual report for 1999: A compilation of control technology determinations. EPA-456/R-99-004. U.S. Environmental Protection Agency, Office of Air Quality Planning and Standards, Research Triangle Park, NC.

Stolarski, R. S. (1985). *A century of nature: Twenty-one discoveries that changed science and the World*. Chapter 18: A hole in the Earth's shield. Chicago: University of Chicago Press.

Striebig, B. A., Schneider, J., Spaeder, T., Mallery, M., and Heinsohn, R. (1997). "Analysis of advanced oxidation processes in a hybrid air pollution control system." *International Chemical Oxidation Association Symposium*, Chemical Oxidation: Technology for the Nineties, Nashville, TN, April 14-17, pp. 260–287.

Striebig, B. A., and Son, H.-K. (2001). "Comparison of water content fluctuations in a unidirectional and a switched-flow biofilter." *Air and Waste Management's 94th Annual Meeting and Exhibition*, Orlando, FL, June 24–28.

Striebig, B. A., Son, H-K., and Regan, R. W. (2001). "Understanding water dynamics in the biofilter." *Biocycle* 42(1):48–50.

Turner, D. B. (1994). *Workbook of atmospheric dispersion estimates: An introduction to dispersion modeling,* 2nd ed. Boca Raton, FL: CRC Press.

U.S. EPA. (1995). *Compilation of air pollutant emission factors*—Volume I: *Stationary point and area sources.* AP-42, 5th ed. U.S. Environmental Protection Agency, Washington, D.C.

U.S. EPA. (1998). National emission standards for hazardous air pollutants: Publicly owned treatment works. 40 CFR Part 63. U.S. Environmental Protection Agency, Washington, D.C.

U.S. EPA. (1999). Smog—Who does it hurt? What you need to know about ozone and your health. Office of Air and Radiation, U.S. Environmental Protection Agency, Washington, D.C.

U.S. EPA. (2001). Latest findings on National Air quality: 2000 Status and Trends. Office of Air Quality, Planning and Standards, U.S. Environmental Protection Agency, Research Triangle Park, NC.

U.S. EPA. (2003a). Latest findings on national air quality: 2002 status and trends. Office of Air Quality, Planning and Standards, U.S. Environmental Protection Agency, Research Triangle Park, NC.

U.S. EPA. (2003b). Ozone—Good up high, bad nearby. Office of Air and Radiation, U.S. Environmental Protection Agency. Washington, D.C.

U.S. EPA. (2005). Guidelines for carcinogen risk assessment, Final Report. Risk Assessment Forum, U.S. Environmental Protection Agency, Washington, D.C. *http://cfpub.epa.gov/ncea/cfm/*

U.S. EPA. (2007a). A plain English guide to the Clean Air Act. Office of Air Quality, Planning and Standards, U.S. Environmental Protection Agency, Research Triangle Park, NC.

U.S. EPA. (2007b). Achievements in stratospheric ozone protection: Progress report. Office of Air and Radiation. U.S. Environmental Protection Agency, Washington, D.C.

U.S. EPA. (2008). Latest findings on national air quality: Status and trends through 2006. Office of Air Quality, Planning and Standards, U.S. Environmental Protection Agency, Research Triangle Park, NC.

U.S. EPA. (2009a). Policy assessment for the review of the particulate matter national ambient air quality standards. Office of Air Quality Planning and Standards, U.S. Environmental Protection Agency, Research Triangle Park, NC.

U.S. EPA. (2009b). Ozone and your health. Office of Air and Radiation. U.S. Environmental Protection Agency, Washington, D.C.

U.S. EPA. (2011a). The benefits and costs of the Clean Air Act from 1990 to 2020: Summary teport. Office of Air and Radiation, U.S. Environmental Protection Agency, Research Triangle Park, NC.

U.S. EPA. (2011b). An overview of methods for EPA's National-Scale Air Toxics Assessment. Office of Air Quality Planning and Standards, U.S. Environmental Protection Agency, Research Triangle Park, NC.

U.S. EPA. (2011c). Policy assessment for the review of the particulate matter national Ambient Air Quality Standards. Office of Air Quality Planning and Standards, U.S. Environmental Protection Agency, Research Triangle Park, NC.

U.S. EPA. (2011d). Compilation of air pollutant emission factors—Update 2011. U.S. Environmental Protection Agency, Washington, D.C. *www.epa.gov/ttn/chief/ap42 /index.html*

U.S. EPA. (2012a). National Ambient Air Quality Standards (NAAQS). Air and Radiation, U.S. Environmental Protection Agency. *www.epa.gov/air/criteria.html*

U.S. EPA. (2012b). Our nation's air: Status and trends through 2010. Office of Air Quality, Planning and Standards, U.S. Environmental Protection Agency, Research Triangle Park, NC.

Wark, K., and Warner, C. F. (1981). *Air pollution: Its origin and control.* New York: HarperCollins.

Watson, A. Y., Bates, R. R., and Kennedy, D. (Eds.) (1988). *Air pollution, the automobile, and public health.* National Academy Press, Washington, D.C.

WEF. (1990). *Air pollutants at municipal wastewater Treatment facilities.* Water Environmental Federation, Alexandria, VA.

WEF. (1994). *Guidance document on control for VOC air emissions from POTWs.* Water Environmental Federation, Alexandria, VA.

WEF. (1995a). *Odor Control in Wastewater Treatment Plants.* Water Environmental Federation, Alexandria, VA.

WEF. (1995b). *Toxic air emissions from wastewater treatment facilities.* Water Environmental Federation, Alexandria, VA.

WHO. (2011). Indoor air pollution and health. Fact Sheet No. 292. World Health Organization.

Wilson, W. E., and Suh, H. H. (1997). "Fine particles and coarse particles: Concentration relationships relevant to epidemiologic studies." *Journal of the Air and Waste Management Association* 47:1238–1249.

Key Concepts

National Ambient Air Quality
 Standards
Prevention of Significant Deterioration
Air Quality Index
Acute health effect
Chronic health effects
Nasopharyngeal region
Trachiobronchial region
Hypoxia
Hypercapnia
Oncogenesis
Inhalable particle

Fine particle
Hazardous air pollutant
Emission factor
Lapse rate
Stability
Effective stack height
Dispersion model
Stratosphere
Troposphere
Stratospheric ozone depletion
Ozone-depleting substance

Active Learning Exercises

ACTIVE LEARNING EXERCISE 5.1: Solid waste combustion case study

Due to the current administration's change in power regulations, it may be cost effective to build another waste-to-energy plant in the Latah Creek Valley. You've been asked to evaluate the proposed facility in terms of potential impact on the air quality in the valley. The facility will duplicate the one currently operating on the west plains of Zinfadell County. You have been provided the following information about the proposed facility and site:

Solid waste loading rate: 800 metric tonnes
Generating capacity: 22,000 kW
Furnace temperatures: 2,500°C
Type: Municipal multi chamber refuse incinerator
Conversion factors: 1 lb_m = 7,000 gr (grains)

What will the uncontrolled emission rate of the following be in **kg/s**?
- Particulates
- SO_2
- CO
- NO_2

a) What removal efficiency is required to meet NSPS standards for particulates for the waste incinerator?
b) Are there any *possible* chronic effects from association with any of the compounds being emitted and, if so, (based on information in your text) what are they?
c) If the particles are all 0.1 μm in size, what part of the respiratory system is most likely to be affected? Reference where you obtain this answer for the Air Quality Board reviewing your work.
d) During a still, cloudy week in February, the volumetric airflow rate through the valley is 1×10^6 m³/s. What will be the steady-state concentration of CO, SO_2, and NO_2 in the valley?
e) Will the ambient air quality in the valley during this week be in compliance with the NAAQS?
f) Based on your answer to part (e), will healthy college-age students living in the valley report any acute symptoms of CO exposure? Reference where you obtained this answer as if the Air Quality Board would be reviewing your work.

Problems

5-1   If a typical adult inhales 3.5 L of air per breath and breathes 15 times per minute, how many grams of oxygen will be inhaled during 24 hours if the density of air is 1.225 kg/m³?

5-2   What is the average molecular weight of "clean air"?

5-3   Research one of the following incidents and list the air pollutants. Describe the possible health effects caused by exposure and describe the pathway for exposure to the pollutants.
a. Cyanide gas exposure from the Union Carbide plant in Bhopal, India
b. Chernobyl nuclear reactor, Ukraine
c. Asbestos and vermiculite mining in Libby, Montana, United States
d. Fukushima Daiichi nuclear power plant, Okuma, Fukushima, Japan

5-4   Is breathing second-hand smoke a voluntary or an involuntary risk?

5-5   Exactly 500.0 ml of CO is mixed with 999,500 ml of air. Calculate the resulting concentration of CO in ppm and mg/m³. Would you expect any health impacts from this air? Explain your answer.

5-6   What is the National Ambient Air Quality Standard for the following compounds expressed in μg/m³?

a. Carbon monoxide, 1-hr standard

b. Carbon monoxide, 8-hr standard

c. Nitrogen dioxide, 1-hr standard

d. Nitrogen dioxide, 8-hr standard

e. Ozone, 8-hr standard

f. Sulfur dioxide, 1-hr standard

g. Sulfur dioxide, 3-hr standard

5-7 Calculate the AQI for a city with a measured 8-hr carbon monoxide concentration of 4 mg/m³. What AQI-based level of health concerns should be reported?

5-8 Calculate the AQI for a city with a measured 1-hr nitrogen dioxide concentration of 0.267 mg/m³. What AQI-based level of health concerns should be reported?

5-9 Calculate the AQI for a city with a measured 8-hr ozone concentration of 0.110 mg/m³. What AQI-based level of health concerns should be reported?

5-10 Calculate the AQI for a city with a measured 24-hr $PM_{2.5}$ concentration of 0.023 mg/m³. What AQI-based level of health concerns should be reported?

a. What are the possible health effects?

b. What could someone do to protect himself or herself?

5-11 Describe the potential health effects from sulfur dioxide exposure to

a. 0.1 ppm_v SO_2 for 8 hr

b. 0.5 ppm_v SO_2 for 8 hr

c. 5.0 ppm_v SO_2 for 8 hr

5-12 Describe the potential health effects from carbon monoxide exposure to

a. 30 ppm_v CO for 12 hr

b. 100 ppm_v CO for 12 hr

c. 500 ppm_v CO for 12 hr

5-13 Identify a city or metropolitan area associated with high concentrations of nitrogen dioxide from Figure 5.16.

a. What is the population of the city or metropolitan area?

b. What is the population density per km² of the city or metropolitan area?

c. What is the population density of automobiles of the city or metropolitan area?

5-14 Identify a city or metropolitan area associated with high concentrations of ozone from Figure 5.20.

a. What is the population of the city or metropolitan area?

b. What is the population density per km² of the city or metropolitan area?

c. What is the population density of automobiles of the city or metropolitan area?

5-15 Identify a city or metropolitan area associated with increased total cancer risk from Figure 5.25.

a. What is the population of the city or metropolitan area?

b. What is the population density per km² of the city or metropolitan area?

c. What is the population density of automobiles of the city or metropolitan area?

5-16 What are the major natural and anthropogenic sources of primary PM less than 2.5 μm and greater than 2.5 μm for the following types of aerosol species?

a. Sulfates

b. Nitrates

     c. Minerals
     d. Ammonium
     e. Organic carbon
     f. Elemental carbon (EC)
     g. Metals
     h. Bioaerosols

5-17 What is the estimated atmospheric half-life and removal process for the following?
     a. Ultrafine particles
     b. Fine (accumulation) particles
     c. Coarse particles

5-18 What size particle (μm) is most likely to be deposited in the pulmonary region of the lungs?

5-19 Describe in what region of the respiratory system the following particles are likely to be deposited and how the particle could be removed from the respiratory system.
     a. A grain of dirt with a diameter of 500 μm
     b. A grain of sand with a diameter of 90 μm
     c. A particle of dust with a diameter of 25 μm
     d. A pollen particle with a diameter of 5 μm
     e. An aerosol with a diameter of 1 μm

5-20 A person smokes one cigarette in 60 sec in a waiting room. The concentration of smoke particles in the inhaled smoke is 10^{15} particles/m^3. The smoke particles have a diameter of 0.2 μm and a density of 0.8 g/ml. If the smoker breathes in air at a rate of 5 L per minute, what is the inhalation rate of smoke particles in mg/m^3 during the time he or she is smoking?

5-21 A miner has a documented diagnosis involving respiratory effects, cancer, and central nervous system impairments.
     a. Atmospheric conditions in his work environment could have what degree of causation for the miner's health diagnosis?
     b. What, if any, significance would there be in linking workplace exposure to causation of the symptoms if this individual smoked two packs of cigarettes a day, every day, for 30 years?

5-22 What gas analysis method(s) would be appropriate for analyzing hydrocarbons in the smokestacks at a petroleum refinery?

5-23 What gas analysis method(s) would be appropriate for analyzing carbon in the smokestacks from a cement kiln?

5-24 There are 40 customers who are heavy smokers in a small pub. Each person smokes one cigarette every 10 minutes. Each cigarette emits 10^{15} particles of smoke. The smoke particles have a diameter of 0.1 μm and a density of 0.8 g/m^3.
     a. What is the mass emission rate of smoke particles into the room in mg/m^3?
     b. If the room measures 5 m wide, 2.5 m high, and 8 m long, and the ventilation rate is 10 m^3 per hour, what is the steady-state smoke concentration in the room?
     c. If the background PM$_{2.5}$ concentration is 0.025 mg/m^3, what total steady-state smoke concentration would the nonsmoking bartender be exposed to during an 8-hr shift?
     d. The causality of smoke of this size has been determined for what potential health outcomes?
     e. Where would these particles be deposited in the respiratory system?

5-25   Find the estimated emissions listed below for a spreader stoker cogeneration facility that burns 10,000 tons of bituminous coal per year with a 3% by-weight sulfur content. Use the EPA's AP-42 (Volume 1, 5th ed.), Chapter 1: External Combustion Sources Emission factors for Bituminous and Sub-bituminous Coal Combustion.
   a. Tons of SOx emitted per year
   b. Tons of NOx emitted per year
   c. Tons of CO emitted per year

5-26   Find the estimated emissions listed below for an uncontrolled (post-NSPS) large wall-fired boiler and natural gas cogeneration facility that burns 500 million standard cubic feet (scf) of natural gas per year. Use the EPA's AP-42 (Volume 1, 5th ed.) Chapter 1: External Combustion Sources Emission Factors for Natural Gas Combustion.
   a. Tons of NOx emitted per year
   b. Tons of CO emitted per year

5-27   The following data are from the U.S. EPA AP-42 emission factors for malt beverages. A closed fermenter vents CO_2 = 2,100, VOC (ethanol) = 2.0, and hydrogen sulfide, H_2S = 0.015. The units for the emission factors in the EPA tables are in lbs of pollutant per 1,000 barrels (bbl). A brewery produces about 7 million barrels of product per year. Estimate the emissions of the following.
   a. Kg of carbon dioxide per year from fermenter venting
   b. Kg of VOCs (ethanol) per year from fermenter venting
   c. Kg of hydrogen sulfide per year from fermenter venting

5-28   The following data are from the U.S. EPA AP-42 emission factors for malt beverages. For a steam-heated brewers grain dryer, CO = 0.22, CO_2 = 53, and VOC (ethanol) = 0.73. The units for the emission factors in the EPA tables are in lbs of pollutant per 1,000 barrels (bbl). A brewery produces about 7 million barrels of product per year. Estimate the emissions of
   a. Kg of carbon monoxide per year from fermenter venting
   b. Kg of carbon dioxide per year from fermenter venting
   c. Kg of VOCs (ethanol) per year from fermenter venting

5-29   How many pounds of the following pollutants are emitted from burning one ton of bituminous coal in a domestic hand-fired furnace?
   a. PM
   b. Sulfur oxides
   c. Carbon monoxide
   d. Hydrocarbons
   e. Nitric oxides
   f. HCl

5-30   Assuming each mole of sulfur in fuel oil (#2) forms 1 mole of SO_2, is more or less sulfur emitted from the energy released from burning 1 ton of #2 fuel oil or from the same energy produced by burning supercritical PC coal with the best available control technology? Reference Appendix D and support your answer with an engineering analysis.

5-31   How much air is needed to burn 2 tons per year of methane generated at a municipal landfill?

5-32   How much air is needed to completely combust a mixture of 20,000 ppm$_v$ butane and 8,000 ppm$_v$ propane?

5-33   An exhaust air stream contains 40,000 ppm methane, 10,000 ppm ethane, and 5% oxygen from a landfill.
  a. How many moles of oxygen are required for combustion?
  b. Does additional air need to be added to the exhaust gas to reach the required amount of oxygen?
  c. If so, what percentage of make-up air must be added?

5-34   A traffic accident causes traffic to come to a halt in an underwater tunnel. The people in their vehicles allow their engines to idle. There are 100 vehicles stopped in the tunnel; assume their vehicle engines emit on average 0.269 g/min of VOC (aldehyde), 3.82 g/min CO, and 0.079 g/min NOx. The tunnel is long and narrow with a height of 10 m, a width of 25 m, and a length of 0.5 km. Assume the ambient air concentrations of CO, VOCs, and NOx are small. The mean air velocity through the tunnel, when traffic stops, is limited to 0.2 m/s.
  a. Assuming traffic remains halted long enough for the concentration of carbon monoxide, nitric oxides, and hydrocarbons in the canyon to reach steady state, what would the expected concentrations of carbon monoxide, nitric oxides and hydrocarbons be in ppm_v in the tunnel?
  b. Do the CO and NOx concentrations exceed NAAQS?
  c. Does the CO concentration exceed the OSHA's permissible exposure limit (PEL)?

5-35   The following data are recorded at meteorological field stations. Determine the environmental lapse rate and stability classification of the atmosphere.
  a. Temperature at ground level is 15.0°C. The temperature measured at 10 m is 15.1°C.
  b. Temperature at ground level is 15.0°C. The temperature measured at 10 m is 14.7°C.
  c. Temperature at ground level is 15.0°C. The temperature measured at 10 m is 15.5°C.
  d. Temperature at ground level is 15.0°C. The temperature measured at 10 m is 14.1°C.

5-36   Determine the Pasquill stability category for the following scenarios and the type of plume to be observed from a smokestack.
  a. An overcast day with leaves rustling.
  b. A sunny afternoon with the solar altitude 75 degrees above the horizon and loose paper blowing on the street.
  c. A sunny afternoon with the solar altitude 75 degrees above the horizon and smoke from a match rises vertically.
  d. A later summer day with broken low clouds and smoke from a match rises vertically.
  e. Nighttime when chimney smoke drifts but does not move wind vanes.
  f. A winter day when chimney smoke drifts but does not move wind vanes.

5-37   The inside diameter of a building exhaust smokestack is 2.0 m. The flow rate of ventilation air emitted through the stack is 10 m³/s at 25°C. The exhaust exits the stack into an atmosphere with a 3 m/s average wind speed and an ambient temperature of 21°C. Determine the effective stack height for a 30-m-high smokestack under the stability conditions calculated in 5.37 as assigned by your instructor.

5-38   The inside diameter of an incinerator's smokestack is 0.5 m. The flow rate of ventilation air emitted through the stack is 0.1 m³/s at 800°C. The exhaust exits the stack into an atmosphere with a 3 m/s average wind speed and a temperature of 23°C at the stack exit. Determine the effective stack height for a 10-m-high smokestack given the following additional information.

a. Ground-level temperature, $T = 23.2$

b. Ground-level temperature, $T = 23.0$

c. Ground-level temperature, $T = 22.0$

5-39 What is the maximum ground-level pollutant concentration (mg/m³) from an elevated source in an urban area with the effective stack height $H = 20$ m, cyanide (HCN) emissions equal to 0.5 g/s, and the average wind velocity = 3 m/s with a stability class C for the atmosphere? Assume $\sigma_z = 100$ m and $\sigma_y = 100$m.

5-40 A 1,530 MW bituminous coal-fired wet-bottom plant emits mercury into the atmosphere. The EPA AP-42 emission factor for the power plant is 6.8 grams of Hg per Terrajoule. The plant operates around the clock, 365 days a year, 24 hours per day. The 20 m³/s (actual) of exhaust from the plant exits the 2-m-diameter stack at 700°C. The stack is 50 m high. The air pressure within the stack is essentially equal to the ambient air pressure of 1 atm. The average ambient temperature measured at the stack height is 20°C. The average temperature measured at 110 m on a nearby radio tower is 18.5°C. The prevailing wind speed is 3 m/s. Determine:

a. The mass emission rate of Hg in units of mg/s

b. The plume rise from the stack in m

c. The maximum ground-level pollutant concentration (mg/m³) from an elevated source in ppm

d. The concentration of mercury 2 km directly downwind from the power plant. Calculate the concentration at ground level for gaseous pollutants that are not absorbed by the ground.

5-41 An 1,880 MW wet-bottom, wall-fired sub-bituminous coal power plant emits sulfur oxides, nitrogen oxides, and carbon monoxide into the atmosphere in a rural area. The plant uses coal with a 4% sulfur content. The plant operates around the clock, 365 days a year, 24 hours per day. The 25 m³/s (actual) of exhaust from the plant exits the 1-m-diameter stack at 800°C. The stack is 25 m high. The air pressure within the stack is essentially equal to the ambient air pressure of 1 atm. The average ambient temperature measure at the stack height is 20°C. The average temperature measured at 110 m on a nearby radio tower is 20.0°C. The prevailing wind speed is 2 m/s. Determine:

a. The AP-42 emission factor for the sulfur oxide, nitrogen oxides, or carbon monoxide (as assigned by your instructor)

b. The mass emission rate of the sulfur oxide, nitrogen oxides, or carbon monoxide (as assigned by your instructor) in units of mg/s

c. The plume rise from the stack in m

5-42 Create a list of chemical and/or materials that have changed in order to comply with the Montreal Protocol.

5-43 Formaldehyde is emitted from a furniture manufacturing plant. The process only operates 12 hours per week over four different days. The exhaust gas temperature is 25°C. You are an engineer for a low-cost furniture manufacturer. The owners are planning to significantly expand their manufacturing facility in a rural area of the northwest. The U.S. EPA has asked them to estimate the plant expansion effects on local air quality. The plant expansion will process 3 million m² of particle board (std, UF) daily.

You, the engineer, are given the following information based on previous sampling events.

- Exhaust gas temperature from the stack is on average 300 K and exits the stack at 145 m/s.
- Typical meteorological conditions were obtained from a local weather station:
  - Strong solar radiation
  - Ambient temperature = 299 K
  - Wind speed at 10 m = 2 m/s
  - Actual lapse rate = −0.015 K/m.
- Both formaldehyde and particulates are a concern. Use emission factors *for the most conservative estimate* of formaldehyde emissions. Also, assume that 1 mole of formaldehyde reacts in the atmosphere to form 1 mole of ozone.
- The data in Table 5.26 were collected from meteorological stack testing,

TABLE 5.26 Emissions from the furniture manufacturer

PM TYPE	PARTICLE SIZE (μm)	EMISSION FACTOR (mg/m²-day)
Chips	> 100	46
Sawdust	1 to 100	18
Fine PM	0.01 to 1	4
Total	0.01 to 100	68

Calculate the emission rate of formaldehyde and total particulate matter (PM) in mg/s.

  a. What region of the respiratory system is most likely to be affected by each type of PM?
     i)   Chips
     ii) Sawdust
     iii) Fine PM
  b. What region is least likely to be affected by each type of PM?
  c. What is the concentration of formaldehyde in units of PPM in the exhaust stack prior to being emitted into the atmosphere?
  d. What is the permissible exposure level (PEL) for formaldehyde?
  e. Does this concentration exceed the PEL for workers in the building?
  f. How high will the plume rise from the stack in meters?
  g. Use Excel to determine the distance in kilometers from the stack will the plume concentration at ground level be greatest (maximum)?
  h. What is the maximum predicted ground-level concentration of formaldehyde and total PM assuming these pollutants are reflected?
  i. The background concentration of PM is 10 mg/m³ and ozone is 45 microgram (µg/m3). Will this location remain in compliance with the United States NAAQS PM and ozone standards based upon the maximum predicted concentration of the pollutants?

CHAPTER 6

The Carbon Cycle and Energy Balances

FIGURE 6.1 One of many cartoons that illustrates the perceived complicated science of climate change.

Source: www.CartoonStock.com

According to a new U.N. report, the global warming outlook is much worse than originally predicted. Which is pretty bad when they originally predicted it would destroy the planet.

—JAY LENO

GOALS

THE EDUCATIONAL GOALS OF THIS CHAPTER are to use basic principles of chemistry, physics, and math to define and solve engineering problems, in particular mass and energy balance problems related to the Earth's carbon flux. This chapter will examine the modern problems associated with fossil fuel use, carbon dioxide concentrations in the atmosphere, and atmospheric climate change. The global context of climate change will be explored, in particular the relationships between climate change and resources.

OBJECTIVES

At the conclusion of this chapter, you should be able to:

- Define and discuss the major repositories and fluxes for carbon.

- Calculate carbon emissions from various fossil fuels.

- Define basic thermodynamic principles such as enthalpy, conduction, convection, and radiation.

- Solve steady-state blackbody radiation problems.

- List and define the major greenhouse gases.

- Describe and calculate changes to the global energy balance.

- Define and discuss the impacts of greenhouse gas and radiative forcing.

- Define and discuss anthropogenic climate change.

- Describe potential environmental and societal impacts related to climate change.

6.1 Introduction

Climate change has been a hit in the media for at least the past two decades. But when did scientists begin to realize the important link between atmospheric gas concentrations and our climate on Earth? Many people believe that the science related to climate change is a relatively new field. In this chapter, we will examine how scientists over 100 years ago developed the fundamental principles that predict how and why the climate changes. We'll also explore the fundamental equations and data that show how climate change is tied to anthropogenic emissions. We'll build on the tools developed in Chapters 1 and 2 to create mathematical models that can be used to predict changes in any system, regardless of whether this is a small laboratory reactor or the entire planet's atmosphere.

Simple but powerful mathematical models will be developed that allow the engineer and scientist to define problems and their solutions. The chemical transformation of fossil fuels through combustion will be evaluated. Mathematical models and fossil-fuel usage data will be used to estimate the amount of carbon dioxide added to the atmosphere from societies' energy consumption. The impact of these global changes will be investigated by developing simple physical thermodynamic models and solving an energy balance or radiation flow from the sun through the atmosphere and to the Earth. The basic chemical, physical, and mathematical models developed in this chapter form the basis for complex climate change models that are used to predict the impact of climate change on the environment, economy, and society.

6.2 Climate Science History

In 1827, Jean-Baptist Fourier recognized that the chemical composition of the atmosphere absorbed energy reflected from the surface of the planet and trapped heat from the sun in a manner similar to a greenhouse. The heat trapped by the gases in the atmosphere create conditions on the surface of the planet that are much more hospitable to human life as we know it than the temperature of the planet without this insulating atmospheric blanket.

John Tyndall measured the effects of carbon dioxide and water vapor on the absorption of energy radiated from the surface of the planet in 1860 and identified the importance of the concentration of these gases on the average surface temperature of the Earth. It wasn't until much later, in 1896, that Svante Arrhenius estimated the effects that increasing carbon dioxide concentrations would have on the Earth's average surface temperature. By balancing the knowledge of the atmosphere's composition and the energy into and out of the atmosphere, these early scientists were able to model and predict the impact of various components of the atmosphere on the greenhouse gas effect and, subsequently, the average surface temperature of the Earth.

Guy Stewart Callendar (1898–1964) was a noted steam and power engineer who was curious about how the climate worked. Callendar (Figure 6.2) examined the work of early scientific studies on the effect of carbon dioxide in the atmosphere and linked increasing carbon dioxide concentrations in the atmosphere to the emissions from fossil fuel use that occurred as a result of the Industrial Revolution. In 1938, Callendar published his paper "The Artificial Production of Carbon Dioxide and Its Influence on Temperature." Callendar's paper linked

FIGURE 6.2 Photo of Guy Stewart Callendar, one of the "fathers" of climate change science.

Source: G.S. Callendar Archive, University of East Anglia

anthropogenic emissions of carbon dioxide and predicted those emissions may impact the Earth's climate. In 1957, Hans Seuss and Roger Revelle referred to the "Callendar effect" as the change in climate brought about by increases in the concentration of atmospheric carbon dioxide, primarily through the processes of combustion. Since that time, scientists around the world have worked with the International Panel on Climate Change and other organizations to examine the fate and transport of carbon dioxide in the atmosphere. These scientists and engineers are trying to create more accurate models to predict how changing carbon dioxide concentration of the Earth's atmosphere will influence the climate.

The **International Panel on Climate Change (IPCC)** is comprised of a group of scientists from many nations on the planet that are intimately involved in better understanding changes to the Earth's climate. The IPCC (2007) statement on the current status of this research was summarized:

> *As climate science and the Earth's climate have continued to evolve over recent decades, increasing evidence of anthropogenic influences on climate change has been found. Correspondingly, the IPCC has made increasingly more definitive statements about human impacts on climate. Debate has stimulated a wide variety of climate change research. The results of this research have refined but not significantly redirected the main scientific conclusions from the sequence of IPCC assessments.*

Figure 6.3 illustrates how much the concentration of carbon dioxide, methane, and nitrous oxide has changed over the last two thousand years.

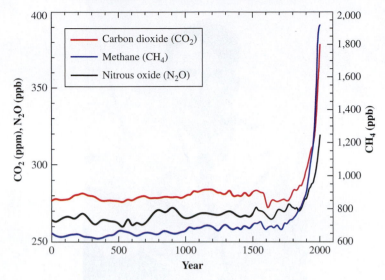

Concentrations of Greenhouse Gases from 0 to 2005

FIGURE 6.3 Carbon dioxide, methane, and nitrous oxide concentrations in the atmosphere.

Source: Based on IPCC (2007) Climate Change 2007: The Physical Science Basis. Contributions of working group I to the Fourth Assessment Report of the intergovernmental Panel on Climate Change [Solomon, S., Qin, D., Manning, M., Chen, Z., Marquis, M., Averyt, K.B., Tignor, M., and Miller, H. L. (eds.)] Cambridge University Press, Cambridge, United Kingdom and New York, NY, USA. 996 pp. FAQ 2.1, Figure 1. Page 135.

EXAMPLE 6.1   How Much Has the Atmosphere Changed?

The volume of the atmosphere at Standard Temperature and Pressure (STP) is approximately 3.94×10^{18} m³. For over a thousand years prior to the Industrial Revolution, the concentration of CO_2 in the atmosphere was approximately 280 ppm. The current (2011) concentration of CO_2 in the atmosphere is approximately 390 ppm. How much (mass) CO_2 has been added to the atmosphere since the Industrial Revolution?

Solution

First, convert units of ppm_v to mg/m³ in air for the concentration of carbon dioxide prior to the Industrial Revolution:

$$c_{280}\left(\frac{mg}{m^3}\right) = \frac{280\,(ppm_v) \cdot 28.96\left(\dfrac{mg}{mmol}\right)}{24.5\left(\dfrac{m^3 - ppm_v}{mmol}\right)} = 331\,\frac{mg}{m^3}$$

Similarly, the approximate present-day concentration of carbon dioxide in mg/m³ is

$$c_{390}\left(\frac{mg}{m^3}\right) = \frac{390\,(ppm_v) \cdot 28.96\left(\dfrac{mg}{mmol}\right)}{24.5\left(\dfrac{m^3 - ppm_v}{mmol}\right)} = 461\,\frac{mg}{m^3}$$

Subtract the pre-Industrial Revolution concentration from the present-day concentration of CO_2 to determine the increase of carbon dioxide in each cubic meter of air:

$$\Delta CO_2 = 461\ \frac{mg}{m^3} - 331\ \frac{mg}{m^3} = +130\ \frac{mg}{m^3}$$

The total mass of the atmosphere is estimated to be 5.14×10^{18} kg. The density of air at STP = 1.293 kg/m^3. If the atmosphere were all compressed and assumed be at conditions equivalent to STP, then

$$m_{CO_2\ added} = 130\ \frac{mg\ of\ CO_2}{m^3} \cdot \frac{kg\ of\ CO_2}{10^6\ mg\ of\ CO_2} \cdot \frac{1\ m^3\ of\ air}{1.293\ kg\ of\ air} \cdot 5.14 \times 10^{18}\ kg$$

air in the atmosphere, and

$$m_{CO_2\ added} = 517 \times 10^{12}\ kg\ of\ CO_2$$

Approximately 517×10^{12} kg of CO_2 have been added to the atmosphere since the Industrial Revolution.

6.3   Carbon Sources and Emissions

Since mass is neither created nor destroyed (except in nuclear reactions, and we will neglect those for the time being), the carbon added to the atmosphere must have come from somewhere. We have already discussed biogeochemical cycles. Just as we have followed water molecules in the hydrologic cycle, we can also trace the movement of carbon atoms from one repository to another.

Carbon concentrations in the atmosphere have increased significantly since the start of the Industrial Revolution. We can examine the possible sources and sinks of carbon in the Earth's atmosphere.

The total U.S. energy use is shown in Figure 6.4. Information about energy use and fossil fuel consumption can be used to calculate the carbon emissions associated

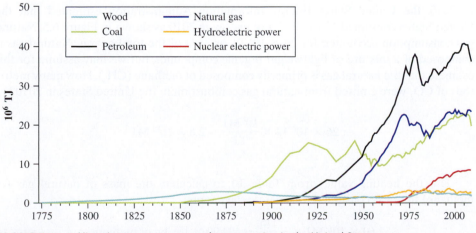

FIGURE 6.4 Historic energy sources and consumption in the United States.

Source: US EIA (2012) Annual Energy Review 2011. U.S. Energy Information Administration. Office of Energy Statistics. U.S. Department of Energy. Washington, D.C. DOE/EIA-0384(2011)

with the conversion of fossil fuels to energy. In general the combustion of fossil fuels can be described by

$$\text{Fuel} + \text{oxygen} \rightarrow \text{energy} + \text{carbon dioxide} + \text{water}$$

Depending on the chemical composition of the fossil fuel, differing amounts of energy and carbon dioxide are produced. Table 6.1 illustrates the differences in typical net heating values for common fuels.

TABLE 6.1 Typical heating values and default carbon dioxide emission factors for common fuels

FUEL	NET HEATING VALUE (MJ/kg)	DEFAULT CARBON DIOXIDE EMISSION FACTORS (KG CO_2/TJ)
Peat	10.4	106,000
Wood, oak	13.3–19.3	95,000–132,000
Wood, pine	14.9–22.3	95,000–132,000
Coal, anthracite	25.8	98,300
Charcoal	26.3	112,000
Coal, bituminous	28.5	94,600
Fuel oil, no, 6 (bunker C)	42.5	73,300
Fuel oil, no 2 (home heating oil)	45.5	77,400
Gasoline (84 octane)	48.1	69,300
Natural gas (density = 0.756 kg/m³)	53.0	56,100
Reference:	Davis and Masten (2009)	IPCC (2006)

EXAMPLE 6.2   Estimating U.S. Contributions to Atmospheric CO_2 from Natural Gas

In 2010, the United States Energy Information Administration estimated that the United States consumed 103×10^6 TJ of energy in 2010 as shown in Figure 6.5. Natural gas consumption accounted for 26×10^6 TJ, or 25% of the total energy. Natural gas is composed of a mixture of lightweight organic compounds, but we may assume for this estimate that the natural gas is primarily composed of methane (CH_4). How many metric tons of CO_2 were emitted from natural gas combustion in the United States in 2010?

$$26 \times 10^6 \text{ TJ} \times \frac{10^6 \text{ MJ}}{\text{TJ}} = 2.6 \times 10^{13} \text{ MJ}$$

Use the heating values from Table 3.1 to calculate the mass of natural gas (as methane) that was consumed to generate the energy:

$$2.60 \times 10^{13} \text{ MJ} \times \frac{1 \text{ kg}}{53.0 \text{ MJ}} = 4.90 \times 10^{11} \text{ kg of methane}$$

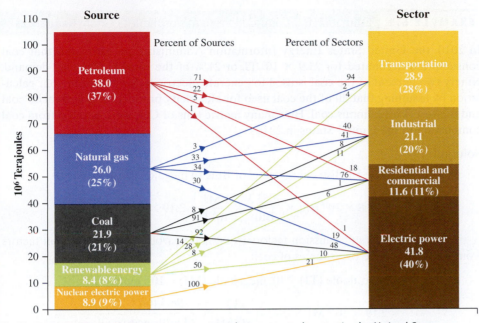

FIGURE 6.5 Primary energy consumption by source and sector in the United States in 2010.

Source: US EIA (2012) Annual Energy Review 2011. U.S. Energy Information Administration. Office of Energy Statistics. U.S. Department of Energy. Washington, D.C. DOE/EIA-0384(2011) http://www.eia.gov/energy_in_brief/major_energy_sources_and_users.cfm

Balance the chemical equation for the combustion of methane to determine the molar ratio of CO_2 to methane:

$$CH_4 + 2O_2 \rightarrow CO_2 + 2H_2O$$

Calculate the molecule weight of methane and carbon dioxide:

$$MW_{methane} = 1 \times 12 + 4 \times 1 = 16 \frac{g}{mole}$$

$$MW_{carbon\ dioxide} = 1 \times 12 + 2 \times 16 = 44 \frac{g}{mole}$$

Convert the mass of methane consumed to the mass of carbon dioxide produced:

$$4.90 \times 10^{11}\,kg\ of\ methane\ as\ CH_4 \times \frac{1{,}000\ g}{1\ kg} \times \frac{1\ mole\ CH_4}{16\ g\ of\ CH_4} \times \frac{1\ mole\ CO_2}{1\ mole\ CH_4}$$

$$\times \frac{44\ g\ CO_2}{1\ mole\ CO_2} \times \frac{ton}{10^6\ g} = 1.35 \times 10^9\ metric\ tons\ of\ CO_2$$

Over 1 billion metric tons of carbon dioxide were emitted by the United States in 2010 from burning natural gas for energy production. This amount of carbon emissions is due solely to the emissions from natural gas. Thus, the total U.S. carbon dioxide emissions in 2010 were more than 4 times the amount calculated here. Almost all of the mass of the carbon is being transferred from its repository where it was stored as natural gas and is being transferred in large amounts into the atmosphere. We will discuss what happens to the carbon dioxide emitted to the atmosphere in Section 6.4.

EXAMPLE 6.3   Estimating U.S. Contributions to Atmospheric CO_2 from Natural Gas

In 2010, the United States Energy Information Administration estimated that coal consumption accounted for 21.9×10^6 TJ, or 21% of the total U.S. energy demand. Ninety two percent of the coal is used to generate electricity. We'll simplify our calculation by assuming that all of the coal used for electricity production is anthracite coal and the rest is bituminous coal. How many metric tons of CO_2 were emitted from coal combustion in the United States in 2010?

Solution

First, convert the English energy units of Btus to megajoules (MJ):

$$20.8 \times 10^{15} \, \text{Btu} \times \frac{1055 \, \text{J}}{\text{Btu}} \times \frac{1 \, \text{MJ}}{10^6 \, \text{J}} = 2.19 \times 10^{13} \, \text{MJ}$$

Next use the default emission factors to estimate carbon dioxide emission factors from Table 6.1 for the two sources of coal:

$$1 \text{ terrajoule (TJ)} = 10^6 \text{ megajoule (MJ)} = 10^{12} \text{ joule (J)}$$

$$2.19 \times 10^{13} \, \text{MJ} \times 0.92 \times \frac{\text{TJ}}{10^6 \, \text{MJ}} \times \frac{98.3 \text{ ton } CO_2}{\text{TJ anthracite coal}}$$

$$= 1.98 \times 10^9 \text{ kg of carbon dioxide from anthracite}$$

$$2.19 \times 10^{13} \, \text{MJ} \times 0.08 \times \frac{\text{TJ}}{10^6 \, \text{MJ}} \times \frac{94.6 \text{ ton } CO_2}{\text{TJ anthracite coal}}$$

$$= 1.66 \times 10^8 \text{ kg of carbon dioxide from bituminous}$$

Total CO_2 emissions from both anthracite and bituminous coal = 2.15×10^9 metric tons.

Over 2 billion metric tons of carbon dioxide were emitted by the United States in 2010 from burning coal, primarily for electricity production. While slightly less coal energy is used, the carbon dioxide emissions associated with coal use are nearly twice the total carbon dioxide emissions from natural gas. This is due to a higher carbon intensity value for coal than for natural gas.

EXAMPLE 6.4   Comparing Carbon Intensities for Fuels

Due to the chemical structure and energy efficiency associated with the chemical energy stored in different fuels, some fuels emit less carbon dioxide per unit energy than others. In the previous example, we compared the total mass of carbon dioxide emitted from natural gas and coal in the United States. Calculate the greater carbon intensity from burning coal and charcoal (fuels more commonly used in low-income countries) by comparing their carbon dioxide emissions to natural gas.

Solution

Carbon intensity of coal to compared to natural gas is

$$\frac{98,300 \, \dfrac{\text{kg}}{\text{TJ}} \text{ anthracite coal}}{56,100 \, \dfrac{\text{kg}}{\text{TJ}} \text{ natural gas}} = 1.75$$

So burning coal will emit 175% of the carbon dioxide if the same amount of energy is derived from natural gas.

Carbon intensity of charcoal to compared to natural gas:

$$\frac{112,000 \; \frac{\text{kg}}{\text{TJ}} \; \text{charcoal}}{56,100 \; \frac{\text{kg}}{\text{TJ}} \; \text{natural gas}} = 2.00$$

Burning charcoal, a practice in many low-income countries with high population growth rates, emits twice the amount of carbon dioxide compared to the amount of carbon dioxide if natural gas had been used to generate the energy.

Carbon dioxide and other gases that increase the energy absorbed by the atmosphere, called greenhouse gases, are emitted from anthropogenic and natural sources. Approximately 26% of greenhouse gas emissions in 2004 were due to energy uses, such as electricity production, heating, and cooling, as shown in Figure 6.6. Industrial emissions were the second highest source of greenhouse gas emissions in 2004 and accounted for 19% of emissions. Carbon dioxide is removed from plant growth, particularly in densely forested areas; the deforestation due to changes in land use for agriculture or the built environment accounted for 17% of the increase in greenhouse gas emissions in 2004. Energy supply, industrial emissions, and deforestation account for about two-thirds of the annual increase in greenhouse gases. Agriculture, transportation, the built environment, water, and wastewater treatment account for the majority of the remainder of annual greenhouse gas emissions.

The sum-total of worldwide emissions of carbon dioxide from fossil fuel use has been compared to changes in atmospheric carbon dioxide levels, and as shown in Figure 6.7 there is a strong correlation between these two values. About 55% of the carbon dioxide emitted stays in the atmosphere, as illustrated in Figure 6.8. Where

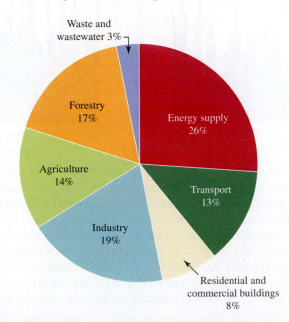

FIGURE 6.6 Percentage of greenhouse gas emissions from important sectors in 2004.

Source: Based on IPCC (2007a) Climate Change 2007: The Physical Science Basis. Contributions of working group I to the Fourth Assessment Report of the intergovernmental Panel on Climate Change [Solomon, S., Qin, D., Manning, M., Chen, Z., Marquis, M., Averyt, K.B., Tignor, M., and Miller, H. L. (eds.)] Cambridge University Press, Cambridge, United Kingdom and New York, NY, USA. pp. 996

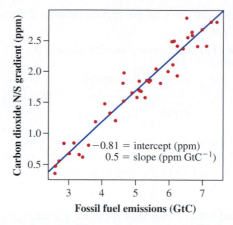

FIGURE 6.7 The difference between CO_2 concentration in the Northern Hemisphere and Southern Hemisphere (*y*-axis), computed as the difference between annual mean concentrations (ppm) at Mauna Loa and the South Pole (Keeling and Whorf, 2005, updated), compared with annual fossil fuel emissions (*x*-axis; GtC; Marland et al., 2006), with a line showing the best fit. The observations show that the north–south difference in CO_2 increases proportionally with fossil fuel use, verifying the global impact of human caused emissions.

Source: Based on IPCC (2007) Climate Change 2007: The Physical Science Basis. Contributions of working group I to the Fourth Assessment Report of the intergovernmental Panel on Climate Change [Solomon, S., Qin, D., Manning, M., Chen, Z., Marquis, M., Averyt, K.B., Tignor, M., and Miller, H. L. (eds.)] Cambridge University Press, Cambridge, United Kingdom and New York, NY, USA. pp. 996. Figure 7.5.

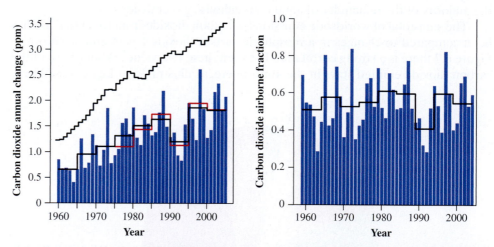

FIGURE 6.8 Changes in global atmospheric CO_2 concentrations. (a) Annual (bars) and five-year mean (lower black line) changes in CO_2 concentrations, from Scripps Institution of Oceanography observations (mean of South Pole and Mauna Loa; Keeling and Whorf, 2005, updated). The upper stepped line shows annual increases that would occur if 100% of fossil fuel emissions (Marland et al., 2006, updated) remained in the atmosphere, and the red line shows five-year mean annual increases from National Oceanic and Atmospheric Administration (NOAA) data (mean of Samoa and Mauna Loa; Tans and Conway, 2005, updated). (b) Fraction of fossil fuel emissions remaining in the atmosphere ("airborne fraction") each year (bars) and five-year means (solid black line) (Scripps data)(mean since 1958 is 0.55). Note the anomalously low airborne fraction in the early 1990s.

Source: Based on IPCC (2007) Climate Change 2007: The Physical Science Basis. Contributions of working group I to the Fourth Assessment Report of the intergovernmental Panel on Climate Change [Solomon, S., Qin, D., Manning, M., Chen, Z., Marquis, M., Averyt, K.B., Tignor, M., and Miller, H. L. (eds.)] Cambridge University Press, Cambridge, United Kingdom and New York, NY, USA. pp. 996. Figure 7.4.

does the "other" 45% of the carbon emitted from burning fossil fuels go? (In 2004, National Geographic published an award-winning article titled, "The Case of the Missing Carbon.") In order to answer this question, we have to look more closely at the biogeochemical cycle for carbon.

6.4 The Carbon Cycle, Carbon Flow Pathways, and Repositories

The atmosphere contains only a small fraction of the world's carbon. Until relatively recently, the atmosphere contained about 597 gigatons (10^9 metric tons, or 10^{12} kilograms) of carbon (abbreviated GtC). Over 165 GtC have been added to the atmosphere since the Industrial Revolution, as illustrated in red in Figure 6.9.

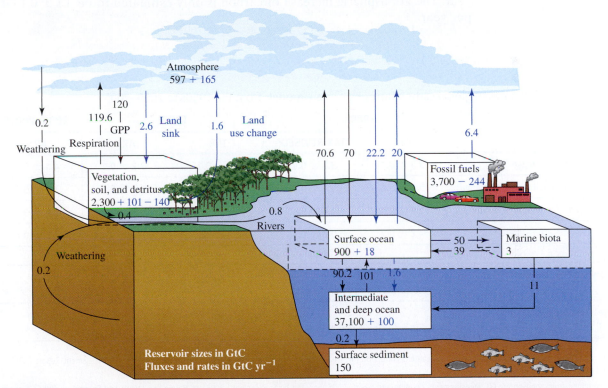

FIGURE 6.9 The global carbon cycle for the 1990s, showing the mean annual fluxes in GtC/year: preindustrial "natural" fluxes in black and "anthropogenic" fluxes in red (modified from Sarmiento and Gruber, 2006, with changes in pool sizes from Sabine et al., 2004a). The net terrestrial loss of −39 GtC is inferred from cumulative fossil fuel emissions minus atmospheric increase minus ocean storage. The loss of −140 GtC from "vegetation, soil, and detritus" represents the cumulative emissions from land-use change (Houghton, 2003) and requires a terrestrial biosphere sink of 101 GtC (in Sabine et al., given only as ranges of −140 to −80 GtC and 61 to 141 GtC, respectively; other uncertainties are given in their Table 1.) Net anthropogenic exchanges with the atmosphere are from Column 5 'AR4' in Table 7.1 of the 2007 IPCC report. Gross fluxes generally have uncertainties of more than 20%, but fractional amounts have been retained to achieve overall balance when including estimates in fractions of GtC/year for riverine transport, weathering, deep ocean burial, and so on. GPP is annual gross (terrestrial) primary production. Atmospheric carbon content and all cumulative fluxes since 1750 are as of the end of 1994.

Source: Based on IPCC (2007) Climate Change 2007: The Physical Science Basis. Contributions of working group I to the Fourth Assessment Report of the intergovernmental Panel on Climate Change [Solomon, S., Qin, D., Manning, M., Chen, Z., Marquis, M., Averyt, K.B., Tignor, M., and Miller, H. L. (eds.)] Cambridge University Press, Cambridge, United Kingdom and New York, NY, USA. pp. 996. Figure 7.3.

While this is a huge mass of carbon, it is small compared to the amount of carbon stored in vegetation, soil, and detritus that contain about 2,300 GtC. It is estimated that there is even more carbon stored as fossil fuels. Far and away the largest carbon reservoirs on the planet are the oceans with approximately 921 GtC stored in the surface water and marine biota and 38,200 GtC stored in the intermediate and deep ocean. The combustion of fossil fuels causes carbon to migrate from fossilized deposits in geologic reservoirs to the atmosphere and oceans See Figure 6.10.

The IPCC Fourth Assessment Report (AR4) has reported the ongoing efforts of several climate models to calculate the net mass flow rate, or flux, from one carbon repository to another. Model results from 1985 to 2005 are shown in Table 6.2. The most recent model, AR4 2000–2005, shows that the net worldwide emissions of carbon are currently estimated to be 7.2 ± 0.3 GtC per year. The atmospheric increase of carbon is only estimated to be 4.1 ± 0.1 GtC per year.

TABLE 6.2 Estimates of global carbon fluxes by climate models from the 1980s to 2000–2005

	1980s		1990s		2000–2005[c]
	TAR	TAR REVISED[a]	TAR	AR4	AR4
Atmospheric increase [b]	3.3 ± 0.1	3.3 ± 0.1	3.2 ± 0.1	3.2 ± 0.1	4.1 ± 0.1
Emissions (fossil + cement)[c]	5.4 ± 0.3	5.4 ± 0.3	6.4 ± 0.4	6.4 ± 0.4	7.2 ± 0.3
Net ocean-to-atmosphere flux[d]	−1.9 ± 0.6	−1.8 ± 0.8	−1.7 ± 0.5	−2.2 ± 0.4	−2.2 ± 0.5
Net land-to-atmosphere flux[e]	−.2 ± 0.7	−0.3 ± 0.9	−1.4 ± 0.7	−1.0 ± 0.6	−0.9 ± 0.6
Partitioned as follows					
Land-use change flux	1.7 (0.6 to 2.5)	1.4 (0.4 to 2.3)	n/a	1.6 (0.5 to 2.7)	n/a
Residual terrestrial sink	−1.9 (−3.8 to −0.3)	−1.7 (−3.4 to 0.2)	n/a	−2.6 (−4.3 to −0.9)	n/a

[a] TAR values revised according to an ocean heat content correction for ocean oxygen fluxes (Bopp et al., 2002) and using the Fourth Assessment Report (AR4) best estimate for the land-use change flux.

[b] Determined from atmospheric CO_2 measurements (Keeling and Whorf, 2005, updated by S. Piper until 2006) at Mauna Loa (19°N) and South Pole (90°S) stations, consistent with the data shown in Figure 6.7, using a conversion factor of 2.12 GtC/yr = 1 ppm.

[c] Fossil fuel and cement emission data are available only until 2003 (Marland et al., 2006). Mean emissions for 2004 and 2005 were extrapolated using energy use data with a trend of 0.2 GtC/yr.

[d] For the 1980s, the ocean-to-atmosphere and land-to-atmosphere fluxes were estimated using atmospheric O_2, N_2, and CO_2 trends, as in the TAR. For the 1990s, the ocean-to-atmosphere flux alone is estimated using ocean observations and model results, giving results identical to the atmospheric O_2:N_2 method (Manning and Keeling, 2006), but with less uncertainty. The net land-to-atmosphere flux then is obtained by subtracting the ocean-to-atmosphere flux from the total sink (and its errors are estimated by propagation). For 2000 to 2005, the change in ocean-to-atmosphere flux was modeled (Le Quéré et. al., 2005) and added to the mean ocean-to-atmosphere flux of the 1990s. The error was estimated based on the quadratic sum of the error of the mean ocean flux during the 1990s and the root mean square of the five-year variability from three inversions and one ocean model presented in Le Quéré et al. (2003).

[e] Balance of emissions due to land-use change and a residual land sink. These two terms cannot be separated based on current observations.

Source: Based on IPCC (2007) Climate Change 2007: The Physical Science Basis. Contributions of working group I to the Fourth Assessment Report of the Intergovernmental Panel on Climate Change [Solomon, S., Qin, D., Manning, M., Chen, Z., Marquis, M., Averyt, K.B., Tignor, M., and Miller, H. L. (eds.)] Cambridge University Press, Cambridge, United Kingdom and New York, NY.

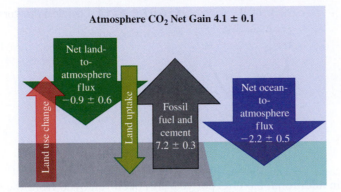

FIGURE 6.10 Schematic of the net carbon fluxes as predicted by the AR4 climate model.

Source: Bradley Striebig.

EXAMPLE 6.5 | Partitioning of Carbon Emissions

What percent of carbon emissions is stored in the atmosphere, oceans, and land?

Solution

To find the percent of carbon emissions that partition to or move into the atmosphere from carbon emissions, divide the atmospheric flux by the total carbon emissions as illustrated in Figure 6.9. Using the data from AR4 2000–2005 yields

$$\frac{4.1}{7.2} \times 100\% = 57\% \text{ of carbon emissions remain in the atmosphere}$$

Repeat these calculations using the data for the carbon that is transported into the oceans and repositories on land:

$$\frac{2.2}{7.2} \times 100\% = 31\% \text{ of carbon emissions migrate into the oceans}$$

$$\frac{0.9}{7.2} \times 100\% = 12\% \text{ of carbon emissions taken up by repositories of carbon on land}$$

Global forests provide the largest land-based reservoir for carbon. Carbon dioxide is converted to cellulose and other plant matter through the process of photosynthesis:

Carbon dioxide + water → plant materials + oxygen

A simple chemical process, whereby carbon dioxide can form glucose, which is a common organic energy-containing compound (Figure 6.11), is represented by the chemical reaction:

$$6CO_2 + 6H_2O \xrightarrow{uv} C_6H_{12}O_6 + 6O_2 \tag{6.6}$$

Photosynthesis changes atmospheric carbon dioxide to plant tissue material, such as cellulose (see Figure 6.11). Photosynthesis and uptake of carbon by the land-based process historically removed about 120 GtC per year, and since the Industrial

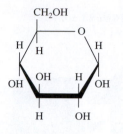

FIGURE 6.11 Chemical structure for glucose.

Source: Bradley Striebig.

Revolution, an additional 2.6 GtC per year has been removed from the atmosphere. However, land-use changes estimated in Table 6.3, mainly deforestation associated with agriculture and cooking fuel harvesting of trees, result in an additional 1.6 GtC added back into the atmosphere. Respiration processes from simple organisms as well as more complex organisms account for the addition of 119.6 GtC per year to the atmosphere. The respiration process is the opposite process of photosynthesis and strongly resembles the combustion process of extracting energy from carbon sources. Animals use sugars, like glucose (see Figure 6.11), carbohydrates and proteins in combination with oxygen to produce energy, cell growth, water, and carbon dioxide via the representative reaction:

$$\text{Glucose} + \text{oxygen} \rightarrow \text{water} + \text{carbon dioxide}$$

$$C_6H_{12}O_6 + O_2 \xrightarrow{\text{energy}} \text{cell mass} + CO_2 + H_2O \tag{6.7}$$

An important aspect of the movement of carbon from one repository to another is how long the carbon will remain in that repository. The **residence time**, t_r, is the term used to describe how long a substance remains as a defined volume, such as the atmosphere, a lake, or a small reactor in a laboratory. The residence time can be determined by dividing the system volume, V, by the volumetric flow rate, Q, of the material, or alternatively, by dividing the mass of material, m, in the control volume by the mass flow rate of material, M, flowing through the volume.

$$t_r = \frac{V}{Q} = \frac{m}{M} \tag{6.8}$$

EXAMPLE 6.6 **Calculating Residence Time**

The carbon flux to the atmosphere is approximately 55.3×10^{12} kg/yr. The approximate steady-state mass of carbon in the atmosphere is 7.35×10^{12} kg. How long, on average, is carbon likely to remain in the atmosphere before cycling through another repository?

Solution

$$t_r = \frac{V}{Q} = \frac{m}{M} = \frac{7.35 \times 10^{12}\,\text{kg}}{55.3 \times 10^{12}\,\text{kg/yr}} = 0.133\,\text{years} \times \frac{365\,\text{days}}{\text{year}} = 48.5\,\text{days}$$

Thus, the average time carbon molecules spend in the atmosphere is about 48.5 days.

TABLE 6.3 Estimates of land-to-atmosphere emissions resulting from land-use changes during the 1980s and 1990s (in Gt/yr). The Fourth Assessment Report (AR4) estimates used in the global carbon budget are shown in bold. Positive values indicate carbon losses from land ecosystems. Uncertainties are reported as ± 1 standard deviation. Numbers in parenthesis are ranges of uncertainty.

	TROPICAL AMERICAS	TROPICAL AFRICA	TROPICAL ASIA	PAN-TROPICAL	NON-TROPICS	TOTAL GLOBE
1990s						
Houghton (2003a)[a]	0.8 ± 0.3	0.4 ± 0.2	1.1 ± 0.5	2.2 ± 0.6	-0.02 ± 0.5	2.2 ± 0.8
DeFries et al. (2002)[b]	0.5	0.1	0.4	1.0	n/a	n/a
	(0.2 to 0.7)	(0.1 to 0.2)	(0.2 to 0.6)	(0.5 to 1.6)	n/a	n/a
Achard et al. (2004)[c]	0.3	0.2	0.4	0.9	n/a	n/a
	(0.3 to 9.4)	(0.2 to 0.4)	(0.4 to 1.1)	(0.5 to 1.4)	n/a	n/a
AR4[d]	0.7	0.3	0.8	1.6	−.02	1.6
	(0.4 to 0.9)	(0.2 to 0.4)	(0.4 to 1.1)	(1.0 to 2.2)	(−0.5 to 0.5)	(0.5 to 2.7)
1980s						
Houghton (2003a)[a]	0.8 ± 0.3	0.3 ± 0.2	0.9 ± 0.5	1.9 ± 0.6	0.06 ± 0.5	2.0 ± 0.8
DeFries et al. (2002)[b]	0.4	0.1	0.2	0.7	n/a	n/a
	(0.2 to 0.5)	(0.08 to 0.14)	(0.1 to 0.3)	(0.4 to 1.0)	n/a	n/a
McGuire et al. (2001)[e]				0.6 to 1.2	−0.1 to +0.4	(0.6 to 11.0)
Jain and Yang (2005)[f]	0.22 to 0.24	0.08 to 0.48	0.58 to 0.34			1.33 to 2.06
TAR[g]						1.7
						(0.6 to 2.5)
AR4[d]	0.6	0.2	0.6	1.3	0.06	1.4
	(0.3 to 0.8)	(0.1 to 0.3)	(0.3 to 0.9)	(0.9 to 1.8)	(−0.4 to +0.6)	(0.4 to 2.3)

[a] His Table 2.

[b] Their Table 3.

[c] Their Table 2 for mean estimates with the range indicated in parentheses corresponding to their reported minimum and maximum estimates.

[d] Best estimate calculated from the mean of Houghton (2003a) and DeFries et al. (2002), the only two studies covering both the 1980s and the 1990s. For non-tropical regions where DeFries et al. have no estimate, Houghton has been used.

[e] Their Table 5: Range is obtained from four terrestrial carbon models.

[f] The range indicated in parentheses corresponds to two simulations using the same model, but forced with different land-cover change data sets from Houghton (2003a) and DeFries et al. (2002).

[g] In the TAR estimate, no values were available for the 1990s.

Source: Based on IPCC (2007) Climate Change 2007: The Physical Science Basis. Contributions of working group I to the Fourth Assessment Report of the Intergovernmental Panel on Climate Change [Solomon, S., Qin, D., Manning, M., Chen, Z., Marquis, M., Averyt, K.B., Tignor, M., and Miller, H. L. (eds.)] Cambridge University Press, Cambridge, United Kingdom and New York, NY, USA.

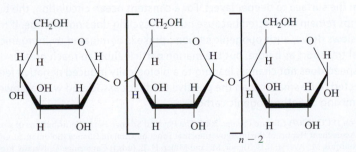

FIGURE 6.12 Chemical structure for cellulose.

Source: Bradley Striebig.

By far the largest repository of carbon in the carbon cycle is the Earth's oceans. Carbon dioxide dissolves into the oceans in accordance with Henry's law, as illustrated in Figure 6.13. Yet, the oceans are so large and massive that the current atmospheric levels of carbon dioxide have not yet reached a steady state, or equilibrium condition, between the amount of carbon dioxide in the atmosphere and the amount of carbon containing compounds in the oceans. However, as the atmospheric levels of carbon dioxide increase, transfer of carbon from the atmosphere to the oceans has also increased by 22.2 GtC per year. The increased ocean temperatures has also increased the amount of carbon dioxide transferred from the

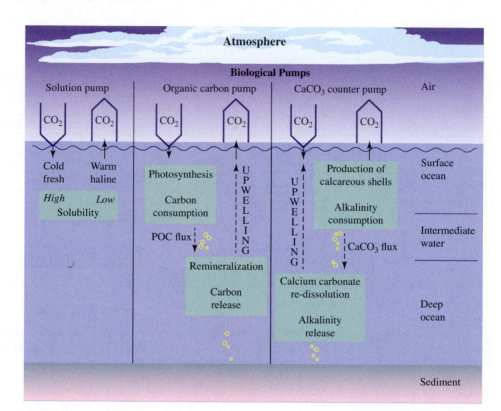

FIGURE 6.13 Three main carbon pumps govern the regulation of natural atmospheric CO_2 changes by the ocean (Heinze et al., 1991): the solubility pump, the organic carbon pump, and the $CaCO_3$ "counter pump." The oceanic uptake of anthropogenic CO_2 is dominated by inorganic carbon uptake at the ocean surface and physical transport of anthropogenic carbon from the surface to deeper layers. For a constant ocean circulation, the biological carbon pumps remain unaffected because nutrient cycling does not change. If the ocean circulation slows down, anthropogenic carbon uptake is dominated by inorganic buffering and physical transport as before, but the marine particle flux can reach greater depths if its sinking speed does not change, leading to a biologically induced negative feedback that is expected to be smaller than the positive feedback associated with a slower physical downward mixing of anthropogenic carbon.

Source: Based on IPCC (2007) Climate Change 2007: The Physical Science Basis. Contributions of working group I to the Fourth Assessment Report of the intergovernmental Panel on Climate Change [Solomon, S., Qin, D., Manning, M., Chen, Z., Marquis, M., Averyt, K.B., Tignor, M., and Miller, H. L. (eds.)] Cambridge University Press, Cambridge, United Kingdom and New York, NY, USA. Figure 7.10 – Redrawn from original citation: Heinze, C., Maier-Reimer, E., and Winn, K. (1991) Glacial pCO2 Reduction by the World Ocean: Experiments With the Hamburg Carbon Cycle Model. Paleoceanography. 6(4):395–430.

oceans to the atmosphere by 20 GtC per year. The net result of this changing, non-equilibrium situation is a net increase in carbon removed by the atmosphere, so that 2.2 GtC per year, or 31% of the carbon emitted, is transferred to the planet's oceans. This increase in the carbon content of the oceans has already had a profound effect by increasing the acidity of the oceans.

In order to understand the environmental impacts of this change to the carbon cycle, we'll evaluate the energy balances that are used to describe the average surface temperature of the planet. The energy balances calculated in the next section of this chapter will allow us to predict a range of potential impacts of our changing climate on our environment.

6.5 Global Energy Balance

The IPCC has summarized some of the key questions people and policymakers have about our changing climate in Figure 6.14. We have already discussed present and past conditions related to carbon dioxide concentrations in the atmosphere, but we have not yet determined how carbon dioxide concentrations are related to climate. To make this connection, we must review the thermodynamic energy balance Callendar studied to link carbon dioxide concentrations in the atmosphere to the changes occurring in our climate.

In order to analyze the effect of atmospheric gas concentrations on the Earth's climate, we must be able to perform an energy balance on the planet. The law of conservation of energy states that energy cannot be created or destroyed. The energy balance within a given control volume is similar in form to that of the general mass balance equation, where

$$\text{Change in energy} = \text{energy inputs} - \text{energy outputs} \qquad (6.9)$$

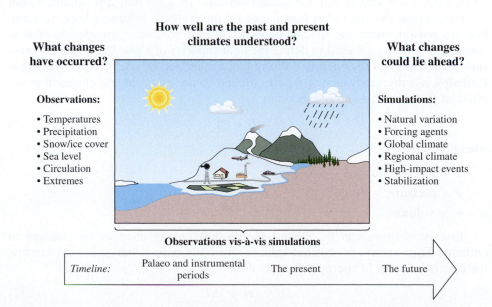

How well are the past and present climates understood?

What changes have occurred?

Observations:

• Temperatures
• Precipitation
• Snow/ice cover
• Sea level
• Circulation
• Extremes

What changes could lie ahead?

Simulations:

• Natural variation
• Forcing agents
• Global climate
• Regional climate
• High-impact events
• Stabilization

Observations vis-à-vis simulations

Timeline: Palaeo and instrumental periods   The present   The future

FIGURE 6.14 Key questions about the climate system and its relationship to humankind.

Source: Based on IPCC (2001) Climate Change 2001: Synthesis Report. A Contribution of Working Groups I, II, and III to the Third Assessment Report of the Integovernmental Panel on Climate Change [Watson, R.T. and the Core Writing Team (eds.)]. Cambridge University Press, Cambridge, United Kingdom, and New York, NY, USA.

This is very similar in form to Equation (6.1), which is the general mass balance equation. The first law of thermodynamics states that energy cannot be created or destroyed, but we can change the form of the energy. (Once again we'll simplify reality slightly by neglecting the processes of nuclear fission and fusion that are reactions that typically only take place in nuclear reactors.) Energy is defined as the capacity to do work. Work is defined as the force acting on a body through a distance. The SI unit of a joule is defined as the constant force of one newton that is applied to a body that is moved one meter in the direction of that force. The first law of thermodynamics relates work and energy by

$$Q_H = U_2 - U_1 + W \tag{6.10}$$

where

Q_H = heat absorbed
U_1 = energy of the system at state 1
U_2 = energy of the system at state 2
W = work

Energy has many different forms, and a complete analysis of general thermodynamics is beyond the scope of this text. However, some basic terms and concepts must be introduced. Energy stored as thermal energy is measured in the system by a calorie. One **calorie** is defined as the amount of energy required to raise one gram of water one degree Celsius from 14.5°C to 15.5°C. The **British thermal unit (Btu)** is defined as the amount of energy required to raise one pound of water one degree Fahrenheit.

The **specific heat** of a substance is the quantity of heat required to increase a unit mass of the substance by one degree with units of energy/(mass-temperature). For most solids and liquids, the specific heat varies only slightly with temperature. However, gases may expand when heated, doing work on the environment. Thus it takes much more energy to raise the temperature of a gas that can expand within the system than the amount of heat needed if the system's volume is kept constant. For this reason, either the specific heat at constant volume, c_v, or specific heat at constant pressure, c_p, is used to define the heat capacity of a gas. The effects of pressure and volume on the energy of the system are related by a term called enthalpy. Enthalpy is a thermodynamic property related to the material and chemical properties of that material and is defined as

$$H = U + PV \tag{6.11}$$

where

U = internal energy
P = pressure
V = volume

Energy changes can be related to internal energy changes or changes in enthalpy. For a system at constant volume that undergoes a change in temperature, the internal change in energy is described by

$$\Delta U = (m)(c_v)\Delta T \tag{6.12}$$

Similarly, for a system at constant pressure that undergoes a change in temperature, the change in enthalpy is given by

$$\Delta H = (m)(c_p)\Delta T \tag{6.13}$$

For most liquids and solids in the natural environment that do not change phase or change substantially in pressure or volume, then, $c_v = c_p = c$ and $\Delta U = \Delta H$. The change in the energy stored within the system due to a change in temperature, if c is assumed constant over the temperature range, is given by

$$\text{Change in stored energy} = (m)(c)\,\Delta T \tag{6.14}$$

When enthalpy changes are considered, then the energy balance equation can be written as

$$\frac{\Delta dH}{dt} = \frac{d(H)_{in}}{dt} + \frac{d(H)_{out}}{dt} \tag{6.15}$$

The energy accumulation within a system may be a result of the transfer of energy to the system by conduction, convection, or radiation. Conduction is the direct physical transfer of heat via molecular diffusion from a substance of greater heat (or energy) coming into direct contact with a substance of less heat (or energy). Convective heat transfer occurs via fluid motion, for example, by air currents or ocean currents, such as those that move heat from the warmer equator toward the cooler polar regions. Radiation is the heat emitted by an object that is absorbed by the surrounding objects and is due to the emission of electromagnetic radiation.

Heat loss through the walls of a home occurs because of a combination of conductive, convective, and radiant energy exchange. The overall heat transfer process can be estimated by

$$q = \frac{A(T_i - T_o)}{R} \tag{6.16}$$

where

q = heat transfer through a surface (energy/time)

A = area of the surface through which the heat transfers

T_i = air temperature within the surface

T_o = air temperature outside the heat transfer surface

R = resistance or R-value [area-degree/energy (typically m²-°C/W or hr-ft²-°F/Btu)].

Radiative heat transfer is related to the frequency and wavelength of the radiation. Incoming solar radiation is divided into several different wavelengths as shown in Figure 6.15. The wavelength and frequency of the radiation are related by the speed of light, which remains constant:

$$\text{Speed of light} = c_o = 2.998 \times 10^8 \text{ m/s} = (\lambda)(v) \tag{6.16}$$

where

λ = wavelength (m)

v = frequency (1/s or hertz)

The energy associated with the radiation at a given frequency is related by Planck's law:

$$E = (h)(v) \tag{6.17}$$

where

h = Planck's constant = 6.6256×10^{-34} J-s.

As shown in Figure 6.15, energy may be radiated from a body, such as the sun or Earth, over a range of wavelengths. A simplifying assumption in thermodynamic analysis is often to model an object as a blackbody, that is, a plain object of simple geometry. The blackbody temperature is defined as the maximum amount of radiation an object or body can emit at a given temperature. The amount of energy at any wavelength is called the spectral emissive power, $E_{b-\lambda}$, and is given by the Planck distribution law:

$$E_{b-\lambda} = \frac{2\pi h c_o^2}{\lambda^5}\left[\exp\left(\frac{h c_o}{\lambda k T}\right) - 1\right]^{-1} \tag{6.18}$$

where

$E_{b-\lambda}$ = spectral emissive power (W/m²-μm)

λ = wavelength of the radiant energy (μm)

k = Boltzmann constant = 1.3805×10^{-23} (J/K)

T = absolute temperature of the radiating body (K)

Integrating over the spectrum of wavelengths for the blackbody yields the Stefan–Boltzmann law of radiation:

$$E_{bb} = \sigma A T^4 \tag{6.19}$$

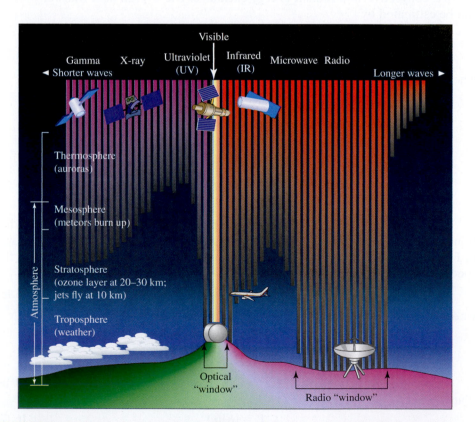

FIGURE 6.15 Types of incoming solar radiation and their penetration into the layers of the Earth's atmosphere.

Source: STScI/JHU/NASA

where

E_{bb} = total blackbody emission rate (W)

σ = Stefan–Boltzmann constant = 5.67×10^{-8} W/(m²-K⁴)

T = absolute temperature (K)

A = surface area of the object (m²)

6.6 Global Energy Balance and Surface Temperature Model

The Earth can be modeled as a simple blackbody to gain an understanding of the principles of how the sun heats the Earth and to estimate the Earth's average surface temperature. The incoming solar radiation is quite constant, with only small variations noted since AD 1600 as shown in Figure 6.16.

Solar irradiation strikes the Earth at a nearly constant energy of 1,365 W/m². Of this radiation, the atmosphere reflects, on average, 30% of the energy back into

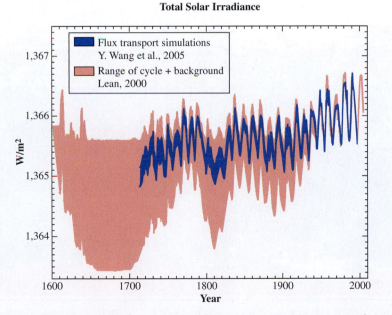

FIGURE 6.16 Reconstruction of total solar irradiance time series starting as early as 1600. The upper envelope of the shaded regions shows irradiance variations arising from the 11-year activity cycle. The lower envelope is the total irradiance reconstructed by Lean (2000), in which the long-term trend was inferred from brightness changes in sun-like stars. In comparison, the recent reconstruction of Y. Wang et al. (2005) is based on solar considerations alone, using a flux transport model.

Source: Based on Forster, P., V. Ramaswamy, P. Artaxo, T. Berntsen, R. Betts, D.W. Fahey, J. Haywood, J. Lean, D.C. Lowe, G. Myhre, J. Nganga, R. Prinn, G. Raga, M. Schulz and R. Van Dorland, 2007: Changes in Atmospheric Constituents and in Radiative Forcing. In: Climate Change 2007: The Physical Science Basis. Contribution of Working Group I to the Fourth Assessment Report of the Intergovernmental Panel on Climate Change [Solomon, S., D. Qin, M. Manning, Z. Chen, M. Marquis, K.B. Averyt, M.Tignor and H.L. Miller (eds.)]. Cambridge University Press, Cambridge, United Kingdom and New York, NY, USA.

space. The percent reflected is called the Earth's albedo, and the albedo is a function of the light reflected or deflected. For instance, ice coverage in the polar regions reflects more than 70% of the sun's energy, while the albedo near the equator is less than 20% (see Figure 6.17).

Based on Planck's law, assuming the Earth is a blackbody that absorbs the solar radiation as shown in Figure 6.18 yields the following equations:

$$\text{Blackbody radiation absorbed by the Earth} = E_{ebb} = S_o \pi r_{e^2} \tag{6.20}$$

$$\text{Energy absorbed by the Earth} = E_{abs} = S_o \pi r_{e^2} (1 - \alpha) \tag{6.21}$$

The Stefan-Boltzmann law can be applied to the Earth, assuming the average temperature of the Earth is nearly constant. Balancing the energy absorbed by the Earth using Equation (6.21) with the Earth's blackbody temperature from Equation (6.19) yields

$$S_o \pi r_{e^2} (1 - \alpha) = \sigma (4\pi r_{e^2}) T_{e^4} \tag{6.22}$$

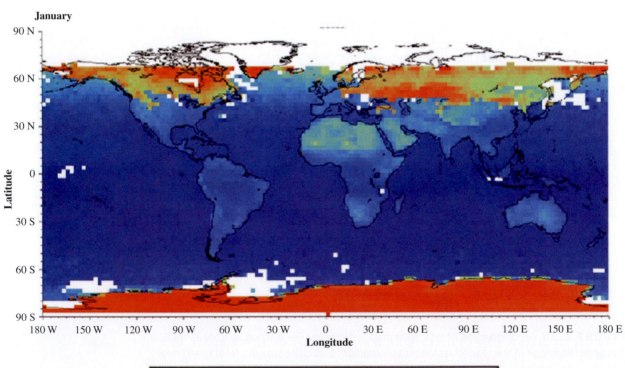

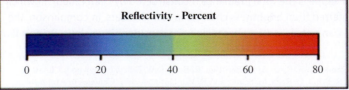

FIGURE 6.17 Surface reflectivity of the Earth for January 1987. Cells with missing data are colored white.

Source: NASA - Earth Radiation Budget Experiment

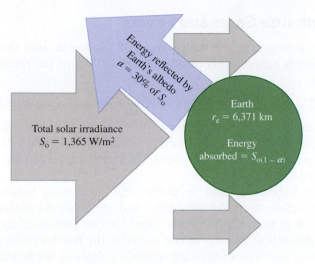

FIGURE 6.18 Model of the Earth as a blackbody radiator in space.

Source: Based on Masters, G.M. and Wendell, E. (2007) Introduction to environmental engineering and science. Pearson Education, Upper Saddle River, NJ, USA.

where T_e = average surface temperature of the Earth (K).

Solving for T_e yields

$$T_e = \left[\frac{S_o(1-\alpha)}{4\sigma}\right]^{\frac{1}{4}} = \left[\frac{1{,}365(1-0.30)}{4(5.67 \times 10^{-8})}\right]^{\frac{1}{4}} = 254\ \text{K} = -18°\text{C} \qquad (6.23)$$

It is obvious that the average surface temperature of the Earth is not −18°C, which would mean a long-lasting ice age. Therefore, we know our model is not valid. What we have neglected in this model is the insulating capacity of the atmosphere. The Earth's atmosphere provides an insulating effect that keeps the Earth much warmer than the predicted −18°C. This insulation performs the same thermodynamic function as the insulation in a home, a coffee cup, or a greenhouse. The magnitude of the insulation effect is a function of the chemical properties that make up the atmosphere.

EXAMPLE 6.8   An Ice Planet without an Atmosphere

If the Earth were completely covered with ice, so that the planet's albedo was 80%, what would be the Earth's blackbody radiation temperature?

Using Equation (6.23) and an albedo, $\alpha = 0.80$, yields

$$T_e = \left[\frac{S_o(1-\alpha)}{4\sigma}\right]^{\frac{1}{4}} = \left[\frac{1{,}365(1-0.80)}{4(5.67 \times 10^{-8})}\right]^{\frac{1}{4}} = 186\ \text{K} = -87°\text{C}$$

This would be a very cold planet indeed without the gases that make up and insulate our atmosphere.

6.7 Greenhouse Gases and Effects

The long-term average surface temperature of the Earth over the last epoch of human evolution is agreed upon to be 15°C. The effect of the greenhouse gases is approximately

$$15°C - T_e = 15°C - (-18°C) = 33°C \text{ effective increase due to the greenhouse gas effect}$$

The **greenhouse gas effect**, illustrated in Figure 6.19, makes the planet much more hospitable by increasing the average surface temperature of the planet by 33°C. The composition of the atmosphere has a profound effect on the magnitude of the greenhouse effect.

The energy balance is used, in conjunction with mass balance information from the carbon cycle, to model the energy and atmosphere interactions. Typically, numerical values in the energy balance equation are presented by averaging the effects over a theoretical average square meter of the Earth's surface. The effective average incoming solar radiation is determined by dividing the incoming irradiation by the surface area of the Earth:

$$\frac{<S_{o\pi}r_e^2}{4\pi r_e^2} = \frac{1,365 \frac{W}{m^2}}{4} = 341 \frac{W}{m^2} \text{ of incoming solar radiation}$$

Approximately 30% of the incoming solar radiation, or 102 W/m², is reflected back into space by the Earth's albedo. In order to maintain a constant temperature on the Earth, 239 W/m² must be radiated back into space, as shown in Figure 6.20.

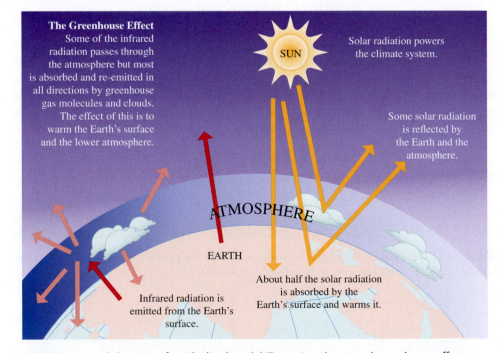

FIGURE 6.19 Schematic of an idealized model illustrating the natural greenhouse effect.

Source: Based on Climate Change 2007: The Physical Science Basis. Contributions of working group I to the Fourth Assessment Report of the intergovernmental Panel on Climate Change [Solomon, S., Qin, D., Manning, M., Chen, Z., Marquis, M., Averyt, K.B., Tignor, M., and Miller, H. L. (eds.)] Cambridge University Press, Cambridge, United Kingdom and New York, NY, USA.

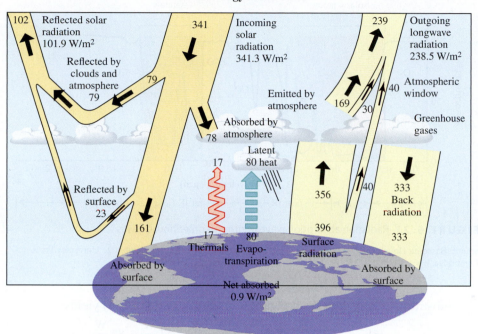

Global Energy Flows W/m²

FIGURE 6.20 An energy balance on the Earth that includes atmospheric insulation effects.

Source: Based on Trenberth, K. E., J. T. Fasullo, and J. Kiehl, 2009: Earth's global energy budget. Bull. Amer. Meteor. Soc., 90, No. 3, 311–324.

The atmosphere absorbs about 33%, or 78 W/m², of the incoming irradiation that is not reflected, and the remaining 67%, or 161 W/m², is absorbed by the surface of the Earth.

The Earth absorbs the incoming solar radiation and, acting as a blackbody, radiates heat back into the atmosphere. Not all components of the atmosphere are capable of trapping the heat from the sun and Earth. The major components of the atmosphere, nitrogen, as N_2 and oxygen in the form of O_2, absorb very little of the radiation from the sun and Earth. The atmosphere's major energy-absorbing compounds are shown in Figure 3.21, and these greenhouse gases absorb various amounts of radiation, depending on the wavelength of the electromagnetic waves. Water vapor, (H_2O) in the atmosphere absorbs radiation with wavelengths near 7 and 20 μm, and it is the most important greenhouse gas due to its fairly high concentration and the wide range of wavelengths over which it absorbs energy. Methane, (CH_4) and nitrous oxide, (N_2O) absorb radiation with wavelengths near 8 μm. Ozone, (O_3) absorbs radiation with wavelengths near 10 μm. Carbon dioxide, (CO_2) absorbs radiation with wavelengths near 15 μm. Radiation with wavelengths near 9 μm and from 10 to 13 μm largely passes through the atmosphere and approximately 40 W/m² is radiated directly from the Earth back into space. The greenhouse gases in the atmosphere retain and radiate a tremendous amount of heat, 324 W/m², of radiation back to Earth.

The net energy balance across the surface of the Earth then can be written as

(Direct energy absorbed from solar radiation) + (energy absorbed from back radiation from the atmosphere) = (energy radiated from the Earth and emitted to space) + (energy radiation from the Earth to the atmosphere) + (thermal convective energy) + (energy lost to evapotranspiration)   (6.24)

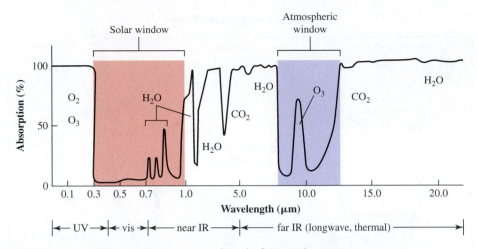

FIGURE 6.21 Radiation absorption wavelengths for greenhouse gases.

Source: Based on Houghton, J.T. (2009) Global Warming: The Complete Briefing. Cambridge University Press. Cambridge, UK.

Substituting the values from Figure 6.20 into Equation (6.24) yields

$$161\,\frac{W}{m^2} + 333\,\frac{W}{m^2} = 40\,\frac{W}{m^2} + 356\,\frac{W}{m^2} + 17\,\frac{W}{m^2} + 80\,\frac{W}{m^2}$$

Greenhouse gases in the Earth's atmosphere effectively act as a temperature regulator for the planet and control the Earth's climate. The radiative forcing for each greenhouse gas can be computed by calculating the amount of energy each gas absorbs in the atmosphere based on the combination of the concentration of the gas in the atmosphere and the ability of the gas to absorb energy over various wavelengths. Radiative forcings for the atmospheric greenhouse gases were calculated for the period between 1860 and 2000. These radiative forcings and uncertainties associated with the calculations are shown in Table 6.4.

TABLE 6.4 Total instantaneous forcing at 200 hPa (W/m²) from AOGCMs and LBL codes in RTMIP (W. D. Collins et al., 2006). Calculations are for cloud-free climatological midlatitude summer conditions. (from 2007 IPCC report)

RADIATIVE SPECIES	CO₂	CO₂	N₂O + CFCs	CH₄ + CFCs	ALL LLGHGs	WATER VAPOR
Forcing[a]	2000–1860	2x–1x	2000–1860	2000–1860	2000–1860	1.2x–1x
AOGCM mean	1.56	4.28	0.47	0.95	2.68	4.82
AOGCM std. dev.	0.23	0.66	0.15	0.30	0.30	0.34
LBL mean	1.69	4.75	0.38	0.73	2.58	5.08
LBL std. dev.	0.02	0.04	0.12	0.12	0.11	0.16

[a] 2000–1860 is the forcing due to an increase in the concentrations of radiative species between 1860 and 2000. 2x–1x are forcings from increases in radiative species by 100% and 20% relative to 1860 concentrations.

Source: Based on IPCC (2007) Climate Change 2007: The Physical Science Basis. Contributions of working group I to the Fourth Assessment Report of the Intergovernmental Panel on Climate Change [Solomon, S., Qin, D., Manning, M., Chen, Z., Marquis, M., Averyt, K.B., Tignor, M., and Miller, H. L. (eds.)] Cambridge University Press, Cambridge, United Kingdom and New York, NY, USA.

Since many of the concentrations of these gases changed during the period of industrialization from 1860 to 2000, scientists have also calculated the net change in the radiative forcing over this time period (see Figure 6.22). Carbon dioxide has had the largest net impact on increasing radiative forcing, and we can also see that there is very little uncertainty associated with this forcing. Simply put, we have added more insulation to our atmosphere, which will in turn make the planet warmer, as carbon dioxide levels and temperatures approach a new equilibrium condition. It is

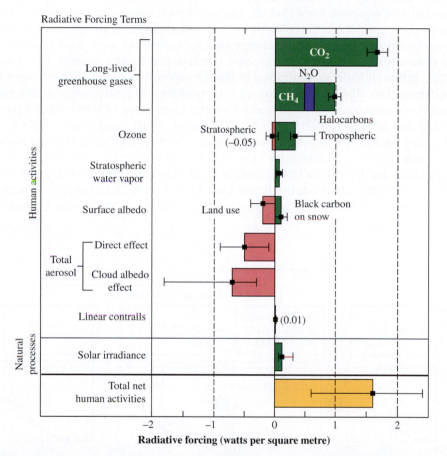

FIGURE 6.22 Summary of the principal components of the radiative forcing of climate change. All these radiative forcings result from one or more factors that affect climate and are associated with human activities or natural processes as discussed in the 2007 IPCC report. The values represent the forcings in 2005 relative to the start of the industrial era (about 1750). Human activities have caused significant changes in long-lived gases, ozone, water vapor, surface albedo, aerosols, and contrails. The only increase in natural forcing of any significance between 1750 and 2005 occurred in solar irradiance. Positive forcings led to warming of climate and negative forcings led to a cooling. The thin black lines attached to each colored bar represent the range of uncertainty for the respective value.

Source: Based on IPCC (2007) Climate Change 2007: The Physical Science Basis. Contributions of working group I to the Fourth Assessment Report of the intergovernmental Panel on Climate Change [Solomon, S., Qin, D., Manning, M., Chen, Z., Marquis, M., Averyt, K.B., Tignor, M., and Miller, H. L. (eds.)] Cambridge University Press, Cambridge, United Kingdom and New York, NY, USA.

important to notice that calculating the exact rise in the average surface temperature of the planet is made more complicated by additional positive and negative forcings from changes in other greenhouse gases and potential changes in the Earth's albedo due to increased cloud cover and aerosol effects.

6.8 Climate Change Projections and Impacts

In this chapter, we've shown how fossil fuel use has resulted in increased carbon dioxide emissions. The basic principle of the law of conservation of mass and the mass balance equation show that the carbon dioxide emitted from fossil fuel use is closely correlated to carbon dioxide concentrations in the atmosphere. The law of conservation of energy and the energy balance equations were used to show that the greenhouse gases in the atmosphere regulate the Earth's climate. The changes in the carbon dioxide concentration in the atmosphere have the greatest impact on our climate, but other factors influence climate change in both positive and negative ways. Basic mass and energy balances form the foundation of sophisticated climate models that are being used to predict the type of climate and climatic effects that will occur on Earth as we approach a new equilibrium condition at some point in the future. Figure 6.23 illustrates how emission projections can be used to model and predict the change in concentration of atmospheric greenhouse gases under different assumptions about the future use of fossil fuels. The change in concentration of the future greenhouse gases will create a change in the energy balance due

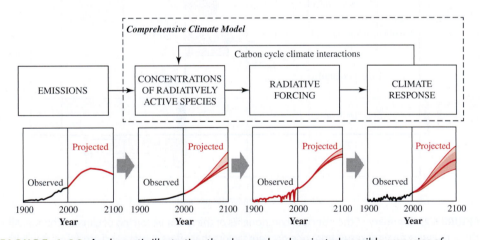

FIGURE 6.23 A schematic illustrating the observed and projected possible scenarios of climate change and how mass and energy fundamentals may be used to predict how the Earth's climate will respond to changing conditions in the atmosphere. Several steps from emissions to climate response contribute to the overall uncertainty of climate model projection. These uncertainties can be quantified through a combined effort of observation, process understanding, a hierarchy of climate models, and ensemble simulations. In comprehensive climate models, physical and chemical representations of processes permit a consistent quantification of uncertainty. Note that the uncertainty associated with the future emissions path is of an entirely different nature.

Source: Based on IPCC (2007) Climate Change 2007: The Physical Science Basis. Contributions of working group I to the Fourth Assessment Report of the intergovernmental Panel on Climate Change [Solomon, S., Qin, D., Manning, M., Chen, Z., Marquis, M., Averyt, K.B., Tignor, M., and Miller, H. L. (eds.)] Cambridge University Press, Cambridge, United Kingdom and New York, NY, USA.

to a change in the radiative forcing that will influence how the Earth's climate will respond over the next century and farther into the future.

The climate models used to make the predictions today started out with many of the exact calculations that have been illustrated in this chapter. The models we have examined are quite simple and take into account only limited variables. Development of a full climate model is well beyond the scope of this textbook and chapter. However, Figures 6.24–6.30 illustrate how climate models have evolved by continuing to include

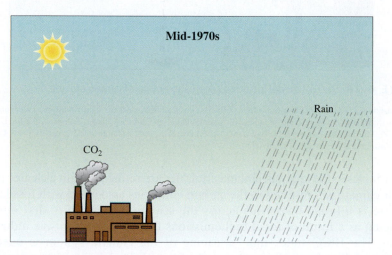

FIGURE 6.24 The complexity and level of detail in models have improved greatly since the early evolution of climate models were developed. In the 1970s, carbon dioxide was modeled in conjunction with models geared toward gaining a greater understanding of the effects of acid rain.

Source: Based on IPCC (2007) Climate Change 2007: The Physical Science Basis. Contributions of working group I to the Fourth Assessment Report of the intergovernmental Panel on Climate Change [Solomon, S., Qin, D., Manning, M., Chen, Z., Marquis, M., Averyt, K.B., Tignor, M., and Miller, H. L. (eds.)] Cambridge University Press, Cambridge, United Kingdom and New York, NY, USA.

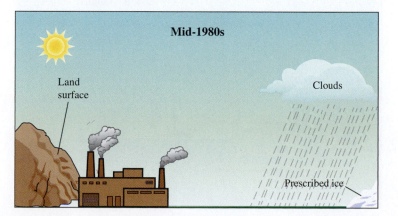

FIGURE 6.25 In the 1980s, carbon dioxide concentrations in the atmosphere were modeled and changes in the Earth's albedo were evaluated.

Source: Based on IPCC (2007) Climate Change 2007: The Physical Science Basis. Contributions of working group I to the Fourth Assessment Report of the intergovernmental Panel on Climate Change [Solomon, S., Qin, D., Manning, M., Chen, Z., Marquis, M., Averyt, K.B., Tignor, M., and Miller, H. L. (eds.)] Cambridge University Press, Cambridge, United Kingdom and New York, NY, USA.

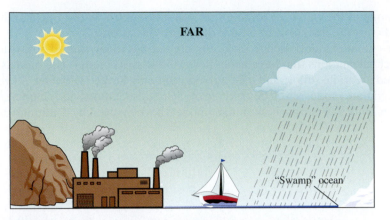

FIGURE 6.26 The FAR model included ocean absorption of carbon dioxide.

Source: Based on IPCC (2007) Climate Change 2007: The Physical Science Basis. Contributions of working group I to the Fourth Assessment Report of the intergovernmental Panel on Climate Change [Solomon, S., Qin, D., Manning, M., Chen, Z., Marquis, M., Averyt, K.B., Tignor, M., and Miller, H. L. (eds.)] Cambridge University Press, Cambridge, United Kingdom and New York, NY, USA.

more variables in the analysis, and uncertainties are refined to gain a clearer picture of how our climate will respond to the carbon emissions associated with fossil fuel use.

There is a broad consensus across the scientific community that our climate is changing as a result of fossil fuel emissions from energy-intensive industrialized societies. The IPCC states that "warming of the climate system is unequivocal, as is now evident from observations of increases in global average temperatures, widespread melting of snow and ice and rising global average sea level."

The IPCC has called for sustainable strategies for development and climate change adaptation and mitigation. The interrelationships between climate,

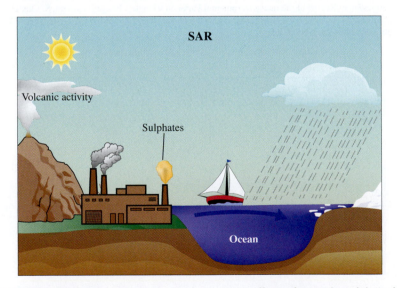

FIGURE 6.27 The SAR model began to investigate the effects of aerosols and their ability to act as a negative radiative forcing.

Source: Based on IPCC (2007) Climate Change 2007: The Physical Science Basis. Contributions of working group I to the Fourth Assessment Report of the intergovernmental Panel on Climate Change [Solomon, S., Qin, D., Manning, M., Chen, Z., Marquis, M., Averyt, K.B., Tignor, M., and Miller, H. L. (eds.)] Cambridge University Press, Cambridge, United Kingdom and New York, NY, USA.

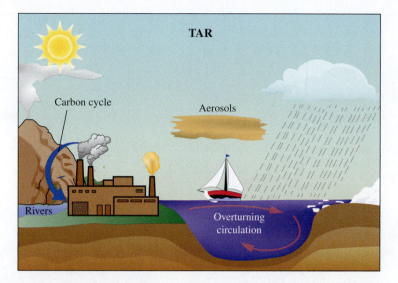

FIGURE 6.28 The TAR model began to combine the previous work evaluated in the FAR and SAR models. This increasingly complex model included most major sources and sinks of carbon and began to evaluate other radiative forcing factors. However, the TAR grid was very large and had difficulty evaluating changes in land use and associated forcings that would occur on small regional scales.

Source: Based on IPCC (2007) Climate Change 2007: The Physical Science Basis. Contributions of working group I to the Fourth Assessment Report of the intergovernmental Panel on Climate Change [Solomon, S., Qin, D., Manning, M., Chen, Z., Marquis, M., Averyt, K.B., Tignor, M., and Miller, H. L. (eds.)] Cambridge University Press, Cambridge, United Kingdom and New York, NY, USA.

societies' activities, adaptation, and mitigation of climate change are illustrated in Figure 6.31.

The 2007 IPCC climate model report (AR4) further concludes that:

Global atmospheric concentrations of CO_2, CH_4 and N_2O have increased markedly as a result of human activities since 1750 and now far exceed

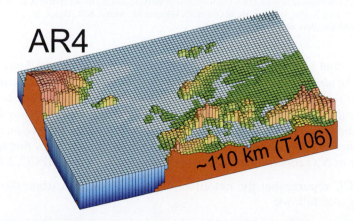

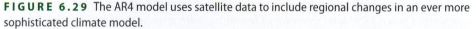

FIGURE 6.29 The AR4 model uses satellite data to include regional changes in an ever more sophisticated climate model.

Source: IPCC (2007) Climate Change 2007: The Physical Science Basis. Contributions of working group I to the Fourth Assessment Report of the intergovernmental Panel on Climate Change [Solomon, S., Qin, D., Manning, M., Chen, Z., Marquis, M., Averyt, K.B., Tignor, M., and Miller, H. L. (eds.)] Cambridge University Press, Cambridge, United Kingdom and New York, NY, USA. 996 pp. Figure 1.4 p113.

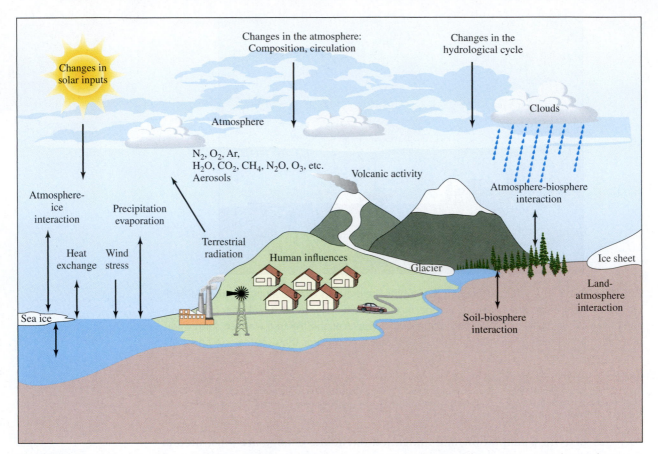

FIGURE 6.30 Current models use high-resolution data to include changes in land use and the most recent understanding of radiative forcing mechanisms to model past, present, and future climates. By replicating climates in the past, these models are calibrated and validated and are able to enhance our understanding of our current climate conditions and predict how the climate will respond to future changes in societies' use of fossil fuels and alternative energy technologies, with a low degree of uncertainty.

Source: Based on IPCC (2007) Climate Change 2007: The Physical Science Basis. Contributions of working group I to the Fourth Assessment Report of the intergovernmental Panel on Climate Change [Solomon, S., Qin, D., Manning, M., Chen, Z., Marquis, M., Averyt, K.B., Tignor, M., and Miller, H. L. (eds.)] Cambridge University Press, Cambridge, United Kingdom and New York, NY, USA.

pre-industrial values determined from ice cores spanning many thousands of years. The atmospheric concentrations of CO_2 and CH_4 in 2005 exceed by far the natural range over the last 650,000 years. Global increases in CO_2 concentrations are due primarily to fossil fuel use, with land-use change providing another significant but smaller contribution. It is very likely that the observed increase in CH_4 concentration is predominantly due to agriculture and fossil fuel use. The increase in N_2O concentration is primarily due to agriculture (IPCC, 2007c).

The IPCC reports that the net effect of the increase in carbon dioxide in the atmosphere is as follows:

There is very high confidence that the global average net effect of human activities since 1750 has been one of warming, with a radiative forcing of $+1.6$ [$+0.6$ to $+2.4$] W/m^2.

The estimated effect of this radiative forcing is a change in the average surface temperature of the Earth, as shown in Figure 6.32. By the year 2100 the Earth's average surface temperature is expected to increase by 1.8 to 4.0 °C. The IPCC

Schematic Framework of Anthropogenic Climate Change Drivers, Impacts, and Responses

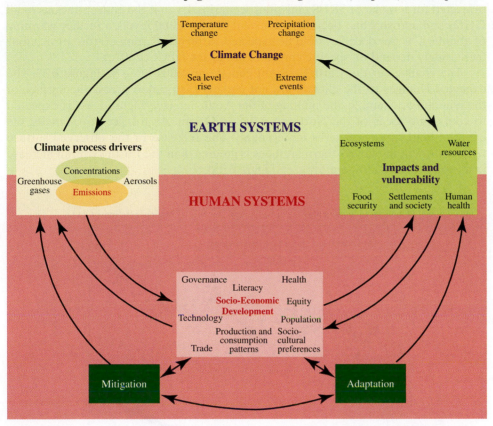

FIGURE 6.31 Schematic illustrating the relationship between sustainability and climate change.

Source: Based on Climate Change 2007: Synthesis Report. Contribution of Working Groups I, II and III to the Fourth Assessment Report of the Intergovernmental Panel on Climate Change, Core Writing Team, Pachauri, R.K. and Reisinger, A. (Eds.), IPCC, Geneva, Switzerland.

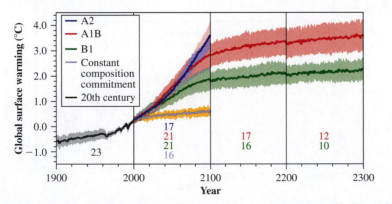

FIGURE 6.32 Multimodel means of surface warming (relative to 1980–1999) for multiple scenarios. Values beyond 2100 are for the stabilization scenarios. Lines show the multimodel means; shading denotes the ±1 standard deviation range of individual model annual means. Uncertainty across scenarios should not be interpreted from this figure.

Source: Based on : IPCC (2007) Climate Change 2007: The Physical Science Basis. Contributions of working group I to the Fourth Assessment Report of the intergovernmental Panel on Climate Change [Solomon, S., Qin, D., Manning, M., Chen, Z., Marquis, M., Averyt, K.B., Tignor, M., and Miller, H. L. (eds.)] Cambridge University Press, Cambridge, United Kingdom and New York, NY, USA.

estimates that "for the next two decades a warming of about 0.2°C per decade is projected for a range of SRES emissions scenarios. Even if the concentrations of all GHGs and aerosols had been kept constant at year 2000 levels, a further warming of about 0.1°C per decade would be expected. Afterwards, temperature projections increasingly depend on specific emissions scenarios (IPCC, 2007c)."

The changes in the Earth's current climate are illustrated in Figure 6.33. The expected changes to the climate are not equally distributed across the surface of the planet. For example, the melting of the Arctic ice (see Figure 6.34) is expected to cool some areas of the Northern Atlantic Ocean. As this ice cover melts, the exposed land surface absorbs more energy. The increase in the observed surface temperature in the far northern latitudes is greater than the Earth's overall increase in average surface temperature (see Figure 6.35).

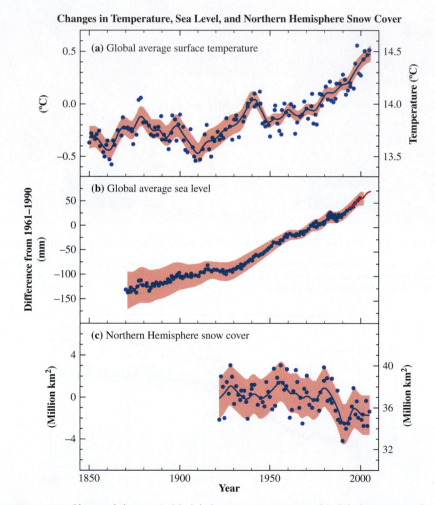

FIGURE 6.33 Observed changes in (a) global average temperature; (b) global average sea level from tide gauge (blue) and satellite (red) data; and (c) Northern Hemisphere snow cover for March–April. All differences are related to the period 1961–1990. Smoothed curves represent decadal averaged values, while circles show yearly values. The shaded areas are the uncertainty intervals estimated from a comprehensive analysis of known uncertainties (a and b) and from the time series (c).

Source: Based on Climate Change 2007: Synthesis Report. Contribution of Working Groups I, II and III to the Fourth Assessment Report of the Intergovernmental Panel on Climate Change, Core Writing Team, Pachauri, R.K. and Reisinger, A. (Eds.), IPCC, Geneva, Switzerland.

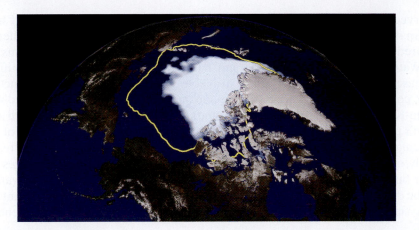

FIGURE 6.34 An image of the Arctic sea ice on September 16, 2012, the day that the National Snow and Ice Data Center announced the minimum reached in 2012. The yellow outline shows the average sea ice minimum from 1979 through 2010. The sea ice is shown with a blue tint.

Source: NASA/Goddard Space Flight Center Scientific Visualization Studio. The Blue Marble data is courtesy of Reto Stockli (NASA/GSFC).

A detailed discussion of the possible effects, adaptation, and mitigation strategies related to climate change can be investigated in a great deal more detail in the IPCC reports. The change in the Earth's average surface temperature is expected to have a profound influence on the environment and human society and also to have economic consequences. The IPCC verdict is that "anthropogenic warming over the last three decades has *likely* had a discernible influence at the global scale on

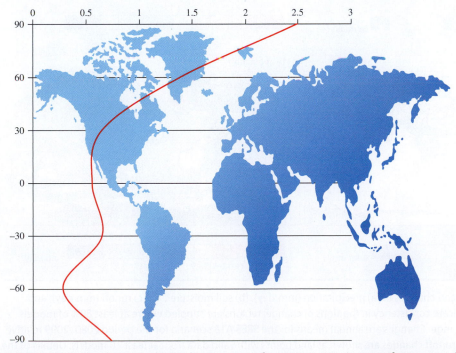

FIGURE 6.35 The change in the Earth's average surface temperature is not uniform across all latitudes. The increase in surface temperature is most dramatic over the far northern latitudes.

Source: Leo Blanchette/Shutterstock.com. Data from UNEP (2011). Keeping Track of Our Changing Environment: From Rio to Rio+20 (1992–2012). Division of Early Warning and Assessment (DEWA), United Nations Environment Programme (UNEP), Nairobi.

observed changes in many physical and biological systems (IPCC, 2007b). Changes in climate that societies have grown accustomed to during the previous millennium are predicted to have significant effects on societies' surrounding ecosystems.

Water resources were discussed in Chapter 3. Climate change may significantly decrease the amount of fresh water available from glaciers and from reduction in snow cover. Particular concern focuses on major mountain ranges, including the Hindu-Kush, Himalaya, and Andes Mountains where more than one-sixth of the world's population currently lives. In addition to drinking water scarcity, lack of water, particularly in drought-prone regions, may have a negative effect on agriculture, water supply, energy production, and health. Figure 6.36 illustrates the likely changes to the water budget parameters that will be driven by climate change. Too much water can also be detrimental as the increased risk of flooding is likely to pose a significant threat to physical infrastructure and water quality. There is a high degree of confidence that

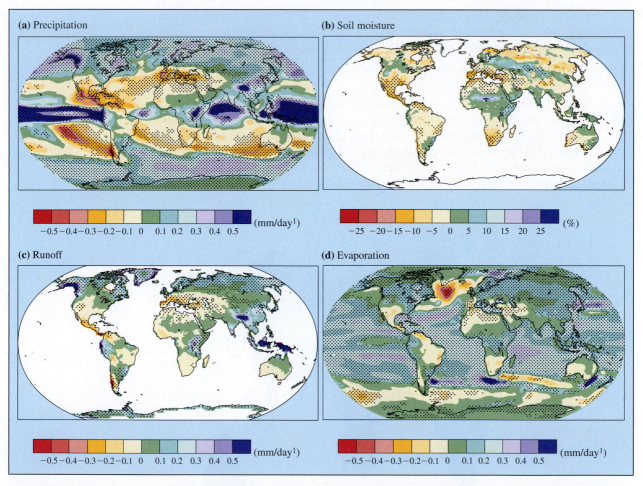

FIGURE 6.36 Multimodel mean changes in (a) precipitation (mm/day), (b) soil moisture (%), (c) runoff (mm/day), and (d) evaporation (mm/day). To indicate consistency in the signs of change, regions are stippled where at least 80% of models agree on the sign of the mean change. Changes are annual means for the SRES A18 scenario for the period 2080–2099 relative to 1980–1999. Soil moisture and runoff changes are shown at land points with valid data from at least 10 models. Details of the method and results for individual models can be found in the IPCC reports.

Source: IPCC (2007) Climate Change 2007: The Physical Science Basis. Contributions of working group I to the Fourth Assessment Report of the inter-governmental Panel on Climate Change [Solomon, S., Qin, D., Manning, M., Chen, Z., Marquis, M., Averyt, K.B., Tignor, M., and Miller, H. L. (eds.)] Cambridge University Press, Cambridge, United Kingdom and New York, NY, USA. 996 pp. Figure 10.12.

the negative consequences of climate change for water resources outweigh the positive changes. Water-scarce areas of the world face the likelihood of less rainfall and higher evaporation rates. Areas such as the southwestern United States could see a drying period even greater in scope and duration than those that may have resulted in the changes associated with the society of Chaco Canyon (see Chapter 1).

The IPCC has reported potential ecosystem impacts of climate change, including the following:

- The resilience of many ecosystems is *likely* to be exceeded in this century by an unprecedented combination of climate change, associated disturbances (e.g., flooding, drought, wildfire, insect infestations, and ocean acidification) and other global change drivers (e.g., land-use change, pollution, fragmentation of natural systems, and overexploitation of resources) (IPCC, 2007b).
- Over the course of this century, net carbon uptake by terrestrial ecosystems is *likely* to peak before mid-century and then weaken or even reverse, thus amplifying climate change (IPCC, 2007b).
- Approximately 20 to 30% of plant and animal species assessed so far are *likely* to be at increased risk of extinction if increases in global average temperature exceed 1.5 to 2.5°C (*medium confidence*) (IPCC, 2007b).
- For increases in global average temperature exceeding 1.5 to 2.5°C and in concomitant atmospheric CO_2 concentrations, major changes are projected in ecosystem structure and function, species' ecological interactions, and shifts in species' geographical ranges, with predominantly negative consequences for biodiversity and ecosystem goods and services, for example, water and food supply (IPCC, 2007b).

The changing climate may also affect human food supplies, health, and the economy. Some of the changes in specific regions may include increases in the yearly growing season, with a potential increase in crop yields in those regions (Figures 6.37 to 6.39). The IPCC reports that food production will be impacted by climate change in the following ways:

- Crop productivity is projected to increase slightly at mid- to high latitudes for local mean temperature increases of up to 1 to 3°C, depending on the crop, and then decrease beyond that in some regions (*medium confidence*) (IPCC, 2007b).

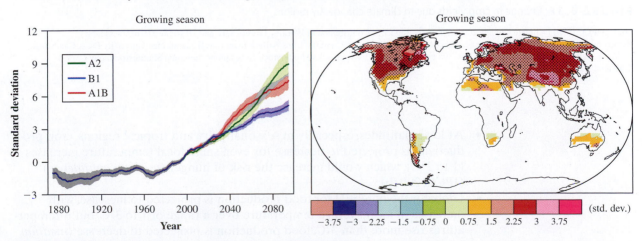

FIGURE 6.37 Impacts of climate change on the growing season.

Source: Based on IPCC (2007) Climate Change 2007: The Physical Science Basis. Contributions of working group I to the Fourth Assessment Report of the intergovernmental Panel on Climate Change [Solomon, S., Qin, D., Manning, M., Chen, Z., Marquis, M., Averyt, K.B., Tignor, M., and Miller, H. L. (eds.)] Cambridge University Press, Cambridge, United Kingdom and New York, NY, USA.

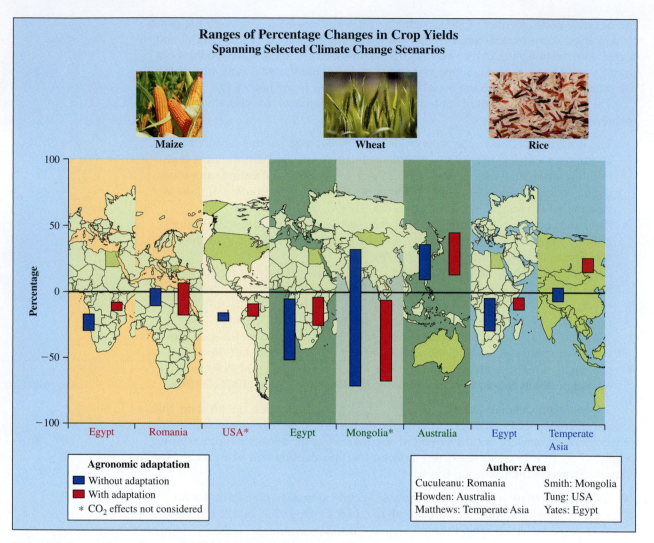

FIGURE 6.38 Change in crop yields due to climate change, by region.

Source: IPCC (2001) Climate Change 2001: Synthesis Report. A Contribution of Working Groups I, II, and III to the Third Assessment Report of the Integovernmental Panel on Climate Change [Watson, R.T. and the Core Writing Team (eds.)]. Cambridge University Press, Cambridge, United Kingdom, and New York, NY, USA, 398 pp. WG2-Figure TS-4; smereka/Shutterstock.com; simone mescolini/Shutterstock.com; Coprid /Shutterstock.com

- At lower latitudes, especially in seasonally dry and tropical regions, crop productivity is projected to decrease for even small local temperature increases (1 to 2°C), which would increase the risk of hunger (*medium confidence*) (IPCC, 2007b).
- Globally, the potential for food production is projected to increase, with increases in local average temperature over a range of 1 to 3°C, but if temperatures rise more than 3°C, food production is projected to decrease (*medium confidence*) (IPCC, 2007b).

Potential Climate Changes Impact

Temperature

Sea level rise

Impacts on...

Health	Agriculture	Forest	Water resources	Coastal areas	Species and natural areas

Health	Agriculture	Forest	Water resources	Coastal areas	Species and natural areas
Weather-related mortality		Forest composition		Erosion of beaches	Loss of habitat and species
Infectious diseases	Crop yields irrigation demands	Geographic range of forest	Water supply Water quality Competition for water	Inundation of coastal lands	Cryosphere: diminishing glaciers
Air-quality respiratory illnesses		Forest health and productivity		Additional costs to protect coastal communities	

FIGURE 6.39 Illustration of the inter-relationship between climate change, environmental impacts, economic impacts, and societal impacts.

Source: Based on Philippe Rekacewicz, UNEP/GRID-Arendal. http://www.grida.no/graphicslib/detail /potential -climate-change-impacts_035a. Data from United States environmental protection agency (EPA); Emiel de Lange /Shutterstock.com; chbaum/Shutterstock.com; Mirexon/Shutterstock.com; Lightspring/Shutterstock.com; SNEHIT /Shutterstock.com; sittitap/Shutterstock.com; KPG_Payless/Shutterstock.com; Vibrant Image Studio/Shutterstock .com; Hung Chung Chih/Shutterstock.com

6.9 Carbon Dioxide Mitigation, Capture, and Sequestration

There are numerous methods under investigation to reduce greenhouse gases in the atmosphere. In order to reduce, or mitigate, the impacts of future carbon dioxide emissions, carbon emissions may be limited by regulations or financial incentives. **Carbon Capture and Sequestration (CCS)** is a three-step process that includes carbon dioxide capture, transport of the captured carbon dioxide, and storage of the carbon dioxide for extended periods of time in biogeochemical repositories other than the atmosphere. The first step in reducing carbon emissions is to separate and capture the carbon either from a carbon-emitting source or from carbon stored in the atmosphere.

Carbon dioxide emitted from combustion processes is at much higher concentrations than carbon dioxide in the atmosphere. The higher the concentration of the carbon dioxide, the easier it is to separate the carbon dioxide from other compounds and capture the carbon. Therefore, most carbon dioxide mitigation, capture, and sequestration methods are targeted toward capture and sequestration of carbon from fossil fuel-burning power plants or other carbon-intensive industries. Several alternative carbon capture processes are illustrated in Figure 6.40, including capturing carbon dioxide after standard combustion, capturing even higher concentration carbon dioxide from combustion with oxygen (Oxyfuel), and separating and capturing carbon from the hydrogen in the fossil fuel.

Pipelines are expected to become the principal form of transporting captured carbon dioxide to geologic storage sinks. A review of the 500 largest sources of carbon dioxide emissions in the United States shows that 95% are within 50 miles of a possible storage site (Dooley, 2006). However, significant work and resources must be allocated to develop the infrastructure required to transport large quantities of captured carbon dioxide.

Once captured and transported, the carbon must be stored or sequestered in such a fashion that it cannot reach the atmosphere for long time periods (hundreds to thousands of years). The carbon may be injected into deep geologic formations, injected into the deep oceans, mineralized (combined with minerals to form an inactive solid), or other industrial uses may be found. The relationships between the

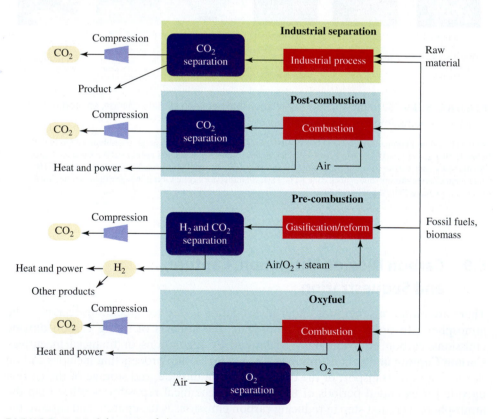

FIGURE 6.40 Schematic of alternative carbon dioxide capture systems.

Source: Based on IPCC (2005) IPCC Special Report on carbon Dioxide Capture and Storage. Prepared by Working Group III of the Intergovernmental Panel on Climate Change [Metz, B., Davidson, O., de Coninck, H.C., Loas, M., and Meyer, L.A. (eds.)] Cambridge University Press, Cambridge, United Kingdom and New York, NY, USA.

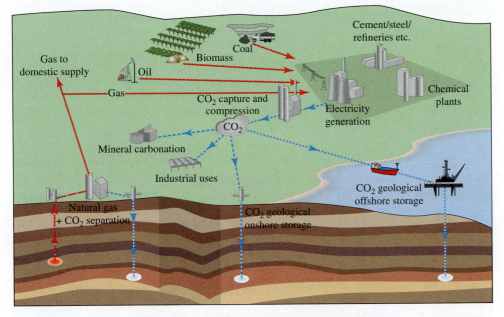

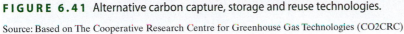

FIGURE 6.41 Alternative carbon capture, storage and reuse technologies.

Source: Based on The Cooperative Research Centre for Greenhouse Gas Technologies (CO2CRC)

fossil fuels source, fossil fuel use, transport of captured CO_2, and possible sequestration processes are illustrated in Figure 6.41.

Carbon storage includes at least three phases of operation. (1) A pre-injection phase for carbon dioxide storage includes extensive geologic site characterization, evaluation of site suitability, and modeling to predict carbon dioxide transport and movement in the subsurface. (2) The operational phase includes injecting the carbon dioxide into a well, monitoring groundwater chemistry, and tracking the subsurface carbon dioxide plume. (3) The final phase of operation should include continuous monitoring of water quality and the carbon dioxide plume near the injection site.

Subsurface geological sequestration, illustrated in Figure 6.42, may include storage in (1) depleted oil and gas reserves, (2) enhanced oil and gas recovery (EOR and EGR) wells, (3a) offshore and (3b) onshore deep saline formations, and (4) use in Enhanced Coal Bed Methane Recovery (ECBMR). Carbon dioxide has been used in EOR operations since the 1970s, so in some ways the theory of deep-well carbon dioxide injection is not new. However, the current number of EOR projects is insufficient to meet the expected demands of future carbon dioxide storage. Nonetheless, the U.S. Department of Energy and International Energy Agency (IEA) estimates that there is enough potential geologic storage in the United States alone to store over 1,000 years of carbon emissions from current coal-fired electricity production in the United States. Potential geologic sequestration capacity worldwide is estimated to be about 2,000 Gt CO_2 (IPCC, 2005). The majority of large point sources of carbon dioxide, worldwide, are located within 300 kilometers of potential geologic storage sites. Although the capacity for geologic sequestration

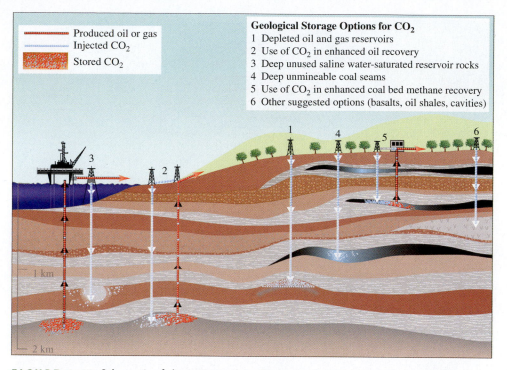

FIGURE 6.42 Schematic of alternative carbon dioxide geological capture and sequestration options.

Source: Based on The Cooperative Research Centre for Greenhouse Gas Technologies (CO$_2$ CRC)

is available, there are several possible mechanisms for carbon dioxide escape or leakage from geologic repositories (illustrated in Figure 6.43). The environmental and human health risks in general and for each individual site must be thoroughly characterized before carbon injection.

Ocean storage of carbon dioxide, illustrated in Figure 6.44, through deep sea injection of carbon produces dense, pressurized liquid carbon dioxide that would form carbon "lakes" that would delay dissolution of the CO$_2$ into the surrounding ocean environment. A smaller portion of large point sources of carbon dioxide is located in close proximity to feasible ocean storage sites. However, alternatives to geologic sequestration may be needed. The effects of carbon storage on ocean pH and deep sea ecosystems are under investigation in several regions.

Current cost estimates for CCS technology range widely. The U.S. Department of Energy (DOE) estimates that CCS technology would cost 60 to 114 U.S. dollars (U.S. DOE, 2010) per metric ton of carbon dioxide, with 70 to 90% of that cost associated with capture and compression processes. Based on 2002 electricity generation costs, this would represent an increase of 0.01 to 0.05 U.S. dollars per kilowatt hour ($/kWh), depending on the fuel, the CSS technology, and the location of the facility (IPCC, 2005). Capturing carbon dioxide requires additional energy, so the total energy produced and fuel use from a CCS-equipped power generation plant is estimated be 10 to 40% more than a comparable non-CCS plant. However, carbon dioxide emissions would

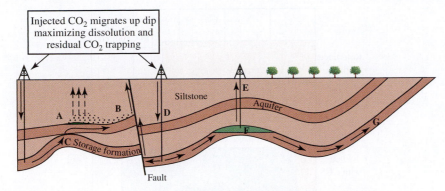

Injected CO_2 migrates up dip maximizing dissolution and residual CO_2 trapping

Potential Escape Mechanisms

A. CO_2 gas pressure exceeds capillary pressure and passes through siltstone	B. Free CO_2 leaks from A into upper aquifer up fault	C. CO_2 escapes through 'gap' in cap rock into higher aquifer	D. Injected CO_2 migrates up dip, increases reservoir pressure and permeability of fault	E. CO_2 escapes via poorly plugged old abandoned well	F. Natural flow dissolves CO_2 at CO_2/water interface and transports it out of closure	G. Dissolved CO_2 escapes to atmosphere or ocean

Remedial Measures

A. *Extract and purify groundwater*	B. *Extract and purify groundwater*	C. *Remove CO_2 and reinject elsewhere*	D. *Lower injection rates or pressures*	E. *Re-plug well with cement*	F. *Intercept and reinject CO_2*	G. *Intercept and reinject CO_2*

FIGURE 6.43 Potential escape mechanisms and remedial measures for geologic sequestration of carbon dioxide.

Source: Based on IPCC (2005) IPCC Special Report on carbon Dioxide Capture and Storage. Prepared by Working Group III of the Intergovernmental Panel on Climate Change [Metz, B., Davidson, O., de Coninck, H.C., Loas, M., and Meyer, L.A. (eds.)] Cambridge University Press, Cambridge, United Kingdom and New York, NY, USA.

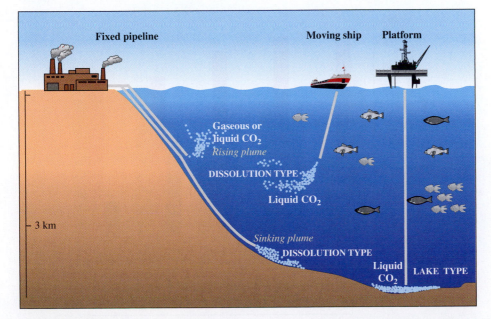

FIGURE 6.44 Schematic of possible carbon dioxide sequestration methods in the oceans.

Source: Based on IPCC (2005) IPCC Special Report on carbon Dioxide Capture and Storage. Prepared by Working Group III of the Intergovernmental Panel on Climate Change [Metz, B., Davidson, O., de Coninck, H.C., Loas, M., and Meyer, L.A. (eds.)] Cambridge University Press, Cambridge, United Kingdom and New York, NY, USA.

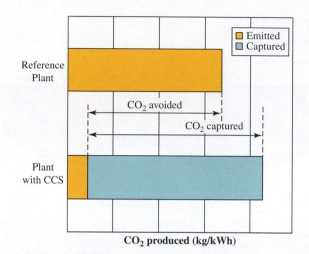

FIGURE 6.45 Comparison of carbon dioxide emissions from reference facilities with and without carbon dioxide capture and storage technology.

Source: Based on IPCC (2005) IPCC Special Report on Carbon Dioxide Capture and Storage. Prepared by Working Group III of the Intergovernmental Panel on Climate Change [Metz, B., Davidson, O., de Coninck, H.C., Loas, M., and Meyer, L.A. (eds.)] Cambridge University Press, Cambridge, United Kingdom and New York, NY, USA.

80 to 90% lower as illustrated qualitatively in Figure 6.45. Depending on the type of carbon capture system implemented at existing coal-powered plants in the United States, the comparable cost of electricity (COE) delivered to the consumer may increase from 40 to 90% (U.S. DOE, 2010), as noted in Figure 6.46 and Tables 6.5 and 6.6.

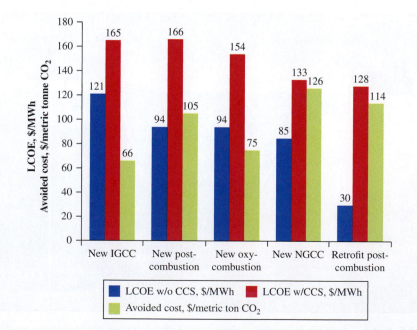

FIGURE 6.46 Expected increases in levelized costs of electricity (LCOE) and the cost associated with carbon dioxide removal from coal-generating power plants in the United States.

Source: DOE/EPA (2010) Report of the Interagency Task Force on Carbon Capture and Storage.

TABLE 6.5 Energy generation costs and carbon dioxide avoidance costs associated with CCS systems. Gas prices are assumed to be 2.8 to 4.4 U.S. dollars/GJ, and coal prices 1 to 1.5 U.S. dollars/GJ.

TYPE OF POWER PLANT WITH CCS	NATURAL GAS COMBINED CYCLE REFERENCE PLANT ($/ METRIC TON CO_2-AVOIDED)	PULVERIZED COAL REFERENCE PLANT ($/ METRIC TON CO_2-AVOIDED)
Power plant with capture and geological storage		
Natural gas combined cycle	40–90	20–60
Pulverized coal	70–270	30–70
Integrated gasification combined cycle	40–220	20–70
Power plant with capture and enhanced oil recovery (EOR)		
Natural gas combined cycle	20–70	0–30
Pulverized coal	50–240	10–40
Integrated gasification combined cycle	20–190	0–40

Source: Based on IPCC (2005) IPCC Special Report on carbon Dioxide Capture and Storage. Prepared by Working Group III of the Intergovernmental Panel on Climate Change [Metz, B., Davidson, O., de Coninck, H.C., Loas, M., and Meyer, L.A. (eds.)] Cambridge University Press, Cambridge, United Kingdom and New York, NY, USA.

TABLE 6.6 2002 Energy generation costs and carbon dioxide avoidance costs associated with CCS systems. Gas prices are assumed to be 2.8 to 4.4 U.S. dollars/GJ, and coal prices 1 to 1.5 U.S. dollars/GJ.

CCS SYSTEM COMPONENT	COST RANGE	REMARKS
Capture from a coal- or gas-fired power plant	15–75 $/metric ton CO_2 net captured	Net cost of captured CO_2 compared to the same plant without capture
Capture from hydrogen and ammonia production or gas processing	5–55 $/metric ton CO_2 net captured	Applies to high-purity sources requiring simple drying and compression
Capture from other industrial sources	25–115 $/metric ton CO_2 net captured	Range reflects use of a number of different technologies and fuels
Transportation	1–8 $/metric ton CO_2 net transported	Per 250 km pipeline or shipping for mass flow rates of 5 (high end) to 40 (low end) Mt CO_2/yr
Geological storage[a]	0.5–8 $/metric ton CO_2 net injected	Excluding potential revenues
Geological storage: monitoring and verification	0.1–0.3 $/metric ton CO_2 net injected	This covers pre-injection, injection, and post-injection monitoring, and depends on the regulatory requirements
Ocean storage	5–30 $/metric ton CO_2 net injected	Including offshore transportation of 100–500 km, excluding monitoring and verification
Mineral carbonization	50–100 $/metric ton CO_2 net mineralized	Range for the best case studied. Includes additional energy use for carbonation

[a] Over the long term, there may be additional costs for remediation and liabilities.

Source: Based on IPCC (2005) IPCC Special Report on carbon Dioxide Capture and Storage. Prepared by Working Group III of the Intergovernmental Panel on Climate Change [Metz, B., Davidson, O., de Coninck, H.C., Loas, M., and Meyer, L.A. (eds.)] Cambridge University Press, Cambridge, United Kingdom and New York, NY, USA.

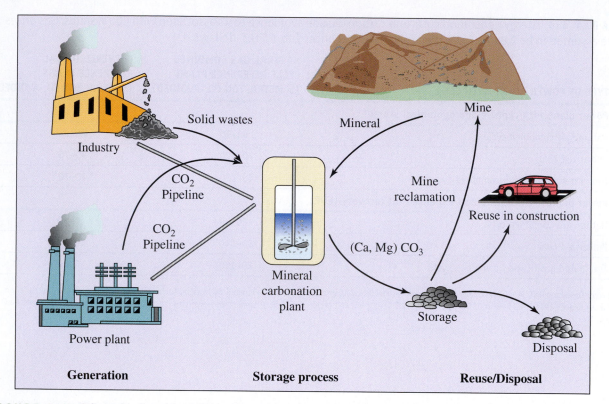

FIGURE 6.47 Schematic of combined CCS energy production and carbon resource recovery options.

Source: Based on IPCC (2005) IPCC Special Report on carbon Dioxide Capture and Storage. Prepared by Working Group III of the Intergovernmental Panel on Climate Change [Metz, B., Davidson, O., de Coninck, H.C., Loas, M., and Meyer, L.A. (eds.)] Cambridge University Press, Cambridge, United Kingdom and New York, NY, USA.

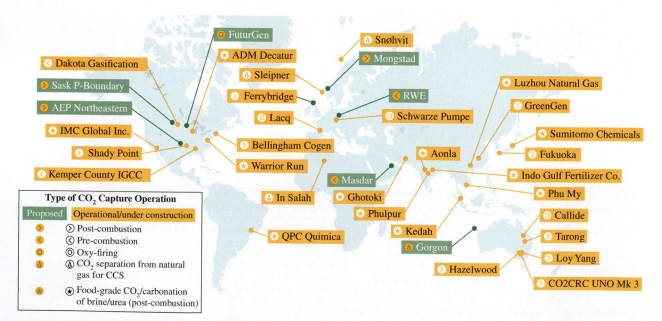

FIGURE 6.48 Demonstration sites under investigation in 2012 for CCS demonstration.

Source: Based on The Cooperative Research Centre for Greenhouse Gas Technologies (CO₂ CRC)

CCS technology may be most cost-effective when integrated into an energy and resource recovery system (see Figure 6.47) based on the principles of industrial ecology that will be discussed in more detail in Chapter 9. The most cost-effective CCS systems based on energy and economic modeling occur in association with energy production. The IPCC estimates that 220 to 2,200 Gt CO_2 from CSS would be needed to stabilize atmospheric greenhouse gas concentrations between 450 and 750 ppm$_v$ until the year 2100. Furthermore, the IPCC (2005, p. 12) states that "*the role of CCS in mitigation portfolios increases over the course of the century, and the inclusion of CCS in a mitigation portfolio is found to reduce the costs of stabilizing CO_2 concentrations by 30% or more.*" CCS technology is already being demonstrated throughout the world in the locations shown in Figure 6.48.

6.10   Summary

Climate change has far-reaching consequences for the environment, society, and economy. These potential consequences and interrelationships are represented in Figure 6.49. As the climate changes, societies must weigh the cost and risks

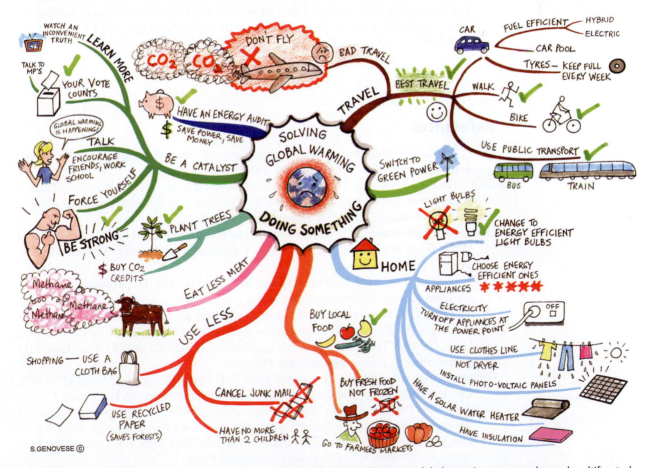

FIGURE 6.49 Action to minimize the impacts associated with climate change, or global warming, are complex and multifaceted. Economic, environmental, and societal impacts from our changing climate are already occurring. Technologic changes and advances, along with behavioral adaptations, may reduce the degree of the negative impacts and future extent of climate change.

Source: Learning Fundamentals

associated with both action and inaction in curbing current levels of carbon dioxide emissions. In this chapter we have examined the fundamental science and theories that explain climate change. The fundamental chemical transformation that illustrates some of the environmental risks associated with a warming climate will be discussed in the next several chapters. The IPCC suggests a strategy of weighing the costs and the threats in order to deal with the changing climate:

> *There is high confidence that neither adaptation nor mitigation alone can avoid all climate change impacts. Adaptation is necessary both in the short term and longer term to address impacts resulting from the warming that would occur even for the lowest stabilization scenarios assessed. There are barriers, limits, and costs that are not fully understood. Adaptation and mitigation can complement each other and together can significantly reduce the risks of climate change. (IPCC 2007b and 2007c)*

Sustainable development can reduce vulnerability to climate change, and climate change could impede nations' abilities to achieve sustainable development pathways. Making development more sustainable can enhance mitigative and adaptive capacities, reduce emissions, and reduce vulnerability, but there may be barriers to implementation. (IPCC 2007b and 2007c)

References

Davis, M.L., and Masten, S.J. (2009). *Principles of Environmental Engineering and Science*, 2nd ed. New York: McGraw-Hill Higher Education.

DOE. (2010a). Cost and Performance Baseline for Fossil Energy Plants. Volume 1: Bituminous Coal and Natural Gas to Electricity. U.S. Department of Energy, National Energy Technology Laboratory.

DOE. (2010b). Advanced Oxycombustion 2015+ Bituminous Coal Fossil Energy Plants, Draft Final Report (Revision 2), U.S. Department of Energy, National Energy Technology Laboratory.

DOE. (2010c). Industrial Carbon Capture and Storage. U.S. Department of Energy, National Energy Technology Laboratory.

DOE/EPA. (2010). Report of the Interagency Task Force on Carbon Capture and Storage.

Dooley, J.J. (2006). Carbon Dioxide Capture and Geologic Storage: A Core Element of a Global Energy Technology Strategy to Address Climate Change. College Park, MD, Global Energy Technology Strategy Program.

Forster, P., Ramaswamy, V. Artaxo, P. Berntsen, T. Betts, R. Fahey, D.W. et al. (2007). Changes in Atmospheric Constituents and in Radiative Forcing. In *Climate Change 2007: The Physical Science Basis. Contribution of Working Group I to the Fourth Assessment Report of the Intergovernmental Panel on Climate Change*. Solomon, S., Qin, D. Manning, M. Chen, Z. Marquis, M. Averyt, K.B. Tignor, M. and Miller H.L. (Eds.). Cambridge, UK: Cambridge University Press.

Heinze, C., Maier-Reimer, E., and Winn, K. (1991). "Glacial pCO_2 Reduction by the world ocean: experiments with the Hamburg carbon cycle model." *Paleoceanography* 6(4):395–430.

Houghton, J.T. (2009). *Global Warming: The Complete Briefing*. Cambridge, UK: Cambridge University Press.

Houghton, J.T., Jenkins, G.T., and Ephraums, J.J. (1990). *Climate Change, the IPCC Scientific Assessment.* Cambridge, UK: Cambridge University Press.

IPCC. (2001). *Climate Change 2001: Synthesis Report.* "A Contribution of Working Groups I, II, and III to the Third Assessment Report of the Intergovernmental Panel on Climate Change." Watson, R.T., and the Core Writing Team (Eds.). Cambridge, UK: Cambridge University Press.

IPCC. (2005). *IPCC Special Report on Carbon Dioxide Capture and Storage.* Prepared by Working Group III of the Intergovernmental Panel on Climate Change. Metz, B., Davidson, O., de Coninck, H.C., Loas, M., and Meyer, L.A. (Eds.) Cambridge, UK: Cambridge University Press.

IPCC. (2006). *IPCC Guidelines for National Greenhouse Gas Inventories.* Prepared by the National Greenhouse Gas Inventories Programme. Eggleston, H. S., Buendia, L., Miwa, K., Ngara, T., and Tanabe, K. (Eds.). Institute for Global Environmental Studies, Kanagwa, Japan.

IPCC. (2007a). *Climate Change 2007: The Physical Science Basis.* "Contributions of Working Group I to the Fourth Assessment Report of the Intergovernmental Panel on Climate Change." Solomon, S., Qin, D., Manning, M., Chen, Z., Marquis, M., Averyt, K. B., et al. (Eds.). Cambridge, UK: Cambridge University Press.

IPCC. (2007b). *Climate Change 2007: Impacts, Adaptations and Vulnerability.* Contributions of Working Group II to the Fourth Assessment Report of the Intergovernmental Panel on Climate Change. Parry, M.L. Canziani, O. F. Palutikof, J.P. van der Linden, P.J. and Hanon C. E. (Eds.). Cambridge, UK: Cambridge University Press.

IPCC. (2007c). *Climate Change 2007: Mitigation.* Contributions of Working Group III to the Fourth Assessment Report of the Intergovernmental Panel on Climate Change. Metz, B.Davidson, O.R. Bosch, P.R. Dave, R. and Meyer L.A. (Eds.). Cambridge, UK: Cambridge University Press.

IPCC. (2012). *Managing the Risks of Extreme Events and Disasters to Advance Climate Change Adaptation.* Field, C.B., Barros, V. Stocker, T.F. Qin, D. Dokken, D.J. Ebi, K.L. et al. (Eds.). Cambridge, UK: Cambridge University Press.

Le Quéré, C., Andres, R.J., Boden, T., Conway, T., Houghton, R.A., House, J. I., et al. (2012). The global carbon budget 1959–2011. Earth System Science Data—Discussions (*manuscript under review*) 5:1107–1157.

Logan, J., and Venezia, J. (2007). Weighing U.S. Energy Options: The WRI Bubble Chart. WRI Policy Note: Energy Security and Climate Change. World Resources Institute. Washington, D.C.

Marion, W., and Wilcox, S. (1992). A Comparison of Data from SOLMET/ERSATZ and the National Solar Radiation Data Base, NREL/TP-463-5118. Golden, CO. National Renewable Energy Laboratory.

Murnane, R.J., Sarmiento, J.L., and Le Quéré, C. (1999). "Spatial distribution of air–sea CO_2 fluxes and the interhemispheric transport of carbon by the oceans." *Global Biogeochemical Cycles* 13(2):287–305.

NSRDB-Vol. 1. (1992). User's Manual—National Solar Radiation Data Base (1961–1990). Version 1.0. Ashville, NC, National Climatic Data Center.

Peters, G., Andrew, R. Boden, T. Canadell, J. Ciais, P. Le Quéré, C. et al. (2012). The challenge to keep global warming below two degrees. Nature Climate Change. Accessed December 2, 2012 at http://bit.ly/Qpt3ub. A pdf can be requested from press@nature.com.

Trenberth, K.E., J. T. Fasullo, and J. Kiehl. (2009). "Earth's global energy budget." *Bull. Amer. Meteor. Soc.* 90(3):311–324.

U.S. EIA. (2010). Annual Energy Review 2009. U.S. Energy Information Administration, Office of Energy Statistics, U.S. Department of Energy, Washington, D.C. DOE/EIA-0384(2009).

U.S EIA. (2012). Annual Energy Review 2011. U.S. Energy Information Administration. Office of Energy Statistics. U.S. Department of Energy, Washington, D.C. DOE/EIA-0384(2011).

Key Concepts

International Panel on
  Climate Change
Heating values
Carbon intensity
Carbon flow and repositories
Blackbody temperature

Stefan–Boltzmann constant
Energy balance
Solar radiation
Albedo
Greenhouse gas
Radiative forcing

Active Learning Exercises

ACTIVE LEARNING EXERCISE 6.1: Draw your personal carbon cycle

Carbon atoms are essential building blocks for all life forms on Earth. The carbon atom shares electrons with other atoms. The ability to share electrons between carbon molecules and other elements makes the carbon atom unique. Carbon can form bonds with hydrogen to form methane (CH_4). Carbon atoms may share atoms with oxygen to form CO_2, CO in the gas phase, and CO_3^{2-} in aqueous solutions. Organic chemistry is the field that studies carbon–carbon bonded elements that are naturally occurring and can be synthesized (or can form) in the laboratory and for industrial use.

The following steps should be followed in order to develop a carbon mass balance equation.

1. Draw a detailed schematic of the major carbon sources and sink for the Earth's atmosphere. You may use the conceptual carbon cycle model shown in Figure 6.50 to start your analysis.
2. Show the system boundaries and define the control volume. Note that the mass of the atmosphere is 5.14×10^{18} kg. For the purposes of our simple atmospheric model, determine the volume of the atmosphere in m^3.
3. Define the variables and list the known data and assumption needed to solve a mass balance equation for carbon in the atmosphere.
4. Create a table that lists each variable in units of
   a) Volume $= m^3$ AND liters
   b) Mass $=$ kg AND g

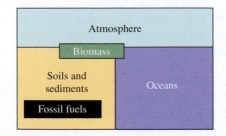

FIGURE 6.50 Basic box model of biogeochemical repositories related to climate modeling. The flux of carbon and numerical values need to be added in order to illustrate variables, quantities, and fluxes associated with the movement of carbon from one repository to another.

Source: Bradley Striebig.

c) Energy = btus AND joules
d) Concentrations in air = ppm_v AND mg/m^3
e) Concentrations in water = ppm_m and mg/L

ACTIVE LEARNING EXERCISE 6.2: Regional changes

The IPCC states that "impacts of climate change will vary regionally. Aggregated and discounted to the present, they are *very likely* to impose net annual costs, which will increase over time as global temperatures increase." Based on the regional differences of climate impacts shown in Figures 6.31 to 6.39, estimate whether the current climate change models would likely cause home values in your area to increase or decrease. Consider the influence of the following parameters in making your decision:

- Water quantity
- Water quality
- Landscaping
- Home heating/cooling costs
- Local food production and prices

Problems

6-1 When did scientists begin to realize the important link between atmospheric gas concentrations and our climate on Earth?

6-2 What changes have occurred in our atmosphere over the past 2000 years?

6-3 How well are past and present climates understood?

6-4 What are the possible effects of a changing climate on economic systems, ecosystems, social structures, and technological challenges that may lie in the future?

6-5 Describe the accomplishments and cite the year the accomplishments were achieved by the following scientists who were studying the way our atmosphere works:
a. Jean-Baptist Fourier
b. John Tyndall
c. Svante Arrhenius
d. Guy Stewart Callendar
e. Hans Seuss and Roger Revelle
f. The International Panel on Climate Change (IPCC)

6-6 In general, the combustion of fossil fuels can be described by what chemical equation?

6-7 Burning coal will emit how much carbon dioxide compared to the same amount of energy derived from natural gas?

6-8 Complete the following table by listing the gigatons of CO_2 stored in the repository or transferred in each transformation pathway listed.

6-9 What percent of carbon emissions is stored in the atmosphere, oceans, and land?

6-10 Carbon dioxide is converted to cellulose and other plant matter through the process of photosynthesis. Write the chemical equation in words for the formation of glucose.

REPOSITORY OR PATHWAY	GIGATONS OF CO_2	REPOSITORY OR PATHWAY	GIGATONS OF CO_2
Atmosphere		Fossil fuels	
Weathering (atmosphere)		Rivers	
Respiration		Surface ocean	
GPP		Marine biota	
Land sink		Intermediate and deep ocean	
Land-use changes		Surface sediment	
Vegetation, soil, and detritus		Weathering (geologic)	

6-11 Writing in your own words and using typical chemical equations, describe the respiration process for animals using sugars (like glucose), carbohydrates, and proteins in combination with oxygen to produce energy, cell growth, water, and carbon dioxide.

6-12 Write the equation for the residence time as a function of the system volume, V, and the volumetric flow rate, Q, of the material or, alternatively, by dividing the mass of material, m, in the control volume by the mass flow rate of material, M, flowing through the volume.

6-13 Define the following terms:
a. Blackbody temperature
b. Calorie
c. Planck's constant
d. Specific heat
e. Speed of light
f. Stefan–Boltzmann constant

6-14 The energy balance within a given control volume can be expressed in words as_____.

6-15 List the variable and the equation that relate the first law of thermodynamics to work and energy.

6-16 List the variable and the equation that define the enthalpy of a system.

6-17 List the variable and the equation that describe heat loss through the boundaries of a system.

6-18 Show the equation and variables associated with the integrated form of the Stefan–Boltzmann law of radiation.

6-19 What are the average and range of incoming solar radiation since AD 1600?

6-20 What are the average and range of the Earth's albedo?

6-21 Write the equation that describes the blackbody radiation absorbed by the Earth.

6-22 Write the equation that describes the energy absorbed by the Earth.

6-23 Write the equation that describes the average surface temperature of the Earth as described by a blackbody.

6-24 What is the effect of the greenhouse gases in degrees Celsius on the Earth's average surface temperature?

6-25 What is the effective average incoming solar radiation determined by dividing the incoming irradiation by the surface area of the Earth?

6-26   The net energy balance across the surface of the Earth can be written as_____.

6-27   What is the radiative forcing parameter for the following atmospheric gases?
a. Carbon dioxide
b. Methane
c. Nitrogen oxide (N_2O)
d. Halocarbons
e. Ozone (stratosphere)
f. Ozone (troposphere)

6-28   Describe how atmospheric models have evolved and what variables have been added in each evolutionary step since 1970.

6-29   Describe in 500 words or less the evidence that is used to support the following statement: *"Warming of the climate system is unequivocal, as is now evident from observations of increases in global average temperatures, widespread melting of snow and ice, and rising global average sea level."*

6-30   What are measurable changes and what magnitude of those changes is expected in the environment due to the enhanced greenhouse effect?

6-31   What are the potential economic and social changes that may result from the environmental changes associated with the enhanced greenhouse effect?

6-32   What engineering and scientific technologies are currently on the horizon that may mitigate the likely effects of the enhanced greenhouse effect? What "science fiction" technologies could you imagine that are not yet understood by today's body of knowledge, but could be imagined to mitigate or reverse the enhanced greenhouse effect?

6-33   World energy production by source is shown in Figure 6.51. Using Excel, plot the data for crude oil, coal, and natural gas for each decade: 1970, 1980, 1990, and 2000. Establish a linear trend line through your graph for each fuel source. Also add a data set for the summation of each source to approximate world energy use of fossil fuels.

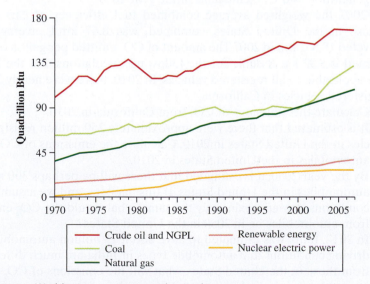

FIGURE 6.51 World primary energy production by source.

Source: US EIA (2012) Annual Energy Review 2011. U.S. Energy Information Administration. Office of Energy Statistics. U.S. Department of Energy. Washington, D.C. DOE/EIA-0384 (2011)

    a. Forecast and extend the x-axis time and the trend line backwards to 1850 and forward to 2050. Show this trend line on a graph.

    b. Show on a sheet of paper an example calculation for CO_2 emissions for the year 1990 for:

       i)   Oil

       ii)   Coal

       iii) Natural gas

       iv) All CO_2 emissions for 1990 from fossil fuel use

6-34 Plot on a graph the CO_2 emissions for each decade from 1850 to 2050 based on your fossil fuel forecasts of Problem 6-33. Show:

    a. Oil

    b. Coal

    c. Natural gas

    d. All CO_2 emissions for 1990 from fossil fuel use

6-35 The sum of all CO_2 emissions from 1850 to 2050, from fossil fuel emissions, is equal to the total area under the graph from Problem 6-33.

    a. Based on this method, how much CO_2 was emitted from fossil fuels between 1850 and 2010?

    b. How does this compare to the atmospheric increase in CO_2 between 1850 and 2010, based on measurements of CO_2 in ppm in the atmosphere between 1850 and 2010?

6-36 Based on the summary statistics of sources for the electricity portfolio within the United States (see Table 6.7), electricity generation results in 0.68956 kg of CO_2 per kWh.

    a. Calculate the metric tons of CO_2 emitted from all energy electricity generation for each year from 2000 to 2010 in Table 6.7.

    b. Plot the net generation from all energy sources and the CO_2 emissions from energy generation for the years 2000 to 2010. Be sure to properly label your graph axis and units.

    c. Fit a curve to the graph and use the curve to show predicted energy generation and CO_2 emissions in the year 2050.

6-37 In 2007, the weighted average combined fuel efficiency of cars and light trucks in the United States, combined, was 8.67 km/L average vehicle traveled 18,861 km in 2007. The amount of CO_2 emitted per gallon of gasoline burned is 8.92 kg. Assume for the following calculations that the 2007 data are applicable to all registered vehicles. In 2010, there were nearly 32 million registered vehicles in California.

    a. Calculate the emissions of CO_2 from California in 2010.

    b. It is estimated that there were approximately 250 million registered vehicles in the United States in 2010. Calculate the emissions of CO_2 from all automobiles in the United States in 2010.

    c. By the year 2050, it is estimated that there will be perhaps 300 million automobiles in the United States. If this happens, and we assume driving conditions stay approximately the same, what would the CO_2 emissions from automobiles be in 2050 in the United States?

    d. In 2012, India was estimated to have about 15 million automobiles. While driving conditions and automobile types in India are much different from those in the United States, what will the emissions of CO_2 be from automobiles in India if the gas mileage and driving patterns were the same as those in the United States?

TABLE 6.7 Summary statistics of energy consumption for the United States, 1999–2012

DESCRIPTION	Net Generation (thousand megawatt hours)											
	2010	2009	2008	2007	2006	2005	2004	2003	2002	2001	2000	1999
Coal (1)	1,847,290	1,755,904	1,985,801	2,016,456	1,990,511	2,012,873	1,978,301	1,973,737	1,933,130	1,903,956	1,966,265	1,881,087
Petroleum (2)	37,061	38,937	46,243	65,739	64,166	122,225	121,145	119,406	94,567	124,880	111,221	118,061
Natural gas (3)	987,697	920,979	882,981	896,590	816,441	760,960	710,100	649,908	691,006	639,129	601,038	556,396
Other gasses (4)	11,313	10,632	11,707	13,453	14,177	13,464	15,252	15,600	11,463	9,039	13,955	14,126
Nuclear	806,968	798,855	806,208	806,425	787,219	781,986	788,528	763,733	780,464	768,826	753,893	728,254
Hydroelectric conventional (5)	260,203	273,445	254,831	247,510	289,246	270,321	268,417	275,806	264,329	216,961	275,573	319,536
Other renewables (6)	167,173	144,279	126,101 [R]	105,238	96,525	87,329	83,067	79,487	79,109	70,769	80,906	79,423
Wind	94,652	73,886	55,363	34,450	26,589	17,811	14,144	11,187	10,354	6,737	5,593	4,488
Solar thermal and photovoltaic	1,212	891	864	612	508	550	575	534	555	543	493	495
Wood and wood-derived fuels (7)	37,172	36,050	37,300	39,014	38,856	38,856	38,117	37,529	38,665	35,200	37,595	37,041
Geothermal	15,219	15,009	14,840 [R]	14,637	14,568	14,692	14,811	14,424	14,491	13,741	14,093	14,827
Other biomass (8)	18,917	18,443	17,734	16,525	16,099	15,420	15,421	15,812	15,044	14,548	23,131	22,572
Pumped storage (9)	-5,501	-4,627	-6,288	-6,896	-6,558	-6,558	-8,488	-8,535	-8,743	-8,823	-5,539	-6,097
Other (10)	12,855	11,928	11,804 [R]	12,231	12,964	12,821	14,232	14,045	13,527	11,906	4,794	4,024
All energy sources	4,125,060	3,950,331	4,119,388	4,156,745	4,064,702	4,055,423	3,970,555	3,883,185	3,858,452	3,736,644	3,802,105	3,694,810

Source: U.S. EIA (2012) Annual Energy Review 2011. U.S. Energy Information Administration. Office of Energy Statistics. U.S. Department of Energy. Washington, DC.

e. By the year 2050, India is expected to be the largest automobile market in the world. Goldman Sachs has estimated that there will be more than 610 million automobiles in India by 2050. If this happens, and we assume driving conditions stay approximately the same, what would India's CO_2 emissions from automobiles be in 2050?

f. Create a bar graph illustrating the number of vehicles and CO_2 emitted from scenarios in parts (a) through (e).

6-38 An average home in the United States consumed about 12,773 kWh of electricity in 2005. The national average carbon dioxide emission rate for electricity generated in 2007 was 1,293 lbs of CO_2 per megawatt-hour. Approximately 7% of the electricity generated is wasted due to losses in the transmission and distribution system.

a. How much CO_2 was generated per year for an average U.S. home due to electricity use?

b. How much CO_2 was emitted from household use of electricity if there were 111.1 million homes in the United States in 2005?

c. How much CO_2 will be emitted from household use of electricity if there are 166 million homes in the United States in 2050?

6-39 An average home in the United States consumed about 1,344 m³ natural gas, 224 liters of liquid petroleum gas, 220 liters of diesel fuel oil, and 3.2 liters of kerosene.

a. How much CO_2 was generated per year for an average U.S. home, due to natural gas usage?

b. Assume that 0.0545 kg of CO_2 are emitted per cubic foot of natural gas burned. How much CO_2 was emitted from household natural gas usage, if there were 111.1 million homes in the United States in 2005?

c. How much CO_2 will be emitted from household natural gas usage, if there are 166 million homes in the United States in 2050?

6-40 The default carbon dioxide emission factor for diesel fuel is 74,100 kg/TJ. Which common transportation fuel (gasoline or diesel) has the greater carbon intensity per unit of energy? What are their respective carbon intensities compared to natural gas?

6-41 The default carbon dioxide emission factor for biodiesel fuel is 70,800 kg/TJ, and the emission factor for diesel fuel is 74,100 kg/TJ.

a. Which diesel fuel, biodiesel or "regular" diesel fuel, has the greater carbon intensity per unit of energy?

b. What are their respective carbon intensities compared to natural gas?

c. If a city were to invest in climate-friendly buses for public transportation, should it choose a diesel-powered bus, a biodiesel-powered bus, or a natural gas-powered bus?

6-42 According to the 1980 TAR report from the IPCC, what percent of carbon emissions is stored in the atmosphere, oceans, and land?

6-43 Cellulose can be represented by the chemical formula $(C_6H_{10}O_5)_n$, where n represents a long number of groups of this formula that can be combined. How many molecules of carbon dioxide would be needed to form one molecule of the base structure of cellulose, $C_6H_{10}O_5$?

6-44 The carbon flux to the oceans is approximately 30.3×10^{12} kg/yr. The approximate steady-state mass of carbon in the oceans is 40×10^{15} kg. How long, on average, is carbon likely to remain in the atmosphere before cycling through another repository? How much longer or shorter is this than the residence time of carbon in the atmosphere?

6-45 The water flux to the atmosphere is approximately 5.18×10^{14} kg/yr. The approximate steady-state mass of water in the atmosphere is 1.3×10^{13} kg. How long, on average, is water likely to remain in the atmosphere before cycling through another repository?

6-46 The water flux to the oceans is approximately 1.54×10^{14} kg/yr. The approximate steady-state mass of water in the oceans is 1.35×10^{18} kg. How long, on average, is water likely to remain in the oceans before cycling through another repository?

6-47 If the Earth were completely covered with water, so that the planet's albedo was 20%, what would be the Earth's blackbody radiation temperature?

6-48 A medium-growth coniferous tree planted in an urban or suburban setting after being raised in a nursery for 1 year can sequester 10.5 kg of CO_2 over a 10-year period of growth. If you lived in an average home in the United States, how many trees would you need to plant to remove the CO_2 you would produce from driving your car, using electricity in your home, and heating your home with natural gas for a period of 10 years?

CHAPTER 7

Models for Sustainable Engineering

FIGURE 7.1 A drive-through tree in Humboldt County, California. A fundamental requirement of sustainability is that people rethink their relationship with the natural world. This giant redwood tree in California was cut to allow a road to pass through it for our recreational entertainment, not out of any real necessity.

Source: Image Source/Getty Images

In the course of history, there comes a time when humanity is called to shift to a new level of consciousness, to reach a higher moral ground. A time when we have to shed our fear and give hope to each other. That time is now.

—WANGARI MAATHAI, WINNER OF THE 2004 NOBEL PEACE PRIZE

GOALS

THE PURPOSE OF THIS CHAPTER is to provide you with a variety of concepts, models, and principles that you can use to analyze how engineering choices can enhance environmental, social, and economic sustainability. You should have a good understanding from Chapters 1 through 6 of the scope of environmental problems and how human agency has contributed to them. Now we begin the process of *doing things differently* so that human technologies and material cultures become inherently more sustainable.

OBJECTIVES

At the conclusion of this chapter, you should be able to:

- Discuss how environmental, social, cultural, and economic considerations shape our understanding of what is meant by the term *sustainable development*.

- Describe the difference between renewable and nonrenewable natural resources, and give examples of each.

- Explain the concept of ecosystem services, and how that relates to environmental sustainability.

- Identify each of the steps of the waste management hierarchy, and give examples of engineering methods and practices that reduce and minimize the waste stream.

- Summarize why strategies such as industrial ecology and life cycle analysis provide more sustainable engineering outcomes.

- Recognize the elements of ecological design, and give multiple examples of how this is used in engineering practice.

- Apply biomimicry as a design strategy.

- Analyze a business decision using the principles of sustainable and green engineering.

- Compare and contrast footprint analysis, waste management, ecological design, and the principles of sustainable engineering as models for better design.

7.1   Introduction

Kenyan Wangari Maathai was the first environmentalist to win the Nobel Peace Prize. Because of Maathai's initiatives in the Green Belt movement and sustainable development, more than 40 million trees have been planted and thousands of African women have been empowered in their communities. Her call to action—to "shift to a new level of consciousness"—refers specifically to our need to change how we understand, think about, and interact with the natural world.

Very simply and quite literally, Earth is our *life support system*. Over its 4.5-billion-year history, it has created environments in which single-celled organisms, tiny sea creatures, carbon-rich fern forests, dinosaurs, and humankind have evolved and thrived. At all times and in all places on our planet, the health and survival of living things depends on complex webs of interdependence with other living systems and Earth's biogeochemical cycles. The emergence of sustainability as a concept and a *mode of thought* requires us to acknowledge that human actions have profound consequences for the natural world and can disrupt the ability of the planet's living systems and physical cycles to support human and other life.

Chapter 7 is a bridge to the second half of your textbook. The first half introduced you to the science of many of the problems that we face as a consequence of our technologies, our production systems, our energy fuels, and our built environment. In this chapter, we will introduce you to ways of thinking that are changing how we engineer our world. We will introduce concepts and models that you can use to develop more sustainable materials, products, and processes as well as think more holistically about engineering problem solving. These include the nature of natural resources, the concepts of waste in ecological and human systems, the basis of ecological design, and the principles of both sustainable and green engineering. In Chapters 8 through 11, you will apply these concepts and models to energy conservation, industrial ecology, life cycle analysis, and green building as methods for enhancing sustainable engineering practices.

Unlike other chapters in this textbook, Chapter 7 does not contain a number of detailed analytical problems for you to work through. Instead, we present four learning activities that encourage you to engage in open-ended problem solving. These problems are designed to help you analyze engineering situations in a broader context. You will not have all of the information that you need, and there is no one "right answer." What you will find is that more sustainable engineering practice involves trade-offs and tensions between environmental, economic, and social values that must be resolved.

7.2   What Do We Mean by Sustainability?

Sustainability is a term that carries a wide variety of meanings depending on context. "Sustainability" to a biologist studying a small wetlands ecosystem is different than "sustainability" for a commercial forester, which is yet again different than "sustainability" to a businessperson. So what are the implications of multiple contexts and meanings for engineers?

The *social* (as opposed to environmental or ecological) concept of sustainability came into widespread use after a 1987 United Nations report entitled *Our Common Future.* Referred to as the Brundtland Report, this study explored the interdependent relationship between the environment and economic development and addressed the consequences of environmental degradation for long-term economic growth and social well-being. The concept that the Brundtland Report put forward was that of **sustainable development**, which the report defined as "development that meets the needs of the present without compromising the ability of future generations to meet their own needs." What is critical about this definition is *time*, the sense that in providing for human societies today, we should not jeopardize the ability of our descendants to provide adequately for themselves. Because human survival and material quality of life depend on the natural world, this requires us to consider the long-term consequences of our impact on the environment for others. This sense of obligation to future generations is referred to as **intergenerational ethics**.

Many individuals and organizations have tried to articulate a clear set of principles to guide practical action, but there is no uniform, standardized approach to sustainable development. Nonetheless, all discussions explore the relationships among the "three pillars" of sustainability—the environment, society, and the economy—with an understanding that:

- The environment and biosphere should be protected *as a dynamic system*.
- Socioeconomic development is necessary for all societies and should be equitable; no one group should bear an unfair burden of environmental harm.
- Economic considerations reflect legitimate concerns about the cost and affordability of solutions to environmental problems, but economic profitability may also be achieved by more environmentally sustainable actions.

What continues to be debated is the relationship between the environment, society, and the economy. Figure 7.2 illustrates two of the most common ways of representing these relationships. In the Venn diagram on the left, sustainable solutions are those that acceptably integrate environmental, societal, and economic goals. This model is known as the *constrained growth* model. This image suggests that there are a broad range of activities in the three pillars that have little bearing on one another, and the model is based on the need for economic growth and expansion. In the image on the right, the nested concentric circles suggest that environmental dynamics represent a defining limit on economic and social activity; this approach is referred to as the *resource maintenance* model. From this perspective, human action, quality of life, and survival are ultimately dependent on and bounded by processes of the natural world.

The value of sustainability concepts is that they encourage us *to think holistically about systems*. In a world in which nature, human well-being, and economic

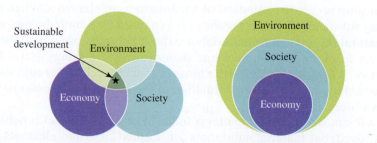

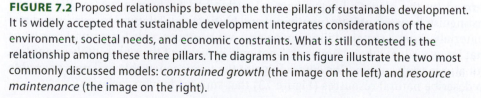

FIGURE 7.2 Proposed relationships between the three pillars of sustainable development. It is widely accepted that sustainable development integrates considerations of the environment, societal needs, and economic constraints. What is still contested is the relationship among these three pillars. The diagrams in this figure illustrate the two most commonly discussed models: *constrained growth* (the image on the left) and *resource maintenance* (the image on the right).

production are interdependent, it is critical that we understand how change in one aspect of these relationships has consequences for the others. In popular terminology, this is referred to as a "butterfly effect," but in systems dynamics language, it represents feedback and system disturbance.

As you will see throughout this chapter, sustainability concepts *do* generate a variety of ideals for practice. These are to minimize our use of raw materials and finite natural resources, to strive for zero waste, to protect ecological processes, and to avoid creating toxic and hazardous substances.

7.3 The Nature of Natural Resources

The first step toward developing new insights into sustainability is understanding the nature of natural resources. From a scientific perspective, Earth is a biophysical system constructed of matter (the elements) and energy flows. From a practical perspective, Earth provides the natural resources on which our survival and material well-being depend. The simple and obvious substances are air and water. But people do not survive like other living things, where existence is based on constant, direct interaction with nature. Instead, human survival is mediated by a technological and material culture: We exploit plant and animal fibers for clothing (cotton, wool, and flax for linen, for example), we construct buildings, we manufacture synthetic substances that are not found naturally in the physical world, we cultivate animals and scientifically hybridize crops for food, we burn fossil fuels, we design complex transportation systems, we have toilets and sanitation, we fabricate metals, we invent chemical pharmaceuticals, and we manufacture a nearly incomprehensible array of consumer goods and products. In an industrialized society, virtually nothing about human existence—except breathing—reflects a "natural" interaction with our environment. Even in hunter-gatherer communities, material culture, technology, and dwellings leave a small ecological footprint. To develop models for sustainable engineering, we need to understand both the finite nature of our material resources and the ecological processes on which our lives depend.

7.3.1 Traditional Concepts of Natural Resources

Laws of physics limit and govern our material world of natural resources. First and foremost, all tangible things that we drink, eat, use, touch, or consume are made of matter. The periodic table identifies the elemental matter on Earth, and all material substances are composed of combinations of elements. When a naturally occurring substance or living organism is exploited by human beings, we refer to it as a **natural resource**. **Raw materials** are those naturally occurring resources that we extract from the environment and then process into things useful for people. As you saw in Chapter 3, the Earth's biogeochemical cycles are responsible for the environmental cycling and transformation of the elements and essential resources (such as water).

As a reminder, elemental matter is finite. Mass on our planet is neither created nor destroyed, but material substances are formed from the elements and transformed by the planet's biogeochemical cycles and by human action. We usually distinguish between **abiotic** (nonliving) and **biotic** (living) natural resources and raw materials. From a sustainability perspective, we need concepts for natural resources that let us evaluate their "finiteness" not only in terms of their absolutely quantities, but also in terms of *time*. We commonly use the terms **renewable** and **nonrenewable** to describe natural resources (Figure 7.3) that support human society.

Nonrenewable Resources				Renewable Resources	
Nonregenerative	Theoretically recoverable	Recyclable	Regenerative	Regenerative	Infinite
Coal Natural gas Oil Fossil water	All elemental minerals and metals Water	Metals Water	Soil Surface water Groundwater	Animals Fish Vegetation/forests Surface water Groundwater	Air Solar energy Wind energy Geothermal energy Tidal/wave energy Water

FIGURE 7.3 A typology of natural resources. This schematic organizes critical natural resources according to their renewability, recyclability, and regenerative properties.

Renewable resources are those that replenish themselves through natural processes. As a general rule, a resource must replenish itself roughly within a human lifetime to be considered renewable. The most commonly mentioned renewable resources are not really tangible material substances, but they result from the physical forces of our planetary system, such as solar energy, wind and tidal power, and air (Earth's atmosphere). What makes these resources unique is that they are virtually infinite—their presence is the result of the mechanics of our solar system and planet, and human use does not diminish their availability. In contrast, many of our other renewable natural resources are biotic, such as animals, fish, plants, and forests. As growing, living things, they have variable rates of regeneration through biological reproduction.

Therefore, these resources are renewable *only to the extent that we do not overharvest them*. If we consistently harvest biotic resources at a rate faster than they regenerate, one of two things will generally happen. The most extreme is that the species is driven to extinction. Alternatively, we can **collapse the population** to the point at which it has a radically diminished abundance and can no longer serve as a natural resource. Classic examples of overharvest and collapse include many fisheries and forests worldwide.

Nonrenewable resources are those that regenerate themselves extremely slowly, if at all. Metals, for example—iron, copper, tin, gold, silver, aluminum, and so forth—are elemental substances that are absolutely finite in the conventional physical and chemical sense. Soil is an amalgam of organic life and minerals and can be regenerated, but depending on climate and location, it requires 100 to 500 years to create 2.5 centimetres topsoil. It took millions of years to form fossil fuels, and as a consequence these natural resources must be considered finite and nonregenerative. The **rare earths** are intriguing nonrenewable resources because they are not particularly rare geologically; instead, geopolitical "scarcity" is one of the forces driving technological innovation to recycle and recover these elements (Box 7.1).

BOX 7.1 How Rare Are Rare Earths?

The rare earths are 17 natural elements with exotic names such as yttrium, neodymium, and samarium. Actually, they aren't particularly rare, and until recently, we weren't especially concerned about their relative scarcity. What is unusual about most rare earths is that they are not found in concentrated deposits, but are instead diffused throughout the Earth's crust. They were labeled rare not because they lacked abundance, but because it is difficult to locate them in economically viable concentrations that can be readily mined.

So what do we use these elements for (Fig. 7.4)? There are a number of rare earth powerhouses, all of which promote a high-tech world. Lanthanum is a critical element in modern battery technology on which electric vehicles like the Toyota Prius depend. Europium was used to create the bright red color in old cathode-ray TV picture tubes, but it is finding new life in white light-emitting diode (LED) light bulbs. Erbium, when added in minute quantities, is needed for lasers and fiber optics. Samarium and neodymium are used to create tiny electromagnets so that we can have miniature consumer electronics. Cerium promotes cleaner air as a component of catalytic converters. The list goes on, and only trace amounts of a rare earth are usually needed to significantly change the properties of a material or its performance.

The rare earth elements are creating a buzz because of geopolitics. Although we use fewer than 136,000 tonnes per year of all rare earth elements, China accounts for nearly 97% of production in the world today and is also the largest consumer of these substances. China has imposed a series of restrictions and export quotas on its rare earth exports, citing

FIGURE 7.4 Your iPod contains dysprosium, neodymium, praseodymium, samarium, and terbium.

Source: mills21/Shutterstock.com

environmental concerns over damage from rare earth mining. But the belief is that China wants to move up the global supply chain by manufacturing more high-value-added products. Stockpiles of rare earths are dwindling, and countries that previously produced them are now re-evaluating old mines.

As industries become concerned over the cost, relative scarcity, and reliability of rare earth supplies, recycling has become more important. Innovations in recycling technology have made recovery of some rare earth elements technically feasible and cost effective in light of rising global prices. Key product classes for recycling and rare earth recovery are fluorescent light bulbs, batteries, magnets, consumer electronics, and computer components.

Because of the finiteness of elemental matter and some raw materials, as well as the extremely slow regenerative times of others, the **recoverability** of nonrenewable resources is of interest. Some raw materials are entirely nonrecoverable because the process of using them prevents us from physically capturing the substance and reusing it in its original form. Fossil fuels are the most extreme example of nonrecoverability because combustion effectively consumes them. All elemental minerals and metals are theoretically recoverable because they are elemental, but we may lack the practical know-how to do so or the required technologies may be prohibitively expensive. Notably, elemental metals are considered to be infinitely **recyclable** because they can be continuously recovered and reused. As you will see in Chapter 8, aluminum recycling not only conserves aluminum as a natural resource, but it significantly lowers the amount of energy required to extract and process aluminum as a usable metal from aluminum oxide.

As you can see in Figure 7.3, water appears in several categories and is a challenging natural resource to classify. Traditionally water has been regarded as an infinitely renewable natural resource because of its biogeochemical cycling, but this perspective has changed. Because of the Earth's hydrological cycle (see Chapter 2), the amount of water in the biosphere stays relatively constant, and human use does not diminish the absolute quantity of water molecules. At the planetary scale, it is appropriate to regard water as an infinite natural resource. At the local scale, however, fresh water in aquifers, groundwater, and surface lakes, rivers, and streams can be either a regenerative renewable or nonrenewable resource depending on climatic and geological conditions.

Fresh water regenerates locally in surface and groundwater as a consequence of regional weather, climate, and hydrologic cycles. As a result, the rate of regeneration can fluctuate, and there is a risk of overabstraction of water—we withdraw it at rates faster than it can replenish. In the case of aquifer **fossil water**, overabstraction will result in a permanent loss of groundwater. This is what is happening to the Ogallala Aquifer in the U.S. Great Plains.

In the case of rivers and streams, overabstraction will reduce flow to a trickle and make water unavailable to people and other living things downstream, as we see with the Colorado River. Overabstraction of river water for agricultural irrigation has also contributed to the virtual disappearance of the Aral Sea, one of the four largest lakes in the world until it was reduced to only one-tenth of its original size by 2010 (Figure 7.5). Sometimes fresh water can be recovered and recycled, but it depends on how it is being used, and it is a costly process.

In sum, a useful way of understanding natural resources with regard to more sustainable engineering is in the context of stocks and flows. **Natural resources management** is the field of study that scientifically explores how to manage regenerative natural resources in a sustainable manner, in which human rates of harvest and extraction are in balance with the biogeochemical regeneration rates for abiotic resources and with the reproduction and population dynamics of biotic systems. As engineers, we can reflect on the absolute amount of resources (their finiteness at any given point in time), the degree and speed with which they can regenerate, and whether or not the resources are recoverable or recyclable in principle. These considerations can, for example, help us decide on materials selection, reevaluate the water intensity of industrial processes, or develop systems for recovering diminishing and costly resources.

7.3.2 Ecosystem Services and Natural Capital

The Earth is a biosphere, and traditional ways of thinking about natural resources as stocks and flows of discrete materials and substances are an incomplete way of understanding the complexity of Earth as a life support system. For example, many fish that we eat are themselves supported by a complex marine ecology in which the fish reproduce only in coastal estuaries, and those estuaries are in turn the product of habitats provided by mangrove trees, diurnal variations in water levels,

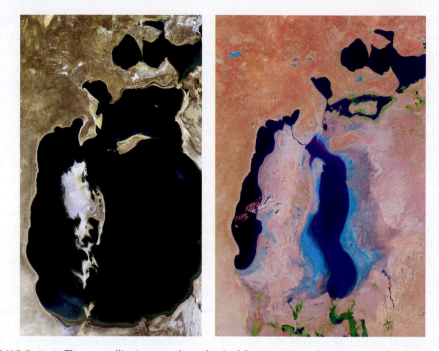

FIGURE 7.5 These satellite images show the Aral Sea in 1989 and 2010. Located in Eurasia, the Aral Sea was previously one of the four largest lakes in the world, but it is now only one-tenth of its original size by volume. Water for irrigation was withdrawn from rivers that fed the lake, to the point that the lake could not replenish itself. Local economies have collapsed, and the region is plagued by environmental degradation.

Source: NASA; NASA

microscopic bacteria, and so on. For many living species, the problem is not that we are overharvesting them, but that we are disturbing, degrading, or even destroying the habitats through which their own life cycles are maintained in delicate balance. Similarly, once deforested, many tropical regions cannot regenerate vegetation. Because of the unique mineral composition of soils in some parts of the tropics, direct sunlight and the loss of organic matter from decomposing vegetation rapidly harden the soil to the point that it cannot naturally regenerate or support the biodiversity of tropical forests.

Human threats to natural resources are therefore not just through the overconsumption of individual species, substances, or materials. By disrupting the ecological balances though which living things directly and indirectly survive, we affect not only the number and diversity of species, but also biogeochemical processes. As a consequence, we must take into account the physical integrity of **ecosystems** themselves. Ecosystems are communities of living organisms, where ecosystem dynamics represent the flow of nutrients and energy between living entities and their abiotic physical environment (water, soil minerals, air, climate). In an ecosystem, the living organisms and the abiotic environment are mutually interdependent. For example, forests require rainfall, but forests themselves create the microclimates in which rainfall can occur. In addition, as you learned in Chapter 3, biological processes are significant contributors to matter cycling on Earth, particularly for carbon and nitrogen.

Two concepts have emerged over the last decade or so to help us better model the usefulness of ecosystems dynamics. These two concepts are **ecosystem services** and **natural capital**. The term *ecosystem services* represents the idea that a wide variety of ecological dynamics support humankind in a way that a simple consideration of natural resource use does not. Traditionally, we have thought of natural resources in terms of their material products, such as oil, metals, food, and potable water. In the ecosystem services model, ecosystems generate tangible and intangible provisioning, regulating, and cultural services for people (Figure 7.6). Provisioning services are analogous to the production of material natural resources such as the food, fresh water, fuels, fibers, minerals, and metals discussed in the previous section. Regulating services are ecosystem processes that moderate harmful impacts on human communities, such as controlling disease, purifying water, and mitigating floods. Regulating services also include processes that are essential for the provisioning of natural resources, such as pollination of food crops and climate dynamics that regulate temperature and rainfall (Figure 7.7). Finally, ecosystems provide services to humans in the form of culture and values, such as an aesthetic appreciation of the natural world, spiritual and religious beliefs, recreational opportunities, and our sense of place in the world. The concept and model of ecosystem services provide us with a more holistic portrait of the natural world, of its functions, and of its benefits to humanity.

Natural capital is a concept developed by Paul Hawken and Amory and Hunter Lovins in their book *Natural Capitalism: Creating the Next Industrial Revolution*. We will provide a simplified version of this concept here, because a background in business or economics is necessary to understand its finer points. Imagine that natural resources and ecosystem services are the Earth's "money in the bank" and that humankind lives off of this money. Ideally, we would protect this investment and its ability to generate interest. That is, we would conserve natural resources by

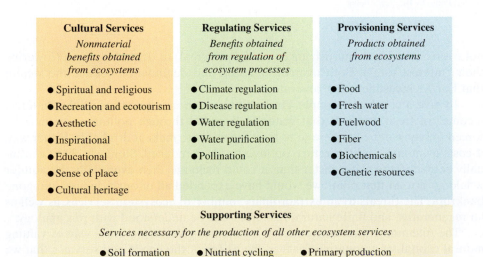

FIGURE 7.6 The scope of ecosystem services. Ecosystems provide a wealth of services to humankind, ranging from traditional raw materials to water purification to our spiritual values. Not all of these services have direct monetary value, and as a consequence, they are hard to protect through market economic systems.

Source: Based on Board of the Millennium Ecosystem Assessment, Millennium Ecosystem Assessment: Ecosystems and Human Well-Being—A Framework for Assessment. Island Press, 2003.

FIGURE 7.7 Ecosystem pollination services. We can buy honey, but the real value of honeybees and other insects is in their ecosystem service as pollinators. Almost all major human food crops depend on pollinators to produce the grains, fruits, nuts, and vegetables of our food supply. From an economic perspective, these pollination services are worth billions of dollars, even though we don't have to pay for them most of the time. Because of shrinking populations of wild pollinators from habitat loss, farmers must now hire commercial pollination services or risk losing their crops. The almond industry in California is particularly dependent on commercial honeybee services.

Source: Photo by Bradley Striebig

not overconsuming them (driving down the principal) or degrading them (lowering their "interest rate"). Both overconsumption and degradation reduce the wealth that Earth's investment can generate.

To give a concrete example, groundwater in an aquifer offers us the "interest income" of fresh water. If that water is degraded through pollution, we may be forced to buy water elsewhere or to treat the water to make it drinkable. Either way, it costs us money, and the return on investment is reduced. Similarly, if we continually extracted the water faster than it could recharge, then eventually the aquifer would go dry. At that point, we would have expended all of our principal and gone bankrupt. The threats to an extraordinary number of ecosystem services, as well as all regenerative and finite natural resources, can be understood with this analogy.

The dilemma is that conventional economics gives us no good way of valuing natural capital, of accounting for it in the prices of the goods and services that we buy and sell, or of incentivizing its stewardship. Natural capitalism as a concept helps us understand why ecosystem services are valuable in economic terms and why they should be protected as investments for the future. As a consequence, it can inspire people and businesses to consider their environmental impacts in a different way and to be willing to conserve resources. In addition, it requires new techniques that will allow us to better represent and incorporate the monetary value of natural capital in our economic systems.

Ecosystem services and natural capital are not simple concepts. They require that we not only understand the relationship between complex ecological processes and human systems, but that we devise methods of better integrating environmental values and costs into our economic decision making. In the sections below, we explore models for applying ideals about sustainable development and natural resource use to engineering design and problem solving. These models include footprint analysis, waste management, ecological design, and sustainable and green engineering.

7.4 Footprint Indicators of Sustainability

Many sustainability metrics have been developed since the 1992 Earth Summit in Rio de Janeiro. These indicators are designed to quantitatively characterize and assess the multiple environmental, social, economic, and cultural dimensions of sustainability. Footprint indicators have become widely accepted as a way of communicating information about the pressures that human activity places on natural resources and ecosystem services. They are increasingly important in the ecological design toolkit, because they provide insights into the availability of natural resources and disruptions to biogeochemical cycles.

7.4.1 Mass Balance and the Footprint Concept

Footprint analysis relies on the mass balance and accounting methods developed in Chapters 2 through 6, a process called a material flow analysis. For example, the water footprint estimates all of the water used by a person, product, or region; the water footprint tries to account for all of the water embodied in an individual's lifestyle, in a product's life cycle, or in a region's overall economic and social systems. Similarly, the carbon footprint uses mass balance and thermodynamics to calculate greenhouse gas emissions from direct or embodied energy use.

Several types of footprint analyses are becoming popular, especially for illustrating the consequences of particular lifestyle choices, predominant technologies, or production systems (Table 7.1). The ecological footprint is the oldest of the indicators and was developed to convey an understanding of overall natural resource sustainability. Designed by William Rees and Mathis Wackernagel in the 1990s, the ecological footprint is usually expressed as the amount of land (hectares) required to support particular patterns of lifestyle and consumption. The carbon footprint was developed during approximately the same timeframe, and it is widely used to calculate the reduction in carbon emissions from specific CO_2 mitigation strategies. The carbon footprint is typically expressed in CO_2 equivalents.

The water footprint was introduced in 2002 to help communicate information about water consumption and quality issues at multiple levels, from personal to national. Water footprints are usually expressed as a water volume. The nitrogen footprint was most recently developed (Leach et al., 2011) as a means of helping people understand their impact on the nitrogen cycle. The nitrogen footprint is measured as kilograms of nitrogen per person per year, and it is used to inform consumers how their food consumption impacts reactive nitrogen cycling.

7.4.2 The Ecological Footprint

People rely on the production of goods, services, infrastructure, and waste absorption provided by natural resources. The ecological footprint is a metric that calculates the mass balance between the demand of human activities and available resources; it

TABLE 7.1 A comparison of sustainability footprint indicators

TYPE OF FOOTPRINT INDICATOR	AUTHORS CITED	YEAR CITED	ISSUE TO CONVEY	INPUTS	OUTPUT
Carbon		~ 2005	Climate change	Greenhouse gas emissions Embodied energy of products from UN COMTRADE Fraction of anthropogenic emissions sequestered by the ocean Rate of carbon uptake per hectare of forestland	CO_2 equivalents
Ecological	Rees (1992); Wackernagel and Rees (1996)	1992, 1996	Resource sustainability	Product harvest Carbon dioxide emitted National average yield of product Market price of product Carbon uptake capacity Area available for given land use	Hectares of land
Nitrogen	Leach, Galloway, Bleeker, Erisman, Kohn, and Kitzes	2011	Food impacts and sustainability	Electricity use Food consumption Food production Sanitation system Heating system Transportation	Kg of reactive Nitrogen per capita per year
Water	Hoekstra	2008	Water scarcity	Evapotranspiration Effective precipitation Environmental flow requirements Crop water requirements Crop yield Water stress coefficient Irrigation schedule Anthropogenic pollutant concentration Natural background concentration Natural assimilation capacity	Green, blue, or gray water volume

Source: Bradley Striebig.

also takes into account the impacts of different prevailing technologies and resource management strategies on the environment (Box 7.2). The ecological footprint relies on the concept of biocapacity—the biologically productive land and sea area available to provide the ecosystem services that humanity consumes (Wackernagel et al., 2002). Ecological footprint calculations are based on international datasets and literature cited in the *Ecological Footprint Atlas* (Ewing et al., 2010a).

Several simplifying assumptions must be made to calculate the ecological footprint. The following assumptions were used in calculating the ecological footprint of nations in 2010 (Ewing et al., 2010b, p. 3):

- *The majority of the resources people consume and the wastes they generate can be quantified and tracked.*
- *An important subset of these resources and waste flows can be measured in terms of the biologically productive area necessary to maintain flows. Resource and waste flows that cannot be measured are excluded from the assessment, leading to a systematic underestimate of humanity's true Ecological Footprint.*
- *By weighting each area in proportion to its bioproductivity, different types of areas can be converted into the common unit of global hectares, hectares with world average bioproductivity.*
- *Because a single global hectare represents a single use, and each global hectare in any given year represents the same amount of bioproductivity, they*

BOX 7.2 **Calculating the Ecological Footprint**

The ecological footprint of production that represents the demand for biocapacity was given in Equation 1.10:

$$EF_{\hat{P}} = \frac{\hat{P}}{Y_N} \cdot YF \cdot EQF \qquad (1.10)$$

where P is either the amount harvested or the carbon dioxide emitted, Y_N is the average yield for the product or carbon dioxide uptake, and YF and EQF are the yield factor and equivalence factor for the type of land use as defined by Ewing et al. (2010a).

Yield factors, YF, are used in the ecological footprint calculations to differentiate between the productivity associated with different types of land use. The yield factor for a particular nation is the ratio of the national average to the world average yields, as illustrated in Table 7.2.

Equivalent area factors, EQF, are used to convert from actual hectares of different types of land use into equivalent *global hectares*, at global average bioproductivity across all land-use types. The equivalence factors are calculated based on the ratio of the world average suitability index for a given land-use type versus the average suitability index for all land-use types. For example, the world average productivity for cropland, as the land-use type, was more than 2.5 times as productive as the average productive of all land on a global basis. Equivalence factors vary by land-use type and year as shown in Table 7.3. Land use globally is divided into the following five categories based on crop productivity (Ewing et al., 2010b):

- Very suitable (VS)—0.9
- Suitable (S)—0.7
- Moderately suitable (MS)—0.5
- Marginally suitable (mS)—0.3
- Not suitable (NS)—0.1

(Continued)

BOX 7.2 Calculating the Ecological Footprint (*Continued*)

TABLE 7.2 Example yield factors for selected countries based on 2007 data

YIELD	CROPLAND	FOREST	GRAZING LAND	FISHING GROUNDS
World Average	**1.0**	**1.0**	**1.0**	**1.0**
Algeria	0.3	0.4	0.7	0.9
Germany	2.2	4.1	2.2	3.0
Hungary	1.1	2.6	1.9	0.0
Japan	1.3	1.4	2.2	0.8
Jordan	1.1	1.5	0.4	0.7
New Zealand	0.7	2.0	2.5	1.0
Zambia	0.2	0.2	1.5	0.0

Source: Based on Ewing, B., A. Reed, A. Galli, J. Kitzes, M. Wackernagel. 2010b. Calculation Methodology for the National Footprint Accounts, 2010 Edition. Oakland: Global Footprint Network.

TABLE 7.3 Equivalence factors for land-use types based on 2007 data

LAND-USE TYPE	EQUIVALENCE FACTOR (GLOBAL HECTARES PER ACTUAL HECTARE OF LAND USE TYPE)
Cropland	2.51
Forest	1.26
Grazing land	0.46
Marine and inland water	0.37
Built-up land	2.51

Source: Based on Ewing, B., A. Reed, A. Galli, J. Kitzes, M. Wackernagel. 2010b. Calculation Methodology for the National Footprint Accounts, 2010 Edition. Oakland: Global Footprint Network.

The ecological footprint associated with consumption is the sum of the production footprint; EF_P; the footprint for imported commodities, EF_I; minus the footprint for exported commodities, EF_E, as shown in Equation 1.11:

$$EF_C = EF_P + EF_I - EF_E \qquad (1.11)$$

A country's biocapacity, BC, for any land type is calculated from

$$BC = A \cdot YF \cdot EQF \qquad (7.1)$$

where A is the actual area in hectares available for a specified land-use type, and YF and EQF are the associated yield factors and equivalence factors for that country's land-use type.

Land-use types are associated with either cropland, grazing land, forestland, fishing grounds, or built-up land as previously illustrated in Figure 1.20 and discussed next. The carbon use is also accounted for, but in the case of the ecological footprint the carbon is accounted for based on the associated area of land required for carbon uptake.

Cropland in the ecological footprint consists of the area required to grow all crop products, including livestock feeds, fish meals, oil crops, and rubber (Ewing et al., 2010b). The footprint associated with each type of crop is calculated as the area of cropland required to produce the reported harvest amount of the crop based on worldwide average yield for that crop.

Grazing land accounts for the land area required for grazing and the additional area required for growing feed crops to support livestock. The total demand for pasture grass, P_{GR}, can be calculated from Equation 7.2:

$$P_{GR} = TFR - F_{Mkt} - F_{Crop} - F_{Res} \qquad (7.2)$$

where TFR is the total feed requirement, F_{Mkt} is the feed from marketed crops, F_{Crop} is the feed from crops grown specifically for fodder, and F_{Res} is feed from crop residues.

Similarly, the footprint associated with fishing grounds is based on the primary production requirement, PPR, or the ratio of the mass of fish harvest to the mass needed to sustain the species (Pauly and Christensen, 1995).

$$PPR = CC \cdot DR \cdot \left(\frac{1}{TE}\right)^{(TL-1)} \qquad (7.3)$$

where CC is the carbon content of wet-weight fish biomass, DR is 1.27 the global average discard rate for bycatch, TE is 0.1 that assumes the transfer efficiency of biomass between trophic levels is 10%, and TL is the trophic level of the particular fish species. The total sustainably harvestable primary production requirement, PP_S, may be calculated from

$$PP_S = \Sigma(Q_{S,i} \cdot PPR_i) \qquad (7.4)$$

where $Q_{S,i}$ is the estimated sustainable catch for a species i. The worldwide average yield, Y_M, can be calculated from the total harvest PP_S per area of fishing along the continental shelf, A_{CS}:

$$Y_M = \frac{PP_S}{A_{CS}} \qquad (7.5)$$

The forestland footprint is a measure of the annual harvest of fuelwood and timber for all forest supply products. The world average yield was reported by Ewing et al. (2010b) to be 1.81 m³ of harvestable wood per hectare per year. The built-up land footprint is calculated based on the area of land covered by human infrastructure, including transportation, housing, industrial structures, and reservoirs for hydroelectric power generation. In 2007, the world's total estimated built-up land area was estimated to be 169.59 million hectares (Ewing et al., 2010b).

The carbon footprint is typically the largest contributing component of the ecological footprint for people living in higher-income countries. The carbon footprint component of the ecological footprint is calculated based on the total land area required to remove carbon dioxide in the atmosphere associated with human activities (such as fossil fuel combustion and changes in land use), as well as natural carbon dioxide emissions in an average year from forest fires, volcanoes, and respiration from animals and microorganisms. Carbon dioxide uptake in ecological footprint calculations consider only forest uptake of carbon dioxide and partitioning of carbon to the oceans as shown in

$$EF_C = \frac{P_C(1 - S_{Ocean})}{Y_C} \cdot EQF \qquad (7.6)$$

where P_C is the annual emissions of carbon dioxide, S_{Ocean} is the fraction of anthropogenic emissions sequestered by oceans in a given year, and Y_C is the annual rate of carbon uptake per hectare of forestland at world average yield.

LEARNING ACTIVITY 7.1 How Big Is Your Footprint?

FIGURE 7.8

Source: hddigital/Shutterstock.com

The ecological footprint assesses how our wealth, lifestyle, production technologies, and resource management affect our consumption of the Earth's ecological services and natural resources. The ecological footprint metric estimates the amount of biologically productive area (land and oceanic) that is required to sustain a particular lifestyle. Although there are criticisms of the ecological footprint methodology as described in the text, it is a useful way to understand how individual consumption affects ecosystems and natural resources.

In this activity, you will use an online calculator to estimate your own ecological footprint. Go to the ecological footprint quiz administered by the Center for Sustainable Economy at *www.myfootprint. org*. Read the webpage "About the Quiz," then take the quiz. Answer the following questions:

1. What exactly does this ecological footprint calculator measure? What is the global average footprint?
2. If everyone in the world ***had your lifestyle***, how many "planet Earths" would be required to sustain the global population?
3. How does your footprint for carbon, food, housing, and goods and services compare to the national average? Why do you think your footprint for these aspects of your lifestyle is above or below average?
4. For which type of biome (ecosystem) do you have the largest footprint? Pasture, forestland, fisheries, or cropland?
5. Based on your footprint analysis, what can you do to *reduce* your ecological footprint?

can be added up to obtain an aggregate indicator of Ecological Footprint or biocapacity.
- Human demand, expressed as the Ecological Footprint, can be directly compared to nature's supply, biocapacity, when both are expressed in global hectares.
- Area demanded can exceed area supplied if demand on an ecosystem exceeds the ecosystems regenerative capacity.

The ecological footprint essentially reports how many hectares of biocapacity are required to support human consumption for a given lifestyle, at a particular level of economic development, and using a prevailing suite of production technologies and natural resource management strategies. The ecological footprint of a person in Japan is therefore likely to be different from that of an individual in the United

States, even though both countries are at similar levels of economic development and enjoy comparable standards of living. Japanese lifestyles (diet, consumerism, housing needs, energy conservation, and so on) are not those of an American, the technological efficiencies of their economies are not the same, nor are their farming and forestry practices identical.

The ecological footprint provides a widely accepted method for assessing the mass balance between human consumption and biocapacity. However, there are several limitations and biases associated with the simplifications and calculations used for this footprint. Human demand on resources is *underestimated* because it does not consider all possible resource impacts; for example, the ecological footprint excludes water consumption, water and air pollution, soil erosion, and the impacts of greenhouse gases (GHGs) other than carbon dioxide. In contrast, biocapacity is likely to be *overestimated* by the method because neither land degradation nor the long-term sustainability of resource extraction (such as phosphorus; see Chapter 4) is considered in footprint calculations.

In sum, footprint indicators provide a measure of mass accounting between the natural world and human consumption. Intrinsic to the footprint concept is the dynamic of material flow, and many footprints directly or indirectly capture critical biogeochemical cycles. As we shall see over and over again, mass balance and the virtually closed-loop biogeochemical cycles of Earth are a defining feature of sustainable engineering as both a design constraint and as an ideal to be imitated.

7.5 Waste Management and Material Life Cycles

Traditional concepts about pollution, toxicity, natural resource abundance, scarcity, recoverability, and recyclability are all integrated by engineering models dealing with waste management and product life cycles. As a consumer, you are already familiar with waste management principles if you know the slogan "Reduce, Reuse, Recycle." If you have ever heard the expressions "cradle-to-grave" or "cradle-to-cradle" in reference to products and manufacturing, then you are also familiar with life cycle concepts. In this section, we will summarize how waste management and product life cycle models represent a more sustainable engineering approach to natural resource use and environmental protection.

7.5.1 The Waste Management Hierarchy

As you might infer from your understanding of biogeochemical cycles, **waste** is a uniquely human concept. In the natural world, all matter has some basic biological or physical purpose, so the outputs from one system are taken up as inputs by another. Elements, matter, and substances are productively used across geographic scales and time. In contrast, complex societies generate high concentrations of human (body) waste as well as material waste—something that has no useful purpose—to dispose. In the case of pollution, this waste is toxic or presents some other form of health and environmental hazard. In the case of natural resources, once they have been embodied in products that we no longer want, they become trash. Unused raw or processed materials in industry become "scrap waste."

Waste management evolved as a field of study and profession from sanitary engineering, which dealt with the public health threats created by urban sewage, garbage, and human excrement. Over time, other types of waste required management as well; today, **waste management** broadly refers to the collection, transport, processing, disposal, and monitoring of waste materials. Such wastes may be liquid, solid, or gaseous, and they may be hazardous or nonhazardous. The purpose of waste management is to minimize the impacts of waste on human health and the environment.

Waste management is guided by a ranking of most preferred to least preferred strategies known as the **waste hierarchy**, or waste pyramid (Figure 7.9). The most preferred action is ideally to prevent waste at the source by using only those natural resources as are absolutely necessary, by reducing packaging, by optimizing production processes so that scrap waste is minimized, and by avoiding materials and processes that will create toxic or hazardous waste. In the waste management hierarchy, such **source reduction** represents the prevention of waste at the beginning of a product's life cycle. Source reduction conserves resources and energy and can reduce pollution and toxicity as well. Source reduction and waste prevention derive from various engineering actions at this stage, including:

- The redesign of products to minimize their materials content, a process known as **dematerialization**.
- Better materials selection and manufacturing processes to minimize scrap waste, pollution, and toxicity.
- Designing more durable consumer products that will last longer and that can be repaired or refurbished more easily for continued use.
- Reusing products that have not reached the end of their useful life.
- **Remanufacturing** products by recovering modules and components for reassembly or reuse.

After source reduction ("reduce and reuse" in Reduce, Reuse, Recycle), waste management becomes a process of diverting waste streams for productive purposes. **Recycling** refers to the activities involved in collecting waste that has been disposed

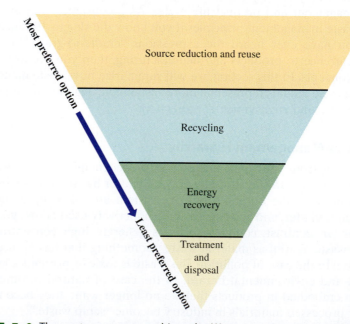

FIGURE 7.9 The waste management hierarchy. Waste management is guided by a ranking of most-to-least preferred actions, where the most preferred is to prevent waste from being created and the least preferred is to arrange for its permanent disposal. "Reduce, Reuse, Recycle" is a slogan modeled on the waste pyramid and encourages consumers and businesses to reduce their waste stream.

of, separating out usable materials, processing these materials into their constituent raw substances, and then using these reconstituted substances as **feedstocks**. Ideally, recycled substances would be used again as a source material for the product from which they came, such as aluminum for beverage cans, glass for containers, and so on. In practice, many recycled substances cannot be readily or cost effectively reconstituted anew, and they are repurposed for alternative uses. When a recycled material or product is used in an application that is of lower quality or has more limited functionality, it is known as **downcycling**. Examples include asphalt roofing shingles that are recycled into roadway materials and recycled office paper that is used for lower quality paperboard.

When material waste cannot be recycled, recovering its embodied energy is the next most desirable step in the waste management hierarchy. This topic is explored in Chapter 8 and generally involves "burning trash" to recover embodied energy as steam or electricity. Least desirable in waste management is the permanent disposal of the waste itself. When useful material and energy can no longer be technically or cost-effectively extracted, there is little recourse but to treat and dispose.

Incentives to practice aggressive waste management are based largely on government policies and economic costs. Government regulations that control pollution and limit the use of toxic substances put pressure on manufacturers for source reduction and can promote recycling. When landfill space is scarce, local municipalities may charge large disposal fees or ban certain kinds of solid waste altogether in an effort to avoid the cost of expanding landfills. This forces large producers of waste to reduce at the source, creatively reuse their waste products, or recycle.

High or rising costs of raw materials and energy also facilitate materials reuse, recovery, and recycling. For example, it is far more economical to recycle aluminum than to create it from raw materials. Successful recycling markets do, however, depend on the demand for the recycled material. Without this demand, there would be no one to whom one could sell recovered materials and products. A lack of recycling markets for particular materials at the regional scale is often one of the largest barriers to effective recycling programs. Recycling therefore relies on a cost-effective business case for the recycled substances or on government policies that create markets because of disposal restrictions.

Today, because of the growing focus on sustainability, the waste management field is moving toward a philosophy of **zero waste**. This notion is based on the principle that there is no waste in natural ecological systems and that human communities should have similarly closed-loop waste systems. This principle is widely practiced in Europe, where laws require producers to take back their packaging and/or products when they have reached their end of life (Box 7.3). The zero waste concept not only encourages a looping waste stream in which waste is processed and reused as an input, but it also places an emphasis on nontoxic materials that can biodegrade naturally once they must be permanently disposed. Biodegradable plastics (polymers) made from corn starch are an example of materials selection that would foster zero waste sustainability.

7.5.2 Engineering Models Based on Waste and Materials Management

Waste management principles are expressed in several different types of engineering models. **Industrial ecology** is a systems approach to industrial processes that models material and energy flows. Its goal is to close the loop in production systems by optimizing processes and finding opportunities in which wastes can be

BOX 7.3 Take Back Programs and Extended Producer Responsibility

Many manufacturers around the world now participate in voluntary and mandatory **take-back programs**. A take-back program is one in which a product (usually postconsumer) can be returned to the manufacturer at the end of its life. The producer then becomes physically responsible for the product and may reprocess it back into a feedstock for production; may refurbish components and modules for remanufacture or reuse; may recycle and reclaim raw materials; or may arrange for final end-of-life disposal.

Take-back programs result from environmental policies and regulations that create the principle of **extended producer responsibility**. Extended producer responsibility makes a manufacturer responsible for the environmental consequences of its product over a greater range of the product's life cycle; it is intended to capture the cost of environmental protection in producer costs and consumer prices.

Extended producer responsibility is a waste management principle that first emerged in Sweden, then Germany, and is now practiced throughout Europe. Scarce landfill space, the presence of potentially hazardous substances in product componentry, and the growing cost to government of waste disposal are strong motives for this type of management innovation. The Green Dot® logo represented in Figure 7.10 is used within the EU to designate products for which there is a take-back packaging program supported by the manufacturer. The Green Dot program is not mandatory; it is one of several strategies that producers can select from to comply with EU regulations.

By shifting the financial and environmental burden of waste disposal back onto manufacturers and consumers,

Der Grüne Punkt –
Duales System Deutschland GmbH

FIGURE 7.10 The Green dot logo indicates a product that has a take back packaging program in the EU.

Source: The Green Dot, the financing symbol of Duales System Deutschland GmbH

economic markets can then work to encourage producers to adopt less environmentally harmful materials and to develop products that can be more readily disassembled. High embodied costs of product recovery and disposal will also signal consumers to buy products that have a smaller environmental footprint. Extended producer responsibility and take-back programs are also intended to stimulate innovation in materials recovery and recycling technology.

There are no federal mandatory take-back programs or requirements in the United States, although more than 20 states have some sort of take-back law with respect to electronics waste, commonly referred to as E-waste. Many large consumer electronics and office machinery companies have well-known take-back initiatives, including Xerox, Dell, Hewlett Packard, IBM, and Sony. Other firms are active, too, in such industries as power tools, automobiles, and cameras.

Through the Carpet America Recovery Effort, the U.S. carpet industry has set a goal of 40% waste diversion for discarded carpeting. Carpet constitutes a significant portion of the U.S. waste stream: About 2.9 million tonnes were landfilled in 2010, which is a little less than 1.5 percent of all municipal waste by weight. For large organizations such as schools and office buildings, the cost of carpet disposal in landfills can be significant. In response to growing pressures for carpet waste diversion to both conserve landfill space and reduce costs to consumers, the U.S. carpet industry has initiated a substantial take-back initiative. More than 50% of all carpeting manufactured in the United States is made of nylon 6 or nylon 6,6; these materials can be economically reused as new feedstock for carpet (as is the case for nylon 6) or downcycled for automobile components and other plastics (nylon 6,6). Most major carpet manufacturers now sponsor take-back programs. However, according to the U.S. Environmental Protection Agency, as of 2010 only 9% of discarded carpeting was being successfully diverted from the waste stream.

reused as inputs by other processes. Industrial ecology derives its analytical concepts and design principles from ecosystem dynamics and biogeochemical cycles; it applies these to manufacturing and fabrication processes. Industrial ecology as an engineering model and method is the basis of Chapter 9 and is explored more thoroughly there.

Many engineering practices based on waste management principles are variations on what we refer to as **life cycle models**. Life cycle models are those that analyze a product from its source raw materials (its "cradle"), through its fabrication and use, to its ultimate end-of-life disposal (its "grave"). Figure 7.11 illustrates the extraction and processing of raw materials for manufactured goods and the multiple pathways that these products can follow for reuse, recycling, remanufacture, and disposal. The architect William McDonough popularized life cycle models with the term **cradle-to-cradle** design in his book *Cradle-to-Cradle: Remaking the Way We Make Things*. The cradle-to-cradle model also derives from the absence of "waste" in natural ecological systems and idealizes closed-loop raw material, product, and waste flows.

In engineering, innovations based on waste management principles and life cycle concepts have already taken place to facilitate waste minimization and to move us toward closed-loop manufacturing systems. These innovations broadly fall under the term **design for environment (DfE)** and are grounded in the idea that environmental protection should be *designed into* products and processes rather than managed as an after-the-fact harm. Major design for environment practices include:

- Dematerialization, which is the reduction in the amount of materials required for a product without changing its functionality.
- Design for recyclability, is used to facilitate materials recovery and reuse.
- Design for disassembly, in which products are designed to readily come apart into their constituent components for reuse or recycling.

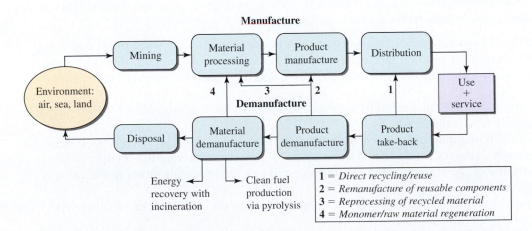

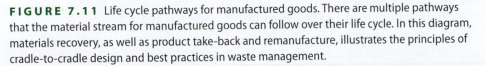

FIGURE 7.11 Life cycle pathways for manufactured goods. There are multiple pathways that the material stream for manufactured goods can follow over their life cycle. In this diagram, materials recovery, as well as product take-back and remanufacture, illustrates the principles of cradle-to-cradle design and best practices in waste management.

Source: Based on Bert Bras, "Incorporating Environmental Issues in Product Design and Realization," *Industry and Environment*, Vol. 20, Nos. 1–2, 1997.

LEARNING ACTIVITY 7.2 Dissecting Postconsumer Waste

How often do you get to break something without getting into trouble? In this learning activity, you will disassemble a consumer product, *but you don't have to put it back together again!* It is a reverse engineering exercise in which your goal is not to figure out how to copy, imitate, or streamline assembly of the product, but to redesign it for lesser environmental impact.

This activity illustrates a variety of the concepts associated with product life cycle and waste management. One of the major challenges in sustainable design is to create products that can be easily disassembled at their end of life, that can have their components reused, that facilitate materials extraction for recycling or be used for reprocessing as feedstock, or that can be disposed of in a manner that mitigates environmental hazard.

Begin by obtaining a small consumer product that you don't mind breaking or that may already not work. Examples include kitchen appliances (toasters, coffee pots, blenders), consumer electronics (cell phones, calculators, game stations, TV sets, radios, stereo equipment, fax machines, alarm clocks, electric staplers), personal hygiene products (electric razors, blow dryers, curling irons), power tools, and toys. You might have something lying around at home that you can use; thrift stores and charity shops are good sources for very cheap consumer products. You'll also need a tool set with an assortment of screwdrivers and pliers.

Clear and set up a workspace, ideally covered with newspaper or other material to protect your work surface and corral small parts. Have a notebook and pencil ready; a digital camera is also useful. Begin by sketching and photographing the product in its fully assembled form, and make notes about how it appears to be put together. Start disassembling the product. As you take it apart, draw a series of sketches and/or take photographs of the units you remove and disassemble. Keep track of the *sequence* in which the item's components and modules come apart, and make sketches of how components and modules fit together. Also keep a record of how the product is physically held together in terms of screws, fasteners, clips, adhesives, soldering, and so on.

Break down your product into the smallest units for which you can safely do so. Keep your materials organized into their dissected groupings to make it easier to identify your product's substructure. Once your item is completely taken apart, reflect on the following:

- Count and describe the discrete modules and components you dissected. At what point could you no longer easily reduce your product into smaller units? What is the relative complexity of disassembly? Consider such factors as time, manual dexterity, and the nonobviousness of how components connect.
- Are there any component units that could be refurbished and used for remanufacture? Which ones? What cannot be easily refurbished? Why?
- Can you readily identify the materials used in your product? Are any of these recyclable? Can these materials be physically extracted, or do they require chemical processing to separate them?
- Can you imagine if any of the materials could be reprocessed as a feedstock for either the same or a different product?
- Can you find any information regarding take-back programs, design for environment, or other types of waste reduction efforts for your product?
- Are there ways in which the design and fabrication of this product could be simplified for easier disassembly and component or materials recovery? What do you suggest, and why?

- Remanufacturing, which is the process of recovering product modules and components, repairing and refurbishing them, and then reusing them again in new production or sales.
- Minimized use of energy, toxic materials, and toxic production processes to limit the environmental release or disposal of contaminants.

Life cycle analysis has emerged as a standard approach to material, energy, and waste flows in product fabrication, and it is an essential engineering tool for designing more sustainable products, processes, and systems. You will have the opportunity to perform simple life cycle analysis in Chapter 10.

7.6 Ecological Design

Life cycle models and waste minimization represent one major category of strategies for more sustainable engineering. These approaches share an intrinsic effort to reduce materials use, waste, pollution, toxicity, and energy consumption. Design decisions are based on careful analysis of material and energy flows and are representative of the ecological dynamics of nutrient flows and the absence of waste in natural systems. These approaches help us shift away from a linear, cradle-to-grave model of resource extraction, processing, and product disposal to a model in which materials, resources, and "waste" products are dynamically cycled and reused as both inputs and outputs in engineered systems. In other words, our technological processes become more "ecological" in their operation. In the idealized closed-loop system, all materials and energy instead flow from "cradle-to-cradle."

Concepts of **ecological design** push us even further by encouraging designers to adapt ecological processes to human constructs. Products, manufacturing processes, energy use, and buildings are all candidates for ecological design, as are infrastructures and even whole communities. Ecological design principles are emerging in engineering, architecture, art, urban planning, and other design endeavors (Box 7.4). The idea is that by "looking to nature" for inspiration, we can take advantage of sustainable forms and processes already successfully established in our environment. In this section we will explore ecological design broadly as well as **biomimicry**, which is a specific design strategy that systematically analyzes natural processes for engineering solutions.

7.6.1 Broad Perspectives on Ecological Design

In art, architecture, urban planning, and industrial design, ecological design can simply refer to the aesthetics of forms that emulate nature. The design problem in this instance is not necessarily functional. Instead, shape is patterned on natural objects and landscapes with the intention of making the design more natural, or organic, in form. The belief is that ecological design evokes an innate human sense of beauty about the natural world and that human structures should be integrated with their landscapes. From this perspective, ecological design is a process of visually modeling nature as a way of appealing to the human senses (Figure 7.13).

In engineering, ecological design refers more specifically to functional processes, and this aspect of ecological design is represented in the professions of architecture, urban planning, and industrial design as well. In the previous section on waste management and life cycle models, several principles of ecological design were discussed—living organisms take only what they need from their environment, waste products are cycled as nutrient or matter inputs to other systems, and as a general rule toxicity is not a characteristic feature of ecological systems. In architecture and urban planning, these ecological dynamics are often applied to buildings

BOX 7.4 What Can We Learn from Termites?

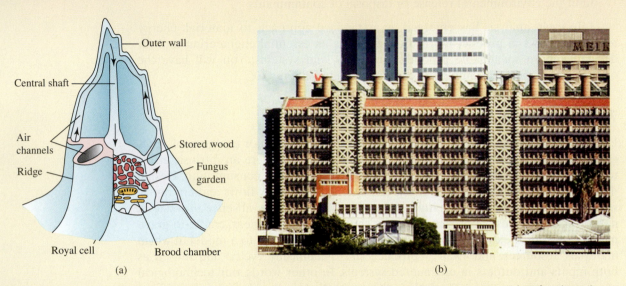

FIGURE 7.12 The Eastgate Centre. The Eastgate Centre in Zimbabwe's capital city of Harare is an exemplar of ecological design. Patterning the natural dynamics of ventilation in (a) termite mounds, Architect Mick Pearce constructed (b) a building of considerable beauty and low environmental impact.

Source: (a) Based on what-when-how, Nervous System (Insects), http://what-when-how.com/insects/nervous-system-insects; (b) Courtesy of Mick Pearce

Architect Mick Pearce's Eastgate Centre in Harare, Zimbabwe (Figure 7.12), is one of the most celebrated examples of ecological design in architecture. The ventilation, heating, and cooling system in this building is inspired by the natural convection processes of *Odontotermes transvaalensis* termite mounds; their structural "chimney" effect draws and moves cool, warm, and fresh air throughout the mound. The process is commonly referred to as **passive heating and cooling**.

Eastgate Centre incorporates both passive and active energy in its ventilation and cooling systems. Electric fans move and exhaust air throughout the building. Natural convection in the chimneys, driven by internal and external air temperature differentials, enhances fan-powered ventilation. Active and passive energy are used to move large volumes of evening air through the building and cool its thermal

mass overnight. Eastgate starts its morning cooled down and able to draw on thermally "stored" night air for daytime cooling. This ecological design has lowered the energy needed for cooling to about 10% of that needed for conventional air conditioning.

The elegance of the design inspiration is deceptively simple. Substantial computer simulation and modeling was required by the engineering firm OveArup in order to achieve the architect's vision and structural features. Parameters for direct sunlight, window operability, and interior light control had to be established for the passive system to work properly. In addition, the feasibility of this design is tied to its geographic location—Zimbabwe's climate is especially conducive to large scale passive cooling. Engineering knowledge and skill is thus an essential element in translating nature's designs to human use.

and sites in order to control water, use locally sourced construction materials, design passive cooling and heating systems, and compost organic wastes.

However, ecological design also goes further by looking to the natural world for pre-existing, more sustainable solutions to design problems that we currently face. Over millions of years of natural selection, living organisms and ecosystems

FIGURE 7.13 Fallingwater. Architect Frank Lloyd Wright pioneered what is referred to as "organic architecture," a style of design in which buildings and their surrounding landscape are perceived as an integrated composition. Fallingwater is a home in Pennsylvania and one of Wright's iconic designs.

Source: Education Images / Contributor/UIG via Getty Images

have evolved highly efficient, nontoxic functions for thriving in their environments. Ecological design looks to nature for answers to common design problems, regardless of whether that problem is confronted by a bacterium or a building. In some instances, ecological design may actually use biological processes to engineer a desired function.

Stormwater management is a good example of this perspective. Stormwater runoff from streets, roadways, parking lots, and grassy hillsides is a significant nonpoint source of water pollution. It is a problem created specifically by our built environment, which causes water to run over hard surfaces like asphalt, concrete, and roof tops, collecting contaminants as it flows. In order to prevent flooding in streets and neighborhoods, stormwater is usually channeled through buried stormwater drains or through open concrete aqueducts. Accumulated debris, garbage, automotive oils, chemical pollutants, animal waste, and other contaminants are thus rapidly directed to the nearest stream, river, or lake.

To avoid such water degradation, stormwater is now frequently managed by containing it locally. Detention ponds halt the flow of contaminants to natural bodies of water by retaining stormwater and letting it evaporate. Ponds are sized to accommodate a defined volume of water from a storm surge (Figure 7.14). Detention ponds are not, however, ecological systems. The water in them is usually so contaminated with chemicals or nutrients that they are not healthy for wildlife or aquatic organisms.

FIGURE 7.14 Stormwater detention pond. A typical stormwater detention pond that is designed to retain stormwater and prevent the flow of contaminants into natural bodies of water.

Source: Photo by Bradley Striebig

The concepts of ecosystem services and ecological design allow us to approach stormwater management very differently. Low-impact development (a topic explored in Chapter 11) is an engineering model in which rainwater runoff is managed at its source by imitating the hydrology of natural systems. Several techniques are used for low-impact development, but one common feature is rain gardens. Water filtration zones are planted in shallow depressions, and storm-water accumulates in these zones to naturally contain and purify the water as it percolates through the soil (Figure 7.15). Rain gardens have the added benefit of

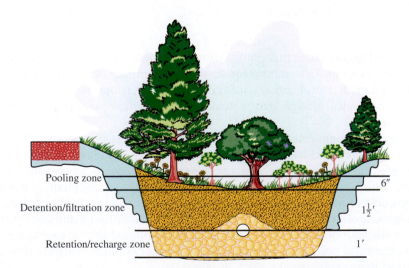

FIGURE 7.15 Rain garden schematic. Rain gardens are modeled after the Earth's own natural water regulation and purification system in which vegetation, topography, and geology all dynamically interact to control the flow of water and clean it of contaminants. Rain gardens are a key design feature of low-impact development that manages stormwater runoff at its source, rather than channeling it to natural bodies of water and polluting them.

Source: Based on City of Des Moines, http://www.dmgov.org/Departments/Parks/PublishingImages/rain_garden.gif

FIGURE 7.16 Constructed wetlands. This constructed wetland is artificial and is basically a scaled-up rain garden. However, at this larger scale, it begins to function as a defined ecosystem with a broader range of dynamic cycles and species.

Source: Photo by Bradley Striebig

being aesthetically beautiful gardens that can become miniature habitats. Rain garden principles can also be scaled up to handle larger volumes of water through bioretention ponds and constructed wetlands, which are themselves complex engineered ecosystems (Figure 7.16). In this manner, low-impact development imitates the ecosystem services of water regulation and purification, while at the same time creating small-scale ecosystems and habitats for wildlife, pollinators, and other living organisms.

Bioremediation is another example in which we use actual biological processes to more sustainably resolve an environmental engineering problem. **Bioremediation** is the process of using microorganisms (bacteria, fungi, protozoa, and so on) to break down contaminated waste into nonhazardous substances. Contaminants are transformed by the organism's growth and digestive processes, producing harmless metabolic wastes. Although not all contaminants may be addressed in this manner, bioremediation offers a cost-effective way to eliminate some types of contaminants, from soil in particular. Living Machines® is a commercial-scale wastewater treatment process that decontaminates sanitation waste at building and community scales. Living machine systems (both commercial and noncommercial) are modeled after tidal wetlands and provide a variety of aesthetic and productive benefits depending on how they are designed. Greenhouses, micro habitats, irrigation water, and passive energy are examples of the productive uses that can be designed into a living machine.

7.6.2 **Biomimicry**

Biomimicry is also a type of ecological design because it looks to nature to solve human design problems. Whereas ecological design is often used to create aesthetic forms in architecture and products, biomimicry is instead an emerging *design discipline* that attempts to systematically structure and analyze the natural world. Biomimicry as a discipline is based on the understanding that living organisms are themselves masterful engineers, having evolved over millions of years through processes of natural selection. What living systems literally embody are successful adaptations to their environment through a vast array of forms and functions. The underlying designs of nature can be the basis for sustainable innovation in energy, transportation, nontoxic materials selection and fabrication, waste processing, building structures, and so on. For example, the Vitalis water bottle (Figure 7.17) is a biomimetic design of pine trees that have spiraled, or helical, fibrous structures. This allows the trees to withstand heavy load and force vertically, horizontally, and diagonally. Vitalis incorporated this ecological design as a grooved geometry in their bottles, making them ultra lightweight yet strong. Unlike the broader approaches of ecological design, biomimetic designs and processes can also be at the nano-, micro-, and molecular scales (Box 7.5, Box 7.6, and Figure 7.18).

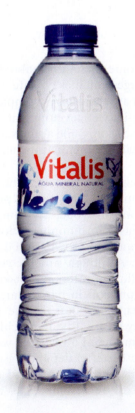

FIGURE 7.17 Vitalis water bottle. The Portuguese brand Vitalis elegantly designed an ultra-light PET (a type of plastic) water bottle. The company was able to reduce materials yet integrate strength by analyzing the spiral growth structures of whitebark pine tree fibers. In 2010, Vitalis reduced its use of raw materials by 227 tonnes through the improved design.

Source: Unicer

BOX 7.5 Sustainable Design Thinking in Action: "A Shark Tale"

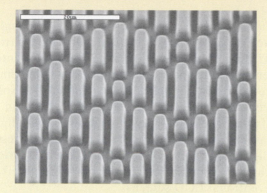

FIGURE 7.18 Sharklet™ technology.

Courtesy of Sharklet Technologies, Inc.

Sharklet Technologies, Inc. invented the first antibacterial material that does not require any chemical, antibiotic, or antiseptic agents. Its antibacterial properties are based exclusively on its material patterning and texture, which inhibits bacterial growth and is biomimetic of sharkskin. Sharklet's use on surfaces and medical devices could inhibit the development of antibiotic resistant bacteria because it does not interfere with the biological processes of the organisms. Below is an excerpt from the company's website that describes the design thinking and process that led to this award-winning innovation. As can be seen from Dr. Brennan's story, good powers of observation and a persistent curiosity can lead to fundamentally new engineering ideas.

Dr. Anthony Brennan, a materials science and engineering professor at the University of Florida, was visiting the U.S. Naval base at Pearl Harbor in Oahu as part of Navy-sponsored research. The U.S. Office of Naval Research solicited Dr. Brennan to find new antifouling strategies to reduce use of toxic antifouling paints and trim costs associated with dry dock and drag.

Dr. Brennan was convinced that using an engineered topography could be a key to new antifouling technologies. Clarity struck as he and several colleagues watched an algae-coated nuclear submarine return to port. Dr. Brennan remarked that the submarine looked like a whale lumbering into the harbor. In turn, he asked which slow moving marine animals don't foul. The only one? The shark.

Dr. Brennan was inspired to take an actual impression of shark skin, or more specifically, its dermal denticles. Examining the impression with scanning electron microscopy, Dr. Brennan confirmed his theory. Shark skin denticles are arranged in a distinct diamond pattern with tiny riblets (Figure 7.18). Dr. Brennan measured the ribs' width-to-height ratios which corresponded to his mathematical model for roughness—one that would discourage microorganisms from settling. The first test of Sharklet™ yielded impressive results. Sharklet reduced green algae settlement by 85% compared to smooth surfaces.

While the U.S. Office of Naval Research continued to fund Dr. Brennan's work for antifouling strategies, new applications for the pattern emerged. Brennan evaluated Sharklet's ability to inhibit the growth of other microorganisms. Sharklet proved to be a mighty defense against bacteria.

Similar to algae, bacteria take root singly or in small groups with the intent to establish large colonies, or biofilms.

Similar to other organisms, bacteria seek the path of least energy resistance. Research results suggest that Sharklet keeps biofilms from forming because the pattern requires too much energy for bacteria to colonize. The consequence is that organisms find another place to grow or simply die from inability to signal to other bacteria.

Dr. Brennan's and Sharklet Technologies' research has demonstrated Sharklet's success in inhibiting the growth *of Staph a., Pseudomonas aeruginosa*, VRE, *E. coli*, MRSA and other bacterial species that cause illness and even death.

Source: Sharklet Technologies, Inc. "Inspired by Nature." Accessed February 2, 2014 at http://sharklet.com/technology.

BOX 7.6 Butterfly Effects

FIGURE 7.19

Source: Photo by Bradley Striebig.

The *butterfly effect* is a popularized social metaphor that helps us understand how small changes in complex systems can have profound impacts in unexpected places. But the butterfly (Figure 7.19) is also a powerful source of engineering innovation through biomimicry.

Nature's beauties are inspiring a host of pragmatic innovations that promise a more sustainable future. One of the most significant biological features of many species of butterfly is that their color *is not* generated by a pigment, but by cellular structures that are arranged as shingled plates at the nanoscale. So what can we learn from this? That color can be created by optical interference. This property was adapted by Qualcomm, an electronics company that invented a display screen using electromechanical manipulation of thin films. The display requires only ambient light; it has no internal light source of its own. The result is highly readable color displays even in bright sunlight, using only 10% of the power required by conventional LCD screens.

Morphotex®, a nanotechnology-based synthetic fiber no longer in production, used the structural chromogenic properties of butterfly color to create color without pigments, potentially reducing water consumption, dyes, and chemical waste. Similarly, ChromaFlair™ is a color-shifting pigment based on structural chromogenics as well. On automobiles, this paint is a rainbow of iridescent color.

Butterfly biomimicry is not all about color aesthetics. Chinese scientists are exploring how to develop more efficient solar energy technology by studying the deep black in butterfly wings, which functions as a heat trap. Parts of butterfly wings are also self-cleaning, inspiring research that will help us develop materials that do not require detergents, surfactants, and large amounts of water to stay clean.

Biomimicry as a design discipline has the potential to transform many aspects of engineering. It has developed a taxonomic classification of eight fundamental functions found in nature from which 30 subgroups and 162 specific functions are derived (Figure 7.20). The challenge over the coming years is to translate and adapt our biological knowledge to the language and processes of engineering. But it also requires that engineers adapt their problem-solving approaches to more effectively study and understand the insights of nature.

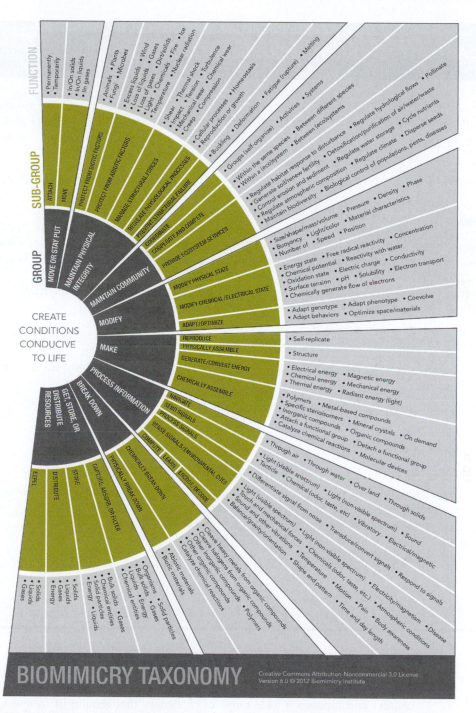

FIGURE 7.20 Biomimicry taxonomy. Biological processes can be standardized into a basic functional typology. These categories guide designers who are looking for ways in which nature has already developed solutions to particular kinds of design "problems."

Source: © Biomimicry Institute, 2012

LEARNING ACTIVITY 7.3 Using the Biomimicry Taxonomy

Biomimetic design requires us to problem-solve a little differently, because we have to align our thoughts with the underlying functions of nature. In this learning activity, you will take a challenge presented by the Biomimicry 3.8 Institute so that you can better understand how the process works.

Design Problem

You are designing a building in an area of low rainfall. You want your building to collect rainwater and store it for future use.

Use the Taxonomy

1. Refer to the taxonomy, and find the verbs. Move away from any predetermined ideas of what you want to design, and think more about what you want your design to do. Try to pull out single functional words in the form of verbs.

2. Go to the online taxonomy at *www.asknature.org/aof/browse*. The questions you might pose through the Search or Browse options might be "How would Nature…"

   • Capture rainwater?
   • Store water?

3. Try a different angle. Some organisms live in areas that don't experience any rain, yet they still get all of the water they need. So other questions to pose might be: "How would Nature…"

   • Capture water?
   • Capture fog?
   • Absorb water?
   • Manage humidity?
   • Move water?

4. Turn the question around. Instead of asking how Nature stores water, you might think about how Nature protects against excess water or keeps water out: "How would Nature…"

   • Remove water?
   • Stay dry?

Source: The Biomimicry 3.8 Institute, AskNature.com, "Biomimicry Taxonomy: Biology Organized by Challenge," www.asknature.org/article/view /biomimicry_taxonomy. [Accessed 5 October 2012]

7.7 Sustainable and Green Engineering

As you can see from the discussions in this textbook, sustainability in an engineering context encompasses a wide range of challenges that require us to manage pollution, diminish toxic and hazardous wastes, conserve natural resources, and analyze the life cycle of material and energy flows. By emulating principles of the natural world, especially ecological systems and organisms, we gain insights into how nature closes a waste loop and handles design problems in efficient, nonharmful ways. As engineers, we are also required to be mindful of economic costs and to minimize risks to people and the environment.

Can we distill from these insights a framework that will systematically guide us as we innovate more sustainable products, materials, and processes? The answer is yes. Two slightly different, but complementary, engineering frameworks have emerged that provide guidance for both the comprehensive process of engineering design as well as more specific design choices. These frameworks are known as **sustainable engineering** and **green engineering**.

Sustainable and green engineering are grounded in a disciplinary shift in which engineering thought and practice actively consider and promote sustainability. Sustainable engineering solutions are those that promote human well-being, protect human health and the environment, and preserve the biosphere. In addition to these criteria, our technologies must also be technically viable and cost effective.

Nine principles of sustainable engineering and social justice were defined by an intensive working session in 2003. At a conference entitled *Green Engineering: Defining the Principles*, 65 professional scientists and engineers derived a set of holistic principles that would inform practice and contribute to a more sustainable society. The principles function as a paradigm for thought and action. In particular, green engineering advocates the philosophy that sustainability can be most effectively achieved, with the greatest impacts, *when it is designed into products and processes in their initial phases of design and development.* The nine principles are (U.S. Environmental Protection Agency, undated, citation a):

1. Engineer processes and products holistically, use systems analysis, and integrate environmental impact assessment tools.
2. Conserve and improve natural ecosystems while protecting human health and well-being.
3. Use life cycle thinking in all engineering activities.
4. Ensure that all material and energy inputs and outputs are as inherently safe and benign as possible.
5. Minimize depletion of natural resources.
6. Strive to prevent waste.
7. Develop and apply engineering solutions, while being cognizant of local geography, aspirations, and cultures.
8. Create engineering solutions beyond current or dominant technologies; improve, innovate, and invent (technologies) to achieve sustainability.
9. Actively engage communities and stakeholders in the development of engineering solutions.

As you might expect from the earlier discussion of sustainable development and the context in which environmental protection must take place, we cannot disregard the real-life constraints on engineering represented by government, business, and society. Cultural needs and expectations shape what is desirable in design, and considerations of cost, safety, and technical performance also inform our technologies. These nine principles of green engineering thus reflect these constraints and are not prescriptive. In other words, they do not give us an exact method for developing a sustainable design. However, they do change the way that we think about the purpose of our work and what we should try to accomplish in our design of materials, products, processes, and systems. These nine principles are analogous to a country's political constitution: They are the guiding rules from which all others derive.

Green engineering is very similar to sustainable engineering, and it also emerged as a model in early 2003. Green engineering is often discussed in the context of chemical engineering because of the significant role that materials and chemical processes play in our industrial world. Nonetheless, its guiding tenets are generalizable to all aspects of engineering and design. Paul Anastas and Julie Zimmerman put forward the 12 principles of green engineering, and as you can see in Box 7.7, many of these are similar, if not identical to, those of sustainable engineering. What

BOX 7.7 The 12 Principles of Green Engineering

1. **Inherent Rather Than Circumstantial**
   Designers need to strive to ensure that all materials and energy inputs and outputs are as inherently nonhazardous as possible.

2. **Prevention Instead of Treatment**
   It is better to prevent waste than to treat or clean up waste after it is formed.

3. **Design for Separation**
   Separation and purification operations should be designed to minimize energy consumption and materials use.

4. **Maximize Efficiency**
   Products, processes, and systems should be designed to maximize mass, energy, space, and time efficiency.

5. **Output-Pulled Versus Input-Pushed**
   Products, processes, and systems should be "output pulled" rather than "input pushed" through the use of energy and materials.

6. **Conserve Complexity**
   Embedded entropy and complexity must be viewed as an investment when making design choices on recycle, reuse, or beneficial disposition.

7. **Durability Rather Than Immortality**
   Targeted durability, not immortality, should be a design goal.

8. **Meet Need, Minimize Excess**
   Design for unnecessary capacity or capability (e.g., "one size fits all") solutions should be considered a design flaw.

9. **Minimize Material Diversity**
   Material diversity in multicomponent products should be minimized to promote disassembly and value retention.

10. **Integrate Local Material and Energy Flows**
    Design of products, processes, and systems must include integration and interconnectivity with available energy and materials flows.

11. **Design for Commercial "Afterlife"**
    Products, processes, and systems should be designed for performance in a commercial "afterlife."

12. **Renewable Rather Than Depleting**
    Material and energy inputs should be renewable rather than depleting.

Source: Paul Anastas and Julie Zimmerman. (2003) "Design Through the Twelve Principles of Green Engineering," *Environmental Science and Technology,* 37 (5), March 1, 2003, pp. 94A–101A.

makes green engineering slightly different from sustainable engineering is that it moves us toward prescriptive action. In other words, the principles give us more direction about how to achieve sustainability through science and engineering, and they provide a framework for evaluating design choices.

Anastas and Zimmerman argue that there are "two fundamental concepts that designers should strive to integrate at every opportunity: life cycle considerations and the first principle of green engineering, inherency" (2003, p. 96A). By inherency, the authors mean that materials and energy use should be as environmentally benign as possible, a principle illustrated well by carbon-dioxide-based dry cleaning methods (Box 7.8). In Chapter 9 you will learn about life cycle considerations in much greater detail, and Chapters 1 through 6 of this textbook reviewed for you the many problems of pollutants, toxicity, and hazardous materials.

Most of the principles in Box 7.7 are readily understandable, but a few require a little more explanation. Principle #5, for example, that products, processes, and systems should be output-pulled and not input-pushed, is not exactly obvious. It derives from the Le Châtelier principle in chemistry which states that systems reestablish their balance when they are induced to make a change. It requires less material and less energy to create system transformations by steadily drawing off small amounts of output from that system (output-pull) instead of driving through large amounts of input (input-push). This principle is fundamental to chemical engineering, but it also "scales up" to larger systems. Just-in-time manufacturing and just-in-time delivery are industrial strategies that represent the Le Châtelier principle for a completely different kind of technological system.

Principle #6 is also not transparent, but it provides useful insights into what to do with a product or material that has reached the end of its useful life. The integrity of items that embody a variety of complex materials, fabrication processes, or structures should be preserved at the end of their lives. This is because creating complex materials, products, and structures is time consuming, costly, usually energy-intensive, and may involve expensive elements such as noble metals and rare earths. Under these circumstances, *complexity should be preserved*.

Rather than breaking these items down into their constituent elements for recovery, recycling, or safe disposal, it would be better to maintain the integrity of the item's complexity and reuse it, as with inkjet cartridges (Figure 7.21). This principle leads us to such engineering practices as remanufacture, in which modules and components of products are reused again. In contrast, aluminum beverage cans are comparatively simple and are better candidates for recycling than reuse. Principle #6 is closely related to principle #11, which is to design for a commercial afterlife. Complex manufactures—such as consumer electronics and office machinery and equipment—often reach the end of their commercial life before they reach the end of their productively useful life. Changing consumer tastes (think about cell phones), leasing arrangements for photocopiers, and rapid technological change (iPads and similar tablet devices) drive the commercial obsolescence of products long before they are no longer physically functional. Design for commercial afterlife complements preserving complexity because it suggests we should anticipate market obsolescence and design-in the reuse of out-of-date componentry for future generations of products.

Finally, principle #10 encourages us to use the materials and energy that are at hand for production, which could be at the molecular, component, production line,

FIGURE 7.21 Inkjet cartridges are a simple example of the principle of preserving complexity. The integrity of these products can be completely preserved by refurbishing and re-inking them, a process that is more cost effective than manufacturing new cartridges. Some manufacturers make cartridges that can be productively re-inked and reused up to 100 times.

Source: Photo by Bradley Striebig

BOX 7.8 How Green Is Your Dry Cleaning?

Dry cleaning, a common way of cleaning delicate fabrics and other textiles, uses a solvent called perchloroethylene, or "perc," to clean clothes. Perc is classified and handled as a hazardous waste in the United States, and the U.S. Environmental Protection Agency regards it as a human carcinogen. California declared perc a toxic chemical in the 1990s and banned its use after 2023.

Finding a less toxic—and ideally nontoxic—way to clean fabrics that water would otherwise damage is a challenge. Micell Technologies developed just such a process, using liquid CO_2 and recoverable/recyclable residues and surfactants. Partnering with a variety of scientific R&D institutes and aided by government research funding, the Micell process uses simple, beverage-grade CO_2. The process eliminates hazardous waste, chemical odors, exposure to harsh solvents, and ozone-depleting substances. And with rare exceptions, it also cleans fabric just as well, if not better. Dry cleaning establishments that use the CO_2 process are award-winning, dedicated advocates.

But out of more than 20,000 dry cleaning enterprises in the United States, fewer than 50 use the eco-friendly CO_2 process. Micell Technologies closed its dry cleaning division and aftermarket product support. The innovation is clearly a success story of green engineering. But why didn't it get adopted more widely in the industry?

The answer lies in cost, marketing, and the size of the machines. The capital cost of a CO_2 system is over $150,000 compared to half that for conventional perc equipment. The machines are larger and have a bigger footprint, an issue for dry cleaners in urban areas where rent and real estate are costly. Lack of appropriate technical support and machinery parts and components may also have been an issue.

FIGURE 7.22 Clothing labels. A variety of symbols indicate the best way to clean our clothing.

Source: bicubic/Shutterstock.com

Although evidence suggests that the CO_2 systems are equally profitable, their marketing may not have been done effectively to illustrate this. There is also the problem that dry cleaning has traditionally been a perc process for 80 years, and change is hard.

New manufacturers that could bring equipment costs down, better after-sales support by service companies, more consumer demand for green dry cleaning, and tighter government regulations on perchloroethylene are all factors that could influence the more widespread adoption of this innovative solution to an inherently unsustainable practice.

facility, or community scale. Principle #10 is an integral strategic feature for closing waste and energy loops and for minimizing the use of natural resources for materials and energy. For example, waste steam can be captured for hot water processes. Iron ore waste at a steel foundry can be sintered on-site and put into the foundry's blast furnace. The waste fryer oil from local restaurants can be the feedstock for biodiesel. Industrial ecology, which is explored in depth in Chapter 9, illustrates how to

LEARNING ACTIVITY 7.4 Which Insulation Is Most Sustainable?

Good product engineering requires that we design with our end users in mind. In this activity, you will integrate principles of green engineering with market analysis to evaluate which type of building insulation is most sustainable for a start-up firm. This activity also illustrates the powerful role that economics plays in getting (or not getting) sustainable products into the marketplace.

Assume that you are the founder of a start-up building materials company committed to green engineering and making sustainable resources readily available to the do-it-yourself homeowner market. You conduct some market research and find that the U.S. building insulation market is projected to reach nearly $9 billion by 2016, with annual increases in sales of nearly 8%. The strongest growth will be in the residential sector, as new housing starts recover from the economic downturn and existing homeowners focus on greater energy conservation. Because of this market data, you decide to expand your product line to include insulation products.

Conduct an Internet search in which you obtain information about fiberglass batt, mineral wool, cellulose, polyurethane spray foam, and soy spray foam insulation. As you search, look for the following types of information:

- Materials content, including any use of toxic or hazardous substances in the manufacturing process.
- Safety and health issues associated with do-it-yourself installation of such products.
- Ease of installation, and flexibility of the product (for example, it can be used in a wide variety of climates or building conditions).
- The cost of the amount of insulation required to achieve roughly R-12 to R-15 for 100 square metres of surface.

The **material safety data sheet** for these products will give you a wide variety of information, as will various websites that provide insulation cost calculators.

Use the principles of sustainable and green engineering to evaluate which of these materials are the most sustainable from a business perspective. Which one do you select to develop for your company? What would have to change (if anything) for your product to be more competitive in the insulation market? What are your next steps in terms of directing your engineering and product design team for new product development?

evaluate industrial operations in order to identify opportunities for exploiting local materials and energy.

7.8   Summary

In this chapter we introduced a number of concepts and models that will help you develop analytical skills for creating more sustainable materials, products, and processes. Natural resources and raw materials are limited substances, and the ecological dynamics of our biosphere support life through a variety of ecosystem services. By both emulating and protecting the resources, biodiversity, and biogeochemical cycles of the Earth, we have a foundation for sustainable engineering and design. Waste management and ecological principles provide us with industrial ecology and life cycle models of zero waste and closed-loop waste systems. Biological organisms and systems hold the keys to addressing complex design problems in ways already demonstrated by nature to be effective. Ecological design at all scales, as well as biomimicry, are emerging as new modes of thought, analysis, and design in the engineering disciplines.

References

Abraham, Martin (Ed.). (2006). *Sustainability Science and Engineering: Defining the Principles*. San Diego: Elsevier.

Anastas, Paul and Zimmerman, Julie. (2003, March 1). "Design through the twelve principles of green engineering." *Environmental Science and Technology* 37(5): 94A–101A.

Childers, Everett. Carbon dioxide. Accessed October 5, 2012 at www.textilecleaning.com /carbon-dioxide.

Ewing, B., et al. (2010a). *Ecological Footprint Atlas*. (2010). Oakland, CA: Global Footprint Network. www.footprint.org/atlas.

Ewing, B., et al. (2010b). (2010b). *Calculation Methodology for the National Footprint Accounts*, 2010 Edition. Oakland, CA: Global Footprint Network.

Fishbein, Bette K. (2000). "Carpet take-back: EPR American style." *Environmental Quality Management* 10(1):25–36.

Hoekstra, A. Y., and Chapagain, A. K. (2008). *Globalization of Water: Sharing the Planet's Freshwater Resources*. Oxford, UK: Blackwell Publishing.

Hoekstra, A., Chapagain, A. K., Aldaya, M. M., and, Mekonnen M. M. (2011). *The Water Footprint Assessment Manual: Setting the Global Standard*. London: Earthscan.

Jones, C. M., and Kammen, D. M. (2011) "Quantifying Carbon Footprint Reduction Opportunities for U.S. Households and Communities: Supporting Materials." Environmental Science and Technology 45(9): pp. 4088–4095. Accessed 1 January 2014 at http://pubs.acs.org/doi/abs/10.1021/es102221h.

Leach, A., et al. (2011). "A nitrogen footprint model to help consumers understand their role in nitrogen losses to the environment." *Environmental Development* **1**(1): 40–66.

McDonough, William and Michael Braungart. (2002). *Cradle-to-Cradle: Remaking the Way We Make Things*. New York: North Point Press.

Moomaw, W., Burgherr P., Heath G., Lenzen M., Nyboer J., and Verbruggen. A. (2011). Annex II: Methodology. In O. Edenhofer, R. Pichs-Madruga, Y. Sokona, K. Seyboth, P. Matschoss, S. Kadner, et al. (Eds.). *IPCC Special Report on Renewable Energy Sources and Climate Change Mitigation*. Cambridge, UK: Cambridge University Press.

Pauly, D., and Christensen., V. (1995). "Primary production required to sustain global fisheries." *Nature* 374:255–257.

Rees, William E. (October 1992). "Ecological footprints and appropriated carrying capacity: what urban economics leaves out." *Environment and Urbanisation* **4**(2): 121–130.

U.S. EPA. (2012). Methodology for Understanding and Reducing a Project's Environmental Footprint. U.S. Environmental Protection Agency, Office of Solid Waste and Emergency Response and Office of Superfund Remediation and Technology Innovation.

U.S. EPA. (a). What is green engineering? Accessed October 5, 2012 at www.epa.gov /oppt/greenengineering/pubs/whats_ge.html.

U.S. EPA. (b). Case study: Liquid carbon dioxide (CO_2) surfactant system for garment care. Accessed 1 January 2014 at http://www.epa.gov/dfe/pubs/garment/lcds /micell.htm

Wackernagel, M. and W. Rees. (1996). *Our Ecological Footprint: Reducing Human Impact on the Earth*. British Columbia: New Society Publishers.

Wackernagel, M., B. Schultz, D. Deumling, A. Callejas Linares, M. Jenkins, V. Kapos, et al. (2002). "Tracking the ecological overshoot of the human economy." *Proc. Natl. Acad. Sci.* 99(14):9266–9271.

World Business Council for Sustainable Development and World Resources institute. (2004). The Greenhouse Gas Protocol: A Corporate Accounting and Reporting Standard, Revised Edition. Geneva, Switzerland and Washington, D.C.

Key Concepts

Sustainable development	Source reduction
Intergenerational ethics	Dematerialization
Natural resource	Rare earths
Raw materials	Remanufacturing
Abiotic	Recycling
Biotic	Downcycling
Renewable resource	Zero waste
Nonrenewable resource	Take-back programs
Population collapse	Extended producer responsibility
Recoverability	Industrial ecology
Recyclable	Life cycle models
Fossil water	Design for environment (DfE)
Natural resource management	Ecological design
Ecosystems	Biomimicry
Ecosystem services	Passive heating and cooling
Feedstocks	Bioremediation
Natural capital	Sustainable engineering
Waste management	Green engineering
Waste hierarchy	Material safety data sheet

Active Learning Exercises

ACTIVE LEARNING EXERCISE 7.1: Your carbon footprint

Use the U.S. EPA individual carbon footprint calculator (*www.epa.gov/climatechange/ghgemissions/ind-calculator.html*) to estimate your carbon footprint and answer the following questions.

- How much carbon dioxide do you emit for your lifestyle in an average year based on the EPA calculator for:
  - Heating
  - Transportation
  - Home energy
  - Consumer waste emission
- Create a pie chart of your emissions for each of the four categories above. What category contributes the greatest amount of GHG emissions? What category contributes the greatest amount of GHG emissions?
- How do your emissions compare to the U.S. average indicated in your summary of estimated savings?
- What would your emission rate be if you reduced emissions associated with electricity use by 75% by using electricity produced from a power generating facility that would adopt carbon capture and sequestration technology?
- How would this impact your total CO_2 emissions?
- What is the difference in your CO_2 emission if you were to recycle zero of your consumer waste compared to if you were to recycle all possible components of your consumer waste?

Problems

7-1 Consider the issues associated with the Ogalalla Aquifer presented in Chapter 3. Using the two images of sustainable development in Figure 7.2, sketch the relationship between the environmental sustainability of the aquifer and sustainable development using aquifer water.

7-2 Can all of the Earth's abiotic substances and biotic species be considered natural resources? Explain why or why not.

7-3 What is the difference between a regenerative renewable resource and a regenerative nonrenewable resource?

7-4 "All renewable resources are the same." Do you agree with this statement? Explain why or why not.

7-5 Classify the following resources as either renewable or nonrenewable.
  a. Biogas
  b. Tin
  c. Fossil water
  d. Tuna
  e. Geothermal energy
  f. Diamonds

7-6 Classify the following as either an infinitely available renewable resource or a regenerative renewable resource.
  a. River water
  b. Solar energy
  c. Wheat
  d. Algae
  e. Wind energy
  f. Bagasse

7-7 Why are all elemental minerals and metals theoretically recoverable?

7-8 What economic and technical factors affect whether a substance is recoverable and/or recyclable?

7-9 Consider the biogeochemical cycle for water (Chapter 3). Why *isn't* water always an infinitely renewable resource?

7-10 Use the second law of thermodynamics to explain why fossil fuels and their energy aren't recoverable once combusted.

7-11 Classify the following by the type of ecosystem service(s) that they represent.
  a. Vegetation growing along a streambed
  b. Sisal production
  c. A scenic mountain vista
  d. Insect pollination
  e. A sacred flower

7-12 A forest provides multiple ecosystem services. Identify and describe five of these services. (*Hint*: Some can be found in this chapter, but you may need to do a quick Internet search.)

7-13 How do the concepts of ecosystem services and natural capital help us better understand the importance of environmental sustainability?

7-14 Identify and briefly explain each of the four steps of the waste management hierarchy. Why is source reduction the most preferred waste management strategy?

7-15 Think about the slogan "Reduce, Reuse, Recycle." Why do you think so much emphasis is placed on recycling in our popular culture, and not on reduce or reuse?

7-16 Give examples of five different engineering methods and practices that reduce and minimize the waste stream.

7-17 Which is better from a sustainability perspective—recycling or downcycling? Why?

7-18 Is it possible to have a truly zero waste production system? Why or why not?

7-19 What is design for environment (DfE)? List and briefly describe five different engineering practices that reflect DfE.

7-20 Explain how industrial ecology and life cycle analysis are related to source reduction in the waste management hierarchy.

7-21 Compare and contrast ecological design as both aesthetic and functional approaches to design. Can you find an example of each in your community or campus?

7-22 Reflect on the principles of sustainable and green engineering. Do you think it's easy to be green? What are common obstacles to greener engineering and design?

7-23 Critique the product that you dissected in Learning Activity 7.2 using the principles of green engineering. Which principles does it reflect? Which principles are violated?

Open-Ended Problems

7-1 A young couple in the city of San Antonio, Texas, wants to design a rainwater harvesting system that will let them water their garden without using the municipal water supply, which comes from the increasingly stressed Edwards Aquifer. The San Antonio climate is mild, so it has two growing seasons per year: spring (February 1 through July 31) and fall (September 1 through December 31). These homeowners would like to have a total supply of 3,700 litres of water for spring and summer and 1,800 litres for fall. Design a rainwater harvest system that meets these goals. Assume that your materials budget is limited to $750. The roof area of the house is 110 square metres.

7-2 You work for a U.S.-based clothing manufacturer, Denim Inc. This company is a profitable niche producer of high-end women's jeans that sell for about $185 per pair. Denim Inc. has become competitive by branding itself as "made-in-America" and by having great-fitting jeans, but believes this strategy will no longer be enough to protect its market share. The company would like to develop a triple bottom-line strategy and has pledged to create a zero waste, take-back program for all of its blue jeans sold in the United States. In other words, these jeans would create no postconsumer waste if everyone who bought them returned them to the manufacturer. Denim Inc. sells about 5 million pairs of jeans in the United States each year. Design the logistics of the take-back program and develop a waste management strategy for these returned jeans so that they create no waste stream.

7-3 Coordinate a "garbage analysis" event with your school's facilities management or sustainability coordinator (if it has one). Select a classroom building of manageable size, gather up all of the waste in the building that has been thrown away in trash cans, and take it outdoors. Sort and separate the trash to see what has been thrown away (make sure to wear gloves when you do this). How much of the trash could have been recycled at your school or in your town? What do the results suggest about recycling awareness and efforts on your campus? (This is a good event for your school newspaper to cover, too; photos and videos can help document your activity.)

7-4 Fleas are amazing insects, including the fact that relative to their body size, they can leap extraordinary distances both horizontally and vertically. They also have unusual sensing mechanisms. Conduct research on fleas and identify all of their biological processes that could be useful from a design perspective. Then classify each of those processes according to the biomimicry taxonomy.

CHAPTER 8

Energy Conservation and Development

FIGURE 8.1 The Earth at night. This satellite image clearly illustrates population density and energy consumption, but it also more subtly highlights regions of the world that experience widespread energy poverty, such as in sub-Saharan Africa and Asia. Notice the low levels of lighting in these regions; more than 95% of the world's population without access to modern energy services lives in sub-Saharan Africa and low-income Asian countries. Nearly 85% of the energy poor live in rural communities.

Source: NASA Earth Observatory/NOAA NGDC

While not exactly a true measure of well being and enlightenment, the amount of power used is a reliable indication of the degree of safety, comfort and convenience, without which the human race would be subject to increasing suffering and want and civilization might perish.
                                                                    —NIKOLA TESLA

GOALS

This chapter will introduce you to the major energy challenges facing the global community today and enable you to understand the impacts of energy use on the environment. Several major strategies for greater energy sustainability will be explored, including efficiency, conservation, energy recovery, renewable energy and alternative fuels, fuel switching, pollution prevention and control, appropriate technology, and distributed energy.

OBJECTIVES

At the conclusion of this chapter, you should be able to:

- Explain the different energy needs of low-income and high-income countries.

- Calculate rates of depletion of finite fossil fuel reserves.

- Estimate energy savings from energy conservation and efficiency improvements.

- Size a photovoltaic array relative to a power load.

- Summarize multiple strategies for achieving greater energy sustainability.

- Explain how water and energy sustainability are interconnected.

8.1 Introduction

As Nikola Tesla suggests, energy is a basic human need, without which such necessities as lighting, cooking, industrial processes, transportation, long-distance communication, refrigeration, and space heating would not be possible. Since the Industrial Revolution, human societies have increasingly met their critical energy requirements with fossil fuels and technologically advanced energy systems such as centralized electric power production, the internal combustion engine, and highly automated thermodynamic steam processes. Access to modern energy services does indeed provide "safety, comfort, and convenience," yet 1.3 billion people live without electricity worldwide, and 2.7 billion use cooking facilities that pollute their homes and threaten their health.

Energy use today is characterized by an extraordinary duality of energy poverty and wealth in the global community. Citizens of affluent nations benefit from virtually universal access to modern energy services, which include direct connections to electricity, clean cooking fuels instead of traditional biomass (dung, grass, wood, peat), and both reliable *and* affordable energy supplies. Homes, hospitals, schools, offices, businesses, and factories in these societies are rarely threatened by a serious disruption of energy production due to shortages of fuels, costly spikes in the cost of energy, or an insufficient delivery and distribution infrastructure. In contrast, low-income countries confront a widespread lack of access to modern energy services at the household level, a significant inequality in rural and urban access to energy supplies, and an inability to provide an adequate amount of energy to their growing populations. Even in capital cities such as Nairobi, Kenya, rolling brownouts occur frequently, and protracted periods of time without electricity are a regular experience.

Energy sustainability presents two significant challenges to the professional engineering community. First and foremost, it reflects the challenge of providing a safe, affordable, abundant, and reliable supply of energy to all. In doing so, we enhance the ability of individuals to meet their basic needs, to improve their health and standard of living, and to achieve their full human potential. Robust energy systems not only increase the well-being of people, but also provide a critical prerequisite to economic development.

The second significant challenge is to supply the world's unmet—and still growing—energy needs while protecting environmental quality and ensuring adequate energy resources for the future. Not only are we trying to manage consumption of the Earth's finite fossil fuels in the face of intensifying demand, but we are also trying to mitigate the impact of energy use on the environment and climate. As you will learn in this chapter, there are many opportunities for new and better engineering design through efficiency and conservation, alternative energy resources, energy recovery, and the development of appropriate technologies.

8.2 Energy and Society

Most discussions of energy begin with a review of energy resources, their availability, their energy content, and so forth. However, energy consumption is always a means to a human end, and more sustainable energy use requires that we appreciate—at a very fundamental level—why people use energy. By understanding when, how, and why people need energy, we can creatively design more sustainable solutions for the future.

Human energy consumption represents **derived demand**. That is, our demand for an energy fuel (such as oil) or an energy-producing technology (such as a wind turbine) results from a primary personal or economic need for a service that energy provides. **Energy services** include lighting, transportation, space heating, refrigeration, telecommunications, plowing fields, industrial manufacturing, materials fabrication, cooking, and so on. Any given energy service is potentially met in a variety of ways. Consider transportation, for example. We can walk, ride a camel, row a boat, or pedal a bicycle to get around, and none of these modes of transportation require the direct consumption of an energy resource (other than food for people and animals to provide their biological energy). Alternatively, we could take a train, drive a car, or fly in a plane, all of which require the supply of an energy fuel to power that form of transport.

These choices represent different degrees of energy intensity, and the technology on which our mode of transportation is based determines the demand for a particular energy resource. For example, commercial aircraft will only fly on jet fuel; the demand for jet fuel is therefore wholly derived from the demand for airline and other aircraft services. Train locomotives, on the other hand, can use coal if they are a steam locomotive, or diesel fuel if they have an internal combustion engine. The demand for trains as a passenger vehicle or to ship freight will therefore result in derived demand for coal and diesel fuel. *In short, the way we engineer a particular technology to provide an energy service is a critical determinant of the type and amount of energy resources that are needed.*

Two key dynamics influence the overall character of energy consumption in a society. One is a process that occurs at the household level, and the other is a feature of the economic system. The **energy ladder** is a model developed by the International Energy Agency that describes how both the quality and quantity of energy used in a household change as incomes rise (Figure 8.2). When incomes are low, only the most

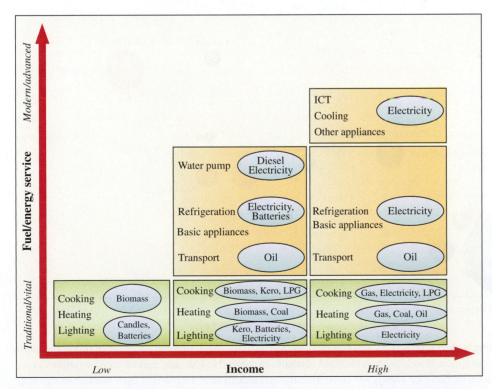

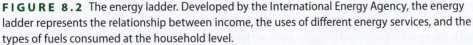

FIGURE 8.2 The energy ladder. Developed by the International Energy Agency, the energy ladder represents the relationship between income, the uses of different energy services, and the types of fuels consumed at the household level.

Source: World Energy Outlook 2002 © OECD/IEA, 2002, fig. 13.1, p. 370.

critical energy needs of the home are met, using fuels that are typically polluting and harmful to human health. As incomes rise, the energy services consumed by the household become more diverse, more modern (derived from electricity or other forms of commercial energy), and more technologically complex. When households "climb" the energy ladder, they benefit from a wider variety of energy services, they transition from low- to high-density fuels, and they increase their overall energy consumption.

Energy intensity reflects the dynamics of a nation's economic system as it industrializes and intensifies its use of energy overall, and it is measured as energy consumed per monetary unit of gross domestic product (for example, as MJ per $1 GDP). Intuitively, a country that has an economy based on subsistence agriculture will require less energy for its economic production than an industrialized nation producing steel, chemicals, and manufactured goods. The most energy-intensive industries include metals, electric power production, pulp and paper, chemical processes, oil and gas refining, and cement manufacturing.

When societies industrialize, their energy intensity goes up, and the industrialization process increases household income through the dynamics of economic development. Economic development therefore intensifies energy use in two ways—through the growing needs of industrial production and through the growing wealth of households and their associated climb up the energy ladder. Figure 8.3 illustrates this relationship between energy consumption and wealth. What is clearly seen in this graph is

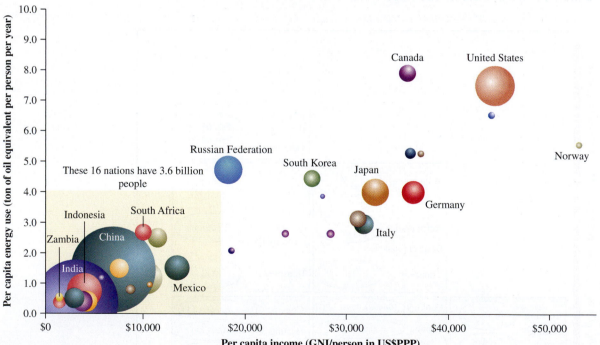

FIGURE 8.3 The relationship between prosperity and energy consumption. This graph demonstrates the dramatically higher levels of per capita energy consumption in wealthy industrialized countries compared to low- and middle-income nations in 2009. The increasing future global demand for energy will come from the less industrialized nations as their economies develop and citizens get wealthier. The great challenge is providing for the growing energy needs of these societies in a sustainable manner. The 16 nations in the bottom left corner of this graph account for 3.6 billion people, which is half of the world's total population. (The population of each country is represented proportionally in this chart by the size of the "bubble.")

Source: Calculated by the author from World Bank indicators

the strong association between per capita income and per capita energy consumption and the strong presence of the industrial and postindustrial economies at the upper end of the income spectrum. Figure 8.3 also illustrates the very large number of countries and people at lower per capita incomes; most of the future growth in world energy consumption will come from these nations as they pursue their economic development and improve their social well-being. Table 8.1 provides key statistics on energy use, population, and gross domestic product (GDP) for different countries.

TABLE 8.1 Energy and economic indicators for selected countries

COUNTRY NAME	TOTAL POPULATION 2011	GDP U.S. dollars, 2009	HUMAN DEVELOPMENT INDEX 2011	TOTAL ENERGY USE, KILOTONNE OIL EQUIVALENT (ktoe) 2007	PER CAPITA ELECTRICITY USE, KILOWATT HOUR (kWh) 2010
World	6,973,738,433	57,933,353,274,622	n/a	11,710,482	2,975
Least developed countries: UN classification	843,773,370	532,595,905,367	n/a	244,464	180
Low income	816,810,477	386,240,724,913	n/a	248,486	242
Middle income	5,021,923,999	16,292,931,520,411	n/a	5,802,331	1,823
OECD members	1,245,198,487	40,919,230,975,511	n/a	5,548,771	8,281
Australia	22,620,600	921,971,672,011	0.929	119,755	10,177
Bolivia	10,088,108	17,339,992,191	0.663	5,371	616
Brazil	196,655,014	1,621,661,507,655	0.718	235,356	2,384
Canada	34,482,779	1,337,577,639,752	0.908	271,731	15,138
Chile	17,269,525	172,590,627,161	0.805	30,565	3,297
China	1,344,130,000	4,991,256,406,735	0.687	2,014,822	2,944
Costa Rica	4,726,575	29,397,499,977	0.744	4,510	1,855
Croatia	4,407,000	62,202,619,240	0.796	9,306	3,813
Germany	81,726,000	3,298,635,952,562	0.905	331,169	7,215
Ghana	24,965,816	25,978,537,279	0.541	9,064	298
Greece	11,304,000	321,016,154,721	0.861	30,217	5,242
Guatemala	14,757,316	37,733,791,089	0.574	8,527	567
Haiti	10,123,787	6,470,254,240	0.454	2,777	24
Iceland	319,000	12,113,097,224	0.898	4,836	51,440
India	1,241,491,960	1,361,056,679,992	0.547	598,801	616
Indonesia	242,325,638	539,579,959,053	0.617	184,681	641
Iran, Islamic Rep.	74,798,599	331,014,973,186	0.707	190,092	2,652
Israel	7,765,700	194,866,363,197	0.888	20,719	6,856
Italy	60,770,000	2,111,148,008,712	0.874	179,599	5,384

(*Continued*)

TABLE 8.1 (Continued)

COUNTRY NAME	TOTAL POPULATION 2011	GDP U.S. dollars, 2009	HUMAN DEVELOPMENT INDEX 2011	TOTAL ENERGY USE, KILOTONNE OIL EQUIVALENT (ktoe) 2007	PER CAPITA ELECTRICITY USE, KILOWATT HOUR (kWh) 2010
Jamaica	2,709,300	12,012,993,251	0.727	4,687	1,222
Japan	127,817,277	5,035,141,567,659	0.901	515,193	8,394
Jordan	6,181,000	23,820,013,059	0.698	7,209	2,226
Kenya	41,609,728	30,580,367,979	0.509	17,178	156
Korea, Rep.	49,779,000	834,060,441,841	0.897	222,147	9,744
Malta	419,000	8,099,400,961	0.832	877	4,151
Mexico	114,793,341	879,703,353,505	0.770	175,937	1,990
Morocco	32,272,974	90,908,402,631	0.582	14,350	781
New Zealand	4,405,200	117,376,308,375	0.908	17,124	9,566
Nicaragua	5,869,859	8,061,961,615	0.589	3,078	473
Norway	4,952,000	374,757,527,038	0.943	27,546	25,175
Philippines	94,852,030	168,333,540,385	0.644	38,514	643
Qatar	1,870,041	97,583,513,671	0.831	20,554	14,997
Russian Federation	141,930,000	1,222,648,134,225	0.755	672,591	6,452
South Africa	50,586,757	283,012,416,483	0.619	137,294	4,803
Spain	46,235,000	1,455,956,368,264	0.878	143,829	6,155
Sweden	9,453,000	405,782,994,635	0.904	50,060	14,939
Thailand	69,518,555	263,711,244,889	0.682	104,886	2,243
Tunisia	10,673,800	43,522,032,141	0.698	9,039	1,350
United Arab Emirates	7,890,924	270,334,929,438	0.846	51,963	11,044
United States	311,591,917	13,898,300,000,000	0.910	2,337,014	13,394
Vietnam	87,840,000	97,180,304,813	0.593	45,777	1,035
Zambia	13,474,959	12,805,029,522	0.430	7,398	623

Note: OECD (Organisation for Economic Co-Operation and Development) member states are: Australia, Austria, Belgium, Canada, Chile, Czech Republic, Denmark, Estonia, Finland, France, Germany, Greece, Hungary, Iceland, Ireland, Israel, Itality, Japan, Korea, Luxembourg, Mexico, Netherlands, New Zealand, Norway, Poland, Portugal, Slovak Republic, Slovenia, Spain, Sweden, Switzerland, Turkey, United Kingdom, and United States.

Source: Based on World Bank, Data Indicators, http://data.worldbank.org/indicator.

EXAMPLE 8.1   Calculating Per Capita Energy Use and Energy Intensity

$$\text{Per capita energy use} = \frac{\text{total energy use}}{\text{total population}} \qquad (8.1)$$

Using values for the United States in Table 8.1:

$$\text{Per capita energy use} = \frac{2{,}337{,}014 \text{ ktoe}}{311{,}591{,}917} = .0075 \text{ ktoe per person per year}$$

$$\text{Energy intensity} = \frac{\text{total energy use}}{\text{total GDP}} \qquad (8.2)$$

Using values for the United States in Table 8.1:

$$\text{Energy intensity} = \frac{2{,}337{,}014 \text{ ktoe}}{\$13{,}898{,}300{,}000{,}000} = 1.68 \text{ ktoe per \$1 GDP per year}$$

Energy intensity is often misinterpreted by students. Correctly, it reflects the amount of energy used to generate one dollar of GDP. We would therefore say that the United States consumed 1.68 ktoe of energy for every dollar of GDP that it generated. (Students often incorrectly state this in the reverse: that it costs one dollar to generate 1.68 ktoe of energy.)

8.2.1 Energy Issues in Low-Income Countries

Low-income countries face a different set of energy issues compared to those of more affluent nations. A compelling problem is **energy poverty** itself, in which a large proportion of households lack access to the **formal energy sector**. Under conditions of energy poverty, families are unable to access (or afford) commercially provided energy, such as electricity or liquid petroleum gas (LPG gas). Instead, households are low on the energy ladder and may rely on biomass that they gather themselves or on limited purchases of kerosene, charcoal, and batteries.

The **social costs** of energy poverty are high. (Private costs refer to the negative effects of a condition on individuals, whereas social costs refer to the negative impacts experienced by society as a whole.) The social costs of energy poverty are reflected by low human productivity, health impacts, and lost social capital. For example, in many countries, gathering fuelwood is women's work, and it may consume several hours a day. The time that women spend gathering fuelwood is time lost to more productive activities, such as running a small business, farming, or tending to other household needs. Traditional biomass is also harmful to human health, especially when burned indoors—respiratory infections and asthma are debilitating. Other health consequences of energy poverty include the inability to purify water or pump potable groundwater, to refrigerate food and protect it from spoiling, and to sterilize medical equipment in community health clinics. Lost social capital includes the inability of children to do their homework or study at night because of a lack of lighting (Figure 8.4). The clean cookstove movement (see Box 8.1) is a global effort to reduce the health, environmental, and economic impacts of traditional biomass and inefficient cookstoves.

Low-income nations are also challenged to provide a reliable electric power infrastructure. Because of its exceptional versatility as an energy resource—it can power an extremely wide variety of technologies and their associated energy services, and it can be shipped over long distances—electricity is the fastest growing form of energy worldwide and the energy form of choice in developing countries (Figure 8.5). Although developing countries have typically electrified all of their major urban areas, rural connectivity may be lacking. In India, even with an intensive electrification effort, about one-third of its rural population still has no access to electric power; in urban areas, the reliability of the system may be in question, with frequent brownouts and blackouts. There may consequently be both an *insufficient amount* of electricity and challenges with *reliability* even when access is not an issue. The result is compromised economic development, in which the scope, scale, and productivity of business and industry is lessened.

FIGURE 8.4 Homemade kerosene lamps. An example of a homemade tin kerosene lamp common in many low-income countries. Such lamps provide light at night in rural areas and during the day in urban dwellings without direct access to sunlight. Kerosene is a costly and precious commodity to those low on the energy ladder, and lack of lighting carries a high social cost. Children cannot study at night, adults cannot extend their work into evening hours, and villages lose the ability to have community meetings and deliberations when the day's work is done.

Source: Photo by Bradley Striebig

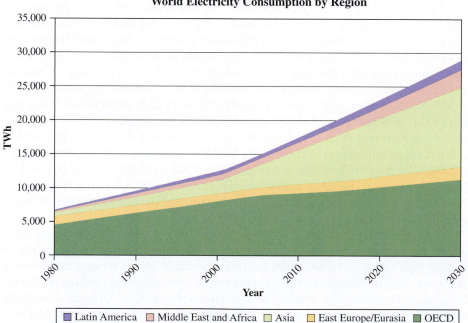

FIGURE 8.5 Past and projected growth in world electricity consumption. Electricity is the fastest growing form of electricity worldwide, and as seen in this figure, most growth in electric power will occur in Asia, the Middle East, and Africa.

Source: Based on World Nuclear Organization, http://www.world-nuclear.org/info/Current-and-Future-Generation/World-Energy-Needs-and-Nuclear-Power/

BOX 8.1 The Global Clean Cookstoves Movement

Nearly one-half of the world's population, about 3 billion people, uses traditional biomass to cook and heat their homes. This unsafe form of energy accounts for the deaths of nearly 2 million people per year (mostly women and children). Emphysema, asthma, pneumonia, cancer, and low birthweights are significant health consequences of indoor fires that burn wood, charcoal, dung, and crop residues; the need for biomass by the world's energy poor is also a notable cause of deforestation and soil erosion. Not only is there a high social cost to the time spent gathering biomass fuels in rural areas, but in urban areas the poor can spend one-fifth or more of their scant incomes on charcoal and wood. Traditional cookstoves and open fires are not only costly in terms of health, the environment, and household economics, but they are extremely inefficient from an engineering standpoint. In the most extreme case of an open fire for cooking, 90% of the heat is lost to the air, with only 10% going to a cooking pot.

The global clean cookstoves movement represents a 30-year initiative to develop more efficient cooking facilities for residents of low-income countries. The World Bank has invested over $160 million in cookstove programs, and in 2010 the United States pledged $50 million to the Global Alliance for Clean Cookstoves, an international public–private partnership dedicated to the adoption of 100 million new cookstoves worldwide by 2020. Engineering an improved cookstove is not trivial because of the wide varieties of fuels, cultural cooking practices, and prepared foods worldwide. In all instances, improved cookstoves are designed to drastically reduce indoor air pollutants,

FIGURE 8.6 Burmese woman with improved cookstove. In Myanmar, almost all of its fuelwood is used for cooking in homes. The improved ceramic cookstove shown here reduces fuelwood consumption for the typical family by more than 40%.

Source: ©FAO/Giuseppe Bizzarri

conserve scarce biomass resources, and protect the environment. As a consequence, several dominant technological designs have emerged to meet the cooking needs of families in Latin America, Africa, Asia, and eastern Europe. These include advanced biomass cookstoves, alcohol and biogas stoves, electric cookers, planchas (flat griddles used in Latin America), propane gas stoves, solar ovens, and "rocket" cookstoves with specialized combustion chambers. However, new technologies are not enough. The affordability of the stoves and fuel is still a challenge. In addition, there is resistance to some types of cookstoves in many communities because of cultural factors regarding the appearance of food and their effects on cooking practices as well.

EXAMPLE 8.2 Calculating Average Annual Rates of Change

When we want to analyze growth trends and patterns, a useful rate to know is the annual average rate of change over a defined period of time. Many real-world phenomena do not increase at a steady rate from year to year. Examples include population, the economy, energy use, and so on. Consider the following data on coal consumption.

	2010	2011	2012	2013
metric tons of coal consumed	100,000	125,000	137,500	145,750
Annual rate of increase	—	25%	10%	6%

Here we see that the increase in coal consumed from year to year is not constant. Often we would like to be able to discuss an average rate, especially when we are working with data that are projections into the future.

To calculate an average rate of change, also known as the compound average growth rate (CAGR), we use

$$CAGR = \left(\frac{Q_n}{Q_0}\right)^{\frac{1}{(t_n - t_0)}} - 1 \qquad (8.3)$$

where

Q_n = quantity at the end of the period

Q_0 = quantity at the beginning of the period

t_n = time at the end of period

t_0 = time at the beginning of the period

Using the data in the table above, the average annual growth rate of coal consumed from 2010 to 2013 is computed as

(a) $t_n - t_0 = 2{,}013 - 2{,}010 = 3$

(b) $CAGR = \left(\frac{145{,}750}{100{,}000}\right)^{1/3} - 1 = .1338 = 13.4\%$

We would therefore state that the average annual growth rate of coal consumption from 2010 to 2013 was 13.4% per year.

8.2.2 Energy Issues in High-Income Countries

While low- and middle-income societies are faced with varying degrees of energy poverty and its consequences, high-income nations such as the United States are challenged by energy security, energy efficiency and conservation, and their obligations to mitigate CO_2 emissions.

Although Americans tend to regard **energy security** in terms of U.S. dependence on foreign oil (Figure 8.7), as a concept energy security more generally refers to the ability of a nation to protect itself from the economic, political, and social disruptions of an interrupted supply of a critical energy resource, of the failure of an important energy infrastructure, or of rapid and steep changes in energy prices. Although such "shocks" do not occur frequently, when they do happen, the results can be devastating. To this end, countries attempt to buffer themselves and create more resilient energy systems in a variety of ways. The three basic strategies are to reduce dependence on fossil fuels, to decentralize electric power production into a more distributed system, and to develop larger domestic supplies of energy fuels.

A second strategic energy goal of high-income countries is to control the growth in demand for energy in their societies. There are three major reasons for this. First, nations not on a growth trajectory are less vulnerable to energy shocks due to supply disruptions, and conservation is one element of an energy security strategy. Second, when

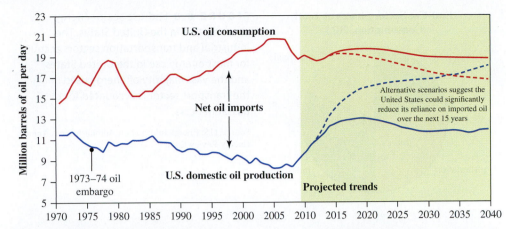

FIGURE 8.7 U.S. oil consumption and imports. President Jimmy Carter implemented policies in the late 1970s to try and limit U.S. oil imports to 50% of total American oil consumption as a way of protecting America from oil shocks. However, oil imports as a share of U.S. consumption increased steadily from 35% in the 1970s to a peak of 60% in 2005. Advances in technology and increases in the value of oil are making it more cost effective to extract "tight" oil deposits, such as in shale. The U.S. Department of Energy projections indicate that over the next 30 years the United States could significantly reduce its net oil imports to a small proportion of total consumption.

Source: Based on U.S. Department of Energy, Annual Energy Outlook 2013.

energy prices are high, a conservation strategy can help contain the economic impacts of escalating costs of fuel. The third reason, of greater importance for Europe than for the United States, is to reduce the greenhouse gas emissions associated with energy production and consumption. Not only have all of the European nations ratified the Kyoto Protocol on climate change (the United States has not), but the European Union also has set greenhouse gas reduction targets for the EU as a whole and for its member states.

8.2.3 Accounting for Societal Energy Consumption

National and international energy agencies have a standardized way of measuring and reporting societal energy consumption. The traditional approach is to account for the shares of total energy consumption accounted for by the four **end-use sectors** composed of the residential, commercial, industrial, and transportation sectors. The residential sector includes private households. The commercial sector represents business, government, and nonprofit enterprises engaged in services such as insurance, real estate, sales, tourism, education, and health care. The industrial sector represents mining, construction, electric power production, agriculture, and manufacturing, whereas the transportation sector involves the movement of passengers as well as freight. Figure 8.8 illustrates the proportion of total energy in the United States accounted for by the end-use sectors. In the United States, a 1% increase in population growth has historically resulted in a 1% increase in total energy use; however, due to aggressive efficiency and conservation measures, as well as the economic recession of 2008, energy growth now increases a little more slowly than U.S. population growth.

In the United States, the end-use sectors have distinctive patterns of fuel needs and uses. Figure 8.9 represents the flow of energy resources to various end-use sectors, and some notable patterns are evident. For example, 94% of all transportation energy comes from petroleum, and 71% of all petroleum used in the United States flows to the transportation sector. This suggests that not only is transport literally

End-Use Sector Shares of Total Consumption, 2011

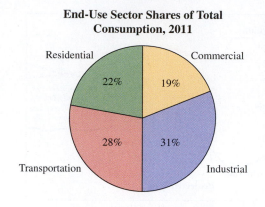

FIGURE 8.8 End-use sector energy consumption in the United States. The industrial and transportation sectors account for most energy use in the United States, and the vast majority of energy used in the transport sector is accounted for by passenger cars.

Source: U.S. Energy Information Administration, Annual Energy Review 2011.

the "driver" of American dependence on oil, but that transportation systems would be most seriously affected by any disruptions to the supply or price of petroleum. In addition, the data in Figure 8.9 calls attention to electric power consumption as an end-use form of energy. Here we see that 40% of all U.S. fuel consumption is for electricity, that 100% of all nuclear energy is for electricity, and that electric power accounts for more than 90% of our coal consumption. As a consequence, finding alternative ways to provide more sustainable electric power production (as well as more sustainable patterns in the demand for electricity) is of growing importance to Americans.

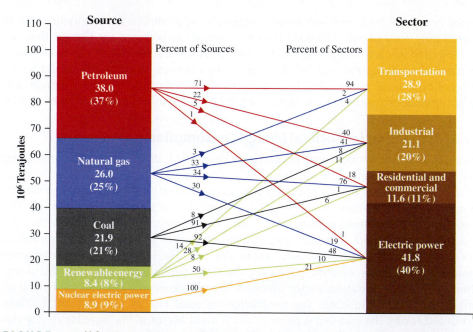

FIGURE 8.9 U.S. primary energy use by source and sector, 2011. This image is one of the most powerful graphical representations of U.S. energy use. The percentages for the bars in each stacked column show the total share of that item. For example, petroleum accounts for 37% of all energy sources, and transportation accounts for 28% of all end-use energy. The outflow arrows on the left indicate the percentage distribution of the resource to the end-use sectors; the inflow lines on the right indicate what percentage of the energy used by each sector comes from a specific source. The energy units in this infographic are million terajoules.

Source: U.S. Department of Energy, Energy Information Administration.

8.3 Energy and the Environment

The Industrial Revolution launched the "modern" energy era. Prior to the first Industrial Revolution of the 1700s, people met their energy needs in simple ways. In the absence of electricity and before steam power, most forms of physical energy were simple machines involving pulleys, wheels, gears, levers, and so on. Beasts of burden and human labor provided the primary forces for drawing carts and coaches, for turning grinding stones, and for lifting heavy objects. Where water or wind resources were abundant, their kinetic energy provided the force for milling grain through windmills and water wheels.

Fuels were low-density in their energy content and were consequently required in very large volumes to be effective. Traditional biomass—such as wood, charcoal, peat, dung, and grass—was the basic fuel for fire and heat. Animal fats and vegetable oils provided low-quality light. Virtually all fuels and lighting polluted indoor air, caused respiratory diseases, and contributed to failing eyesight. Localized deforestation due to overharvest of fuelwood and timber led not only to episodic energy crises in several parts of the world, but also resulted in desertification of vast regions of the Middle East. Indeed, it was the energy crisis associated with chronic wood shortages in Great Britain from about 1550 to 1700 that led to the transition from wood to coal as a primary fuel; this transition to coal was accompanied by significant innovations in manufacturing and mining, which then ultimately contributed to the Industrial Revolution itself. The basic impetus to the invention of mass production systems in textiles and manufactures, as well as to the development of the steam engine, was a national energy crisis!

The Industrial Revolution was a watershed in the intersection between human energy use and the environment because of the transition from low-energy biomass to high-density fossil fuels (Table 8.2). Today, as the demand for fossil fuels continues to grow, concerns about the availability of these fuels—oil in particular—has sharpened our focus on finite and renewable energy resources. The extraction, processing, and combustion of fossil fuels is also a major cause of air pollution and environmental degradation, and as we now know, burning fossil fuels for energy is the primary cause of anthropogenic CO_2 emissions.

TABLE 8.2 Energy density of common fuels

FUEL	ENERGY DENSITY (MJ/kg)
Peat	10.4
Wood, oak	13.3–19.3
Wood, pine	14.9–22.3
Coal, anthracite	25.8
Charcoal	26.3
Coal, bituminous	28.5
Fuel oil, no. 6 (bunker C)	42.5
Fuel oil, no. 2 (home heating oil)	45.5
Gasoline (84 octane)	48.1
Kerosene	45.6
Natural gas (density = 0.756 kg/m³)	53.0

Source: Based on Davis and Masten, p. 293.

Discussions of fossil fuel energy and resources can be confusing because data are often presented in terms of energy content, energy density, physical volume, and/or mass. Complicating matters is the fact that non-SI and nonmetric units are frequently used to analyze fossil energy data. Examples include a barrel of oil, a ton of oil equivalent, a therm of natural gas, a quad of energy consumed, short tons of coal, and so on. Table 8.3 provides the most commonly used quantities of energy as well as the English units that are conventionally used in the United States and sometimes elsewhere. *We cannot caution you enough* to take extra care in understanding the quantities and units of energy data that are used outside of the scientific context, especially in government and industry reports. Whereas scientific analyses rather reliably use SI units, U.S. government, industry, and nongovernmental organization (NGO) global energy analyses may use a variety of SI, metric, and alternative (but common) measurements.

TABLE 8.3 Energy conversion units

COMMONLY USED QUANTITIES OF ENERGY AND ENERGY CONTENT

1 British thermal unit (Btu)	The amount of energy required to heat one pound mass of water by one degree Fahrenheit; 1 Btu = 1055 joules (J)
1 megajoule (MJ)	948 Btu
1 kilowatt hour (kWh)	3,412 Btu
1 horsepower (HP)	746 watts
1 barrel of crude oil (bbl)	42 U.S. gallons = 5,100,000 Btu
1 ton of oil equivalent (toe)	41.868 GJ
1 therm	100 ft³ natural gas (ccf) = 103,100 Btu
1 gallon of gasoline	125,000 Btu
1 gallon of #2 fuel oil	138,690 Btu
1 gallon of LP gas	95,000 Btu
1 ton of coal	25,000,000 Btu
1 ft³ natural gas	1,031 Btu

CONVENTIONAL ENGLISH UNITS FOR PRESENTING ENERGY DATA, ESPECIALLY IN THE UNITED STATES

MMBtu	1 million Btu (Note that in the SI System, M = 1,000,000)
1 Quad	10^{15} Btu (a quadrillion Btu)
ccf	100 ft³
mcf	1,000 ft³
ton	2,000 lb (also known as a short ton)
long ton	2,240 lb

EXAMPLE 8.3   Calculating Fuel Equivalencies

The significance of fossil fuels as an energy source can best be understood by comparing them to the fuel that they replaced—wood. While we are concerned about the climate impacts of CO_2 emissions today, it is important to remember that the fossil fuels replaced wood as our primary energy source and that energy crises due to the lack of wood occurred throughout human history. Widespread deforestation and other environmental degradation occurred in many parts of the world because wood was the sole source of heat energy for local societies.

1. How much red oak is required to provide the equivalent amount of energy as a ton of bituminous coal, if 1 kg of red oak has an energy density of 14.9 MJ?

   Using the equivalencies in Tables 8.2 and 8.3,

   $$1 \text{ ton of bituminous coal} = 2{,}000 \text{ lb} \times \frac{.454 \text{ kg}}{1 \text{ lb}} = 908 \text{ kg} \times \frac{28.5 \text{ MJ}}{1 \text{ kg}} = 25{,}878 \text{ MJ}$$

   The energy density of red oak is 14.9 MJ/kg. To continue using the given equivalencies,

   $$25{,}878 \text{ MJ of coal} = \frac{1 \text{ kg red oak}}{14.9 \text{ MJ}} = 1{,}736.8 \text{ kg red oak}$$

   From these calculations, we see that it takes 1,737 kg of red oak to provide the same amount of energy as 1 ton (908 kg) of bituminous coal; we need almost double by mass of oak.

2. In addition to the amount of the resource, it would be interesting to know how much space is required to store our fuel. So, how much space would 1,737 kg of red oak take up compared to 1 ton of coal? A ton of coal by volume is about 38.5 ft³. (This is a cube a little over 1 meter in length per side). By volume, 20 kg of red oak is equivalent to 1 ft³. To calculate,

   $$1{,}737 \text{ kg red oak} = \frac{1 \text{ ft}^3}{20 \text{ kg}} = 86.85 \text{ ft}^3$$

   By volume, red oak thus requires more than double the space as its equivalent amount of coal, which as given is 38.5 ft³.

8.3.1 Energy and Natural Resources

The Earth's natural energy resources are generally categorized as **finite** and **renewable**. Finite resources—in particular the fossil fuels and uranium—are those that took millions of years to form and can only be regenerated on a geological timescale, if at all (Box 8.2). Renewable natural resources are those that are replenishable

BOX 8.2 How Long Did It Take for Fossil Fuels to Form?

The history of the Earth represents deep time and is measured on a geological timescale. Key changes in our planet's climate, atmospheric content, living species, and landscape are demarcated by chronological eons and eras that last millions of years. The three fossil fuels (coal, oil, and natural gas) were created over millions of years by the transformation of highly concentrated amounts of organic matter. Coal formations are about 300 million years old, and the median age of oil fields is about 65 million years, although large oil fields in Siberia have been dated at 650 million years. The conditions required to replicate large-scale fossil fuel formation are not generally present on Earth today.

Fossil fuels originated during the Carboniferous Era, which did not have the biological decomposition of the carbon cycle (see Chapter 3) of our current geological era. As a consequence, organic matter today is cycled before dense amounts of carbon can accumulate to be transformed by fossilization into a fuel.

If we converted the planet's 4.5 billion year history to a 24-hour clock, then biologically modern humans have existed on Earth less than 3 seconds of this 24-hour period. In turn, the Industrial Revolution represents only 6/1,000 of a second of that time. As a species, we have consumed, in a fraction of a second, finite resources that took "hours" to form.

within a human lifetime and include resources such as biomass and surface waters. **Infinitely renewable resources** are those that derive from the physical workings of our planet and solar system and are inexhaustible in their supply. These would include, for example, solar, wind, tidal, and geothermal sources of energy. Figure 8.10 demonstrates the contributions of different finite and renewable energy resources to total world energy consumption today; renewables account for less than one-fifth of world energy use, and many of these resources (such as biomass) are being used unsustainably.

The terms *occurrences*, *resources*, and *reserves* are used to discuss and analyze the available supply and recoverability of an energy fuel (Figure 8.11). An occurrence is simply the physical presence of a substance in a form that we would recognize. **Resources** are a subset of occurrences and reflect deposits of sufficient quantity that they may be recovered, although we might not be able to do so profitably. Resources include deposits that have not yet been discovered, so they are also speculative; Resource estimates try to account for the available supply of a substance at present and in the future. A **reserve** is the most reliable calculation of the availability of a substance. It is a measure that reflects (1) a high degree of certainty about the location and amount of occurrences, (2) occurrences that are recoverable using existing technologies, and (3) economically profitable recovery of the substance.

No standardized classification system exists for all energy substances. Each resource (such as coal, uranium, gas, and petroleum) has evolved its own terminology and methods for quantifying the amount and availability of a fuel. This is due to the very different geophysical properties and context of each substance. However, to be counted as a proven or recoverable reserve, the geological deposit of a resource has to be both technically recoverable and economically cost effective to extract, regardless of the type of resource. Table 8.4 provides estimates of global coal, oil, and natural gas reserves.

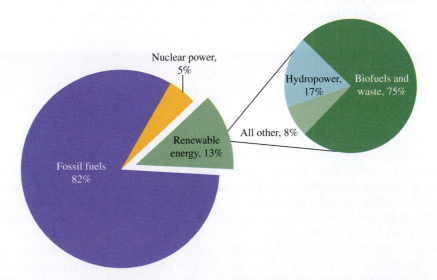

FIGURE 8.10 Total world primary energy production by source, 2011. Total world primary energy production is still dominated heavily by the fossil fuels. The role of renewable resources is increasing, but solar, wind, and geothermal energy still represent only about 1% of total primary production and about 8% of the total renewable energy supply.

Source: Based on International Energy Agency, Key World Energy Statistics 2013.

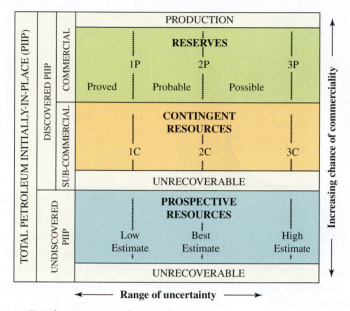

FIGURE 8.11 Classification system for petroleum reserves. The Society of Petroleum Engineers developed the classification system for estimating petroleum reserves, which are classified as either proved, probable, or possible. Note that undiscovered deposits are referred to as "prospective *resources*." Even though estimates are made of the amounts of these resources, they are not counted in estimates of the available supply of petroleum. Proved reserves are those that are technically and economically recoverable and form the basis of petroleum supply and depletion rates.

Source: Based on Society of Petroleum Engineers, Petroleum Resource Management System 2007. http://www.spe.org/industry /docs/Petroleum_Resources_Management_System_2007.pdf#redirected_from=/industry/reserves/prms.php

TABLE 8.4 Estimates of global coal, oil, and natural gas reserves

RESOURCE	RESERVE	RESERVE-TO-PRODUCTION RATIO
Liquid fuels	1,471.2 billion barrels of proved reserves (2011)	47.0 years*
Natural gas	190 trillion cubic meters of proven reserves (2011)	60.2 years
Coal (all types)	860 billion metric tons of recoverable reserves (2008)	126.3 years

*Estimated by author from 2008 production levels and 2011 reserves. The U.S. Department of Energy cautions against calculating reserve-to-production ratios for liquid fuels because of the volume of proven resources that become available each year.

Source: Based on U.S. Department of Energy, Energy Information Agency, International Energy Outlook 2011.

Table 8.4 also presents the **reserve-to-production ratio**. This number represents the amount of time it would take to deplete current proved reserves if the current yearly *amount* of consumption (production) was held constant. In reality, new proven reserves are added each year as resources become scientifically proven and as it becomes more cost effective or technically possible to extract resources.

The **Hubbert Curve** (Figure 8.12) is a concept that lets us analyze the different rates of growth in demand for a fossil fuel and the rate at which new reserves become available. The Hubbert Curve is often mistakenly interpreted as a model of available oil, coal, or gas reserves and as a representation of the depletion rate of

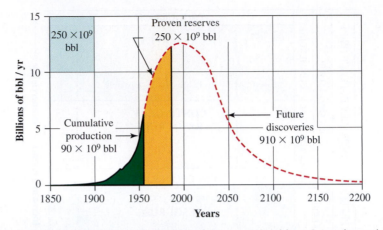

FIGURE 8.12 The Hubbert Curve. This image of the original Hubbert Curve shows the time at which the *rate of increase* in the production (or supply) of a resource reaches its peak. This is not the same thing as the rate of depletion or absolute depletion of the resource. The peak of the curve is a signal that our ability to supply the resource at a pace that can keep up with growing demand has reached its maximum limit.

Source: Based on Hubbert (1956).

the **stock** of those reserves. The amount of a resource—its reserve—represents the available stock of the resource.

The Hubbert Curve most intensively discussed is **peak oil**, which represents not the stock of oil, but the **flow rate** of oil into the economy. Very simply, the Hubbert model explains why the *rate* of flow (basically the same concept as the rate of extraction or the rate of production) of a finite resource cannot increase indefinitely. The rate of increase in production must ultimately slow down, stabilize, and then decrease. The timing of the peak, which is the point at which the maximum rate of increase in productive output occurs, matters. After that point in time, if demand for the resource is still growing at an increasing rate, then supply cannot keep up with demand, prices will increase, and demand will have to change. Peak oil is not a prediction that we are running out of oil, but an indication that major economic, social, and technological changes in the supply and demand for oil and its energy services are imminent. In the United States, we see the political stress of peak oil dynamics in the controversy about whether or not to drill for oil in the Arctic National Wildlife Refuge (Box 8.3).

Methane hydrate, which is a form of natural gas trapped in ocean sediments and Arctic permafrost, is a substance under intense investigation as a potential energy source. Methane hydrates are believed capable of supplying anywhere from 350 to 3,500 years of energy; however, we do not yet know how to reliably characterize this resource geologically, how to extract and process it, or what such costs would involve. Consequently, we are currently unable to estimate methane hydrate *reserves* that are available on Earth and the amount of energy that they could supply.

Uranium is an unusual energy resource because it is regarded by some to be a renewable, and not finite, resource. Uranium is a metal found in ore deposits and must undergo significant physical and chemical change in order to become a usable fuel for thermoelectric power production. Because spent nuclear fuel can be reprocessed and used again, many regard it as a renewable resource. However,

BOX 8.3 To Drill or Not to Drill?

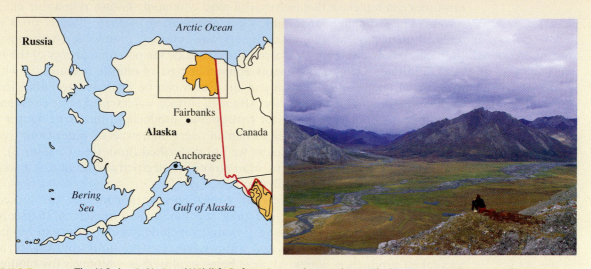

FIGURE 8.13 The U.S. Arctic National Wildlife Refuge. Located in northeast Alaska, ANWR is the source of political controversy because of its ecological significance and the presence of oil.

Source: U.S. Fish and Wildlife Service; Steven Chase, U.S. Fish and Wildlife Service.

The U.S. Arctic National Wildlife Refuge (ANWR) is managed by the U.S. Fish and Wildlife Service and is a federally protected conservation area in northeastern Alaska. Comprising 7.7 million hectares, ANWR is a pristine wilderness that lies north of the Arctic Circle (Figure 8.13); its landscape and habitats represent all six of the ecological regions of the circumpolar north. Biodiversity within the Reserve is considerable, and it hosts hundreds of bird species, major mammals (including the charismatic polar bears, arctic fox, and caribou), and fragile ecosystems. ANWR is valued aesthetically as an undisturbed wilderness and for its role in supporting unique ecological diversity.

At issue in ANWR are 608,000 hectares known as Section 1002. This tract of land has been exempted by Congress from full wilderness protection because of its potential for oil development. On the North Slope of the Brooks Range on the Beaufort Sea, Section 1002 is near the proven (and rich) commercial Alaskan oil fields. Its potential for oil and its extraordinary beauty have made ANWR a political hotspot, reflecting competing interests in the environment, energy security, and economic development. When the price of oil and gasoline go up, the focus on ANWR is renewed. At present, no drilling is allowed in Section 1002, but that could change with an act of Congress. The environmental impacts of different drilling strategies are highly contested.

Oil in ANWR is a prospective resource and not a proven reserve, and estimated amounts are based on a U.S. Geological Survey (USGS) study in 1998. The USGS estimates that there are 5.7 to 16 billion barrels of technically recoverable oil in ANWR, which represents less than 15% of all technically recoverable, undiscovered, prospective resources of the United States. The U.S. Department of Energy indicates that the potential oil in ANWR is equivalent to about two years or less of U.S. petroleum consumption and would have negligible effects on U.S. oil imports. The Department of Energy's analysis indicates that ANWR oil would reduce the cost of oil by 41 cents to $1.44 a barrel, depending on the amount available. The price of oil has fluctuated in recent years between $50 and $130 per barrel.

there are physical limits to the number of times the fuel can be reprocessed economically, so it is best to regard uranium as a recoverable/recyclable *finite* natural resource.

In contrast to finite energy resources that have a fixed stock, renewable energy resources can replenish themselves. The most commonly known renewable energy resources are infinitely renewable—our exploitation of these "fuels" does not diminish their physical abundance in any way. Infinitely renewable energy results from nature's own geophysical forces. These include the potential energy of the sun, wind, and tides. Geothermal energy is another type of infinite resource, in which the natural temperature gradients of the Earth can be exploited technologically for heating, cooling, and electricity. Although surface waters used to be considered an infinite renewable resource for hydropower production, this is no longer so. Localized climate change, as well as human overwithdrawals of water from rivers, have changed hydrological cycles as well as watershed flows (see Chapter 2). Of the infinitely renewable resources, wind power for electricity is the fastest growing energy application worldwide (Figure 8.14).

Other important renewable energy resources are not infinite, but are considered to be renewable because they regenerate and replenish themselves generally within a human lifetime. Examples include a wide variety of biomass fuels such as cultivated grasses and willows, bagasse (the pulp from sugarcane), and forest products residues. Algae is being explored as a biomass for biodiesel because of its high oil content. Indeed, biofuels are becoming increasingly important, and include methane from landfill decomposition (referred to as biogas) and ethanol as well as biodiesel. Surface water for hydropower is also a regenerative renewable resource.

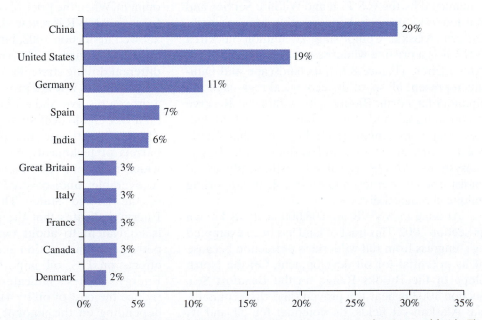

FIGURE 8.14 Wind power is the fastest growing renewable energy application worldwide. The 10 countries with the largest installed capacity and their shares of total installed wind capacity are presented here. These 10 nations account for 85% of all wind energy systems in the world.

Source: Based on Global Wind Energy Council, Global Wind Statistics 2013, http://www.gwec.net/wp-content /uploads/2014/02/GWEC-PRstats-2013_EN.pdf

EXAMPLE 8.4 How Much Usable Energy Can We Get from the Sun?

The amount of usable energy that we can get from a renewable resource varies by site, climate, and technology. In order to reliably estimate the productive energy that we can get for electricity, heat, or other work, we have to conduct what is called a "resource characterization." Solar energy is one of the easier renewable resources to evaluate because solar radiance is a relatively predictable quantity and has been mapped for all parts of the globe. The challenge is in accounting for the intermittency of solar energy—it varies by *diurnal phase* because of day and night, by season due to the angle of the Earth, and by climate because of variations in precipitation and cloudy days.

The electric power output of a solar photovoltaic panel can be calculated using the concept of a **peak sun-hour** (or just "sun-hours") to account for intermittency. Throughout the world, solar energy is at its most intense at solar noon in the summertime, when the sun is perpendicular to a surface. A peak sun-hour is therefore equivalent to the amount of solar energy over a period of 1 hour at solar noon in the summer, where this energy is striking a 1-square-meter area. This amount of energy is equivalent to 1,000 watt-hours. By adding up all of the available solar radiation over the course of a day and throughout the year, we can construct the average annual peak sun-hours for every location around the world. This average basically tells us how many hours of maximally productive sunlight is available to a solar panel each day at a specific location.

Solar PV panels are rated by their theoretically maximum output capacity. A 2-kW PV panel would produce 2 kW of electricity at its maximum output, which would be during peak solar radiance. A 2-kW PV panel exposed to peak solar radiance for one hour would therefore generate 2 kWh of electricity.

If you know the average number of peak sun-hours in a location, you can then readily estimate the average output of a solar panel for a year:

Solar PV output = rated capacity of the panel × average daily peak sun hours × 365

(8.4)

A 500-W PV panel in San Antonio, Texas, where the average daily peak sun-hours are 5.3, would produce 967.25 kWh per year, as

$$\text{PV output} = 500 \text{ W} \times \frac{1 \text{ kW}}{1,000 \text{ W}} \times \frac{5.3 \text{ hours}}{\text{day}} \times \frac{365 \text{ days}}{\text{year}} = 967.25 \text{ kWh}$$

8.3.2 Energy and Environmental Degradation

All energy use has environmental consequences. Even renewable energy, which comes from "clean" natural resources, cannot be harnessed for human use without environmental impacts. For example, hydroelectric dams have had a significant impact on the ecosystems of the U.S. Pacific Northwest because they interfere with salmon spawning; some solar panels contain heavy metals; and biomass creates air contaminants when burned. Not only must we consider the energy fuel used, but the technologies that are used to convert that fuel into usable forms of energy. Very simply, when we convert natural energy resources into usable forms of human energy for transportation, electricity, building heat, and so on, we degrade landscapes, disrupt wildlife, and generate a wide variety of pollutants and wastes that must be disposed of. As seen in Chapters 5 and 6, pollution from fossil fuels is considerable, and the negative impacts of fossil energy were known long before CO_2 emissions and climate change became an issue (Figure 8.15).

FIGURE 8.15 Smog in Donora, Pennsylvania, 1948. This image shows the factory town of Donora, Pennsylvania, at mid-day in 1948 during a smog event. The air pollution from local mills and metal furnaces was trapped in the atmosphere, killing two dozen people and making thousands ill. Similar events were experienced in London, England, where the often-chronicled "London fog" was not really fog at all, but a blanket of air pollutants. In both countries, these events led to the development of the first clean air laws.

Source: Pittsburgh Post-Gazette

In order to obtain energy for people, the Earth's natural energy resources must be converted from a raw material to a usable form of energy in a process that we can think of as the life cycle for that fuel. As explained in Chapter 7, the life cycle of a material or product involves all the stages of its production process, including resource extraction, material processing, fabrication, shipment, and disposal. A life cycle assessment evaluates the critical environmental impacts of that product throughout its life cycle.

For the fossil fuels and uranium, significant environmental contamination is associated with the life cycles of these fuels (Figure 8.16). The extraction and refining of coal and uranium creates ore and mine wastes that can acidify water supplies and degrade soils, landscapes, and habitats. Coal mining that involves mountaintop removal has become a politically contentious issue in eastern coal regions of the United States; oil spills continue to shock and disturb communities and nations; and fuel refining complexes emit noxious fumes. In nations such as Nigeria, violent conflict has broken out because of the environmental degradation created by global oil corporations. The generation of solid hazardous waste throughout the life cycles of all the fossil fuels and uranium is an inescapable feature of these energy resources.

The combustion of fossil fuels also results in a wide variety of air pollutants, including sulfur dioxide (a health threat as well as a precursor to acid rain), nitrogen oxides (the precursors to smog and associated respiratory diseases), particulate

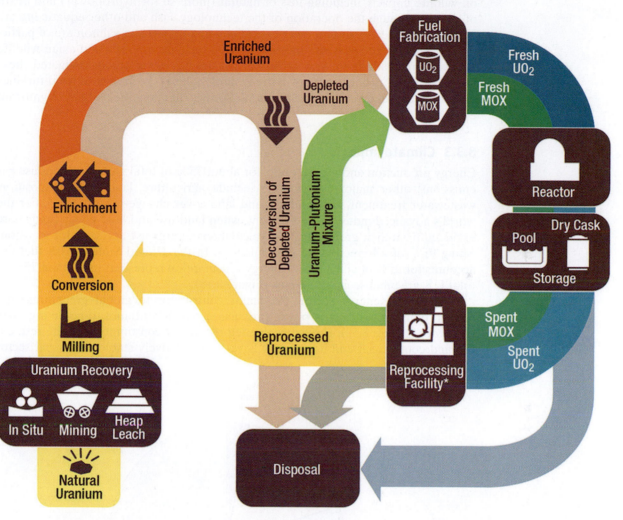

The Nuclear Fuel Cycle

FIGURE 8.16 The life cycle of uranium. Nuclear power is often considered clean energy because nuclear power plants emit no air pollutants. However, when the entire fuel cycle of uranium is taken into account, a different picture of nuclear power emerges. To provide fuel for electricity, raw uranium ore is processed into enriched uranium, used in nuclear reactors until it is spent, and then disposed of. Radioactive and highly toxic waste is created at all stages of uranium fuel fabrication and use, presenting challenges for safe disposal throughout the process.

Reprocessing of spent nuclear fuel including MOX is not practiced in the U.S.
Note: The NRC has no regulatory role in mining uranium.
Source: U.S. Nuclear Regulatory Commission, http://www.nrc.gov/materials/fuel-cycle-fac/stages-fuel-cycle.html

matter (a contributor to heart disease), mercury (a heavy metal with neurological effects), and carbon monoxide (a poisonous gas). These critical pollutants are all regulated in the United States and Europe in order to achieve safer levels of emissions, but this is not necessarily so in all countries.

As previously mentioned, even renewable energy has its consequences. With respect to solar photovoltaic panels, much of the concern is over the materials content and manufacturing of the panels as well as their safe end-of-life disposal. Issues associated with wind energy and hydroelectric power focus more intensively on wildlife impacts, including loss of habitat (more so for hydropower) and death of wildlife through the operation of the technology. Fish and other aquatic life are primarily affected by hydropower operations, and impacts on salmon are of particular concern. In the past, the impacts of wind turbines on birds and avian wildlife received great attention; new technological designs have largely mitigated these impacts. Of great concern today in the United States is the impact of wind turbines on bat populations, a rather uncharismatic species that nonetheless plays significant ecological roles.

8.3.3 Climate Impacts

Energy production and use accounts for about 75% of total global greenhouse gas emissions; other major contributors include agriculture, landfill decomposition, wastewater treatment, and land-use and land-cover change (Figure 8.17). For the world's advanced industrialized nations, when land-use and land-cover change are excluded from their greenhouse gas calculations, energy use accounts for an astonishing 91% of all greenhouse gas emissions. In the United States, fossil fuel use accounts for 95% of all CO_2, and energy consumption represents more than 75% of total U.S. greenhouse gas emissions (Figure 8.18).

Within the energy sector, the major contributor to global greenhouse gas (GHG) emissions is electric power production, which continues to be on a growth trajectory (Figure 8.19). Emissions created by energy consumption in the residential and commercial end-use sectors have been relatively stable; industrial sector

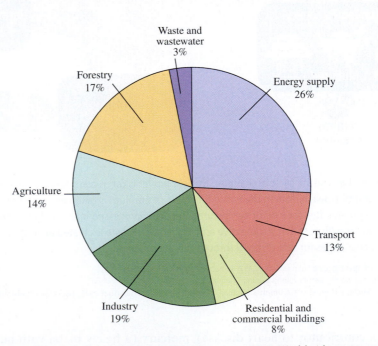

FIGURE 8.17 Global greenhouse gas emissions by source, 2004. Worldwide, energy use accounts for the majority of all greenhouse gas emissions. Virtually all of the emissions from the transport, energy supply, industry, and buildings sectors result from energy consumption and production in those sectors.

Source: U.S. Environmental Protection Agency, http://www.epa.gov/climatechange/ghgemissions/global.html

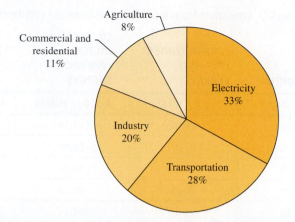

FIGURE 8.18 Sources of U.S. 2011 greenhouse gas emissions. In the United States, energy consumption for electric power production, transportation, homes, and businesses accounts for most of U.S. greenhouse gas emissions, about 75% of the total. A portion of industrial emissions are also due to direct energy use by industry for heat and steam.

Source: U.S. Environmental Projection Agency, http://www.epa.gov/climatechange/ghgemissions/sources.html

emissions fluctuate with global economic cycles and are not rapidly growing. Road transportation is on a strong upward growth trend; controlling CO_2 emissions in the transportation and electric power sectors is therefore critical to achieving the international community's goals for mitigating climate change. Because of the chemical character of fuels and the nature of combustion technology, fossil fuels emit different levels of CO_2. Table 8.5 lists the most commonly burned fuels, their energy density, and the amount of CO_2 emitted in kilograms per million BTU of fuel burned for the United States; such CO_2 emissions factors are usually calculated for countries or regions because of geographic variations in fuel chemical composition and the nature of combustion technologies used.

Carbon footprint techniques are used to calculate both the GHG emissions for different activities and the emissions that are avoided by different carbon mitigation efforts. Institutions and policy makers may use the carbon footprint to track changes in greenhouse gas emissions and potential climate impacts over time. Typically an

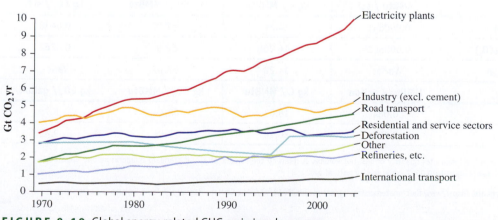

FIGURE 8.19 Global energy-related GHG emissions by source.

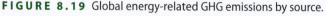

Source: Based on Intergovernmental Panel on Climate Change, 2007 Climate Inventory.

TABLE 8.5 U.S. default factors for calculating CO_2 emissions from fossil fuel and biomass combustion

FUEL TYPE	HEAT CONTENT	CARBON CONTENT (PER UNIT ENERGY)	CO_2 EMISSION FACTOR (PER UNIT ENERGY)	CO_2 EMISSION FACTOR (PER UNIT MASS OR VOLUME)
Coal and Coke	**MMBtu / short ton**	**kg C / MMBtu**	**kg CO_2 / MMBtu**	**kg CO_2 / short ton**
Anthracite	25.09	28.24	103.54	2,597.82
Bituminous	24.93	25.47	93.40	2,328.46
Subbituminous	17.25	26.46	97.02	1,673.60
Lignite	14.21	26.28	96.36	1,369.28
Coke	24.80	27.83	102.04	2,530.59
Mixed electric utility/electric power	19.73	25.74	94.38	1,862.12
Mixed industrial sector	22.35	25.61	93.91	2,098.89
Natural Gas	**Btu / scf**	**kg C / MMBtu**	**kg CO_2 / MMBtu**	**kg CO_2 / scf**
U.S. Weighted Average	1,028	14.46	53.02	0.0545
Petroleum Products	**MMBtu / gallon**	**kg C / MMBtu**	**kg CO_2 / MMBtu**	**kg CO_2 / gallon**
Distillate fuel oil no. 1	0.139	19.98	73.25	10.18
Distillate fuel oil no. 2	0.138	20.17	73.96	10.21
Residual fuel oil no. 6	0.150	20.48	75.10	11.27
Kerosene	0.135	20.51	75.20	10.15
LPG	0.092	17.18	62.98	5.79
Propane (liquid)	0.091	16.76	61.46	5.59
Motor gasoline	0.125	19.15	70.22	8.78
Aviation gasoline	0.120	18.89	69.25	8.31
Kerosene type jet fuel	0.135	19.70	72.22	9.75
Asphalt and road oil	0.158	20.55	75.36	11.91
Crude oil	0.138	20.32	74.49	10.28
Biomass Fuels—Gaseous	**MMBtu / scf**	**kg C / MMBtu**	**kg CO_2 / MMBtu**	**kg CO_2 / scf**
Biogas (captured methane)	0.000841	14.20	52.07	0.0438
Landfill gas (50% CH_4/50%CO_2)	0.0005025	14.20	52.07	0.0262
Wastewater treatment biogas	Varies	14.20	52.07	Varies
Biomass Fuels—Liquid	**MMBtu / gallon**	**kg C / MMBtu**	**kg CO_2 / MMBtu**	**kg CO_2 / gallon**
Ethanol (100%)	0.084	18.67	68.44	5.75
Biodiesel (100%)	0.128	20.14	73.84	9.45

Note: scf denotes "standard cubic feet." scf measures quantity of gas, not volume. It refers to the amount of gas at standard temperature and pressure. One standard cubic foot of natural gas is equal to 0.0027 cubic feet of natural gas.

Source: Based on 2012 Climate Registry, http://www.theclimateregistry.org/downloads/2012/01/2012-Climate-Registry-Default-Emissions-Factors.pdf

institution or organization chooses a baseline year for emissions and calculation of their carbon footprint. The potential climate benefits of investments in resource conservation and green technologies can be tracked over time. For instance, if companies shift to a renewable energy portfolio, they can calculate the potential impact of these changes and decisions such as investing in forest protection, planting trees to increase carbon sequestration, or investing in renewable energy. Box 8.4 and Examples 8.5, 8.6, and 8.7 illustrate various techniques for calculating carbon footprints and offsets.

BOX 8.4 Calculating Carbon Footprints

The carbon footprint is the sum of all greenhouse gas (GHG) emissions associated with an individual, organization, or product expressed in CO_2 equivalence. Greenhouse gas emissions other than carbon dioxide are expressed in terms of their CO_2 equivalence based on their 100-year Global Warming Potential.

GHG emissions are calculated using standards determined from approved inventories such as the Global Greenhouse Gas Protocol (World Business Council for Sustainable Development and World Resources Institute, 2004) and the U.S. EPA Environmental Footprint Methodology. The carbon footprint should include direct emissions (scope 1), indirect emissions related to electricity consumption (scope 2), and indirect emissions associated with other products and materials (scope 3), including emissions associated with transportation, waste disposal, and consumer goods as illustrated in Figure 8.20. Data for global GHG emissions and production of goods can be found from various sources, including the Food and Agriculture Organization of the UN (FAOSTAT), the UN Commodity Trade Statistics Database (UN Comtrade), the International Energy Agency Statistics (IEA), and the Ecological Footprint Atlas. Table 8.6 provides examples of several types of industrial production and other facilities for which specific tools are used to calculate GHG emissions.

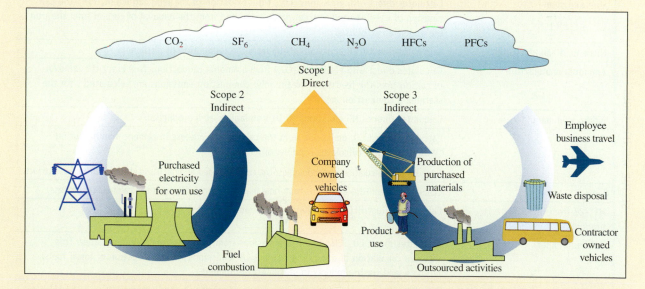

FIGURE 8.20 Direct and indirect GHG emissions for a value chain. This illustration shows how GHG emissions occur from the direct and indirect activities related to producing goods and services.

Source: Based on World Business Council for Sustainable Development and World Resources Institute (2004) The Greenhouse Gas Protocol: A Corporate Accounting and Reporting Standard, Revised Edition. (Washington, DC: World Resources Institute), Figure 3, p. 26.

(Continued)

BOX 8.4 Calculating Carbon Footprints (*Continued*)

TABLE 8.6 Greenhouse gas (GHG) emission calculation tools available online for calculating standardized emission rates for GHGs from various sectors

	CALCULATION TOOLS	MAIN FEATURES
CROSS SECTOR TOOLS	Stationary combustion	Calculates direct and indirect CO_2 emissions from fuel combustion stationary equipment
		Provides two options for allocating GHG emissions from a co-generation facility
		Provides default fuel and national average electricity emission factors
	Mobile combustion	Calculates direct and indirect CO_2 emissions from fuel combustion in mobile sources
		Provides calculations and emission factors for road, air, water, and rail transport
	Hydrofluorocarbons (HFCs) from air conditioning and refrigeration use	Calculates direct HFC emissions during manufacture, use, and disposal of refrigeration and air-conditioning equipment in commercial applications
		Provides three calculation methodologies: a sales-based approach; a life cycle-,stage-based approach; and an emission factor-based approach
	Measurement and estimation uncertainty for GHG emissions	Introduces the fundamentals of uncertainty analysis and quantification
		Calculates statistical parameter uncertainties due to random errors related to calculation of GHG emissions
		Automates the aggregation steps involved in developing a basic uncertainty assessment for GHG inventory data
SECTOR SPECIFIC TOOLS	Aluminum and other nonferrous metals production	Calculates direct GHG emissions from aluminum production (CO_2 from anode oxidation, PFC emissions from the "anode effect," and SF_6 used in nonferrous metals production as a cover gas)
	Iron and steel	Calculates direct GHG emissions (CO_2) from oxidation of the reducing agent, from the calcination of the flux used in steel production, and from the removal of carbon from the iron ore and scrap steel used
	Nitric acid manufacturing	Calculates direct GHG emissions (N_2O) from the production of nitric acid
	Ammonia manufacture	Calculates direct GHG emissions (CO_2) from ammonia production. This is for the removal of carbon from the feedstock stream only; combustion emissions are calculated with the stationary combustion module
	Acidic acid manufacture	Calculates direct GHG emissions (N_2O) from adipic acid production
	Cement	Calculates direct CO_2 emissions from the calcination process in cement manufacturing (WBCSD tool also calculates combustion emissions)
		Provides two calculation methodologies: the cement-based approach and the clinker-based approach
	Lime	Calculates direct GHG emissions from lime manufacturing (CO_2 from the calcination process)
	HFC-23 from HCFC-22 production	Calculates direct HFC-23 emissions from production of HCFC-22
	Pulp and paper	Calculates direct CO_2, CH_4, and N_2O emissions from production of pulp and paper. This includes calculation of direct and indirect CO_2 emissions from combustion of fossil fuels, biofuels, and waste products in stationary equipment
	Semiconductor wafer production	Calculates PFC emission from the production of semiconductor wafers
	Guide for small office-based organizations	Calculates direct CO_2 emissions from fuel use, indirect CO_2 emissions from electricity consumption, and other indirect CO_2 emissions from business travel and commuting

Source: Based on World Business Council for Sustainable Development and World Resources Institute (2004) The Greenhouse Gas Protocol: A Corporate Accounting and Reporting Standard, Revised Edition. (Washington, DC: World Resources Institute).

EXAMPLE 8.5 Calculating the Carbon Footprint of an American Household

The U.S. EPA (2013) reported that, on average, an American home consumed the following energy resources in 2010. What is the simplified carbon footprint for average home energy use in the United States?

ENERGY SOURCE	AMOUNT CONSUMED
Delivered electricity	11,319 kWh
Natural gas	66,000 ft³
Gasoline	464 gallons
Fuel oil	551 gallons
Kerosene	108 gallons
Auto transportation (gasoline)	2 vehicles at 11,493 miles each
Auto fuel economy	21.5 mpg

The EPA uses a simplified method of estimating an individual's greenhouse gas emissions in their Household Carbon Footprint Calculator (*www.epa.gov/climate-change/ghgemissions/ind-calculator.html*). The carbon dioxide equivalent emission rates are estimated from the conversion factors in Table 8.7.

TABLE 8.7 USEPA carbon dioxide emission factors for household resource consumption

SOURCE	SOURCE UNITS	CO$_2$ CONVERSION FACTOR	CONVERSION FACTOR UNITS
Electricity	kWh	7.0555×10^{-4}	Metric tons CO$_2$/kWh
Coal	Railcar	232.74	Metric tons CO$_2$/90.89 metric ton railcar
Natural gas	ft³	5.44×10^{-5}	Metric tons CO$_2$/ft³
Oil	Barrel	0.43	Metric tons CO$_2$/barrel
Gasoline	Barrel	0.2913	Metric tons CO$_2$/barrel
Kerosene	Barrel	0.42631	Metric tons CO$_2$/barrel
Propane cylinders used for home barbecues	Cylinder	0.024	Metric tons CO$_2$/cylinder
Uptake by trees	Urban tree planted	0.039	Metric tons CO$_2$/tree
U.S. forests storing carbon for one year	1 acre of average U.S. forest	1.22	Metric tons CO$_2$/acre-year

Source: Based on US EPA. Clean Energy Calculations and References. Accessed May 15, 2013. (www.epa.gov/cleanenergy/energy-resources/refs.html)

To answer this question, the individual GHG emissions for each category are calculated and then summed to estimate the total carbon footprint for an average U.S. household.

Step 1: Calculate CO_2 emissions for each energy source using the emissions factors in Table 8.7.

Electricity usage:

$$11{,}319 \text{ kWh} \times 7.0555 \times 10^{-4} \text{ tons } \frac{CO_2}{\text{kWh}} = 7.99 \text{ metric tons } CO_2$$

Natural gas usage:

$$66{,}000 \text{ ft}^3 \times 5.44 \times 10^{-5} \text{ tons } \frac{CO_2}{\text{ft}^3} = 3.59 \text{ metric tons } CO_2$$

Home gasoline use:

$$464 \text{ gallons} \times \frac{\text{barrel}}{42 \text{ gallons}} \times 0.2913 \text{ tons } \frac{CO_2}{\text{barrel}} = 2.42 \text{ metric tons } CO_2$$

Home fuel oil use:

$$551 \text{ gallons} \times \frac{\text{barrel}}{42 \text{ gallons}} \times 0.43 \text{ tons } \frac{CO_2}{\text{barrel}} = 5.64 \text{ metric tons } CO_2$$

Home kerosene use:

$$108 \text{ gallons} \times \frac{\text{barrel}}{42 \text{ gallons}} \times 0.42631 \text{ tons } \frac{CO_2}{\text{barrel}} = 1.10 \text{ metric tons } CO_2$$

Gasoline for automobile transportation:

$$2 \text{ cars} \times \frac{11{,}319 \text{ miles}}{\text{car}} \times \frac{\text{gallon}}{21.5 \text{ miles}} \times \frac{\text{barrel}}{42 \text{ gallons}} \times 0.2913 \text{ tons } \frac{CO_2}{\text{barrel}} = 5.58 \text{ metric tons } CO_2$$

Step 2: Sum the emissions from all energy sources.

The simplified carbon footprint for the average U.S. home in 2010 is:

$$(7.99 + 3.59 + 2.42 + 5.64 + 1.10 + 5.58) \text{ metric tons } CO_2 = 26.32 \text{ metric tons } CO_2$$

EXAMPLE 8.6 **Estimating Avoided CO_2 Emissions**

We can calculate the CO_2 emissions avoided by using cleaner energy resources. The solar PV panel in Example 8.4 generates 967.25 kWh of electricity per year, carbon-free. How much CO_2 do we avoid by using the solar panel?

Step 1: Convert energy equivalences.

To estimate the avoided CO_2, we refer to the emissions factors in Table 8.5. From the available data, we see that we need to convert our kilowatt hour output into million Btu (MMBtu). Using the energy equivalencies provided in Table 8.3:

$$\text{Output of PV panel} = 967.25 \text{ kWh} \times \frac{3{,}412 \text{ Btu}}{1 \text{ kWh}} \times \frac{1 \text{ MMBtu}}{1{,}000{,}000 \text{ Btu}} = 3.3 \text{ MMBtu}$$

Step 2: Account for source energy losses.

Electric power plants, as **source energy**, are not 100% efficient. There are conversion losses in the generation and transmission of electricity. To get electricity to the end user as **site energy**, we will have to produce more than 100% of the site energy at the power plant. In the United States, conventional thermoelectric power plants powered by fossil fuels are only 34% efficient; that is, 66% of the heat content of the fuel is lost in the process of burning it, converting it to electricity, and distributing it to the end users. We therefore have to multiply our site energy by a factor of 3 to arrive at the source energy:

$$\text{Source energy} = 3.3 \text{ MMBtu} \times 3 = 9.9 \text{ MMBtu}$$

Step 3: Calculate the avoided emissions.

Referring to Table 8.5, we see that CO_2 emissions for a coal-fired electric power plant are 94.38 kg per MMBtu burned.

$$\text{Avoided } CO_2 = 9.9 \text{ MMBtu} \times \frac{94.38 \text{ kg}}{1 \text{ MMBtu}} = 934.4 \text{ kg } CO_2$$

Our solar panel therefore avoids 934.4 kg per year of CO_2.

EXAMPLE 8.7   Estimating a Portfolio of Carbon Offsets

By building on Examples 8.5 and 8.6, we can now estimate the reduction in GHGs from a portfolio of carbon mitigation initiatives. If electricity for the average U.S. home shifted to 50% fossil energy, 15% solar PV, 15% wind, and 20% nuclear energy and we also invested in 0.5 hectares of newly planted trees, what are the yearly offsets of carbon emissions?

Step 1: Calculate the emissions reductions from lower fossil fuel electricity.

In Example 8.5, we see that the total CO_2 emissions from electricity is 7.99 metric tons of CO_2 per year, which we assume to be entirely based on fossil energy. A 50% reduction in this emission rate would be 7.99/2, or 4.00 metric tons of CO_2.

Step 2: Calculate GHG emissions for non-fossil energy resources.

Renewable and nuclear energy resources do emit some GHGs as part of their electric power generation process or as a consequence of their embodied energy. We can calculate the rate at which GHGs are emitted from nonfossil energy resources by drawing on their life cycle GHGs averaged over their lifetime power output. This estimate yields GHG emissions produced per kilowatt hour generated (Table 8.8).

TABLE 8.8 Estimates of life cycle GHG emissions per kilowatt hour generated from renewable energy resources

ENERGY SOURCE	g CO_2 EQUIVALENTS/kWh	ENERGY SOURCE	g CO_2 EQUIVALENTS/kWh
Biopower	18	Wind energy	12
Solar (photovoltaic)	46	Nuclear energy	16
Solar (concentrating solar power)	22	Natural gas	469
Geothermal energy	45	Oil	840
Hydropower	4	Coal	1001
Ocean energy	8		

Source: Based on Moomaw, W., and others (2011), Table A.II.4.

Using data from Table 8.8:

Solar (photovoltaic) emissions:

$$11{,}319 \text{ kWh} \times 0.15 \times 46 \text{ g} \frac{CO_2}{\text{kWh}} \times \frac{1 \text{ metric ton}}{10^6 \text{ grams}} = 0.0781 \text{ metric tons } CO_2$$

Wind energy emissions:

$$11{,}319 \text{ kWh} \times 0.15 \times 12 \text{ g} \frac{CO_2}{\text{kWh}} \times \frac{1 \text{ metric ton}}{10^6 \text{ grams}} = 0.0204 \text{ metric tons } CO_2$$

Nuclear energy carbon equivalent emissions:

$$11{,}319 \text{ kWh} \times 0.20 \times 16 \text{ g} \frac{CO_2}{\text{kWh}} \times \frac{1 \text{ metric ton}}{10^6 \text{ grams}} = 0.0362 \text{ metric tons } CO_2$$

Annual carbon emissions from the nonfossil electricity generation are

$$C_{\text{nonfossil}} = (0.0781 + 0.0204 + 0.0362) \text{ metric tons } CO_2 = .1347 \text{ metric tons } CO_2$$

Step 3: Calculate total offsets.

The total carbon offset, including the offsets associated with tree planting, is

$$C_{\text{offset}} = \text{original GHG emissions} - (\text{fossil emissions} + \text{nonfossil emissions})$$
$$+ \text{ other offsets}$$

Given that 0.5 hectares of newly planted trees sequesters 1.22 metric tons of CO_2 per year,

$$C_{\text{offset}} = 7.99 - (4.00 + .1347) + 1.22 = 5.075 \text{ metric tons } CO_2 \text{ per year}$$

Each year, just over 6 metric tons of CO_2 would be avoided compared to previous emission profiles, or a 76% reduction in electricity-related GHGs.

8.4 Direct and Embodied Energy

Efforts to make our energy more sustainable focus on two kinds of energy: **direct energy** and **embodied energy**. Direct energy provides an energy service that we need. Examples include electricity for lighting, natural gas for industrial processes, and diesel for transportation. More sustainable direct energy results from conservation, efficiency, and substituting renewable or less polluting fuels for fossil fuels.

Embodied energy, also referred to as indirect energy, represents all of the energy required to fabricate, package, ship, and sell a product, and it is accounted for using life cycle analysis. This sort of analysis allows us to think more holistically about energy conservation by considering where and how products—over most of their life cycle—consume energy. Embodied energy also allows us to think about the role of recycling in energy conservation (Box 8.5), and our ability to recover the energy embedded in the materials of our manufactured products. Waste-to-energy facilities, for example, burn trash to make process steam or electricity.

Laws of physics govern the potential contributions of embodied energy to sustainability. Newton's first law of thermodynamics postulates that all energy is conserved in a closed system; however, the second law, because of principles of entropy, tells us that not all energy can be recovered because of the irreversibility of some

BOX 8.5 The Value of Recycling Aluminum

FIGURE 8.21

Source: Mikael Damkier/Shutterstock.com

Commercial aluminum is refined from aluminum oxide found in bauxite ore, and it is an extraordinarily energy-intensive process. Although a known substance since 1800, aluminum could not be mass produced because of both technological and energy-based barriers to the extraction of aluminum oxide from bauxite and to refining the metal itself. The engineering and economic breakthroughs arrived first with the invention of the electric arc furnace and second with the development of inexpensive hydroelectric power. Commercial aluminum manufacturing began in 1886 with hydroelectricity powered by Niagara Falls.

Metals are important from a sustainability perspective for two reasons. First, as natural resources, all metals are theoretically recoverable in terms of principles of physics and chemistry. As a consequence, it is indeed possible (theoretically) to have a closed-loop materials cycle for metals and metal products where no metallic waste is ever generated. Indeed, nonferrous metals are regarded as being infinitely recyclable without any serious degradation of materials quality or properties.

Second, metal manufacturing is one of the most energy-intensive industrial processes because most metals must be extracted from low-grade ores and also refined by sustained, intense heat. They should therefore be a prime candidate for energy conservation. Regrettably, there are both technical and economic barriers to fully recovering all metals used in human societies and closing the materials loop for these substances. Where technically possible and economically feasible, metal recycling is critically important. Aluminum, copper, tin, and steel have strong recycling markets, as do precious metals (gold, silver, platinum).

Aluminum is a fascinating metal for the scope of materials recovery that occurs worldwide, and because recycling aluminum requires only 5% of the energy needed to process aluminum from raw ore, a 95% energy savings. According to Alcoa, "almost 75% of all the primary aluminum ever produced since 1888 is still in productive use." And, in such industries as automobiles, construction, and electric power systems, the aluminum recycling rates are near 90%, and the U.S. aluminum recycling industry contributes $77 billion per year to the U.S. economy.

But the real story is in aluminum beverage cans (Figure 8.21). Beverage cans are an infinitely recyclable packaging material; in the United States, the recycled content of a can is 68%. Beverage cans are the most highly recycled container and are "the only packaging material that covers the cost of its own collection and reprocessing." The life of a recycled can is a mere 60 days from the time it is collected as scrap to the time it is back on the shelf as a beverage. Can recycling in a number of countries such as Brazil, Japan, and Switzerland exceeds 90%; Germany and Finland recycle cans at a rate over 95%. There is room for improvement in the United States, where we recycle about 60% of our aluminum cans. The aluminum industry has set a goal of 75% recycling by 2015.

processes. For example, once burned, the energy contained in a fossil fuel cannot be *fully* recovered and reconstituted to be exploited again.

The concept of embodied energy also enables us to consider the overall productivity of life cycle energy for an energy source. **Energy return on investment (EROI)** measures the usable energy from a source relative to the amount of energy required to produce the resource. The resulting EROI ratio should be greater than 1;

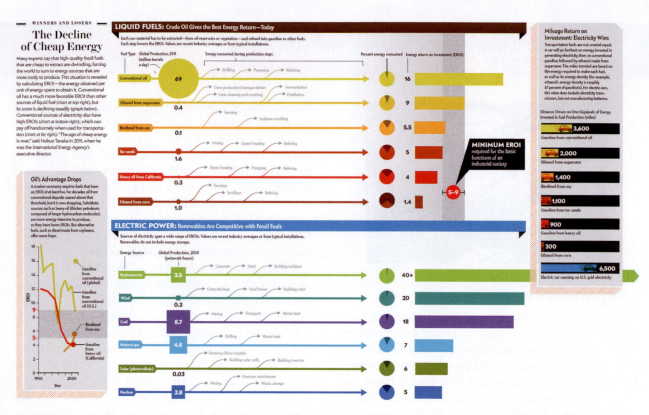

FIGURE 8.22 Energy return on investment for various fuels. Petroleum still provides the highest EROI. The amount of energy required to extract, process, and distribute petroleum returns more usable energy than any other source, with an EROI of 69.

Source: Mason Inman, "The True Cost of Fossil Fuels," Scientific American 308, 58–61. Reproduced with permission. Copyright © 2003 Scientific American, a division of Nature America, Inc. All rights reserved.

otherwise the energy expended to obtain a fuel or technology would be more than that created by the source over its life time. As seen in Figure 8.22, the popular impression that it takes more energy to manufacture a solar panel than it produces is not well founded.

8.5   Opportunities for Energy Sustainability

By integrating our understanding of human energy needs with (a) the nature of energy resources and their environmental impacts and (b) the concepts of direct and embodied energy, we now have the pieces we need to begin understanding the opportunities for more sustainable energy production and consumption. These opportunities are considerable and encompass conservation, efficiency, the development of renewable resources, energy recovery, fuel switching, and pollution prevention and control.

8.5.1 Energy Conservation and Curtailment

In common everyday language, energy conservation simply means some initiative to save energy and could refer to a wide variety of practices from better efficiency to turning off the lights. We therefore try to distinguish between **energy efficiency** and **energy curtailment**. The term *energy curtailment* represents efforts to reduce the *demand* for energy by changing people's need for energy services. We can think of curtailment as ways to prevent energy from being used when it is not needed (such as turning out the lights when a room is empty) or measures that prevent the loss of energy (such as better insulation in a building).

Many people confuse curtailment with energy efficiency. Efficiency refers to the performance of a piece of equipment that uses energy or transforms it from one state to another. It is important to keep these two concepts distinct from both an engineering design and an energy management perspective because they lead to very different opportunities for improved sustainability. Curtailment measures range from simple human behaviors such as taking a bus instead of driving a car to complex building automation systems that dynamically adjust building temperatures, lighting levels, and air flow based on time of day, building occupancy, and changes in weather.

Behavioral changes are especially attractive forms of energy conservation because they can be accomplished with little or no cost. At the residential scale, for example, energy conservation can be achieved in myriad ways that involve nothing more than changes in personal habit, such as washing laundry in cold water, driving less, turning off lights, lowering thermostats when homes are unoccupied, and so on. However, these changes can be remarkably difficult to achieve because they require that the average person be aware of ways to conserve energy, have the motivation to do so, and then actually undertake a change in habit. In the United States, this is complicated by popular perceptions that conservation implies deprivation, discomfort, or inconvenience. As a consequence, unless energy prices are extremely high or people have strong values about energy conservation, it is difficult to motivate people to change their energy habits.

However, for the commercial and industrial sectors, conserving energy and saving money are often synonymous. Here, because of the scale of buildings, technologies, and industrial operations, energy conservation is more commonly achieved by using technological innovations that regulate energy usage through advanced automation and control systems. One example would be lighting controls that regulate the amount of artificial lighting in a room relative to the available daylight coming in through windows. Another would be carbon dioxide sensors that regulate air-handling equipment based on a room's occupancy rather than a set rate of fresh air replacement every hour.

Energy conservation is not just for wealthy industrialized countries. For example, in Box 8.1 on cookstoves, a family that switches from open-fire cooking to a cookstove is practicing energy conservation because the family changed its need for energy and the associated demand for fuelwood. In general, when households, businesses, or industry replace less efficient technology with more efficient, this is considered a conservation measure. For example, replacing incandescent light bulbs with compact fluorescent light bulbs in a home is a conservation *behavior*—the household has reduced its demand for the energy service of lighting by switching to a more efficient technology.

EXAMPLE 8.8   Estimating Energy Conservation

It can be relatively straightforward to estimate energy conservation for a variety of energy systems and to calculate how much curtailment (behavioral change) is required to equal an improvement in energy efficiency. For example, suppose you are thinking about buying a new car. Your current automobile has a fuel economy of 8.5 kilometres per litre (km/l), and you are thinking about a new vehicle that has a rating 13.6 km/l. If you kept your existing car, how many fewer miles would you have to drive per year to achieve the same level of energy conservation as the new car? Assume you currently drive 36,000 kilometres per year.

Step 1: Calculate your current annual fuel usage.

$$\text{Total fuel usage} = \frac{36{,}000 \text{ kilometers}}{\text{year}} \times \frac{1 \text{ liter}}{8.5 \text{ kilometers}} = 4{,}235 \text{ liters per year}$$

Step 2: Calculate your annual fuel usage with the new car.

$$\text{Total fuel usage} = \frac{36{,}000 \text{ kilometers}}{\text{year}} \times \frac{1 \text{ liter}}{13.6 \text{ kilometers}} = 2{,}647 \text{ liters per year}$$

Step 3: Calculate the equivalent miles driven on the fuel savings for your current automobile.

$$\text{Equivalent miles driven} = \frac{(4{,}235 - 2{,}647) \text{ liters}}{\text{year}} \times \frac{8.5 \text{ kilometers}}{\text{liter}} = 13{,}498 \text{ kilometers per year}$$

Based on these calculations, we see that buying a new car would save 1,588 liters of fuel per year, the equivalent of driving 13,498 kilometers less in your existing vehicle. To achieve the equivalent fuel economy of buying a new car, you would need to reduce your kilometers traveled by 37.5%.

8.5.2 Energy Efficiency

Energy efficiency refers to the productivity of technologies that convert an energy input into a more useful or usable form. In physics, this useful or usable form is commonly referred to as "work," or mechanical energy. However, usable energy often takes the form of heat, mechanical work, electric power, light, and cooling.

Energy efficiency therefore represents the conversion efficiency (and associated losses) of transforming energy from one form to another. It is represented by the Greek letter eta (η) and the equation

$$\eta = \text{efficiency} = \frac{\text{energy out}}{\text{energy in}} = \frac{\text{work done}}{\text{energy in}}$$

Energy efficiency is the ratio of the amount of energy created by a piece of machinery or equipment relative to the amount of energy that was put into the system. Another way of saying this is that energy efficiency is the amount of work done relative to the amount of energy used. Efficiency can be gained in one of two ways: (1) by creating more work with the same amount of energy or (2) by obtaining the same amount of work with less energy.

Many types of products, machines, and equipment can be understood as energy conversion devices. These include anything that generates electricity—from steam turbines to photovoltaic cells—as well as combustion engines, motors, and boilers. Appliances and many consumer goods are also energy conversion devices, including refrigerators, light bulbs, air conditioners, and hot water heaters. In addition to specific pieces of equipment, we can also think about the energy efficiency of complex systems. For example, a conventional thermoelectric power plant is only able to convert about one-third of its fuel input into usable electricity. Losses occur mostly in the power plant itself because of the thermodynamics of the process; lesser amounts are lost in the transmission of electric power (Figure 8.23).

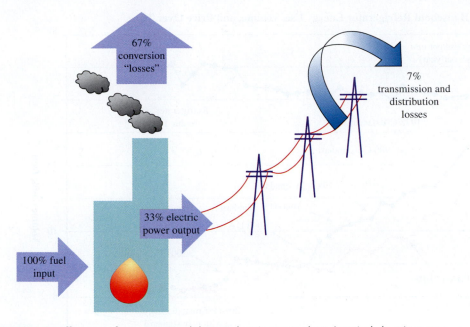

67% conversion "losses"

7% transmission and distribution losses

33% electric power output

100% fuel input

FIGURE 8.23 Efficiency of a conventional thermoelectric power plant. A typical electric power plant burns a fossil fuel (or uses nuclear energy) to create steam for conversion into electricity. The thermodynamic and electromechanical conversion "losses" are high; only about one-third of the fuel energy ultimately becomes electricity. There are also slight transmission losses when electricity is moved through power lines.

Source: Maria Papadakis

Energy efficiency improvements are a significant way of reducing energy consumption. For example, the energy efficiency of American refrigerators has increased significantly largely due to government regulation. As seen in Figure 8.24, the average energy use of a refrigerator has declined four-fold since the late 1970s, even as the average size of refrigerators has increased. The invention of new, high-efficiency lighting—compact fluorescent bulbs, mercury vapor lights, and LEDs—also represents significant gains in lighting efficiency. In the United States, the federal government sets energy efficiency standards for a wide variety of residential and commercial appliances and equipment, and the Energy Star™ program enables certification and labeling of consumer products that exceed the government's minimum requirements or that may not have mandated standards.

A major engineering challenge today is represented by the physical limits to efficiency for heat engines. A heat engine is any device or system that converts thermal energy to mechanical work and includes the internal combustion engine in your car as well as a traditional steam engine. The **Carnot Limit** to the efficiency of heat engines was deduced by Nicolas Carnot in the early 1800s, and it results from the temperature difference between the source heat in a system and the cooler temperature at which the heat is discharged. The Carnot Limit tells us that for any particular heat engine, there will be a maximum efficiency that can physically be achieved. Once the Carnot Limit is reached, the only way to improve overall system performance is to capture and utilize waste heat in some way. The Carnot Limit of electric power plant steam generators is about 51% today; gas-fired turbines have achieved efficiencies over 40%. The thermal efficiency of a typical automobile engine is under 20%, compared to the Carnot theoretical limit of 37%.

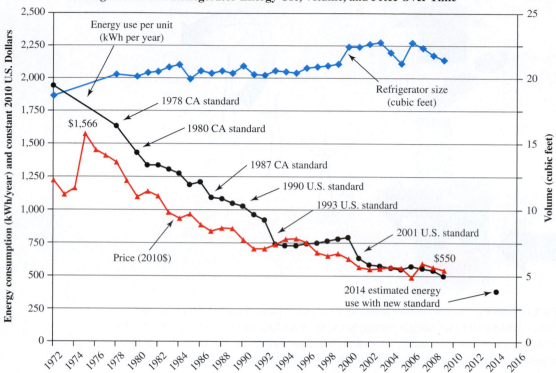

FIGURE 8.24 U.S. refrigerator energy use over time. In the United States, the energy use of refrigerators has been declining sharply because of government standards. Energy use has declined four-fold even as the average size of refrigerators has continued to gradually increase.

Source: Based on Appliance Standards Awareness Project, http://www.appliance-standards.org/sites/default/files/Refrigerator%20Graph_July_2011.PDF.

8.5.3 Energy Recovery

The recovery of energy otherwise lost as waste represents notable opportunities for energy sustainability. **Energy recovery** can exploit waste heat from industrial processes as well as the energy embodied in solid waste (such as garbage). **Waste-to-energy** facilities burn municipal solid waste to generate steam or electricity; not only do these energy plants use solid waste as a fuel, but they have the added benefit of conserving landfill space (Figure 8.25).

The exploitation of usable waste heat is referred to as **combined heat and power (CHP)** (Box 8.6). CHP, also known as cogeneration, most commonly uses the waste steam or hot gasses from electric power production for hot water, heating, and cooling in buildings. Less commonly, it uses the waste heat from high-temperature industrial processes to generate electricity. Integrated gasification combined cycle (IGCC) power plants use gasified coal and CHP to reduce emissions (which is why they are often referred to as a "clean coal" technology) and raise the thermal efficiency of the facility to about 60%.

Another type of energy recovery is the exploitation of biogas. When organic matter breaks down, it produces methane (natural gas) as a by-product of anaerobic decomposition. Three major sources of biogas are landfills, wastewater treatment plants, and cow manure. Landfill gas can be tapped and piped; biogas

FIGURE 8.25 **A waste-to-energy facility.** This resource recovery facility in Harrisonburg, Virginia, burns up to 180 metric tons of municipal waste per day and generates 25,800 kilograms of steam all of which is used on the James Madison University campus for heating and cooling. Not only is this considered to be a renewable energy resource, but it has the added benefit of reducing the volume of landfill waste by 90%.

Source: Photo by Bradley Striebig

BOX 8.6 Combined Heat and Power

Combined heat and power (CHP) systems, also commonly referred to as **cogeneration**, are unsung heroes in the world of sustainable energy and deserve more popular attention. Rather than directly consuming an energy fuel to provide energy, CHP systems exploit waste heat to provide cooling and heating or to generate electricity.

There are two basic types of CHP systems. **Topping cycle plants** use waste steam and heat from electric power production to create steam or hot water that is then used for heating and cooling a building. Because electricity is generated by both steam and gas turbines, the design of CHP systems will vary depending on the generating system (Figure 8.26). The capacity of topping cycle CHP is extraordinary. For example, Consolidated Edison of New York operates the largest district steam system in the United States and supplies 1,800 major customers (such as the United Nations and the Empire State Building) with hot water, heating, and cooling via a network of 105 miles of lines and pipes. Amazingly, 45% of Con Edison's district steam is CHP from its own electric power plants and those of the Brooklyn Navy Yard. By installing CHP in electric power plants, the overall energy efficiency of these plants increases from 45 to 80% because waste energy is put to productive use.

The other type of CHP is less common and is referred to as a **bottoming cycle plant**. Here, the waste heat from industrial furnaces and other high-temperature manufacturing processes (metals, glassmaking) is captured to generate electricity that is typically used within the facility itself. These types of cogeneration systems are much less common than topping cycle plants.

According to the U.S. Combined Heat and Power Association, CHP systems currently:

- Generate nearly 12% of electricity in the United States
- Save $5 billion per year in customers' energy costs

(Continued)

BOX 8.6 Combined Heat and Power (*Continued*)

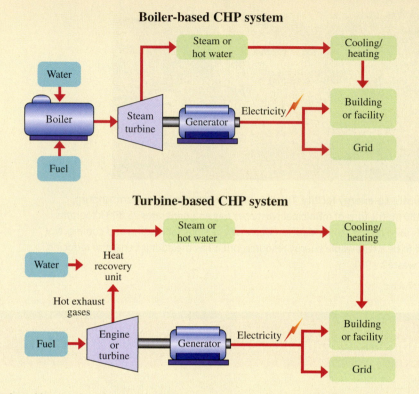

Boiler-based CHP system

Turbine-based CHP system

FIGURE 8.26 Combined heat and power plants. CHP systems use waste heat for productive purposes. These diagrams illustrate two different types of topping cycle plants. The boiler-based system is a conventional steam turbine in which a fuel is combusted to heat water into steam. Turbine-based systems burn a fuel, and it is the combustion exhaust gas that turns the electric generator, not steam.

Source: U.S. Environmental Protection Agency, www.epa.gov/chp/basic/

- Offset 1.4 million megajoules per year of primary energy production from a fuel source
- Avoid the emissions of 362,800 metric tons of NOx, 816,300 metric tons of SO$_2$, and 35 million metric tons of atmospheric carbon equivalent per year

There is room for expanded cogeneration capability in the United States. The Combined Heat and Power Association estimates that there is still about 140 gigawatts (GW) of CHP that is technically possible, although not necessarily economically profitable without public policy assistance. Currently, there is a little more than 80 GW of installed cogeneration capacity in the United States.

The U.S. military is a leading adopter of CHP systems. In 2012, the Marine Corps Air Ground Combat Center (in California) and the 82nd Airborne at Fort Bragg (North Carolina) won the EPA's CHP Energy Star™ award. At the Air Ground Combat Center, the new CHP system improves plant efficiency to 64%, saves nearly $6 million per year in energy costs, and annually avoids CO$_2$ emissions that would be equivalent to the electricity used by 2,400 homes. At Fort Bragg, the CHP system saves $1 million each year and avoids CO$_2$ emissions equivalent to the electricity used by 1,500 residences.

from wastewater and cow manure is obtained from anaerobic digesters. In all instances, biogas can be substituted for natural gas and is considered to be a renewable resource.

8.5.3 Renewable Energy and Alternative Fuels

Renewable energy enhances energy sustainability by relying on natural resources that are regenerative and not finite, and they have the added benefit of being largely nonpolluting as direct energy sources. The most commonly discussed renewable energy resources are solar, wind, water, and tides. At present, these renewable energy resources are exploited primarily for electricity: Wind power, hydropower, and tidal systems are used exclusively for electric power production, although solar energy is used for electricity from photovoltaics as well as for hot water production from solar thermal systems. Major challenges with electric power from renewable energy sources include the intermittency of the energy—it is not produced in continuous amounts due to daily and seasonal variations (e.g., the sun does not shine at night)—and the fact that the energy output is low relative to the energy loads and economic costs. For example, a 1-kilowatt solar photovoltaic system or wind turbine will cost several thousand dollars in the United States, but the amount of electricity they could generate at peak output is equivalent to a typical hair dryer. These systems therefore need to be quite large to supply even a small proportion of the electricity needed by a typical household.

Geothermal energy is exploited as a source of renewable energy in two ways. First, because of the natural and stable temperature gradients of the Earth, heating and cooling systems can be preheated or prechilled by circulating their heat-exchanging fluids underground. **Ground source heat pumps** (Figure 8.27) are an example of this technology. **Geothermal power** refers to the generation of electricity from geothermal energy. This requires intense heat or steam, and such plants are typically found near active volcanoes, tectonic plates, and natural geysers. There is about 10.7 GW of installed electric power capacity in the world; the United States has 77 geothermal power plants and the most installed geothermal capacity, but in Iceland, about one-third of its electric power is generated from geothermal energy.

As a combustion fuel, biomass is also a form of renewable energy, although most traditional biomass is considered unsustainable for both its negative health consequences and overall environmental impacts. Fuelwood harvesting is a significant contributor to deforestation in some nations, and destruction of peat bogs affects wetlands ecology. More sustainable and commercially viable amounts of biomass are usually waste products from another industry or deliberately cultivated fuels. For example, forest products residues (sawdust, wood chips, logging chaff) are burned for electric power production in the United States; **bagasse**—the waste fiber from sugarcane—is widely used in countries with large sugarcane operations. Certain species of willows and grasses are cultivated agriculturally as biomass fuels.

In the transportation sector, ethanol and biodiesel are alternative fuels that are also considered to be renewable. Both fuels are blended to varying degrees with petroleum. Ethanol, which is obtained by fermentation, is commonly formulated as 85% ethanol, 15% gasoline (E85) and 10% ethanol, 90% gasoline (E10, commonly known as gasohol). While E10 can be used by most engines, E85 requires a flex-fuel vehicle. Biodiesel results from the chemical conversion of animal fats and vegetable oils and is used in diesel engines.

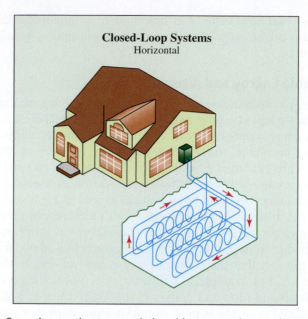

FIGURE 8.27 Ground source heat pump. A closed-loop ground source heat pump cycles compressed fluid through buried piping. During the summer, the heat pump takes advantage of geothermal energy by precooling compressor fluids for air conditioning; in the winter, the compressor fluid is prewarmed by the ground, which is warmer than air temperatures.

Source: U.S. Department of Energy, http://energy.gov/sites/prod/files/styles/large/public/closed_loop_system_horiz.gif?itok=A8hCygmB

The sustainability of both types of fuels is not without criticism, especially for their impacts on agriculture and forestry. Ethanol requires large-scale grain production (especially corn) and has been criticized for its impact on crop prices and for the perpetuation of large-scale monoculture and its associated environmental impacts. Worldwide, palm oil is a major source of plant oil for biodiesel production; the widespread cultivation of palm oil plantations is a contributing factor in tropical deforestation as well as large-scale destruction of peat bogs in Indonesia.

Fuel cells are a technology often considered as renewable energy because they operate on hydrogen as a fuel. Fuel cells create electricity through a chemical process involving hydrogen ions, and they have efficiencies as high as 60%. Because their by-products are water and heat, their "emissions" have negligible environmental impacts. In addition, if the waste heat is recovered with CHP, the overall efficiency of a fuel cell system can exceed 90%. Although many fuel cells are now in commercial use—especially in the transportation sector and as small-scale power plants for buildings—there are still many obstacles to the widespread development and adoption of this technology. Costs are high, and we do not have an effective way of sustainably producing large quantities of hydrogen as a fuel. Many systems use natural gas as the fuel source of hydrogen.

EXAMPLE 8.9 Estimating the Energy Savings of Ground Source Heat Pumps

The energy performance of heating and cooling equipment is measured in two ways. For cooling systems, the rating is called an **energy efficiency ratio (EER)** and is calculated as

$$EER = \frac{\text{Btu of cooling output}}{\text{watt} - \text{hour of electric input}} \qquad (8.5)$$

EER units are expressed in Btus per watt-hour.

Heating energy efficiency is represented as the **coefficient of performance (COP)** and is calculated as

$$\text{COP} = \frac{\text{power output in watts}}{\text{power input in watts}} = \frac{\text{EER}}{\dfrac{3.412 \text{ Btu}}{\text{watt} - \text{hour}}} \tag{8.6}$$

The COP is a dimensionless ratio and can be understood as a measure of instantaneous work output. Note that to convert the COP to an EER, 3.412 Btu/Wh is applied as a straight conversion factor. (Recall from Table 8.3 that there are 3,412 Btu in a kWh.)

Both the EER and the COP represent the same ratio, which is the amount of work output relative to one unit of electricity input (power). A COP of 3.8, for example, means that a heating system produces 3.8 times more energy in heat than it consumes in electricity. An EER of 11 means that a watt-hour of electricity produces 11 Btu of cooling output.

Suppose that the COP for a conventional heat pump is 2.4 and 3.5 for a ground source heat pump. If a building requires 43.3 million Btu per year for heat, how much electricity does the geothermal system save?

Step 1: Convert the annual heating load to kilowatt-hours.

This question requires us to express the energy savings in units of electricity, which is commonly expressed in kilowatt hours (kWh). Using the equivalencies from Table 8.3:

$$\text{Annual heating load} = 43{,}300{,}000 \text{ Btu} \times \frac{1 \text{ kWh}}{3{,}412 \text{ Btu}} = 12{,}690.5 \text{ kWh}$$

Step 2: Calculate the input energy required for each system.

Because the COP is dimensionless, it represents a straightforward input/output ratio. Manipulating Equation 8.6 above,

$$\text{Power input} = \frac{\text{power output}}{\text{COP}}$$

We can thus calculate the energy input requirements for each system as

$$\text{Power input, conventional heat pump} = \frac{12{,}690.5 \text{ kWh}}{2.4} = 5{,}287.7 \text{ kWh}$$

$$\text{Power input, geothermal heat pump} = \frac{12{,}690.5 \text{ kWh}}{3.5} = 3{,}625.9 \text{ kWh}$$

Step 3: Compute the net energy savings.

$$\text{Net energy savings} = 5287.7 - 3{,}625.9 \text{ kWh} = 1{,}661.8 \text{ kWh}$$

The geothermal heat pump saves us 1,661.8 kWh per year for this particular heating load. This is equivalent to 31% less input energy than the conventional heat pump. We can therefore say that the ground source heat pump is almost one-third more efficient than the conventional system.

8.5.5 Fuel Switching

Fuel switching is not a widely discussed sustainable energy option, but it is none-theless an important strategy for mitigating carbon dioxide. Fuel switching typically involves the substitution of natural gas for coal or oil because systems that use nat-ural gas operate at higher efficiencies and the fuel burns more cleanly. Although it does not significantly reduce our use of fossil fuels, it does lessen the environmental impacts of that fuel use. Fuel switching is increasingly common in electric power production, in which coal-based thermoelectric steam generation is replaced by combined cycle systems that have two thermodynamic power generation cycles. The first uses gas turbines, and the second uses the hot exhaust of the turbine through CHP to generate a steam cycle. Motor vehicles that operate on compressed natural gas are also a form of fuel switching.

A type of fuel switching was popularized by the financial magnate T. Boone Pickens in the late 2000s. The **Pickens Plan** proposed that the United States invest heavily in wind farms for electric power production, enabling the natural gas used for electric power to be shifted to the transportation sector for use in compressed natural gas vehicles. The result of the Pickens Plan would be reduced petroleum usage for transportation by substituting natural gas. Others have pointed out that this proposal would also facilitate the use of electric vehicles that are grid-charged by a renewable energy source.

8.5.6 Pollution Prevention and Control

The field of environmental engineering has historically focused on pollution prevention and control, and it represents the oldest efforts to make industrial pro-cesses more sustainable. The purpose of pollution prevention and control is—as its name quite literally implies—to limit the amount of pollutants entering the envi-ronment. This may be accomplished in one of three primary ways: (1) re-designing systems to reduce the amount of pollutants generated by a process, (2) physically "capturing" a pollutant for disposal through a formal waste management system, and (3) treating waste to reduce its impacts before it is disposed of in the open environment.

Pollution prevention and control efforts often result from governmental regu-lation. Because of the harmful effects of pollution on human health in particular, governments regulate critical pollutants by establishing levels of contaminants that are generally considered "safe" for society and that minimize health risks. Examples of such pollutants are particulate matter, lead, carbon monoxide, nitrogen oxides, and sulfur dioxide. These have traditionally been controlled by what is known as **command and control regulation**, in which the government requires industries to obtain permits, pay fees, and/or install best available technologies to achieve the desired limits or controls on pollution.

With respect to energy, several examples illustrate the principles of pollution prevention and control. In the transportation sector, government air quality reg-ulations on permissible levels of nitrogen oxides from automobile tailpipe emis-sions resulted in the re-engineering of combustion processes (e.g., fuel injection systems) as well as the development of catalytic converters by auto manufacturers. Localized air quality issues also resulted in the development of reformulated gasoline, including unleaded gas. In the electric power sector, coal-fired power plants capture their fly ash with electrostatic precipitators; fly ash that is not used as a raw material for other products is usually stored wet in containment ponds to prevent the ash from entering the air or water. Sulfur dioxide is largely controlled by flue gas desulfurization technology. Today, carbon capture and storage represents an emerging suite of pollution prevention and control technologies to mitigate CO_2 emissions.

8.6 Appropriate Technology, Scale, and Distributed Energy

Energy sustainability is most often discussed in terms of renewable energy, conservation, efficiency, and better energy recovery. These traditional approaches were summarized in Section 8.5. However, extraordinary opportunities for more sustainable energy use can be found in rethinking the scale of energy technologies and in being mindful that providing energy is about creatively meeting people's *needs* for an *energy service*. Two technological concepts allow us to consider these opportunities: appropriate technology and distributed energy. Both require that we explicitly consider the way in which geographic scale and technological scale are interrelated. A promising future for more sustainable energy can be found when we provide small amounts of energy at the location where it is needed: technologies that are inappropriate for large, centralized power production because of their low energy output may be more than adequate when they are matched to loads of similar size at the point of end use. In addition, for areas that are remote from formal energy supplies, we can also design systems that use the energy that is locally available.

8.6.1 Appropriate Technology

Appropriate technology is an imprecise term that is used to generally describe a technology that meets the needs of a local community, is ideally locally produced from local materials, and is simple enough that it can be operated and maintained by community members themselves. *Appropriate* thus means appropriate to the context of the needs, incomes, geography, skills, and resources of the people whom the technology is to serve. This approach, first advocated in the 1970s by E.O. Schumacher in his book *Small Is Beautiful*, can provide much-needed technology for socioeconomic development. With respect to energy, appropriate technology typically refers to small-scale energy systems that do not require fossil fuels and that meet critical social needs.

Appropriate technologies are, almost by definition, not energy intensive and not especially reliant on the formal energy sector. The case study on cookstoves discussed earlier is one such example, where cookstove design is based on local cooking cultures and techniques; materials are locally sourced, produced, and of natural materials, and they do not require fossil fuels for their operation. Another example is the Electrical Conductivity kit (EC-kit), which is a water quality test kit that can be incubated in the field by human body heat. Unlike other tests for drinking water contaminated by coliform bacteria, these kits do not require a sophisticated testing lab, electricity, or an incubator to test for contamination—just a human body! Other examples are micro- and pico-hydropower systems, which provide electricity to rural areas (Box 8.7). Designing appropriate technologies can be challenging, and first design and installs may not necessarily be successful. The PlayPump is an example of a highly promising technology that has not yet met expectations (Box 8.8).

BOX 8.7 The Power of Microhydro

No other energy technology demonstrates the sustainability implications of technological scale better than hydropower. The Three Gorges Dam in China is the world's largest hydropower station with 22,500 MW of generating capacity, yet around the world hundreds of thousands of pico-hydro turbines provide much-needed electricity with a capacity of less than 5 kW. Large-scale, centralized hydropower has been the norm historically, but as greater attention is given to energy poverty, distributed generation, and appropriate technology, more and more micro-hydropower is coming online. Although there is not an internationally standardized terminology for hydropower, small hydropower usually has a generating capacity of 10 MW or less. Mini hydropower is less than 1 MW, micro hydro is less than 100 kW, and pico-hydro is less than 5 kW. Most pico-hydro systems are between 200 W and 1 kW! Figure 8.28 illustrates the different types of pico-hydro turbines and their suitability for different head heights and water flow conditions.

The power of pico-hydro should not be underestimated. For the energy poor in regions of the world without access to electricity, even 200 W can power a few precious light bulbs and a radio. In a World Bank community pico-hydro pilot program in Ecuador, participants reported better health (less vision and eye irritation), an ability to use community radio and television for several more hours per day, better educational experiences as children could study at night, improved community cohesiveness because of an ability to confer at night and celebrate local events, a greater sense of personal efficacy (community members maintain the systems), financial savings because of no longer

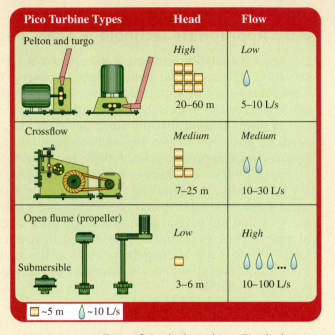

FIGURE 8.28 Types of pico-hydro turbines. Pico-hydro technology is designed to extract the potential energy of water under conditions of either low head height or low flow. Different types of turbines have been engineered depending on whether head height or flow are the most constraining factors.

Source: Based on Energypedia, https://energypedia.info/wiki/Pico_Hydro_Power

needing to purchase candles, batteries, or generator fuel, and a greater tendency for men to stay home during the evenings and weekends. While these are seemingly small changes by wealthy nation standards, they are transformative developments at the smaller scale. The World Bank estimates that there is the potential for 4 million low-head pico-hydro systems worldwide.

8.6.2 Distributed Energy

Some of our most energy-intensive technical systems use fossil fuels to produce energy in large volumes that is then shipped over long distances to the ultimate customer. Electric power systems fundamentally operate in this way, where electric power is produced on a mega scale at centralized power plants, then transmitted and distributed over hundreds of miles from the source. Baseload electric power

BOX 8.8 Why Wasn't the PlayPump a Success?

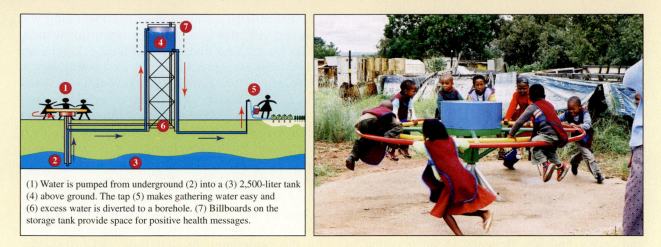

(1) Water is pumped from underground (2) into a (3) 2,500-liter tank (4) above ground. The tap (5) makes gathering water easy and (6) excess water is diverted to a borehole. (7) Billboards on the storage tank provide space for positive health messages.

FIGURE 8.29 PlayPump as an energy source. The PlayPump was intended as an appropriate technology designed to provide energy for water pumping. Mechanical energy provided by children playing on a merry-go-round enabled communities without electricity or money for diesel generators to pump potable groundwater. The technology generated considerable enthusiasm, but it has had a problematic history.

Source: Courtesy of Playpumps

The PlayPump was an exciting, sensational technology introduced in southern Africa in 1994. A water pumping system disguised as a children's merry-go-round (or "roundabout"), the PlayPump extracted and stored large quantities of potable water for communities. As seen in Figure 8.29, the mechanical energy provided by the merry-go-round pumped water to a gravity-based storage and tap system. It seemed like a terrific and astonishingly simple solution to the water crisis in Africa—children's playtime at school (Figure 8.29) could be used to provide much needed fresh water to a village, eliminating the drudgery involved in this task for local women. In many locations, the PlayPump was a viable source of water. However, within a few short years, PlayPump deployment halted and the technology came under severe criticism as an ill-conceived "technological fix." The history of this technology is both predictable and regrettable, and reminds us that there is no quick substitute for the process of human-centered design and careful field testing.

The initial success of the PlayPump attracted widespread media attention, including highly publicized donations by celebrities and the U.S. government. Rapid deployment of the pumps followed as production scaled up and as donors hoped to see rapid results of their charity. However, when PlayPumps spread across a wide variety of communities, faults quickly emerged. Technically, the pumps did not always work as designed because of hydrological conditions. In other words, the mechanical design was fine as long as specific site requirements were met. In addition, there were a large number of mechanical failures without adequate technical assistance or spare parts; failures also tended to occur in areas difficult to reach and repair below ground. Very simply, the pumps broke and could not be fixed. Regrettably, many communities lost their only source of water when this happened, because the PlayPump was either installed in the existing borehole or the old water supply was dismantled. Early

(Continued)

BOX 8.8 Why Wasn't the PlayPump a Success? (*Continued*)

PlayPump installations therefore violated two long understood rules about appropriate technology: (1) make sure it is repairable by local residents and that spare parts are available, and (2) leave working systems in place, especially if they meet critical community needs.

Other issues emerged *socially* and *culturally*. First, not all villages had enough children to operate the pump, or the children got bored easily and did not play long enough to pump an adequate supply of water. The job defaulted to women, who were physically challenged by the equipment. (Consider that elderly or pregnant women would have to run or sit on a merry-go-round.) The PlayPump is large and heavy, and was designed to be operated by a group of children running all out; an individual woman cannot adequately power the machine. Culturally, the role of children also shifted, and they became viewed as laborers in some places. Because the system was usually installed at schools to enable an opportunity to play, children's class time was reportedly diverted to water production. *Safety* also became an issue. *Economically*, the PlayPump was also a financial burden on communities because of shared costs, choices of materials for the water tank, and the renting of heavy machinery involved in its installation.

Eventually, installation of the PlayPump was banned in Malawi. In 2010, the nonprofit responsible for making and distributing the PlayPump, PlayPumps International, ceased operation and transferred its technology and inventory to Water for People. Water for People is an aid organization with a successful track record in water projects, and it quickly conducted a technology assessment of the PlayPump in partnership with its major donors. (**Technology assessment** is the process of analyzing the technical, social, cultural, safety, economic, environmental, reliability, usability, and infrastructure dynamics of a technology.) A number of design changes were made to the equipment, its materials, and installation techniques. New requirements were put in place for appropriately siting the pump in a manner that clearly made the technology more appropriate to social, cultural, and environmental circumstances. However, by this time more than 1,000 PlayPumps had been installed, and reports from the field were highly damaging. While not a fatally flawed concept, the PlayPump could not overcome its first set of limitations or the unwillingness of the development community to pursue it further. It has, for all practical purposes, disappeared.

plants typically generate 1,000 to 1,500 megawatts (MW) of power with fossil and nuclear energy, and there are really no substitute fuels or technologies that can produce electricity at this density on a 24/7 basis. District steam systems work in a similar manner, where steam is manufactured at a centralized plant and then piped to facilities for building heat and hot water.

The alternative to highly centralized power production is distributed energy, where small amounts of energy are generated at multiple sites throughout a connected system. Distributed energy systems have many advantages. One is that they are better able to use technologies that do not produce large amounts of energy but do produce energy from intermittent renewable sources or at higher levels of efficiency from fossil fuels. While solar panels, wind turbines, gas microturbines, CHP systems, and so on simply cannot be used for large-scale, remote, centralized power output, they can successfully supplement the energy provided by a centralized system.

Technical and economic challenges confront our ability to implement distributed energy systems on a widespread basis. First, our technical ability to have

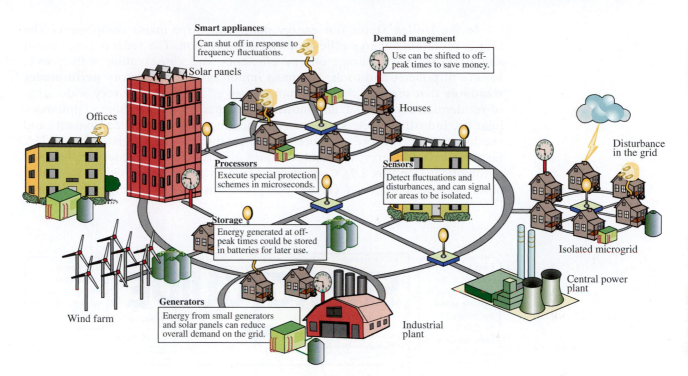

FIGURE 8.30 Smart grid connectivity. Electric power systems of the future will integrate centralized power production with small-scale distributed generation throughout the system. In addition to being able to regulate diffuse sources of energy, a smart grid also has storage capacity, the ability to regulate appliances and energy-consuming equipment at the source, and is more resistant to disturbances in the grid itself.

Source: Based on Smart Grid 2030™ Research Associates. http://www.smartgrid2030.com/wp-content/uploads/2009/10/SG-Nature.jpg

a "smart grid" that can dynamically regulate and control the electricity flowing from hundreds of diffused sources is not in place, and major R&D efforts are underway to create the electronics and communication technologies required by such a network (Figure 8.30). Second, with respect to consumers in the United States and elsewhere, small-scale sources are still not cost competitive with conventional, fossil-fuel-based electric power generation. In addition, some electric power utilities are resistant to distributed generation because it can undermine their own system's profitability. As a consequence, there is not yet a strong demand for the smaller technologies that are required for a robust distributed energy system.

8.7   Energy Policy

All governments strive to promote a safe, affordable, reliable, and abundant supply of energy through their public policies. Many different policy strategies and tools are used to achieve these goals and vary depending on the energy needs of the country. For low- and middle-income countries, their initiatives focus heavily on providing and managing an adequate energy infrastructure. In industrialized and high-income nations, the focus is often on their global energy security and on reducing the environmental impacts of energy, especially greenhouse gases.

In the United States, our energy policy has three major components. The first is promoting energy efficiency and conservation. The federal government and some states encourage energy efficiency and conservation with a wide variety of policies and tools. The most important are mandatory **performance standards** that regulate the minimum energy efficiencies of a very wide array of residential, commercial, and industrial equipment, such as home appliances, lighting, industrial boilers, commercial heating and cooling equipment, and vending machines. There are more than 30 product classes regulated by the U.S. Department of Energy; the United States also regulates the fuel efficiency of automobiles (see Box 8.9). Consumer awareness programs provide information about product energy use (Figure 8.31) and ways to save energy; the Energy

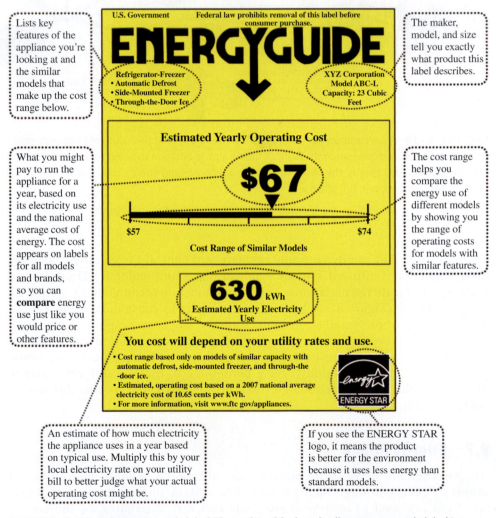

FIGURE 8.31 U.S. energy guide label. The striking black and yellow energy guide label is required by law and is regulated by the U.S. Federal Trade Commission. Its purpose is to promote energy efficiency by allowing consumers to compare the energy costs of products as part of their purchasing decision.

Source: US Federal Trade Commission

BOX 8.9 Legislating Fuel Economy

Fuel economy (also known as gas mileage) is an important form of energy conservation. Fuel efficient vehicles can be the result of market forces, voluntary targets set by government or industry, and/or mandatory performance standards required by law. The fuel economy of European and Japanese cars has historically been higher than that of the United States largely because of market forces, higher fuel prices, and the "dieselization" of auto fleets (diesel engines achieve better fuel efficiency). Europe has not had mandatory fuel efficiency standards, only voluntary targets; the European Union instead started regulating automotive CO_2 emissions in 2009, which has a direct impact on fuel performance. Japanese fuel efficiency standards were voluntary until the early 2000s but are now mandatory.

The United States has the most long-standing legally required standards for fuel economy. The Corporate Average Fuel Economy (CAFE) standards are government regulations of the gas mileage of automobiles and light trucks. These standards were originally enacted to conserve gasoline in the transportation sector after the Arab Oil Embargo from 1973 to 1974. At that time, the average fuel economy of American cars was only 5.5 kilometers per liter (km/L), and the original goal of the CAFE standards was to enhance U.S. energy security by doubling the fuel economy of new cars. Originally set at 18 mpg for passenger cars, the fuel economy standards rose to 11.7 km/L in 1985. They remained unchanged for 20 years until the Energy Independence and Security Act of 2007 made significant changes in the way in which the United States designs and implements its fuel economy standards and programs. A study by the National Academy of Sciences estimated that in the absence of the CAFE standards, fuel consumption in the United States would have been 14% higher in the early 2000s than it actually was.

The current law requires that light-duty vehicles (passenger cars and light-duty trucks) achieve a fleetwide fuel economy of 15 km/L by 2020. In the past, cars and trucks had separate gas mileage standards, but now they have been combined into a single fleet. In addition, the older CAFE standards did

FIGURE 8.32

Source: Olaru Radian-Alexandru/Shutterstock.com

not apply to individual model cars: The standard was a sales-weighted average for an auto manufacturer's entire fleet. Now, there are standards that apply to individual vehicles based on their footprint (wheelbase by track width) in addition to the fleetwide average of 15 km/L. The Obama Administration successfully negotiated a new set of standards for automakers—set to go into effect in 2017—that will increase fleetwide average fuel performance to 23 km/L by 2025. This will undoubtedly be achieved by the greater use of hybrid and alternative fuel vehicles.

In addition to the CAFE standards, there is a "gas guzzler tax" applied to passenger cars (excluding SUVs, trucks, and minivans) that do not meet a minimum fuel economy of 10 km/L; the worse the gas mileage, the larger the tax penalty. CAFE standards apply to both domestic and imported cars; failure to meet the standard results in a fine on the automaker. Almost all fines paid in the United States under the CAFE system involve European luxury cars.

A study published by the United Nations carefully standardized automobile fuel economy in different countries in order to compare them. In 2012, U.S. fuel economy standards were 11.7 km/L compared to about 20 km/L in Europe and Japan and 16 km/L in China. The EU emissions standards will result in a fuel economy of 28 km/L by 2020.

Star™ program is a **voluntary certification** initiative in which private businesses apply to the government to get their products endorsed by the government as being extraordinarily energy efficient. The government also funds research and development programs to enhance the performance of energy technologies, from batteries to coal-fired power plants. A variety of tax credits are used to encourage the purchase of energy conserving equipment by homeowners, businesses, and industry.

Second, U.S. policy also promotes the development and use of alternatives to fossil fuels, including renewable energy, alternative fuels, and **emerging technologies**. R&D programs, producer incentives, and consumer tax credits promote renewables and alternative fuels. Emerging technologies include the "next generation" of energy supply and require significant investments in R&D to make them commercially viable energy systems. Examples include hydrogen fuel cells and research into methane hydrates. At the state level, **renewable portfolio standards** are an important tool for promoting the use of renewable energy for electric power generation. Renewable portfolio systems establish a minimum baseline for the amount of electricity that utilities must provide from renewable sources.

Finally, U.S. energy policy is also concerned with energy security and a reliable supply of energy. Here a variety of policies—particularly tax incentives—promote gas and oil exploration and recovery to increase the domestic reserves of fossil fuels in the United States. In addition, considerable efforts are going into the design, development, and deployment of a smart grid to drastically improve the ability of the U.S. electric power transmission and distribution infrastructure. A smart grid would enable widespread distributed generation from renewable energy; highly automated energy demand management systems, such as the ability of electric power utilities to remotely control appliances in homes and equipment in businesses and industry; and a more reliable system that prevents power outages due to system instability.

8.8 The Water–Energy Nexus

An increasingly important sustainability issue is the interdependence between water resources and energy production and consumption. This relationship is referred to as the **water–energy nexus**, and it reflects generally the need for water to generate electricity as well as the energy-intensity of water supply systems, domestic water use, and wastewater treatment.

In the United States, nearly as much fresh water is withdrawn from the environment for thermoelectric power production (to both generate and cool steam) as is withdrawn for agricultural irrigation and livestock. About 85% of U.S. electricity is generated by thermoelectric power production, which involves producing steam to power turbines. A typical thermoelectric power plant will withdraw 378 to 946 million liters of water per day (Torcellini, Long, and Judkoff, 2003). While most of these withdrawals are discharged back into surface waters as condensate, evaporative losses are considerable and represent about 1.8 L per kilowatt hour of electricity generation. Because hydropower often relies on large reservoirs from dammed rivers, evaporative losses represent 68 L per kilowatt hour generated by a hydroelectric power plant. (About 7% of U.S. electricity is produced from hydropower.) The weighted national average evaporation for U.S. electricity is 7.6 liters of water (7.6 L) per kilowatt hour (Torcellini, Long, and Judkoff, 2003). Hydrofracking, an increasingly controversial extraction process for natural gas, requires a significant amount of water for drilling and fracturing shale, and has raised concerns about both the depletion and contamination of aquifers and other groundwater resources.

Areas where there is either physical or economic water scarcity are limited in their ability to install conventional centralized electric power systems, which can be a constraint on development. Climate change also threatens electric power production worldwide. Diminished water flows reduce the potential for hydropower, which is an important renewable source of electricity. Likewise, intense or protracted drought can result in lower electricity output from conventional thermoelectric power plants and may even result in facilities being taken offline; this was one consequence of an extensive drought in the southeastern United States during 2007 to 2009.

Water supply and production systems are also energy-intensive. Energy is required to pump fresh water from remote locations to drinking water treatment plants, to treat water and make it potable, and to distribute water to consumers. In California, which relies heavily on water transported over long distances from out of state, almost 20% of the state's electricity consumption is for water-related purposes (California Energy Commission, 2005). In terms of sanitation, wastewater treatment facilities are one of the most energy-intensive industrial activities and are necessary to discharge wastewater safely back into the environment. The EPA (2014) reports that drinking water and wastewater systems account for 3 to 4% of all energy used in the United States and that the proportion of drinking water costs accounted for by energy use can be as high as 40% (and is expected to increase by 20% over the next 15 years). Regions of the world where fresh water is scarce often rely on reverse osmosis water desalination technology, itself a highly energy-intensive process.

Sustainable development in the context of the water–energy nexus relates to both the environmental dynamics of water scarcity and the demand for electricity. Electric power production both depends on and affects the availability of freshwater; limited—or low-quality—fresh water requires increasing energy inputs to obtain, treat, and deliver high-quality drinking water, including water desalination in some areas. As the global population continues to urbanize and produce large volumes of wastewater, energy-intensive water treatment processes are required to protect surface water from sewage and other types of contamination. Finding opportunities to (1) improve energy efficiency and conservation in public water systems, (2) conserve water in power plants, and (3) conserve water at the household scale are increasingly important to protect both our natural resources and human well-being.

8.9 Summary

The world's dependence on fossil fuels has created a host of environmental problems, and as finite resources they cannot be used indefinitely into the future. Our challenge today is to provide for the unmet needs of the world's energy poor and to maintain safe, reliable, abundant, and affordable energy systems for all. To accomplish this goal, we must engineer or re-engineer our energy technologies in a variety of ways. First, we can diversify our resources by substituting cleaner and more renewable forms of energy for electrical, mechanical, and thermal energy. Second, we can enhance energy conservation by making our systems more efficient, by recovering otherwise wasted energy, and by inventing devices and materials that curtail the amount of energy required. Third, we can rethink the scale and distribution of our energy technologies to take better advantage of small-scale energy generation.

References

California Energy Commission. (2005, November). *Integrated Energy Policy Report*, CEC-100-2005-007-CMF. (Sacramento, CA: California Energy Commission.

Hubbert, M. King. (1956, March). Nuclear Energy and the Fossil Fuels. Paper presented at the spring meeting of the Southern District Division of Production, American Petroleum Institute, San Antonio, TX.

Moomaw, W., et al. (2011). "Annex II: Methodology". In O. Edenhofer and others, *IPCC Special Report on Renewable Energy Sources and Climate Change Mitigation*. Cambridge, UK: Cambridge University Press.

Torcellini, P., Long, N., and Judkoff, R. (2003, December). Consumptive Water Use for U.S. Power Production. National Renewable Energy Laboratory Technical Report NREL/TP-550-33905 (Golden, CO: National Renewable Energy Laboratory).

U.S. EPA. (2013). Clean Energy Calculations and References, Accessed May 15, 2013, at www.epa.gov/cleanenergy/energy-resources/refs.html.

U.S. EPA. (2014). *Water: Sustainable Infrastructure*. Energy Efficiency for Water and Wastewater Facilities. Accessed March 17, 2014, at http://water.epa.gov/infrasturce/sustain/energy efficiency.cfm.

World Business Council for Sustainable Development and World Resources Institute. (2004). *The Greenhouse Gas Protocol: A Corporate Accounting and Reporting Standard, Revised Edition*. Washington, D.C.; World Resources Institute.

Key Concepts

Derived demand
Energy services
Energy ladder
Energy intensity
Social costs
Energy security
End use sectors
Infinitely renewable resources
Reserve
Proven reserve
Methane hydrate
Hubbert Curve
Peak oil
Direct energy
Embodied energy
Energy return on investment
Energy efficiency
Energy curtailment
Carnot Limit
Energy recovery
Waste-to-energy

Combined heat and power
Topping cycle plant
Ground source heat pump
Geothermal power
Bagasse
Pickens Plan
Command and control regulation
Distributed energy
Performance standards
Voluntary certification
Emerging technologies
Renewable portfolio standards
Reserve-to-production ratio
Peak sun-hour
Source energy
Site energy
EER
COP
Technology assessment
Water–energy nexus

Active Learning Exercises

ACTIVE LEARNING EXERCISE 8.1: Calculate national energy savings

This activity requires a number of computational steps to come up with the most compelling statistics, but it is readily doable. It will allow you to practice the full sequence of conversions, equivalences, and estimates of energy output and consumption that would be commonly done in a professional setting.

Assume that you work as an energy analyst for the U.S. secretary of energy. She wants to provide a compelling fact about energy conservation to really grab the audience's attention in a speech she is giving. The point that she wants to stress is that small actions conducted by many people can add up to significant energy savings and carbon mitigation. She wants to tell people how much energy the United States would save if every home replaced one incandescent light bulb with a CFL.

Use the following "givens," as well as other information provided in this chapter, to calculate four different statistics for the secretary that reflect national energy savings and CO_2 mitigation. At least one of these statistics must be the number of baseload electric power plants that can be shut down by switching bulbs.

- There are 120 million households in the United States.
- A "high-use" bulb is one that burns for at least 4 hours per day.
- A 13-watt CFL is equivalent to a 60-watt incandescent bulb.
- The average baseload power plant has a rated output capacity of 800 MW.
- The typical baseload electric power plant in the United States uses a mixture of different kinds of coal.

ACTIVE LEARNING EXERCISE 8.2: Conduct a personal energy audit

Designing any type of energy system requires that you size the system to meet the energy "load," or amount of energy required. This exercise will give you practice estimating your personal weekly electricity load. You can do this for yourself regardless of where you live, which could be a dormitory room, an apartment, a shared house, or your family home.

1. Begin by listing all devices and equipment that you personally use that require electricity. You would include, for example, lights, computers, game systems, washing machines, and so on. If you cook on an electric stove, you would include that. If someone else prepares your meals, do not count it. Set up a table or form like the sample below to record your device data and energy calculations.
2. Some challenging items to estimate are shared or communal equipment, such as refrigerators, hot water systems, and space heating and cooling machinery. What are various ways of estimating your portion of these loads?
3. Determine how much energy is used by an item when it is on. This will be a power rating in watts or kilowatts. The power rating is often on a label attached to the device, and it may be specified in watts, amps, voltage, or volt-amps. If you cannot find it, a list of power ratings for a wide variety of consumer and household appliances can be found quickly through an Internet search. Fill in this information in the accompanying table as well.
4. Estimate how much time you use each item over the course of a week, and complete the table accordingly. When you are done itemizing, add the total watt-hours of electricity you use per week and convert it to kilowatt hours.

Sample

ELECTRICAL APPLIANCE, DEVICE, OR PIECE OF EQUIPMENT	POWER RATING, IN WATTS (Watts = Volts × Amps)	TIME USED PER WEEK, IN HOURS OR FRACTIONS OF	WATT-HOURS USED PER WEEK (Power × Hours Used)
Desk lamp	75 watts	28 hours	2,100
Hair dryer	1,100 watts	1 hour	1,100
Laptop computer	40 watts	42 hours	1,680
Item.........			
Item.........			
Total, in kWh			Divide total by 1,000 to convert to kWh

After you have completed your audit, answer the following questions:

1. What surprised you about your energy use?
2. What did you find most difficult to estimate? Why?
3. What are some opportunities for energy conservation that you can implement into your lifestyle right now?

Problems

8-1 List three different kinds of energy services, and identify at least two ways that each of these energy needs can be met.

8-2 Explain how the technologies that we design to satisfy energy needs affect the type and amount of energy resources that we use.

8-3 What is the energy ladder?

8-4 Calculate per capita energy use and energy intensity for the following countries:
  a. China
  b. Australia
  c. Mexico
  d. Kenya
  e. Germany

8-5 Plot per capita energy use for the five countries in Problem 8-4 on the *y*-axis, and energy intensity on the *x*-axis. Does the position of the country data points make sense? Explain why or why not.

8-6 What is energy poverty, and why is it a problem?

8-7 True or false: Petroleum is the fastest growing form of energy worldwide.

8-8 Calculate aggregate per capita electricity use for the least developed nations as well as low-income, middle-income, and OECD nations (see Table 8.1). What do these variations suggest about the capacity of these nations for economic development?

8-9 What is the average annual rate of depletion of global reserves for liquid fuels, natural gas, and coal assuming no new reserves and no increase in annual production/consumption levels?

8-10 What are the three basic strategies that high-income nations can use to improve their energy security?

8-11 What are the differences between the residential, commercial, and industrial end-use sectors?

8-12 What is the distinction between source energy and site energy?

8-13 Consider the data presented in Figure 8.9.

a. Which end-use sectors do you think are most vulnerable to an energy shock? Why?

b. Which fuel has the most diversified applications? Why?

8-14 Consider the energy content of the different fossil fuels and their EROI. Discuss why it will be difficult to shift the global energy system away from these resources.

8-15 Calculate the energy content of the following in MJ.

a. 10 tons of coal

b. 2 barrels of oil

c. 4,000 kWh

d. A quad of natural gas

8-16 Which has more energy in MJ: 1 toe or 1 bbl?

8-17 Why does a liter of gasoline have slightly more energy than a gallon of crude oil?

8-18 Consider an electric power plant that generates 5,256,000 MWh of electricity per year. How much of the following resources are required to provide this much power? Report your results in metric tons.

a. Fuel oil #6

b. Natural gas

c. A coal blend composed of 60% bituminous coal and 40% anthracite coal

8-19 What is the distinction between a reserve and a resource?

8-20 Refer to the data in Table 8.4 to answer the following questions.

a. Calculate the amount of liquid fuels, natural gas, and coal consumed globally in the base year given.

b. Assume that coal consumption is not flat, but growing at a rate of 4% a year. What would the reserve-to-production ratio be for this growth rate?

c. What would the annual rate of increase in natural gas reserves need to be in order to have a reserve-to-production ratio of 125 years? (Assume consumption does not increase.)

d. Assume that liquid fuels consumption is increasing by 6.3% a year. What would the annual rate of increase in reserves need to be in order to have a reserve-to-production ratio of 75 years?

8-21 Why is the Hubbert Curve for peak oil not a model of the rate of oil depletion?

8-22 Make the argument that uranium is (a) a finite resource, (b) a renewable resource, and (c) and an infinitely renewable resource.

8-23 The following is a table that lists major U.S. cities and their available sun-hours.

a. How much electricity does a 4-kW solar PV system in Chicago generate in a year?

b. How large would a PV system need to be in Fairbanks, New York City, and Miami to generate the same amount of electricity in a year as a 2-kW system in Phoenix?

c. Miami is further south than Phoenix, so in principle it should have more sun-hours per year than Phoenix. What are some reasons why it does not?

CITY	SUN-HOURS	CITY	SUN-HOURS
Fairbanks, Alaska	3.99	Portland, Oregon	4.51
Phoenix, Arizona	6.58	Chicago, Illinois	3.14
Los Angeles, California	5.62	Miami, Florida	5.62
New York City	4.08	San Antonio, Texas	5.30

8-24 Countries in certain kinds of climates do not have the four seasons of winter, spring, summer, and fall. Instead, they have relatively little temperature variation, but "wet seasons" and "dry seasons" that can last for several weeks each. What are the implications of this climate pattern if you wanted to design and install?

    a. A solar hot water heating system

    b. A solar PV system that is independent of the electric power grid (e.g., it stores all of its excess electricity in batteries)

8-25 Rank-order the following sectors in terms of the amount of greenhouse gasses they generate, from highest to lowest: waste, industrial process, agriculture, international transport, and primary energy production.

8-26 True or false: In the United States, the transportation sector generates more greenhouse gas than the electric power sector.

8-27 True or false: Globally, deforestation contributes more to net greenhouse gas emissions than do the residential and commercial building sectors.

8-28 What is EROI, and why is it a helpful energy metric?

8-29 What are the energy conservation gains from the following?

    a. Switching from a single 60-watt incandescent bulb that burns for 6 hours per day to a 13-watt CFL.

    b. Using pure biodiesel instead of motor gasoline for an automobile that travels 60,000 kilometers per year with a fuel economy of 15 km/l.

    c. Replacing a heat pump with a COP of 3.2 with a heat pump with a COP of 2.6 if the heating load is 43.3 million Btu per year.

    d. Replacing an air conditioner with an EER of 11 with a unit that has an EER of 14 if the cooling load is 22 million Btu per year.

    e. Upgrading a natural gas furnace from one with 88% efficiency to one with 92% efficiency if the heating load is 58,000 megajoules per year.

8-30 Estimate the avoided yearly CO_2 emissions from the following.

    a. Switching from a single 60-watt incandescent bulb that burns for 6 hours per day to a 13-watt CFL.

    b. Using pure biodiesel instead of motor gasoline for an automobile that travels 60,000 kilometers per year with a fuel economy of 15km/l.

    c. Replacing an air conditioner with an EER of 11 with a unit that has an EER of 14 if the cooling load is 22 million Btu per year.

    d. Upgrading a natural gas furnace from one with 88% efficiency to one with 92% efficiency if the heating load is 58,000 megajoules per year.

8-31 What are the two different kinds of CHP systems?

8-32 Define and discuss the features of distributed energy.

8-33 What conditions need to be met in order for an appropriate technology to be successful and sustainable over the long run?

8-34 Which aspects of the water–energy nexus are most important where you live?

Open-Ended Design Problem

Design a solar PV system (with battery storage) for a rural village in the Andes Mountains of Bolivia. The community would like to have five days of backup power. This village needs power to meet the following energy needs.

8-1 A community center that can burn three lights for 6 hours each evening.

8-2 A ¼ HP water pump that can supply the village with 400 litres of fresh water daily.

8-3 A radio that will be turned on for 3 hours per day.

8-4 A small refrigerator to store vaccines.

Industrial Ecology

FIGURE 9.1 Environmental impacts are caused by a variety of factors including the products we consume, the quantity we consume, how we make them, and how we dispose of them.

Source: Andy Singer Cartoons http://www.andysinger.com/

The individual serves the industrial system. . . by consuming its products.

—JOHN KENNETH GALBRAITH

GOALS

THE GOAL OF THIS CHAPTER is to introduce the basic principles of industrial ecology and to help students understand the influence of social, economic, and technological factors on environmental impacts by exploring the relationships between industrial and ecological systems.

OBJECTIVES

At the conclusion of this chapter, you should be able to:

- Identify and use the IPAT equation in its variant forms.

- Describe the contributing factors to environmental impact.

- Use equivalence factors to evaluate the contribution of chemical species to

different categories of environmental impact.

- Define the similarities and differences between natural ecological systems and industrial ecological systems.

9.1   Introduction

Ecology is the scientific field devoted to the study of the relationships that organisms have with each other and the natural environment in which they exist. In previous chapters, we have learned that a significant part of sustainable development efforts is concerned with minimizing the environmental impacts associated with activities. We have also learned about the concept of carrying capacity that specifies the sustainable rates of resource consumption and waste discharge within an ecosystem. In ecology studies, both resource consumption and waste discharge are generally linked to the population of organisms in a particular ecosystem. This has led some researchers to conclude that the primary driver of environmental impacts leading to unsustainable development is population growth. However, others have identified various other factors that contribute to environmental impacts. Current industrial systems rely on the extraction, processing, manufacturing, transportation, and distribution of natural resources in the form of products. These different processes and stages rely on technologies that generate waste and emissions. Understanding the interrelationship between people, products, technology, and the environment is necessary for developing and implementing sustainable practices.

9.2   The IPAT Equation

In the early 1970s, environmentalist Paul Ehrlich suggested that environmental impact from human activities was the result of three contributing factors and proposed a conceptualized mathematical formula to represent this concept. This formula has become known as the **IPAT equation**:

$$I = P \times A \times T \tag{9.1}$$

where

I = Environmental Impact

P = Population

A = Affluence

T = Technology

The IPAT equation is often used as a starting point to study the relationships between population, economics, and technological development and how they contribute to environmental impact. However, one of the major challenges of the IPAT equation as a usable mathematical formula is finding quantitative parameters to represent each of the constituent variables. While population is relatively easy to quantify and estimate (as previously discussed in Chapter 1), the terms *impact*, *affluence*, and *technology* require better definition usually within varied contexts. We will discuss the terms before returning to the IPAT equation.

9.2.1 Impact

Environmental impact can be defined in terms of the carrying capacity of the planet. Anything that contributes to diminishing the sustainable rate of resource consumption and waste discharge can be classified as an **impact**. While the Earth has natural mechanisms that replenish most resources and manage most waste discharges, these processes typically occur at very slow rates compared to the rates at which current impacts occur. Current rates of most environmental impacts can be directly linked to human-made developmental activities. Chapters 4 and 5 deal with these concepts.

Impacts can be classified based on the degree of severity and scale of the impact. Critical impacts include global climate change, loss of water quality and quantity, depletion of fossil fuel resources, loss of biodiversity, land degradation, and stratospheric ozone depletion. Critical impacts can potentially lead to irreversible consequences. Significant but noncritical impacts include depletion of nonfossil fuel resources, acid precipitation, smog, and aesthetic degradation. Less significant impacts include radionuclide contamination, depletion of landfill capacity, thermal pollution, oil spills, and odors.

Impacts may also be classified based on the scale of the area directly affected. Hence, global scale concerns like global climate change, ozone layer depletion, and loss of habitat and biodiversity ultimately affect the entire planet regardless of where the source of the impacts originate.

Regional scale concerns have more limited impacts, and these include soil degradation, acid precipitation, and changes in surface water chemistry leading to consequences such as acidification and eutrophication. The impacts associated with the use of herbicides and pesticides may also be included in this category.

The impacts of local scale concerns tend to be localized to the areas of origin of the impacts. These include groundwater pollution, oil spills, hazardous waste sites, and photochemical smog.

Most of these impacts are the direct or indirect results of technological developments and advancements in society. Understanding these classifications and consequently applying them to preventive efforts is central to industrial ecology studies.

9.2.2 Population

Population has been previously covered in Chapter 1, but we will review it again in this chapter in the context of industrial ecology and the IPAT equation. **Population** refers to the number of individual organisms present in a particular ecological system. Since organisms consume resources and generate waste as part of their metabolic activities, the population of an ecological system has a direct impact on the sustainability of that system. As described in Chapter 1, population growth follows an exponential mathematical relationship, which can be represented by

$$P(t) = P_0 e^{r(t - t_0)} \tag{9.2}$$

where

$P(t)$ = the projected population at a time t

P_0 = the initial population at initial time t_0

r = the population growth rate, usually measured per 1,000 people per year

Equation (9.2) shows that a positive value of r will lead to population growth, while a negative value is required for the total population to decline with time. A value of zero for r means there will be no change in total population over time.

For a given area or region of study, the effective growth rate, r, is calculated from four factors:

- The rate of birth, r_b
- The rate of death, r_d

- The rate of immigration into the area, r_i
- The rate of emigration from the area, r_e

$$r = r_b + r_i - r_d - r_e \qquad (9.3)$$

If we take the entire planet as our area of interest, we can conclude that the global population growth rate is not affected by immigration and emigration from Earth (not yet anyway). Therefore, the birth rate, r_b, and the death rate, r_d, are the contributing factors to population growth.

To reduce the rate of population growth or even reverse the growth trend, the rate of birth has to decrease or the rate of death has to increase, or a combination of both. However, advances in health care, public health education and policies, as well as modern methods of resolving social and political conflicts all contribute to increases in the average global life expectancy. In general, people are living longer today than in previous generations. As shown in Table 9.1, data from the Population Division of the United Nations, Department of Economic and Social Affairs, indicates that the global death rate, measured in deaths per 1,000 persons per year, is currently at a historical low, with projections that the death rate may start to rise in the next decade. These estimates were calculated from information collected for

TABLE 9.1 World historical and predicted crude death rates. Medium variant projection.

YEAR(S)	CRUDE DEATH RATE (per 1,000 per year)	YEAR(S)	CRUDE DEATH RATE (per 1,000 per year)
1950–1955	18.713	2030–2035	8.728
1955–1960	17.135	2035–2040	9.112
1960–1965	16.122	2040–2045	9.514
1965–1970	12.964	2045–2050	9.896
1970–1975	11.751	2050–2055	10.239
1975–1980	10.636	2055–2060	10.553
1980–1985	10.097	2060–2065	10.834
1985–1990	9.554	2065–2070	11.066
1990–1995	9.245	2070–2075	11.256
1995–2000	8.974	2075–2080	11.393
2000–2005	8.689	2080–2085	11.463
2005–2010	8.390	2085–2090	11.479
2010–2015	8.194	2090–2095	11.492
2015–2020	8.159		
2020–2025	8.239		
2025–2030	8.426		

Source: Based on UNdata, http://data.un.org/

"every country in the world, including estimates and projections of 60 demographic indicators such as infant mortality rates and life expectancy." These values are for medium variant projections that balance between the upper and lower bound projection values.

Intentionally increasing the death rate as a means of reducing the global population growth rate is clearly morally and ethically unacceptable. It is also unarguable that one of the primary roles of engineering in society is to improve the quality of life, thereby increasing life expectancy. So we are left with developing methods that will lead to reductions in the birth rate.

In his controversial article written in 1968 titled "The Tragedy of the Commons," Garret Hardin, an ecologist, proposed the theory that, if unrestricted, most people will seek to maximize the benefit to themselves with little regard for the ultimate consequences for the common good, thereby leading to a tragedy that befalls all. He concluded that overpopulation would lead to a limitation of access to the finite Earth resources through overconsumption and environmental degradation and is thus an example of a tragedy of the commons. He insisted that "freedom to breed will bring ruin to all." Though many have agreed and many have disagreed with his suppositions, one of the most controversial points he advocated was the use of governmental action and the mechanisms of state to limit individual rights to procreate.

The social, cultural, economic, and religious dispositions that inform attitudes toward procreation as well as the basic principles of personal liberty promoted by many nations in the modern world make Garret Hardin's proposal unpopular and impractical.

While governmental action by way of taxation have been proposed, like the one instituted in the one-child policy of China, this has also been controversial, with the current tax laws in many countries designed to benefit families with more children rather than penalize them. Hence, current efforts to curb population growth by reducing the birth rate are limited to programs of education, enlightenment, and provision of access to contraception.

Even with reductions in the growth rate, in the foreseeable future it is projected that the overall global population growth rate, r, will remain positive, leading to an overall increase in global population over time.

9.2.3 Affluence

Affluence is a bit more difficult to define and quantify. In general, it refers to a measure of the quality of life of individual members of the society. It represents the complex link between the economic well-being of the society as a whole and the consumption patterns of the average member of that society. One indicator of affluence is the gross domestic product (GDP), which is an estimate of the market value of all goods and services produced. The GDP can be measured on a per capita basis, as illustrated in Figure 9.2. Figure 9.2 illustrates the disproportion of economic means allocated on a per capita basis, but GDP alone is insufficient to adequately describe the broader term, *affluence*. The easiest way to think about affluence is to consider the access to and consumption of resources by members of the society. Figure 9.3 illustrates the correlation between GDP, population, and consumption. Consumption refers to the selection, use, reuse, maintenance, repair, and disposal of goods and services (D. Leslie 2009). It has also been defined as the "human transformations of materials and energy" (Myers, 1997). Consumption becomes

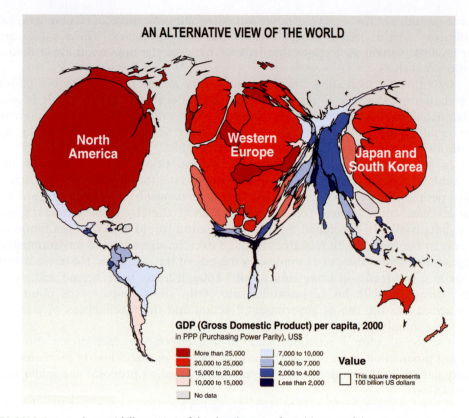

FIGURE 9.2 A carotid illustration of the distribution of wealth sizes of the countries according to their relative financial status based on GDP per capita.

Source: GRID-Arendal, World economy cartogram, http://www.grida.no/graphicslib/detail/world-economy-cartogram_1551

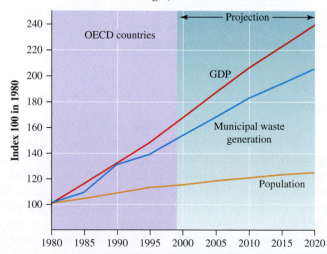

FIGURE 9.3 The correlation between GDP, population, and consumption shows that according to various scenarios, the GDP will most likely continue for the next decades—but at a slower pace for those countries that can afford advanced waste management strategies.

Source: Based on GRID-Arendal, The richer we get, the more we discard - human consumption, waste and living standards, 2005, http://www.grida.no/graphicslib/detail/the-richer-we-get-the-more-we-discard-human-consumption-waste-and-living-standards_5bcc

a problem when it makes materials and energy less available for future use and threatens human health and welfare.

Professor Thomas Princen of the University of Michigan has identified three layers of consumption:

- **Background Consumption**: This term refers to the normal biological functioning of all organisms to meet physical and/or psychological needs in order to survive and reproduce. The total impact of background consumption is a function of the aggregate consumption of the total population. In this case, population is the significant driver of impact.
- **Overconsumption**: This term refers to a level of consumption based on individual and collective choices that undermine a species' own life support system. In this instance, individual behavior may seem rational and conform to societal norms or dictates, yet the aggregate effect is injurious to the collective.
- **Misconsumption**: Misconsumption is when the individual consumes in a way that undermines his or her own well-being even if there are no aggregate effects on the collective.

People sometimes mistake overconsumption for misconsumption and vice versa. Consumption trends are often correlated with economic metrics, and more affluent societies tend to consume more resources per capita than less affluent ones.

The observable trend is that consumption per capita is increasing over time. For example, evidence suggests that Americans today consume more natural resources and artifacts per capita than they did in the past (Putnam and Allshouse, 1999), as was shown in Figure 3.2. The per capita purchasing power and income of individuals in less affluent nations are also on the rise as many nations aspire to Western standards, leading to more overall consumption (see Figures 9.4 and 9.5). There are many potential indicators of consumption, including the consumption of energy, which is used for manufacturing, home necessities, and electricity production. Figure 9.4 shows the global disparity of energy consumption—with North America and northern Europe having very high rates of energy consumption per capita, due in part to the combination of lifestyle and climate factors. Figure 9.5 also shows that energy consumption is a very good indicator of the number of personal computers owned.

Social scientists have proposed various theories on why people consume beyond what they need for survival. Reasons include status, pleasure, display, convenience, and the result of marketing (Wilk, 2002). It can be said that satisfaction now depends less on what a person possesses or has access to, in an absolute sense, than on attaining aspirations developed out of social expectations. In the past, consumption beyond what was necessary was discouraged within religious, legal, and cultural frameworks in most societies. However, those deterrents are largely absent in modern societies. In fact, it can be said that there now exists a cultural and economic imperative to increase consumption. Aggressive consumerism has sometimes been correlated with patriotism, with the vindication being that it drives economic prosperity, creates jobs and better wages, and enhances overall productivity.

The problem, as noted by anthropologist Willett Kempton, is that the sense of well-being will always be relative, generating an unending spiral of increasing consumption (Kempton, 2001). So the implication of population growth for the carrying capacity of Earth is magnified by the fact that the average person today

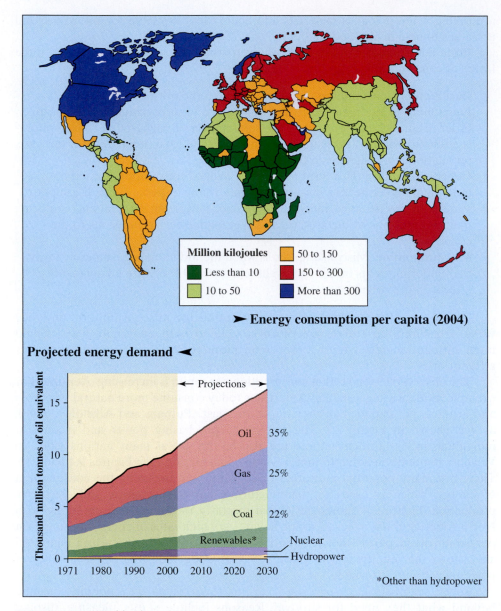

FIGURE 9.4 Worldwide energy consumption per capita.

Source: Based on GRID-Arendal, Energy consumption per capita (2004), 2006, http://www.grida.no/graphicslib/detail/energy-consumption-per-capita-2004_5dca

is consuming significantly more than the average person of a few decades ago. Moreover, the overconsumption of goods leads to increased waste production. Of the billion personal computers in the world, all will sooner or later end up as waste. The computer waste or "e-waste" is composed of a mixture of metals, plastic, glass, and electronic boards (see Figure 9.6). The higher GDP per capita countries in Europe and North America often ship their e-waste to East Asian countries with fewer regulations and policies for disposal of these hazardous materials (Figure 9.7).

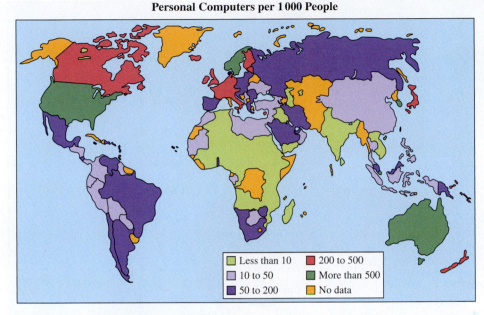

Personal Computers per 1000 People

Legend:
- Less than 10
- 10 to 50
- 50 to 200
- 200 to 500
- More than 500
- No data

FIGURE 9.5 The worldwide distribution of over a billion personal computers in the world. In developed countries these have an average lifespan of only 2 years. In the United States alone there are over 300 million obsolete computers.

Source: Based on GRID-Arendal, Personal computers per 1000 people, 2005, http://www.grida.no/graphicslib/detail /personal-computers-per-1000-people_e1e7

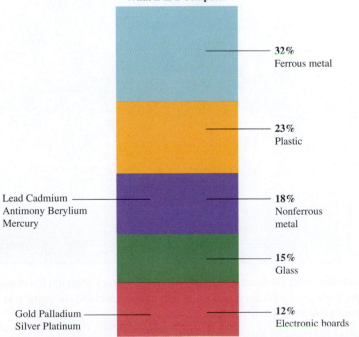

What is in a Computer

- **32%** Ferrous metal
- **23%** Plastic
- Lead Cadmium Antimony Berylium Mercury — **18%** Nonferrous metal
- **15%** Glass
- Gold Palladium Silver Platinum — **12%** Electronic boards

FIGURE 9.6 A computer is composed of a mixture of metals, glass, plastics, and electronic boards. Only about 50% of the computer can be recycled. Of the remaining nonrecyclable wastes, lead, mercury, and cadmium represent significant hazards.

Source: Based on GRID-Arendal, What is in a computer, 2005, http://www.grida.no/graphicslib/detail/what-is-in-a -computer_134a

FIGURE 9.7 e-waste recycling involves the major producers in North America and Europe shipping the obsolete products to Asia, eastern Europe, and Africa. However, there are significant economic, environmental, and societal impacts associated with the processing and often unregulated disposal of these waste products.

Source: Based on GRID-Arendal, http://www.grida.no/graphicslib/detail/who-gets-the-trash_10

TABLE 9.2 Consequences of technology-based solutions to technology-caused problems

YESTERDAY'S NEED	YESTERDAY'S SOLUTION	TODAY'S PROBLEM
Nontoxic, nonflammable refrigerants	Chlorofluorocarbons	Ozone hole
Automobile engine knock	Tetraethyl lead	Lead in air and soil
Locusts, malaria	DDT	Adverse effects on birds, mammals
Fertilizer to aid food production	Nitrogen and phosphorous fertilizer	Lake and estuary eutrophication

9.2.4 Technology

Technology refers to all the artifacts of human development as well as the processes and systems that contribute to these artifacts. It can be said that observed trends in consumption are driven directly or indirectly by achievements made in technological development. The role of technology in facilitating social and economic development is not as contentious as the question of whether technology is to be held responsible for unsustainable trends and environmental impacts. Societies that have harnessed the benefits of technological advancements have ultimately become more socioeconomically prosperous.

In the past, the development and deployment of technology was driven primarily by need and effectiveness in dealing with immediate problems, with little or no regard accorded to potential downstream consequences and implications. Some of those technology-based solutions did consequently lead to modern-day problems and concerns (Table 9.2). For this reason, technological solutions may only lead to more unforeseen challenges. So while technology fosters positive trends in social and economic development, unmanaged it could also nullify those achievements by its potential impacts. In industrial ecology, we study industrial and technological systems and the relationships these systems have with the environment in order to minimize the overall impacts.

9.2.5 Application of the IPAT Equation

The IPAT equation [Equation (9.1)] implies that impacts occur owing to a combination of technological and socioeconomic factors and provides a basis for evaluating the parameters of an acceptable technological solution. One useful feature of the IPAT equation is that it can be written in many variants to represent the same relationship in different contexts and you can develop your own variant to suit your purpose.

Consider one variant of the IPAT equation:

Total impact =

$$\text{number of people} \times \frac{\text{number of units of technology}}{\text{person}} \times \frac{\text{impact}}{\text{unit of technology}} \qquad (9.4)$$

The number of people represents the population factor, P.

The number of units of technology per person represents the affluence factor, A.

And the impact per unit of technology represents the technology factor, T.

Using the IPAT equation, we can estimate how the values of the contributing factors affect the total impact as well as what changes can be made to minimize impact. Based on our previous discussions on the rising trends in population (P) and consumption (A), to keep the current total impact value unchanged requires significant reduction in the value of the technology factor (T). And to reverse the trend and bring the impact values to naturally manageable equilibrium values, even more significant reductions must be achieved.

For example, consider the impact associated with the combustion of gasoline in automobiles. The overall impact is related to the technology of the internal combustion engine, which burns gasoline and emits carbon dioxide, contributing to global climate change, but it is also a function of how many people use this technology, that is, how many cars are driven and how much gasoline is consumed per car—that is, average distance traveled.

This relationship may be represented by an equation like this

$$\text{Total } CO_2 \text{ emmitted} = \text{no. of automobiles in use}$$
$$\times \text{ average distance traveled per automobile}$$
$$\times CO_2 \text{ emitted per distance traveled} \qquad (9.5)$$

To reduce the total CO_2 emitted from gasoline powered automobiles, one could

- Reduce the total number of automobiles in use
- Reduce the average distance traveled per automobile
- Reduce the CO_2 emitted per distance traveled

The U.S. Department of Transportation study of travel trends in the United States shows that since 1970 the increase in vehicle miles traveled has far outpaced population growth (Figure 9.8). This trend aligns closely with economic trends as seen with the GDP. The average annual driving distance in the United States increased by nearly 300% in the three decades from 1970 to 2000. A similar study by the Department of Energy shows that the number of vehicles in the United States is growing faster than the population and that the percentage of households with three or more vehicles has increased significantly over the past 50 years (Table 9.3).

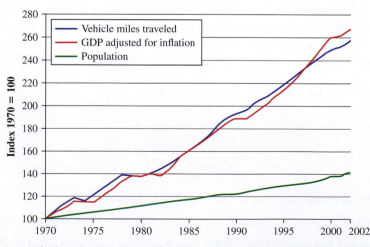

FIGURE 9.8 Vehicle miles traveled as a function of population and GDP in the United States.

Source: U.S. Department of Transportation, Transportation Air Quality Facts and Figures January 2006, Travel Trends, http://www.fhwa.dot.gov/environment/air_quality/publications/fact_book/page06.cfm

TABLE 9.3 Household vehicle ownership

YEAR	NO VEHICLES	ONE VEHICLE	TWO VEHICLES	THREE OR MORE VEHICLES
1960	21.53%	56.94%	19.00%	2.53%
1970	17.47%	47.71%	29.32%	5.51%
1980	12.92%	35.53%	34.02%	17.52%
1990	11.53%	33.74%	37.35%	17.33%
2000	9.35%	33.79%	38.55%	18.31%
2009	8.90%	33.69%	37.56%	19.85%

Source: Based on 1960–2009 Census. U. S. Department of Transportation, Volpe National Transportation Systems Center, Journey-to-Work Trends in the United States and its Major Metropolitan Area, 1960–1990, Cambridge, MA, 1994, p. 2–2. 2000 data – U.S. Bureau of the Census, American Fact Finder, factfinder.census.gov, Table QT-04, August 2001. (Additional resources: www.census.gov) 2009 data – U.S. Bureau of the Census, American Community Survey, Table DP-4, 2009. http://cta.ornl.gov/data/chapter8.shtml

For example, the total CO_2 emitted represents the impact, I; the number of automobiles in use represents the population, P; the average distance traveled per automobile represents a measure of affluence, A; and the CO_2 emitted per distance traveled represents the technology, T. In this case, it gives an indication of the efficiency of the engine. Based on the observed trends (Figures 9.9 and 9.10) and the fact that options 1 and 2 are related to socioeconomic factors previously discussed, a technological approach to reducing total CO_2 emissions would be to reduce the CO_2 emitted per distance traveled. More efficient engines will have less CO_2 emitted per distance traveled by requiring less gasoline to travel.

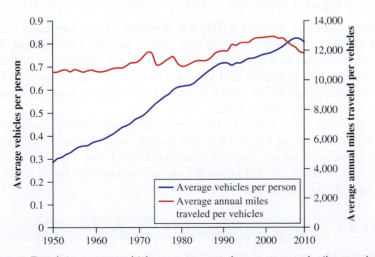

FIGURE 9.9 Trends in average vehicles per person and average annual miles traveled per vehicle in the United States.

Sources: Resident population and civilian employed persons – U.S. Department of Commerce, Bureau of the Census, Statistical Abstract of the United States–2011, Washington, DC, 2011, tables 2, 59, 601, and annual. (Additional resources: www.census.gov). Licensed drivers and vehicle-miles – U.S. Department of Transportation, Federal Highway Administration, Highway Statistics 2009, Tables DL-1C and VM-1, and annual. (Additional resources: www.fhwa.dot .gov) http://cta.ornl.gov/data/chapter8.shtml

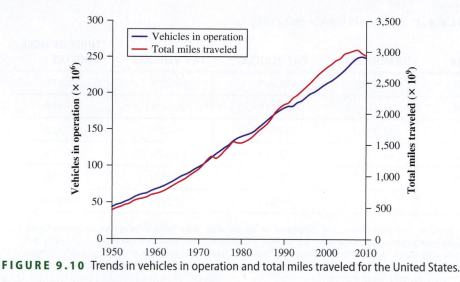

FIGURE 9.10 Trends in vehicles in operation and total miles traveled for the United States.

Sources: Resident population and civilian employed persons – U.S. Department of Commerce, Bureau of the Census, Statistical Abstract of the United States–2011, Washington, DC, 2011, tables 2, 59, 601, and annual. (Additional resources: www.census.gov) and Licensed drivers and vehicle-miles – U.S. Department of Transportation, Federal Highway Administration, Highway Statistics 2009, Tables DL-1C and VM-1, and annual. (Additional resources: www.fhwa.dot.gov) http://cta.ornl.gov/data/chapter8.shtml

The IPAT equation allows us to calculate what rate of reduction of CO_2 emissions per distance traveled we would need to achieve to balance the effect of the increasing number of automobiles in use and average distance traveled per automobile.

Another variant of the IPAT equation states

$$\text{Total impact} = \text{population} \times \$\text{GDP per capita} \times \text{impact per }\$\text{GDP} \qquad (9.6)$$

The gross domestic product, GDP, is a measure of the monetary value of all goods and services produced by a country over a specific time period. The GDP per capita of a country is the total GDP divided by the population of the country. It is a measure of the affluence and standard of living of the country. While some countries may have high GDP values, their GDP per capita may be comparatively low because of high population values.

The impact per GDP refers to the impact of all activities associated with the generation of one unit of GDP. This may include manufacturing, transportation, packaging, sales, and so on. The value of the impact per GDP of a country is often seen as a measure of that country's environmental efficiency. More efficient countries have lower impact per GDP values.

EXAMPLE 9.1   Using the IPAT Equation

The population of a country is 256 million, and the annual GDP per capita is estimated at \$49,389 per person. If the total CO_2 emission of the country is estimated at 156 teragram (Tg) for the year, calculate the value of the CO_2 emitted per dollar of GDP.

Using the IPAT equation in the form

$$\text{Total }CO_2\text{ emitted} = \text{population} \times \$\text{GDP per capita} \times CO_2\text{ emitted per }\$\text{GDP}$$

or

$$\text{Total } CO_2 \text{ emitted} = \text{population} \times \frac{\$GDP}{\text{population}} \times \frac{CO_2 \text{ emitted}}{\$GDP}$$

$$156 \times 10^{12}g = 256 \times 10^6 \text{ people} \times \frac{\$49{,}389}{\text{person}} \times \frac{CO_2 \text{ emitted}}{\$GDP}$$

$$\frac{CO_2 \text{ emitted}}{\$GDP} = \frac{156 \times 10^{12}\, g}{256 \times 10^6 \text{ people} \times \dfrac{\$49{,}389}{\text{person}}} = 12.34 \; \frac{g\, CO_2}{\$}$$

EXAMPLE 9.2   Using the IPAT Equation

If the population of a country increased by 8% in the same time that the GDP per capita increased by 5%, what percentage change in the emissions per dollar of GDP would be required to maintain the total emissions at the original values?

Solution

Using the IPAT equation

$$I = P \cdot A \cdot T$$

$$\text{Impact} = \text{population} \times \text{affluence} \times \text{technology}$$

$$\text{Total emissions} = \text{population} \times \frac{\$GDP}{\text{population}} \times \frac{\text{emissions}}{\$GDP}$$

Let I_i, P_i, A_i, and T_i represent the initial impact, population, affluence, and technology factors, respectively, and I_f, P_f, A_f, and T_f represent the final impact, population, affluence, and technology factors, respectively.

If the final impacts are not to be greater than the initial impacts, then

$$I_i = I_f$$

then substituting from Equation (9.1) yields

$$P_i \times A_i \times T_i = P_f \times A_f \times T_f$$

From the information given with a population increase of 8%,

$$P_f = P_i + \left(\frac{8}{100} \times P_i\right) = 1.08 P_i$$

GDP per capita, the measure of affluence increased by 5% is

$$A_f = A_i + \left(\frac{5}{100} \times A_i\right) = 1.05 A_i$$

Substituting the relationships between final and initial population and affluence into the previous equation yields

$$P_i \times A_i \times T_i = 1.08 P_i \times 1.05 A_i \times T_f$$

Solving for the technology terms, T:

$$T_i = 1.134 T_f$$

$$\frac{T_f}{T_i} = \frac{1}{1.134} = 0.88 \quad or \quad T_f = 0.88 T_i$$

The technology factor defined by the emissions per dollar of GDP needs to be reduced by 12%.

9.3 Resource Allocation

One of the core motivations for sustainable development is the principle that humanity in the current generation should be able to pursue progress and development in a manner that does not jeopardize the ability of future generations to do the same. Entrenched in this notion is the principle that has come to be known as **intergenerational equity**. Along with this principle is the supposition that achieving true progress and development would require equitable distribution of available natural resources among people in every generation. Social and economic justice have already been identified as integral to human development. This principle is defined as intragenerational equity. Communities advancing toward sustainable development must develop and implement principles for the planning and equitable management of natural resources.

9.3.1 Intergenerational Equity

In industrial ecology, the stock of a resource refers to the quantity of that resource that is available within a system over a period of time. The total stock of available natural capital comprises renewable resources that can be regenerated like forests and fish stocks; the natural system's capacity to absorb waste discharges and convert them into useful or benign products; and nonrenewable resources that are not regenerated once extracted—for example, minerals and fossil fuels.

If renewable resources are not consumed at a rate greater than their rate of regeneration and if waste discharges are maintained at or below the capacity of the natural system, the viability of the first two items of natural capital can, in theory, be maintained over generations.

Maintaining the viability of nonrenewable resources however, poses a challenge. Nonrenewable resources are by definition exhaustible, and the notion of one generation being able to pass on sufficient quantities to the succeeding generations appears counterintuitive. Added to this challenge is the lack of knowledge about what successive generations would require for their own development. What one generation demands may not be valued by another. One proposed solution to this problem is to steadily substitute nonrenewable resources with renewable alternatives while keeping consumption rates at or below regeneration rates. The consumption of nonrenewable resources could then be increased only if there is a danger of exceeding the consumption/regeneration rates of renewable resources.

9.3.2 Intragenerational Equity

Sustainability often takes on different meanings between developed and developing countries. While developed countries focus on economic development while simultaneously preserving the natural environment, developing countries tend to concentrate on increasing the number of people with access to adequate shelter, potable water, effective waste disposal, employment, and food. It is therefore not strange that developing countries will often mimic the path of developed countries to achieving this end. The notion exists that developed countries have reached their levels of economic prosperity by uneven distribution and utilization of global natural resources. Models show that if the living standards of the average citizen of developed nations are applied to the entire earthly population, the resources from a multiple number of earths would be required. For example, it is estimated that if everyone in the world consumed resources at the rate the average American does, we would need four Earths to meet the needs of everyone (Source: *www.footprintnetwork.org*). However, if everyone adopted the current resource consumption pattern of the average Nigerian, 0.8 earths would be sufficient. Intragenerational equity implies an equitable distribution of global resources for the population within a generation.

9.3.3 Implications for Technology

Resource allocation is a method of allocating available resources in a sustainable way that has been proposed to advance intergenerational and intragenerational equity in sustainable development and planning. It starts with identifying the total stock of a particular resource that is available to be used sustainably over a specific time period and distributing this resource equally among the population. The resulting value indicates the allocation to individual members of the population and puts a limit on how much of that resource individuals can sustainably use over specific time periods.

For example, if we estimate the maximum sustainable amount of greenhouse gases that can be emitted into the atmosphere without causing impacts on climate and other natural processes, then we can divide this number by the global population to establish the allowable emissions that each person in the world is allocated. This emissions per capita value would be the total from various activities such as driving, air travel, buying manufactured goods, food (especially meats), and burning fuel to heat houses using electricity.

While this approach may not be realistic nor in itself sustainable because of the limitations it places on individual desires for affluence and other social factors, it is nonetheless a starting point in the use of resources for development of technology.

One potential application is in estimating how much of a particular resource should be included in the sustainable design of a technological artifact to be used over a specified period. This could also be applied to environmental quality by determining the allocation of sustainable rate of waste discharge per capita and incorporating this limit in processes that generate environmental emissions in the production of items of technology.

9.4   Technology, Emissions, and Impacts

Modern societies engage in various technology-dependent activities such as crop production, animal husbandry, transportation, energy production and distribution, and industrial manufacturing. These activities usually result in impacts to the environment in one form or another. Three main types of impacts can be identified.

- **Depletion**—The removal of resources from the environment
- **Pollution**—The emission of waste and dispersion of environmentally harmful materials
- **Disturbances**—Detrimental changes to the environment

Sustainable engineering challenges us to develop alternatives that will substitute current technologies and the associated emissions and impacts. However, to do this, we need to be able to quantify impacts in order to properly evaluate if the alternatives we propose are truly better. Correlating emission sources, emitted species, and the resultant impacts is necessary in quantifying the collective impact due to technology use.

A single emission source could result in multiple impacts, while multiple emission sources could lead to a single impact. So, for example, a global scale impact like climate change has been linked to the emission of multiple chemical species such as carbon dioxide, methane, nitrous oxide, and chlorofluorocarbons (Table 9.4). These chemicals are emitted in varying quantities and contribute to this impact in varying degrees. Moreover, the amount of time that each of these impact-causing chemicals are active in the environment also varies widely.

Since emissions occur from various different sources at different times and since the quantity of each chemical species emitted can vary from source to source and from time to time, common indices called impact potentials were developed to help quantify the total impact anticipated from various sources. This helps us compare these impacts even if the specific chemicals emitted may be different. Next, we will discuss these impact potentials to better understand what they mean and how they can be used.

TABLE 9.4 Emissions from some technology-based activities.

ACTIONS	PRIMARY EMISSIONS
Biomass combustion for energy demand Land clearing for agriculture, grazing, construction	CO_2, CO, NOx, methane, hydrocarbons, particulates
Crop production	CO_2 from soil organic matter, methane from some crops, N_2O, nitrates, and phosphates from fertilizer, herbicides, pesticides
Animal husbandry	Methane, ammonia, hydrogen sulfide

Carbon-based energy production and use
 	
    Coal
 	
    Oil
 	
    Gas
	Mining/production:
	
	    Drilling waste, CO_2, hydrocarbons, methane
	
	combustion:
	
	    CO_2, CO, hydrocarbons, NOx, SO_2, particulates

Incineration:	CO_2, HCl, metals, dioxins
Landfills:	Methane, water contaminants
Industrial manufacturing	Chlorofluorocarbons, CO_2, particulates, hydrocarbons, metals

Source: Bradley Striebig.

9.4.1 Greenhouse Gases and Global Warming Potential (GWP)

Consider this: The greenhouse effect occurs because of the presence of greenhouse gases (GHG) in the atmosphere. The molecular vibrations of greenhouse gases occur at frequencies that enable them to absorb and emit infrared radiation. The most abundant GHGs in the atmosphere are water, H_2O (yes, water is a greenhouse gas!); carbon dioxide, CO_2; methane, CH_4; nitrous oxide, N_2O; and ozone, O_3 as discussed in Chapter 6. Other relevant GHGs include sulfur hexafluoride and halocarbons like hydrofluorocarbons (HFCs), and perfluorocarbons (PFCs).

The greenhouse gases come from both natural and anthropogenic sources. N_2O is emitted from the combustion of fuels and more importantly from fertilizers. Halocarbons are industrial products, such as chlorofluorocarbons (CFCs) and hydrochlorofluorocarbons (HCFCs), that are used as refrigerants and solvents. Methane is emitted from natural gas releases, coal mining, sewage treatment plants, landfills, cows, rice paddies, and so on. The major source for carbon dioxide is burning fossil fuel. Another major contribution to the build-up of carbon dioxide in the atmosphere is change in land use, due mostly to tropical deforestation.

Greenhouse gases vary in their capacity to trap infrared radiation. For example, one molecule of methane is 26 times more effective at IR absorption than one molecule of CO_2. For this reason, the concept of **radiative forcing** was developed and is used as a quantitative comparison of the strength of different greenhouse gases in their contribution to global warming. Radiative forcing is a measure of the influence a substance has in altering the balance of incoming and outgoing energy in the Earth's atmosphere. It is the difference between the energy received from the sun and the energy reflected back into space caused by a substance and measured in Watts per square meter (W/m^2) of Earth surface. See Table 9.5.

The contribution of any greenhouse gas to global warming also depends on its lifetime in the atmosphere. Some compounds are unstable and highly reactive, so they have short lifetimes. Others are stable and unreactive, so they stay in the atmosphere for longer periods of time. Typically, the rate of decay is proportional to the concentration, C, and follows a first-order kinetic relationship:

$$\frac{dC}{dt} = -kC \tag{9.7}$$

$$\frac{C}{C_0} = e^{-kt} \tag{9.8}$$

where

$$k = \frac{1}{\tau} \tag{9.9}$$

and τ is the atmospheric lifetime.

To quantify the contribution of GHGs to global warming, a common index was developed called the **global warming potential (GWP)**. CO_2, being the most significant contributor to anthropogenic radiative forcing, is used as the benchmark. The relative effectiveness of a unit mass of greenhouse gas when compared to an equivalent unit mass of CO_2 is its global warming potential (GWP) and is calculated from

$$GWP_i = \frac{\int_0^{ITH} a_i C_i \, dt}{\int_0^{ITH} a_{CO_2} C_{CO_2} \, dt} \tag{9.10}$$

where a_i and a_{CO_2} are the radiative efficiencies of substance i and CO_2, respectively.

TABLE 9.5 Greenhouse gases with significant contributions to radiative forcing

GREENHOUSE GAS	AVERAGE ATMOSPHERIC CONCENTRATION	RADIATIVE FORCING (W/m²)
CO_2	379 ± 0.65 ppm	1.66
CH_4	$1,774 \pm 1.8$ ppb	0.48
N_2O	319 ± 0.12 ppb	0.16
CFC-11	251 ± 0.36 ppt	0.063
CFC-12	538 ± 0.18 ppt	0.17
CFC-113	79 ± 0.064 ppt	0.024
HCFC-22	169 ± 1.0 ppt	0.033
HCFC-141b	18 ± 0.068 ppt	0.0025
HCFC-142b	15 ± 0.13 ppt	0.0031
CH_3CCl3	19 ± 0.47 ppt	0.0011
CCl_4	93 ± 0.17 ppt	0.012
HFC-125	3.7 ± 0.10 ppt	0.0009
HFC-134a	35 ± 0.73 ppt	0.0055
HFC-152a	3.9 ± 0.11 ppt	0.0004
HFC-23	18 ± 0.12 ppt	0.0033
SF_6	5.6 ± 0.038 ppt	0.0029
CF_4 (PFC-14)	74 ± 1.6 ppt	0.0034
C_2F_6 (PFC-116)	2.9 ± 0.025 ppt	0.0008

Source: Based on IPCC Fourth Assessment Report, 2007, Chapter 2 http://www.ipcc.ch/publications_and_data/ar4/wg1/en /ch2s2-3.html#table-2-1

The radiative efficiency is the radiative forcing that occurs due to a unit increase in the atmospheric abundance of a substance, that is, radiative forcing $(W \cdot m^{-2})$/unit change in concentration C_i and C_{CO_2} are the concentrations of substance i and CO_2, respectively.

As we have previously discussed, the concentration of substances in the environment change over time at different rates. Unstable and reactive substances will decay at faster rates and thus have shorter residence times in the environment regardless of the value of their radiative forcing efficiencies. Stable substances, on the other hand, will stay longer in the environment, causing impacts for longer periods.

For example, methane remains in the atmosphere for approximately a decade, while CO_2 has a variable lifetime, with about 50% being removed from the atmosphere within a century and about 20% remaining for thousands of years. Consequently, in the near term a given mass of methane has a greater contribution to global warming than the equivalent mass of CO_2 released at the same time, but after 100 years the methane will have long been removed from the atmosphere and the CO_2 will still be present continuously, contributing to global warming. For this reason, the impact of greenhouse gases in the atmosphere is also dependent on the time that has passed since the gasses were emitted. The **integrated time horizon (ITH)** allows us to calculate the relative impact of substances over different time periods. The integrated time horizon is typically chosen as 20, 100, and 500 years to model short-term, medium-term, and long-term global warming potential values.

From Equation (9.9), we can see that that the GWP of CO_2 will be equal to 1. Carbon dioxide (CO_2), methane (CH_4), and nitrous oxide (N_2O) are released in large volumes but have relatively low global warming potentials. Other gases, such as perfluorocarbons, are released in much smaller volumes but have much higher global warming potentials.

The product of the global warming potential for a greenhouse gas and the emitted mass gives equivalent mass of CO_2 that would result in the same radiative forcing effect.

$$\text{Equivalent mass } CO_2 = GWP_i \times m_i \qquad (9.11)$$

where m_i is the emitted mass of the greenhouse gas.

For emissions containing multiple greenhouse gases, the equivalent CO_2 values of each GHG may be summed to obtain the **total equivalent CO_2**.

$$\text{Total equivalent mass } CO_2 = \sum_i (GWP_i \times m_i) \qquad (9.12)$$

The total equivalent CO_2 allows us to compare the contributions to global warming by two or more systems that may or may not have identical emitted species in different quantities.

EXAMPLE 9.3 Calculating Equivalent CO_2 using Global Warming Potentials

Calculate the equivalent CO_2 for 40 kg of methane.

Solution

Using Equation 9.11,

$$\text{Equivalent mass } CO_2 = GWP_i \times m_i$$

The GWP value for methane obtained from Table 9.6 using the 100-year time horizon is 25.

$$\text{Equivalent mass } CO_2 = 25 \times 40 \text{ kg} = 1,000 \text{ kg equivalent } CO_2$$

TABLE 9.6 Global warming potential for selected greenhouse gases

INDUSTRIAL DESIGNATION OR COMMON NAME	CHEMICAL FORMULA	APPROX. LIFETIME (years)	GLOBAL WARMING POTENTIAL FOR GIVEN TIME HORIZON		
			20-yr	100-yr	500-yr
Carbon dioxide	CO_2	Variable	1	1	1
Methane	CH_4	12	72	25	7.6
Nitrous oxide	N_2O	114	289	298	153
CFC-11	CCL_3F	45	6,730	4,750	1,620
CFC-12	CCL_2F_2	100	11,000	10,900	5,200

(Continued)

TABLE 9.6 *(Continued)*

INDUSTRIAL DESIGNATION OR COMMON NAME	CHEMICAL FORMULA	APPROX. LIFETIME (years)	GLOBAL WARMING POTENTIAL FOR GIVEN TIME HORIZON		
			20-yr	100-yr	500-yr
CFC-13	$CClF_3$	640	10,800	14,400	16,400
CFC-113	CCl_2FCClF_2	85	6,540	6,130	2,700
CFC-114	$CClF_2CClF_2$	300	8,040	10,000	8,730
CFC-115	$CClF_2CF_3$	1,700	5,310	7,370	9,990
Halon-1301	$CBrF_3$	65	8,480	7,140	2,760
Halon-1211	$CBrClF_2$	16	4,750	1,890	575
Halon-2402	$CBrF_2CBrF_2$	20	3,680	1,640	503
Carbon tetrachloride	CCl_4	26	2,700	1,400	435
Methyl bromide	CH_3Br	0.7	17	5	1
Methyl chloroform	CH_3CCl_3	5	506	146	45
HCFC-22	$CHClF_2$	12	5,160	1,810	549
HCFC-123	$CHCl_2CF_3$	1.3	273	77	24
HCFC-124	$CHClFCF_3$	5.8	2,070	609	185
HCFC-141b	CH_3CCl_2F	9.3	2,250	725	220
HCFC-142b	CH_3CClF_2	17.9	5,490	2,310	705
HCFC-225ca	$CHCl_2CF_2CF_3$	1.9	429	122	37
HCFC-225cb	$CHClFCF_2CClF_2$	5.8	2,030	595	181
HFC-23	CHF_3	270	12,000	14,800	12,200
HFC-32	CH_2F_2	4.9	2,330	675	205
HFC-125	CHF_2CF_3	29	6,350	3,500	1,100
HFC-134a	CH_2FCF_3	14	3,830	1,430	435
HFC-143a	CH_3CF_3	52	5,890	4,470	1,590
HFC-152a	CH_3CHF_2	1.4	437	124	38
HFC-227ea	CF_3CHFCF_3	34.2	5,310	3,220	1,040
HFC-236fa	$CF_3CH_2CF_3$	240	8,100	9,810	7,660
HFC-245fa	$CHF_2CH_2CF_3$	7.6	3,380	1030	314
HFC-365mfc	$CH_3CF_2CH_2CF_3$	8.6	2,520	794	241
HFC-43-10mee	$CF_3CHFCHFCF_2CF_3$	15.9	4,140	1,640	500
Sulfur hexafluoride	SF_6	3,200	16,300	22,800	32,600
Nitrogen trifluoride	NF_3	740	12,300	17,200	20,700
PFC-14	CF_4	50,000	5,210	7,390	11,200
PFC-116	C_2F_6	10,000	8,630	12,200	18,200

(Continued)

TABLE 9.6 (*Continued*)

INDUSTRIAL DESIGNATION OR COMMON NAME	CHEMICAL FORMULA	APPROX. LIFETIME (years)	GLOBAL WARMING POTENTIAL FOR GIVEN TIME HORIZON		
			20-yr	100-yr	500-yr
PFC-218		2,600	6,310	8,830	12,500
PFC-318		3,200	7,310	10,300	14,700
PFC-3-1-10		2,600	6,330	8,860	12,500
PFC-4-1-12		4,100	6,510	9,160	13,300
PFC-5-1-14		3,200	6,600	9,300	13,300
PFC-9-1-18		>1,000	>5,500	>7,500	>9,500
Trifluoromethyl sulfur pentafluoride		800	13,200	17,700	21,200
HFE-125		136	13,800	14,900	8,490
HFE-134		26	12,200	6,320	1,960
HFE-143a		4.3	2,630	756	230
HCFE-235da2		2.6	1,230	350	106
HFE-245cb2		5.1	2,440	708	215
HFE-245fa2		4.9	2,280	659	200
HFE-254cb2		2.6	1,260	359	109
HFE-347mcc3		5.2	1,980	575	175
HFE-347pcf2		7.1	1,900	580	175
HFE-356pcc3		0.33	386	110	33
HFE-449sl (HFE-7100)		3.8	1,040	297	90
HFE-569sf2 (HFE-7200)		0.77	207	59	18
HFE-43-10pccc124 (H-Galden 1040x)		6.3	6,320	1,870	569
HFE-236ca12 (HG-10)		12.1	8,000	2,800	860
HFE-338pcc13 (HG-01)		6.2	5,100	1,500	460
PFPMIE		800	7,620	10,300	12,400
Dimethylether		0.015	1	1	≪1
Methylene chloride		0.38	31	8.7	2.7
Methyl chloride		1	45	13	4

Source: Based on IPCC Fourth Assessment Report, 2007, Chapter 2 http://www.ipcc.ch/publications_and_data/ar4/wg1/en/ch2s2-10-2.html#table-2-14

> **EXAMPLE 9.4** Calculating Equivalent CO_2 using Global Warming Potentials
>
> Consider two process systems with the following emission values
>
> | SPECIES | SYSTEM A | SYSTEM B |
> |---------|----------|----------|
> | CO_2 | 22 kg | 18 kg |
> | CH_4 | 3 kg | 5 kg |
> | N_2O | 1.2 kg | 1.5 kg |
>
> Which system has the greater equivalent CO_2 value?
>
> Solution
>
> | SPECIES | GWP | SYSTEM A | EQUIVALENT CO_2 | SYSTEM B | EQUIVALENT CO_2 |
> |---------|-----|----------|-------------------|----------|-------------------|
> | CO_2 | 1 | 22 kg | 22 kg | 18 kg | 18 kg |
> | CH_4 | 25 | 3 kg | 75 kg | 5 kg | 125 kg |
> | N_2O | 298 | 1.2 kg | 357.6 kg | 1.5 kg | 447 kg |
> | | | | Σ454.6 kg | | Σ590 kg |
>
> Use Equation (9.11) and Table 9.6 for the GWP values for the 100-year time horizon.
>
> System B has the greater equivalent CO_2 and hence makes the greater contribution to global warming.

9.4.2 Ozone Depletion

Ozone is the triatomic molecule of oxygen atoms and exists unevenly distributed in the atmosphere. The presence of ozone (O_3) results in different impacts, depending on which layer of the atmosphere it exists in. Ozone in the troposphere, which is the layer closest to the Earth's surface, and acts as an undesirable pollutant contributing to photochemical smog. However, ozone in the stratosphere, which is the layer just above the troposphere, is a necessary constituent forming a layer that absorbs harmful ultraviolet radiation, as discussed in Chapter 6. The depletion of the ozone in the stratosphere is a significant global concern. Excessive exposure to ultraviolet radiation is harmful to animals and vegetation. The ozone layer is a naturally formed protective shield for life on Earth that has been affected by human-made ozone-depleting substances (ODS).

Chlorofluorocarbons (CFCs) and halogenated compounds have been identified as contributing significantly to the destruction of stratospheric ozone. CFCs are highly stable chemical structures composed of carbon, chlorine, and fluorine. One important example is trichlorofluoromethane, CCl3F or CFC-11.

At high altitudes, in the presence of solar radiation, CFCs dissociate to produce chlorine atoms that act as catalysts in the conversion of ozone to oxygen. The reactions can be represented thus:

$$CFC \xrightarrow{\text{Solar radition}} Cl + \text{other products}$$

$$Cl + O_3 \rightarrow ClO + O_2$$

$$ClO + O \rightarrow Cl + O_2$$

$$\text{Net:} \quad \overline{O_3 + O \rightarrow 2O_2}$$

From the net reaction above, we can see that the chlorine atom is not used up in the reaction and remains in the atmosphere, catalyzing the conversion of more ozone to oxygen. Other halogenated ozone-depleting substances employ similar mechanisms.

Just as with the greenhouse gases, a common index has been developed to quantify the relative contribution of ozone-depleting substances (ODS) to the depletion of stratospheric ozone. This index is called the **ozone depletion potential (ODP)** and uses CFC-11 as a benchmark (Table 9.7). The ODP is the effectiveness in the degradation of stratospheric ozone by the emission of a unit mass of an ozone-depleting substance relative to the depletion caused by CFC-11.

$$\text{Equivalent mass of CFC-11} = ODP_i \times m_i \qquad (9.12)$$

And the same rule applies to emissions containing multiple ODPs:

$$\text{Total equivalent mass CFC-11} = \sum_i (ODP_i \times m_i) \qquad (9.13)$$

TABLE 9.7 Ozone depletion potential of some ozone-depleting substances

CHEMICAL NAME	ODP	CHEMICAL NAME	ODP
CFC-11 (CCl_3F) Trichlorofluoromethane	1	$C_2H_2FBr_3$	0.1–1.1
CFC-12 (CCl_2F_2) Dichlorodifluoromethane	1	$C_2H_2F_2Br_2$	0.2–1.5
CFC-113 ($C_2F_3Cl_3$) 1,1,2-Trichlorotrifluoroethane	0.8	$C_2H_2F_3Br$	0.7–1.6
CFC-114 ($C_2F_4Cl_2$) Dichlorotetrafluoroethane	1	$C_2H_3FBr_2$	0.1–1.7
CFC-115 (C_2F_5Cl) Monochloropentafluoroethane	0.6	$C_2H_3F_2Br$	0.2–1.1
Halon 1211 (CF_2ClBr) Bromochlorodifluoromethane	3	C_2H_4FBr	0.07–0.1
Halon 1301 (CF_3Br) Bromotrifluoromethane	10	C_3HFBr_6	0.3–1.5
Halon 2402 ($C_2F_4Br_2$) Dibromotetrafluoroethane	6	$C_3HF_2Br_5$	0.2–1.9
CFC-13 (CF_3Cl) Chlorotrifluoromethane	1	$C_3HF_3Br_4$	0.3–1.8
CFC-111 (C_2FCl_5) Pentachlorofluoroethane	1	$C_3HF_4Br_3$	0.5–2.2

(Continued)

TABLE 9.7 *(Continued)*

CHEMICAL NAME	ODP	CHEMICAL NAME	ODP
CFC-112 ($C_2F_2Cl_4$) Tetrachlorodifluoroethane	1	$C_3HF_5Br_2$	0.9−2.0
CFC-211 (C_3FCl_7) Heptachlorofluoropropane	1	C_3HF_6Br	0.7−3.3
CFC-212 ($C_3F_2Cl_6$) Hexachlorodifluoropropane	1	$C_3H_2F_2Br_4$	0.2−2.1
CFC-213 ($C_3F_3Cl_5$) Pentachlorotrifluoropropane	1	$C_3H_2F_3Br_3$	0.2−5.6
CFC-214 ($C_3F_4Cl_4$) Tetrachlorotetrafluoropropane	1	$C_3H_2F_4Br_2$	0.3−7.5
CFC-215 ($C_3F_5Cl_3$) Trichloropentafluoropropane	1	$C_3H_2F_5Br$	0.9−14
CFC-216 ($C_3F_6Cl_2$) Dichlorohexafluoropropane	1	$C_3H_3FBr_4$	0.08−1.9
CFC-217 (C_3F_7Cl) Chloroheptafluoropropane	1	$C_3H_3F_2Br_3$	0.1−3.1
CCl_4 Carbon tetrachloride	1.1	$C_3H_3F_3Br_2$	0.1−2.5
Methyl chloroform ($C_2H_3Cl_3$) 1,1,1-trichloroethane	0.1	$C_3H_3F_4Br$	0.3−4.4
Methyl bromide (CH_3Br)	0.7	$C_3H_4FBr_3$	0.03−0.3
$CHFBr_2$	1	$C_3H_4F_2Br_2$	0.1−1.0
CH_2FBr	0.73	$C_3H_4F_3Br$	0.07−0.8
C_2HFBr_4	0.3−0.8	$C_3H_5FBr_2$	0.04−0.4
$C_2HF_2Br_3$	0.5−1.8	$C_3H_5F_2Br$	0.07−0.8
$C_2HF_3Br_2$	0.4−1.6	C_3H_6FBr	0.02−0.7
C_2HF_4Br	0.7−1.2	CH_2BrCl Chlorobromomethane	0.12

Source: Based on EPA, Ozone-depleting Substances, http://www.epa.gov/Ozone/science/ods/index.html

9.4.3 Other Impact Categories

Similar equivalent impact factors have been developed for some other impact categories.

Photochemical Ozone Creation Potential (POCP)

While ozone in the stratosphere is essential for reducing the amount of ultraviolet radiation that falls on Earth and its depletion is a global-scale concern, tropospheric ozone or ground-level ozone (smog) is a regional-scale concern. It is caused by the degradation of volatile organic compounds (VOCs) in the presence of nitrogen oxides

TABLE 9.8 Photochemical ozone creation potential of selected compounds

CHEMICAL	POCP	CHEMICAL	POCP	CHEMICAL	POCP
Acetylene	0.168	Ethanol	0.268	Propane	0.420
Acetaldehyde	0.527	Ethylene	1.000	Propene	1.030
Acetone	0.178	Formaldehyde	0.421	Toluene	0.563
Benzene	0.189	Methane	0.007	o-xylene	0.666
Ethane	0.082	Methanol	0.123		

Source: Based on Heijungs, R. et al; Environmental Life Cycle Assessment of Products: Guide, Ed. CML (Center of Environmental Science), Leiden 1992 http://www.ziegel.at/gbc-ziegelhandbuch/eng/umwelt/wirkkatvoc.htm

(NOx) and ultraviolet radiation. Tropospheric ozone causes damage to vegetation at low concentrations and is hazardous to human health at high concentrations.

The **photochemical ozone creation potential (POCP)** is an index used to quantify the contribution of chemical species to the creation of tropospheric ozone (Table 9.8). Similar to the global warming potential (GWP), it is calculated using the primary contributing substance as the reference. In this case, ethylene is used and has a POCP value of 1. Other substances that contribute to photochemical ozone formation can then be converted to equivalent ethylene values using

$$\text{Equivalent mass of ethylene} = POCP_i \times m_i \tag{9.14}$$

Acidification Potential (AP)

Another regional-scale concern is acid precipitation. "Normal rain" is usually acidic with a pH of around 5.6 due to the natural presence of CO_2 in the atmosphere. However, the presence of other acidifying substances in the atmosphere leads to precipitation with lower pH values. Falling acid precipitation causes damage to soils, surface waters, and plants. Acidic rain changes the chemistry of soils and affects the way plants take up nutrients. Low pH values can also affect fish in surface waters.

The acidification potential is the index that was developed to quantify the contribution to acid precipitation and uses sulfur dioxide (SO_2) as a benchmark (Table 9.9). This is determined by

$$\text{Equivalent mass of } SO_2 = AP_i \times m_i \tag{9.15}$$

TABLE 9.9 Acidification potential of some compounds

CHEMICAL	ACIDIFICATION POTENTIAL
SO_2	1.00
NO	1.07
N_2O	0.70
NOx	0.70
NH_3	1.88
HCl	0.88
HF	1.60

Source: Based on Heijungs, R. et al; Environmental Life Cycle Assessment of Products: Guide, Ed. CML (Center of Environmental Science), Leiden 1992 http://www.ziegel.at/gbc-ziegelhandbuch/eng/umwelt/wirkkatap.htm

TABLE 9.10 Eutrophication potential of some substances

SUBSTANCE	FORMULA	EP
Nitrogen oxide	NO	0.200
Nitrogen dioxide	NO_2	0.130
Nitrogen oxides	NOx	0.130
Ammonium	NH^{4+}	0.330
Nitrogen	N	0.420
Phosphate	PO_4^{3-}	1.000
Phosphorus	P	3.060
Chemical oxygen demand (COD)		0.022

Source: Based on Heijungs, R. et al; Environmental Life Cycle Assessment of Products: Guide, Ed. CML (Center of Environmental Science), Leiden 1992 http://www.ziegel.at/gbc-ziegelhandbuch/eng/umwelt/wirkkatnp.htm

Eutrophication Potential (EP)

Eutrophication is the enrichment of nutrients in soils and water associated with excessive levels of nitrogen and phosphorous containing compounds as discussed in Chapter 4. Current agricultural practices involve the application of fertilizers to soils to boost crop production. Fertilizers add nutrients to soils, but excess nutrients in soil can also get washed into surface water bodies. In water it leads to accelerated algae and phytoplankton growth that slowly leads to depletion of dissolved oxygen values, ultimately choking life in the water body. Eutrophication can also be caused by the industrial emission of nutrient rich wastewater.

The eutrophication potential (Table 9.10) was developed to quantify the contribution to eutrophication of water bodies using phosphate (PO_4^{3-}) as a benchmark similar to previously discussed indices.

$$\text{Equivalent mass of } PO_4^{3-} = EP_i \times m_i \qquad (9.16)$$

We can use these impact potential values to compare the impacts that may occur even if the actual emitted species are not the same. It also allows us to a have a quantitative basis for decision making and choice.

EXAMPLE 9.5 Using Multiple Impact Potentials to Compare Systems

The table shows the emissions from two systems. Compare systems 1 and 2 on the basis of their contribution to global warming, acidification, and photochemical ozone creation.

CHEMICAL EMISSION	SYSTEM 1 kg	SYSTEM 2 kg
Methane (CH_4)	5.0	21.0
Nitrous oxide (N_2O)	0.2	0.1

Solution

To calculate the contribution to global warming, we will use Equation (9.11) and data for a 100-year time horizon from Table 9.6.

CHEMICAL EMISSION	GWP	SYSTEM 1 kg	EQUIVALENT CO$_2$ kg	SYSTEM 2 kg	EQUIVALENT CO$_2$ kg
Methane (CH$_4$)	25	5.0	125.0	21.0	525.0
Nitrous oxide (N$_2$O)	298	0.2	59.6	0.1	29.8
			Σ184.6		**Σ554.8**

To calculate the contribution to acidification, we will use Equation (9.15) and data from Table 9.9.

CHEMICAL EMISSION	AP	SYSTEM 1 kg	EQUIVALENT SO$_2$ kg	SYSTEM 2 kg	EQUIVALENT SO$_2$ kg
Methane (CH$_4$)	0.007	5.0	0.00	21.0	0.00
Nitrous oxide (N$_2$O)	0.700	0.2	0.14	0.1	0.07
			Σ0.14		Σ0.07

To calculate the contribution to photochemical ozone creation, we will use Equation (9.14) and data from Table 9.8.

CHEMICAL EMISSION	POCP	SYSTEM 1 kg	EQUIVALENT SO$_2$ kg	SYSTEM 2 kg	EQUIVALENT SO$_2$ kg
Methane (CH$_4$)	0.007	5.0	0.035	21.0	0.147
Nitrous oxide (N$_2$O)	0	0.2	0	0.1	0
			Σ0.035		**Σ0.147**

System 1 contributes more to acidification, while system 2 contributes more to global warming and photochemical ozone creation.

9.5 Risk Assessment and Analysis

The physical flow of energy and the mass of chemicals in the environment is related to the physical and chemical properties of those chemicals and other properties, such as temperature and pressure in the environment. An evaluation of chemical concentration in a given environment is insufficient without an attempt to understand the risk associated with the changing levels of a given chemical in the environment. A sustainable system and approach to development seeks a balance between the risks industrialized chemical changes pose to the economy, environment, and social systems. Many measures of risk may affect human health, behavior, and communities, as illustrated in the diagram shown in Figure 9.11. Critical resources such as air, water, and food are the main subject of this section; however, risks to societal systems may be important in managing climate change adaptation and mitigation scenarios as discussed in Chapter 6.

There are over 60,000 chemicals and millions more mixtures in industrial use, with more than 1000 new chemicals synthesized each year. The ability to manage the risk associated with existing industrial processes and new processes is daunting, and the inadequacy of data to estimate the risk of new chemicals is a concern. It is the goal of scientists, engineers, policy makers, and regulatory agencies to work in concert to detect and mitigate the environmental and human health risks of

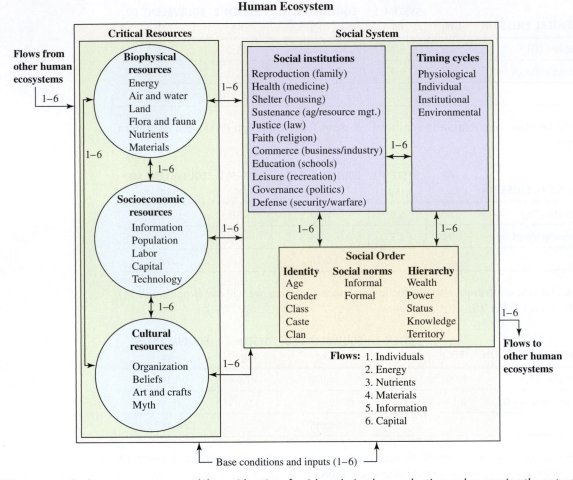

FIGURE 9.11 The human ecosystem model, considerations for risk analysis when evaluating and comparing the potential risks between industrialized processes.

Source: "The human ecosystem Part I: The human ecosystem as an organizing oncept in ecosystem management," Machlis, G.E., Force, J. E., and Burch, W.R. Jr., Jul 1, 1997, Society & Natural resources: An International Journal. 10(4):347–367, Copyright © 1997 Routledge, reprinted by permission of the publisher (Taylor & Francis Ltd, http://www.tandf.co.uk/journals)

exposure to existing and potential contaminants. Figure 9.12 illustrates the methods for conveying risk levels among different professions in order to manage existing and potential risks. In this section, human health risks associated with environmental and workplace exposure procedures will be demonstrated.

Typically, a human health risk assessment includes four steps (illustrated in Figure 9.13):

- Hazard identification
- Dose–response assessment
- Exposure assessment
- Risk characterization

9.5.1 Hazard Identification and Levels of Risk

A **hazard** is anything with the potential to produce conditions that endanger safety and health, but does not, by itself, express the likelihood that a dangerous incident will occur. Risk represents the number of incidents per time of exposure or

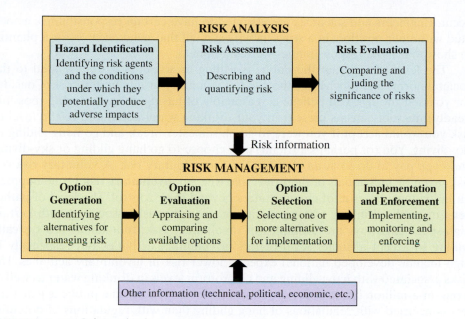

FIGURE 9.12 Relationship between risk analysis, risk assessment, risk management, and professional occupations.

Source: Based on Covello and Merkhofer. Risk Assessment Methods: Approaches for addressing health and environmental risks. Plenum Publishing Corp.

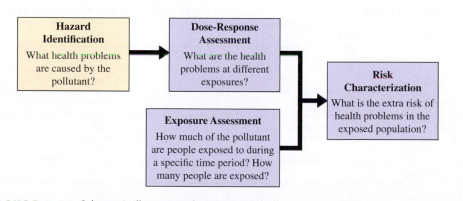

FIGURE 9.13 Schematic illustrating the steps in the risk-assessment process.

Source: USEPA (1998) Guidelines for Ecological Risk Assessment. Risk Assessment Forum, U.S. Environmental Protection Agency. Washington, D.C. U.S.A. EPA/630/R-95/002F. Exhibit 1.2.

probability that an incident related to a hazard will occur. The engineer or scientist who ignores the risks and hazards associated with a process risks the possibility of being liable for damages associated with their work if damages are perceived to have occurred. **Liability** is a legal concept associated with a wrongful or injurious act for which a civil court action occurs. There are two types of liability: negligence and strict liability. Negligent liability may occur if an engineer or a scientist is careless or does not adhere to professional standards that results in some harm or damages that occur. Engineers and scientists are assumed to have certain responsibilities associated with their position; in a strict liability suit, all that need occur is for a plaintiff to show that those responsibilities were not upheld.

The level of risk that the public is willing to accept is closely related to the societal norms and to whether the risk is a voluntary or involuntary risk. Consider for yourself the risk you believe is reasonable of having disease-causing (possibly deadly) microorganisms present in your drinking water compared to the level of risk you would accept if you were to go on semester break and go hang gliding or sky-diving. You (or perhaps a friend) may choose to go hang gliding or sky-diving, and even pay to do so at an elevated level of voluntary risk. A **voluntary risk** is a risk taken by individuals of their own free will. An **involuntary risk** is one imposed on individuals because of circumstances beyond their control. The level of pathogens in public drinking water supplies or even the risk associated with certain high-speed highways are examples of involuntary risk. Individuals are typically willing to accept a higher level of risk in voluntary activities but (especially in high-income developed nations) expect a lower risk in involuntary activities. The risks associated with hang gliding and chloroform levels in drinking water as well as a one-in-a-million lifetime risk are shown in Table 9.11, but the public is generally less concerned with regulations of hang gliding than with regulations of potential carcinogens in the water supply.

TABLE 9.11 Annual mortality risk associated with giving birth in selected countries, disease, physical activities, and environmental exposures

ACTIVITY/EXPOSURE	ANNUAL RISK (DEATHS PER 100,000 PERSONS AT RISK)
Maternal mortality, Somalia	1,000
Maternal mortality, Benin	350
Smoking, all causes	284
Maternal mortality, India	200
Cancer	186
Heart disease	180
Hang gliding	179
Parachuting	175
Pedestrian transportation	142
Motorcyclist	131
Boxing	45
Coal mining	29
Agriculture	29
Maternal mortality, China	37
Maternal mortality, United States	21
Motor vehicle accidents	12

(Continued)

TABLE 9.11 (*Continued*)

ACTIVITY/EXPOSURE	ANNUAL RISK (DEATHS PER 100,000 PERSONS AT RISK)
Death due to firearms	10
Unintentional fall deaths	8
Maternal mortality, Australia	7
Mountain hiking	6
Cataclysmic storm	3
Scuba diving	3
American football	2
Contact with hornets, wasps, and bees	1
Earthquakes and other earth movements	1
Chlorinated drinking water (chloroform)	0.8
4 tbsp peanut butter per day (aflatoxin)	0.8
Lightning strike	0.7
3 oz charcoal-broiled steak per pay (PAHs)	0.5
Flooding	0.2
Skiing	0.1

Source: Bradley Striebig.

In the United States, many risk-based measures of public protection, such as regulations set for environmental pollutants and workforce exposure to chemicals, are designed to reduce the level of risk associated with a given workplace or environmental exposure below a one-in-a-million lifetime risk. This risk level is comparable to some of the activities shown in Table 9.12. The ability to calculate this risk is subject to many interpretations in toxicity and exposure data. Certain assumptions must also be made to determine the appropriate risk. For example, an environmentally related risk calculation might assume that a person lives in the same home location for 70 years and is exposed to a given environmental substance present in the home or local environment 24 hours per day, 365 days per year for 70 years. Exposure in the workforce might be calculated based on a 30-year career, with exposure 8 hours per day for 250 days per year. Risk factors need to be estimated, and these estimates may be used to make decisions about publicly acceptable levels of risk compared to publicly accepted levels of regulation.

Government agencies in many countries are responsible for creating regulations that minimize the risks, particularly involuntary risks, associated with transportation systems, workplace exposure, air quality, and water quality. **Risk management** is the process of identifying, selecting, and implementing appropriate actions, controls, or regulations to reduce risk for a group of people. Scientists and engineers develop the scientific comparisons risk managers use to relate the public acceptance of risk to the acceptance of government regulations and restrictions (see Figure 9.14). **Risk assessment** is the process of estimating the spectrum and frequency of negative impacts based on the numerical values associated with exposure for certain activities. Risk assessment requires adequate knowledge of the actual amount of contaminants present, the chemical and physical

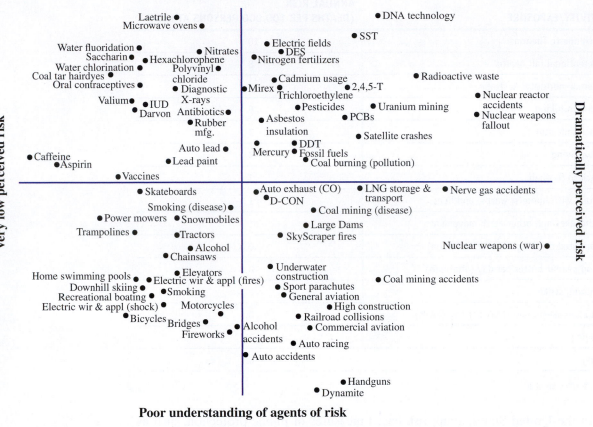

FIGURE 9.14 The public's perception of risk may differ dramatically from that of someone trained in risk assessment. Public perception may be related to the degree of the perceived risk and the understanding of the perceived risk.

Source: Reprinted with permission from Stern, P.C. and Fineberg, H.V. (Eds.), Understanding Risk: Informing Decisions in a Democratic Society. Committee on Risk Characterization, Commission on Behavioral and Social Sciences and Education, Figure 2.4, page 62, National Research Council, 1996, by the National Academy of Sciences, Courtesy of the National Academies Press, Washington, D.C.

form in which the contaminant is present, and the methods by which the contaminant will be transported to an individual that will result in an exposure pathway as illustrated in Figure 9.15. Toxicity tests have been developed to enhance the understanding of how organism and ecosystems respond to contaminants.

The response of an organism to a contaminant may be classified as an acute or chronic effect. An acute effect is a response to short-term hazards that result in immediate symptoms. Chronic effects are long-term effects that result from an exposure, and these effects may or may not disappear when the hazard is removed. Some types of contaminants with long-term chronic effects include:

- **Carcinogens**—contaminants that may cause cancer
- **Teratogens**—contaminants that may cause reproductive system problems
- **Neurotoxins**—contaminants that may cause impaired brain or nervous system function
- **Mutagens**—contaminants that may cause birth defects

TABLE 9.12 Activities that increase mortality risk by one-in-a-million

ACTIVITY	TYPE OF RISK
Smoking 1.4 cigarettes	Cancer, heart disease
Drinking 0.5 liters of wine	Cirrhosis of the liver
Spending 1 hour in a coal mine	Black lung disease
Spending 3 hours in a coal mine	Accident
Living 2 days in New York or Boston	Air pollution
Drinking Miami water for 1 year	Cancer from chloroform exposure
Traveling 300 miles by car	Accident
Flying 1,000 miles by jet	Accident
Flying 6,000 miles by jet	Accident
Traveling 10 miles by bicycle	Accident
Traveling 6 miles by canoe	Accident
Living 2 summers in Denver (vs. sea level)	Cancer from cosmic radiation
Living 2 months with a cigarette smoker	Cancer, heart disease
Eating 40 tablespoons of peanut butter	Cancer from aflatoxin
Eating 100 charcoal-broiled steaks	Cancer from benzopyrene
Living 50 years within 5 miles of a nuclear reactor	Accident releasing radiation

Source: Based on Richard Wilson. Analyzing the Daily Risks of Life. Technology Review. February, 1979. Pages 41–46.

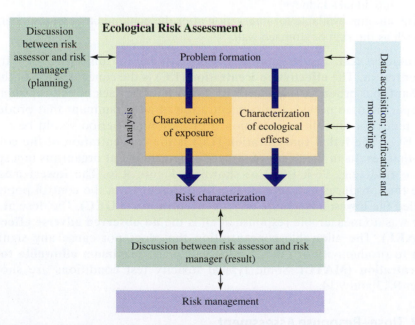

FIGURE 9.15 The framework for ecological risk assessment.

Source: Based on USEPA (1992) Framework for Ecological Risk Assessment. Risk Assessment Forum, U.S. Environmental Protection Agency. Washington, D.C. U.S.A. EPA/630/R-92/001. Exhibit 1.2.

- Disorders of the pulmonary, cardiovascular, and immune system
- Disorders of the skeletal system
- Disorders of the skin and mucous membrane hypersensitivity

Toxicology is the study of the adverse effects of contaminants on biological systems. Toxicity tests typically involve animal (often microorganisms) experiments, large populations, and extrapolation to broader ecological response or human effects. Toxicity tests may be used to comply with government water quality standards and to identify the toxicity of specific substances. It has been found that the toxicity of effluent discharges of wastewater correlate well with toxicity measurements for similar contaminants in receiving waters and ecological community responses for many toxicity measurement procedures.

Toxicity tests may be classified according to the duration of the test, the method of adding the contaminant to the population being studied, and the purpose of the test. Acute toxicity tests typically provide a response shortly after exposure (in 48 to 96 hours). Chronic toxicity tests typically require a time period equal to at least one-tenth the lifespan of the test population. The exposure time, in which the population is exposed to a specific contaminant, may be different from total time required for the procedure. The toxicity tests may be carried out *in vitro* (using part of an organism's cells in glass or test tubes) or *in vivo* (in living organisms or using the whole organism).

The **dose** of a contaminant is defined as the amount of a chemical received by an individual that can interact with an individual's biological functions and receptors. The dose typically has units measured in mg of contaminant per kg of the subject's body mass (mg/kg). The dose may be any one of these.

- The amount of a contaminant that is administered to the individual
- The amount administered to a specific location of an individual (for example, the individual's kidneys)
- The amount available for interaction after the contaminant crosses a barrier such as the skin or stomach wall.

Toxicity test results may be reported as an inhibiting or effective concentration. The **effective concentration (EC)** is the concentration of the contaminant that produces a specified response (such as lack of cell reproduction) in a specific time period. The concentration of a contaminant that produced a 50% reduction in cell reproduction over a 96-hour period would be called a 96-h EC_{50}. The **lethal concentration (LC)** is the concentration of the contaminant that results in the death of a specific percentage of organisms in a specific time period (e.g., 96-h LC_{50}) as shown in Figure 9.16. The lowest measured value that produces a result statistically different from the control population is called the **lowest observable effect concentration (LOEC)**. The dose at which there was no measurable response at all is the **no observed adverse effect level (NOAEL)**. The allowable concentration that does not cause any significant harm to productivity or other uses is called the **maximum allowable toxicant concentration (MATC)**. Some typical toxicity test conditions are shown in Tables 9.13 and 9.14.

9.5.2 Dose–Response Assessment

A plot of the measured response versus the dose administered is called the **dose–response curve**, as shown in Figure 9.16. Three generic compounds, *A*, *B*, and *C*, are shown in Figure 9.16. The measured toxicity of compound *A* is less

TABLE 9.13 Typical toxicity test conditions for freshwater species

SPECIES/COMMON NAME	DURATION	ENDPOINTS
Cladoceran (*Ceriodaphnia dubia*)	Approx. 7 days	Survival, reproduction
Flathead minnow (*Pimephales promelas*)	7 days	Larval growth
	9 days	Embryo–larval survival, percent hatch, percent abnormality
Freshwater algae (*Selastrum capriocomutum*)	4 days	Growth

TABLE 9.14 Typical toxicity test conditions for marine and estuarine species

SPECIES/COMMON NAME	DURATION	ENDPOINTS
Sea urchin (*Arbacia puntulata*)	1.5 hour	Fertilization
Red macroalgae (*Champia parvula*)	7-9 day	Cystocarp production (fertilization)
Mysid (*Mysidopsis bahia*)	7 day	Growth survival, fecundity
Sheephead minnow (*Caprinodon variegatus*)	7 day	Larval growth
	9 day	Embryo–larval survival, percent hatch, percent abnormality
Inland silverside (*Menidia beryijina*)	7 day	Larval growth, survival

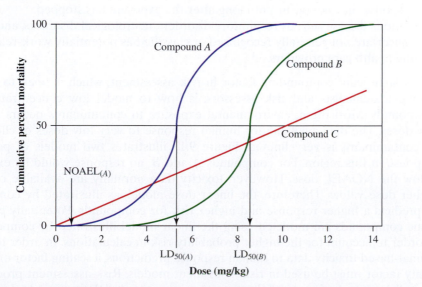

FIGURE 9.16 Theoretical dose–response acute toxicity curve for three generic compounds, A, B, and C.

Source: Based on Masters, G.M. and Ela, W. P. (2008) Introduction to Environmental Engineering and Science, Third Edition. Pearson Education, Inc. Upper Saddle River, NJ. Figure 4.11.

than compound B at all doses. However, notice that the slopes of compounds A and C are significantly different and that compound A shows a lower toxicity than compound C at doses less than 4 mg/kg, even though compound A has a lower LD_{50}, but the toxicity of compound A becomes greater than compound C at higher doses. Thus, it is important to note that there is no standardized toxicity value or procedure that is suitable for all applications. Toxicity is a relative term. The shape of the response curve and the actual response to different contaminants may vary greatly among species. Also, individual organ toxicity may vary as well; for example, a compound may profoundly affect the lungs, but have no effect on the nervous system.

Epidemiology is the study of the cause of disease and adverse responses. Epidemiological work may be based on historical data and toxicity data. Epidemiological studies may often be complicated by a lack of information about the duration and exposure to a contaminant, the change in exposure as individuals and populations change locations, long lag times, and multiple pathways of exposure to multiple contaminants. The lack of available data and data variability create a high level of uncertainty in the results of epidemiological and hazard assessment. However, historical examples of a few specific environmental exposure cases have led to several models that are useful to estimate human risk based on exposure data and toxicological data.

Extrapolation of toxicity data to human response is quite difficult. Hazard identification or assessment involves predicting adverse human responses and severity of the response to contaminants. Human-based risk assessment for environmentally caused illness is confounded by several factors, including:

- Environmental illness may be indistinguishable from other disease.
- The relationship between a response (such as cancer) and the environment is not always recognizably different from other populations and environments.
- Chronic effects may be complicated by long latency periods, and the negative response may not occur until long after the exposure has stopped.
- Serious disease such as respiratory disorders, neurological disorders, and cancer are not generally recognized or classified as potentially work-related by the health care providers.

A significant confounding factor in risk assessment, which is open to debate among toxicologists and risk assessors, is how to model low concentrations of compounds. Most often, environmental exposure to contaminants occurs at very low doses. The data available for human response to very low doses of chemicals or contaminants is very limited. Figure 9.17 illustrates two models of potential response in this region. For compounds A and B, no response would be expected below the NOAEL dose. However, toxicity data normally are available only for higher dose values. Therefore, the linear dose model, as illustrated by compound C, predicts a higher response and higher risk for compounds. Potentially carcinogenic compounds are modeled using the linear model illustrated by compound C in order to account for the higher probability risk in calculations. In order to apply animal-based toxicity data to human response predictions, a scaling factor or uncertainty factor must be used in risk-assessment models. Risk-assessment procedures typically use the body mass of the test animal compared to the average body mass of a human adult (typically considered to be 70 kg), to relate toxicity data from animal tests to predict human risk. A total exposure assessment considers the cumulative risk of all exposure pathways, summarized in Table 9.15.

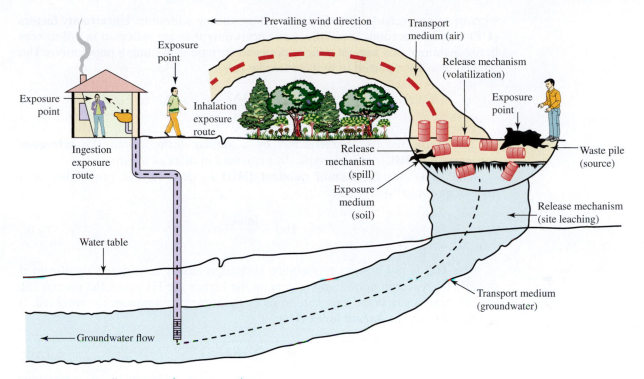

FIGURE 9.17 Illustration of exposure pathways.

Source: Based on U.S. EPA (1989) Risk Assessment Guidance for Superfund: Volume I Human Health Evaluation Manual (Part A). Office of Emergency and Remedial Response. U.S. Environmental Protection Agency. Washington, D.C. U.S.A.

The U.S. Environmental Protection Agency (EPA) uses a nonlinear model for noncarcinogenic risk assessment. This method is based on the assumption that there is a NOAEL value for which no toxic effects of exposure exist. The **reference dose (RfD)** is defined as an estimate (with uncertainty spanning perhaps an order of magnitude) of a daily oral exposure by a human population (including sensitive groups, such as asthmatics, children, or elderly populations) that is likely to be

TABLE 9.15 Routes of exposure from environmental contaminants (adapted from Davis, *Principles of Environmental Engineering and Science*.)

ENVIRONMENTAL SOURCE	ROUTES OF EXPOSURE
Groundwater	Ingestion, dermal contact, inhalation during showering
Surface water	Ingestion, dermal contact, inhalation during showering
Sediments	Ingestion
Air	Inhalation of gases and particles
Soil and dust	Incidental ingestion (especially children), dermal contact
Food	Ingestion

without an appreciable risk of adverse effects during a lifetime. **Uncertainty factors (UF)** take into account variability and uncertainty that are reflected in differences between animal and human toxicity data and variations in human populations. The RfD is typically reported in units of (mg/kg-day).

$$RfD = \frac{NOAEL}{UF} \tag{9.17}$$

Inhalation risk is characterized by a similar term, called the **reference concentration (RfC)**, which is typically expressed in units of mg/m^3.

The EPA uses the **hazard quotient (HQ)** to describe the probability of a noncarcinogenic toxicity risk:

$$HQ = \frac{intake}{RfD} \tag{9.18}$$

The HQ is not a strict probability. Instead, a risk is considered significant if the HQ is greater than one, and in general the higher the HI value, the greater the risk. If there is suspected exposure to more than one contaminant, the total risk is estimated to by the **hazard index (HI)**:

$$HI = \sum HQ_{ij} \tag{9.19}$$

The EPA has created an online **integrated risk information system (IRIS)** that provides a database for human health assessment data (*www.epa.gov/iris/index.html*). Through the IRIS Program, the EPA provides science-based human health assessment data to support the agency's regulatory activities. The IRIS database is web accessible and contains information, including NOAEL levels, RfDs, slope factors, and carcinogenic health risk information on more than 550 chemical substances.

The EPA relates the low-dose response for carcinogenic compounds to human response in risk assessment models with the slope factor (also sometime called the potency factor). The EPA method is referred to as the **reasonable maximum exposure (RME)** technique. The slope factor assumes a linear dose–response relationship and typically has units of risk per unit dose (kg-day/mg):

$$SF = \frac{incremental\ lifetime\ cancer\ risk}{chronic\ daily\ intake \left(\dfrac{mg}{kg - day} \right)} \tag{9.20}$$

The total cancer risk is determined by the sum of all risk associated with each potential pathway of concern (i.e., inhalation, ingestion, and dermal exposure). The single-exposure pathway cancer risk is determined by multiplying the **chronic daily intake (CDI)** by the slope factor (SF) as:

$$Cancer\ risk = CDI \times SF \tag{9.21}$$

The CDI is a composite estimate of various exposure scenarios and assumptions used to estimate contaminant intake, as shown in Table 9.16. The CDI has units of mg/kg-day.

TABLE 9.16 Intake due to exposure routes for various pathways

ROUTE	INTAKE EQUATION	VARIABLES	UNITS	EQ. #
Ingestion in drinking water	$$CDI = \frac{(C_w)(IR)(EF)(ED)}{(BW)(AT)}$$	C_w = concentration in water	mg/L	
		IR = ingestion rate	L/day	
		EF = exposure frequency	Days, years, or events	
		ED = exposure duration	Years	
		BW = body weight	kg	
		AT = averaging time	Days	
Ingestion while swimming	$$CDI = \frac{(C_w)(CR)(ET)(EF)(ED)}{(BW)(AT)}$$	CR = contact rate	L/h	
		ET = exposure time	h/event	
Dermal contact with water	$$AD = \frac{(C_w)(SA)(P_C)(ET)(EF)(ED)(CF_w)}{(BW)(AT)}$$	AD = absorbed dose	$\frac{mg}{kg\|day}$	
		SA = skin surface area for contact	cm^2	
		P_C = chemical specific dermal permeability constant	cm/h	
		CF = conversion factor for water	10^{-3} L/cm^3	
Ingestion of chemicals in soil	$$CDI = \frac{(C_S)(IR)(CF_S)(FI)(EF)(ED)}{(BW)(AT)}$$	C_S = concentration in soil	mg/kg	
		CF_s = conversion factor for soil	10^{-6} kg/mg	
		FI = fraction ingested		
Dermal contact with soil	$$AD = \frac{(C_S)(SA)(AF)(AB_S)(EF)(ED)(CF_S)}{(BW)(AT)}$$	AF = soil-to-skin adherence factor	mg/cm^2	
		AB_s = absorption factor for soil contaminants		
Inhalation of airborne (vapor phase) chemicals	$$CDI = \frac{(C_A)(IR_A)(ET)(EF)(ED)}{(BW)(AT)}$$	C_A = concentration in air	mg/m^3	
		IR_A = inhalation rate in air	m^3/h	
Ingestion of contaminated foods	$$CDI = \frac{(CF_S)(IR_F)(FI)(EF)(ED)}{(BW)(AT)}$$	IR_F = ingestion rate of food	kg/meal	

Source: Based on USEPA (1989) Risk Assessment Guidance for Superfund: Volume I Human Health Evaluation Manual (Part A). Office of Emergency and Remedial Response. U.S. Environmental Protection Agency. Washington, D.C. U.S.A.

9.5.3 Exposure Assessment

The process of determining the amount of time and contact to a particulate contaminant to a study population is called **exposure assessment**. The EPA defines exposure as "contact between an agent and the visible exterior of a person (e.g. skin and openings into the body)." The size, nature, and types of human populations exposed to a contaminant and the uncertainties in exposure should be thoroughly documented as part of the exposure assessment as shown in the Appendix A. The total exposure to a study population may be estimated from measured concentrations in the environment, consideration of models of chemical transport and fate, and estimates of human intake over time. The exposure assessment must also consider both the pathway of exposure (how a contaminant moves though the environment to contact the study population) and the route of exposure (how the contaminant enters an individual's body).

Both in human health assessment and toxicity evaluations for other species, bioaccumulation of contaminants in the environment is an important consideration in risk evaluation. A **bioaccumulation factor (BCF)** with units of L/kg is used to estimate the equilibrium concentration in fish (in mg/kg) from a given concentration of a contaminant in water (in mg/L):

$$C_{fish} \frac{mg}{kg} = C_w \frac{mg}{L} \times BCF \frac{L}{kg} \tag{9.22}$$

The bioaccumulation factor characterizes the tendency of some chemicals to partition to the fatty tissue found in fish. Some examples of bioaccumulation factors are shown in Table 9.17.

TABLE 9.17 Bioaccumulation factors for selected compounds

COMPOUND	BIOCONCENTRATION FACTOR (L/kg)
4,4—Methylene dianiline	11.1
Arsenic	4
Cadmium	366
Chromium	2
Dioxins and furans	19,000
Hexachlorocyclohexanes	456
Hexochlorobenzene	13,130
Lead	155
Mercury (inorganic)	5,000
PAH as benzo[a]pyrene	583
Polychlorinated biphenyls	99,667
Diethylhexylphthalate	483.1

Source: Based on Marty, M.A. and Blaisdell, R. J. (Eds.) (2000) Air Toxics Hot Spots Program Risk Assessment Guidelines: Part IV technical Support Document for Exposure Assessment and Stochastic Analysis. California Office of Environmental Health and Hazard Assessment.

9.5.4 Risk Characterization

Risk assessment is a complex process with a high level of uncertainty that should be considered as part of the analysis. Data for risk assessments should be collaboratively gathered and reviewed in order to make quantitative decisions about risk.

Detailed documentation of the risk assessment is essential; each component of the risk assessment (e.g., hazard assessment, dose–response assessment, exposure assessment) should have an individual risk characterization that summarizes the key findings, assumptions, limitations, and uncertainties. The objective of the risk characterization is to inform the risk manager and others about the rationale and approach for conducting the risk assessment. The risk characterization should be transparent, clear, and consistent and should show that reasonable assumptions and judgment have been used in the assessment.

The EPA's risk characterization policy calls for conducting risk characterizations in a manner consistent with the following principles.

- **Transparency**—The characterization should fully and explicitly disclose the risk assessment methods, default assumptions, logic, rationale, extrapolations, uncertainties, and overall strength of each step in the assessment.
- **Clarity**—The products from the risk assessment should be readily understood by readers inside and outside of the risk assessment process. Documents should be concise, free of jargon, and should use understandable tables, graphs, and equations as needed.
- **Consistent**—The risk assessment should be conducted and presented in a manner which is consistent with EPA policy and should be consistent with other risk characterizations of similar scope prepared across programs within the EPA.
- **Reasonable**—The risk assessment should be based on sound judgment with methods and assumptions consistent with the current state-of-the-science and conveyed in a manner that is complete and balanced, informative.

These four principles—transparency, clarity, consistency, and reasonableness—are referred to collectively as TCCR.

The **Toxic Substance Control Act (TOSCA)** in the United States catalogues over 56,000 substances that identify the properties of these substances that require special handling or safety precautions. The EPA is charged with evaluating new substances and predicting potential release points, estimating potential exposure, and mitigating exposure if there is an elevated environmental or human health risk. The agency has identified six areas of investigation where environmental or human health risks are potentially of concern.

- Drinking water disinfection
- Particulate matter inhalation
- Endocrine disruptors
- Improved ecosystem risk assessment
- Improved health risk assessment
- Pollution prevention and new technologies

It is the responsibility of engineers and scientists to design products, processes, and chemicals that are not intentionally defective or harmful. Engineers, scientists, and industry may be liable for exposure that results in adverse impacts to the environment or human health. The physical and chemical properties of a substance determine the likely fate and method of transport in the environment. Toxicity measures help provide knowledge for the handling and risks associated with chemicals and contaminants. Knowledge of the concentration, duration, and level of toxicity are important parameters for understanding the relative and quantitative risks associated with potential emissions or exposure. Risk-assessment tools are useful for estimating or predicting environmental and human health risks associated with various processes. These risks contain significant uncertainties that should be communicated to the public policy makers and designers.

9.6   Industrial Metabolism

We have explored the role of technology in impacts as well as specific ways to quantify these impacts. However, sustainable engineering requires a change in the way technology is used and, in general, a minimization of impacts from industrial processes.

Industrial ecology proposes the development of technological systems that have minimal impacts on the surrounding "natural system" by mimicking the way nature handles materials and energy transformations.

"The concept requires that an industrial system be viewed not in isolation from its surrounding systems but in concert with them" (Graedel and Allenby, 2003).

Since biological organisms tend to adopt metabolic patterns that keep them in balance with their environment, we can use the analogue of biological systems to study and improve the relationship of industrial systems with the environment.

To this end, we can assume that the analogue of the biological organism is the industrial process or processes that produce a particular product or family of products.

Both natural and industrial systems have cycles of energy and nutrients or materials. Natural systems utilize a complex system of feedback mechanisms that initiate appropriate responses when certain limits are reached. This is what industrial ecology aims to do for industrial systems.

Three types of ecosystems, as follows, have been identified based on the way materials and energy flow through them.

9.6.1   Type I System

This is a linear and open system relying completely on materials and energy from outside the system (Figure 9.18). Materials and energy enter and leave either as products or wastes. No recycle or reuse occurs within this system. The supply of materials and energy is typically not infinite and the capacity for the surrounding system to absorb waste is also finite, so the system is ultimately unsustainable.

9.6.2   Type II System

In a type II system (Figure 9.19), a quasi-cyclic materials flow takes place. Some material is recycled or reused within the systems. The rest leaves the system as waste,

FIGURE 9.18 A type I system.

Source: Based on Braden R. Allenby, "Industrial Ecology: The Materials Scientist in an Environmentally Constrained World," MRS Bulletin 17, no. 3 (March 1992): 46–51.

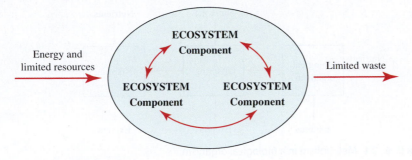

FIGURE 9.19 A type II system.

Source: Based on Braden R. Allenby, "Industrial Ecology: The Materials Scientist in an Environmentally Constrained World," MRS Bulletin 17, no. 3 (March 1992): 46–51.

so some resources are still required externally. This represents the current state of industrial systems where some recycle and reuse occurs.

9.6.3 Type III System

The type III system refers to a completely integrated closed system in which only some energy crosses its boundary (Figure 9.20). All materials are recycled and reused within the system. This represents an ideal sustainable state and the ultimate goal of industrial ecology where the effluents from one process serve as the raw material for another.

If we think of human beings as existing in an ecosystem with Earth's natural systems, we can infer that a type I system with its infinite resource source and infinite waste sink is not representative of our interaction with Earth's system.

A type III system represents an ideal that can be aspired to but will likely take a very long time to achieve given the current state of technology.

A type II system is, however, feasible with changes to our industrial process systems' overall consumption patterns. The objective should be to significantly reduce the extraction of new resources from the Earth's natural system, maximizing the recycling/reuse of resources, and thereby significantly reducing the amount of waste that is ultimately generated. This model can be sustainable if the extraction rates are limited to natural rates of replenishment and the waste discharge rates are limited to the natural rates of waste absorption.

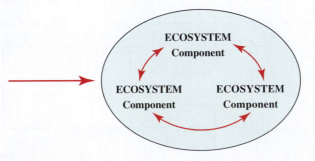

FIGURE 9.20 A type III system.

Source: Based on Braden R. Allenby, "Industrial Ecology: The Materials Scientist in an Environmentally Constrained World," MRS Bulletin 17, no. 3 (March 1992): 46–51.

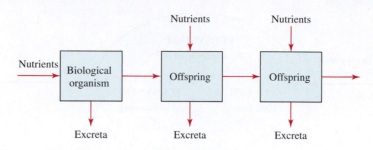

FIGURE 9.21 Metabolism in a biological organism

Source: Bradley Striebig.

9.6.4 Biological Metabolism

In the metabolic process of biological organisms as shown in Figure 9.21, the organisms take in nutrients from the environment and utilize them to grow and reproduce while generating waste. Ultimately, the nutrients taken from the environment end up back in the environment in a useful and available form, coming from the waste generated over the lifetime of the organism and the eventual death of the organism.

9.6.5 Industrial Metabolism

The industrial metabolic process, as shown in Figure 9.22, consumes materials and energy from the environment in the form of natural resources to make products. Emissions contain materials and energy in forms that are not easily retrievable. The use and disposal of the products also consume energy and generate emissions. A sustainable industrial ecological system will minimize the emissions generated and increase the recovery of materials and energy within the system.

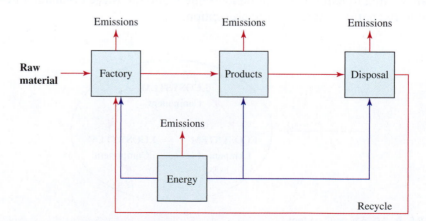

FIGURE 9.22 Metabolism in an industrial organism.

Source: Bradley Striebig.

BOX 9.1 Case Study: Kalundborg Eco-Industrial Park

One of the best-known examples is the Kalundborg eco-industrial park in Denmark. This park was one of the first models for industrial symbiosis and was a result of the collaboration of participating businesses; these businesses have all reported significant achievements in environmental performance. Four main industries—a coal-fired power plant, a refinery, a pharmaceuticals and enzymes maker, and a plasterboard manufacturer—as well as the municipal government and a few smaller businesses accept each other's wastes—in the process turning them into useful inputs for their respective processes.

The power company supplies residual steam to the refinery and, in exchange, receives refinery gas that used to be flared as waste. The power plant burns the refinery gas to generate electricity and steam. It sends excess steam to a fish farm that it operates, to a district heating system serving 3,500 homes, and to the pharmaceuticals plant. Sludge from the fish farm and pharmaceutical processes becomes fertilizer for nearby farms. The power plant sends fly ash to a cement company, while gypsum produced by the power plant's desulfurization process goes to a company that produces gypsum wallboard. The refinery removes sulfur from its natural gas and sells it to a sulfuric acid manufacturer.

Some of the reported results include:

- Annual CO_2 emission is reduced by 240,000 tons
- 3 million m^3 of water is saved through recycling and reuse
- 30,000 tons of straw are converted to 5.4 million liters of ethanol
- 150,000 tons of yeast replaces 70% of soy protein in traditional feed mix for more than 800,000 pigs
- Recycling of 150,000 tons of gypsum from desulphurization of flue gas (SO_2) replaces the import of natural gypsum ($CaSO_4$)

9.7 Eco-Industrial Parks

One of the ways that the objectives of industrial ecology can be achieved is by the development of eco-industrial parks (Figure 9.23). An **eco-industrial park** is a community of manufacturing and service businesses that collaborate in the management of environmental and resource issues for improved environmental and economic performance. A symbiotic relationship is developed between participants wherein materials, energy, and waste resources are exchanged, thereby minimizing the need for a new influx of materials into the park and minimizing total waste generated by the park. By adopting principles of industrial ecology and pollution prevention, this model can provide one or more of the following benefits.

- Reduction in the use of virgin materials
- Reduction in pollution
- Increased energy efficiency
- Reduction in the volume of waste products requiring disposal
- Increase in the amount and types of process outputs that have market value

9.8 Summary

Industry ecology is a concept that utilizes ecological principles, such as interconnected webs of material and waste flows, in much the same way that a food web is conceived in ecology. This represents an extension of the biomimicry

Industrial Ecosystem at Kalundborg, Denmark

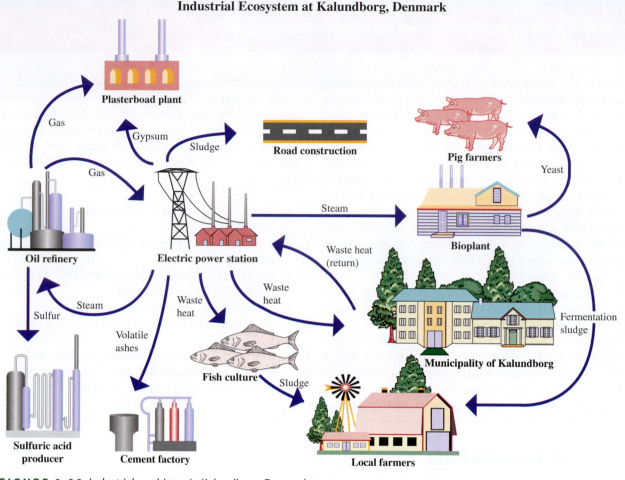

FIGURE 9.23 Industrial symbioses in Kalundborg, Denmark.

Source: Based on Pollution Issues, Industrial Ecology, http://www.pollutionissues.com/Ho-Li/Industrial-Ecology.html

principles discussed in Chapter 7, but it applies these principles to a system of material use, manufacturing, and waste disposal. Industrial ecology utilizes economic, environmental, and societal factors, like those considered in the IPAT equation, to help find innovative methods to better utilize resources, reuse waste products, and minimize the risk associated with waste disposal. Because these systems can be complex and extensive, computer based tools and large databases may be required to fully utilize the power associated with industrial ecology concepts. Life cycle analysis and assessment tools, discussed in Chapter 10, will relate data associated with material extraction, use, and disposal to potential positive and negative impacts.

To properly develop and model an eco-industrial park, we need to quantify the various materials and energy flow streams. We also need to ensure that in trying to reduce the total quantities of materials used and waste emitted, we do not unknowingly replace one type of emission that has minimal impact for another with more significant impacts. To do this, we can employ the methods of life cycle analysis, which we will cover in Chapter 10.

References

Allenby, Braden R. (1992, March). "Industrial ecology: The materials scientist in an environmentally constrained world." *MRS Bulletin* 17(3):46–51.

ATDSR (Agency for Toxic Substances and Disease Registry). (2009). Case Studies in Environmental Medicine: Toxicity of Polycyclic Aromatic Hydrocarbons (PAHs). U.S. Department of Health and Human Services. Washington, D.C.

Bull, P. J., Brooke, R. K., Cocker, J., Jone, K., and Warren, N. (2005, November). "An occupational hygiene investigation of exposure to scrylamide and the role for urinary S-carboxyethyl-cysteine (CEC) as a biological marker." *Ann Occup Hyg* 49 (8):683–690.

CDC (Centers for Disease Control and Prevention). (2009). National Hospital Ambulatory Medical Care Survey: 2009 Emergency Department Summary Tables. U.S. Department of Health and Human Services, Washington D.C.

Covello, V.T., and Merkhofer, M.W., *Risk Assessment Methods: Approaches for Addressing Health and Environmental Risks*, New York, Plenum Publishing Corp.

Davis, M.L., and Masten, S.J., (2008). *Principles of Environmental Engineering and Science*, 2nd edition New York McGraw-Hill.

Graedel, T. E., and B. R. Allenby. (2003). *Industrial Ecology*. 2nd edition. Upper Saddle River, NJ: Pearson Education, Prentice Hall.

Hardin, G. (1968). "The Tragedy of the Commons." *Science* 162 (3859): 1243–1248.

Heran, M. (2012). Deaths: Leading Causes for 2009. National Vital Statistics Report 60(7). U.S. National Vital Statistics Systems. National Center for Health Statistics, Centers for Disease Control and Prevention, U.S. Department of Health and Human Services, Washington, D.C.

Kempton, W., (2001). "Cognitive anthropology and the environment." In *New directions in anthropology and environment.* (Crumley C. L. (ed.), pp. 49–71. Walnut Creek, CA: AltaMira Press.

Leslie, D., (1997). "Consumption, In *International Encyclopedia of Human Geography*, Rob Kitchin and Nigel Thrift, (Eds.). Elsevier, Oxford, 2009, pp. 268–274.

Machlis, G. E., Force, J. E., and Burch, W. R., Jr. "The human ecosystem. Part I: The human ecosystem as an organizing concept in ecosystem management." *Society and Natural Resources: An International Journal* 10(4):347–367.

Marty, M. A., and Blaisdell, R. J. (Eds.). (2000). Air Toxics Hot Spots Program Risk Assessment Guidelines: Part IV. Technical Support Document for Exposure Assessment and Stochastic Analysis. California Office of Environemntal Health and Hazard Assessment.

MOE (Ministry of the Environment). (1997). Scientific Criteria Document for Multimedia Standards Development, Polycyclic Aromatic Hydrocarbons (PAH). Part 1: Hazard Identification and Dose–Response Assessment. Ministry of Environment, Toronto, Ontario.

Myers, N., (1997). "Consumption in relation to population, environment and development." *Environmentalist*, Volume 17, Issue 1 pp 33–44.

Putnam, J.J., and Allshouse, J., (1999). "Food consumption, prices and expenditures, 1970–1997." U.S. Department of Agriculture, *Statistical Bulletin* No. 965.

Stern, P. C. and Fineburg, H. V. (Eds.). (1996). Understanding Risk: Informing Decisions in a Democratic Society. Committee on Risk Characterization, Commission on Behavioral and Social Sciences and Education. National Research Council. National Academy Press, Washington, D.C.

Thomas P. (December 1999). "Consumption and environment: some conceptual issues." *Ecological Economics*, Volume 31, Issue 3, pp. 347–363.

Thun, M. J., Osorlo, A. M., Schober, S., Hannon, W. H., Lewis, B., and Halperin, W. (1989). "Nephropathy in cadmium workers: Assessment of risk from airborne occupational exposure to cadmium." *Br J Ind Med* 46:689–697.

U.S. Department of Health and Human Services. (2010). A Report of the Surgeon General: How Tobacco Smoke Causes Disease: What It Means to You. U.S. Department of Health and Human Services, Centers for Disease Control and Prevention, National Center for Chronic Disease Prevention and Health Promotion, Office on Smoking and Health, Washington, D.C.

U.S. EPA. (1989). Risk Assessment Guidance for Superfund: Volume I Human Health Evaluation Manual (Part A). Office of Emergency and Remedial Response, U.S. Environmental Protection Agency, Washington, D.C.

U.S. EPA. (1998). Guidelines for Ecological Risk Assessment. Risk Assessment Forum, U.S. Environmental Protection Agency, Washington, D.C.

U.S. EPA, (2000). Technical Support Document for Exposure Assessment and Stochastic Analysis. Risk Assessment Forum, U.S. Environmental Protection Agency, Washington, D.C.

U.S. EPA. (2005a). Guidelines for Carcinogen Risk Assessment. Risk Assessment Forum, U.S. Environmental Protection Agency, Washington, D.C.

U.S. EPA. (2005b). Supplemental Guidance for Assessing Susceptibility from Early-Life Exposure to Carcinogens. Risk Assessment Forum, U.S. Environmental Protection Agency, Washington, D.C.

U.S. EPA. (2006). A Framework for Assessing Health Risks of Environmental Exposures to Children. National Center for Environmental Assessment, Office of Research and Development, U.S. Environmental Protection Agency, Washington, D.C.

U.S. EPA. (2010). Valuing Mortality Risk Reductions for Environmental Policy: A White Paper. National Center for Environmental Economics, U.S. Environmental Protection Agency, Washington, D.C.

U.S. EPA. (2011). Exposure Factors Handbook: 2011 Edition. National Center for Environmental Assessment, Office of Research and Development, U.S. Environmental Protection Agency, Washington, D.C.

U.S. EPA. (2012). Microbial Risk Assessment Guideline: Pathogenic Microorganisms with Focus on Food and Water. Interagency Microbiological Risk Asessment Guideline Workgroup, U.S. Environmental Protection Agency. Washington, D.C.

Wilk, R., (2002). "Consumption, human needs, and global environmental change." *Global Environmental Change,* 12 pp. 5–13.

Wilson, R. (1979, February). "Analyzing the daily risks of life." *Technology Review,* pp. 41–46.

Windsor, J. S., Firth, P. G., Grocott, M. P., Rodway, G. W., and Montgomery, H. E. (2009). "Mountain mortality: A review of deaths that occur during recreational activities in the mountains." *Postgrad Med J.* 85:316–321.

Key Concepts

Ecology	Technology
IPAT equation	Intergenerational equity
Impact	Intragenerational equity
Population	Global warming potential **(GWP)**
Affluence	Integrated time horizon **(ITH)**
Background consumption	Total equivalent CO_2
Overconsumption	Ozone depletion potential **(ODP)**
Misconsumption	Acidification potential **(AP)**

Eutrophication potential **(EP)**
Hazard
Liability
Voluntary risk
Involuntary risk
Risk management
Risk assessment
Carcinogens
Teratogens
Neurotoxins
Mutagens
Effective concentration **(EC)**
Lethal concentration **(LC)**
Lowest observable effect concentration **(LOEC)**
No observable effect concentration
Maximum allowable toxicant concentration **(MATC)**

Epidemiology
Reference dose **(RfD)**
Uncertainty factor **(UF)**
Reference concentration **(RfC)**
Hazard index **(HI)**
Integrated Risk Information System **(IRIS)**
Chromic daily intake **(CDI)**
Exposure assessment
Bioaccumulation factor **(BCF)**
Toxic Substance Control Act **(TOSCA)**
Industrial ecology
Eco-industrial park
Reasonable Maximum Exposure **(RME)**
Hazard quotient **(HQ)**

Active Learning Exercises

ACTIVE LEARNING EXERCISE 9.1: Tragedy of the Commons

Read and analyze Garrett Hardin's essay "The Tragedy of the Commons." Do you agree with his premises and conclusions? What are your thoughts on the implications for freedom, the role of governmental authorities, and the "public good"? Be prepared to discuss your conclusions.

ACTIVE LEARNING EXERCISE 9.2: The other greenhouse gas

If water is a greenhouse gas, how come we are not concerned about the quantity of water in the atmosphere?

ACTIVE LEARNING EXERCISE 9.3: Ecological system models in society

Discuss your view on a society modeled after a type III ecological system. What would this society be like? What advancements in science and technology will be required? What changes to current social and economic models will be necessary?

ACTIVE LEARNING EXERCISE 9.4: Industrial symbiosis

Can you think of any potential environmentally beneficial industrial symbiotic relationships that can be developed in your community?

Problems

9-1 Describe how affluence contributes to environmental impact.

9-2 If the population of a country in 2010 was 72 million and the projected *exponential rate* of increase is 6.3 per 1,000, what total percentage reduction of environmental impact per GDP will be required by 2050 to keep the environmental impact at the 2010 levels if the GDP per capita is predicted to increase at a rate of 5% per year between 2010 and 2050?

9-3 In 2012, the population of a country was 65 million and the projected **exponential** rate of increase is 5.2 per 1,000. If the environmental impact per GDP is reduced by 15% by 2040, how will this affect the GDP per capita assuming the total environmental impact in 2040 is maintained at the 2010 levels?

9-4 The annual exponential population growth rate is estimated to be 11.9 per 1000 persons from 2012 to 2030 and 12.1 per 1,000 persons from 2030 to 2060. If the GDP per capita is estimated to increase at a yearly rate of 2.5% from 2012 to 2030 and 1.5% from 2030 to 2060, calculate the annual rate of emissions reductions from 2012 to 2060 to keep impact levels in 2060 at the 2012 values.

9-5 Compare the contribution to global warming in millions of metric tons of carbon dioxide equivalent of the two systems below. Based on your results, which process is environmentally preferred and why?

GAS	EMISSIONS FROM SYSTEM 1 (kg/year)	EMISSIONS FROM SYSTEM 2 (kg/year)
Carbon dioxide	5.00E + 09	3.00E + 09
Methane	3.00E + 09	5.00E + 09
Nickel	0.00E + 00	1.20E + 14
Nitrous oxide	3.00E + 09	3.00E + 09
CF_4	5.60E + 08	0.00E + 00
C_2F_6	0.00E + 00	3.00E + 09

9-6 Compare systems 1 and 2 on the basis of their contribution to **global warming**, **acidification**, and **photochemical oxidant formation** using equivalence factors (*perform Internet searches to find values you may not have and provide a reference*). State any assumptions made.

CHEMICAL EMISSION	SYSTEM 1 (kg/day)	SYSTEM 2 (kg/day)
Methane (CH_4)	8	14
Sulfur dioxide (SO_2)	0.2	0.9
Ammonium (NH_3)	0	4
1,1,1 trichloroethane	70	5
Benzene	5	67
Nitrogen dioxide (NO_2)	24	0

9-7 Determine the hazard quotient for an adult exposure to acrylamide in the air during a 30-year industrial career. Use the U.S. EPA IRIS website to find relevant risk assessment parameters.

9-8 Determine the hazard quotient for an adult exposure to cadmium in the air during a 30-year industrial career. Use the U.S. EPA IRIS website to find relevant risk assessment parameters.

9-9   Determine the hazard quotient for an adult exposure to 0.1 mg/m³ of acrylamide in the air of a polymer production facility during a 30-year industrial career. Use the U.S. EPA IRIS website to find relevant risk assessment parameters.

9-10   Determine the hazard quotient for an adult exposure to 1.5 mg/m³ of cadmium in the air at a smelting facility during a 30-year industrial career. Use the U.S. EPA IRIS website to find relevant risk assessment parameters.

9-11   A family living in a home is exposed to 0.008 mg/L of 1,4-dioxane in their drinking water. Calculate the lifetime risk for
   a. Adult male
   b. Adult female
   c. Child

9-12   A female mechanic working in an industrial facility is exposed to 18 mg/m³ of benzene in the air at the workplace. Calculate the lifetime cancer risk for this exposure.

CHAPTER 10

Life Cycle Analysis

FIGURE 10.1 Do you ever wonder what the true impacts of the choices you make are?

Source: www.CartoonStock.com.

For the first time in the history of the world, every human being is now subjected to contact with dangerous chemicals, from the moment of conception until death.

—RACHEL CARSON, 1962

GOALS

THE GOAL OF THIS CHAPTER IS to introduce the concept of product life cycle thinking as well as present elementary principles of life cycle assessment methodologies.

OBJECTIVES

At the conclusion of this chapter, you should be able to:

- Develop an understanding of the concept and framework of life cycle assessment.

- Understand the role of life cycle assessments (LCA) in industrial ecology and sustainability.

- Describe the components and steps utilized to calculate a water footprint for a product.

- Develop frameworks for conceptualizing complex materials balance problems.

- Model material and energy system cycles.

- Appreciate the limitations of LCA.

10.1 Introduction

In Chapter 9, we learned about the relationship between technology and environmental impacts. We also learned about the concept of industrial metabolism, which is the process by which industrial systems consume materials and energy from the environment to make products. In addition to these products, a by-product of industrial metabolism is waste in the form of environmental emissions. One of the objectives of sustainable development is to supply society's needs with minimal harm to the environment and the ability of future generations to meet their needs. The current needs of society are met in a wide variety of ways due to advancements in technology. It is a major challenge to determine which of various options are the least harmful and thus most sustainable.

In the past couple of decades, studies have shown that emissions produced during industrial and technological manufacturing are just a small percentage of the overall emissions that result from technological products. The extraction of raw materials, production, transportation, the use and the eventual disposal of products—all result in emissions and various impacts that have to be considered in the calculating the full impact of technology on the environment. There is a need for methods that can assess the sustainability of different technologies or options over the entire life of a product. The life cycle of a product is defined as "consecutive and interlinked stages of a product system, from raw material acquisition or generation from natural resources to final disposal" (ISO 14040). This has led to the use of the life cycle assessment (LCA), which has been defined as "the compilation and evaluation of the inputs and outputs and the potential environmental impacts of a product system throughout its life cycle" (ISO 14040, ISO 14044). In this context, product is defined in a broad sense to include physical goods as well as services.

As an example, your textbooks may be printed on paper that was produced from pulp that was extracted from wood obtained from trees using a chemical process. After use, these textbooks may be disposed of in various ways, including recycling, incineration, or burial in landfills. In addition, each of these stages requires energy drawn from various sources. Every stage in the life of your textbooks contributes to the overall impact associated with the production and use of textbooks. Life cycle assessments allow us to compare the impacts associated with different products or processes to determine which have the least impacts. It may also allow us to determine which stage in the overall life cycle of products contributes the most to the overall impact.

The Food and Agriculture Organization (FAO) of the United Nations predicts that global meat consumption will approximately double between 2001 and 2050. Consider this: An average of 18% of global greenhouse gas emissions from all human activities comes from livestock production (FAO). Transportation, which includes emissions from fuel combustion, accounts for approximately 14%. The equivalent of up to 36.4 kg of CO_2 are emitted in the production of 1 kg of beef without considering the emissions from transportation and managing infrastructure (Ogino et al.). The livestock sector emits 37% of all anthropogenic methane and accounts for 65% of anthropogenic nitrous oxide (NOx), mostly coming from manure. Methane and nitrous oxide have global warming potentials (GWP) of 25 and 298, respectively, as illustrated in Figure 10.2. Life cycle assessment studies can help us determine which processes contribute the most to these emission numbers.

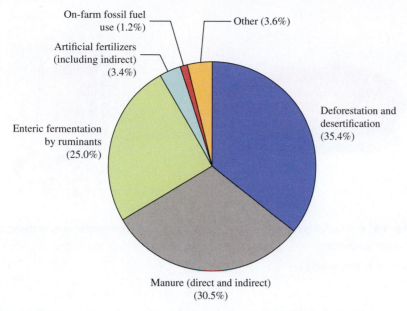

FIGURE 10.2 Proportion of greenhouse gas emissions from different parts of livestock production.

Source: Based on McMichael et al., Adapted from the Food and Agriculture Organization.

10.2   Life Cycle Thinking

In Chapter 9, we established that the primary goal of industrial ecology is the development of technological systems that simulate the way materials and energy are handled within natural systems, leading to minimal impacts on the surrounding systems. This goal requires comprehensive tracking of the way materials and energy flow through the product system—from the extraction of virgin raw material from the environment through the various stages of processing, refinement, packaging, sales, and use all the way to the ultimate disposal back into the environment.

Let us consider products we use and conveniences that you have become accustomed to. Do you ever think of the various stages and processes that were required to get these to you? For example, consider what it takes to make a peanut butter sandwich. You need just two slices of bread and some peanut butter as illustrated in Figure 10.3.

FIGURE 10.3 Ingredients to make a peanut butter sandwich.

Source: Photo by Bradley Striebig.

FIGURE 10.4 A loaf of bread and a jar of peanut butter used to manufacture, or make, a peanut butter sandwich.

Source: Photo by Bradley Striebig.

But bread usually isn't made in individual slices, so your slices will likely come from a whole loaf and peanut butter will likely come from a jar, both of which have greater mass than the individual sandwich, as shown in Figure 10.4.

Even though sliced bread and peanut butter are readily available in supermarkets and stores, they have to be made first. Bread is made using flour, water, and yeast. Flour is obtained from the milling of cereal grains, typically wheat. Wheat has to be grown and requires land for planting, water, and sources of plant nutrients, as shown in Figure 10.5.

Similarly, the peanut butter is made from crushing and grinding roasted peanuts that also have to be grown requiring land, water, and plant nutrients, as illustrated in Figure 10.6.

FIGURE 10.5 Simplified life cycle of bread.

Sources: Photo by Bradley Striebig, Andy Sacks/Photographer's Choice/Getty Images.

FIGURE 10.6 Simplified life cycle of peanut butter.

Sources: Photo by Bradley Striebig, Rick Rudnicki/Lonely Planet Images/Getty Images, nanao wagatsuma/Moment /Getty Images, Peeter Viisimaa/Vetta/Getty Images.

Note that, in this example, we have ignored crucial stages and processes that include packaging and transportation in addition to the processes involved in the production of the other ingredients. However, we can appreciate the concept that all products come to us through processes that we often do not think about while we use the products.

Consider the following requirements for manufacturing a peanut butter sandwich (National Peanut Board, 2014).

- It takes about 540 peanuts to make a 12-ounce jar of peanut butter.
- There are enough peanuts in one acre to make 30,000 peanut butter sandwiches.
- The average American consumes more than six pounds of peanuts and peanut butter products each year.
- The average child will eat 1,500 peanut butter and jelly sandwiches before he/she graduates high school.
- Americans consume on average over 1.5 billion pounds of peanut butter and peanut products each year.
- Peanut butter is consumed in 90% of U.S. households.
- Americans eat enough peanut butter in a year to make more than 10 billion peanut butter and jelly sandwiches.
- The amount of peanut butter eaten in a year could wrap the Earth in a ribbon of 18-ounce peanut butter jars one and one-third times.
- Peanuts account for two-thirds of all snack nuts consumed in the United States.
- Peanut butter is the leading use of peanuts in the United States.

Each of the process steps considered in making peanut butter involves the use of materials and energy and may generate wastes, as illustrated in

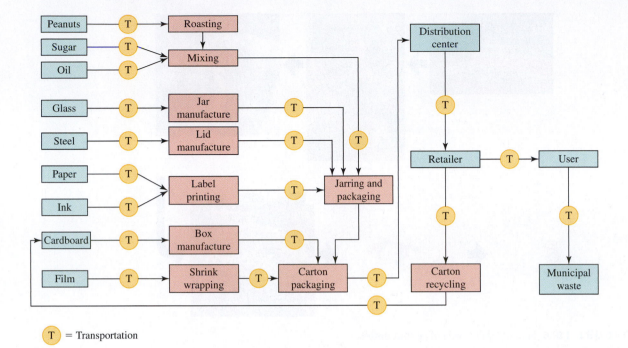

$\widehat{T}$ = Transportation

FIGURE 10.7 The life cycle of packaged peanut butter.

Source: Bradley Striebig.

Figure 10.7. We must also consider the potential impacts associated with eating a peanut butter sandwich, possibly relating to the use of fertilizers, herbicides, and pesticides in crop production. Consideration of unintended consequences and impacts associated with products and manufacturing are presented in the cartoon presented in Figure 10.8, which illustrates the impacts associated with using ethanol as a gasoline additive.

Figure 10.9 shows a simple schematic of the inputs and outputs from a single stage in a product life cycle system. This is the basic building block of life cycle assessments. Figure 10.10 shows the connected multiple stages in the life cycle of a product.

Electrical energy is generated from other forms of energy in facilities called power plants and must be transmitted through high-voltage wires to local utility companies before being transmitted to useful electricity for work in our homes, schools, and offices. Carbon-based forms of chemical energy such as coal, petroleum, and natural gas are combusted to generate heat, which in turn generates electrical energy. The combustion of these fossil fuels generates CO_2 that has been linked to global climate change, as discussed in Chapters 6 and 8. This means that when generated form fossil fuels, every unit of electrical energy we use has a direct impact on the global climate. Companies may market "zero emissions" electrical processes as illustrated by the cartoon in Figure 10.11, but this is deceptive marketing since the production of electricity has measurable environmental impacts.

In technological systems, all raw materials are initially extracted from the environment in one form or another. The environment is often called the cradle; the source of all materials. These materials are processed in various industrial stages

FIGURE 10.8 Cartoon showing the complexity of choices that become apparent from a life cycle perspective.

Source: Andy Singer Cartoons http://www.andysinger.com/.

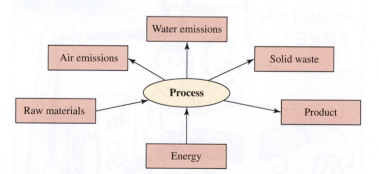

FIGURE 10.9 Simple schematic of a single stage in a product.

Source: Bradley Striebig.

and processes called gates and are ultimately disposed into the environment. The methods and location of waste disposal are often described as the grave for the waste material. Thus, the environment is both the cradle and the grave. The changes that occur to materials from cradle to grave and the implications of extraction, processing, and disposal can be evaluated using **life cycle assessment (LCA)**.

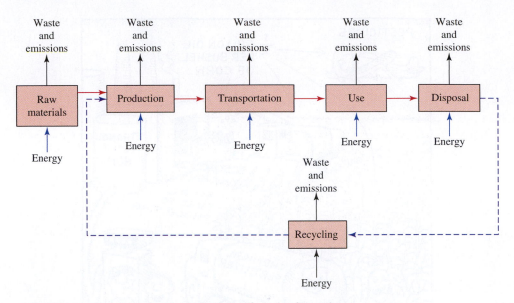

FIGURE 10.10 A simple schematic of the life cycle of a product.

Source: Bradley Striebig.

FIGURE 10.11 A cartoon showing how life cycle assessments may reveal unapparent impacts.

Source: Andy Singer Cartoons http://www.andysinger.com/.

LCA is an objective process used to quantitatively evaluate the environmental burdens associated with a product, process, or activity throughout its entire life cycle (through the cradle, through all the gates, to the grave). It utilizes various mass and energy balance protocols as well as environmental impact evaluation techniques described in Chapters 3 through 8 to model the associated impacts across every stage of the life of a product. This may include the impacts associated with processing materials as well as the impacts associated with subsidiary actions. For example, in the extraction of natural resources, LCA includes the impacts associated with the extracted materials as well as the impact associated with the extraction process.

Consider the cutting of trees to make wood products. The impacts associated with this activity include the environmental impacts from the loss of the trees as well as from the logging technology itself. Depending on the objective for which an LCA is performed, the scope may range from

1. Raw materials extraction to the disposal of finished goods (i.e., cradle-to-grave)
2. Raw materials extraction to finished goods (i.e., cradle-to-gate)
3. One processing stage to another (i.e., gate-to-gate)

10.3 Life Cycle Assessment Framework

An LCA contains three phases: goal and scope definition, inventory analysis, and impact analysis, and each phase is followed by interpretation according to the generally agreed structure shown in Figure 10.12.

10.3.1 Goal and Scope Definition

The goal specifies the reasons for carrying out the LCA. The goal is important because the parameters to be used in the assessment are usually dependent on what the intended objectives are. The goal also specifies the intended application of the LCA as well as the intended audience.

The scope helps to establish the system boundaries and the limits of the LCA. For example, if an LCA is to be performed to compare products, it is usually not the products themselves that are the basis of the comparison but the

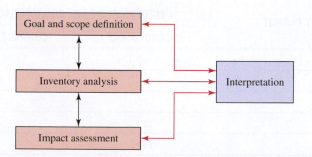

FIGURE 10.12 The LCA Framework.

Source: Based on ISO 14040:2006, Environmental management, Life cycle assessment – Principles and framework.

functions they provide. So a useful term called the **functional unit** is defined in the scope.

The functional unit is used to establish a basis for comparison of two products by identifying a common function and how each product achieves that function over its life. The functional unit is quantitative and measurable, and a careful and proper identification of the functional unit is crucial, particularly if the LCA is used for comparisons.

For example, if an LCA is to be performed comparing reusable plastic cups and disposable paper cups, it would be incorrect to compare the impacts associated with one paper cup and those associated with one plastic cup since the plastic cup will continue to perform its function after the paper cup is discarded. Rather, it will be more appropriate to determine how many times one plastic cup will be used before disposal and to calculate the equivalent number of paper cups to fulfill the same function. So the comparison will be of the impacts associated with one plastic cup and the functional equivalent number of paper cups.

Another example involves comparing the life cycle impact of incandescent bulbs and compact fluorescent lamps. One fluorescent lamp uses significantly less energy than an incandescent bulb to produce the same amount of visible light. Since the function of the bulb and the fluorescent lamp is to produce light, it would be incorrect to compare a 40 W incandescent bulb to a 40 W fluorescent lamp. From Table 10.1, we see that a 9 W fluorescent lamp produces approximately the same amount of light as a 40 W bulb. So the functional unit would be the amount of visible light required, and then LCA would be performed comparing the number of incandescent light bulbs and compact fluorescent lamps required to provide the required amount of light. However, a fluorescent lamp also lasts significantly longer than an incandescent bulb, so one would have to use multiple incandescent bulbs over the lifetime of one fluorescent lamp. Incandescent light bulbs are typically rated to last an average of 1,000 hours, while compact fluorescent lamps are rated to last up to 8,000 hours. This means that, in addition to comparing the two choices based on the amount of light that is needed, the functional unit will also include the length of time that the light will be provided.

TABLE 10.1 Energy consumption of incandescent bulbs and compact fluorescent lamps

MINIMUM LIGHT OUTPUT (Lumens)	ELECTRICAL POWER CONSUMPTION (Watts)	
	INCANDESCENT	COMPACT FLUORESCENT
450	40	9–13
800	60	13–15
1,100	75	18–25
1,600	100	23–30
2,600	150	30–52

Source: USEPA, https://www.energystar.gov/index.cfm?c=cfls.pr_cfls_lumens.

So, consider that our functional unit is ***450 lumens of light for 5,000 hours***. For this we would compare the life cycle impact of one 9 W compact fluorescent lamp to five 40 W incandescent light bulbs.

10.3.2 Inventory Analysis

Inventory analysis involves determining the quantitative values of the materials and the energy inputs and outputs of all process stages within the life cycle. This includes

- Raw materials/energy needs
- Manufacturing processes
- Transportation, storage, and distribution requirements
- Use and reuse
- Recycle and end-of-life scenarios such as incineration and landfilling

Inventory analysis is usually initiated with a flowchart or process tree identifying the relevant stages and their interrelations. Relevant materials and energy data are then collected for each process stage using materials and energy balance protocols to account for unknown values. Standardized units are used to make calculations and comparisons easy. It is also in this phase that the system boundaries necessary to meet the predetermined goal and scope of the LCA are determined. The system boundaries are the limits placed on data collection. For example, if the goal of the LCA is a comparative study of two products for which some life cycle stages are the same with the same materials and energy input–output values, then the system boundary may be drawn to exclude the data related to those stages. Including these would only burden the LCA without providing additional information.

10.3.3 Impact Assessment

Impact assessment entails determining the environmental relevance of all the inputs and outputs of each stage in the life cycle. This includes the environmental impacts associated with the production, use, and disposal of the products. Relevant impact categories are selected such as degradation of ecological systems, depletion of natural resources, and impacts on human health and welfare. For example, if we conducted the LCA of paper-based textbooks, examples of ecological systems degradation can include the impact of cutting trees as well as the impact of chemicals discharged during pulping on water quality. Trees as natural resources are renewable but not replaceable rapidly enough to immediately offset the impact of logging. The impact of the chemicals used on the health of the populations directly connected to the paper processing industries will also be assessed.

10.3.4 Interpretation

LCA results present the opportunities to reduce or mitigate identified environmental impacts arising from the manufacture, use, and disposal of products. The product design may be changed; the materials used may be replaced with less impactful materials; and the entire industrial process may even be changed based on the results of an LCA. The LCA represents the most objective tool currently available to inform decisions on the environmental sustainability of products and processes.

10.4   Materials Flow Analysis

Materials flow analysis is a method that can be used to develop the inventory for an LCA. It is used to evaluate the metabolism of anthropogenic systems. It is a systematic assessment of the flows of materials within the boundaries of a system in space and time. Materials flow analysis relies on the law of conservation of mass and is performed by simple mass balances comparing inputs, outputs, depletions, and accumulations within systems and subsystems. Some relevant terminologies that are used in materials flow analysis are

- *Materials:* Substances as wells as goods.
- *Process:* The transport, transformation, or storage of materials.
- *Reservoir:* A system that holds materials. It can be a source, from which materials come, or a sink, into which materials go, or both.
- *Stocks:* The quantity of materials held in reservoirs within a system.
- *Flows/Fluxes:* The ratio of the mass per time that passes through a system boundary is termed the *flow* while the *flux* is the flow per cross section.

Consider the reservoir system shown in Figure 10.13, where

$\dot{m}_{i,1}$ and $\dot{m}_{i,2}$ are mass flows into the reservoir;

$\dot{m}_{o,1}$ and $\dot{m}_{o,2}$ are mass flows out of the reservoir;

Δm_s is the change in quantity of stock in the reservoir; and

$\dot{m}_s$ is the rate of change of stock in the reservoir.

Let us recall the general mass balance equation:

$$\text{Input} - \text{output} = \text{accumulation} \tag{10.1}$$

For a steady-state system, accumulation will be zero and

$$\text{Input} = \text{output} \tag{10.2}$$

but steady-state systems are rare.

Solving the mass balance for the reservoir:

$$\sum \dot{m}_i - \sum \dot{m}_o = \dot{m}_s \tag{10.3}$$

By definition,

$$\dot{m}_s = \frac{\Delta m_s}{dt} \tag{10.4}$$

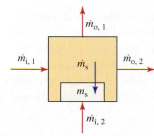

FIGURE 10.13 A materials reservoir system showing input and output flows.

Source: Bradley Striebig.

Combining the two equations

$$\Delta m_s = \left(\sum \dot{m}_i - \sum \dot{m}_o\right) dt \tag{10.5}$$

The change in the stock in a reservoir is given by the expression

$$\Delta m_s = \int_{t_1}^{t_2}\left(\sum \dot{m}_i - \sum \dot{m}_o\right) dt \tag{10.6}$$

10.4.1 Efficiencies in Mass Flow Systems

One of the applications of mass flow analysis in industrial systems is the ability to track how much of the original materials going into the process actually end up as part of the desired product stream. Applying the principles of conservation of mass to the system shown in Figure 10.10, we can see that only a portion of the raw materials entering the system becomes the product, while the rest ends up in the waste streams. The efficiency of the process is defined as the ratio of the mass in the desirable product and the total mass entering the system. For a 100% efficient system, no emissions are generated, and all of the raw material is converted to useful product. This is probably not a realistic goal for industrial systems. A more feasible aspiration is for industrial systems to mimic a type II biological ecosystem as defined in Chapter 9, in which limited waste is generated across the entire industrial systems.

Figure 10.14 represents a typical industrial materials flow system of a substance where

m_E, is the mass of the substance in raw materials extracted and processed

m_M, is the mass of the substance in materials sent to product manufacturing

m_P, is the mass of the substance in the consumer product

m_D, is the mass of the substance recovered after consumer use

m_R, is the mass of the substance in recycled material

w_E, is the mass of the substance in the waste from extraction and processing

w_M, is the mass of the substance in the waste from product manufacturing

w_P, is the mass of the substance of the unrecovered waste after consumer use

w_R, is the mass of the substance of the waste from recycling

w_T, is the total mass of the substance in all waste streams

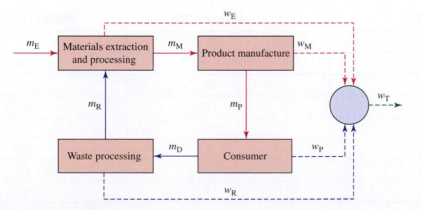

FIGURE 10.14 Materials flow system of a substance in an industrial system.

Source: Bradley Striebig.

The efficiencies of each stage within the system can be defined as

$$\text{Extraction efficiency: } E_{\text{eff}} = \frac{m_M}{m_E + m_R} = \frac{m_M}{m_M + w_E} \tag{10.7}$$

$$\text{Manufacturing efficiency: } M_{\text{eff}} = \frac{m_P}{m_M} = \frac{m_P}{m_P + w_M} \tag{10.8}$$

$$\text{Recovery efficiency: } D_{\text{eff}} = \frac{m_D}{m_P} = \frac{m_D}{m_P + w_P} \tag{10.9}$$

$$\text{Recycling efficiency: } R_{\text{eff}} = \frac{m_R}{m_D} = \frac{m_R}{m_R + w_R} \tag{10.10}$$

In each stage, reduction of the mass in the waste stream increases the efficiency of that stage.

The system efficiency can be calculated as

$$S_{\text{eff}} = E_{\text{eff}} \times M_{\text{eff}} \times D_{\text{eff}} \times R_{\text{eff}} \tag{10.11}$$

$$S_{\text{eff}} = \frac{m_M}{m_E + m_R} \times \frac{m_P}{m_M} \times \frac{m_D}{m_P} \times \frac{m_R}{m_D} = \frac{m_R}{m_E + m_R} \tag{10.12}$$

The most efficient system will be the one in which the total waste emitted from the system is minimal.

The *reuse factor* is defined as the amount of recycled material that can be recovered and used in the process to make a product. This can calculated as

$$\psi = \frac{m_R}{m_M} \tag{10.13}$$

Because of the large values of the masses that are sometimes encountered in materials flow analysis, unit prefixes listed in Table 10.2 are used to make assessments easier.

TABLE 10.2 Unit prefixes used in materials flow analysis

PREFIX	ABBREVIATION	MULTIPLIER VALUE
kilo	k	10^3
mega	M	10^6
giga	G	10^9
tera	T	10^{12}
peta	P	10^{15}
exa	E	10^{18}
zetta	Z	10^{21}
yotta	Y	10^{24}

Source: Bradley Striebig.

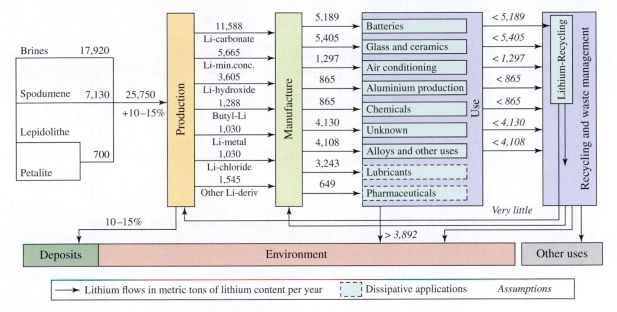

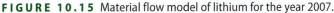

FIGURE 10.15 Material flow model of lithium for the year 2007.

Source: Reprinted from Resources, Conservation and Recycling, Volume 63, Saskia Ziemann, Marcel Weil, Liselotte Schebek, "Tracing the fate of lithium—The development of a material flow model," Pages 26–34, Copyright 2012, with permission from Elsevier.

10.4.2 Constructing a Materials Flow System

The boundaries of a materials flow system can be drawn across different spatial levels, from a single industrial facility to national and even global levels. Figure 10.15 shows the global material flow of lithium for 2007.

All materials flow analyses entail the construction of a materials budget, which requires:

1. Determination of the material(s) to be evaluated.
2. Identifying all relevant reservoirs and flow streams.
3. Quantifying the contents of the reservoir and the magnitudes of the flows. This information is obtained from data if available, taking measurements where possible, and finally making estimations if necessary.

EXAMPLE 10.1 Material Analysis

The following questions refer to the materials flow system shown in Figure 10.16.

1. Calculate the values of the

   i.   Extraction efficiency
   ii. Manufacturing efficiency
   iii. Recovery efficiency
   iv. Recycling efficiency
   v.   System efficiency
   vi. Reuse factor

2. Which subsystem is the least efficient?

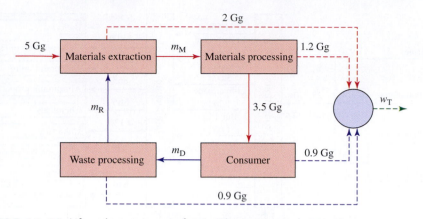

FIGURE 10.16 Life cycle assessment of material extraction and processing.

Source: Bradley Striebig.

Solution

1. The best approach is to conduct a mass balance across each subsystem.

   The mass balance for a steady-state system is

   $$\text{Input} - \text{output}$$

   The materials extraction stage is illustrated in Figure 10.17.

   $$5 \text{ Gg} + m_R = 2 \text{ Gg} + m_M$$

   There are two unknowns in this subsystem that cannot be evaluated until we solve the mass balance for other subsystems.

   For the materials processing subsystem shown in Figure 10.18,

   $$m_M = 3.5 \text{ Gg} + 1.2 \text{ Gg} = 4.7 \text{ Gg}$$

   We may now return to the previous equation and substitute for m_M, which yields

   $$5 \text{ Gg} + m_R = 2 \text{ Gg} + 4.7 \text{ Gg} = 6.7 \text{ Gg}$$
   $$m_R = 6.7 \text{ Gg} - 5 \text{ Gg} = 1.7 \text{ Gg}$$

   For the consumer subsystem shown in Figure 10.19,

   $$3.5 \text{ Gg} = m_D + 0.9 \text{ Gg}$$
   $$m_D = 3.5 \text{ Gg} - 0.9 \text{ Gg} = 2.6 \text{ Gg}$$

   We do not need to evaluate the final subsystem (waste processing) since the values of all unknown mass flows have been found, where

   $$m_M = 4.7 \text{ Gg}$$
   $$m_D = 2.6 \text{ Gg}$$
   $$m_R = 1.7 \text{ Gg}$$

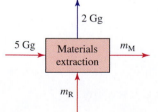

FIGURE 10.17 Materials extraction component of the example LCA.

Source: Bradley Striebig.

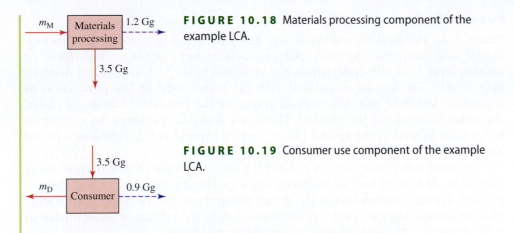

m_M Materials processing 1.2 Gg

3.5 Gg

FIGURE 10.18 Materials processing component of the example LCA.

3.5 Gg

m_D Consumer 0.9 Gg

FIGURE 10.19 Consumer use component of the example LCA.

2. We can now calculate the values of the efficiencies.

   i. Extraction efficiency

$$E_\text{eff} = \frac{m_\text{M}}{5\ \text{Gg} + m_\text{R}} = \frac{4.7\ \text{Gg}}{5\ \text{Gg} + 1.7\ \text{Gg}} = \frac{4.7\ \text{Gg}}{6.7\ \text{Gg}} = 0.70$$

   ii. Manufacturing efficiency

$$M_\text{eff} = \frac{3.5\ \text{Gg}}{m_\text{M}} = \frac{3.5\ \text{Gg}}{4.7\ \text{Gg}} = 0.74$$

   iii. Recovery efficiency

$$D_\text{eff} = \frac{m_\text{D}}{3.5\ \text{Gg}} = \frac{2.6\ \text{Gg}}{3.5\ \text{Gg}} = 0.74$$

   iv. Recycling efficiency

$$R_\text{eff} = \frac{m_\text{R}}{m_\text{D}} = \frac{1.7\ \text{Gg}}{2.6\ \text{Gg}} = 0.65$$

   v. System efficiency

$$S_\text{eff} = E_\text{eff} \times M_\text{eff} \times D_\text{eff} \times R_\text{eff}$$

$$S_\text{eff} = 0.70 \times 0.74 \times 0.74 \times 0.65 = 0.25$$

   vi. Reuse factor

$$\psi = \frac{m_\text{R}}{m_\text{M}} = \frac{1.7\ \text{Gg}}{4.7\ \text{Gg}} = 0.36$$

Note that the *least efficient* subsystem is the waste processing subsystem.

Embodied Water

Water is as essential to industrial and manufacturing systems as it is to biological and ecological systems. All products require water in one form or another over their life cycle, with some products having a higher water demand than others. The impact associated with the water used in the production of a product, summed over the various stages of the production chain, is called the water footprint of the product. The water footprint captures the volume of water used as well as the spatial (where) and temporal (when) attributes of the water used.

The total water footprint of a product is a measure of the volume of fresh water consumed directly as well as indirectly when producing, distributing, and using a product. Direct consumption is the actual physical volume of water used during these processes, while indirect consumption refers to the volume of water related to auxiliary processes along the production and supply chain.

The water footprint is composed of three components; blue water, green water, and gray water, and their interactions are illustrated in Figure 10.20. Blue water is composed of traditional surface and groundwater storage sources (discussed previously in Chapter 3) along the supply chain of a product. Green water is water that may be captured, such as in rainwater containment systems, which does not become runoff. Crops that do not use irrigation rely primarily on green water, and thus the primary losses of green water are associated with the evapotranspiration (ET) from growing crops or forest production. Gray water is defined as the volume of fresh water that is required to assimilate the load of pollutants, given natural background concentrations and existing ambient water standards (Hoekstra et al., 2011), and is similar to the U.S. EPA Total Maximum Daily Load (TMDL) described in Chapter 4.

The calculated water footprint is dependent on how the inventory boundaries of a process are determined. The inventory boundary forms the divisions between what steps are included and what steps are excluded in the analysis. The scope of the inventory boundaries that are considered must be explicitly stated when

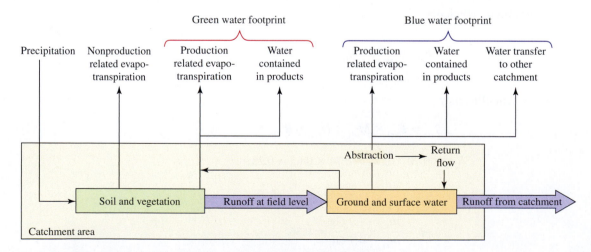

FIGURE 10.20 Relationships between blue water, green water, and hydrologic processes.

Source: Based on Hoesktra, A, A.K. Chapagain, M. M. Aldaya, M. M. Mekonnen. 2011. The Water Footprint Assessment Manual: Setting the Global Standard. Earthscan, Ltd., London, UK.

performing water footprint analyses and may include consideration of the following points.

- What components, blue water, green water, or gray water, should be considered?
- How far back along the supply chain are contributions to the water footprint significant?
- What are the appropriate geographic and time boundaries for the analysis?
- How small must the time-steps for the water analysis be (typically days or months)?
- The water footprint for consumers and businesses must consider which direct and/or indirect contributions should be included.
- Water footprint analysis on a national scale should consider not only national consumption, but also the water footprint associated with import and export of goods and services.

Ideally, the water footprint should include all significant sources that contribute to the water footprint, including all major direct and indirect contributions. For example, if consumers included only direct contributions to their food consumption, the most significant portion of their water footprint would be omitted. Also, the indirect water footprint contributions from most businesses' supply chain is greater than that associated with their own direct operations. In general, agricultural ingredients are often a major contributor for products and processes, as an estimated 85% of humanity's water footprint is from the agricultural sector (Chapagain and Hoekstra, 2008). Industrial contributions to a water footprint are likely to be important, especially to the gray water footprint, if they contribute to water pollution. Generally, transportation and energy contributions are small, except when the energy source or transportation source uses biofuels or hydropower. The time scale of the analysis and data (days, months, years, etc.) may affect the outcome of the analysis, especially concerning water hotspots.

For example, cotton is used to manufacture various items of clothing. Over the entire production chain, the production of 1 kg of cotton requires approximately

TABLE 10.3 Types and scale of water footprint calculations

BOUNDARY	DESCRIPTION	UNIT
Product (good or service)	The aggregate of the water footprints of the various process steps relevant in the production of the product	Volume/time
Consumer	A function of the water footprints of the various products consumed by the consumer	Volume/product unit
Business (producer)	Is equal to the sum of the water footprints of the products that the producer or business delivers	Volume/time
Geographic area	Is equal to the sum of the water footprints of all processes taking place in that area	Volume/time
Humanity	The sum of the water footprints of all consumers of the world, which is equal to the sum of the water footprints of all final consumer goods and services consumed annually and also equal to the sum of all water-consuming or polluting processes in the world	Volume/time

Source: Based on Hoesktra, A, A.K. Chapagain, M. M. Aldaya, M. M. Mekonnen. 2011. The Water Footprint Assessment Manual: Setting the Global Standard. Earthscan, Ltd., London, UK.

1,100 liters of water, and the average t-shirt that weighs 250 g requires approximately 2,700 liters of water.

Water Footprint of a Process

The blue water footprint of a process is calculated from

$$WF_{proc,blue} \frac{\text{volume}}{\text{time}} = \text{blue water evaporation} + \text{blue water incorporation}$$
$$+ \text{ lost return flow} \tag{10.14}$$

The green water footprint occurs from rainwater that is used for a process that comes from precipitation that does not recharge groundwater. Since this water does not require well-drilling, treatment, or transport, the cost of green water is typically far less than that for blue water. The green water footprint for a process step is determined from

$$WF_{proc,green} \frac{\text{volume}}{\text{time}} = \text{green water evaporation} + \text{green water incorporation} \tag{10.15}$$

The gray water footprint is calculated by dividing the pollutant mass-loading rate, L, by the difference between the ambient water quality standard for that pollutant, c_{max} and its natural concentration in the receiving body of water, $c_{nat,}$ shown in Equation (10.16) where

$$WF_{proc,gray} \frac{\text{volume}}{\text{time}} = \frac{L}{c_{max} - c_{min}} \tag{10.16}$$

The mass-loading rate of a pollutant associated with the discharge of a point source process, L_{pts}, is calculated by Equation (10.17) and is affected by any treatment process that occurs prior to the discharge.

$$L_{pts} = Q_{eff}c_{eff} - Q_{abstr}c_{abstr} \tag{10.17}$$

If there is no appreciable water consumption by the process, the volumetric flow rate of the effluent, Q_{eff}, and the water abstracted from the source, Q_{abstr}, are equal to one another—so that Equations (7.9) and (7.10) can be combined and simplified to derive Equation (10.18). Notice that the ratio of the difference between the concentration of the pollutant in the effluent and that in the water source compared to the difference between the maximum allowable concentration and the naturally occurring background level is equivalent to the gray water dilution factor, DF. The dilution factor represents the number of times that the effluent volume must be diluted in order to arrive at the maximum allowable discharge level (Hoekstra et al., 2011).

$$WF_{proce,pts} = \left\{ \frac{c_{eff} - c_{act}}{c_{max} - c_{nat}} \right\} Q_{eff} = \{DF\}Q_{effl} \tag{10.18}$$

If the process gray water is treated so that the concentration of the effluent is equal to the concentration in the source water, the gray water footprint is negligible. If the effluent concentration is equal to the regulatory limit of the maximum degradable concentration and if the source water concentration and natural background concentration are equivalent, then the dilution factor, DF, is equal to one and the gray water footprint is equal to the volumetric flow rate of the process effluent. If the maximum concentration is either equal to or less than the natural background concentration (for a particularly toxic nondegradable pollutant), then the gray water footprint becomes infinitely large.

The mass flow rate of a nonpoint-source pollutant, such as fertilizer on a field, is more difficult to calculate directly than that of a definable point-source pollutant, as discussed in Chapter 4. Partitioning factors discussed in Chapter 2 are used to estimate the fraction of a given pollutant from the chemicals from the source area. The nonpoint-source mass-loading rate, L_{nps}, can be estimated from the unitless fraction of the pollutant that may migrate, α, multiplied by the mass application rate of the pollutant (such as a fertilizer), M_{appl}, with units of mass/time.

$$L_{nps} = \alpha \times M_{appl} \tag{10.19}$$

The complete water footprint is the sum of the blue, green, and gray water footprints:

$$WF_{proc} = WF_{proc,green} + WF_{proc,blue} + WF_{proc,gray} \tag{10.20}$$

The water footprint of growing a crop or tree is a process that is dependent on the crop water use (CWU), which has units of cubic meters per hectare and is defined by

$$CWU = 10 \times \sum_{d=1}^{lgp} ET \tag{10.21}$$

where ET is the evapotranspiration rate from the crop in units of mm/day. The total water used for the crop is summed over the first day of planting until harvest, which is the length of the growing period (lgp). Several sources can be used to provide the evapotranspiration rate for crops or trees, and the UN's Food and Agriculture Organization CROPWAT model is one source that may provide data for a wide variety of crops. The blue and green water components of the water footprint from the process of growing a crop is calculated from the CWU divided by the yield of the crop given in

$$WF_{blue\ or\ green} = \frac{CWU}{Y} \tag{10.22}$$

Water pollution from crops generally comes from nonpoint-source pollutants that may include nitrogen and phosphorus from fertilizers, pesticides, and insecticides. The gray water footprint is calculated using Equations (10.16) and (10.19), and dividing by the crop yield:

$$WF_{crop,gray} = \left(\frac{\alpha \times M_{appl}}{c_{max} - c_{min}} \right) \frac{1}{Y} \tag{10.23}$$

Water Footprint of a Product

The water footprint of a product is the sum of all the direct and indirect fresh water used to produce the product. The total water footprint consists of all the water used in the steps in the production chain as illustrated in Figure 10.21. The product water footprint, $WF_{prod,p}$, is calculated from the sum of all the water from

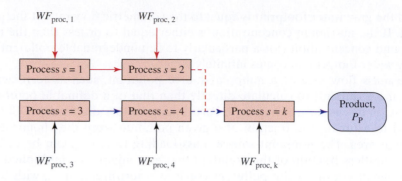

FIGURE 10.21 Schematic that represents the product water footprint for a unit value of product, P_P, and the chain of supply of the water footprints from process 1 to process k associated with the process steps to produce the product.

Source: Based on Hoesktra, A, A.K. Chapagain, M. M. Aldaya, M. M. Mekonnen. 2011. The Water Footprint Assessment Manual: Setting the Global Standard. Earthscan, Ltd., London, UK.

the process steps, $WF_{proc,s}$, divided by the number of units of product produced, P_P, as

$$WF_{prod,p} = \frac{\sum_{s=1}^{k} WF_{proc,s}}{P_P} \qquad (10.24)$$

The ultimate sustainability associated with the water footprint must consider both a geographic perspective and a holistic social justice perspective. Geographically, the water footprint within a particular catchment basin provides a natural boundary for analysis. If the water footprint is greater than available water resources, the water consumptive rate is unsustainable, and measures should be put in place to mitigate water use as discussed in Chapters 3 and 4. The water footprint of processes, products, and consumers within the geographical region can help identify which components create the greatest water demand and may provide information that can produce the greatest water saving and least costs for mitigation strategies. Holistically, the water footprint of humanity can be compared to the total resources that the Earth can sustainably support.

Water hotspots can be identified when and where the water footprint exceeds the sustainable supply. A hotspot may be a particular time period, usually during a dry season, in a specific catchment basin when the water does not meet environmental demands and water quality standards, or because the water cannot be equitably and economically shared. The hotspots create problems associated with water scarcity, such as water pollution, decreased human health, or conflict.

Where hotspots occur, processes and products may be studied to identify an appropriate response to mitigate the water hotspots as illustrated in the example shown in Table 10.4. Products or processes can be identified that may reduce the water footprint by determining the processes involved in each product and if the contribution to the water footprint is significant. If the water footprint is significant and the water footprint for production can be avoided or reduced, policy or practices can be implemented to reduce water consumption in this area.

The water footprint is a single multiparameter indicator of sustainable water use. However, the water footprint cannot, in and of itself, serve as a metric for all water-related issues. For instance, the water infrastructure does not address issues

TABLE 10.4 A theoretical example of using the water footprint analysis to identify practical responses to unsustainable water consumption

DATA DERIVED FROM THE PRODUCT WATER FOOTPRINT ACCOUNT			IS THE PROCESS LOCATED IN A CATCHMENT HOTSPOT?	CHECK THE SUSTAINABILITY OF THE WATER FOOTPRINT OF THE PROCESS ITSELF	CONCLUSION		CHECK RELEVANCE FROM PRODUCT	CHECK WHETHER RESPONSE IS REQUIRED
Process Step	Catchment in Which the Process Is Located	Water Footprint (m^3 per unit of final product)		Can the Water Footprint Be Reduced or Avoided?	Is this a Sustainable Component in the Product Water Footprint?	Fraction of the Product Water Footprint That Is not Sustainable	Share Above Threshold of 1%	Is This a Priority Component?
1	A	45	No	No	Yes		Yes	No
	B	35	Yes	Yes	No	35%	Yes	**Yes**
2	A	10	No	No	Yes		Yes	No
3	C	6	No	No	Yes		Yes	No
	D	2	Yes	No	No	2%	Yes	**Yes**
	E	1.1	No	Yes	No	1.1%	Yes	**Yes**
4	F	0.5	Yes	No	No	0.5%	No	No
5	A	0.3	No	No	Yes		No	No
6	A	0.1	No	Yes	No	0.1%	No	No
Total		100						

Source: Based on Hoesktra, A, A.K. Chapagain, M. M. Aldaya, M. M. Mekonnen. 2011. The Water Footprint Assessment Manual: Setting the Global Standard. Earthscan, Ltd., London, UK.

associated with development (discussed in more detail in Chapter 11), flooding, and lack of infrastructure for economically disadvantaged communities. The water footprint does not yield very useful results or inform the discussion for particular harmful pollutants or pollutants that do not degrade in the environment, and thus, it does not address broad environmental issues beyond water scarcity. The water footprint may inform portions of life cycle analysis. The water footprint method can provide data for economic and environmental measures in LCA computations, such as those discussed in Table 10.5 that have not previously adequately accounted for water resources in the assessment methodologies.

The water footprint, or embodied water in the production of a cotton shirt illustrated in Figure 10.22, includes all the blue, green, and gray water utilized to produce the product. For example, the production of 1 kg of cotton requires approximately 1,100 liters of water, and the average t-shirt that weighs 250 g requires approximately 2,700 liters of water (Chapagain et al., 2006). Of this total water volume, 45% is irrigation water consumed by the cotton plant; 41% is rainwater evaporated from the cotton field during the growing period; and 14% is water required to dilute the wastewater flows that result from the use of fertilizers in the field and the use of chemicals in the textile industry.

TABLE 10.5 Outcomes associated with water footprint sustainability and how they may be able to inform life cycle assessments

WATER FOOTPRINT ASSESSMENT PHASE	OUTCOME	PHYSICAL MEANING	RESOLUTION	LCA PHASE
Product water footprint accounting	Green, blue, and gray water footprints (volumetric)	Water volume consumed or polluted per unit volume of product	Spatiotemporally explicit	Life cycle inventory
Product water footprint sustainability assessment	An evaluation of the sustainability of a green, blue, and gray product water footprint from an environmental, social, and economic perspective	Various measurable impact variables	Spatiotemporally explicit	Life cycle impact assessment
Aggregation of selected information from the water footprint sustainability assessment	Aggregated water footprint impact indices	None	Nonspatiotemporally explicit	

Source: Based on Hoesktra, A, A.K. Chapagain, M. M. Aldaya, M. M. Mekonnen. 2011. The Water Footprint Assessment Manual: Setting the Global Standard. Earthscan, Ltd., London, UK.

We can use embodied water calculations to determine and compare the water footprints of different products. A study commissioned by the United Nations Educational, Scientific and Cultural Organization (UNESCO) found significant variations in the water footprints of food products as shown in Tables 10.6 and 10.7 (Mekonnen and Hoekstra, 2010).

Consider that 1 kg of tomatoes has a water footprint of 180 liters. The world average water footprint of pure chocolate is 2,400 liters for a 100-gram bar. The average water footprint of a beef cow raised in an industrial beef production system

FIGURE 10.22 The material for cotton shirts comes from the cotton plant, which requires water for growth.

Sources: Jonathon Pryor/Shutterstock.com, By shibu bhattacharjee/Moment/Getty Images, urfin/Shutterstock.com.

TABLE 10.6 Global average water footprint of 14 primary crop categories

PRIMARY CROP CATEGORY	WATER FOOTPRINT (m³/ton)			
	Green	Blue	Grey	Total
Sugar crops	130	52	15	197
Fodder crops	207	27	20	253
Vegetables	194	43	85	322
Roots and tubers	327	16	43	387
Fruits	727	147	93	967
Cereals	1,232	228	184	1,644
Oil crops	2,023	220	121	2,364
Tobacco	2,021	205	700	2,925
Fibers, vegetable origin	3,375	163	300	3,837
Pulses	3,180	141	734	4,055
Spices	5,872	744	432	7,048
Nuts	7,016	1,367	680	9,063
Rubber, gum, waxes	12,964	361	422	13,748
Stimulants	13,731	252	460	14,443

Source: Based on M.M. Mekonnen and A.Y. Hoekstra, The green, blue and grey water footprint of crops and derived crop products, Volume 1: Main Report, Value of Water Research Report Series No. 47, UNESCO-IHE Institute for Water Education, Delft, The Netherlands, December 2010.

TABLE 10.7 Water footprint of some crop products

VEGETABLE OILS	WATER FOOTPRINT (m³/ton)	FRUITS	WATER FOOTPRINT (m³/ton)
Maize oil	2,600	Watermelon	235
Cottonseed oil	3,800	Pineapple	255
Soybean oil	4,200	Papaya	460
Rapeseed oil	4,300	Orange	560
Palm oil	5,000	Banana	790
Sunflower oil	6,800	Apple	820
Groundnut oil	7,500	Peach	910
Linseed oil	9,400	Pear	920
Olive oil	14,500	Apricot	1,300
Castor oil	24,700	Plums	2,200
JUICES		Dates	2,300
Tomato juice	270	Grapes	2,400
Grapefruit juice	675	Figs	3,350
Orange juice	1,000	**ALCOHOLIC BEVERAGES**	
Apple juice	1,100	Beer	300
Pineapple juice	1,300	Wine	870

Source: Based on M.M. Mekonnen and A.Y. Hoekstra, The green, blue and grey water footprint of crops and derived crop products, Volume 1: Main Report, Value of Water Research Report Series No. 47, UNESCO-IHE Institute for Water Education, Delft, The Netherlands, December 2010.

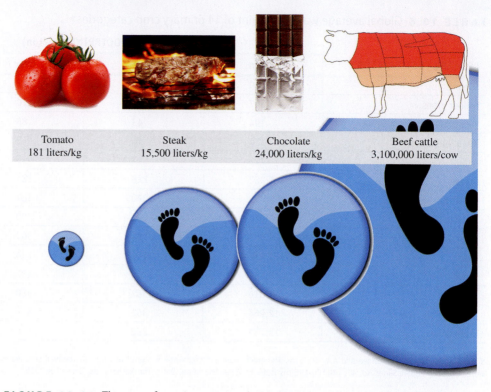

| Tomato | Steak | Chocolate | Beef cattle |
| 181 liters/kg | 15,500 liters/kg | 24,000 liters/kg | 3,100,000 liters/cow |

FIGURE 10.23 The water footprint associated with food types (from Hoeskstra, 2010).

Source: ievgen sosnytskyi/Shutterstock.com, Shira Raz/Shutterstock.com, windu/Shutterstock.com, life_is_fantastic /Shutterstock.com, VectorARA/Shutterstock.com.

is 3,100,000 liters!!! For a typical industrial beef production system, 99% of this water goes toward the production of animal feed. while the rest is used for drinking and other services related to animal production. As a result, 1 kg of steak produced in an industrial system requires 15,500 liters of water, as illustrated in Figure 10.23. The water footprint of beef can vary significantly, depending on the composition and method of production of the animal feed.

The Water Footprint of a Consumer

A consumer's water footprint is the sum of the water footprints of all goods and services consumed. Only a very small fraction (~3%) of the water footprint of the average consumer is from the direct use of water. The rest of the water footprint comes from other goods, services, and consumer products that are used. Global import and export contributes to the fact that many consumers today have water footprints outside the immediacy of consumption. The magnitude of the calculated water footprint of a product is dependent on the availability of water in the location from which it is made. This implies that a product made and imported from a region of water scarcity will have a higher water footprint than a comparable product made and imported from a region with abundant water. Having access to this information allows the consumer to make informed choices in reducing their water footprint. It should, however, be noted that other factors such as economic impact need to be considered.

The Water Footprint of a Business

The water footprint of a business is the total volume of water that is used directly and indirectly to run and support a business. It comprises two components:

- Direct water used to produce and manufacture the product and for supporting activities
- Indirect water use that comes from the producer's supply chain

Since it is possible for a product manufacturing business to lower its direct water footprint but still carry a total water footprint due to the indirect water use from its supply chain, it has become essential for producers to set standards for their suppliers.

10.5 Embodied Energy

Energy is the ability to do work, and it is an essential need for a functional industrial system. The sustainability objective is to reduce and minimize the energy requirements of product manufacture. Energy flows in industrial systems are also tracked closely with materials and water flows because they are often intrinsically linked. One-way energy flows are assessed with the concept of energy efficiency.

Whenever energy from any source is converted into work, some of the energy input is dissipated to the surroundings. The energy of this system is defined as the ratio of the useful work obtained and the total energy input.

$$\text{Efficiency} = \frac{\text{useful work output}}{\text{total energy input}} \tag{10.25}$$

For example, approximately 25% of the energy derived from combustion of gasoline is used to propel a car. The rest is dissipated to the surroundings as heat. Thus, the efficiency of an average automobile is 0.25.

Similar to the concept of embodied water, the total amount of energy required in producing a product or service across the product chain is called the embodied energy. A significant amount of energy used to manufacture and use products is hidden and not always accounted for by manufacturers and consumers.

Consider the materials processing system shown in Figure 10.24. If E_e, E_m, and E_p are the energies consumed per unit mass (J/kg) during the materials extraction, materials processing, and product manufacture stages, respectively, the energy used in each stage of the process is obtained from the product of the energy consumed per unit mass and the total mass processed in that stage.

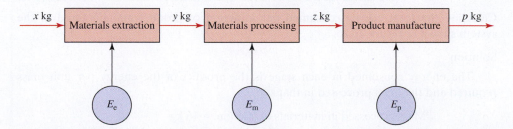

FIGURE 10.24 Example of a materials and energy flow system in product manufacture.

Source: Bradley Striebig.

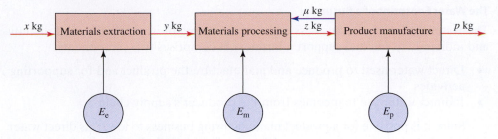

FIGURE 10.25 Example of a materials and energy flow system in product manufacture with a recycle flow stream.

Source: Bradley Striebig

The total energy (in joules) used in the manufacture of p kg of product is then given by the sum the energies consumed across all stages.

$$E_{Total} = (x \times E_e) + (y \times E_m) + (z \times E_p) \tag{10.26}$$

The energy used in the manufacture of a unit mass of product can then be calculated from

$$\alpha = \frac{E_{Total}}{p} \tag{10.27}$$

If some of the materials processed in the product manufacture stage are recycled back into the materials processing stage as shown in Figure 10.25, the energy flow balance can be recalculated as

$$E_{Total} = (x \times E_e) + ((y + \mu) \times E_m) + (z \times E_p) \tag{10.28}$$

Since

$$z = y + \mu \tag{10.29}$$

Then

$$E_{Total} = (x \times E_e) + ((y + \mu) \times E_m) + ((y + \mu) \times E_p) \tag{10.30}$$

$$E_{Total} = (x \times E_e) + ((y + \mu) \times (E_m + E_p)) \tag{10.31}$$

EXAMPLE 10.2 **Embodied Energy Analysis**

Calculate the total energy consumed per kilogram of product manufactured in the system shown in Figure 10.26.

Solution

The energy consumed in each stage is the product of the energy per unit mass required and the mass processed in that stage.

Mass processed in materials extraction = 16 kg

Mass processed in materials processing = 12 kg + 2 kg = 14 kg

Mass processed in product manufacture = 10 kg

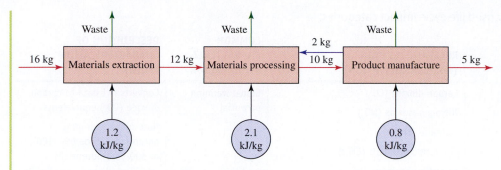

FIGURE 10.26 Flow diagram for Example 10.2.

Source: Bradley Striebig

Note that the mass processed in each stage is equal to the total mass input with no regard to how much ends up in useful product or waste.

$$E_{\text{Total}} = \left(16 \text{ kg} \times 1.2 \frac{\text{kJ}}{\text{kg}}\right) + \left(14 \text{ kg} \times 2.1 \frac{\text{kJ}}{\text{kg}}\right) + \left(10 \text{ kg} \times 0.8 \frac{\text{kJ}}{\text{kg}}\right)$$

$$E_{\text{Total}} = 19.2 \text{ kJ} + 29.4 \text{ kJ} + 8 \text{ kJ} = 56.6 \text{ kJ}$$

Energy consumed per unit mass of product

$$\alpha = \frac{56.6 \text{ kJ}}{5 \text{ kg}} = 11.32 \text{ kJ/kg}$$

10.6 Impact Assessment

Emissions to the environment and the extraction and use of natural resources are called environmental stresses. What are the consequences of the emissions and the use of resources that are quantified in the inventory analysis? The impact assessment phase translates these into the relevant impact categories. We do this because the actual impacts categories are easier to relate to and understand. Impact assessment also makes results more comparable to guide decision making. For example, several gases contribute to the greenhouse effect. Simply knowing the quantities of these gases that may be emitted from a system is not sufficient, but evaluating how these emissions contribute to global temperature increase, climate change, sea level rise, biodiversity changes, and food availability makes it easier to address specific concerns.

Impact assessment requires three elements and can include four optional ones. These are as follows.

Mandatory Elements

1. *Impact Category Definition*

    Impact categories represent environmental issues of concern, some of which are shown in Table 10.8. Different LCA practitioners use various models to make these assessments, depending on the context and specific impacts considered to be of concern. Defining the impact categories of an LCA involves specifying which impacts are relevant to the goal and scope. Not all impact categories

TABLE 10.8 Commonly used life cycle impact categories.

IMPACT CATEGORY	SCALE	RELEVANT LCI DATA (i.e., CLASSIFICATION)	COMMON CHARACTERIZATION FACTOR	DESCRIPTION OF CHARACTERIZATION FACTOR
Global warming	Global	Carbon dioxide (CO_2) Nitrogen sioxide (NO_2) Methane (CH_4) Chlorofluorocarbons (CFCs) Hydroflourocarbons (HCFCs) Methyl bromide (CH_3Br)	Global warming potential	Converts LCI data to carbon dioxide (CO_2) equivalents Note global warming potentials can be 50-, 100-, or 500-year potentials
Stratospheric ozone depletion	Global	Chlorofluorocarbons (CFCs) Hydroflourocarbons (HCFCs) Halons Methyl Bromide (CH_3Br)	Ozone depletion potential	Converts LCI data to trichlorofluoromethane (CFC-11) equivalents
Acidification	Regional Local	Sulfur oxides (SOx) Nitrogen oxides (NOx) Hydrochloric acid (HCl) Hydroflouric acid (HF) Ammonia (NH_4)	Acidification potential	Convert LCI data to hydrogen (H+) ion equivalents
Eutrophication	Local	Phosphate (PO_4) Nitrogen oxide (NO) Nitrogen dioxide (NO_2) Nitrates Ammonia (NH_4)	Eutrophication potential	Convert LCI data to phosphate (PO_4) ion equivalents
Photochemical smog	Local	Nonmethane hydrocarbon (NMHC)	Photochemical oxidant creation potential	Convert LCI data to ethane (C_2H_6) ion equivalents
Terrestrial toxicity	Local	Toxic chemicals with a reported lethal concentration to rodents	LC_{50}	Convert LC_{50} data to equivalents
Aquatic toxicity	Local	Toxic chemicals with a reported lethal concentration to fish	LC_{50}	Convert LC_{50} data to equivalents
Human health	Global Regional Local	Total releases to air, water, and soil	LC_{50}	Convert LC_{50} data to equivalents
Resource depletion	Global Regional Local	Quantity of minerals used Quantity of fossil fuels used	Resource depletion potential	Convert LCI data to a ratio of quantity of resource used versus quantity of resource left in a reserve
Land use	Global Regional Local	Quantity disposed of in a landfill	Solid waste	Convert mass of solid waste into volume using an estimated density

Source: Based on U.S. Environmental Protection Agency and Science Applications International Corporation. *LCAccess—LCA 101*. 2001. Exhibit 4-1, Retrieved from http://www.epa.gov/ORD/NRMRL/lcaccess/lca101.htm.

are of relevance in every environment of concern. The Dutch guide to LCA (Guinée et al., 2001) distinguishes between three sets of impact categories:

- Group A: Baseline impact categories
  These are typically found in almost all LCA studies and include the following.

  - Depletion of abiotic resources
  - Impacts of land use (land competition)
  - Climate change
  - Stratospheric ozone depletion
  - Human toxicity
  - Ecotoxicity (freshwater aquatic ecotoxicity, marine aquatic ecotoxicity, and terrestrial ecotoxicity)
  - Photooxidant formation
  - Acidification
  - Eutrophication

- Group B: Study specific impact categories
  These may be included depending on the goal and scope of the LCA and on whether data is available.

  - Impacts of land use (loss of biodiversity and life support function)
  - Ecotoxicity (freshwater sediment ecotoxicity, marine sediment ecotoxicity)
  - Impacts of ionizing radiation
  - Odor (air)
  - Noise
  - Waste heat
  - Casualties

- Group C: Other impact categories
  These are impact categories for which research is still ongoing and require more studies before being included.

  - Depletion of biotic resources
  - Desiccation
  - Odor (water)

2. *Classification*
   Classification is the process of assigning emissions, wastes, and resource use to relevant impact categories. This process relies on established cause and effect relationships, also known as stressor–impact relationships. As we have previously discovered, multiple sources can contribute to one type of impact (e.g., CO_2 and CH_4 contribute to global climate change), and a single source can contribute to multiple impacts (e.g., Nox contributes to acidification as well as eutrophication).

3. *Characterization*
   Characterization is quantitatively determining the impacts arising from the environmental stresses determined in the inventory stage. Different methods are used to characterize different impact categories. Some impacts are calculated using equivalence factors and impact potential values. For example, the global warming contribution of emissions can easily be calculated using the equivalent CO_2 values discussed in Chapter 9.

Optional Elements

4. *Normalization*

   Some LCAs include a normalization step. This step involves dividing the impact values obtained from characterization by reference values in order to determine the relative magnitude and significance of the impacts caused by the system under study. The reference values are dependent on the context in which the comparison is being made. For example, global reference values will be different from regional reference values.

5. *Grouping*

   This element refers to sorting the impact results from characterization into different defined groups for analysis and presentation. For example, impacts can be sorted under global-, regional-, or local-scale concerns or high-, medium-, and low-priority impacts.

6. *Weighting*

   In this step, the relative importance of one impact is weighted against other impacts. Weighting factors are assigned in a qualitative or quantitative process that is based on perceived or actual relative importance. Weighted factors can be subjective, and there is often no agreement values used, which can sometimes be challenging for engineers who tend to favor objective metrics. It is, however, acceptable to use different values as long as they are appropriately documented.

7. *Data Quality Analysis*

   Life cycle assessments involve the collection and processing of large amounts of data that may come from a wide variety of sources. Sometimes, these data may have been estimated or extracted from literature sources that may not adequately represent the system being studied. As a result, there may be consequent uncertainty in the data and in the results. Data quality analysis is a series of statistical methods used to test the validity of the data and results obtained.

BOX 10.1 Case Study: Life Cycle Assessment of Supermarket Shopping Bags in the UK

A study commissioned by the Environment Agency, which is the leading public body charged with protecting and improving the environment in England and Wales, assessed the life cycle environmental impacts of the production, use, and disposal of different shopping bags for the UK in 2006.

The report considered only the types of shopping bags available from UK supermarkets and did not consider the ability and willingness of consumers to change behavior nor the potential economic impacts of each choice on UK businesses.

The following types of shopping bags were studied and compared.

- A conventional, lightweight carrier made from high-density polyethylene (HDPE)

- A lightweight HDPE carrier with a prodegradant additive designed to break down the plastic into smaller pieces
- A biodegradable carrier made from a starch-polyester (biopolymer) blend
- A paper carrier
- A "bag for life" made from low-density polyethylene (LDPE)
- A heavier, more durable bag, often with stiffening inserts made from nonwoven polypropylene (PP)
- A cotton bag

Functional Unit

Each of these shopping bags is designed for a different number of uses. Those intended for longer multiple use need more resources in their production and are therefore likely to produce greater environmental impacts if compared on a bag-for-bag basis. A comparison of life cycle environmental impacts should be based on a comparable function (or "functional unit") to allow a fair comparison of the results. The researchers decided to compare the impacts from the number of bags required carrying one month's shopping in 2006/2007. The shopping bags studied are, however, also of different volumes, weights, and qualities, so the researchers conducted a survey that found that over a four-week period supermarket shoppers purchased an average of 446 items. So they defined their functional unit as

Carrying one month's shopping (483 items) from the supermarket to the home in the UK in 2006/07.

The researchers used the conventional high-density polyethylene bag (HDPE) as the reference. They then calculated how many times each of the other types of bags would have to be used to reduce its contribution to global warming to below that for conventional HDPE bags. They considered different scenarios that included 100% secondary reuse of the HDPE bags and 40% reuse as well as no secondary reuse; the results are shown in Table 10.9.

The bags were also compared for other impacts, including resource depletion, acidification, eutrophication, human toxicity, freshwater aquatic ecotoxicity, marine aquatic ecotoxicity, terrestrial ecotoxicity, and photochemical oxidation (smog formation).

Since different raw materials, manufacturing processes, and disposal methods are applicable to the different bags being compared, it was necessary for the researchers to define an appropriate systems boundary to encompass all these life cycle stages. The system boundary used in the study is shown in Figure 10.27. A cradle-gate model was used that incorporated consideration of impacts of sub-processes such as extraction, transport, packaging, use, reuse, recycling, and disposal as applicable to each type of bag.

The study found the following results.

- The conventional HDPE bag had the lowest environmental impacts of the lightweight bags in eight of nine impact categories. The bag performed well because it was the lightest bag considered.

TABLE 10.9 Number of times primary use required to bring the global warming potential of other reusable bags below that of HDPE bags with and without secondary reuse

TYPE OF BAG	HDPE BAG (NO SECONDARY REUSE)	HDPE BAG (40.3% REUSED AS BIN LINERS)	HDPE BAG (100% REUSED AS BIN LINERS)	HDPE BAG (USED 3 TIMES)
Paper bag	3	4	7	9
LDPE bag	4	5	9	12
Nonwoven PP bag	11	14	26	33
Cotton bag	131	173	327	393

Source: Based on Environment Agency, Evidence: Life cycle assessment of supermarket carrier bags: a review of the bags available in 2006, Report: SC030148, www.environment-agency.gov.uk, 2011.

(Continued)

BOX 10.1 Case Study: Life Cycle Assessment of Supermarket Shopping Bags in the UK (*Continued*)

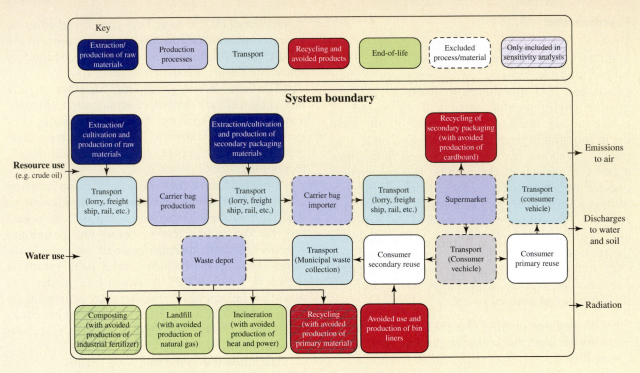

FIGURE 10.27 System boundary applied in the study.

Source: Based on Environment Agency, Evidence: Life cycle assessment of supermarket carrier bags: a review of the bags available in 2006, Report: SC030148, www.environment-agency.gov.uk, 2011.

- The starch-polyester (biopolymer) bag had the highest impact in seven of the nine impact categories considered. This was partially due to it having approximately twice the weight of the conventional HDPE bags and partially due to the high impacts of raw material production, transport, and the generation of methane from landfill at disposal.
- The environmental impact of all types of shopping bags is dominated by resource use and production stages. Transport, secondary packaging, and end-of-life management generally had a minimal influence on their performance.

- Whatever type of bag is used, the key to reducing the impacts is to reuse it as many times as possible. Where reuse for shopping is not practicable, other reuse (e.g., to replace bin-liners) is beneficial.
- The reuse of conventional HDPE and other lightweight carrier bags for shopping and/or as bin-liners is pivotal to their environmental performance, and reuse as bin-liners produced greater benefits than recycling bags.
- Starch-polyester (biopolymer) blend bags have a higher global warming potential and abiotic depletion than conventional polymer

bags, due both to the increased weight of material in a bag and higher material production impacts.

- The paper, LDPE, nonwoven PP, and cotton bags should be reused at least 3, 4, 11, and 131 times, respectively, to ensure that they have lower global warming potential than conventional HDPE carrier bags that are not reused. The number of times each would have to be reused when different proportions of conventional (HDPE) carrier bags are reused are shown in Table 10.9.
- Recycling or composting generally produces only a small reduction in global warming potential and abiotic depletion.

Figure 10.28 shows the comparison of all the bags based on their contribution to global warming (measured as equivalent CO_2 emitted over the life cycle). The results not only show that the polypropylene bag has the largest impact in this category but that a significant portion of this impact from extraction and production of raw materials for all bags.

While the HDPE bag has the lowest impact across all impact categories, Figure 10.29 shows that this impacts can be significantly reduced with increased secondary reuse. Secondary reuse is, however, a function of consumer habits and choices which need to be considered in impact reduction decisions.

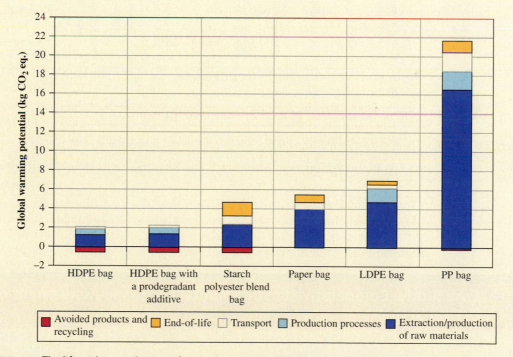

FIGURE 10.28 The life cycle contribution of each shopping bag to global warming.

Source: Based on Environment Agency, Evidence: Life cycle assessment of supermarket carrier bags: a review of the bags available in 2006, Report: SC030148, www.environment-agency.gov.uk, 2011.

(Continued)

BOX 10.1 Case Study: Life Cycle Assessment of Supermarket Shopping Bags in the UK (*Continued*)

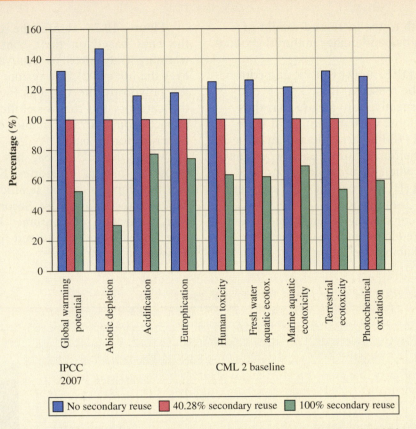

FIGURE 10.29 The influence of secondary reuse on the lifecycle impacts of the conventional HDPE bag.

Source: Based on Environment Agency, Evidence: Life cycle assessment of supermarket carrier bags: a review of the bags available in 2006, Report: SC030148, www.environment-agency.gov.uk, 2011.

BOX 10.2 Case Study: Comparative Environmental Life Cycle Assessment of Hand Drying Systems

Excel Dryer, Inc., are manufacturers of American-made hand dryers for schools, hospitals, restaurants, and many other commercial facilities. Their hand dryers, brand named XLERATOR®, feature patented technology that is marketed to dry hands in 10 to 15 seconds, use 80% less energy than conventional dryers, and offer 95% cost savings over paper towels. The company believed that, while a comparison to other hot-air hand dryers in the use phase of the product life cycle was easy, it was also necessary to know how their product compared with conventional hand dryers in other stages of the life cycle considering a complete set of environmental performance metrics. So the company commissioned Quantis, which is a consulting firm specializing in life cycle assessments, to compare the life cycle impacts of several systems for drying hands in public restrooms: the XLERATOR dryer, a conventional electric dryer, and paper towels containing between 0% and 100% recycled content.

Functional Unit

The primary function of the products that were studied is to dry hands after washing in a public restroom. The functional unit provides a basis for comparing all life cycle components on a common basis: namely, the amount of that component required to fulfill the described function. The researchers assumed that the lifetime of a single installation of an electric dryer is 10 years, and on the basis of approximately 500 uses per week, they defined their functional unit as

Drying 260,000 pairs of hands

The systems boundaries for all three choices included raw materials extraction and processing, production and assembly of subcomponents, consumer use, and disposal scenarios. Figure 10.30 shows the systems boundary used for the XLERATOR system, Figure 10.31 shows the systems boundary

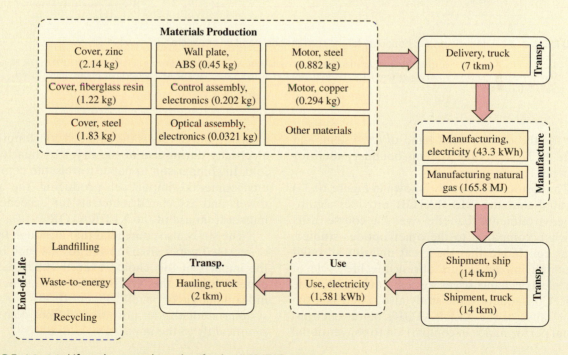

FIGURE 10.30 Life cycle system boundary for the XLERATOR system.

Source: Based on Excel Dryer, Inc., Comparative Environmental Life Cycle Assessment of Hand Drying Systems: The XLERATOR Hand Dryer, Conventional Hand Dryers and Paper Towel Systems, http://www.exceldryer.com/PDFs/LCAFinal9-091.pdf, prepared by Quantis (www.quantis-intl.com).

(Continued)

BOX 10.2 Case Study: Comparative Environmental Life Cycle Assessment of Hand Drying Systems (*Continued*)

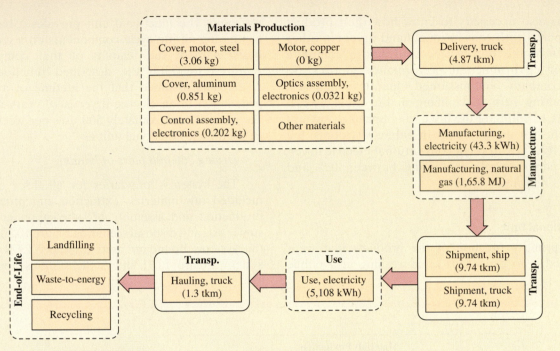

FIGURE 10.31 Life cycle system boundary for the conventional hand dryer system.

Source: Based on Excel Dryer, Inc., Comparative Environmental Life Cycle Assessment of Hand Drying Systems: The XLERATOR Hand Dryer, Conventional Hand Dryers and Paper Towel Systems, http://www.exceldryer.com/PDFs/LCAFinal9-091.pdf, prepared by Quantis (www.quantis-intl.com).

used for the conventional hand dryer system, and Figure 10.32 shows the systems boundary used for the paper towel option.

The results of the study (shown in Figure 10.33) indicated that the XLERATOR provides significant environmental benefits over the course of its life when compared to the other options studied. It was concluded that the great majority of environmental impact occurs during the life cycle of both electric hand dryer systems during the use phase. Consequently, the study also stated that the major cause of the XLERATOR's better environmental performance in comparison to conventional

electric hand dryers is the reduced consumption of electricity during use of the dryer by nearly fourfold. In comparison to paper towels, the combined environmental impact of producing the paper towels and associated materials far exceeded the impact from use of the XLERATOR.

The study also concluded that although the use of recycled paper fibers in the towels may reduce the impacts of this choice, even if 100% recycled content is used, the XLERATOR maintains a significant margin of benefit (Figure 10.34). For all the systems studied, transportation did not contribute significantly to the impacts.

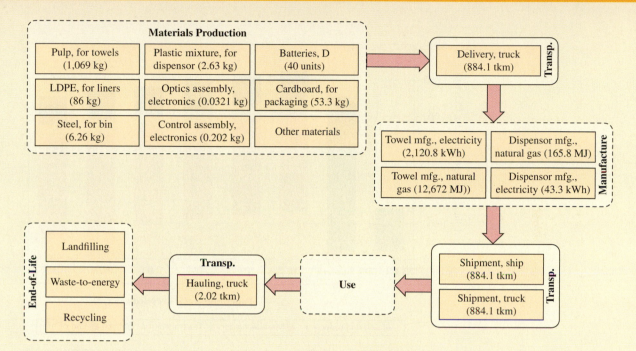

FIGURE 10.32 Life cycle system boundary for the paper towels system.

Source: Based on Excel Dryer, Inc., Comparative Environmental Life Cycle Assessment of Hand Drying Systems: The XLERATOR Hand Dryer, Conventional Hand Dryers and Paper Towel Systems, http://www.exceldryer.com/PDFs/LCAFinal9-091.pdf, prepared by Quantis (www.quantis-intl.com).

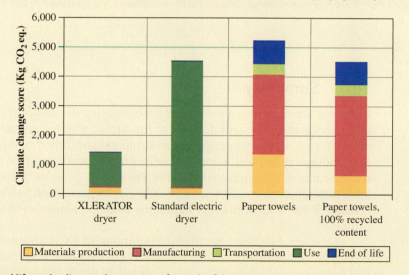

FIGURE 10.33 Total life cycle climate change score for each of the systems.

Source: Based on Excel Dryer, Inc., Comparative Environmental Life Cycle Assessment of Hand Drying Systems: The XLERATOR Hand Dryer, Conventional Hand Dryers and Paper Towel Systems, http://www.exceldryer.com/PDFs/LCAFinal9-091.pdf, prepared by Quantis (www.quantis-intl.com).

(Continued)

BOX 10.2   **Case Study: Comparative Environmental Life Cycle Assessment of Hand Drying Systems** (*Continued*)

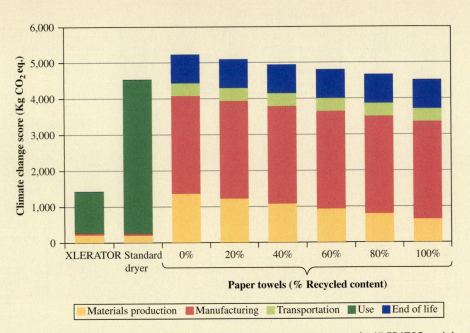

FIGURE 10.34 Climate change score for paper towels of varying recycled content in comparison to the XLERATOR and the conventional electric hand dryer.

Source: Based on Excel Dryer, Inc., Comparative Environmental Life Cycle Assessment of Hand Drying Systems: The XLERATOR Hand Dryer, Conventional Hand Dryers and Paper Towel Systems, http://www.exceldryer.com/PDFs/LCAFinal9-091.pdf, prepared by Quantis (www.quantis-intl.com).

10.7   Summary

Life cycle assessment is a useful tool for evaluating the comprehensive impacts from the manufacture, use, and disposal of products. LCA techniques rely on compiling databases of relevant energy, material inputs, and environmental impacts; evaluating potential impacts associated with process waste; and providing a simplified scale for interpreting modeling results. The results of an LCA can illustrate comparative differences in environmental and health impacts, such as whether product *A* or product *B* has a greater likely impact on climate change, as shown in the previous examples.

There are, however, significant limitations to LCA results that the user must recognize and interpret. Comprehensive LCAs require large amounts of resources and time. Accurate data collection is central to a reliable assessment, and the value of an LCA is only as good as the data used. LCAs are usually performed with truncated boundaries to limit the amount of extraneous data implying a compromise for practicality. While LCAs offer insights into the environmental performance of products, they do not provide information on cost effectiveness, product efficiency, or any social considerations. LCAs produce good global-scale impact results, though

TABLE 10.10 Some currently available LCA software

TOOL	WEBSITE	DEVELOPERS
ECO-it	www.pre-sustainability.com	PRé Consultants
eiolca.net	www.eiolca.net	Carnegie Mellon University Green Design Institute
GaBi	www.gabi-software.com	PE-International
GREET	www.greet.es.anl.gov	Argonne National Laboratory
IDEMAT	www.idemat.nl	Delft Univ. of Technology
SimaPro	www.pre-sustainability.com	PRé Consultants
SULCA	www.vtt.fi/research/technology/	VTT Technical Research Centre of Finland
Umberto	www.ifu.com/software/umberto-e/	ifu Hamburg GmbH

Source: Bradley Striebig.

assessments of local and contextual impacts are still challenging. In general, LCAs are considered to be decision support tools that are used in addition to other points of consideration. The simplified LCA modeling results are suitable for relative comparisons of environmental and health impacts, but they do not indicate actual measurable response or the risk levels associated with the environmental and health risks. For quantifiable estimations of environmental impacts, the environmental models presented in Chapters 4 and 5 should be used, and detailed risk analysis procedures are required to quantify risk and exposure to specific emissions or scenarios.

The U.S. EPA and ISO standard 14043 recommend three key steps in interpreting the results of an LCA:

1. Identification of the significant issues based on the LCA
2. Evaluations that consider completeness, sensitivity, and consistency checks
3. Conclusions, recommendations, and reporting

Due to the large quantities of data that have to be collected and processed for any LCA, computer software are usually used to perform most LCAs. Many institutions and companies have developed their own software using data specific to their needs and concerns. There are, however, commercially available software that utilize comprehensive databases of materials, energy, and emissions inventories as well as multiple impact assessment methods. Some software are used to simply model the *Inventory Analysis* phase of the LCA, while others are useful for more complete LCAs that include *Impact Assessment* and *Interpretation* of results. Table 10.10 shows examples of some currently available LCA software and their developers.

References

Chapagain, A. K. and Hoekstra, A. Y. (2008). "The global component of freshwater demand and supply: An assessment of virtual water flows between nations as a result of trade in agricultural and industrial products," *Water International* 33(1):19–32.

Chapagain, A. K., Hoekstra, A. Y., Savenije, H. H. G., and Gautam, R. (2006). "The water footprint of cotton consumption: An assessment of the impact of worldwide consumption of cotton products on the water resources in the cotton producing countries," *Ecological Economics* 60(1):186–203.

Environment Agency. (2011). Evidence: Life Cycle Assessment of supermarket carrier bags: A review of the bags available in 2006. Report: SC030148, www.environment-agency.gov.uk.

Excel Dryer, Inc. Comparative Environmental Life Cycle Assessment of Hand Drying Systems: The XLERATOR Hand Dryer, Conventional Hand Dryers and Paper Towel Systems (www.exceldryer.com/PDFs/LCAFinal9-091.pdf), prepared by Quantis (www.quantis-intl.com).

FAO. (2010). Greenhouse gas emissions from the dairy sector, a Life Cycle Assessment. Food and Agriculture Organization of the United Nations, Rome, Italy.

Guinée, J. B., Gorrée, M., Heijungs, R., Huppes, G., Kleijn, R., Koning, A. de, et al. (2002). *Handbook on Life Cycle Assessment*. Operational guide to the ISO standards. I: LCA in perspective. IIa: Guide. IIb: Operational annex. III: Scientific background. Dordrecht: Kluwer Academic Publishers.

Hoekstra, A. Y. (2010). "The water footprint of animal products". In D'Silva, J. and Webster, J. (Eds.). *The meat crisis: Developing more sustainable production and consumption. London*, UK: Earthscan, pp. 22–33.

Hoesktra, A., Chapagain, A. K., Aldaya, M. M., and Mekonnen, M. M. (2011). *The Water Footprint Assessment Manual: Setting the Global Standard*. London, UK: Earthscan.

International Organization for Standardization. Environmental Management—Life Cycle Assessment—Principles and Framework, 2006.

Mekonnen, M. M. and Hoekstra, A. Y. (2010, December). The green, blue and grey water footprint of crops and derived crop products. Volume 1: Main Report, Value of Water Research Report Series No. 47. UNESCO-IHE Institute for Water Education, Delft, The Netherlands.

McMichael, A. J., Powles, J. W., Butler, C. D., and Uauy, R. (2007). "Food, livestock production, energy, climate change, and health," *The Lancet* 370, 1253–1263.

National Peanut Board. Accessed 2014 at www.nationalpeanutboard.org/classroom-funfacts.php.

Ogino, A., Orito, H., Shimada, K. and Hirooka, H. (2007), "Evaluating environmental impacts of the Japanese beef cow–calf system by the life cycle assessment method". *Animal Science Journal* 78:424–432.

Ziemann, Saskia, Weil, Marcel, and Schebek Liselotte. (2012, June). "Tracing the fate of lithium–The development of a material flow model." *Resources, Conservation and Recycling* 63:26–34.

U.S. Environmental Protection Agency and Science Applications International Corporation. *LCAccess—LCA 101*. 2001. Exhibit 4-1, Retrieved from www.epa.gov/ORD/NRMRL/lcaccess/lca101.htm.

U.S. EPA. Accessed 2014 at www.energystar.gov/index.cfm?c=cfls.pr_cfls_lumens.

Key Concepts

Life cycle analysis	Reuse factor
Functional unit	Embodied water
Inventory analysis	Water footprint
Impact assessment	Blue water
Materials	Green water
Process	Gray water
Reservoir	Embodied energy
Stocks	Functional unit
Flows/fluxes	Life cycle assessment

Active Learning Exercises

ACTIVE LEARNING EXERCISE 10.1: **Life Cycle Thinking**

Pick a typical household product and construct the life cycle of the product from raw material extraction to disposal. How is this life cycle different from a comparable product that fulfills the same function?

ACTIVE LEARNING EXERCISE 10.2: **Choices we make and their trade-offs**

Discuss your thoughts on the cartoon shown in Figure 10.35.

FIGURE 10.35 Active learning 10.2 illustration.

Source: Andy Singer Cartoons http://www.andysinger.com/.

ACTIVE LEARNING EXERCISE 10.3: **Determining appropriate functional units**

Identify three sets of two or more products on which you would like to perform comparative LCAs. Determine the functional unit of each set.

ACTIVE LEARNING EXERCISE 10.4: **How would you quantify impacts?**

Identify the major sources of impact in the life cycle of your school. Which phase contributes most to the impact?

Problems

10-1 Explain the concept of
a. Embodied water
b. Embodied energy

10-2 Identify the sources that could contribute to the embodied water of an apple just purchased from a grocery store.

10-3 What is the purpose of inventory analysis in conducting a life cycle assessment?

10-4 Figure 10.36 shows the flows of a material through a product life cycle. All symbols on the diagram refer to total mass flows:
- O is locally sourced raw material.
- I_1 is imported raw material.
- W_p is waste from raw material processing.
- R_m is recycled material from manufacturing.
- I_2 is imported finished product.
- W_c is irretrievable waste from consumers.
- R_w is recovered and reused material from waste processing.

Assume that O has a flow of 3500 kg/hour. Calculate the extraction and recovery efficiencies (defined as product outflow/total inflow). W_c is 7.5% of the product stream; P_5, R_m, and R_w each represent 5% of their respective product streams (P_3 and P_6); the total imported material ($I_1 + I_2$) is 60% of locally sourced raw material O, I_1 is three-fold larger than I_2, and P_6 is 25% larger than O.

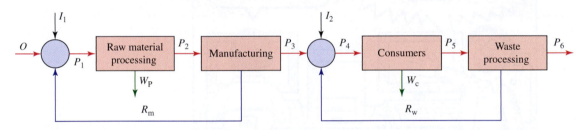

FIGURE 10.36 Illustration for Problem 10-4.

Source: Bradley Striebig.

10-5 Calculate the total energy consumed per kg of Al cans produced if 23% of total production is recycled from Figure 10.37.

10-6 If the materials efficiencies processing, sheet production, and production of ingots are 85%, 92%, and 98%, respectively, calculate new total energy consumed per kg of Al cans for the system shown in Figure 10.37.

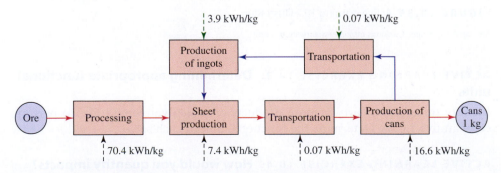

FIGURE 10.37 Illustration for Problems 10-5 and 10-6.

Source: Bradley Striebig.

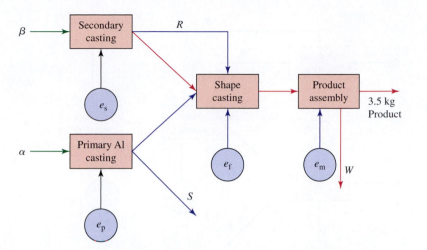

FIGURE 10.38 Illustration for Problem 10-7.

Source: Bradley Striebig.

10-7 For the aluminum production system described in Figure 10.38, calculate the following.
  a. Amount of aluminum input to "primary aluminum casting" (α in kg of aluminum)
  b. Amount of aluminum input into "secondary aluminum casting" (β, in kg aluminum)
  c. Total energy consumed.
  d. If the waste, W, from the product assembly is sent to shape casting to be reused, recalculate all three values for parts (a), (b), and (c).

Energy Flow Information:

  - $e_p = 2.0$ kWh/kg
  - $e_s = 0.50$ kWh/kg
  - $e_f = 0.75$ kWh/kg
  - $e_m = 0.25$ kWh/kg

Mass Flow Information
  - R is scrap, S is slag and, W is waste from the assembly.
  - Fraction of material input into shape casting that becomes scrap = 0.25.
  - Fraction of primary aluminum α that is lost to slag = 0.3.
  - Fraction of the material in product assembly that becomes waste = 0.08.
  - 60% of material in shape casting comes from primary aluminum casting and the remaining 40% from secondary aluminum casting.

10-8 For the aluminum production system described in Figure 10.39, calculate the following.
  a. the amount of aluminum input to "primary aluminum casting" (α in kg of aluminum);
  b. the amount of aluminum input into "secondary aluminum casting" (β, in kg aluminum); and
  c. the total energy consumed.
  d. If the waste, W, from the product assembly is sent to shape casting to be reused, and the percentage ratios from primary casting and

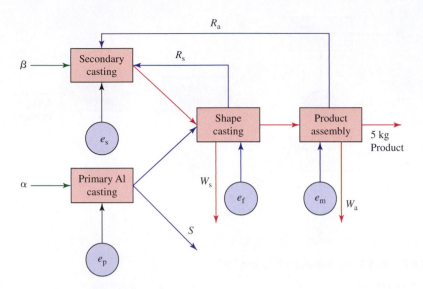

FIGURE 10.39 Illustration for Problem 10-8.

Source: Bradley Striebig.

secondary casting still remain 60:40, recalculate all three values in parts (a), through (c).

Energy Flow Information:
- e_p = 1.8 kWh/kg
- e_s = 0.75 kWh/kg
- e_f = 1.3 kWh/kg
- e_m = 0.25 kWh/kg

Mass Flow Information
- R_s and R_a are scrap flows from shape casting and product assembly, respectively.
- W_s and W_a are waste flows from shape casting and product assembly, respectively
- S is slag.
- Fraction of material input into shape casting that becomes scrap = 0.25.
- Fraction of material input into product assembly that becomes scrap = 0.15.
- Fraction of primary aluminum α that is lost to slag = 0.3.
- Fraction of the material in product assembly that becomes waste = 0.08.
- Fraction of material in shape casting that becomes waste = 0.03.
- 60% of material in shape casting comes from primary aluminum casting and the remaining 40% from secondary aluminum casting.

10-9 Consider the materials flowchart shown in the Figure 10.40. The wastes from primary production and manufacturing are 17% and 24% of their respective stocks. The mass of material from secondary production is 23% of the mass from primary production. Calculate the masses of materials going into primary and secondary production, that is, Ω and Ψ.

10-10 Figure 10.41 shows the flows of two materials α and β used in the manufacture of product $\alpha\beta$. Product $\alpha\beta$ contains 60% α and 40% β by mass. Calculate the mass values of flow streams A, B, C, D, E, F, G, and H, given $F = C$.

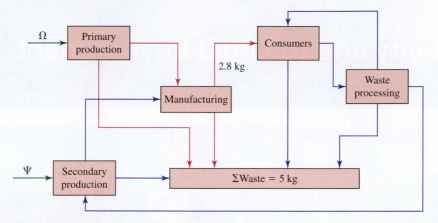

FIGURE 10.40 Illustration for Problem 10-9.

Source: Bradley Striebig.

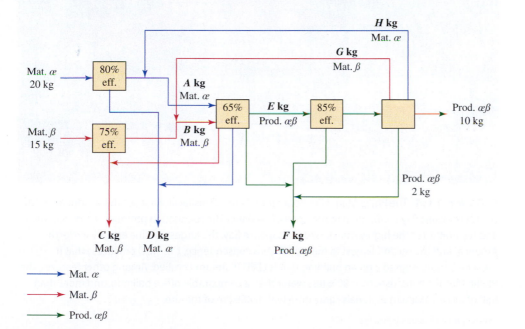

— Mat. *æ*

— Mat. *ß*

— Prod. *æß*

FIGURE 10.41 Illustration for Problem 10-10.

Source: Bradley Striebig.

10-11   Determine the functional unit for comparing the following:
  a. Different brands of clothes washing machine
  b. Wall paint of different qualities
  c. Various goods transportation systems
  d. Cell phones
  e. Air conditioning systems

10-12   Construct the life cycle flowcharts and system boundaries for the LCA of
  a. A car
  b. A mall
  c. Farm-raised fish
  d. A Coal-fired electric power plant

CHAPTER 11

Sustainability and the Built Environment

FIGURE 11.1 The Philip Merrill Environmental Center. This facility houses the headquarters of the Chesapeake Bay Foundation in Annapolis, Maryland. The foundation promotes conservation and restoration of the highly threatened Chesapeake Bay, the largest marine estuary in North America (and the second largest in the world). As a consequence, this NGO practices what it preaches, and designed a green building that is LEED Platinum certified. Among other features, the Phillip Merrill Center uses over 90% less water than a comparable office building, and more than half of its construction materials came from within 480 km of the site.

Source: Photo by Bradley Striebig.

You can design and create, and build the most wonderful place in the world. But it takes people to make a dream a reality.

—WALT DISNEY

GOALS

This chapter introduces you to the concepts of the built environment, low-impact development, conservation design, and green building. We begin with a short overview of land-use and land-cover change and the relationship between urban planning and sustainable development. Sustainable land-use planning principles are introduced through low-impact development and conservation design strategies, which are then linked to green building rating systems and the interactions between structures and their sites. The chapter concludes with building energy conservation. Understanding the impact of design choices on green buildings and rating systems will enhance your ability to use the techniques, skills, and modern engineering tools necessary to communicate principles of sustainable design for the built environment.

OBJECTIVES

At the conclusion of this chapter, you should be able to:

- Define and discuss:
  - Low-impact development
  - Smart growth
  - Green building
  - Conservation design
  - Erosion control
- Compare green building rating systems.
- Describe patterns and trends in land development.
- Define and utilize the concept of building envelope to determine allowable development lot size.

- Use conservation design principles to analyze the differences in housing density patterns for traditional yield and conservation-based development.
- Estimate erosion rates and conceptualize erosion control processes.
- Summarize rating systems that are applicable to buildings and residential developments and describe how those rating systems promote the principles of sustainability.
- Explain energy conservation strategies in green building, and link these to building energy codes and building energy rating systems.

11.1 Introduction

Some of the most significant impacts that humankind has on the natural world can be attributed to our **built environment**. The built environment represents all human-made structures, engineered alterations to land forms, and constructions of every sort. The built environment includes not only our homes and buildings, but also roads, dams, parks, bridges, recreational spaces, flood control systems, and the full landscape of our cities.

The built environment involves different types of land use, and we refer to such human activity and its impact on the natural world as **land-use and land-cover change**. Over time, because of land-use and land-cover change, we have altered local climates through deforestation, reduced the ability of watersheds to filter and purify water, and degraded ecosystems. Indeed, the greatest threat to biodiversity and species loss in the world has been—and continues to be—the loss of natural habitats resulting from the built environment.

The sustainability of human structures is intensely important because of the scale of urbanization in the world today. More than half the population in more developed regions already lives in urban areas; between 2011 and 2050, almost all population growth will be absorbed by urban regions worldwide. The United Nations estimates that the world's cities will grow by 72% during this time, increasing their populations by a total of 2.7 billion people. Urban communities provide spaces for people to live, work, and play, and require extensive construction of buildings, roads, energy infrastructures, communication networks, water purification and distribution systems, sewage treatment plants, and sanitation and waste management facilities. Very simply, the spatial expansion of cities puts tremendous pressure on environmental resources.

As engineers, our challenge is to find ways that will make the urban environment safer and more comfortable while minimizing economic costs and environmental impacts. In the past few decades, the professions of urban and environmental planning, architecture, and engineering have developed strategies such as "environmentally sensitive development," "smart growth," "green building," and "low-impact development" to manage human land use and natural resources more sustainably.

In this chapter, we will focus on the basics of sustainability and the built environment by exploring land-use planning generally and environmentally sensitive development and green building specifically. We will start at the largest spatial scales by reviewing the nature of land-use and land-cover change; then examine how urban planning, low-impact development, conservation design, and green building are changing the way we think about, design, and build our communities.

11.2 Land-Use and Land-Cover Change

To understand our discussion of environmentally sensitive development later in this chapter, we need to briefly introduce the concept of land cover and changing human land uses. Scientists classify the Earth's surface into basic categories of land cover that represent vegetation, ecosystems, and types of human development. Figure 11.2 illustrates one typology commonly used for land-cover analysis.

Because the Earth's surface provides many important environmental functions—including ecosystem services for humans, habitats that support biodiversity, and biogeochemical cycling of matter—the loss of critical land cover can affect these environmental functions in ways that increase economic costs and threats to human life. Land-use and land-cover changes are now monitored intensely, especially with new techniques in satellite imagery and remote sensing.

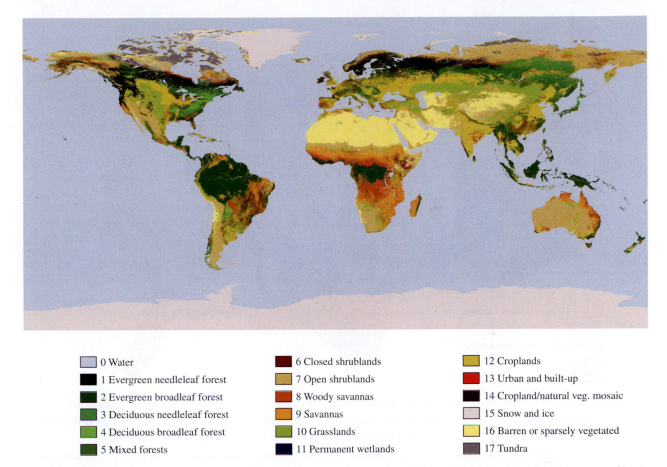

0 Water	6 Closed shrublands	12 Croplands
1 Evergreen needleleaf forest	7 Open shrublands	13 Urban and built-up
2 Evergreen broadleaf forest	8 Woody savannas	14 Cropland/natural veg. mosaic
3 Deciduous needleleaf forest	9 Savannas	15 Snow and ice
4 Deciduous broadleaf forest	10 Grasslands	16 Barren or sparsely vegetated
5 Mixed forests	11 Permanent wetlands	17 Tundra

FIGURE 11.2 Global land-cover patterns. This map illustrates the ways in which we classify Earth's surface into types of land cover. Common categories include surface water (water, snow, ice), types of vegetation (savannah, grassland), ecological systems (forests, permanent wetlands), and human development (cropland, urban development). At the global scale of this map, it is hard to see the scope of the human built environment.

Source: NASA, "Satellites and the City," http://www.nasa.gov/images/content/121557main_landCover.jpg

Of particular interest to us is the scope of land change due to **development**. In the context that is used here, *development* means something different than the way it was introduced earlier in our textbook, which was with respect to improvements in human individual, social, and economic well-being. In the context of land-cover change and land-use planning, development refers to tracts of land that have been newly built on or transformed in some other way for human use. Development represents the conversion of open land and natural areas to residential neighborhoods, airports, industrial parks, hydroelectric dams, shopping centers, parks, and so on. Development reflects the inherent urbanization of communities and is driven most substantially by population growth and economic industrialization.

Figure 11.3 illustrates the scope of development in the United States from 1973 to 2000. As can be seen by this map, the rate of development is not evenly distributed, but it varies considerably by region. In some places, developed land cover increased by more than 80%, notably in the Midwest (Indiana and Ohio),

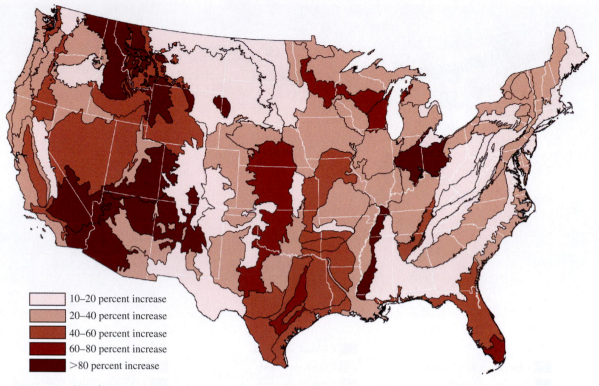

FIGURE 11.3 Net change in U.S. developed land cover, 1973–2000. Developed land cover increased throughout the United States, as it is strongly associated with population growth. Areas of rapid growth and urbanization are most dramatically illustrated by the darkest regions in which land developed for the built environment increased by more than 80%.

Source: Sleeter, et al., "Land-cover change in the conterminous United States from 1973 to 2000," Global Environmental Change 23 (2013) 733–748, p. 744.

in the Pacific Northwest (northern Washington and Idaho), and in a vast area of the southwestern United States covering portions of California, Nevada, Utah, Colorado, Arizona, and New Mexico.

Urbanization and land-cover change is of particular concern for two reasons. First, it can create unsustainable environmental change by altering local climates, stimulating the loss of economically useful natural resources, and otherwise affecting a myriad of ecosystem services. Deforestation, the loss of wetlands, and the destruction of critical habitats are especially problematic. Second, the built environment can also put people in harm's way (Figure 11.4). Destructive forest fires, mudslides, and flooding are examples of hazards to human life and property that can occur when development locates communities in areas naturally prone to fire, unstable geology, or floods.

11.3   Land-Use Planning and Its Role in Sustainable Development

Land-use planning is a form of public policy that governs *how* land is used by community residents within a defined geographical area, such as a state, province, county, district, or city. In theory, land-use planning attempts to manage human activities through laws and regulations designed to make the most efficient use of land, natural resources, and water. In practice, this is not always the case. As we will see, many land-use regulations result in an inefficient use of space and degradation of the environment.

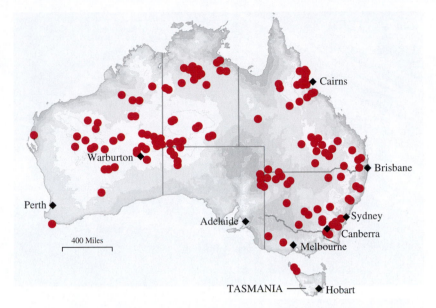

FIGURE 11.4 Brushfire outbreaks in Australia. The worst season for brushfires occurred in 2013 in Australia, where this map shows fire outbreaks over a 12-hour period. Homes were destroyed and many towns had to evacuate. Urbanization and spreading settlements exposed thousands of people to dangerous conditions in regions naturally prone to fire.

Source: Based on The Guardian, "Australian heatwave puts south-east on alert as wildfires burn out of control," 8 January 2013

Land-use planning is often referred to as urban planning when it focuses on cities and large, complex metropolitan areas. It is widely practiced in high-income nations; middle- and low-income countries face many challenges in designing and enforcing land-use and urban planning laws. Land use in these countries tends to be governed by national efforts to protect natural resources, such as forest policies or the establishment of parks and wildlife preserves. Because of budget constraints, low-income countries also often struggle to successfully implement and enforce conservation laws as well.

In the United States, settlement and land-use patterns over the past 75 years have resulted in an increasingly unsustainable form of development known as **urban sprawl**. After World War II, the United States began to suburbanize, particularly because of the "baby boom" and the rise of the automobile culture. As land-use patterns, sprawl and suburbia are characterized by low-density housing that spreads along major streets and arterial roadways. Cities lost their **mixed-use** urban centers in which homes, shops, businesses, and recreational spaces all coexisted with relatively high levels of **population density** (the number of people living in a square mile or square kilometer). Figure 11.5 illustrates the contrast between urban mixed-use space and suburbs in the United States.

The benefits associated with urban sprawl are typically related to less expensive and larger single-family housing and smaller, more manageable political jurisdictions. However, sprawl results in longer work commutes, greater conversion of agricultural land and open spaces, worsening water quality, and higher levels of material usage and cost. Sprawl also resulted in the **hollowing out** of many American cities, making them much less livable spaces. As people migrated from mixed-use urban neighborhoods to the suburbs, amenities declined. Urban centers became shopping districts at best—and sites of decline and decay at worst.

FIGURE 11.5 Mixed-use urban spaces and suburban neighborhoods. U.S. cities began to spread in a new pattern of development after World War II. Historically characterized by mixed-use spaces in which residential neighborhoods, offices, and shops were commingled with one another, urban areas spread out and transformed into highly segregated land-use patterns. Residential neighborhoods were isolated from businesses and shops and became less dense in terms of housing. This overall pattern of spread is generally referred to as sprawl.

Source: trekandshoot/Shutterstock.com; Courtesy of Turner Construction Company and Robert Benson Photography

11.3.1 Zoning and Land-Use Planning

Zoning is the primary tool of land-use planning, and its principal purpose is to define the legally permitted uses of land. In the United States, zoning is often practiced as **single-use zoning**. Community activities are highly segregated from one another and only one type of land use is allowed in a designated zone. Residential neighborhoods are usually isolated from farms and commercial districts, for example, and are often allocated to very large tracts of land. Indeed, the widespread practice of single-use zoning is one of the reasons sprawl emerged as such a basic pattern of the American urban landscape. The most commonly used zoning designations are residential, commercial, industrial, agricultural, mixed use, and recreational spaces (Figure 11.6).

With respect to residential areas, zoning laws also define **housing density**, which specifies the number and type of dwellings allowed, typically per acre of proposed residential space. Residential housing density is commonly classified as low-density residential (LDR), medium-density residential (MDR), and high-density residential (HDR). In the U.S., LDR housing typically ranges from 0.2 to 6 single- or two-family housing units per 0.5 hectares and MDR, from 5 to 15 single- and multifamily units per 0.5 hectares. HDR usually reflects large, multifamily structures like apartment complexes and usually has a density of more than 15 units per acre. A variety of zoning requirements for Spokane County, Washington are shown in Table 11.1.

Zoning ordinances codify the rules and regulations that govern the allowed uses for buildings or dwellings in an area. Ordinances are often extremely detailed, and they can specify building placement, structure heights, parking requirements, signage, landscaping requirements, and other performance standards that affect building engineering and design. Examples from Marion County, Indiana, zoning laws illustrate this point (Table 11.2).

Agricultural District

Light Industrial Park

General Business

Parks and Recreation

Industrial Zone

Flood Plain District

Mixed Use Redevelopment Zone

Planned Commercial Center

Residential Multi Family

Residential Single Family

FIGURE 11.6 Single-use zoning patterns. In the United States, single-use zoning is a common practice in which different types of land uses are segregated from one another. This zoning map clearly shows how residential, business, and town centers are distinctly separated. Single-use zoning is a common policy in suburban communities and has been a contributing factor to sprawl.

Source: Serdar Tibet/Shutterstock.com

TABLE 11.1 Density definitions for Spokane County, Washington

	LOW-DENSITY RESIDENTIAL	LOW-DENSITY RESIDENTIAL PLUS	MEDIUM-DENSITY RESIDENTIAL	HIGH-DENSITY RESIDENTIAL
Density (units/acre)	1–6	1	6–15	>15
Max building coverage (%)	55	55	65	70
Max height (ft)	35	35	40	50
Minimum lot area (ft^2)				
Permitted uses	6,000	6,000	6,000	6,000
Single family	5,000	43,560	4,200	1,600
Duplex	10,000	n/a	8,400	3,200
Min. frontage (ft)				
Permitted uses	50	60	60	60
Single family	50	90	50	20
Duplex	50	n/a	50	40
Min. setback (ft) Front (garage) Side		15 (20) 5	15 (20) 5 + 1 for each additional foot of structure height over 25 feet to a maximum of 15 feet.	
Rear setback	5 + 1 for each additional foot of structure height over 25 feet to a maximum of 15 feet.			

Source: Based on Spokane County Public Works Department (2004) Zoning Code, Spokane County, Washington. 1026 West Broadway Ave. Spokane, WA, USA.

Zoning provisions can affect the sustainability of the built environment by influencing the land area required for construction and the amount of building material needed to comply with zoning regulations. For example, the zoning provisions discussed previously—housing density, the requirements for open space specified in Table 11.2, and single-use zoning—all contribute to less compact, more spread out, developments. Other zoning provisions that can affect sustainable land-use practices include requirements for the building envelope, setback, the angle of bulk plane, and floor-to-area ratios.

The buildable area, or **building envelope**, is the region of a parcel of land that may have a structure legally built upon it, as shown in Figure 11.7. Four square lots are illustrated in this figure, with two lots on either side of the street. In addition to the lots, there is a **right-of-way** composed of the street itself and a narrow strip of land (commonly called an **easement**) adjacent to either side of the street; the easement is often preserved for utilities and other types of infrastructure. No permanent building structure is allowed within the right-of-way. Zoning laws regarding the size of roads and easement requirements can aggravate sprawl by requiring larger land areas to achieve desired lot sizes and roadways.

It is also typical in the United States to have a required distance—called a **setback**—between a structure and an adjacent feature such as a property line, street, another structure, or a stream. Figure 11.7 also illustrates setback requirements for a residential dwelling. As seen in the image, the building structure must be located completely within the buildable area (building envelope) established by the setback distances and size of the right of way. As with single-use zoning,

TABLE 11.2 Marion County standards of lot size, buildable area, and setbacks for single- and two-family residential structures

DISTRICT	MINIMUM LOT AREA		MINIMUM LOT WIDTH		MINIMUM STREET FRONTAGE	MINIMUM REAR YARD	MINIMUM SIDE YARD		MINIMUM OPEN SPACE	MINIMUM MAIN FLOOR AREA	
	Single Family	Two Family	Single Family	Two Family			One side	Aggregate		One story	Above One story
D-A	3 acres		250 ft		125 ft	75 ft	30 ft	75 ft	85%	1,200 sq ft	800 sq ft
D-S	1 acre		150 ft		75 ft	25 ft	15 ft	35 ft	85%	1,200 sq ft	800 sq ft
D-1	24,000 sq ft		90 ft		45 ft	25 ft	8 ft	22 ft	80%	1,200 sq ft	800 sq ft
D-2	15,000 sq ft	20,000 sq ft	80 ft	120 ft	40 ft	25 ft	7 ft	19 ft	75%	1,200 sq ft	800 sq ft
D-3	10,0000 sq ft	15,000 sq ft	70 ft	105 ft	35 ft	20 ft	6 ft	16 ft	70%	1,200 sq ft	800 sq ft
D-4	7,200 sq ft	10,000 sq ft	60 ft	90 ft	30 ft	20 ft	5 ft	13 ft	65%	900 sq ft	660 sq ft
D-5	5,000 sq ft	9,000 sq ft	50 ft	90 ft	25 ft	20 ft	4 ft	10 ft	65%	900 sq ft	660 sq ft
D-5II	3,200 sq ft	7,600 sq ft	40 ft	80 ft	25 ft	10 ft	3 ft	10 ft	65%	900 sq ft	660 sq ft
D-8	n/a		30 ft		30 ft	15 ft	4 ft	10 ft	55%	900 sq ft	660 sq ft
D-12	9,000 sq ft		70 ft		35 ft	20 ft	4 ft	10 ft	65	900 sq ft	660 sq ft

Note: A maximum height of 35 ft for primary buildings and 20 ft for accessory buildings applies to D2, D3, D4, D5, D8, and D12 zoning districts.

Source: Based on Marion County Zoning Ordinance, Chapter 731: Dwelling Districts (DDZO) Zoning Ordinance & Subdivision Control Ordinance, http://www.indy.gov/eGov/City/DMD/Planning/Zoning/Pages/municode.aspx.

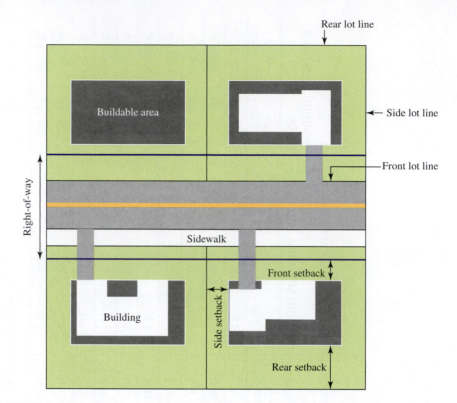

FIGURE 11.7 Buildable area and setbacks for residential structures. Zoning laws usually specify the area of a parcel of land that can be built upon and the minimum distances from property lines (setbacks) required for construction. While these requirements are often for safety and to minimize nuisances to neighbors, setbacks and buildable area requirements can also contribute to sprawl by requiring more space than is necessary.

Source: Bradley Striebig.

U.S. residential setback requirements have also been criticized for contributing to urban sprawl because they encourage larger lot sizes and lower density development.

Zoning ordinances can also specify a maximum building height (Figure 11.8) and may define the **angle of bulk plane** in an attempt to prevent buildings from literally overshadowing smaller, neighboring structures. The angle of bulk plane is typically determined by calculating the angle from the edge of the lot to be developed to the steepest angle of the proposed building, as illustrated in Figure 11.9. Depending on permitted building heights and angle of bulk plan requirements, zoning laws can also result in lower-density residential areas.

Zoning codes may also specify other key building parameters such as the floor-to-area ratio. The **floor area ratio (FAR)** is the ratio of the building's footprint (floor area) to the area of the lot on which the building is to be constructed. Buildings cannot exceed the specified FAR, although multiple building configurations are possible. An example of three methods to comply with a FAR equal to one is illustrated in Figure 11.10. Note that requirements for the floor area ratio will have an impact on the amount of building material required to construct a building.

To sum, land-use zoning designations and ordinances define the basic manner in which land can be used for the built environment, and their provisions can affect

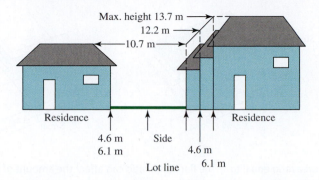

FIGURE 11.8 Maximum building height for residential structures. Zoning ordinances usually restrict the heights of structures. This diagram illustrates how even small variations in height can affect the bulk and visual appearance of a building.

Source: Bradley Striebig.

sustainable land use in important ways. Permitted uses, housing density, building envelopes, setbacks, widths of right of ways, open space requirements, floor area ratios, angle of bulk plane, and many other specifications are *necessary regulations* for safety, convenience, and pleasing spaces. But individually and collectively, these factors affect the location and arrangement of our structures, which ultimately affects land cover and its environmental benefits.

Some zoning is more conducive to a sustainable built environment than others; today, new planning models and design principles are emerging to better balance development and environmental protection. **New Urbanism** is a movement within

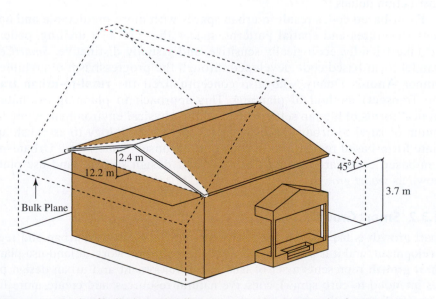

FIGURE 11.9 Angle of bulk plane. This illustration provides an example of a structure encroaching upon the angle of bulk plane.

Source: Based on Boulder Revised Code Home, Chapter 9-7: Form and Bulk Standards, http://www.colocode.com/boulder2/chapter9-7.htm

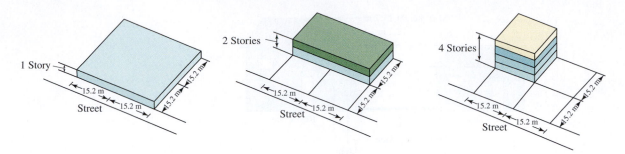

FIGURE 11.10 Configurations for a floor area ratio equal to 1. The floor area ratio can affect the amount of land used for a structure as well as the amount of building materials needed.

Source: Based on City of Boulder. Understanding Density and Floor Area Ratio: Examples of Residential, Mixed Use, Commercial and Industrial Developments in Boulder. http://www.bouldercolorado.gov/files/PDS/planning%20and%20zoning/density_floorarearatio.pdf

the field of urban design that provides alternatives to traditional land-use planning and its extensive use of single-use zoning. New Urbanism focuses on walkable communities, open space, diverse neighborhoods, and architectural styles that reflect local history, culture, and climate. In the United States, **form-based codes** are emerging as part of New Urbanism strategies and have increased in popularity since the early 2000s. Major metropolitan areas like Baltimore, Dallas, Denver, Miami, and Nashville have adopted form-based codes that often use neighborhoods as the primary geographic planning unit. Form-based codes rely on legal regulations controlling the location, size, and appearance of buildings, and also implement standards for public spaces. Streets and squares, for example, will contain design requirements that emphasize safety, green space, building placement, and other construction details.

Form-based codes result in urban spaces with more predictable and harmonious structures and spatial patterns, spaces that are also inviting, pedestrian and bike-friendly, ecologically sensitive, and visually distinctive. *SmartCode* is a model form-based code developed through the progressivism of architect and planner Andrés Duany, who also conceptualized the **rural-to-urban transect** (the Transect) method of planning. This approach to planning evaluates the physical forms of human settlements and the physical environment along a continuum of rural-to-urban spaces and uses this methodology to establish appropriate form-based regulations in each of the transect zones. New Urbanism and form-based codes are also part of a broader design shift in land use planning known as *smart growth*.

11.3.2 Smart Growth

Smart growth is likewise an alternative strategy to traditional urban and regional development, and it is practiced within the larger framework of land-use planning. **Smart growth** represents a set of land-use development and urban design principles intended to curb sprawl, conserve natural resources, and create more livable cities. (In Europe, the smart growth concept goes by the term *Compact City*.) It advocates **urban intensification**—higher levels of housing density in combination with mixed-use zoning, the greater availability of green space and natural areas, and enhanced mass transit. The Smart Growth Network, a coalition of organizations concerned about sustainable development, distilled 10 basic principles to manage growth in the built environment:

1. Mix land uses.
2. Take advantage of compact building design.
3. Create a range of housing opportunities and choices.
4. Create walkable neighborhoods.
5. Foster distinctive, attractive communities with a strong sense of place.
6. Preserve open space, farmland, natural beauty, and critical environmental areas.
7. Strengthen and direct development toward existing communities.
8. Provide a variety of transportation choices.
9. Make development decisions predictable, fair, and cost effective.
10. Encourage community and stakeholder collaboration in development decisions

Smart growth requires a wide variety of urban planning, landscape design, and community development strategies. Three prominent land-use techniques include (1) urban infill, (2) brownfield redevelopment, and (3) cluster developments.

Urban infill is a strategy in which abandoned, derelict, unused, or under-utilized space in urban areas is redeveloped. The infill land use may be purely residential or commercial, but ideally it has a mix of housing options as well as parks, commercial, and business space. The cost economics of urban infill is often problematic, as it can be much more expensive to redevelop existing urban spaces than to simply build on the suburban periphery (another contributing factor to sprawl). Urban infill is most widely practiced in cities that have an **urban growth boundary** (a zoning designation that legally limits the geographic spread of growth and new development) or physical barriers to expansion such as mountains or existing dense urban development. Portland, Oregon, is a city renowned for its urban growth boundary and successful infill projects; however, escalating housing prices and unfavorable patterns of land values have become of growing concern in the region. Most recently, **urban agriculture** has emerged as a type of infill land use, where small farms and community gardens are interspersed in a city's neighborhoods. There are more than 700 urban farms in New York City; derelict neighborhoods in Detroit, Michigan, are being revitalized by a growing community movement to provide healthy local food, connect residents to nature, and put abandoned land to productive use.

Brownfield redevelopment is closely related to the concept of urban infill, as it also involves the redevelopment of an existing space. In the United States, a **brownfield** has traditionally represented an inactive industrial site, usually with some degree of real or perceived environmental/chemical contamination. Environmental remediation may be required to effectively reuse the site for some types of land use; brownfields are nonetheless potentially attractive sites for growth because they can involve large acreages in desirable locations. Atlantic Station in Atlanta, Georgia, is an award-winning, mixed-use brownfield redevelopment project that successfully repurposed 56 hectares in midtown Atlanta (Figure 11.11).

Cluster developments are subdivisions or building tracts that are designed to concentrate housing and other structures in one area, protecting much of the landscape from development. In a cluster development, the overall zoned density of land does not change, but the building envelope, setback requirements, and housing density are

FIGURE 11.11 Atlantic Station in Atlanta, Georgia, is an example of a substantial urban brownfield *and* infill redevelopment initiative, the largest in the United States. It provides mixed use, green spaces, and other amenities to make the neighborhood a livable community for its residents.

Source: Sean Pavone/Shutterstock.com

adjusted to create an alternative arrangement of buildings (Figure 11.12). Ideally, cluster developments contain outdoor recreational areas and different types of housing options to accommodate a range of resident ages, incomes, and lifestyles. They should also be highly walkable with sidewalks, low-traffic areas, and pedestrian ways. What makes cluster developments special from a planning perspective is that they usually violate several aspects of local zoning laws and require special regulations to permit their use.

Smart growth policies and principles are fundamentally designed to plan for population growth and development of the built environment in ways that reduce traffic congestion, provide a wider variety of housing options at reasonable costs, and increase convenience and quality of life for local residents. Smart growth also purposefully protects the natural environment. Burchell and Mukherji (2003) reported that managed growth may decrease land consumption by 21% compared to conventional methods, conserving hundreds of thousands of hectares. Significant savings can be realized for infrastructure and utility costs as well (Table 11.3) because higher density growth reduces overall demand on public services and infrastructure.

To sum, smart growth strategies involve urban intensification by redeveloping unused (or underused) land for productive purposes and by creating more compact, higher-density neighborhoods at the urban/suburban interface. When combined with other design techniques for walkability, greenspace, mixed use, and so on, smart growth can result in cities with pleasing amenities, greater conveniences, and more local employment opportunities. Smart growth does, however, require a major rethinking of zoning laws, and it reflects an intensive effort at progressive urban design (Box 11.1).

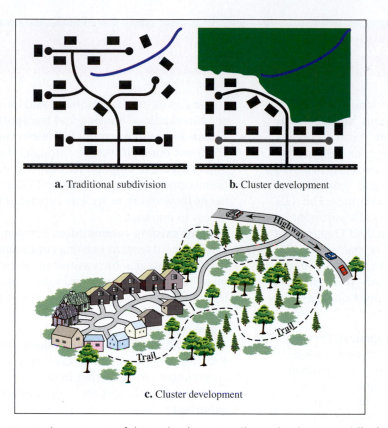

a. Traditional subdivision

b. Cluster development

c. Cluster development

FIGURE 11.12 Arrangement of cluster developments. Cluster developments differ from traditional suburban neighborhoods and subdivisions in the arrangement of their homes. Images (a) and (b) both have the exact same number of housing units, but the cluster development in part (b) arranges the dwellings more compactly, opening up a green space. As seen in part (c) this green space can be used for community recreation.

Source: Based on Triangle J Council of Governments, Green Growth for Clean Water (Research Triangle Park, NC, 2012), pp. 17, 29. http://www.tjcog.org/Data/Sites/1/media/regional-planning/di-partnership/greengrowth_mar2012_final.pdf

TABLE 11.3 Projected water and sewer infrastructure under conventional development and managed growth scenarios in the United States, 2000–2025

REGION	TOTAL WATER AND SEWER DEMAND, gal/d (millions)			TOTAL WATER AND SEWER LATERALS (thousands)			TOTAL INFRASTRUCTURE COSTS, $ (millions)		
	Conventional Development	Managed Growth	Demand Savings	Conventional Development	Managed Growth	Lateral Savings	Conventional Development	Managed Growth	Cost Savings
Northeast	1,451	1,444	7	3,406	3,068	338	16,015	14,751	1,264
Midwest	2,935	2,915	21	7,110	6,604	505	30,393	28,839	1,556
South	7,942	7,870	72	21,243	19,116	2,126	84,573	79,026	5,547
West	5,794	5,737	56	14,108	12,456	1,652	58,786	54,544	4,242
Total	18,121	17,965	156	45,867	41,245	4,621	189,767	177,160	12,609

Source: Based on Burchell and Mukherji (2003) Conventional Development Versus Managed Growth: The Costs for Sprawl. American Journal of Public Health 93(9):1534–1540.

BOX 11.1 Design Values for Sustainable Communities

Smart growth as an approach to land use is effective as a sustainable development strategy only if our communities become more "livable" at the same time—no one wants to live in a "concrete jungle." The Smart Growth Network articulated 10 design principles for managed growth that reflect the need for livable communities. The U.S. Environmental Protection Agency, Department of Housing and Urban Development, and Department of Transportation also created a partnership to promote sustainable communities. These agencies elaborated the smart growth principles by developing the following "Livability Principles" for communities:

- **Provide more transportation choices**. Develop safe, reliable, and economical transportation choices to decrease household transportation costs, reduce our nation's dependence on foreign oil, improve air quality, reduce greenhouse gas emissions, and promote public health.
- **Promote equitable, affordable housing**. Expand location- and energy-efficient housing choices for people of all ages, incomes, races, and

ethnicities to increase mobility and lower the combined cost of housing and transportation.
- **Enhance economic competitiveness**. Improve economic competitiveness through reliable and timely access to employment centers, educational opportunities, services, and other basic needs by workers, as well as expanded business access to markets.
- **Support existing communities**. Focus new development toward existing communities—through strategies such as transit-oriented, mixed-use development and land recycling—to increase community revitalization and the efficiency of public works investments and to safeguard rural landscapes.
- **Value communities and neighborhoods**. Enhance the unique characteristics of all communities by investing in healthy, safe, and walkable neighborhoods—rural, urban, or suburban.

Source: US Environmental Protection Agency, HUD-DOT-EPA Partnership for Sustainable Communities, "Livability Principles," http://www.epa.gov /smartgrowth/partnership/#livabilityprinciples

Higher densities of buildings and people tend to minimize urban sprawl and reduce the environmental impact associated with transportation to and from homes and the workplace. Higher-density areas also have a significantly lower carbon footprint. One life cycle analysis in Toronto, Canada, found that neighborhoods with 148 dwellings per hectare had two-thirds less carbon emissions than neighborhoods with a density of 20 units per hectare. Higher-density residential areas simply require less transportation, lower energy costs in building operations, and fewer materials for infrastructure. Nevertheless, high-density development does come with challenges. It may stress water, sanitation and waste management systems. It can also be more costly to implement, and as seen in some areas, increase the cost of housing. Urban intensification is also not necessarily easy to "retrofit" into existing communities because large, multi-story buildings may negatively affect their surroundings in important ways.

11.4 Environmentally Sensitive Design

Concepts such as smart growth and livable cities provide important guidelines for land-use *planning* that promote more sustainable land uses. These concepts also help engineers and other professionals reconceptualize how we think about space, the built environment, and the landscape. An important remaining consideration is how we actually go about *physically designing* our spaces to achieve the balance between environmental conservation, social need, and economic opportunity that sustainable development requires.

Environmentally sensitive design (also known as environmentally sensitive development) is a term that generically describes a broad range of design strategies and on-the-ground techniques to integrate both development and environmental conservation. Environmentally sensitive design is very different from a conventional **environmental impact assessment**, which is an analysis that simply catalogues the expected environmental impacts a project will have at a site and indicates how those impacts might be mitigated. Environmental impact assessments *are not* design strategies. The Maryland Office of Planning stated the foundational design principle of environmentally sensitive development quite succinctly, which is to "let the land shape your plan." The goal of environmentally sensitive design is to site and construct the built environment in a manner that protects sensitive areas and minimizes the degradation associated with development.

No matter where we live, certain types of "sensitive" landscape features consistently require conservation and protection because of their profound role in ecosystem services and biogeochemical cycling. These include, for example:

- Forest cover
- Floodplains
- Streams and their riparian buffers
- Wetlands
- Steep slopes
- Habitats of threatened, rare, and endangered species

Nonsensitive features of the landscape have important environmental roles as well. For example, a meadowland provides habitat for native pollinators and small mammals, contributing to the overall biological complexity of a region. Large expanses of undeveloped or agricultural land create wildlife corridors, regulate the local climate, support local food systems, and offer recreational opportunities. Environmentally sensitive design therefore focuses on protecting sensitive areas, conserving open space, and maintaining the integrity of the environmental and social functions supported by the landscape.

In addition to the loss of sensitive landscape features, open space, and habitats, the most significant land-cover impacts of the built environment are associated with the disturbance of soil and the disruption of a location's natural hydrology. Site development commonly involves excavating and re-grading land, laying down large areas of impervious surfaces, redirecting or eliminating natural bodies of water, and constructing stormwater systems to manage rainwater runoff. (**Impervious surfaces** are those areas in the urban landscape that do not absorb water or allow water to infiltrate into the ground, such as roads, parking areas, and rooftops.)

As a consequence, wetlands have been eliminated at an astounding rate in the United States, and it is likewise common to bury small surface streams or channelize them into engineered drainage systems. Resculpted land, vast areas of impervious material, and the disturbance of surface waters result in the loss of productive soil, sedimentation of streams and rivers, accelerated stormwater runoff, increased non-point-source water pollution, declining biodiversity, and the destruction of critical water habitats. Figure 11.13 illustrates some of these problems by showing how development can affect the hydrologic dynamics of a forested site.

The social and environmental problems associated with stormwater runoff are particularly significant in urban and suburban areas. Rainfall rapidly moves across impervious roadways, parking lots, and rooftops (gathering contaminants as it goes); it also tends to flow in sheets across the sloped lawns and grassy areas

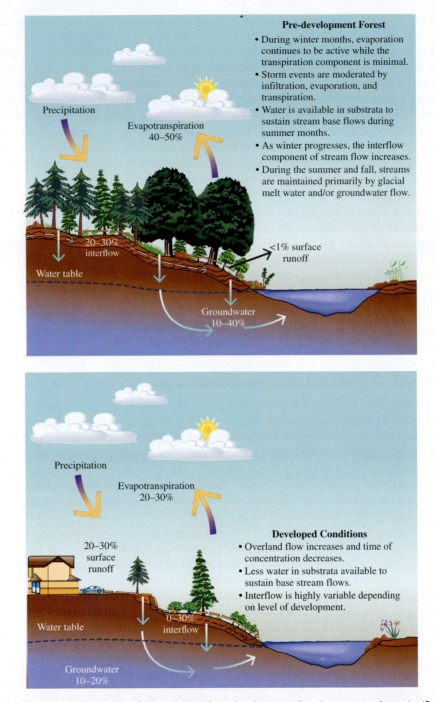

Pre-development Forest
- During winter months, evaporation continues to be active while the transpiration component is minimal.
- Storm events are moderated by infiltration, evaporation, and transpiration.
- Water is available in substrata to sustain stream base flows during summer months.
- As winter progresses, the interflow component of stream flow increases.
- During the summer and fall, streams are maintained primarily by glacial melt water and/or groundwater flow.

Precipitation
Evapotranspiration 40–50%
20–30% interflow
Water table
<1% surface runoff
Groundwater 10–40%

Precipitation
Evapotranspiration 20–30%

Developed Conditions
- Overland flow increases and time of concentration decreases.
- Less water in substrata available to sustain base stream flows.
- Interflow is highly variable depending on level of development.

20–30% surface runoff
Water table
0–30% interflow
Groundwater 10–20%

FIGURE 11.13 Disruption of site hydrology from development. Development can have significant consequences for hydrologic cycles. The image at the top shows how this site's natural topography and vegetation regulate the level of the water table, the volume of groundwater, the rate of interflow, the amount of surface water runoff, and degree of evapotranspiration. After development, the reduced forest cover diminishes evapotranspiration, and the accelerated runoff reduces groundwater recharge. Over time the environmental dynamics of this site will be significantly changed in a manner that will alter its microclimate, degrade its water flow and quality, and diminish its ability to support complex vegetation.

Source: Based on Hinman, C. (2005) Low Impact development: Technical Guidance Manual for Puget Sound. Puget Sound Action Team, Olympia, WA, USA.

common in suburbia. Without stormwater management systems, localized flooding would result, which is clearly a socially undesirable condition. However, traditional stormwater systems direct runoff to local water bodies in a manner and at rates that would not occur naturally. Water contamination, downstream flooding, and degraded stream ecology are the unintended consequences of these systems (Box 11.2). In some places, runoff may not be directed to a structured drainage

BOX 11.2 Impervious Surfaces and the Ecological Integrity of Streams

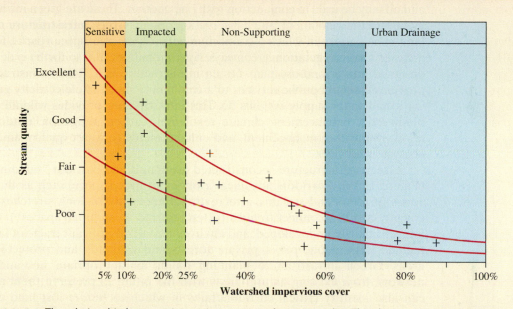

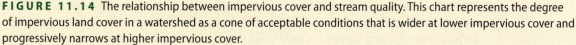

FIGURE 11.14 The relationship between impervious cover and stream quality. This chart represents the degree of impervious land cover in a watershed as a cone of acceptable conditions that is wider at lower impervious cover and progressively narrows at higher impervious cover.

Source: Based on Schueler, T. and Fraley-McNeal, L (2008) The Impervious Cover Model revisited: a review of recent ICM research. Presented at the Symposium on Urbanization and Stream Ecology, May 23 and 24, 2008. http://chesapeakestormwater.net/2009/11/the-reformulated-impervious-cover-model/

Streams in a watershed with less than 10% impervious surface can maintain natural levels of biodiversity and experience little harm to their normal stream channel and habitats. Streams become significantly affected by development when impervious areas represent 11 to 25% of the watershed, as illustrated in Figure 11.14. When an impervious surface exceeds about 25% of the land area, streams, waterways, and their surrounding riparian habitat become irreparably damaged. As a consequence, they cannot function as a healthy ecosystem or as an effective hydrologic channel in the watershed. Streambank erosion, high water velocity during storm events, increased sedimentation, low dissolved oxygen levels,

and nutrient loading are some of the consequences of extensive impervious surfaces that *are not* mitigated by environmentally sensitive design. This has two important results. First, it substantially diminishes a stream's biodiversity and ecology, which has a key role in the surrounding ecosystem and for maintaining water quality. Second, after repeated storm events, streambank erosion and sedimentation tend to "straighten" a stream, which increases water velocity. The result is downstream flooding during heavy rain and the rapid movement of contaminants into larger bodies of water within and outside of the local watershed. Over time, vital streams effectively become ecologically sterile drainage pipes.

system and erodes soil in problematic ways. Mudslides can occur, for example, when sloped land has not been adequately protected from erosion due to runoff from the built environment.

By "letting the land shape the plan," environmentally sensitive design does several things. First, it requires that *landscape features* be the primary determinant of how a site or region is developed. Second, it protects critical environmental features by literally designing around them and disturbing them minimally, if at all. Third, it mitigates stormwater flow and erosion by containing and processing runoff as close to its source as possible and by imitating natural processes.

Three complementary environmentally sensitive design paradigms are emerging, and all may be used in conjunction with one another. These are green infrastructure, low-impact development, and conservation design. **Green infrastructure** refers to a landscape pattern in which a connective network of greenspace (parks, forestland, areas of native vegetation, greenways, riparian buffers, and so forth) is deliberately designed into a land-use plan. Green infrastructure is a *living* infrastructure that provides essential public services to a community much like electricity grids, road-ways, and water supply systems do. This infrastructure provides wildlife corridors, biodiversity enhancement, climate regulation, carbon mitigation, outdoor recre-ation, stormwater management, and enhancements to water quality, among other ecosystem services.

Green infrastructure is constructed using hub-and-corridor "building blocks" (Figure 11.15). Corridors are linear ribbons of green space, such as the riparian zones of rivers and streams, a recreational greenway, or long stretches of forest cover. Corridors connect to hubs (often referred to as "open spaces"), which are large, primarily undeveloped, and environmentally significant tracts of land. Hubs include large public parks, private forests and wetlands, and protected natural areas. Green infrastructure planning builds an appropriate hub-and-corridor network from existing natural areas with the intent to preserve them as such. It can also identify critical corridors/hubs in which to *restore* lost land cover and ecological function. In addition, green infrastructure planning can artificially create hubs and corridors through careful environmental analysis and landscape architecture. Green infrastructure concepts work at all land-use scales, from the suburban backyard to a single city or county to large regions. Most major metro-politan areas in the United States have, or are in the process of, developing green infrastructure master plans; this form of planning is rapidly taking hold in both theory and practice.

The other two environmentally sensitive design strategies—low-impact devel-opment and conservation design—are discussed separately in their own sections. Both strategies are required to implement green infrastructure goals, although it is common to have low-impact development and conservation design *without* a green infrastructure land-use plan. Each of these also represents site-specific techniques and methods that you should be familiar with as engineers because they are rapidly becoming a suite of best management practices in a number of professions.

11.4.1 Low-Impact Development

Because of stormwater management issues in particular, low-impact develop-ment (LID) has emerged as a strategy to *avoid* runoff when possible and to *minimize* the disruption to natural hydrologic processes from development. The goal of LID is to mimic the natural hydrologic characteristics of a site, so that it

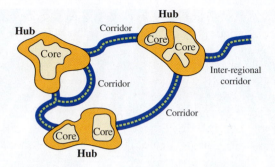

FIGURE 11.15 Interconnected hubs and corridors are the basic design elements for green infrastructure. Hubs are open spaces and usually environmentally important; corridors provide pathways for wildlife and may also protect rivers and streams.

Source: EPA, "Going Green" in the Mid-Atlantic, Natural Infrastructure, http://www.epa.gov/region03/green /infrastructure.html

is soundly within the environmentally sensitive design philosophy of letting the land shape the plan. Although LID is often used to casually describe a variety of sophisticated design strategies, it is most properly understood as a set of **best management practices (BMPs)** to manage runoff at a site. BMPs are techniques based on demonstrated, successful engineering practice; they are measures that consistently achieve the desired results in a cost-effective manner. LID BMPs are now mandatory in the land-use, transportation, and water quality laws of many states and localities.

LID uses water management systems that retain, treat, and filter runoff as close to its source as possible. This enhances water quality and also provides the benefit of soil conservation. LID BMPs are designed to maintain predevelopment water infiltration rates and to minimize downstream sediment loading, streambank erosion, and flooding. LID works in part by diverting and slowing stormwater runoff to natural bodies of water. A large number of techniques and practices are commonly used in LID depending on the nature and location of a project; several examples can serve to illustrate the scale and scope of LID efforts.

At the residential scale, gutters and downspouts traditionally move rainwater off of a rooftop and away from the foundation of a house, where water can be structurally damaging. However, downspouts are commonly channeled to the street, where rain then moves rapidly as stormwater runoff. Discharging downspouts to a lawn is better but may still result in stormwater flow; LID promotes the use of rain gardens (as discussed in Chapter 7) as a simple way of capturing roof water and allowing it to infiltrate naturally on site. Impervious surfaces can be reduced or avoided at the residential scale by using **pervious pavers** instead of concrete or asphalt for hardscapes (Figure 11.16).

At commercial scales or in the urban environment, a variety of more durable permeable paving materials are available to reduce runoff from the larger and more heavily trafficked parking, walking, and road areas. Unavoidable runoff from these surfaces (as well as rooftops) can be managed through a number of LID techniques, many of which involve bioretention and filtration. For example, surface runoff and downspouts can be directed to **vegetated swales**, which are basically a bigger,

FIGURE 11.16 Pervious pavers allow grass to grow through open spaces. The grass allows rainfall to infiltrate the soil naturally and reduces the amount of stormwater flowing into the watershed. Pervious pavers work well in low traffic areas like home driveways and carparks.

Source: mycteria/Shutterstock.com

engineered version of residential rain gardens. Vegetated swales are most commonly used as bioretention areas to hold and filter runoff for groundwater recharge (Figure 11.18). However, they can also be used as drainage *channels*, filtering runoff while also slowing and directing it to a body of surface water (Figure 11.19). Biorentention and filtration can be implemented at very small scales, such as with the use of tree boxes in stormwater drains (Figure 11.20).

Bioretention systems preserve stream health by cleaning, reducing, and slowing runoff. Stream health is further protected through a number of additional LID techniques. One is simply not to bury or otherwise channel a stream itself, something that was done with surprising frequency in conventional development. (Restoring a buried stream in an urban area to natural conditions is referred to as **stream daylighting**.) Another technique is to use a **riparian buffer zone**. Although riparian buffer zones are commonly discussed in the context of agriculture, they are also used in LID. These zones offer significant protection of surface streams by first keeping built structures at an extended distance from the water. Second, riparian buffer zones characteristically grow a mixed variety of grasses, plants, shrubs, and trees adjacent to the stream channel (Figure 11.20); the wider the buffer, the greater the variety of environmental benefits that result, including flood control and wildlife habitat. Streams that have lost their natural riparian buffers can be restored by planting appropriate new vegetation.

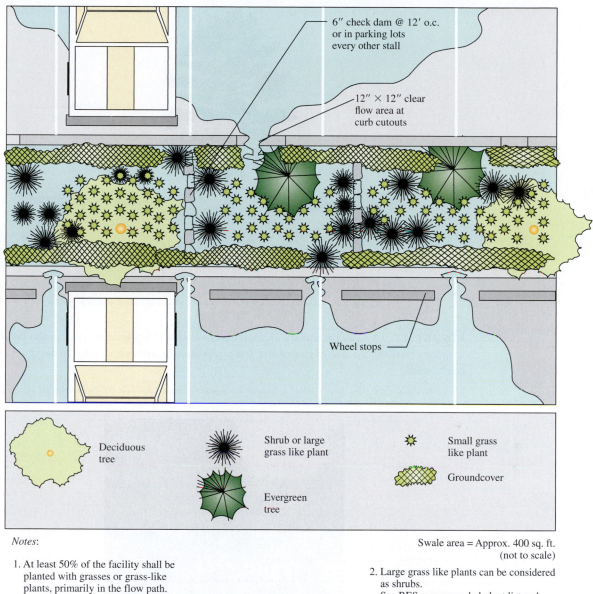

FIGURE 11.17 Bioretention systems for stormwater management. Bioretention areas adjacent to parking lots and roadways are commonly referred to as vegetated swales. These engineered biological systems are shallow depressions in the ground constructed to collect and filter stormwater, both cleaning it and allowing it to either recharge naturally as groundwater or slowly make its way to a body of surface water. This diagram shows a bioretention zone in a parking lot.

Source: Based on City of Portland. (2004) Stormwater Management Manual: Revision #3. Environmental Services, Clean River Works, City of Portland, OR, USA.

Although low-impact development is most commonly discussed in terms of stormwater prevention, management, and control, it does have a notable focus on water conservation as well. In particular, LID advocates reducing the demands for fresh water created by new development. Consequently, LID water conservation strategies include rainwater catchment methods that allow harvested rainwater to be used for landscape and garden irrigation, and potentially for indoor toilet flushing.

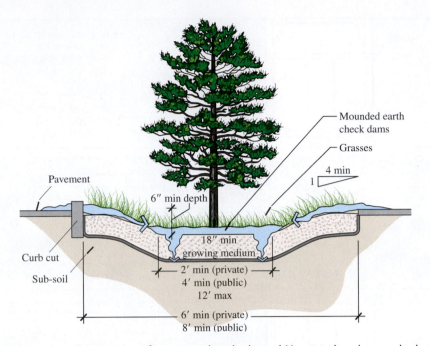

FIGURE 11.18 Cross section of a vegetated swale channel. Vegetated swales can also be used to *move* or *redirect* the flow of stormwater. This profile view illustrates how they are designed and constructed.

Source: Based on City of Portland. (2004) Stormwater Management Manual: Revision #3. Environmental Services, Clean River Works, City of Portland, OR, USA.

FIGURE 11.19 Tree box biofiltration system. Storm drains are unavoidable in urban areas where large areas of impervious surfaces are needed, such as for roadways and parking lots. Runoff can be slowed and cleaned by directing it to a drainage system that contains tree boxes, where the soil and plant roots act as biofilters.

Source: Maria Papadakis

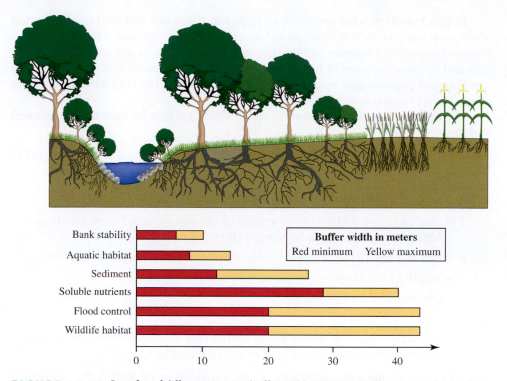

FIGURE 11.20 Benefits of different riparian buffer widths. Riparian buffers are used to protect and restore stream health in agricultural settings (as seen in this image) as well as in low-impact development. As buffers get wider, more environmental benefits result.

Source: Based on http://www.buffer.forestry.iastate.edu/Photogallery/illustrations/Images/Riparian-Widths_sld.jpg

Cisterns are commonly used to capture rainwater in the large volumes required for these applications. LID water conservation can also involve onsite water recycling, such as with **gray water** use for irrigation and indoor toilet flushing. Gray water is the wastewater from activities such as bathing, dishwashing, and laundry; because it is not contaminated with human body waste (known as sewage), it can be safely reused/recycled for nonpotable needs. In sum, low-impact development BMPs are most commonly used in and applied to the landscape, but they can also involve the design of buildings and building water systems as well.

11.4.2 Erosion Challenges

Productive topsoil is a nonrenewable resource that can take hundreds of years to form. It is critical not just to provide our food, but to sustain the plants and photosynthesis that form the basis of life on Earth. In the United States, soils are estimated to be eroding at a rate 17 times faster than they are produced; in Asia, Africa, and South America, the rate of loss is estimated at 30 times the rate of production. A good deal of this soil loss occurs because of agriculture and its associated practices; nonetheless, it is important that development of the built environment not worsen these conditions.

Fluvial (water) erosion is caused when runoff carries away soil particles, and it takes place in four typical forms: (1) sheet, which occurs evenly over the slope of a land surface; (2) rill, when the rainfall collects into very small but well-defined temporary channels; (3) gully, when rills enlarge; and (4) streambank, when the stream channel widens or collapses. Fluvial erosion of all types may be caused by the increased runoff from development.

Eolian (wind) erosion occurs when high winds remove soil, and it is worsened when soils are dry or not covered by vegetation. Development can trigger wind erosion temporarily because of the earth moving involved in the construction process and by creating areas of uncovered top soil. Development can also exacerbate wind erosion permanently by changing the topography of the land and by removing natural wind breaks, such as dense vegetation.

The **universal soil loss equation** is one method that may be used to estimate soil loss through erosion:

$$A = R \times K \times L \times S \times C \times P \tag{11.1}$$

where

A = annual soil loss due to erosion (mg/ha-yr)

R = rainfall factor, which is the measure of energy expected from raindrops hitting the soil during a 30-minute storm

K = soil erodibility factor based on the ratio of silt, sand, and organic matter

L = length of the slope

S = gradient of the slope

C = soil cover factor

P = a "support practice" factor based on the type of agricultural practice. For construction practices, this is assumed to be equal to one.

Typical values for the soil cover factor, C, are provided in Table 11.4. The remaining terms of the universal soil loss equation can be determined from site-specific maps and in soil maps from organizations like the United States Department of Agriculture, the United States Geological Survey, and local soil conservation agencies.

Development most commonly contributes to eroded soil because of disturbances to the natural hydrologic conditions of a location, including the loss of natural vegetative land cover. LID best management practices for stormwater management consequently go far in conserving soil lost through water erosion. With respect to wind erosion, hedgerows, tree rows, and fencerows may be used as windbreaks to avoid conditions that create a permanent risk of soil erosion. A design with diverse plantings of various heights and densities will not only buffer winds but also provide habitat.

Environmentally sensitive design is especially concerned with steep slopes, one of several sensitive landscape features that in principle should either not be developed or else should be aggressively protected from erosion and instability. Steep slopes may be geologically stabilized through a variety of retaining wall systems (Figure 11.21). Soil erosion may be prevented with fabric-based methods that use plastic netting, wood mesh, nylon mesh, or paper twine to hold soils, seeds, and new vegetation in place. Synthetic **geotextile** and turf reinforcement materials can be used on very steep slopes and will lock soil in place but allow water to permeate

TABLE 11.4 Typical soil cover factors used in the universal soil loss equation to estimate erosion

COVER TYPE	NO COVER	STRAW OF HAY TIE DOWN	CRUSHED STONE COVER	WOOD CHIPS	SOD OR LAWN
Cover factor:	1.0	0.02–0.06	0.02–0.08	0.02–0.08	0.01

Source: Bradley Striebig.

a. Common Types of Retaining Structures

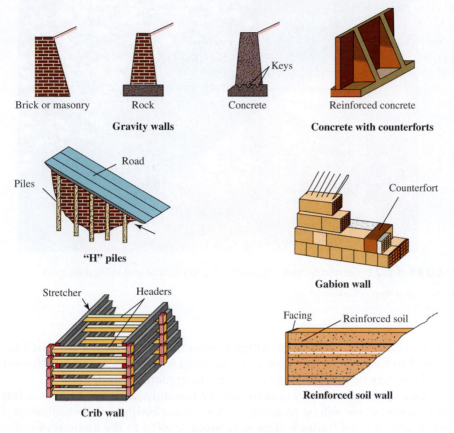

Brick or masonry   Rock   Concrete — Keys   Reinforced concrete

Gravity walls   **Concrete with counterforts**

Road
Piles
"H" piles

Counterfort
Gabion wall

Stretcher   Headers
Facing — Reinforced soil
Reinforced soil wall

Crib wall

b. Typical Rock Wall Construction

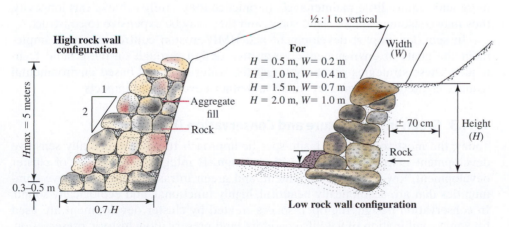

High rock wall configuration

$H_{max} = 5$ meters

1
2

Aggregate fill
Rock

0.3–0.5 m

0.7 H

½ : 1 to vertical

Width (W)

For
$H = 0.5$ m, $W = 0.2$ m
$H = 1.0$ m, $W = 0.4$ m
$H = 1.5$ m, $W = 0.7$ m
$H = 2.0$ m, $W = 1.0$ m

± 70 cm

Rock

Height (H)

Low rock wall configuration

Low-volume roads BMPs: 110

FIGURE 11.21 Common retaining systems for steep slopes. Development that occurs on or near steep slopes requires that the earth itself be held in place by a substantial retaining structure.

Source: Keller, G. and Sherar, J. (2003) Low Volume Roads Engineering: Best Management Practices Field Guide. USDA Forest Service/USAID, Washington, DC, USA. 182p. Figure 11.4.

FIGURE 11.22 Geotextile materials used for erosion control and soil stabilization.

Source: Eugene Sergeev/Shutterstock.com

(Figure 11.22). Geotextiles are synthetic materials purposefully designed to protect soil and manage water flow in a variety of ways; they are straightforward to install but may be unsightly and expensive to implement and repair.

Some development may nonetheless permanently generate soil and sediments. Long-term controls will be required for these sites; BMPs include settlement, infiltration, and wetland basins and must be sized relative to the amount of estimated sedimentation (see Box 11.3). Permanent controls are designed to last 50 years or more and require little maintenance. In practice they rarely achieve this longevity, they may consume valuable land space, and they may be expensive to construct.

In sum, low-impact development and BMP erosion control measures implement the goals of environmentally sensitive design through on-the-ground techniques. These strategies primarily target the water- and soil-based environmental damage created by stormwater runoff and other development impacts.

11.4.3 Green Infrastructure and Conservation Design

Today, the most sophisticated, site-specific approach to environmentally sensitive development is known as **conservation design**. It integrates principles of cluster development, low-impact development, and green infrastructure to achieve communities that are aesthetically beautiful, highly functional, and ecologically sound. In conservation design, the open spaces created by cluster development are used for some combination of wildlife corridors, land preservation, historic preservation, recreation, and water resources protection. It is a holistic approach to environmentally sensitive design that is most commonly used to develop large tracts of land into residential neighborhoods or mixed-use communities. For localities that have formally adopted a green infrastructure plan, conservation design is an essential tool for implementing that plan in new developments.

Conservation design provides many environmental advantages compared to traditionally developed properties. In addition to avoiding the development of sensitive landscape features, this approach can also protect upland buffers, preserve

BOX 11. 3 **Stokes Settling Law**

The design and maintenance of sediment control systems for runoff management requires knowledge of the sedimentation loss rate, runoff velocity, and estimated particle size distribution and weight. Sediments may be removed from runoff when the downward velocity of the particle is greater than the runoff velocity. **Stokes Settling law** [(see Equations (5.35) and (5.38)] may be used to estimate particle removal as a function of the runoff flow velocity and expected residence time in the sedimentation removal device. The settling velocity for a single particle may be estimated by

$$V_s = \frac{2g\left(\frac{\rho_s}{\rho_f} - 1\right)r^2}{9\eta_f} \qquad (11.2)$$

where

V_s = settling velocity (m/s)

g = gravitational constant (m/s^2)

ρ_s = density of the particle (kg/m^3)

ρ_f = density of the fluid (kg/m^3)

r = particle radius (m)

η_f = kinematic viscosity of the fluid (m^2/s)

The settling time, t_s, of a particle is given by the relationship between the depth, h, and settling velocity:

$$t_s = \frac{h}{V_s} \qquad (11.3)$$

habitat and biodiversity, reduce erosion and runoff, and improve groundwater infiltration. Research has also shown that real estate values for homes on smaller lots designed using conservation principles not only sell more quickly, but appreciate in value faster than comparable homes in traditionally designed neighborhoods.

In practice, conservation design literally starts with letting the land shape the plan. Landscape architect Randall Arendt developed a conservation design methodology that is supported by the Natural Lands Trust, the American Planning Association, and the American Society of Landscape Architects. It consists of four steps:

1. Identify conservation areas from resource maps
2. Locate the housing sites
3. Design street alignments and trails
4. Draw lot lines

After the first step, *identify conservation areas from resource maps*, the design process becomes progressively more detailed by establishing (a) areas of the landscape for conservation, (b) the buildable areas of the site, and (c) the methods for water conservation. Figure 11.23 illustrates how, by overlaying land use and resource maps, areas of highest conservation value can be identified. The buildable area will be sited in a manner that does not disturb the conservation zones.

Figure 11.23 also illustrates the contrast between conventional site development and conservation design in terms of step 2, *locate the housing sites*. In image (b), only the wetland (a primary area of conservation) was protected. The rest of the site is fully apportioned to building lots, a strategy that is typical of conventional developments. Secondary conservation features, such as the natural drainage of the property, are built over. The conservation approach of image (c) contains the exact same

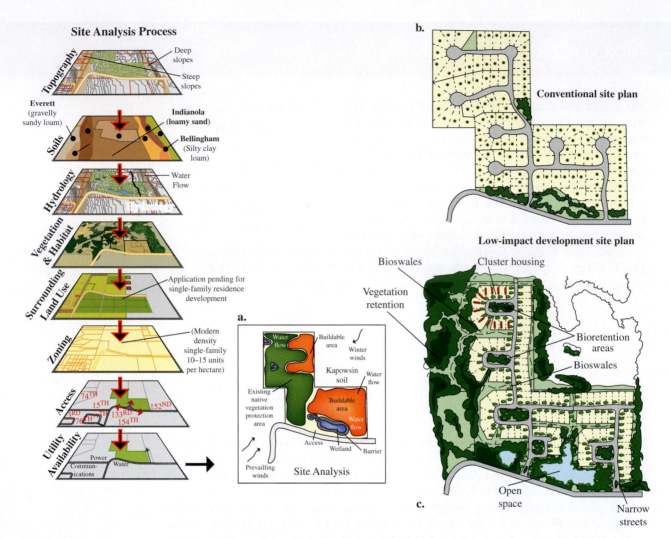

FIGURE 11.23 Resource mapping and conservation design. In conservation design, resource mapping is the critical first step. By evaluating landscape features, geology, and land cover, natural drainage flows can be identified as well as areas for conservation. (a) Resource mapping results in a preliminary site analysis; the buildable area is shown in brown, the conservation area in green, and a wetland in blue. Image (b) demonstrates how this site would be developed using conventional siting approaches; image (c) represents a conservation design strategy. The total housing density of both (b) and (c) is the same, but the conservation approach incorporates cluster development, LID techniques, and densely vegetated greenspace.

Source: Based on Hinman, C. (2005) Low Impact development: Technical Guidance Manual for Puget Sound. (Olympia, WA: Puget Sound Action Team, 2005, PSAT 05-03. pp. 24, 47).

number of housing units as in image (b), making it a **density-neutral** design. But the homes are clustered into more compact arrangements through smaller lot sizes and narrower streets, enabling protection of both the wetland and the natural drainage. The conservation approach also opens spaces for low-impact development stormwater management strategies—vegetated swales are distributed throughout the development, contributing to the green infrastructure of the site. These conservation areas can reduce erosion and sediment control costs for the developer.

Figure 11.24 illustrates all four steps of the conservation design method. Image (a) represents the conservation areas identified through resource mapping. Highly sensitive features of the landscape, such as steep slopes, wetlands, and hydric soils are designated as primary conservation areas. As part of step 1, image (b) diagrams

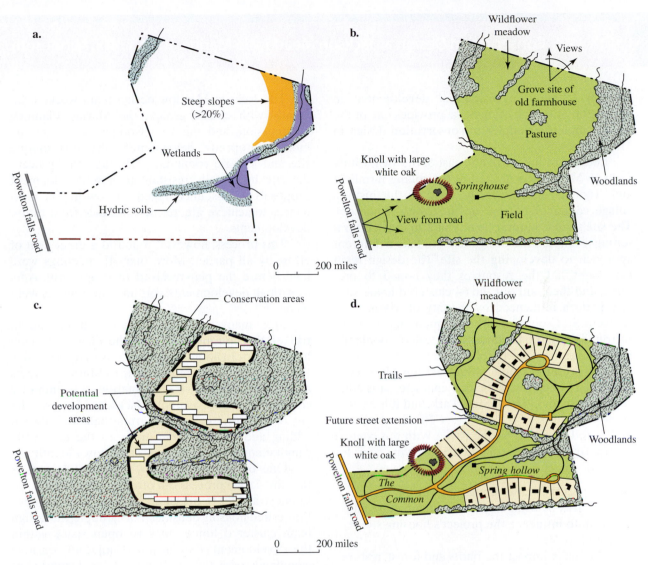

FIGURE 11.24 Stages of conservation design. This series of illustrations demonstrates how conservation design progresses from landscape resource mapping to complete site plans.

Source: Based on Arendt, R. G. (1996) Conservation Design for Subdivisions: A Practical Guide to Creating Open Space Networks. Island Press.

secondary features of the landscape to be incorporated into the plan, such as attractive views, a majestic old tree, woodlands, and a wildflower meadow. Image (c) represents step 2, locating the housing sites and buildable area of the property. In image (c), the buildable area stays clear of all of the primary and secondary areas of conservation, places the housing at a distance from the nuisance of the road, and orients the homes toward attractive views. Finally, image (d) completes the plan with steps 3 and 4: *designing the street alignment and trails and drawing in the lot lines.*

Because density-neutral conservation design strategies usually do not comply with local zoning law—they often have smaller lot sizes and setbacks, larger building envelopes, and narrower streets, for example—they are simply not possible in most locations without a special review and permitting process from local planning authorities. The case study of Jackson Meadow in Minnesota (Box 11.4) shows how one community negotiated a conservation development in a highly sensitive landscape.

BOX 11.4 Conservation Design at Jackson Meadow in Marine on St. Croix, Minnesota

Jackson Meadow, a residential development in Marine on St. Croix, Minnesota, provides one of the most complete examples of conservation design in Minnesota.

Located to the north and east of the Twin Cities, Jackson Meadow is nestled in the rolling farmland and wooded bluffs directly west of the historic village center of Marine on St. Croix, Minnesota. The site's great natural beauty and unique cultural setting prompted the design team to take a different approach to developing the site. The design team first identified the resources they hoped to preserve, and then laid out the 64 clustered home sites in a pattern influenced by the city of Marine and adjacent St. Croix river. The unique home sites are surrounded by 100 hectares of protected woodlands, restored prairies, and farmland.

The preserved land within Jackson Meadow is part of a river city with historic character; it is adjacent to William O'Brien State Park; and it is located along the bluffs of the St. Croix River, which is a designated Wild and Scenic River. With this context in mind, the project team had the following goals for the development:

- Respect the sense of place of Marine on St. Croix by using its historic architecture and form to influence the project's housing styles and layout.
- Minimally impact the bluffs and forest, restore native vegetation, and maintain an agricultural buffer around the development.
- Provide an extensive pedestrian trail system and other amenities.
- Use innovative stormwater and wastewater treatment techniques to minimize water resource impacts.

The Jackson Meadow design team worked diligently with citizen groups, the Marine Planning Commission, and the City Council to design the most appropriate development for this unique site and historic community. Through this process, the city–historically resistant to new development proposals – was encouraged to rewrite its ordinances to achieve greater performance from future developments.

This process, however, required a great deal of effort by all parties. More than 40 meetings were held before the plat received final approval. And, as with all developments, certain compromises were made.

Like many new developments in the urban fringes, the housing is beyond the means of many homebuyers. In addition, though sensitive to the site, the project still adds more traffic to Marine's streets and furthers Marine's transformation to a commuter community. However, with Jackson Meadow, the city of Marine struck an important balance between adding additional housing units to the city while simultaneously preserving its small-town identity.

One of Jackson Meadow's greatest lessons is that the design process can often combine many different strategies to preserve open space and achieve the desired housing densities. For example, the design team clustered homes to save open space within the development, they acquired adjacent sensitive woodlands with the assistance of the Department of Natural Resources and they employed a unique variation of Transfer of Development Rights (TDR) to preserve adjacent farmland.

Source: Based on Minnesota Land Trust, "Jackson Meadow," Conservation Design Portfolio Case Study 1. http://www.buildwise.org/library/design /sustainable-development/ jackson meadow.pdf.

In sum, environmentally sensitive design moves land-use planning and sustainable development a large step forward in its use of green infrastructure, low-impact development, and conservation design. These practices are far from standard, however, and it will take years of effort to change zoning laws, professional practice, and public perception to make them so.

11.5 Green Building

Sustainability of the built environment can be substantially advanced by changing the way we develop land. However, we also have to consider the environmental impacts of building structures and systems. Green building has emerged as a concept that can inform a more sustainable approach to building design, construction, and operation.

Green building is defined by the U.S. Office of the Federal Environmental Executive as "the practice of (1) increasing the efficiency with which buildings and their sites use energy, water, and materials, and (2) reducing building impacts through better siting, design, construction, operation, maintenance, and removal—the complete building cycle." Similarly, the U.S. Environmental Protection Agency defines green building as "the practice of creating structures and using processes that are environmentally responsible and resource-efficient throughout a building's life-cycle from siting to design, construction, operation, maintenance, renovation and deconstruction. The practice expands and complements the classical building design concerns of economy, utility, durability, and comfort."

Green building consequently embodies three concepts that have been previously explored in our text—environmental footprints, life cycles, and industrial ecology. First, green building reduces the environmental footprint of a building on its site by protecting as many of the natural features and functions of the location as possible. Second, we can think of a building as a product with its own life cycle. As with the life cycle principles that you saw in Chapters 7 and 10, the goal of green building is to conserve natural resources and minimize waste over the material life of the building. Third, the building itself is a production system. Buildings dynamically provide services for their occupants such as water, sewage disposal, light, fresh air, heat, cooling, and electricity. In this sense, the principles of industrial ecology can be applied to optimize the "production processes" of the building. It is not inconceivable that buildings could become nearly closed-loop systems in the future.

Green building is not the same thing as the ecological design discussed in Chapter 7, although ecological design can significantly enhance the green properties of a building. As we saw with the termite biomimicry used by Eastgate House in Zimbabwe, ecological principles can dramatically reduce the energy required by a building. Ecological design is also concerned with aesthetics in a way that is not represented in the practicalities of designing and building structures with improved environmental outcomes.

The fundamental challenge with green building is to meaningfully differentiate a building that is "green" from a conventional structure. Establishing a set of green characteristics and performance standards that can be applied to the actual practice of designing and constructing buildings is no easy task. The fundamental question is, *what actually makes a green building objectively more sustainable than one that is not?*

There is no uniform law or government-mandated "code" that defines what a green building is, or requires that new construction be built as such. What has evolved over time is **endorsement labeling**, a concept that is discussed in more detail in Chapter 12. Endorsement labeling is a type of voluntary certification in which an independent, not-for-profit, third party verifies that a product meets certain criteria in its construction, material content, performance, or operation. An example of such a third party is Underwriters Laboratories, a product safety certification organization more than 100 years old, with operations in over 40 countries.

Two voluntary green building endorsement/certification systems are most commonly used in the architecture and construction industries today, and both were first introduced in the 1990s. The **Leadership in Energy and Environmental Design (LEED)** program is used for commercial buildings, residential homes, and even entire neighborhoods. LEED was developed by the U.S. Green Building Council and has been adapted for use in a number of countries. In Europe, the **Building Research Establishment Environmental Assessment Method (BREEAM)** is well established in the UK and commonly used in Germany, the Netherlands, Norway, Spain, and Sweden. BREEAM similarly certifies a wide variety of buildings and communities. Lesser known green building rating systems include EarthCraft in the southeastern United States, Green Globes in North America, Green Star in Australia and New Zealand, the Comprehensive Assessment System for Built Environment Efficiency (CASBEE) in Japan, and LEED India, which offers a Green Factory certification.

All of these programs are based on recommended best practices, and buildings typically earn "credits" toward their certification for the practices or materials that they adopt. In sharp contrast, the Living Building Challenge is the only holistic performance-based green building certification system in the world, and it is inarguably the most sustainable set of green building criteria today. Unlike credit-based methods, the **Living Building Challenge** has mandatory requirements for certification, such as prohibitions against the use of particular materials and the requirement for projects to have "net zero energy" (which we discuss later in this chapter). The great variety of building-related NGOs in the world—as well as differences in geography and goals for sustainable development—make it difficult for a global green building standard to emerge. The Living Building Challenge is perhaps the most applicable globally, since it is a standards-based certification that can be implemented in many contexts.

A recent analysis by the *Architects Journal* in the United Kingdom suggests that LEED may be emerging as a preferred system for sustainable building certification programs; it dominates in many countries and regions compared to BREEAM, except for the United Kingdom (Figure 11.25). Nonetheless, what is also clear from Figure 11.25 is that very few projects get *formally certified* by a third party as a

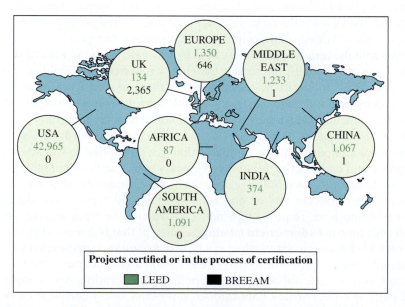

FIGURE 11.25 LEED and BREEAM projects internationally. A recent study in the United Kingdom suggests that LEED is a more common green building certification than BREEAM in many countries and regions.

Source: Based on Laura Mark, "LEED outstrips BREEAM across the globe—including Europe," Architects Journal, 28 February 2013, http://www.architectsjournal.co.uk/news/leed-outstrips-breeam-across-the-globe-including-europe/8643464.article

green building. This underscores the point that green building criteria are rigorous, *holistic* building requirements that may not be easily or inexpensively achieved. More commonly, new buildings are likely to adopt a smaller subset of green building features such as water conservation, renewable energy, or locally sourced building materials. The National Association of Home Builders consequently estimates that 40 to 50% of homes being built in the United States after 2012 will incorporate at least some aspect of more sustainable design. Tax breaks, energy-efficient mortgage programs, utility rebates, and state and local financial incentives are all mechanisms used today to encourage the use of more sustainable construction practices and building systems.

11.5.1 The LEED Rating and Certification System

The LEED rating system is sponsored by the U.S. Green Building Council. Different certification systems are available for commercial buildings (offices, libraries, churches, hotels), retail spaces (shops, banks, restaurants, big box stores), K-12 schools, health care facilities, residential homes, and neighborhood communities. The rating systems undergo periodic revision to reflect emerging best practices and to incorporate new techniques in the certification process. LEED recognizes that sustainable building criteria may vary regionally because of different environmental opportunities and challenges; regional councils consequently designate regionally specific green criteria that projects can earn bonus points for achieving.

All LEED certifications use the same approach. Each rating system has 100 base points, plus 10 bonus points for innovation in design and for regional priorities. A project is credited for the different green features it incorporates from the ratings list, and each feature earns a specific number of points. To be LEED-certified, a building must earn a minimum number of total points. A basic LEED certification requires a minimum of 40 points. Buildings that exceed the minimum standard can be rated as silver, gold, or platinum, where each successive tier requires more points and has increasingly more favorable environmental benefits (Table 11.5). Box 11.5 summarizes key green features incorporated in a LEED-Platinum university residence hall.

All of the rating systems also share five fundamental categories through which certification credits can be earned:

1. *Sustainable sites* criteria minimize a building's impact on ecosystems and water resources.
2. *Water efficiency* criteria enhance water conservation within the building and for its outdoor landscapes.
3. *Energy and atmosphere* criteria reduce energy consumption and promote the use of renewable energy.

TABLE 11.5 LEED 2009 Building Certification Levels

LEED RATING	EARNED POINTS
Certified	40–49
Silver	50–59
Gold	60–79
Platinum	80 points and over

Source: Based on US Green Building Council

BOX 11.5   A LEED-Platinum Residence Hall Renovation

FIGURE 11.26 Wayland Hall at James Madison University. The back side of Wayland Hall hosts native plantings and a bicycle shelter, both of which are seen in this photo. There is also a 38 m³ below-ground cistern that harvests rainwater for use in the building's toilet system. All of these features helped the building earn a LEED Platinum rating.

Photo by Scott Smith

Wayland Hall (Figure 11.26) is a dormitory on the campus of James Madison University in Harrisonburg, Virginia. At the time it was renovated, Wayland was one of only four full-scale LEED platinum residence halls in the country, and the first completely renovated residence hall to achieve platinum status under the USGBC's new construction and major renovation guidelines. The $11.6 million renovation transformed the 41,000-square-foot space into an innovative new living–learning community dedicated to the visual and performing arts. The renovated building includes a gallery, music practice rooms, an art studio, and a performance and exhibition room. All aspects of the renovation, including an ambitious reconfiguration of the bedroom spaces, are designed to encourage interaction, promote sustainable living, and expose students to the discipline and joy of the arts.

A variety of rigorous design strategies contribute to an expected 38% reduction in energy consumption and savings of over 5 million liters of water each year. Energy conservation features include a heat recovery unit that captures heat from exhaust air, window sensors that cut off heating and cooling when windows are open, drain water heat recovery, and a ground source heat pump. Notable contributions to water conservation include systems that collect, filter, and cycle rainwater for toilet flushing.

Site improvements along and behind the residence hall replace a parking lot with a series of landscaped terraces. The new design reduces impervious site cover, improves pedestrian connectivity, and provides new opportunities for residence life to extend outside. In all, the site design creates a stormwater management strategy that is as environmentally friendly as it is beautiful.

Source: Based on VMDO Architects, "James Madison University's Wayland Hall Renovation," http://www.vmdo.com/project.php?ID=26

4. *Materials and resources* criteria foster the use of sustainable building materials and reduced waste.
5. *Indoor environmental quality* criteria enhance the comfort of occupants with access to natural daylight, outdoor views, and better indoor air quality.

Depending on the rating system, the criteria within the categories may be slightly different, and additional categories may be included. For example, "Green Infrastructure and Buildings" is a credit category unique to the LEED Neighborhood Development certification.

Figure 11.27 lists the rating criteria and points for LEED New Construction and Major Renovations, a certification for commercial buildings such as offices, libraries, and hotels. The point system allows a building to earn credits for many of the smart growth, conservation design, and low-impact development principles that have already been introduced in this chapter. Under the Sustainable Sites criteria we see—for example—brownfield redevelopment, maximized open space, habitat restoration, and enhanced stormwater management. The point system also illustrates that green building principles extend beyond land-use criteria to include many other aspects of a building's materials, construction, operation, and comfort levels. Notably, there are prerequisites for several categories that require compliance with existing federal laws, industry standards, or conservation best practices. For example, all LEED-certified buildings must, because of the prerequisites, use 20% less water than they would under conventional design and comply with erosion and sedimentation laws.

The BREEAM rating system and certification operates in a similar manner to that of LEED, although the specifics are different. BREEAM likewise requires earning points, or credits, in a number of categories, which include management, health and well-being, energy, transport, water, materials, waste, land use and ecology, and pollution. BREEAM ratings range from 1 to 5 stars in contrast to the precious metal rankings of LEED.

11.5.2 Green Building and Land-Use Planning

Green building rating systems have the capacity to substantially enhance more sustainable land-use and urban planning. Both BREEAM and LEED certify neighborhoods and communities using a breadth of sustainable development criteria that promote conservation design and low-impact development. New developments and urban infill projects are both captured by the criteria, which can be used as guides for overall site planning design.

From a planning perspective, community green building rating systems can also significantly reduce demands on the urban infrastructure for water use, wastewater treatment, and solid waste disposal (Figure 11.28). Reduced energy consumption also enhances a smaller carbon footprint, an important outcome for meeting carbon mitigation goals. Boulder, Colorado, is a city that has aggressively incorporated mandatory green building requirements into its zoning law (Box 11.6).

In addition, the U.S. Environmental Protection Agency (EPA) provides a variety of resources related to green communities. For example, the *Sustainable Design and Green Building Toolkit for Local Governments* offers guidance to local governments about how to evaluate and revise existing codes and ordinances to support environmentally, economically, and socially sustainable communities. The Toolkit provides a checklist of key principles in categories that mirror those of LEED and BREEAM: Sustainable Sites and Responsible Land Use Development; Materials and Resource Conservation; Energy Conservation and Atmospheric Quality; Water Efficiency, Conservation, and Management; and Indoor Environmental Air Quality.

Sustainable sites — **26 possible points**

☑ Prerequisite 1	Construction activity pollution prevention	Required
☐ Credit 1	Site selection	1
☐ Credit 2	Development density and community connectivity	5
☐ Credit 3	Brownfield redevelopment	1
☐ Credit 4.1	Alternative transportation—public transportation access	6
☐ Credit 4.2	Alternative transportation—bicycle storage and changing rooms	1
☐ Credit 4.3	Alternative transportation—low-emitting and fuel-efficient vehicles	3
☐ Credit 4.4	Alternative transportation—parking capacity	2
☐ Credit 5.1	Site development—protect or restore habitat	1
☐ Credit 5.2	Site development—maximize open space	1
☐ Credit 6.1	Stormwater design—quantity control	1
☐ Credit 6.2	Stormwater design—quality control	1
☐ Credit 7.1	Heat island effect—nonroof	1
☐ Credit 7.2	Heat island effect—roof	1
☐ Credit 8	Light pollution reduction	1

Water efficiency — **10 possible points**

☑ Prerequisite 1	Water use reduction	Required
☐ Credit 1	Water efficient landscaping	2–4
☐ Credit 2	Innovative wastewater technologies	2
☐ Credit 3	Water use reduction	2–4

Energy and atmosphere — **35 possible points**

☑ Prerequisite 1	Fundamental commissioning of building energy systems	Required
☑ Prerequisite 2	Minimum energy performance	Required
☑ Prerequisite 3	Fundamental refrigerant management	Required
☐ Credit 1	Optimize energy performance	1–19
☐ Credit 2	On-site renewable energy	1–7
☐ Credit 3	Enhanced commissioning	2
☐ Credit 4	Enhanced refrigerant management	2
☐ Credit 5	Measurement and verification	3
☐ Credit 6	Green power	2

Materials and resources — **14 possible points**

☑ Prerequisite 1	Storage and collection of recyclables	Required
☐ Credit 1.1	Building reuse—maintain existing walls, floors, and roof	1–3
☐ Credit 1.2	Building reuse—maintain existing interior nonstructural elements	1
☐ Credit 2	Construction waste management	1–2
☐ Credit 3	Materials reuse	1–2
☐ Credit 4	Recycled content	1–2
☐ Credit 5	Regional materials	1–2
☐ Credit 6	Rapidly renewable materials	1
☐ Credit 7	Certified wood	1

Indoor environmental quality — **15 possible points**

☑ Prerequisite 1	Minimum indoor air quality performance	Required
☑ Prerequisite 1	Environmental tobacco smoke (ETS) control	Required
☐ Credit 1	Outdoor air delivery monitoring	1
☐ Credit 2	Increased ventilation	1
☐ Credit 3.1	Construction indoor air quality management plan—during construction	1
☐ Credit 3.2	Construction indoor air quality management plan—before occupancy	1
☐ Credit 4.1	Low-emitting materials—adhesives and sealants	1
☐ Credit 4.2	Low-emitting materials—paints and coatings	1
☐ Credit 4.3	Low-emitting materials—flooring systems	1
☐ Credit 4.4	Low-emitting materials—composite wood and agrifiber products	1
☐ Credit 5	Indoor chemical and pollutant source control	1
☐ Credit 6.1	Controllability of systems—lighting	1
☐ Credit 6.2	Controllability of systems—thermal comfort	1
☐ Credit 7.1	Thermal comfort—design	1
☐ Credit 7.2	Thermal comfort—verification	1
☐ Credit 8.1	Daylight and views—daylight	1
☐ Credit 8.2	Daylight and views—views	1

Innovation in design — **6 possible points**

☐ Credit 1	Innovation in design	1–5
☐ Credit 2	LEED accredited professional	1

Regional priority — **4 possible points**

☐ Credit 1	Regional priority	1–4

FIGURE 11.27 LEED 2009 rating criteria for new construction and major renovations.

Source: Based on US Green Building Council

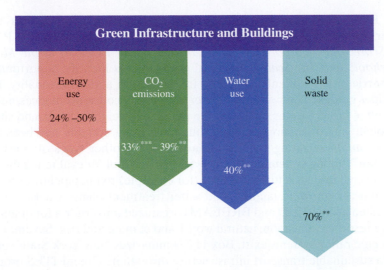

FIGURE 11.28 Environmental benefits from green infrastructure and buildings. LEED has a rating system for entire neighborhoods. When green infrastructure and green buildings are combined at this scale, the environmental benefits can be considerable.

Source: Based on USGBC (2009) LEED 2009 for Neighborhood Development Rating System. U.S. Green Building Council. Washington D.C.

BOX 11.6 Boulder, Colorado: Rocky Mountain Green

The city of Boulder, Colorado, is nestled in the foothills of the Rocky Mountains. It has a long-standing reputation as an environmentally progressive community. Boulder was an early pioneer of green building rating systems and is notably one of the few places in the world with rigorous, *legally mandated* green construction requirements. Boulder's green building initiatives started in the early 1980s with water and energy conservation provisions. In 1996, Boulder introduced Green Points, a rating system that scores green building attributes. Green Point performance criteria are now written into Boulder's building code. Very simply, the city will not issue a construction building permit for a project unless it earns the required number of points in the Green Point system, many of which are physically verified by building inspectors. Projects can earn points for energy efficiency, practices like xeriscaping (landscaping to save water and grow native plants), managing stormwater runoff, reusing existing buildings, and effective waste management

FIGURE 11.29

Source: © Steve Krull City Town & Street Scene Images / Alamy

during the construction phase. The Green Points program is mandatory for all new residential construction and remodeling/addition activities over 47 m^2.

The EPA also offers evaluation tools that enable communities to evaluate their green infrastructure and its effects on water quality. The EPA *Water Quality Scorecard: Incorporating Green Infrastructure Practices at the Municipal, Neighborhood, and Site Scale* helps local governments identify opportunities and remove barriers that will enhance green infrastructure and water quality. The tool also provides guidance on revising and creating zoning codes, ordinances, and incentives to better protect water quality at the municipal, neighborhood, and site levels.

Although our discussion of green building rating systems has been focused literally on *buildings*, rating systems are available for other types of constructions. The **Envision™ rating system** provides an assessment tool for evaluating the sustainability of all civil infrastructure, including (for example) roads, pipelines, power lines, telecommunication towers, landfills, and water treatment plants. Envision is a credit-based system (like LEED and BREEAM) organized into criteria for quality of life, leadership, resource allocation, natural world, and climate and risk. Several transportation-specific rating systems exist; Box 11.7 summarizes New York State's efforts to promote a sustainable transport infrastructure through its GreenLITES program.

BOX 11.7 GreenLITES and Transportation Sustainability

FIGURE 11.30

Source: N.Y. Department of Transportation, https://www.dot.ny.gov/programs/greenlites/repository/Evergreen.jpg

Transportation infrastructure is a significant form of land use and has major impacts on air quality, water quality, and natural resource consumption. Transportation agencies recognize these problems and have responded with the green transportation rating systems modeled after LEED. The New York State Department of Transportation designed GreenLITES (Leadership in Transportation and Environmental Sustainability), which is the first U.S. rating system for transportation projects.

GreenLITES helps New York measure performance, recognize good practices, and identify areas for improvement. The tool reflects a points system in five categories—sustainable sites, water quality, materials and resources, energy and atmosphere, and innovation/unlisted. Because GreenLITES extends beyond just road projects, there are a large number of credit opportunities (more than 200), but not all credits are necessarily relevant or available for a particular project. The rating system classifies successful projects as Certified, Silver, Gold, or Evergreen depending on the total number of points earned.

The first GreenLITES awards honored four Evergreen projects: three highway projects and one greenway/multiuse trail. A notable award was for the reconstruction of three miles of New York State Route 30. This highway segment is located along the Adirondack Trail Scenic Byway and is bordered by environmentally sensitive wetlands, a forest preserve, and Tupper Lake. The project included multiple environmentally and socially sustainable design elements including:

- A wider shoulder along the side of the road that balances the needs of the traveling public and of environmentally sensitive Adirondack ecosystems.
- Fencing installed along wetland borders to protect turtles from the trafficway.
- Relocated utility lines underground to enhance scenic views along the highway and at four designated overlooks.
- A closed storm drainage system to capture sediment and reduce flow of pollutants into the lake.

(Continued)

Green transportation rating systems modeled after GreenLITES include the Illinois Livable and Sustainable Transportation (I-LAST) Rating System and the adaptation by Pennsylvania Turnpike Commission.

Two additional green transportation rating systems are of interest because of their unique features. The BE²ST rating system, created in partnership with the University of Wisconsin, uses qualitative measures to first screen road projects and then rate projects with quantitative measures. BE²ST is also unique because it incorporates both environmental life cycle assessment and economic life cycle cost analysis. STEED (Sustainable Transportation Environmental Engineering and Design) was created by the private consulting firm Lochner Engineering and is a checklist for roadway projects that can track how the sustainability of a project changes throughout the engineering design cycle.

11.6 Energy Use and Buildings

Green building rating systems like LEED and BREEAM are holistic approaches to the built environment, incorporating the sustainability of a building on its site, the surrounding landscape, water usage, material inputs, energy consumption, and interior comfort. As seen earlier, few projects make the effort to become certified as a green building relative to the total amount of new construction that takes place each year.

However, because of the extraordinary environmental, social, and economic issues associated with climate change, considerable effort is made to reduce the impacts of building energy use as a specific component of green building. According to the International Energy Agency, about 40% of all primary energy in the world's industrialized countries is consumed by buildings; this represents space heating, water heating, and other energy production within the building itself. It does not include electricity, which arrives at buildings as a secondary form of energy from power plants. In the United States, it is estimated that buildings account for 70% of the country's electricity use. Because the vast majority of energy produced and used in the world derives from fossil fuels, building energy efficiency and conservation is critical to sustainable development in the built environment.

Residential, commercial, and office buildings all use energy for the same basic energy services, often called *loads* when referring to the amount of energy each of these needs consume. Common energy needs include space heating, cooling, hot water production, ventilation, lighting, refrigeration, and electrical appliances and devices. What varies among types of buildings is how much energy is accounted for by different loads. For example, about half of the energy used by commercial buildings in the United States is for heating, cooling, and lighting. In contrast, these three loads account for almost two-thirds of residential energy use in the United States (Figure 11.31). Reducing building energy use requires that we understand how different kinds of structures use energy so that we can implement effective strategies for energy efficiency and conservation.

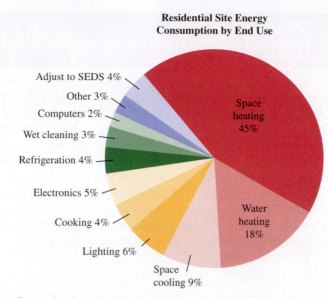

FIGURE 11.31 Energy loads in the U.S. residential sector. U.S. residential energy consumption is heavily driven by the need for space heating and cooling, which accounts for just over half of home energy use nationally. Water heating is significant as well, and these three energy services combined represent almost three-fourths of household energy use.

Source: US Department of Energy, http://buildingsdatabook.eren.doe.gov/ChapterIntro2.aspx

11.6.1 Strategies for Building Energy Conservation

Improved energy use in buildings is largely accomplished in four ways. First, we can conserve energy by virtue of a building's architectural design. For example, Eastgate House in Zimbabwe was designed for natural convection like that of termite mounds, significantly reducing its energy needs for cooling. Similarly, the use of skylights and careful placement of windows provide natural sunlight during the day, reducing the need for artificial light. This design strategy is known as **daylighting** and uses a variety of products and architectural features. **Passive solar design** is likewise a form of architecture that takes advantage of solar energy to help heat homes. Design strategies can be simple or extraordinarily innovative and complex; Figure 11.32 shows 1 Bligh Street, a skyscraper in Australia that has a double-skin glass facade that provides insulation, ventilation, and daylighting for the building.

Second, energy conservation can be achieved by our choice of construction materials and the structure of a building's exterior envelope. Because space heating and cooling represent such significant energy loads for buildings, constructing a tight thermal layer will moderate heat loss and gain between the indoors and outdoors. Double- and triple-paned windows, careful weatherstripping, and substantial insulation in walls, attics, and foundations are examples of ways in which we conserve building energy through the envelope. **Structural insulated panels (SIPs)** are an advanced building product that significantly lower energy use in homes and small commercial buildings. SIPs are used instead of framing lumber and plywood for the sheathing of a house (Figure 11.33) and provide higher insulating values for a building compared to conventional construction. Although SIPs cost more from a material standpoint, they shorten construction time considerably and thus save on labor costs.

FIGURE 11.32 1 Bligh Street, Sydney, Australia. This skyscraper in Sydney has a double skin glass facade. Two layers of glass about 0.6 m apart comprise the exterior envelope of the building, providing natural daylight and insulation for the interior spaces. An innovative cooling system is incorporated into the glass skin as well. 1 Bligh Street reportedly has 42% less CO_2 emissions than an office complex of similar size.

Source: © Martin Cameron / Alamy

Third, energy savings can be realized by selecting highly energy-efficient equipment, appliances, and devices in the building. It is especially critical to optimize heating and cooling given their building energy loads, as well as lighting in commercial structures. Systems requiring motors for fans or pumps—such as air handling and water supply—also need to be considered. Motors

FIGURE 11.33 Structural insulated panels (SIPs) are construction panels that contain a sandwiched layer of highly insulating foam board. They interconnect at their edges, creating a snug joint and reducing air infiltration. SIPs replace standard framing construction techniques for the exterior envelope of a building, and when installed correctly, they substantially increase the thermal barrier. SIPs are often needed to achieve the energy performance targets required by various energy efficiency certification programs.

Source: © Streeter Photography / Alamy

with large horsepower and long **duty cycles** can consume a tremendous amount of energy. (Duty cycle refers to the length of time a device is operating over a defined period.) The efficiency of equipment can vary notably; engineers and designers must carefully select the systems that create the major energy loads in a building.

Efficiency in buildings is enhanced by dynamic controls. For example, occupancy and photo sensors adjust lighting by turning lights off when rooms are vacant and by actively adjusting lighting levels during the daytime relative to the amount of available daylight in a room. CO_2 sensors allow air-handling equipment to ventilate a room relative to the actual air quality, instead of replacing large volumes of air in rooms that may only be minimally occupied. Heat recovery systems are also ways to achieve greater system efficiencies within a building; heat recovery techniques are increasingly used for water heating, space heating and cooling, and combined heat and power systems. Microturbine technology in particular is significantly increasing the opportunity for small, on-site combined heat and power systems, which is an asset for large office buildings and multifamily housing in particular.

Finally, the carbon footprint of building energy can be reduced by using renewable energy. Photovoltaic panels can actively provide on-site electricity, as will wind turbines if the wind resource is substantial enough. Solar hot water heating and solar air heating systems (Figure 11.34) are both highly cost-effective

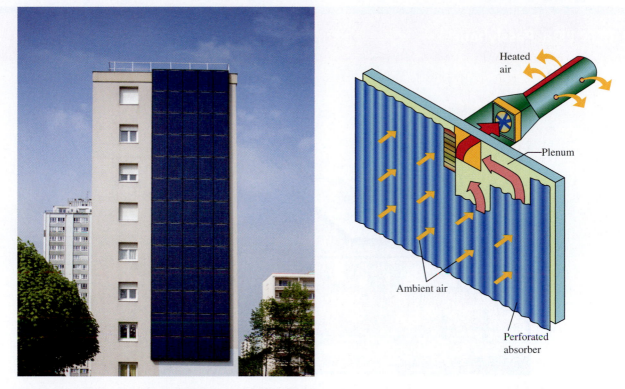

Heated air

Plenum

Ambient air

Perforated absorber

FIGURE 11.34 Solar air heating systems, such as the plenum on this apartment building, use passive solar heating and active ducting to provide space conditioning inside buildings.

Sources: Solar wall: Hadrian/Shutterstock.com; Solar Air Heating Systems: Based on http://www.ecobuildtrends.com/2011/02/solar-air-heating.html

solar technologies. When renewable energy technologies are used to substitute, in part, for a building's conventional materials, we refer to it as a **building integrated system**. The solar air system in Figure 11.34 is an example of a building-integrated heating system.

Combinations of these four energy conservation strategies can achieve dramatic reductions. The Passivehaus concept (Box 11.8) is a good case in point, as these buildings rely on passive solar gain, the body heat of their occupants, and the heat generated from equipment and appliances to provide virtually all of the space heating that a Passivhaus structure requires. However, energy conservation measures *do* interact with one another, and their interactive effects represent an instance in which "one plus one does not equal two." If we implement three different energy efficient technologies in a building, each of which individually might contribute a 10% energy savings, the net savings from all three combined will not be 30%. For example, we can use daylighting techniques to reduce the demands for electricity for lighting during the daytime, but more (or larger) windows will also generate more heat loss during cooler seasons, thus requiring more energy for heating. As a consequence, the total benefit from multiple energy efficiency and conservation efforts must be carefully modeled to understand how much energy will be saved overall. Many commercial software

BOX 11.8 Passivhaus

FIGURE 11.35 Although Passivhaus buildings are often strikingly modern in design, they can also be constructed in traditional architectural styles, such as this row of terraced houses in Houghton-le-Spring, United Kingdom.

Source: © Ashley Cooper pics / Alamy

Passivhaus construction is a voluntary, certified low-energy building that has been implemented most widely in Germany and Scandinavia. More than 20,000 Passivhaus structures—including homes, schools, and offices—have been built in these countries. A Passivhaus building (Figure 11.35) has the rather startling characteristic of needing no major active heating system—passive solar gain, as well as the heat generated by a building's occupants and appliances, are all that a structure constructed to Passivhaus standards usually requires for comfortable space heating during most of its heating season. High insulation values, passive solar design, very low air infiltration, and other properties help a Passivhaus achieve this level of heating performance. Because of other energy efficiency measures incorporated into the building, Passivhaus buildings use 80 to 90% less energy than their conventional equivalents.

products are available to make such calculations in the design stage for the building. Robust, but preliminary, estimates can be made with free software such as DOE-2, eQuest, and RETScreen.

11.6.2 The Role of Energy Building Codes

The minimum energy efficiency performance of residential and commercial buildings can be legally mandated by building codes in the United States. These codes set the baseline energy efficiency and consumption that buildings must achieve in order to comply with local construction laws. Different requirements will apply for commercial and residential structures. Codes affect the building envelope, energy-consuming equipment, and the energy consumption of a building's overall systems, such as ventilation. Some countries may have national building codes; in the United States, individual states and localities are legally responsible for establishing and enforcing code requirements for building construction and energy performance. The vast majority of states in the United States have energy conservation codes on the books and also enforce them; as of 2012, only 10 states had either no codes at all or limited codes and enforcement.

Two different nonprofit organizations shape energy codes in the United States. Energy codes are highly technical, time consuming to develop, and usually outside the scope of the expertise of government entities. By designing detailed model codes, third-party organizations help to improve the level of energy efficiency in American buildings and also help to harmonize (make equivalent) energy conservation requirements around the country.

The International Code Council develops a variety of model building codes for state and local government. The Council sponsors the model **International Energy Conservation Code (IECC)** that covers both residential and commercial structures. Most states and/or communities in the United States use the IECC model code in their own ordinances. The IECC is updated about every three years through a public process to account for improvements in technology, best practice construction techniques, and climate change.

The other building energy standards organization is the **American Society of Heating, Refrigeration, and Air Conditioning Engineers** (**ASHRAE**). ASHRAE produces the ANSI/ASHRAE/IESNA Standard 90.1; *Energy Standard for Buildings except Low-Rise Residential Buildings*. This energy conservation standard applies to all commercial buildings except multifamily residences that are three stories high or less. State and local governments can also adopt the ASHRAE 90.1 standard for their energy commercial building code; because the IECC also uses ASHRAE 90.1 by reference, compliance with ASHRAE 90.1 means that a building is also in compliance with the IECC for commercial structures. Like the IECC, ASHRAE 90.1 is revised about every three years. LEED certified commercial buildings must, at a minimum, comply with ASHRAE 90.1 standards.

Building energy codes vary to allow for differences in climate and the cost effectiveness of different energy conservation measures. **Heating degree days** is the metric used by the IECC to distinguish broad and narrow geographic variations in climate; many aspects of the energy code will vary based on the designated climate zone (Figure 11.36). Degree days represent an estimate of the amount of energy required to heat or cool a building; heating degree days are calculated by summing the extent to which the average outdoor temperature deviates by 1 degree from a baseline temperature, below which it is assumed a building will need heat. In the United States, this baseline temperature is 18°C. To illustrate, a winter day with an average daily temperature of −2°C will have 37 heating degree days for that specific day. Climate zones are calculated by adding up the total number of heating degree days for the entire year, which can be a few hundred days in a warm climate or thousands in a cold area.

11.6.3 "Beyond Code" Energy Rating Systems and Models

Building energy codes simply set the *minimum* energy conservation characteristics of a building. As with green building rating systems, it is also possible to certify that

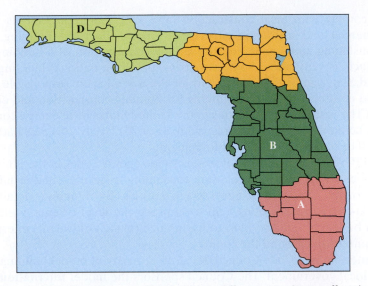

FIGURE 11.36 IECC climate zones in Florida. Subtle differences in climate affect the International Energy Code. This map shows the four climate zones in Florida designated by the 2003 IECC. These variations contribute to different details in the actual requirements of the code, ranging from window selection to insulation levels and equipment efficiency.

Source: Based on IECC 2003

a building exhibits a level of energy efficiency that is significantly greater than that required by building energy codes. These building energy rating systems are *not* the same as LEED or BREEAM, because they evaluate only the energy characteristics of buildings, not their entire potential environmental impacts. Such rating systems are often referred to as **beyond code programs**. There are several third-party building energy efficiency rating systems; the most widely known program is that of the Energy Star building certification system sponsored by the U.S. Department of Energy. It is a surprise to many people that the Energy Star endorsement labeling program not only certifies appliances and equipment, but also residential and commercial buildings.

The Residential Energy Services Network (RESNET) in the United States has also developed a standardized scoring tool to rate a home's overall energy consumption, known as the Home Energy Rating System (HERS). RESNET conducts highly detailed analyses of a residence's energy conservation and standardizes the resulting evaluation to a 100-point rating index (Figure 11.37). The HERS score allows consumers to compare the energy properties of different homes, and may also enable them to qualify for a more favorable loan known as an energy efficient mortgage.

Internationally, the Passivhaus (Box 11.8) is a formally certified structure with respect to energy conservation; the Passive House Institute sets the standards that are used worldwide for both residential and nonresidential structures. Passivhaus standards govern the requirements for heating systems within a building, as well as limit the building's total energy use. Passivhaus structures fall into a category known as *ultra-low energy buildings* or *nearly zero energy buildings*. An actual **net zero building** exhibits no *net* energy consumption and therefore also has a net zero carbon footprint. Also known as zero energy buildings, these structures typically

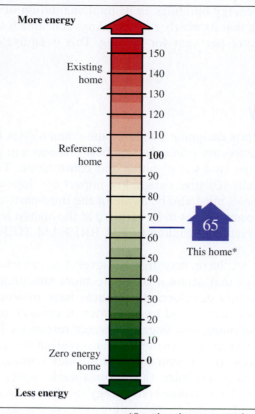

More energy

Existing home

Reference home

This home*

Zero energy home

Less energy

*Sample rating representation.

FIGURE 11.37 The Home Energy Rating System Scale. The HERS index evaluates the energy performance of a home and converts it to a standardized index. This allows people to meaningfully compare housing choices.

Source: Based on RESNET, Home Energy Rating, http://www.resnet.us/energy-rating

integrate passive solar design, solar thermal hot water heating, highly energy efficient technologies, and active solar PV or wind energy systems that generate clean electricity.

The net zero building principle is that over the course of a year, a building will generate as much carbon-free energy as it uses. As a consequence, the building in theory has no net impact on energy resources or CO_2 levels. However, there is ambiguity in the net zero building concept, because a building can be net zero at its site or with respect to its source emissions. Net zero site energy use represents a building that generates as much energy as it actually consumes as an end use on its site. Zero net source energy, however, means that a building would have to compensate for the emissions required to generate the energy used by a building at its source, such as the electric power plant from which a structure would get its electricity when the building's own power system does not generate enough. The United States now requires that all new federal buildings which begin their design process in 2020 or later must be net zero buildings. The European Union has similarly directed its member states to require that all new buildings after 2020 must be nearly zero energy buildings; in its implementation of this EU directive, Germany is requiring that its nearly zero buildings consume less than 15 kilowatt hours per square meter per year for heating. This is equivalent to Passivhaus standards.

11.7 Summary

Engineers and architects designing buildings and communities in the future must understand the contemporary principles and best management practices for more environmentally benign land-use planning and construction. This includes smart growth, environmentally sensitive design, low-impact development, and conservation design. Professionals must also be aware of the third-party rating systems that allow structures and communities to be certified at the highest levels of energy and environmental conservation, including LEED, BREEAM, HERS, Passivhaus, and net zero buildings.

In this chapter, we have investigated several approaches to land development and buildings that attempt to create more sustainable infrastructure. Traditional infrastructure development patterns have produced low-efficiency patterns of development that lead to high rates of erosion, urban sprawl, inefficient energy consumption, and degraded water resources. Development patterns that utilize smart growth planning, conservation design principles, best management practices for erosion and stormwater management, and green building techniques will minimize the unsustainable water resource impacts discussed in Chapters 2 and 4, conserve energy (as discussed in Chapter 8), and mitigate the associated emissions from energy production and use (Chapter 3). The industrial ecology principles and life cycle assessment tools described in Chapters 9 and 10 can help us assess not only the low-impact development techniques described in this chapter, but they can also help us rethink the ways buildings use energy.

Reducing the land-, resource- and energy-intensive nature of new urban development may have significant economic, environmental, and social benefits for a planet already witnessing a significant shift in climate and weather patterns. Future infrastructure investments will need to be more sustainable and more resilient to meet the demands of growing populations and limited available resources. Green building standards, conservation design principles, and energy efficient construction all involve good engineering practices that reflect the following:

- Identifying and providing buffers and setbacks associated with environmentally sensitive areas
- Limiting disturbances and development in floodplains
- Clustering development to minimize impervious surfaces
- Incorporating open spaces and low-impact development into site design
- Achieving the highest levels of energy conservation in new construction

Key Concepts

Built environment	Urban sprawl
Land-use and land-cover change	Mixed use
Development	Population density

Hollowing out
Single-use zoning
Housing density
Zoning ordinances
Building envelope
Right-of-way
Easement
Setback
Angle-of-bulk plane
Floor area ratio (FAR)
New Urbanism
Smart growth
Urban intensification
Urban infill
Urban growth boundary
Form-based codes
Brownfield
Cluster development
Environmentally sensitive design
Environmental impact assessment
Impervious surfaces
Green infrastructure
Best Management Practices (BMPs)
Pervious pavers
Vegetated swales
Stream daylighting
Riparian buffer zone

Gray water
Geotextile
Stokes Settling law
Conservation design
Density neutral
Endorsement labeling
Daylighting
Passive solar design
Structural insulated panels (STPs)
Duty cycle
Building integrated system
International Energy Conservation Code (IECC)
Leadership in Energy and Environmental Design (LEED)
Building Research Establishment Environmental Assessment Method (BREEAM)
Living Building Challenge
Heating degree day
Beyond code program
Net zero building
Envision rating system
American Society of Heating, Refrigeration and Air Conditioning Engineers (ASHRAE)

References

Arendt, R. G. (1996). *Conservation Design for Subdivisions: A Practical Guide to Creating Open Space Networks.* Washington, D.C. Island Press.

Beaulieu, K. M., Bell, A. H., and Coles, J. F. (2012). "Variability in Stream Chemistry in Relation to Urban Development and Biological Conditions in Seven Metropolitan Areas of the United States, 1999–2004." National Water-Quality Assessment Program, U.S. Geological Survey. *Scientific Investigations Report* 2012-5170.

BRE Global. (2011). BREEAM®: The world's foremost environmental assessment method and rating system or buildings. Bucknalls Lane, Watford, UK. www.breeam.org.org

Burchell, R. W., and Mukhrji, S. (2003). "Conventional development versus managed growth: The costs for sprawl." *American Journal of Public Health* 93(9): 1534–1540.

Cassidy, R., Wright, G., Flynn, L., Barista, D., and Zissman, M. (2003). Building Design & Construction White Paper on Sustainability. Building Design and Construction. A Reed Business Information publication. Oak Brook, IL.

Center for Watershed Protection. (1998). *Better Site Design: A Handbook for Changing Development Rules in Your Community.* Ellicott City, MD.

Center for Watershed Protection. (2000). *Maryland Stormwater Design Manual.* Volumes I & II. Water Management Administration, Maryland Department of the Environment, Baltimore.

City of Boulder. Understanding Density and Floor Area Ration: Examples of Residential, Mixed Use, Commercial and Industrial Developments in Boulder. Accessed 1 January 2014 at http://www.lcog.org/documents/meetings/coburgurb/0309/Attachment4.pdf

City of Portland. (2004). *Stormwater Management Manual*: Revision #3. Environmental Services, Clean River Works, Portland, OR.

Fairfax County. (2012). Zoning Ordinance. www.fairfaxcounty.gov/dpz/zoningordinance.

Frischknecht, R., Jungbluth, N., Althaus, H.-J., Doka, G., Heck, T., Hellweg, S., et al. (2007). Overview and Methodology. Ecoinvent report No. 1. Swiss Centre for Life Cycle Inventories, Dübendorf.

GSA Public Building Service. (2008). Assessing green building performance. A post occupancy evaluation of 12 GSA buildings. Washington D.C. U.S. General Services Administration

Hinman, C. (2005). *Low Impact Development: Technical Guidance Manual for Puget Sound.* Puget Sound Action Team, Olympia, WA.

Hirschman, D. J., and Kosco, J. (2008). Managing Stormwater in Your Community: A Guide for Building and Effective Post-Construction Program. Ellicott City, MD.

Homer, C. H., Fry, J. A., and Barnes, C. A. (2012). The National Land Cover Database. U.S. Geological Survey Fact Sheet 2012-3020.

Howard, J. L. The Federal Commitment to Green Building: Experiences and Expectations. Federal Environmental Executive. Accessed 1 January 2014 at http://www.epa.gov/pdf/2010_fed_gb_report.pdf

Kats, G. (2003). The Costs and Financial Benefits of Green Building. A report to California's Sustainable Building Task Force.

Keller, G., and Sherar, J. (2003). Low Volume Roads Engineering: Best Management Practices Field Guide. USDA Forest Service/USAID, Washington, D.C.

Krekeler, P., and Kolb, E. (2010, May 29). NYSDOT (New York Department of Transportation), GreenLITES Program. EPA Webinar Series.

Lewis, P. G. and Neiman, M. (2002). Cities under Pressure: Local Growth Controls and Residential Development Policy. Public Policy Institute of California.

Maryland Office of Planning (1995). Achieving Environmentally Sensitive Design in Growth Areas Through Flexible and Innovative Regulations. Baltimore: Maryland Office of Planning.

Marion County, Indiana. (2010) Marion County Zoning Ordinance, Chapter 731: Dwelling Districts (DDZO) Zoning Ordinance & Subdivision Control Ordinance. www.indy.gov/eGov/City/DMD/Planning/Zoning/Pages/municode.aspx.

McCuen, R. H. (2004). *Hydrologic Analysis and Design*, 3rd ed. Upper Saddle River, NJ: Pearson-Prentice Hall.

Norman, J., MacLean, H., and Kennedy, C. (2006). "Comparing high and low residential density: Life-cycle analysis of energy use and greenhouse gas emissions." *J. Urban Plann. Dev.*, 132(1), 10–21.

Schueler, T. R., Fraley-McNeal, L., and Cappiella, K. (2009, April). "Is impervious cover still important? Review of recent research." *Journal of Hydrologic Engineering*, 309–315.

Spokane County Public Works Department. (2004). Zoning Code. Spokane County, Spokane, WA.

State of Florida. (2009). *Drainage Handbook: Open Channel*. Office of Design, Drainage Section. Department of Transportation, Tallahassee, FL.

Thomas, J. V. (2009). *Residential Construction Trends in America's Metropolitan Regions*. Development, Community, and Environment Division, U.S. Environmental Protection Agency.

Turner, C., and Frankel, M. (2008). Energy Performance of LEED for New Construction Buildings. Final Report.

USDA. (1986). *Urban Hydrology for Small Watersheds: TR-55*. 2nd ed. Conservation Engineering Division, Natural Resources Conservation Service, United States Department of Agriculture.

USDA. (2009). *2007 National Resources Inventory—Development*, Natural Resources Conservation Service, Washington, D.C., and Center for Survey Statistics and Methodology, Iowa State University, Ames, Iowa.

U.S. EPA. (2009). Essential Smart Growth Fixes for Urban and Suburban Zoning Codes. U.S. Environmental Protection Agency.

U.S. Department of Energy. (2010). Building Energy Codes 101: An Introduction. Department of Energy, Office of Energy Efficiency and Renewable Energy, Washington, D.C.

U.S. GBC. (2009). LEED 2009 for Neighborhood Development Rating System. U.S. Green Building Council, Washington, D.C.

U.S. GBC. (2013). LEED for Neighborhood Development and Historic Preservation. U.S. Green Building Council, Washington, D.C.

Wildlife Conservation Society. (2013). "Conservation development has some developers thinking—and seeing—green." *ScienceDaily*. Retrieved March 11, 2013, from www.sciencedaily.com/releases/2013/03/130305130449.htm.

Active Learning Exercises

ACTIVE LEARNING EXERCISE 11.1: Assess a comprehensive plan

Explore the planning requirements in your community by looking at its comprehensive plan for future land use and current zoning laws regarding new residential construction. Also explore whether there is a green infrastructure plan. Analyze and discuss how well your community may practice or advocate smart growth and environmentally sensitive development.

ACTIVE LEARNING EXERCISE 11.2: Debate green building rating systems

Organize your class into teams, and debate which green building rating system is better: LEED, BREEAM, or the Living Building Challenge.

ACTIVE LEARNING EXERCISE 11.3: Design beyond code

Assume that you are planning to build an 170 m² home where you are currently located and that you would like this home to be "beyond code." Develop a set of design and construction criteria that you will use for your new house.

ACTIVE LEARNING EXERCISE 11.4: Sketch out green infrastructure

Obtain an aerial photograph of the region in which your college campus is located. Sketch a potential green infrastructure system for this area.

ACTIVE LEARNING EXERCISE 11.5: Compare and contrast LEED

Compare and contrast LEED green building criteria with the Envision™ sustainable infrastructure rating system. How are these similar and different? Does one system do better than another in capturing what we mean by "sustainability" in their rating criteria? Explain your answer.

Problems

11-1 Identify and explain three to four factors that contribute to sprawl.

11-2 Identify and explain three to four strategies intended to counteract sprawl.

11-3 Why are sprawl and other forms of low-density development unsustainable?

11-4 How do form-based codes differ from traditional zoning codes?

11-5 List the 10 smart growth principles.

11-6 Why might communities be resistant to implementing smart growth land-use practices?

11-7 Which of the following statements about cluster developments is FALSE?
a) They provide for more open green space.
b) They have a lower housing density than a conventional subdivision.
c) They don't necessarily implement the features of low-impact development.
d) They often require a change in local zoning codes.

11-8 What does "let the land shape your plan" mean?

11-9 What are the benefits of large expanses of undeveloped or agricultural land?

11-10 Summarize the environmental impacts and consequences of the built environment for land cover and water quality.

11-11 A watershed with 40% impervious cover would probably have what kind of stream quality?

11-12 Explain the two basic design features of green infrastructure.

11-13 Identify and explain five different BMPs for low-impact development.

11-14 How is low-impact development different from conservation design?

11-15 What is green building?

11-16 The Seattle Library was an innovative green building project launched in 1998. The building design included considerations for sustainable design as described in the report entitled "Libraries For All," available on the Seattle Library website (*www.spl.org/Documents/about/libraries_for_all_report.pdf*). Describe how the design of the **Seattle Central Library** incorporates sustainable design elements in each of the following areas.
a. Size and placement
b. Building/material choices
c. Landscape and ecology
d. Energy
e. Transport
f. Water

11-17 You are designing a building on 0.4 acre. It will be three stories high, and each story will have about 12,000 ft². Determine the development density descriptor from Table 11.1 or 11.2 in Spokane County or Marion County (as selected by your instructor) for the proposed building.

11-18 Explain what type of development—one house per acre with 20% impervious cover or eight houses per acre with 65% impervious cover—generates more stormwater and why.

11-19 Research the local development planning policies and codes for your city or county, or use the 2008 Countywide Planning Policies for Spokane County, (*www.spokanecounty.org/BP/data/Documents/CWPP/cwpp.pdf*) as assigned by your instructor.
a. Do the Planning Policies encourage open space preservation in concept? Explain.

FIGURE 11.38 Illustration for problem 11.25.

Source: USDA – Farm Service Agency

     b. Do the Planning Policies encourage open space preservation in practice? Explain.

     c. Do the Planning Policies follow conservation design theories? Which procedures are utilized and which are not in the growth management act?

11-20 Draw a building envelope for low-density residential zoning described in Table 11.1 with six homes per acre

     a. What is the minimum lot size?

     b. What is the maximum square footage of the home

     c. For a home with a three-car garage, what would be the remaining square footage of the first floor?

11-21 How do conventional buildings and green buildings differ?

11-22 What are the LEED rating categories and possible points? Which rating categories have greater weights (e.g., more credits are possible)? Why might this be so? Do you think the categories are unbalanced? Why or why not?

11-23 Refer to the Fossil Ridge High School LEED summary on the textbook website.
   a. Estimate and assign LEED points for the Fossil Ridge High School design.
   b. In what areas of design did "green building" increase costs the most?
   c. In what areas of design did "green building" result in the greatest potential savings?

11-24 Use Google Earth to select a potential residential development site in your area or the aerial photo shown in Figure 11.38 or 11.39. Draw a yield plan if the property is divided into
   a. 0.25 hectare lots
   b. 0.5 hectare lots
   c. 2 hectare estates

11-25 From the aerial map or photo developed in Problem 11-24, determine the following
   a. Identify primary conservation areas and briefly label and explain them.
   b. Identify secondary conservation areas and briefly label and explain them.
   c. Identify potential development areas and briefly label and explain them.
   d. Locate potential housing sites.
   e. Locate transportation routes.
   f. Identify lots using conservation design principles and maintaining 30% open space.
   g. Does your conservation design look different from the yield plan in Problem 11-24?
   h. Compare and contrast the advantages and disadvantages of the yield plan (Choose parts (a), (b), or (c) from Problem 11-24) and the conservation base design.

11-26 What are the four basic ways we can improve energy use in a building?

11-27 Refer to Figure 11.31. Based on this graph, a 10% reduction in which category of energy end use would yield the greatest energy savings?

11-28 Refer to Figure 11.31. Discuss how you might reduce energy consumption for refrigeration, lighting, cooling, heating, and water heating through various green building strategies.

11-29 What is a heating degree day, and how is it relevant to building codes?

11-30 Discuss how the function of a building and the behavior of the occupants might influence the appropriate exterior design temperature of the following.
   a. Office building
   b. Gymnasium
   c. Hospital

11-31 Lighting can account for up to what percentage of a building's electricity usage?

11-32 How much energy could be saved by replacing five 60-W incandescent bulbs each of which burns 4 hours per day with 15-W compact fluorescent light bulbs?

Open-Ended Design Problems

11-1 Consider a home that uses the equivalent of 16,000 kWh per year for the following energy loads.
   • Domestic hot water: 3,500 kWh
   • Refrigeration: 1,300 kWh

FIGURE 11.39 Illustration for Design Problem 11-3.

Source: USDA – Farm Service Agency

- Lighting: 1,500 kWh
- Heating: 4,800 kWh
- Cooling: 2,400 kWh

Make this house a net zero home.

11-2 Visually evaluate the broader neighborhood, community, and region in which your college campus is located using Google Earth, Google Maps, or other aerial imagery. Consult the SmartCode document, as well as other resources on the rural-to-urban transect and form-based codes. Select an area with a spatial scale of at least 2 km² develop a form-based code for it, and include visual diagrams as reference. How does your code compare to the existing comprehensive plan, zoning, and ordinances for this location?

11-3 The site shown in Figure 11.39 is an area northwest of Harrisonburg, Virginia (38°27"48.13" N, 78°54"25.33"W), that is slated for development. The area for development is 160 hectares. A competitor in the land development field has told the land owner that he is willing to pay $50,000 per hectare if the land can be rezoned to 0.1 hectare lots. He expects to sell homes on the property for four times the land cost and make 25% of total sales to be profit/revenue. However, the Sierra Club, local home owners, and other interested parties have told the land owner that they would sue and seek a court injunction to prevent development if the zoning were changed. They say that at least 4 hectare lots are required under the current zoning code. They are interested mainly in protecting the (real) high value green corridor and habitat that make up 20% of the property. This habitat is used by a wide variety of migratory birds, including some potentially endangered species. A friend of the land owner is also an acquaintance of yours and is familiar with the work you've done in green building. You've been requested to provide an alternative plan and an estimate for the price per hectare your company would be willing to pay. As part of your presentation, be sure to address the following issues.

a. Yield under existing zoning codes and those proposed by the investor
b. Definition of a sustainable design
c. Conservation areas
d. Potential differences in a "green design" approach compared to traditional zoning and development

FIGURE 11.40 Illustration for Design Problem 11-4.

Source: USDA – Farm Service Agency

FIGURE 11.41 Illustration for Design Problem 11-4.

Source: USDA – Farm Service Agency

FIGURE 11.42 Illustration for Design Problem 11-4.

Source: USDA – Farm Service Agency

   e. Definitions of green building characteristics, including identifying those most relevant to this site
   f. Explanation of how you would incorporate LEED practices into your development plan and construction methods
   g. Estimate of your lot size
   h. Relation of your lot size to current zoning codes and explanation of your proposal
   i. Estimated cost(s) of homes on the proposed property
   j. Estimate of amount your company would be willing to pay per hectare

11-4 The site shown in Figures 11.40, 11.41, and 11.42 is in Washington State, and we will assume that the area is slated for development. The area for development is 160 hectares. Suppose the Sierra Club, local home owners, and other interested parties have told the land owner they would sue and seek a court injunction to prevent development if the zoning were changed from agricultural to mixed-use residential. Pretend that at least 4 hectare lots are required under the current zoning code. They are interested mainly in protecting the (real) high value green corridor and wetlands that make up 20% of the property. These wetlands are used by a wide variety of migratory birds, including, potentially some endangered species. Your company has been selected over competitors to develop this site based on your preliminary submittal. You've been requested to provide more development details in order to ensure the environmental impacts due to development are minimized. You may incorporate previous suggestions into your final plans along with the technical details listed below. As part of your presentation, be sure to address the following.
   a. Provide yield under existing zoning codes and LDR-P.
   b. Explain conservation areas.
   c. Cite potential differences in a "green design" approach.
   d. Explain how you would incorporate LEED practices into your development plan and construction methods. Estimate the categories and points associated with the LEED certification that you will try to attain.
   e. Estimate your lot size, relation of your lot size to current zoning codes, and detail your proposal.
   f. On a separate diagram, note the following for the roadways designed for the development
      i. Label the roadway type.
      ii. Determine traffic volume (ADT) on each road.
      iii. Provide a schematic detail of each cross-section.
      iv. Discuss how your transportation plans fits into the LEED analysis.
   g. Provide an erosion and sedimentation control plan.
      i. Label the phases of development.
      ii. Determine temporary and permanent structures.
      iii. Provide a schematic detail of each type of control method.
      iv. Provide a maintenance schedule and identify who is responsible for maintenance.
      v. Discuss how your erosion and sedimentation plan fits into the LEED analysis.

h. Provide a water (water, wastewater, and stormwater) management plan.
  i. Determine temporary and permanent structures.
  ii. Provide a schematic detail of each type of control method.
  iii. Provide a maintenance schedule and identify who is responsible for maintenance.
  iv. Discuss how your water management plan fits into the LEED analysis.
i. Estimate the cost(s) of finished homes (property and built home) on the proposed property.
j. Estimate the amount your company would be willing to pay per hectare.

CHAPTER 12

Challenges and Opportunities for Sustainability in Practice

FIGURE 12.1 Buckminster Fuller was an American designer most widely known for his invention of the geodesic dome. He was also a profound humanist and advocated the need for sustainable development in the early 1920s, long before it became a compelling goal of the global community. He dedicated his life to solving problems related to energy use, affordable housing, poverty, efficient transportation, and environmental degradation. A philosopher and futurist, Fuller called our planet "Spaceship Earth," but, he noted regrettably, it "did not come with an instruction manual."

Source: John Loengard/Contributor/The LIFE Images Collection/Getty Images

To change something, build a new model that makes the existing model obsolete.

—BUCKMINSTER FULLER

GOALS

The purpose of Chapter 12 is to provide you with a greater context for the social and economic factors that shape the widespread acceptance and use of more sustainably designed products, processes, techniques, and behaviors. In this chapter we explain the dynamics of the adoption and diffusion of innovations, the economic concepts that help us understand why achieving greater environmental sustainability can be a challenge, and the role of governmental policymaking in surmounting those obstacles.

OBJECTIVES

By the end of this chapter, you should be able to:

- Summarize and analyze the social, cultural, technical, and economic factors that affect the potential adoption and diffusion of an innovation.

- Conduct a simple benefit–cost analysis comparing two investment options.

- Explain the concept of an externality and give examples of public policies designed to address them.

- Research and summarize the social and environmental accounting of a company.

- Identify and give examples of government policy tools that help to overcome barriers to the adoption and diffusion of innovations.

- Summarize the issues associated with social justice and sustainability in wealthy industrialized nations.

12.1   Introduction

To achieve a more sustainable world, people must change their ideas, their behavior, and their vision of the future. It isn't enough to wish or hope for things to be different. Action is required. Yet in spite of the sometimes grim state of the environment described in the media and elsewhere, there is reason to be encouraged. As we have seen from the past several chapters, sustainable development is *actionable*. There are *workable solutions* available now to prevent us from repeating the environmental harms of the past and to enable us to live more lightly on the planet. The human community has a surprising array of models, concepts, tools, land-use approaches, technologies, conservation behaviors, construction techniques, and analytical methods to put the principles of sustainable development into practice. We just need to use them more extensively and continue to build our repertoire of strategies.

But people can be resistant to change, and some features of our social and economic systems make implementing sustainable development difficult. To create compelling change, Buckminster Fuller advised us to design in a manner that makes the old way of doing things obsolete. His approach to design science captured the principles of sustainable development decades before concerns about sustainability focused the world's attention. Meeting basic human needs, cooperative and collaborative problem solving, doing more with less, and conserving natural resources are central to Fuller's approach to successful change through design.

To conclude our textbook, we consequently want to explore some of the issues that affect how people, communities, and society respond to products and techniques engineered to be more sustainable. Many of the important influences relate to human psychology and sociology. These are beyond the scope of our text, however; we will instead examine factors that you could meaningfully address as engineers in the design process. In particular, we will focus on dynamics that affect the diffusion and adoption of innovations, the nature of economic decision making, and the role of government in offsetting social and economic constraints. We also address issues of social justice in industrialized countries.

12.2   The Diffusion and Adoption of Innovations

Chapters 7 through 11 highlight a wide array of approaches that can result in a more sustainable society generally and a more sustainable engineering practice specifically. Ecological design, industrial ecology, life cycle analysis, technologies for energy conservation, green building, conservation design, low-impact development, and smart growth represent proven opportunities to reduce human impacts on the environment. These approaches are not widely practiced even though most are relatively well known. We have to ask ourselves why this is so.

Environmental sustainability and sustainable development represent radical new ways of thinking both socially and economically. Acting on these goals means changing our values, our lifestyles, our methods of economic accounting and valuation, and the criteria we use for making decisions about what we buy, use, build, and make. Sustainability has achieved a great deal of influence today as a driving force for social change through a process that is referred to as a **social movement**. Social movements occur when new ways of thinking become institutionalized in a society and are broadly advocated by groups that have political and social power. There is no doubt that a social movement is occurring. But sustainability has not yet become a cultural standard against which we weigh decisions about what we do. It will take time for it to become thoroughly established and internalized as a broad societal goal and common habit; understanding these psychological and sociological processes is unfortunately beyond the scope of our text.

However, it *is* possible to discuss why people and organizations may or may not use the sustainability practices discussed in Chapters 7 through 11. The term used to describe when, why, and how we do things differently—whether in our behavior or in the types of products and technologies that we buy—is known as the **adoption and diffusion of innovations**. Home owners who buy a CFL light bulb for the first time are innovating—they are adopting a new technology for energy conservation in their house. Companies that begin to practice life cycle assessment are innovating—they are newly using a method to reduce waste. Local governments that implement green building ordinances are innovating—they have adopted new practices. When many individuals and organizations adopt a specific innovation (such as CFL lighting), the innovation begins to diffuse—or spread—throughout society. By anticipating obstacles to the adoption and diffusion of an innovation, *we can proactively address them in engineering design*.

The degree and rate at which new innovations are adopted and diffused throughout a society can be modeled by a logistics S-curve. For any given innovation, there will be some defined group of adopters (people or organizations) who could potentially use it; this is represented by the *y*-axis in Figure 12.2. To illustrate, if three-fourths of all U.S. households purchased energy-saving CFL light bulbs, then we would say that CFLs as an innovation have diffused to 75% of all potential adopters.

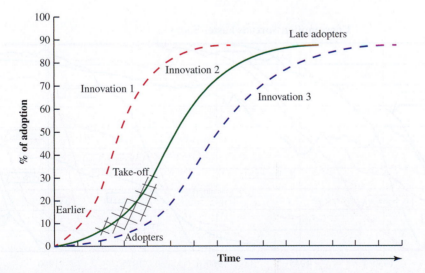

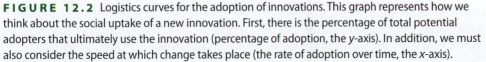

FIGURE 12.2 Logistics curves for the adoption of innovations. This graph represents how we think about the social uptake of a new innovation. First, there is the percentage of total potential adopters that ultimately use the innovation (percentage of adoption, the *y*-axis). In addition, we must also consider the speed at which change takes place (the rate of adoption over time, the *x*-axis).

Source: Based on Rogers (1963)

A critical issue obviously is *how many* potential adopters have actually decided to use an innovation, whether it is a practice, product, or process. Very simply, it does not matter how terrific or wonderful an innovation might be ***if nobody uses it***. Without adoption and diffusion, society simply cannot benefit from the existence of something new.

The rate of innovation matters as well, and the logistics curve also models how fast an innovation might be accepted. As you can see from the x-axis in Figure 12.2, in the early stages of an innovation, there are only a few people or organizations who use it (the earlier adopters); then the rate of adoption accelerates as more people become aware of the innovation and its benefits (the take-off period). The rate of adoption slows down over time because there are fewer and fewer adopters who could potentially use the innovation. Finally, the rate flattens out and plateaus with the "late adopters." Figure 12.3 shows the diffusion curves for a variety of consumer electronics and home appliances in the United States as a way of illustrating these concepts with real innovations.

Broadly speaking, the widespread social diffusion of innovations requires that four conditions be met: that potential users are *aware* of an innovation, that they have an *ability* to acquire and use it, that they *accept* that it will improve their lives in some way, and that they then choose to *adopt* the innovation in practice. These conditions are relevant in all societies, including low-income communities where appropriate technologies are essential. The story of the Jaipur foot (Box 12.1) illustrates how a remarkable innovation languished until a dedicated government official created a way to mass produce and distribute it free of charge to those most in need. In doing so, he eliminated the diffusion barrier to people's ability to acquire the innovation. Another award-winning appropriate technology, the Chotukool, is confronting adoption challenges, even though its inventor has carefully worked out both its manufacture and distribution. The Chotukool is a

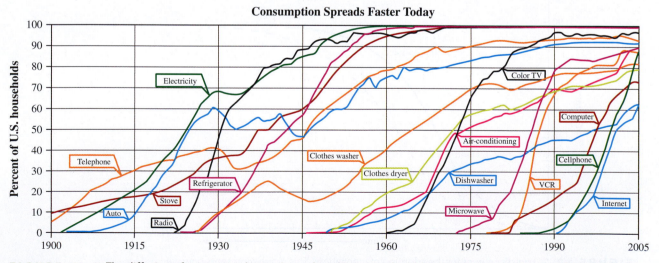

FIGURE 12.3 The diffusion of consumer electronics and appliances in the United States. As seen here, each innovation achieves different levels of total adoption (compare radios to the Internet, for example). In addition, the speed at which the diffusion occurs is also variable; VCRs achieved 70% adoption within a few years compared to decades for washing machines.

Source: Based on Nicholas Felton, The New York Times, 10 February 2008.

BOX 12.1 Diffusing the Jaipur Foot

The Jaipur foot is a prosthetic limb celebrated as a model of design for extreme affordability. The prosthesis was invented in Jaipur, India, in 1968 as a result of the collaboration between orthopedic surgeon Dr. Pramod Karan Sethi and master craftsman Pandit Ram Chandra Sharma. The foot is an extraordinary biomechanical invention, because unlike traditional prosthetics, it allows people to walk, climb, squat, and sit cross-legged; it more closely mimics the human foot than any other device. In addition, it can be custom made and fit to its owner in a few hours, and only costs about US $45 to make. The cost of a western prosthesis would be well over $10,000.

Although the Jaipur foot is of extremely low cost, only about 60 of these prosthetics were fitted between 1968 and 1975. Significant barriers to its diffusion existed. Not only was its cost still out of reach to India's poor—those most in need of the device—but there was no ability to readily make, fit, and distribute it to the tens of thousands of amputees who wanted it.

This excerpt from an article by D. V. Sridharan in the online magazine Good News India *explains how the Jaipur foot overcame obstacles to its diffusion through the efforts of Devendra Raj Mehta. Within 6 years Mehta's non-profit organization, the Bhagwan Mahaveer Viklang Sahayata Samiti (BMVSS), was fitting almost 3,000 Jaipur limbs per year; within 10 years, that number had risen to 6,000. Today BMVSS has fitted more than a half a million of these prosthetics free of charge, and currently fits about 20,000 artificial limbs using the Jaipur foot technology a year, including in countries with significant human limb loss because of war and land mines. The most current and detailed information on Jaipur foot technology, manufacturing, and distribution may be found at the Bhagwan Mahaveer Viklang Sahayata Samiti website, www.jaipurfoot. org.*

After a severe traffic accident in 1969, Mehta received physical therapy in Jaipur, the capital of Rajasthan in Northern India. The Jaipur foot had just been invented around that time, and Mehta was deeply affected by the large number of maimed and poor people coming to Sawai Man Singh (SMS) Hospital trying to get a foot.

"He recalled that experience years later. He was back in government service and Principal Secretary to the Chief Minister of Rajasthan. 1975 was the 2,500th birth year of Mahavira, the founder of Jainism, which the Indian government wanted all states to celebrate fittingly. Mehta suggested rescuing the Jaipur foot from its neglect and delivering it to the needy.

Technology by itself cannot bring about change, however elegant it may be. The Jaipur foot was no doubt a technical winner, but what made it languish, unavailable to the needy? *There were no processes to manage its delivery in the thousands.*

"The first thing I wanted changed was the approach to visitors," says Mehta. "It had to become human." It was common for the maimed to arrive at SMS Hospital and be made to wait for days for mere registration." When BMVSS began operations, the first practice he put in place was that registration must be done on arrival, round the clock. Then the patient is given food and a bed. He and a caretaker are hosted till his limb is custom fitted. And he walks out upright in dignity, with return fare in hand. The whole service is free.

There is an ordered assembly line approach to fabrication. The amputee's stump is covered with a knitted sock and a plaster mould is made. From this socket a plug is made which is an exact replica of the limb. High Density Polyethylene Pipe [HDPE] is warmed and stretched over the plug. A vulcanised rubber foot is attached and suitable straps are provided to fasten the limb to the body. Most of the time, fitment is on the same day and comfort with using it is achieved in hours.

The Jaipur foot costs about Rs.1,300 to make. It can be made by a team of five with average skills, which can produce 15 limbs a day. It's a limb that is suited for Indian use. Wearing it, one can work in the fields, run, pedal a bicycle, squat on the floor or even climb a tree. In contrast, a prosthesis from the West is mostly cosmetic.

(Continued)

BOX 12.1 Diffusing the Jaipur Foot (*Continued*)

Pandit Ram Chandra Sharma's original Jaipur foot was in wood and then aluminum. Later, while visiting a pipe factory in Hyderabad, he was struck by a narration of high density polyethylene's virtues: light weight, low cost, mouldability and strength. Today, body-colored HDPE pipe is manufactured specially for BMVSS use.

To scale up delivery, BMVSS conceived fitment camps. These are held all over the country. Schedules are published at their site. Local chapters screen and line up people in need of prosthesis. Expert team leaves Jaipur complete with equipment and materials. There over several days, limbs are fitted practically at victims' door-steps."

Source: Based on D. V. Sridharan, "Jaipur foot: the real story," Good News India (undated, circa 2005).

low-cost source of refrigeration for rural Indian families without access to electricity (about 400 million people according to the World Bank); in the first two years of its market introduction, only about 15,000 coolers had been sold relative to millions of potential adopters. It remains to be seen whether the Chotukool will experience a diffusion take-off phase.

In industrialized countries, an extraordinary number of factors can influence the awareness, ability to use, acceptance, and ultimate adoption of an innovation (Figure 12.4). The slow acceptance of compact fluorescent light bulbs (CFLs) readily illustrates problems that new innovations confront. Although CFLs are far more energy efficient—for example, a CFL requires only 13 watts of electricity to produce the same amount of light as a 60-watt incandescent—they are not widely seen in the United States (Figure 12.5). One problem is their price; early on, CFLs could cost several dollars compared to an average of 25 to 50 cents for a conventional light bulb. "Sticker shock" was not the only problem that CFLs faced with consumers. Poor light rendering—colors in the lighting spectrum that are visually harsh—was a frequent complaint, as was the slow illumination of these bulbs. Because CFLs are fluorescent, they do not achieve full brightness immediately. Older bulbs could take several minutes to completely light a space, a condition that presents a safety hazard in some locations and is not culturally acceptable to people accustomed to instant light at the flip of a switch. The spiral shape of CFLs also prevented people from using them with existing lamps, because clip-on lamp shades could not be readily attached to the bulb. CFLs therefore confronted several common challenges to the market acceptance of a new innovation.

- They seemed prohibitively expensive compared to the common light bulb, and consumers could not easily calculate the energy savings from these bulbs relative to their up-front cost.
- Lighting quality was poor both in terms of the color spectrum and the length of time required to achieve full brightness.
- Bulbs were incompatible with existing lighting equipment, which would require that consumers buy new lamps and lampshades.

Adopter's Human Resources	Adopter's Organizational Structure	Adopter's Organizational Culture and Decision Process
■ Motivation ■ Skills ■ Technical knowledge resources ■ Specialization and professionalism ■ Commitment ■ Managerial attitudes and support	■ Size and resources ■ Centralization ■ Complexity ■ Communication/ administrative intensity ■ Formalization ■ Flexibility	■ Cooperation and openness ■ Organizational support for innovation ■ Technology champions ■ Innovation proneness ■ Orientation (outward v. inward) ■ Organizational position and role of decision maker
Adopter's Market Context	**Industry Characteristics**	**Communication Channels and Social Networks**
■ Unionization ■ Competitive strategy ■ Location ■ Growth strategy ■ Knowledge of competitors' behavior ■ Market scope	■ Heterogeneity ■ Concentration ■ Regionalization ■ Government regulation ■ Wage rates ■ Growth rate ■ Inter-firm competitiveness	■ Boundary spanners ■ Word-of-mouth ■ Opinion leaders ■ Professional and trade associations ■ Word-of-mouth ■ Informal and indirect links
Technical Attributes of the Innovation	**Economic Attributes of the Innovation**	**Supplier/Vender Characteristics**
■ Complexity-crudeness ■ Communicability ■ Divisibility ■ Type of innovation (process or product) ■ Complementarities required ■ Relative improvements in old technologies ■ Compatibility (values and practice) ■ Learning by doing ■ Relation to innovator product class schemas ■ Radical v. incremental ■ High, medium, and low tech	■ Continuing cost ■ Initial cost ■ Expectations about future prices ■ Expectations about future tech trajectory of innovation ■ Labor saving v. materials saving ■ Scale neutral v. lumpy ■ Uncertainty/risk ■ Profitability ■ Start-up investment ■ Time savings ■ Rate of recovery of cost	■ Technical capabilities and support ■ Public relations ■ Expertise in monitoring deployment ■ Communications skills

FIGURE 12.4 Factors affecting the adoption of an innovation. For businesses, a large number of factors affect their awareness, ability to use, acceptance of, and ultimate adoption of an innovation. Drivers range from the communication and social networks in which they are engaged to industrial structure and their competitive environment, to the technical characteristics of the innovation.

Source: Based on Koebel, Papadakis, Hudson, and Cavell (2003).

Although CFL technology has improved over the past decade and achieved stronger market acceptance, new innovations may ultimately surpass these bulbs. As seen in Figure 12.6, both halogen and LED lighting are projected to dominate the U.S. market by 2020.

Although many of these diffusion factors are beyond your ability to influence as an engineer, you do have the ability to address several important considerations, including quality, user needs, cost factors, and compatibility with existing equipment and technologies (among others). You can engage in people-centered and participatory design for all types of problems (not just those of the poor), which will likely increase the potential for your innovations becoming widely adopted. This is

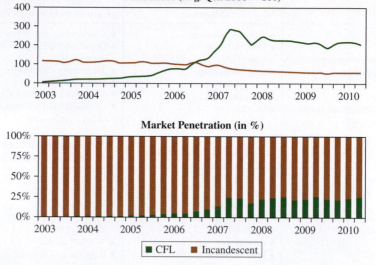

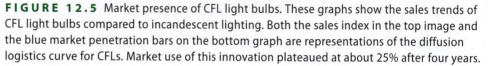

FIGURE 12.5 Market presence of CFL light bulbs. These graphs show the sales trends of CFL light bulbs compared to incandescent lighting. Both the sales index in the top image and the blue market penetration bars on the bottom graph are representations of the diffusion logistics curve for CFLs. Market use of this innovation plateaued at about 25% after four years.

Source: Based on NEMA, http://www.nema.org/News/Pages/CFL-Lamp-Index-Gives-Back-Prior-Gains-for-Second -Quarter-2010.aspx

because people-centered and participatory design are (1) grounded in user-defined needs assessment, (2) sensitive to the values and preferences of users, (3) shaped by an awareness of the economic constraints of potential adopters, and (4) informed by the requirements for technical compatibility of the innovation. The relatively slow adoption and diffusion of solar thermal technologies (Box 12.2) illustrates the

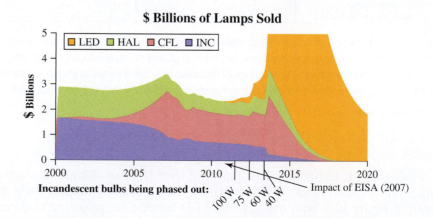

FIGURE 12.6 Diffusion of different lighting technologies. After setting the standard for 100 years of electric lighting, incandescent bulbs (in blue, above) are being challenged by alternative lighting choices. New government lighting standards make it hard for incandescent bulbs to achieve the mandated levels of efficiency; CFL and halogen technology will probably function as transitional innovations as we move to LEDs.

Source: Based on Jason Knott, October 15, 2012, "LEDs Will Eliminate All Other Lamp Sources in 6 Years," CEPro.

many challenges that our global society faces in encouraging a rapid uptake of more environmentally sustainable technologies.

In sum, there are many reasons why an innovation that can enhance environmental sustainability may not be widely accepted or used. Awareness, costs, cultural factors, values, access, and usability all shape people's willingness to try something new. Sustainable development requires that we understand the deeply rooted social, organizational, cultural, and economic contexts that frame the suitability of innovations for individuals, communities, and businesses and their ultimate willingness and ability to change. However, there is no doubt that cost considerations are foremost among many considerations and that cost differences are distorted because we do not account for the environmental harm created

BOX 12.2 Barriers to the Diffusion of Solar Thermal Technologies

Although most people think of photovoltaic electricity when they think of solar energy, the sun can meet an extraordinary amount of human energy needs for *heat*. A variety of solar thermal technologies and systems are available for hot water production and space heating and cooling. Passive solar architecture, solar hot air collectors, and thermal storage systems are proven systems that still are not widely used. Industrial operations that can be supplemented by solar thermal energy include hot water production, drying methods, and a variety of chemical processes, among others. Barriers to the development and widespread adoption of these technologies include technical, economic, legal, cultural, and behavioral obstacles.

For example, technical problems with residential solar hot water systems occurred early in their product life and tended to involve their inability to store hot water for long periods of time or have backup sources of water heating. Consumers were regularly disappointed by the technology, creating a poor reputation in the market for what is now a very reliable form of hot water production and storage. Early technical problems were the result of engineering design that did not account for system complexity and poor installation of even relatively simple systems. In Europe, continued challenges include the inability of some domestic appliances, such as dishwashers and washing machines, to use water from multiple water lines for hot and cold water. Industrial solar thermal technologies confront a wide array of technical challenges that can only be met with further research and development (OECD, 2006).

Almost all solar thermal technologies face economic challenges. The competitiveness of solar thermal products is a function of the available sunshine in a region, the local cost of conventional energy, and the cost of conventional heating alternatives. Not only installation costs, but also the energy costs of alternative equipment, must be considered. The OECD reports that in Greece, with a sunny Mediterranean climate, a solar hot water system for a family would cost about 700 euros (OECD, 2006). In Germany, with less sunshine and freezing temperatures, an appropriate comparable system would cost 4,500 euros. Although the hot water needs of the German family could clearly be met with a solar heating system, the operating cost is equivalent to as much as €0.16 to €0.42 per kilowatt hour (The average cost of electricity in Germany is about €0.25).

Legal, cultural, and behavioral barriers also exist. Passive solar architecture does not cost appreciably more than a conventional building and results in significant energy savings; solar hot air collectors are likewise highly cost effective. Lack of awareness may be one reason these technologies are overlooked in new construction. Better explanations include a

(Continued)

BOX 12.2 Barriers to the Diffusion of Solar Thermal Technologies (*Continued*)

general lack of willingness to innovate in the construction industry, the inability to create standardized building designs using these systems because of the need for careful site orientation, and a local workforce without the requisite trade skills. Landlords will have little incentive to adopt innovative residential solar thermal technologies if they cannot recoup their investments through higher rents. In some U.S. neighborhoods, solar hot water systems cannot be installed because they are unsightly, a practice that is known as "restrictive covenant."

Cultural cooking norms are an important reason why solar cookers and ovens have met with widespread disinterest around the world, in spite of the fact that these resources require no fuel, emit no smoke, and may enable women to spend less time attending to meal preparation. Because solar cookers generally cannot brown food, they result in visually unappetizing meals for cultures that rely on frying (such as in Asia). Women who prepare meals in a communal cooking environment, like some cultures in West Africa, are mystified by a cookstove that a woman would use by herself. In industrialized countries, solar cookers are not useful in the evening when most women arrive home from work. For families that live in dense urban areas with multistory dwellings, they simply do not have access to the amount of sunshine needed for solar cooking.

LEARNING ACTIVITY 12.1 The Diffusion and Adoption of a Packing Peanut Innovation

FIGURE 12.7

Source: Mike Flippo/Shutterstock.com

Packing peanuts, those ubiquitous little Styrofoam™ nuggets that protect products during shipment and storage, are actually made from a variety of materials and come in a range of colors that signify their material properties. According to the radio program *Living on Earth*, Americans dispose of 8.6 million tonnes (8.6 billion kilograms!) of packing peanuts each year. Because this material is largely inert and takes centuries to break down in a landfill, the latest innovation is a completely biodegradable peanut that dissolves in water.

In this learning activity, you should first conduct Internet research on the different kinds of packing peanuts on the market today and their product attributes (appearance, performance characteristics, material content, cost, and so on). If you are a manufacturer of peanuts, what are you most likely to consider in the design and manufacture of this product? Would this change if you conducted a life cycle analysis, or if you considered the fundamental principles of green engineering? (*Hint*: You should research the term bioplastics to get at this quickly.)

If you are a large-scale shipper of products that require peanuts for safe transport and to protect your goods from damage, what factors are you likely to consider as a bulk purchaser of peanuts? If you are a large-scale retailer of products packaged in peanuts that you store in inventory, what are you likely to consider about choice of packing material for the products that you buy?

After reflecting on and analyzing these different perspectives, what factors do you think constitute the major opportunities and obstacles to the widespread diffusion of the biodegradable peanut as an innovation? Who is the actual "adopter" in this context? Do you think it likely that biodegradable materials will come to dominate the peanut market? Be able to discuss why or why not. If you think not, what circumstances might create the conditions necessary for high rates of market adoption of this innovation?

by the traditional way of doing things. Nonpoint-source water pollution, climate change, and the impacts of land-use and land-cover change all have associated social costs that are not reflected in the market prices of goods that embody these environmental problems.

12.3 The Economics of Sustainability

Sustainability presents many challenges to us from an economic perspective. In particular, there are four types of economic issues that we see routinely in environmental problems and that affect our ability to develop more sustainable solutions.

1. The fundamental affordability of greener products, technologies, and systems.
2. The opportunity cost of money and the way individuals and businesses make decisions about how to invest and spend their money.
3. The problem of externalities and the fact that the burden of environmental degradation is not generally reflected in the prices of the goods and services that we buy.
4. The difficulty in establishing a monetary value for a clean, healthy environment and sustainable business practices.

Engineers are often frustrated by these economic factors, especially when they have designed a product or process that reflects a high degree of technical efficiency or reduces impacts on the environment. By better understanding the economic logic at work in human decision making, you will be better able to anticipate how the successful adoption of your design might be affected.

12.3.1 The Fundamental Affordability of Greener Goods and Services

Affordability is an obvious issue when we are talking about the ability of the world's poor—especially those living in $1- or $2-dollar-a-day poverty—to purchase simple life improvements that also reduce their impact on the environment. As seen in Figure 12.8, in 2005 nearly half of the global population lived at a poverty level equivalent to $2.50 per day, and 20% (that's *over a billion people*) lived on less than $1.25. When scarce household cash needs to be protected for medicine, food, or other critical necessities, many families simply cannot afford to invest in more sustainable life improvements. The biomass cookstove discussed in Chapter 8 is an excellent example of this key point, since it reduces the need for fuel wood, improves health, and enables women to spend their time more productively. Yet it is also beyond the means of many families, even if costs of $20 to $50 seem trivial to us. Most of us have no trouble comprehending the basic fact that if you *have no money*, you cannot afford to buy something even if it is of great benefit to you.

Paradoxically, our altruistic impulses to simply give the poor what they need—subsidized cookstoves, solar-powered electricity, water sanitation systems, and so on—have been shown time and again not to work. The failure of such charitable initiatives was touched on previously in this chapter, and it deserves emphasis again here. People are most likely to accept, adopt, and use innovations when their needs are truly being met, when they have made a financial investment by paying

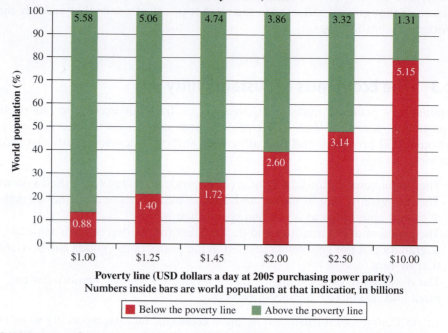

FIGURE 12.8 Percentage of the global population at different poverty levels. This graph illustrates the percentage of the world's population living at different levels of poverty, ranging from $1 per day to $10 per day. Forty percent of Earth's population live on $2 per day or less, and 80% live on less than $10 per day. Even accounting for substantial differences in the cost of living around the world, these are dire levels of poverty in which people's ability to afford more sustainable lifestyles and livelihoods is radically compromised.

Source: Based on World Population Balance, http://www.worldpopulationbalance.org/content/details-maps-and-graphs

something toward the cost of the innovation, and when there is a local capacity for operation, maintenance, and repair. In economic terms, people *value what they are willing to pay for*. As a consequence, human-centered engineering design must be accompanied by an economic program that will enable people to contribute in some manner to the purchase of goods and services that will make them better off. This is why microcredit is often critical to the success of sustainability initiatives such as those provided by Grameen Shakti for solar PV panels (see Chapter 1) and Agua del Pueblo for community water systems (Box 12.3). Not only does it increase the likelihood that an innovation will be used, but it helps build the long-term financial and institutional capacity of the community. Enabling people to pay at least some token amount toward these goods and services also protects their dignity and self-esteem and enhances their sense of being able to provide for their families.

Affordability in the context of the world's wealthier nations is a bit more perplexing, since our purchasing power is significantly greater than that of people in developing nations. Nonetheless, ability to pay is also a consideration because more sustainable products, goods, and services often carry a **green premium**. That is, they

BOX 12.3 Agua del Pueblo (*The People's Water*)

Agua del Pueblo (AdP, "the people's water") is an NGO that helps rural Guatemalan communities design, build, finance, and operate their own community water supply systems. Founded in 1972, the organization has enabled hundreds of villages to develop a secure supply of potable water. In the process of doing so, these communities have improved their health, expanded economic opportunities, enhanced the status of women, improved the social capital of their villages, and uplifted the technical skill sets of virtually all of the adults living in the village.

In addition to the diseases and illnesses described in Chapter 4, the lack of potable water has high **social costs** in Guatemala. The majority of the rural families in Guatemala have no water in their homes, and women must make numerous daily trips to a river, lake, well, or mudhole. They go to wash clothing, to obtain water for household use, and to bathe. Women must carry a two- or three-gallon jug home from the water source several times a day and must make several trips a week to this source to wash clothing. In one community served by Agua del Pueblo, women spent as many as 45 working-days per year in transit just to obtain the use of water. This is an incredible loss of time that could be spent more productively (the social cost), even within the scope of traditional gender roles. With more time available to them, women could pay more attention to performing household sanitary measures (improving family health), preparing better meals (improving family nutrition), or producing such salable items as weaving or pottery (improving family income).

In this context, a community water supply system is an innovation that faces many obstacles and barriers to its adoption and long-term viability. The cost of the system, and the ability to repair and maintain it over time, are the two biggest challenges to potable water supply in rural Guatemala. The subsistence communities and villages of Guatemala are small, men have many demands on their time, and few (if any) adults have the required technical skills to design or manage such systems.

FIGURE 12.9 Community water systems reduce the amount of physical labor and time spent collecting potable water by women and girls in rural Guatemala.

Source: Milosz_M/Shutterstock.com

An early pioneer of reproducible methods for the adoption of appropriate technology, AdP addresses these challenges in a manner that can be easily copied from place to place in Guatemala, enabling the "uptake" of water supply systems to occur much more easily. The AdP strategy is anchored in community-based decision making as well as the training of paraprofessional local water supply technicians.

In order to receive technical support and financial assistance from AdP, a village must elect a committee to oversee community input into the project, because participation is required for all stages of the project—planning, financing, design, construction, education, and

(Continued)

BOX 12.3 Agua del Pueblo (*The People's Water*) (*Continued*)

maintenance. The village is responsible for deciding, with technical guidance, the nature of the water supply system to be constructed. Choices include, for example, whether to have water taps in individual homes or in central strategic locations throughout the community. Local residents build the water system (using volunteer labor and labor-intensive construction techniques) and supply locally available construction materials. This not only motivates a strong sense of community ownership of the system, but it also lowers the costs of outside inputs. AdP educates a cadre of regional and local rural water supply technicians to design and supervise the construction of the systems, as well as train villagers to perform the basic system maintenance and repair. Villagers (both men and women) commonly rotate responsibility for monitoring and fixing problems with the water supply.

The local community must also contribute to the cost of the system, even if it is very modest. A small cash downpayment is made, and AdP provides some matching funds as well as low-cost loans. By being required to pay for a valuable service, the community not only demonstrates its commitment to the system, but it develops what the World Bank

refers to as a "habit of payment for other worthwhile goods" (World Bank, 1975).

The results of AdP's efforts for sustainable development have been striking. Hundreds of communities have chosen to adopt water systems (involving thousands of miles of ditches and small pipelines) in spite of the costs and demanding physical labor. In addition to the health benefits of clean water, the projects have dramatically increased the social capacities of their communities. Elected village committees for the water projects evolved into robust political systems for many other community development initiatives. These include schools, the construction of better homes, the implementation of adult education programs, and empowerment in negotiating with plantation land owners and district politicians. Employment opportunities increased because of the enhanced technical skills of the populace. The status of women increased with their greater involvement in both the technical management of the water system and community decision making. The microfinancing models were extended by villages to other projects, such as schools, community centers, as well as for buying materials to build better homes.

are more costly than "normal" products and services. For example, consider an energy efficient refrigerator that costs $250 more than its less efficient counterpart, or a new automobile with better pollution control technology that costs thousands of dollars more than an older, more polluting, used car. If people cannot afford the marginally higher cost of the green premium, they will not make the more environmentally sustainable purchase. We cannot escape the fact that people must live within their means.

This sense of the relative trade-off in what we get for our money is understood in economic terms as **opportunity cost**. If we need a car but can only spend so much money on it because we also need to pay rent and buy food, then there are opportunity costs between rent, food, and a more fuel efficient car. The more income we have, the more we can afford a wide variety of goods and services, and the opportunity cost of money becomes a more subtle and complex decision that is not based on the absolute affordability of a product or service, but on our needs, wants, and desires. Should I buy the more energy efficient refrigerator, use the extra $250 to go on a weekend holiday, or save the money for other needs at a later date? This relative weighing of opportunity cost is what drives essentially all economic decisions for consumers, businesses, industry, and governments.

12.3.2 The Opportunity Cost of Money

No one has an infinite amount of money. We must make choices about what to buy, how to save, and alternative investments. **Opportunity cost** is an economic concept that represents what we gain (or lose) by choosing one option over another. For example, by choosing to go to college, you gave up the opportunity to work and begin earning an income right away. The opportunity cost of this decision is the amount of money you could have earned by working for four years; however, you willingly incurred the cost of a university education on the assumption that your lifetime earnings with an advanced degree would exceed the money that you would make working immediately.

Virtually all economic decision making has inherent opportunity costs associated with it. As consumers, we aren't even consciously aware of how often we consider these trade-offs. Whenever you quickly choose the cheapest box of cereal on the shelf, you have made a decision about the opportunity cost of your money: You would rather have the benefit of the money you did not spend instead of the additional tastiness or quality that the more expensive cereal might offer.

Opportunity cost is an important consideration in our analyses of why more sustainable products, services, buildings, and industrial practices do not rapidly take hold. Compared to conventional choices, the green premium of environmentally friendly goods and services acts as an adoption barrier for people and organizations for which the opportunity cost of the "greener" purchase is high. The implication for engineering design is rather straightforward: try to make more sustainable products and processes no more costly than their conventional alternatives; better still, make them less expensive.

If a green premium is unavoidable, one of several things must happen in order to motivate people to spend more money up-front and change their perception of opportunity cost. One option is to appeal to environmental values, such as through green marketing in which products may be labeled as "eco-friendly." Another example is to get a product formally certified by an independent third party as meeting specific sustainability criteria, a process known as **endorsement labeling** (Box 12.4). LEED buildings (Chapter 11) and Energy Star products (Chapter 8) are examples of endorsement labeling that protect consumers from **greenwashing**, which is a form of deceptive environmental advertising, marketing, and public relations. It also enables businesses and industry to more easily comply with their own environmental management goals.

Although the initial cost of sustainable products and systems may be higher, they can save money in the long run. A second way in which initial opportunity costs may be ovecome is therefore to demonstrate the long-term benefits of the more sustainable choice. Known broadly as benefit–cost analysis, there are many techniques for comparing the long-term advantages and disadvantages of different economic choices. Commonly used methods are life cycle analysis and return-on-investment ratios. Example 12.1 demonstrates the technique of life cycle costing and return-on-investment ratios with respect to household light bulbs. It shows that the more expensive energy-saving bulb not only pays for itself quickly in energy savings (a concept known as **payback period**) but generates long-term

BOX 12.4 The Benefits of Endorsement Labeling

Endorsement labeling, commonly known as a green certification, has emerged as a way to help consumers and businesses make informed choices and buy more sustainable products. Endorsement labeling requires that an independent third party develop standards and protocols for determining whether or not a product or process has achieved the required sustainability characteristics.

Green certifications are not required by government. Rather, manufacturers and producers voluntarily submit their products for certification (endorsement) as a way of distinguishing themselves in the marketplace. The certification may be done by the third-party institution itself (such as Energy Star and Green Seal) or by those trained and licensed to do so independently (such as Passivhaus and the HERS home energy rating programs).

You have seen examples of endorsement labeling previously. Chapter 8 introduced you to Energy Star, which certifies that equipment and appliances have a level of energy efficiency well above that required by law. Chapter 11 reviewed a number of green certification systems for green building, including LEED, BREEAM, Earthcraft, Passivhaus, Energy Star, and net zero building. Other green certification programs exist, including Green Seal (cleaning products in the United States) and Green Dot (packaging in the European Union). The Forest Stewardship Council is an international NGO which certifies that timber and lumber products are harvested from sustainably managed forests. Endorsement labeling programs can therefore operate effectively at national, regional, and global scales.

Endorsement labeling/green certification has many benefits. First, it provides an independent standard about what makes a product or service more sustainable, and it avoids many problems associated with corporate greenwashing. Second, certified

FIGURE 12.10 The Green Seal endorsement logo. Green Seal is a U.S. nonprofit that has been providing third-party certifications of environmentally safe cleaning products since 1989.

Source: www.greenseal.org

products are the only ones legally allowed to exhibit the endorser's logo, such as the Green Seal emblem illustrated in Figure 12.10. This enables products to legitimately "brand" themselves as green, giving manufacturers a competitive edge in the marketplace.

Third, it allows consumers to overcome information asymmetries. Individuals and small businesses often do not have the time or ability to evaluate the environmental implications of their purchasing decisions, and a third-party endorsement enables people to more easily make choices based on their environmental values. Over time, the expectation is that strong consumer preferences for endorsed products will "pull" sustainability into the marketplace by creating competitive pressures on all competitors to be more green.

Fourth, companies that want their products to be green can use third-party certifications as a way of guiding product attributes and performance specifications. Finally, because endorsement labeling is voluntary, it reduces the burden on government to make and enforce regulations.

financial savings as well. Benefit–cost analysis works well for analyzing many kinds of sustainability choices, in particular for energy and water conservation efforts, waste management, and improved materials selection. Box 12.5 illustrates how benefit–cost techniques were used to compare traditional stormwater management controls with low-impact development BMPs.

BOX 12.5 Cost–Benefit Analysis for an Urban Stormwater Control Project

The city of Caldwell, Idaho, incorporated low-impact development techniques into a daylighting project for its local Indian Creek and a downtown redevelopment project. This is a brief case summary of how planners incorporated cost–benefit analysis to compare conventional stormwater management options with low-impact development (LID) best management practices.

Indian Creek is an excellent focus stream because it is a high-priority subwatershed to the lower Boise River and is contaminated by sediments, bacteria, and phosphorus. The project recognized that the existing collection systems could be upgraded to mitigate the discharge of untreated stormwater into a highly visible, restored stream. The LIFE™ model was used to determine a cost–benefit ratio. LIFE is a physically based, continuous simulation tool that represents the stormwater and runoff processes that occur within bioretention facilities, vegetated swales, green roofs, infiltration devices, and other LID controls.

Indian Creek meanders through farmland, residential, and industrial areas before it enters downtown Caldwell, Idaho, where it joins the lower Boise River. As Caldwell grew, it buried Indian Creek beneath asphalt and concrete. The daylighting project provides important habitat improvement along Indian Creek and is one component of a larger effort aimed at creating an attractive core area for the revitalization of downtown Caldwell. This work builds on existing efforts in the Indian Creek watershed, including a 2002 redevelopment design charette and an existing "Urban Ecology Design Manual for the Lower Boise River." This project recognized that the existing collection systems could be upgraded to mitigate the discharge of untreated stormwater into a highly visible, restored stream.

Final consideration for selected sites was a favorable cost–benefit ratio as predicted by the LIFE model. LIFE also accounts for runoff generated from all categories of land cover, including roadways, landscaping, and buildings over a variety of land uses and soil types, for new development and redevelopment. The model was used to estimate improvements in stormwater runoff quality resulting from application of LID technologies versus traditional stormwater control methods. Basins that drain to critical pinch points in the existing creek were delineated based on historic storm sewer mapping and "ground-truthed" (validated by on-site visual inspection) with recent field visits. In addition, 28.5 hectares in the downtown core were delineated by land use. Results of this land-use characterization are.

- Impervious –75%
  - Sidewalk –26%
  - Parking –22%
  - Roof –16%
  - Road –11%
- Pervious –25%

This information, as well as localized data on rainfall, runoff coefficients, and soil properties, was input into the model. Results of the modeling predicted the following runoff improvements.

- If all sidewalks were converted to permeable pavers, annual stormwater flows would be reduced by 13.2%.
- If half of all parking areas were retrofitted to include bioretention cells, annual stormwater flows would be reduced by 31.6%.
- If all roofs were retrofitted to discharge rainwater into underground swales, annual stormwater runoff would be reduced by 16.3%. If all roofs were also converted to green roofs, then an additional reduction of 0.7% of annual stormwater flows could be achieved.

These reductions in flows would correspond to a reduction in sediment discharge of 2,771 kg a year (34% reduction versus traditional stormwater controls). These reductions in flows would correspond to a reduction in phosphorus discharge of 15 kg a year (33% reduction versus traditional stormwater controls).

BOX 12.5 Cost–Benefit Analysis for an Urban Stormwater Control Project (*Continued*)

TABLE 12.1 Summary of Load Removal Predictions at Demonstration Site

	SEDIMENT	PHOSPHORUS
Traditional upstream technologies	266 kg/yr	0.86 kg/yr
LID technologies	1,089 kg/yr	6 kg/yr

Source: Excerpted from Sherrill Doran and Dennis Cannon, "Cost-Benefit Analysis of Urban Storm-water Retrofits and Stream Daylighting Using Low Impact Development Technologies." Water Environment Federation Technical Exhibit and Conference, October 2006. ©Water Environment Federation.

On a site-specific basis, LIFE predicted increased improvements due to the LID stormwater design as compared to tradition upstream technologies (Table 12.1). From a cost–benefit perspective, traditional stormwater controls have a cost of $8,500, with a removal efficiency of 5% (the resulting cost–benefit ratio is $1,700 per % of load removal). In contrast, the LID technologies cost more ($20,648), but the removal efficiency is much higher (32%) for a resulting cost–benefit ratio of $645 per % of load removal. On a cost–benefit basis, the improvements in LID certainly suggest a better return on investment.

Yet even though it is a valuable tool, benefit–cost analysis still may not result in a decision to purchase (or invest in) the more sustainable option, for several reasons. First, consumers and companies may truly not be able to afford the more costly product up-front. Second, they may not have the time or skill to evaluate the long-term economic benefits of their purchase, which is a problem known as **information asymmetries** that is especially characteristic of households. (Not many people would be able to carefully evaluate the long-term energy savings from a more efficient refrigerator, for example, while they are standing in a store trying to decide which model to buy.) Small businesses likewise suffer from information asymmetries for many of the same reasons as households and individuals. Third, businesses and industry are under pressure to compete and make profits. Consequently, the payback periods for more sustainable options may be too long, or alternative and less sustainable choices may generate larger returns on investment.

In sum, the nature of opportunity costs drives economic decision making, and rational economic choices may not necessarily be wise environmental decisions. Options for making our choices more sustainable—by achieving both environmental goals *and* economic goals—include designing "greener" products and processes in order to be cost-competitive with conventional choices, designing products and processes in a manner that may qualify for an endorsement label, and making appropriate use of benefit–cost analysis to demonstrate the long-term advantages of a sustainable product or technique.

EXAMPLE 12.1	How Much Is That Light Bulb?!

As discussed earlier in this chapter, the up-front cost of CFL light bulbs is a disincentive to many consumers. However, these bulbs pay for themselves many times over in electricity savings. Life-cycle costing (LCC) is one type of analysis that can be used to compare the economic costs of different purchasing decisions. The choice between a CFL and an incandescent bulb readily illustrates the basic technique.

Life-cycle costing sums all expenses associated with owning and operating a device over its lifetime. These costs include the initial purchase and installation price, money spent for maintenance and repair, replacement of components, costs of operation, and any income received for salvage or sale of used equipment at the end of its life cycle.

FIGURE 12.11

Source: Somchai Som/Shutterstock.com

In this example, we will assume that our choice is between a 60-W incandescent light bulb and a 13 W CFL. These bulbs provide equivalent light output in lumens, so the CFL provides the same amount of lighting but uses 78% less energy. The cost of the CFL is $2.25, and the incandescent is $0.50. A CFL bulb will last for 10,000 hours of operation, but the incandescent will last only 1,000 hours. The electricity rate is $0.13/kWh.

The following table is a simple life-cycle cost table of this problem. A life-cycle cost analysis must specify the time period, known as the life cycle period, over which we will conduct our analysis. Because the CFL is a longer-lasting bulb, we will choose 10,000 hours as our life cycle period.

	INCANDESCENT	CFL
Initial purchase cost	$0.50	$2.25
Maintenance and repair	—	—
Replacement costs	$4.50	—
Energy use	$78.00	$16.90
Salvage or resale	—	—
Total life cycle cost	$83.00	$19.15

This table illustrates several basic life-cycle costing elements. We see the initial purchase cost of the bulbs, and we note that there is no maintenance or repair required for light bulbs over their life-cycle period (if a bulb breaks, we cannot fix it!). However, because the incandescent bulb burns out after 1,000 hours, we will have to replace it nine times to achieve the equivalent amount of lighting that our single, 10,000-hour CFL provides. Replacement costs are $4.50 (9 bulbs × $0.50 per bulb).

Energy use is where our CFL bulb really starts to shine. The wattage difference adds up to dramatically higher operating costs for the incandescent bulb:

$$60 \text{ W} \times 10{,}000 \text{ hours} \times \frac{\$0.13}{\text{kWh}} \times \frac{1 \text{ kW}}{1{,}000 \text{ W}} = \$83.00 \qquad (12.1)$$

If we repeat the same calculation for the CFL, we see that total energy costs for the CFL are only $19.15. Finally, neither type of light bulb has a salvage or resale value because once they are burned out they have no useful life left and no material value as scrap.

When we add up the initial purchase cost, replacement expenses, and energy costs over the full 10,000-hour life cycle of the project, we see that the total life cycle cost of using an incandescent bulb is $83.00, whereas it is only $19.15 for the CFL. The CFL bulb creates a *net savings* (or net benefit) of $63.85 (the difference between $83.00 and $19.15).

We can use the same information to calculate a return-on-investment (ROI) ratio. ROI is generally calculated as

$$\text{ROI} = \frac{(\text{benefit from the investment} - \text{cost of the investment})}{\text{cost of the investment}} \tag{12.2}$$

Using Equation (12.2) and the net savings already calculated, the ROI for our CFL is

$$\text{ROI} = \frac{\$63.85}{\$2.25} = 28.38$$

This result means that every $1 invested in our CFL yields a return of $28.38.

Unfortunately, we are still confronted by a major challenge when it comes to more effectively demonstrating—in economic terms—the value of sustainable design. That limitation is in the great difficulty of capturing the real environmental costs of pollution and environmental degradation in the prices of the goods and services that we sell. If this were to happen, environmentally harmful products and services would ultimately be far more expensive than their more benign counterparts. This difficulty is associated with an economic concept known as externalities.

12.3.3 The Problem of Externalities

As you learned early on in this textbook, pollution and environmental degradation are the result of industrial activity, transportation, energy production, and so on. In economic terms, these harmful environmental impacts represent a **negative externality**—a cost (or harm) experienced by someone other than the buyers and sellers of the good or service that resulted in the externality. For example, sulfur dioxide is a major emission from electric power production, and it contributes to respiratory disease and acid rain. The people experiencing the health effects and consequences of acid rain are often different from those buying electricity.

The price of electricity to the consumer simply reflects the costs to the utility of making the energy (including any pollution control measures) as well as a modest rate of profit. However, the social burden of externalities from electric power production is not captured in electricity rates: This price does not reflect the health expenses for people made sick by SO_2 or the economic losses to fisheries and forests because of acid rain. Negative externalities are a challenging problem because they are literally *outside of the market*. This means that they cannot be represented intrinsically through the dynamics of supply and demand that create prices in market transactions. If the social costs of negative externalities could be reflected in market prices, higher prices would be a disincentive for pollution. This, in turn, would create opportunities for correcting the problem. To continue our example, if externalities could be effectively captured in price signals, renewable energy would be much more cost-competitive with fossil fuel electricity because the price of electricity generated by fossil fuels would be considerably higher if social health costs and economic losses to others were included.

Because economic markets cannot fix the problem of negative externalities, they are one of several types of problems referred to as a **market failure**. Government regulations are often used to protect society from negative externalities and other kinds of market failures. Environmental protection laws that limit pollutants are a way of reducing contaminants to less risky levels. For example, in the 1970s, the U.S. Clean Air Act limited SO_2 emissions from new electric power plants to 0.5 kilograms of SO_2 per gigajoule of fuel burned. This type of emission standard is referred to as **command and control regulation** because it represents laws that control industrial behavior through statements about what is permissible activity and what is illegal. Command and control regulation is accomplished through a variety of policy tools and mechanisms, including permits, licenses, fees, emission standards, technology standards, and penalties for noncompliance. Today, most laws governing industrial air and water pollution are command and control regulations.

Market-based incentives are an alternative to command and control regulation and are a relatively new economic mechanism for mitigating environmental and public health externalities. As just mentioned, U.S. command and control regulations for sulfur dioxide required new power plants to limit their emissions to 0.5 kilograms of SO_2 per gigajoule. In 1990, the Clean Air Act replaced this regulation with a **cap-and-trade** market mechanism. The U.S. Acid Rain Program is now widely regarded as the most successful market-based pollution control mechanism in the world (Box 12.6).

Cap-and-trade systems can be successful only under particular circumstances—they are not necessarily effective for all pollutants. These systems must have an actual cap that is either permanent or tightened over time; allowance trading mechanisms must be in place and easily used with clear rules; and measurement and verification methods must be robust and strictly enforced. The concept has been extended to other pollutants; the European Union inaugurated the European Union Emissions Trading System in 2005 with a carbon cap-and-trade system designed to reduce greenhouse gases by 20% in 15 years. The United States does not have a governmental cap-and-trade system for carbon, and the one voluntary scheme initiated by the private sector, the Chicago Climate Exchange, failed in 2010.

Instead, markets for **renewable energy credits (RECs)** are emerging in the United States as a viable alternative for carbon mitigation. In these markets, 1 REC represents 1 megawatt of electric power generated by a clean energy source. RECs are used by utilities to demonstrate the amount of electricity sourced from renewable energy, and they are also used by consumers who want to purchase green electric power. As with cap-and-trade systems, REC markets help provide a price signal regarding the desirability of clean energy and enable it to literally be bought and sold.

In sum, environmental externalities reflect the indirect social costs and harms created by economic activity, and are outside the ability of private markets to correct through the classic price dynamics of supply and demand. As a consequence, government policy is required to protect society and the environment,

BOX 12.6 The U.S. Sulfur Dioxide Cap-and-Trade Program

The U.S. Acid Rain Program uses an innovative market-based system to reduce pollution in a manner that is as economically efficient as possible. The cap-and-trade process works by establishing a permanent, maximum yearly limit on total SO_2 emissions for the United States, referred to as a "cap." SO_2 polluters, basically electric power plants, are then given a number of "allowances," where one allowance is equal to one ton of emitted SO_2. The total number of allowances is equivalent to the national cap on SO_2. At the end of the year, the power plants must produce an allowance for each ton of SO_2 that they emit. If a power plant emits more SO_2 than it has allowances for, it is penalized with a hefty fine.

What makes this a market-based incentive is that the allowances are *tradeable*—they can be bought and sold. If a polluter emits less SO_2 than it has allowances for, it can sell its extra allowances or "bank" them for the future (Figure 12.12). Polluters that discharge more SO_2 than they have allowances for must purchase allowances to make up the difference. Over time, it will be more cost effective for a polluter to begin reducing emissions than to pay for needed allowances. Today, U.S. SO_2 emissions are 40% lower than they were in 1980.

Capping and Trading Emissions: The Concept

Before the program

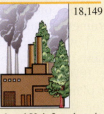

With no reductions required, Unit 1 and Unit 2 each emits 18,149 tonnes a year.

The "CAP"

The cap requires a 50 percent cut in emissions—e.g., from 18,149 to 9,074 tonnes.

Emissions trading under the CAP

If Unit 1 can efficiently reduce 13,612 tonnes of emissions and Unit 2 can only efficiently reduce 4,537 tonnes, trading allows each unit to act optimally while ensuring achievement of the overall environmental goal. Unit 1 can hold on to (and "bank") its excess allowances or can sell them to Unit 2, whereas Unit 2 must acquire allowances from Unit 1 or from another source in the program.

FIGURE 12.12 Cap-and-trade basics. This graphic shows how the U.S. SO_2 cap-and-trade system works to reduce emissions through market mechanisms. U.S. SO_2 emissions declined by 64% after the onset of this program; by 2009, actual SO_2 emission levels for all sources regulated under the Acid Rain Program were well below the national limit set by the EPA.

Source: Based on US EPA, Clearing the Air: Facts About Capping and Trading Emissions (May 2002).

and this protection usually takes the form of command and control regulation. New efforts to develop market-based incentives through cap-and-trade programs have been shown to be successful and are useful because they allow businesses to mitigate pollution in a manner that is most economically effective for themselves. When rigorous caps are set and the rules of cap-and-trade are enforced, the desired level of environmental quality is achieved through the most efficient economic means. Many environmental regulations dealing with externalities have profound implications for engineering design, because the technological systems and processes resulting in the externalities must comply with government standards or target limits. As businesses place more emphasis on pollution prevention (rather than control), greater opportunities arise to practice the methods of industrial ecology and life cycle analysis described in Chapters 9 and 10.

12.3.4 The Difficulty of Environmental Valuation

Environmental sustainability challenges traditional economic approaches in yet another way, which is how we monetarily value and account for the impacts of our actions on natural resources and a clean environment. We see this with the emergence of the ecosystem services concept (see Chapter 7) as well as sustainable development's "three pillars" of society–environment–economy.

Several important changes in economic culture are taking place throughout the world today in order to accommodate the growing emphasis on **social and environmental accounting** by business enterprises. Generally, these terms refer to a formal assessment of how business activity benefits people and the environment. In business, two terms are used somewhat synonymously to reflect social and environmental accounting: **corporate social responsibility** and **triple bottom line**. The British visionary John Elkington has been an advocate of corporate social responsibility for nearly three decades; he also coined the term *triple bottom line*. Both notions stress the idea that business accountability to investors and society involves more than just the traditional financial reporting of profits and losses. Instead, firms also build (or destroy) social and environmental capital, which they can and should account for as standard practice through their corporate reporting systems. In this manner, businesses can more realistically reflect how their operations holistically affect social, environmental, and economic conditions. Triple bottom line thus represents business outcomes in which "people, planet, and profit" benefit simultaneously.

Social and environmental accounting requires new methods, performance measures, and metrics. The ecological footprint, which we discussed in Chapter 7, is one such indicator, as is the United Nations' System of Integrated Environmental and Economic Accounting for a country's macroeconomy. Most initiatives to create social and environmental accounting techniques are global in nature because of the scope of international trade and commerce and concerns about social justice. Businesses around the world—especially transnational corporations—should ideally use common terminology and methods for assessing the social and environmental outcomes of their business practices. Examples include the Global Reporting Initiative standardized sustainability measurement reporting system and the ISO 14000 and ISO 26000 standards from the **International Organization for Standardization (ISO)** (Box 12.7). As yet, however, there are no standard, universally adopted protocols.

BOX 12.7 What Is the International Organization for Standardization (ISO)?

FIGURE 12.13

Source: The International Organization for Standards, ISO

ISO is an international NGO with the primary mission of developing international standards governing most aspects of technology, manufacturing, and business management operations. ISO members are national standards bodies from most of the world's countries. Historically, ISO focused on standards related to manufactured products to facilitate international trade; standards addressed issues of quality, reliability, component substitutability, and so forth. Today, ISO promotes standards that enable the global community to have highly consistent modes of practice and technologies worldwide; ISO has developed more than 19,000 standards. Two ISO programs are of particular relevance to sustainable development: ISO 14000 on environmental management and ISO 26000 on social responsibility.

The ISO 14000 series or "family" of standards helps companies reduce their environmental impacts and comply with environmental laws. These standards govern processes and practices, not products themselves. It is therefore known as a **management system standard**. ISO 14000 does not set environmental performance targets for companies, but rather provides protocols, best practices, and guidance that enable a company to systematically set environmental goals, work toward achieving them, and monitor performance against goals and plans. A primary focus of ISO 14000 is the development of an environmental management system that is a comprehensive way of collecting and analyzing environmentally related data for a company, including tracking performance targets. In addition to environmental management systems, the ISO 14000 family includes standards for the use of life cycle analysis and environmental communication. Companies can get certified as ISO 14000 compliant; third parties validate and certify compliance against ISO requirements. Therefore ISO sets the standards but does not provide the certification services.

ISO 26000 on social responsibility was introduced in 2010 and provides guidance to all types of organizations regarding their obligations to people and the environment. Seven core areas are addressed: organizational governance, human rights, labor practices, the environment, fair operating practices, consumer issues, and community involvement and development. ISO 26000 is not a set of standards for which companies may be certified as ISO compliant; it is not a management system standard. Rather, it is a guidance document intended to help create an internationally standardized and common understanding of terms, obligations, and principles for practice.

As with social and environmental accounting, valuation of ecosystem services challenges traditional economic/financial concepts and methods. Because most ecosystem services are not directly bought and sold, there is no market and no price for them. Assigning economic value becomes a complex effort and is based on methods of **nonmarket valuation** (assigning monetary value that is not based on the price of actually buying and selling the good or service in question). With individual services such as purified water, valuation is relatively straightforward: It is estimated from the cost of operating water treatment and supply facilities and/or the supply of potable water. The process becomes more difficult once we allow for complexity of ecological systems, such as the role of healthy forests in preventing soil erosion or water contamination.

LEARNING ACTIVITY 12.2 Who's Green, and Why?

A number of media organizations regularly issue lists of green companies. *Newsweek* magazine in particular provides an annual ranking of American and global companies that achieve high ratings with respect to their environmental impacts and social responsibility. Because the ability to be green varies by type of industry, *Newsweek* provides ratings for 19 different industrial and service sectors.

Below is a list of *Newsweek's* most highly ranked green businesses in each of these 19 sectors for 2012. Select one of these companies and investigate why it is considered exceptional within its class. The company's annual reports and corporate website, online articles, and the *Newsweek* ratings (see *www.thedailybeast.com/newsweek/2012/10/22 /newsweek-green-rankings-2012-u-s-500-list.html*) should get you all of the information you need. Is your company ISO 14000 compliant? Has it won awards? What does it actually *do* to deserve its rating? Once you have completed your work,

prepare a brief digital presentation summarizing your findings and analysis.

- Baxter International (health care)
- Boeing
- Coca-Cola
- Cummins
- Hartford Financial Services Group
- Hess Energy
- Hewlett Packard
- IBM
- Kimberly Clark
- Manpower
- McGraw-Hill
- Northeast Utilities
- Office Depot
- Praxair
- Sprint Nextel
- The Gap
- United Parcel Service
- Vornado Realty Trust
- Wyndham Worldwide

12.4 The Role of Government

Governments do have a significant role to play in creating opportunities for sustainable development or removing the obstacles and barriers that exist. However, the role of government in fostering sustainability varies worldwide, simply because each nation defines the proper and legitimate role of government based on its own society, culture, and needs. As a consequence, the scope and range of governmental behavior is considerable as we reflect on the breadth of global well-being and circumstance.

In low-income nations, governmental policies and actions (often in partnership with nongovernmental organizations) may focus on achieving several of the millennium development goals of the United Nations (Figure 12.14). For nations that have large numbers of people struggling to meet basic needs, sustainable development is a matter of meeting these needs while preserving and protecting the environment. As discussed earlier in this chapter, the role of appropriate technology at appropriate scales is critical in providing for human material quality of life. Pollution prevention and control regulations often are not as critical as those for education, public health, water supplies, and economic growth generally. Natural resource conservation policies related to forestry and agriculture are also areas of policymaking common in low-income nations, and they are usually designed to prevent environmental degradation.

In high-income nations, public policy has historically focused on environmental protection and natural resource conservation. Clean air and water laws are well over

FIGURE 12.14 Icons of the Millennium Development Goals. The Millennium Development Goals are designed to provide the foundation for sustainable societies in low-income nations. They emphasize social, economic, and cultural well-being and quality of life in a manner that is environmentally sustainable. Policymaking in many nations is explicitly about trying to achieve these goals.

Source: UNDP Brazil

60 years old in the United States and parts of Europe; sewage and sanitation regulations date back 100 years or more. Forestry and agriculture policies are complex and vary considerably by nation. Even though Europe and the United States share a common history in terms of their environmental movements of the 20th century; after the 1973 oil embargo and oil shock of the 1970s, their policies began to diverge significantly. Europe's policy evolution—as reflected by many individual nations as well as the EU—has been decidedly toward laws and guidance that promote sustainable development. The EU is therefore considerably more aggressive toward carbon and climate change mitigation; many European nations also regulate land use and urban growth more proactively than the United States.

In almost all instances, government policies are intended to correct market failures of one kind or another. Market failures contribute not just to negative externalities, but to underinvestment in a range of socially and economically desirable activities such as R&D, technological innovation, and more energy-efficient products. Governments therefore have a role to play not just for traditional environmental protection, but to stimulate both innovation and the social adoption of more sustainable products, processes, and behaviors. The tools and techniques that government laws and policies commonly use to stimulate both innovation and adoption include the following.

- *Tax incentives*—Tax credits can be used to offset the higher cost of more sustainable technologies, such as solar and wind energy systems for homes and business. R&D tax credits facilitate innovation in the corporate sector.
- *Standards*—Performance standards are widely used to achieve higher levels of energy efficiency in products, machinery, and equipment. Examples include fuel economy and carbon emission standards for automobiles, and energy efficiency standards for heating and cooling equipment. We would also include building energy codes in this category.
- *Mandatory Best Practice Requirements*—Government regulations can mandate that best practices be used for pollution prevention and control, as well as other forms of environmental conservation. Examples include state and local governments that mandate low-impact development BMPs for stormwater management.

- *Research and Development*—Government can itself invest in and conduct R&D through national research laboratories, or it can give grants to universities and private companies to develop innovative new products and processes.

In general, governments in western, capitalist societies are reluctant to force people to behave in certain ways or to directly manipulate market prices. As a consequence, many public policies work indirectly to provide incentives for changing the social and economic conditions that can contribute to sustainable development.

12.5 Social Justice and Sustainability in Wealthy Countries

The concept of sustainable development explicitly stresses equity, quality of life, and social justice as part of our global transformation toward environmental stewardship. Much of our textbook has focused on and given examples for low-income countries—for obvious reasons.

But wealthy industrialized nations confront challenges to social justice and sustainability as well. In the United States particularly, the poor and minority communities may live at the margins of the "sustainability movement" in a variety of ways. An affordable, "green" quality of life is beyond the reach of many Americans, 15% of whom lived in poverty in 2012 (U.S. Bureau of the Census, 2014). This is equivalent to more than 46 million individuals; notably, more than 20% of U.S. children under the age of 18 live in poverty. Low-income families must often live in substandard housing that potentially presents many environmental hazards, including lead paint, poor indoor air quality, insect and rodent infestations, and contaminated water from aging plumbing systems. Children are especially vulnerable to the environmental conditions of poverty; the incidence of asthma among poor children is notably higher than among the general population, as is asthma among low-income African-Americans (National Center for Health Statistics, 2012). The cause of these higher rates of disease reflects a portfolio of problems, including cockroaches, inner city or industrial neighborhood air quality, mold, mice, and poor air ventilation. The two U.S. zipcodes with the highest number of children living in poverty had asthma incidence rates over 40% (NBC News, 2014).

High household energy costs due to old appliances, inadequate heating equipment, and poorly insulated buildings are also common. Higher proportions of disposable income may consequently go to paying utility bills; the federal government provided nearly $3 billion in assistance through the Low Income Home Energy Assistance Program in 2013 to help families meet their basic energy needs. Very simply, low-income and poor families in the United States are commonly subject to some of the oldest or most poorly maintained housing stock in America and have little influence over their landlords. The green building innovations explored in Chapter 11 and the energy savings measures noted in Chapter 8 are far beyond the means of many Americans. In addition, developers are unlikely to build more sustainable low-income housing simply because they cannot recover the higher construction costs through rent.

Environmental justice is also a compelling issue in the United States, and it reflects a type of *spatial discrimination*. Many low-income, minority, and tribal communities face environmental health risks such as contaminated drinking water and exposure to toxic chemicals, or they confront environmental nuisances such as noxious odors from landfills and factories. In Appalachia, mountaintop removal—a technique used in coal mining—subjects communities to air pollution, degraded water resources, and constant dynamite explosions. The nature of such discrimination is complex. These communities often lack the political power to resist siting unwanted land uses or to compel environmental clean-up. In other instances, people must live in areas with environmental degradation and contamination simply because it is more affordable.

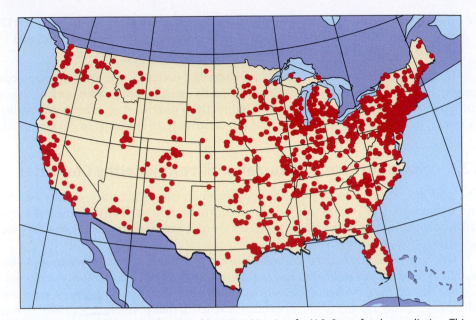

FIGURE 12.15 Location of National Priorities List sites for U.S. Superfund remediation. This map shows the location of hazardous waste sites that qualify for the U.S. Superfund. It can take years for remediation to begin, and many more communities technically qualify for Superfund than are actually on the National Priorities List.

Source: Based on National Atlas, nationalatlas.gov

United States laws help only to a limited extent. The National Environmental Policy Act (NEPA) requires the federal government to address environmental justice considerations for the actions of its own agencies. NEPA does not apply to private organizations, however, and it is often criticized for being weakly enforced at the federal level. The U.S. Comprehensive Environmental Response, Compensation, and Liability Act of 1980, commonly known as **Superfund**, provides funding to clean up the most contaminated sites in the United States; there are over 1,000 sites on the National Priorities List (Figure 12.15). The Superfund list is not exhaustive, however. Many other communities technically qualify but are not put on the priority list for remediation. With respect to low-income housing, state and local public housing authorities are often large landlords, but struggle to upgrade the quality of housing because of budgetary constraints. Overall, weak enforcement of existing laws, as well as limited policy protection generally, characterize the politics of environmental justice. Lack of affordability and access to legal representation likewise prohibits communities from seeking and receiving help through the legal system.

12.6 Summary

Achieving sustainable development requires that we more widely adopt and diffuse innovations that enhance sustainability. Yet there are many obstacles to doing so. By understanding the dynamics of the adoption and diffusion process, we can engage in better engineering design to minimize them as much as possible. In addition, a variety of economic conditions act as barriers to the adoption of green innovations. These conditions can be addressed with a variety of techniques, including endorsement labeling, cost–benefit analysis, implementation of corporate social responsibility and environmental accounting systems, and selective use of governmental policy tools.

References

Clemens, B., Karp, A., and Papadakis, M. (2002). "The people's water: Technology transfer and community empowerment in Guatemala." In M. De Laet (Ed.), *Research in Science and Technology Studies: Knowledge and technology transfer* (pp. 103–125). Amsterdam Boston: JAI.

DiSalvo, Carl, Clement, A., et al. (2012). "Communities: Participatory Design For, With, and By Communities." *Routledge International Handbook of Participatory Design.* Jesper Simonsen and T. Robertson (Eds.). New York: Routledge.

EIA. (2009). *International Energy Outlook 2009.* Washington, D.C.: U.S. Department of Energy.

Forsyth, T. (2005). *Encyclopedia of International Development.* New York: Routledge.

IEA. (2011). Energy for all: Financing access for the poor. Special early excerpt of the World Energy Outlook 2011, International Energy Agency (IEA) and OECD.

Koebel, C. Theodore, Papadakis, Maria, Hudson, Ed., and Cavell, Marilyn. (2003). *The Diffusion of Innovation in the Residential Building Industry.* Prepared for U.S. Department of Housing and Urban Development, Office of Policy Development and Research; Center for Housing Research, Virginia Polytechnic Institute and State University, Blacksburg, VA.

Living on Earth. (2012). Mushroom Packaging. Radio Program Air Date March 23, 2012, Public Radio International. Accessed June 20, 2013, at www.loe.org/shows/segments.html?programID=12-P13-00012&segmentID=4.

Makuch, Karen E., and R. Pereira. (2012). *Environmental and Energy Law.* Malden, MA: Blackwell.

National Center for Health Statistics. (2012, May). *Trends in Asthma Prevalence, Health Care Use, and Mortality in the United States, 2001–2010.* National Center for Health Statistics Data Brief No. 94. Washington, D.C.: U.S. Department of Health and Human Services.

National Research Council. (2010). *Advancing the Science of Climate Change, America's Climate Choices,* Panel on Advancing the Science of Climate Change. National Research Council, The National Academies Press.

NBC News. (2014). *In Plain Sight.* Mold, Mice, and Zipcodes: Inside the Childhood Asthma Epidemic. Accessed 6 April 6, 2014, at http://inplainsight.nbcnews.com/_news/2014/01/03/22149240-mold-mice-and-zip-codes-inside-the-childhood-asthma-epidemic?lite.

Philibert, Cédric. (2006). *Barriers to Technology Diffusion: The Case of Solar Thermal Technologies.* Paris: OECD.

Rogers, Everett M. (1963). *Diffusion of Innovations,* 3rd. ed. New York: Free Press.

Schumacher, E. F. (1973). *Small Is Beautiful: Economics as If People Mattered.*[London: Blond and Briggs.]

Simonsen, Jesper, and Robertson, T. (Eds.) (2012). Preface. *Routledge International Handbook of Participatory Design.* New York: Routledge.

UN. (2012). Glossary of climate change acronyms. Retrieved October 32, 2012, from http://unfccc.int/essential_background/glossary/items/3666.php.

UNDP. (2009). Human Development Report 2009: Overcoming Barriers: Human Mobility and Development. New York, N.Y.: UNDP.

U.S. Bureau of the Census. (2014). Poverty Highlights. Accessed April 6, 2014, at www.census.gov/hhes/www/poverty/about/overview/index.html.

World Bank. (1975). Issues in Village Water Supply. Public Utilities Department, Report No. 793.

World Health Organization. (2006). Fuel for Life: Household Energy and Health. Geneva: World Health Organization.

Key Concepts

Social movement
Adoption and diffusion of innovations
Social costs
Green premium
Opportunity cost
Endorsement labeling
Greenwashing
Payback period
Information asymmetries
Negative externality
Command and control regulation

Market-based incentives
Cap-and-trade program
Renewable energy credits
Social and environmental accounting
Corporate social responsibility
Triple bottom line
International Organization for
　Standardization (ISO)
Nonmarket valuation
Market failure
Management system standard

Problems

12-1　Identify three different, *real* organizations that have made sustainability an explicit consideration in their operations.

12-2　The author Thomas Friedman is credited with saying that "pessimists are usually right and optimists are usually wrong but all the great changes have been accomplished by optimists." Do you consider yourself an optimist or a pessimist about society's ability to accomplish change that leads to more sustainable development? Why?

12-3　Explain the concept of an opportunity cost and how that affects the decision by individuals, organizations, and businesses to buy (or invest in) more sustainable products and practices.

12-4　What are the four ways that we can offset opportunity costs for "green" decision making?

12-5　List five benefits of endorsement labeling/green certification.

12-6　Identify and explain three reasons why the poor don't necessarily use more sustainable products that are charitably given to them.

12-7　Why are environmental externalities so difficult to remedy using traditional market dynamics and financial criteria?

12-8　Give an example of a specific, real government policy for each of the following that promotes the adoption and diffusion of more sustainable products and techniques:
　a. Tax incentives
　b. Standards
　c. Best management practices
　d. Research and development

12-9　Explain three reasons why a company would not select an investment that would improve its environmental performance and save it money in the long run.

12-10　Voluntary compliance with third-party standards is an increasingly common way that companies are creating social and environmental accountability. Why are these standards usually developed for management systems and not specific products?

12-11　Refer to Chapter 4 and the discussion of ceramic water filters in Benin (see Tables 4.4 to 4.6). Develop two different benefit cost analyses for a household as follows.
　a. Construct a life cycle cost table comparing the ceramic water filter to purchased Pur™ filtered water. What is the total life cycle cost of each option, and what is the return on investment ratio for the ceramic filter?

b. What is the benefit cost of each system expressed as the cost per % reduction in coliform bacteria load?

Open-Ended Problems

12-1 Below are a number of technical innovations (products, methods, or techniques) that enhance sustainability. Select one to research and analyze with respect to (a) how it might balance the three pillars of sustainable development and (b) the barriers and obstacles that it confronts with respect to its adoption and diffusion.
- Hydrogen fuel cells
- Biomimetic design
- Solar ovens
- Combined heat and power systems
- Soy-based building insulation
- Biodegradable plastic
- Green roofs
- Ceramic water filters for low-income countries
- Solar walls
- Wind turbines rated at 5 kW or less
- Biodiesel

12-2 Many efforts have been made to apply the success of the U.S. cap-and-trade program for sulfur dioxide to other environmental problems. Below are four efforts that have been introduced as an actual cap-and-trade system or as a credit trading scheme. Select one, analyze its social, economic, and environmental contexts, and critique the basic principles of the trading or credit scheme designed to mitigate this problem. If you do not think a trading program will be successful for the activity you selected, explain why and defend your position.
- Nutrient trading (in watersheds)
- Carbon emissions
- Mitigation banking (of wetlands)
- Regional NOx trading

12-3 You work for a company that wants to use the Global Reporting Initiative's G4 Standard to develop its first Sustainability Report. It has assigned you to a team in order to summarize the criteria and the process, and to develop a 12-month project timeline in which the report can be completed. Prepare a 15-minute presentation of your team's summary and recommended project plan

12-4 Consumer electronics and e-waste are emerging as an extraordinary problem in industrialized countries, especially the United States. Reflect on insights you gained in Chapters 7, 9, and 10. Add to this concepts related to the adoption and diffusion of innovation from this chapter. Develop a solution to the e-waste problem in your state that incorporates all of these insights as well as government policies to overcome market failures.

12-5 Identify the active Superfund sites in your state. Explore and analyze the original source of contamination of these sites, and explain the nature of the contamination, hazard, and risk confronting communities located there. Conduct a socioeconomic and demographic analysis of these sites across the state. What are the environmental justice issues associated with them?

Conversion Factors

CONVERSIONS BETWEEN U.S. CUSTOMARY UNITS AND SI UNITS

U.S. Customary unit		Times conversion factor		Equals SI unit	
		Accurate	**Practical**		
Acceleration (linear)					
foot per second squared	ft/s^2	0.3048*	0.305	meter per second squared	m/s^2
inch per second squared	$in./s^2$	0.0254*	0.0254	meter per second squared	m/s^2
Area					
circular mil	cmil	0.0005067	0.0005	square millimeter	mm^2
square foot	ft^2	0.09290304*	0.0929	square meter	m^2
square inch	$in.^2$	645.16*	645	square millimeter	mm^2
Density (mass)					
slug per cubic foot	$slug/ft^3$	515.379	515	kilogram per cubic meter	kg/m^3
Density (weight)					
pound per cubic foot	lb/ft^3	157.087	157	newton per cubic meter	N/m^3
pound per cubic inch	$lb/in.^3$	271.447	271	kilonewton per cubic meter	kN/m^3
Energy; work					
foot-pound	ft-lb	1.35582	1.36	joule (N·m)	J
inch-pound	in.-lb	0.112985	0.113	joule	J
kilowatt-hour	kWh	3.6*	3.6	megajoule	MJ
British thermal unit	Btu	1055.06	1055	joule	J
Force					
pound	lb	4.44822	4.45	newton ($kg·m/s^2$)	N
kip (1000 pounds)	k	4.44822	4.45	kilonewton	kN
Force per unit length					
pound per foot	lb/ft	14.5939	14.6	newton per meter	N/m
pound per inch	lb/in.	175.127	175	newton per meter	N/m
kip per foot	k/ft	14.5939	14.6	kilonewton per meter	kN/m
kip per inch	k/in.	175.127	175	kilonewton per meter	kN/m
Length					
foot	ft	0.3048*	0.305	meter	m
inch	in.	25.4*	25.4	millimeter	mm
mile	mi	1.609344*	1.61	kilometer	km
Mass					
slug	$lb-s^2/ft$	14.5939	14.6	kilogram	kg
Moment of a force; torque					
pound-foot	lb-ft	1.35582	1.36	newton meter	N·m
pound-inch	lb-in.	0.112985	0.113	newton meter	N·m
kip-foot	k-ft	1.35582	1.36	kilonewton meter	kN·m
kip-inch	k-in.	0.112985	0.113	kilonewton meter	kN·m

CONVERSIONS BETWEEN U.S. CUSTOMARY UNITS AND SI UNITS (Continued)

U.S. Customary unit		Times conversion factor		Equals SI unit	
		Accurate	Practical		
Moment of inertia (area)					
inch to fourth power	in.^{4}	416,231	416,000	millimeter to fourth power	mm^4
inch to fourth power	in.^{4}	0.416231×10^{-6}	0.416×10^{-6}	meter to fourth power	m^4
Moment of inertia (mass)					
slug foot squared	slug-ft^2	1.35582	1.36	kilogram meter squared	kg·m^2
Power					
foot-pound per second	ft-lb/s	1.35582	1.36	watt (J/s or N·m/s)	W
foot-pound per minute	ft-lb/min	0.0225970	0.0226	watt	W
horsepower (550 ft-lb/s)	hp	745.701	746	watt	W
Pressure; stress					
pound per square foot	psf	47.8803	47.9	pascal (N/m^2)	Pa
pound per square inch	psi	6894.76	6890	pascal	Pa
kip per square foot	ksf	47.8803	47.9	kilopascal	kPa
kip per square inch	ksi	6.89476	6.89	megapascal	MPa
Section modulus					
inch to third power	in.^{3}	16,387.1	16,400	millimeter to third power	mm^3
inch to third power	in.^{3}	16.3871×10^{-6}	16.4×10^{-6}	meter to third power	m^3
Velocity (linear)					
foot per second	ft/s	0.3048*	0.305	meter per second	m/s
inch per second	in./s	0.0254*	0.0254	meter per second	m/s
mile per hour	mph	0.44704*	0.447	meter per second	m/s
mile per hour	mph	1.609344*	1.61	kilometer per hour	km/h
Volume					
cubic foot	ft^3	0.0283168	0.0283	cubic meter	m^3
cubic inch	in.^{3}	16.3871×10^{-6}	16.4×10^{-6}	cubic meter	m^3
cubic inch	in.^{3}	16.3871	16.4	cubic centimeter (cc)	cm^3
gallon (231 in.^{3})	gal.	3.78541	3.79	liter	L
gallon (231 in.^{3})	gal.	0.00378541	0.00379	cubic meter	m^3

*An asterisk denotes an *exact* conversion factor

Note: To convert from SI units to USCS units, *divide* by the conversion factor

Temperature Conversion Formulas

$$T(°C) = \frac{5}{9}[T(°F) - 32] = T(K) - 273.15$$

$$T(K) = \frac{5}{9}[T(°F) - 32] + 273.15 = T(°C) + 273.15$$

$$T(°F) = \frac{9}{5}T(°C) + 32 = \frac{9}{5}T(K) - 459.67$$

PRINCIPAL UNITS USED IN MECHANICS

Quantity	International System (SI)			U.S. Customary System (USCS)		
	Unit	Symbol	Formula	Unit	Symbol	Formula
Acceleration (angular)	radian per second squared		rad/s^2	radian per second squared		rad/s^2
Acceleration (linear)	meter per second squared		m/s^2	foot per second squared		ft/s^2
Area	square meter		m^2	square foot		ft^2
Density (mass) (Specific mass)	kilogram per cubic meter		kg/m^3	slug per cubic foot		$slug/ft^3$
Density (weight) (Specific weight)	newton per cubic meter		N/m^3	pound per cubic foot	pcf	lb/ft^3
Energy; work	joule	J	$N \cdot m$	foot-pound		ft-lb
Force	newton	N	$kg \cdot m/s^2$	pound	lb	(base unit)
Force per unit length (Intensity of force)	newton per meter		N/m	pound per foot		lb/ft
Frequency	hertz	Hz	s^{-1}	hertz	Hz	s^{-1}
Length	meter	m	(base unit)	foot	ft	(base unit)
Mass	kilogram	kg	(base unit)	slug		$lb\text{-}s^2/ft$
Moment of a force; torque	newton meter		$N \cdot m$	pound-foot		lb-ft
Moment of inertia (area)	meter to fourth power		m^4	inch to fourth power		$in.^4$
Moment of inertia (mass)	kilogram meter squared		$kg \cdot m^2$	slug foot squared		$slug\text{-}ft^2$
Power	watt	W	J/s $(N \cdot m/s)$	foot-pound per second		ft-lb/s
Pressure	pascal	Pa	N/m^2	pound per square foot	psf	lb/ft^2
Section modulus	meter to third power		m^3	inch to third power		$in.^3$
Stress	pascal	Pa	N/m^2	pound per square inch	psi	$lb/in.^2$
Time	second	s	(base unit)	second	s	(base unit)
Velocity (angular)	radian per second		rad/s	radian per second		rad/s
Velocity (linear)	meter per second		m/s	foot per second	fps	ft/s
Volume (liquids)	liter	L	$10^{-3} m^3$	gallon	gal.	$231\ in.^3$
Volume (solids)	cubic meter		m^3	cubic foot	cf	ft^3

SELECTED PHYSICAL PROPERTIES

Property	SI	USCS
Water (fresh)		
weight density	9.81 kN/m^3	62.4 lb/ft^3
mass density	1000 kg/m^3	1.94 slugs/ft^3
Sea water		
weight density	10.0 kN/m^3	63.8 lb/ft^3
mass density	1020 kg/m^3	1.98 slugs/ft^3
Aluminum (structural alloys)		
weight density	28 kN/m^3	175 lb/ft^3
mass density	2800 kg/m^3	5.4 slugs/ft^3
Steel		
weight density	77.0 kN/m^3	490 lb/ft^3
mass density	7850 kg/m^3	15.2 slugs/ft^3
Reinforced concrete		
weight density	24 kN/m^3	150 lb/ft^3
mass density	2400 kg/m^3	4.7 slugs/ft^3
Atmospheric pressure (sea level)		
Recommended value	101 kPa	14.7 psi
Standard international value	101.325 kPa	14.6959 psi
Acceleration of gravity		
(sea level, approx. 45° latitude)		
Recommended value	9.81 m/s^2	32.2 ft/s^2
Standard international value	9.80665 m/s^2	32.1740 ft/s^2

SI PREFIXES

Prefix	Symbol	Multiplication factor	
tera	T	10^{12}	= 1 000 000 000 000
giga	G	10^{9}	= 1 000 000 000
mega	M	10^{6}	= 1 000 000
kilo	k	10^{3}	= 1 000
hecto	h	10^{2}	= 100
deka	da	10^{1}	= 10
deci	d	10^{-1}	= 0.1
centi	c	10^{-2}	= 0.01
milli	m	10^{-3}	= 0.001
micro	μ	10^{-6}	= 0.000 001
nano	n	10^{-9}	= 0.000 000 001
pico	p	10^{-12}	= 0.000 000 000 001

Note: The use of the prefixes hecto, deka, deci, and centi is not recommended in SI.

Earth and Environmental Physical and Chemical Data

Mass = 5.9726 10^{24} kg
Volume = 108.321 10^{10} km³
Equatorial radius = 6,378.1 km
Polar radius = 6,356.8 km
Volumetric mean radius = 6,371.0 km
Mean density = 5,514 kg/m³
Surface gravity = 9.798 m/s²
Bond albedo = 0.306
Visual geometric albedo = 0.367
Solar irradiance = 1,367.6 W/m²
Black-body temperature = 254.3 K

Terrestrial Atmosphere

Surface atmospheric pressure = 1,014 mb
Surface atmospheric density = 1.217 kg/m³
Atmospheric scale height = 8.5 km
Mass of atmosphere = 5.1×10^{18} kg
Mass of hydrosphere (oceans) = 1.4×10^{21} kg
Mass of stratosphere = 0.5×10^{18} kg
Mass of water in the atmosphere = 1.3×10^{16} kg
Average temperature = 288 K (15°C)
Diurnal temperature range = 283 K to 293 K (10 to 20°C)
Wind speeds = 0 to 100 m/s
Mean molecular weight of the dry atmosphere = 28.97 g/mole

Source: NASA.gov: Earth Fact Sheet

WATER SOURCE	WATER VOLUME, IN CUBIC MILES	WATER VOLUME, IN CUBIC KILOMETERS	PERCENT OF FRESH WATER	PERCENT OF TOTAL WATER
Oceans, seas, & bays	321,000,000	1,338,000,000	–	96.5
Ice caps, glaciers, & permanent snow	5,773,000	24,064,000	68.7	1.74
Groundwater	5,614,000	23,400,000	–	1.69
Fresh	2,526,000	10,530,000	30.1	0.76
Saline	3,088,000	12,870,000	–	0.93
Soil moisture	3,959	16,500	0.05	0.001
Ground ice & permafrost	71,970	300,000	0.86	0.022

(Continued)

WATER SOURCE	WATER VOLUME, IN CUBIC MILES	WATER VOLUME, IN CUBIC KILOMETERS	PERCENT OF FRESH WATER	PERCENT OF TOTAL WATER
Lakes	42,320	176,400	–	0.013
Fresh	21,830	91,000	0.26	0.007
Saline	20,490	85,400	–	0.006
Atmosphere	3,095	12,900	0.04	0.001
Swamp water	2,752	11,470	0.03	0.0008
Rivers	509	2,120	0.006	0.0002
Biological water	269	1,120	0.003	0.0001

Source: Based on Igor Shiklomanov's chapter "World fresh water resources" in Peter H. Gleick (editor), 1993, Water in Crisis: A Guide to the World's Fresh Water Resources (Oxford University Press, New York).

PHYSICAL CONSTANTS FOR DRY AIR AT STP*

CONSTANT	VALUE
Average molecular weight	28.96
Specific heat	
At constant pressure	1,004.2 J/kg °C
At constant volume	719.6 J/kg °C
Density	1.293 kg/m^3
Viscosity	1.72 x 10^{-4} poise
Coefficient of heat conductivity	0.0209 W/m °C
Speed of sound in air	331.4 m/sec

*Standard temperature and pressure, denoted STP, is a temperature of 0°C and a pressure of 1 atm.

AREA, BIOMASS, AND PRODUCTIVITY OF ECOSYSTEM TYPES

ECOSYSTEM TYPE*	AREA (10^{12} m^2)	MEAN PLANT BIOMASS [kg(C)/m^2]	AVERAGE NET PRIMARY PRODUCTIVITY [kg(C)/m^2/yr]
Tropical forests	24.5	18.8	0.83
Temperate forests	12.0	14.6	0.56
Boreal forests	12.0	9.0	0.36
Woodland and scrubland	8.0	2.7	0.27
Savanna	15.0	1.8	0.32
Grassland	9.0	0.7	0.23
Tundra and alpine meadow	8.0	0.3	0.065
Desert scrub	18.0	0.3	0.032
Rock, ice, and sand	24.0	0.01	0.015

(Continued)

ECOSYSTEM TYPE*	AREA (10^{12} m^2)	MEAN PLANT BIOMASS [kg(C)/m^2]	AVERAGE NET PRIMARY PRODUCTIVITY [kg(C)/m^2/yr]
Cultivated land	14.0	0.5	0.29
Swamp and marsh	2.0	6.8	1.13
Lake and stream	2.5	0.01	0.23
Open ocean	332.0	0.0014	0.057
Upwelling zones	0.4	0.01	0.23
Continental shelf	26.6	0.005	0.16
Algae bed and reef	0.6	0.9	0.90
Estuaries	1.4	0.45	0.81

*For a description of each of the major types of ecosystems (deserts, boreal forests, estuaries, etc.) see Whittaker (1970), Whittaker and Likens (1973), and Ehrlich et al. (1977).

APPENDIX C

Sustainability Indicators

TABLE C.1 A comparison of sustainability footprint indicators

TYPE OF FOOT-PRINT INDICATOR	AUTHORS CITED	YEAR CITED	ISSUE TO CONVEY	INPUTS	OUTPUT
Carbon	Hogevold	~ 2003	Climate change	Greenhouse gas emissions Embodied energy of products from UN COMTRADE Fraction of anthropogenic emissions sequestered by the ocean Rate of carbon uptake per hectare of forest land	CO_2 equivalents
Ecological	Rees and Wackernagel	1992	Resource sustainability	Product harvest Carbon dioxide emitted National average yield of product Market price of product Carbon uptake capacity Area available for given land use	Hectares of land
Nitrogen	Leach, Galloway, Bleeker, Erisman, Kohn, and Kitzes	2012	Food impacts and sustainability	Electricity use Food consumption Food production Sanitation system Heating system Transportation	Kg of reactive Nitrogen per capita per year
Water	Hoekstra	2002	Water scarcity	Evapotranspiration Effective precipitation Environmental flow requirements Crop water requirements Crop yield Water stress coefficient Irrigation schedule Anthropogenic pollutant concentration Natural background concentration Natural assimilation capacity	Green, blue, or gray water volume

Source: Hoekstra, A.Y. and Hung, P.Q. (2002). "Virtual water trade: A quantification of virtual water flows between nations in relation to international crop trade" Value of Water research Report Series No. 11. UNESCO-IHE Institute for Water Education, Delft, the Netherlands.

Høgevold, N.M. (2003). "A corporate effort towards a sustainable business model: a case study from the Norwegian furniture industry" International Journal of Operations and Production Management 23(4):392-400.

Leach, A. M., Galloway, J. N., Bleeker, A., Erisman, J. W., Kohn, R. and Kitzes, J. (2012). "A nitrogen footprint model to help consumers understand their role in nitrogen losses to the environment." Environmental Development 1(1):40-66.

Rees, W. E., and Wackernagel, M. (1992). "Ecological Footprints and Appropriated Carrying Capacity: Measuring the Natural Capital Requirements of the Human Economy." Presented to the Conference of the International Society of Ecological Economics, Stockholm, 1992.

TABLE C.2 U.S. SEPA carbon dioxide emission factors for household resource consumption

SOURCE	SOURCE UNITS	CO$_2$ CONVERSION FACTOR	CONVERSION FACTOR UNITS
Electricity	kWh	7.0555×10^{-4}	Metric tons CO$_2$/kWh
Coal	Railcar	232.74	Metric tons CO$_2$/90.89 metric ton railcar
Natural gas	ft^3	5.44×10^{-5}	Metric tons CO$_2$/ ft^3
Oil	Barrel	0.43	Metric tons CO$_2$/barrel
Gasoline	Barrel	0.2913	Metric tons CO$_2$/ barrel
Kerosene	Barrel	0.42631	Metric tons CO$_2$/barrel
Propane cylinders used for home barbecues	Cylinder	0.024	Metric tons CO$_2$/cylinder
Uptake by trees	Urban tree planted	0.039	Metric tons CO$_2$/tree
U.S. forests storing carbon for one year	1 acre of average U.S. forest	1.22	Metric tons CO$_2$/acre-year

Source: Based on USEPA. Clean Energy Calculations and References. Accessed May 15, 2013. (http://www.epa.gov/cleanenergy/energy-resources/refs.html).

TABLE C.3 IPCC 50th percentile carbon emission factors for conventional and renewable energy sources in [g CO$_2$ equivalent/kWh]

ENERGY SOURCE	g CO$_2$ EQUIVALENTS/ kWh	ENERGY SOURCE	g CO$_2$ EQUIVALENTS/ kWh
Biopower	18	Wind Energy	12
Solar (Photovoltaic)	46	Nuclear energy	16
Solar (Concentrating solar power)	22	Natural gas	469
Geothermal energy	45	Oil	840
Hydropower	4	Coal	1,001
Ocean energy	8		

Source: Based on Moomaw, W., P. Burgherr, G. Heath, M. Lenzen, J. Nyboer, A. Verbruggen, 2011: Annex II: Methodology. In IPCC Special Report on Renewable Energy Sources and Climate Change Mitigation [O. Edenhofer, R. Pichs-Madruga, Y. Sokona, K. Seyboth, P. Matschoss, S. Kadner, T. Zwickel, P. Eickemeier, G. Hansen, S. Schlömer, C. von Stechow (eds)], Cambridge University Press, Cambridge, United Kingdom and New York, NY,USA.

TABLE C.4 Example yield factors for selected countries based upon 2007 data

YIELD	CROPLAND	FOREST	GRAZING LAND	FISHING GROUNDS
World Average	**1.0**	**1.0**	**1.0**	**1.0**
Algeria	0.3	0.4	0.7	0.9
Germany	2.2	4.1	2.2	3.0
Hungary	1.1	2.6	1.9	0.0
Japan	1.3	1.4	2.2	0.8
Jordan	1.1	1.5	0.4	0.7
New Zealand	0.7	2.0	2.5	1.0
Zambia	0.2	0.2	1.5	0.0

Source: Based on Ewing, B., A. Reed, A. Galli, J. Kitzes, M. Wackernagel. 2010b. Calculation Methodology for the National Footprint Accounts, 2010 Edition. Oakland: Global Footprint Network.

TABLE C.5 Equivalence factors for land-use types based upon 2007 data

LAND-USE TYPE	EQUIVALENCE FACTOR (GLOBAL HECTARES PER ACTUAL HECTARE OF LAND-USE TYPE)
Cropland	2.51
Forest	1.26
Grazing land	0.46
Marine and inland water	0.37
Built-up land	2.51

Source: Based on Ewing, B., A. Reed, A. Galli, J. Kitzes, M. Wackernagel. 2010b. Calculation Methodology for the National Footprint Accounts, 2010 Edition. Oakland: Global Footprint Network.

TABLE C.6 Types and scale of water footprint calculations

BOUNDARY	DESCRIPTION	UNIT
Product (good or service)	The aggregate of the water footprints of the various process steps relevant in the production of the product	Volume/time
Consumer	A function of the water footprints of the various products consumed by the consumer	Volume/product unit
Business (producer)	Is equal to the sum of the water footprints of the products that the producer or business delivers	Volume/time
Geographic area	Is equal to the sum of the water footprints of all processes taking place in that area	Volume/time
Humanity	The sum of the water footprints of all consumers of the world, which is equal to the sum of the water footprints of all final consumer goods and services consumed annually and also equal to the sum of all water-consuming or polluting processes in the world	Volume/time

Source: Based on Hoesktra, A, A.K. Chapagain, M. M. Aldaya, M. M. Mekonnen. 2011. The Water Footprint Assessment Manual: Setting the Global Standard. Earthscan, Ltd., London, UK.

TABLE C.7 Outcomes associated with water footprint sustainability and how they may be able to inform life cycle assessments

WATER FOOTPRINT ASSESSMENT PHASE	OUTCOME	PHYSICAL MEANING	RESOLUTION	LCA PHASE
Product water footprint accounting	Green, blue, and gray water footprints (volumetric)	Water volume consumed or polluted per unit volume of product	Spatiotemporally explicit	Life-cycle inventory
Product water footprint sustainability assessment	An evaluation of the sustainability of a green, blue and gray product water footprint from an environmental, social and economic perspective	Various measurable impact variables	Spatiotemporally explicit	Life-cycle impact assessment
Aggregation of selected information from the water footprint sustainability assessment	Aggregated water footprint impact indices	None	Nonspatiotemporally explicit	

Source: Based on Hoesktra, A, A.K. Chapagain, M. M. Aldaya, M. M. Mekonnen. 2011. The Water Footprint Assessment Manual: Setting the Global Standard. Earthscan, Ltd., London, UK.

TABLE C.8 Defining global warming potential

GHG	GWP (CO_2e)
Carbon dioxide (CO_2)	1
Methane (CH_4)	21
Nitrous oxide (N_2O)	310
Carbon tetrachloride (CCl_4)	1,400
1, 1, 1-Trichloroethane (CH_3CCl_3, methyl chloroform)	146
Bromomethane (CH_3Br)	5
Chloromethane (CH_3Cl)	13
Methylene chloride (CH_2Cl_2)	8.7
CFC-11 (CCl_3F, Freon-11)	3,800

Source: Based on Intergovernmental Panel on Climate Change Assessment Report 4, Chapter 2, Changes in Atmospheric Constituents and in Radiative Forcing (www.ipcc.ch), which is referenced by U.S. EPA at www.epa.gov/climatechange.

TABLE C.9 Densities of common materials

MATERIAL	DENSITY
Refined	
Cement	94 lbs/ft^3
GAC	30 lbs/ft^3
HDPE	59.6 lbs/ft^3
Lime (hydrated)	30 lbs/ft^3
PVC	87.36 lbs/ft^3
Steel	490 lbs/ft^3
Unrefined	
Asphalt	1.95 tons/cy
Concrete	1.95 tons/cy
Compact clay	1.5 tons/cy
Mulch/compost	0.4 tons/cy
Sand, gravel, soil	1.5 tons/cy

ft^3 = cubic foot

tons = short tons (2,000 pounds)

cy = cubic yard

Source: Based on U.S. EPA (2012) Methodology for Understanding and Reducing a Project's Environmental Footprint. U.S. Environmental Protection Agency, Ofice of Solid Waste and Emergency Response and Office of Superfund Remediation and Technology Innovation.

TABLE C.10 Approximate material and water content of aqueous chemical solutions

CHEMICAL SOLUTION	SPECIFIC GRAVITY	DENSITY OF SOLUTION (lbs/gal)	MASS OF CHEMICAL PER GALLON OF SOLUTION (lbs/gal)	VOLUME OF WATER PER GALLON OF SOLUTION (gal/gal)
Hydrochloric acid (37%)	1.19	9.92	3.67	0.75
Sulfuric acid (98%)	1.84	15.30	15.00	0.04
Sodium hydroxide (20%)	1.22	10.20	2.03	0.98
Sodium hydroxide (50%)	1.53	12.80	6.40	0.77
Hydrogen peroxide (30%)	1.11	9.26	2.78	0.78
Hydrogen peroxide (50%)	1.19	9.92	4.96	0.59
Ferric chloride (37%)	1.40	11.70	4.33	0.88
Sequestering agent (assume 40% solution)	1.20	10.00	4.00	0.72

Source: Based on U.S. EPA (2012) Methodology for Understanding and Reducing a Project's Environmental Footprint. U.S. Environmental Protection Agency, Ofice of Solid Waste and Emergency Response and Office of Superfund Remediation and Technology Innovation.

TABLE C.11 Approximate content of concrete, asphalt, and steel

| | | POUNDS PER CUBIC FOOT OF MATERIAL | | | |
| | | REFINED MATERIAL | | UNREFINED MATERIAL | |
MATERIAL	DENSITY (lbs/ft³)	TOTAL	RECYCLED/REUSED CONTENT	TOTAL	RECYCLED/REUSED CONTENT
Concrete	145	22	0	115	0
Fly ash concrete	145	22	4	115	0
Asphalt	145	7	1.4	138	28
Steel	490	490	270	0	0

Source: Based on Portland Cement Association (www.cement.org), the Federal Highways Administration Materials Group: http://www.fhwa.dot.gov/infrastructure/materialsgrp/flyash.htm, National Asphalt Pavement Association (www.hotmix.org), Steel Takes LEED® with Recycled Content, American Iron and Steel Institute, November 2009.

TABLE C.12 Estimating electricity use for typical remediation components

ITEM	CALCULATION FOR ESTIMATING ELECTRICITY USE
Small motors (< 1 HP) (e.g., for pumps, blowers, mixers)	$kWh = \dfrac{HP \times L_M}{n_m} \times 0.746 \times hours \ (n_m = 0.65, L = 80\%)$
Large motors (≤ HP) (e.g., for pumps, blowers, mixers)	$kWh = \dfrac{HP \times L_M}{n_m} \times 0.746 \times hours \ (n_m = 0.65, L = 80\%)$
Items with known electrical ratings (e.g., kW)	$kWh = kW \times hours$
Interpreting VFD settings	$kWh = \dfrac{HP \times L_V^3}{n_m \times n_V} \times 0.746 \times hours$

kW = kilowatts of electric power

kWh = kilowatt – hours of electricity

HP = horsepower

L_M = percent of motor full load

L_V = percent of VFD full load (or speed in Hertz divided by 60 Hertz)

n_m = motor efficient (typically 60% for less than 1 HP to 85% for 15 HP or greater)

n_V = VFD efficiency (typically 75% for less 50% load to 93% for more than 90% load)

hours = hours of operation over time frame of project

0.746 = conversion of HP to kW

VFD = variable frequency drive

Source: Based on U.S. EPA (2012) Methodology for Understanding and Reducing a Project's Environmental Footprint. U.S. Environmental Protection Agency, Office of Solid Waste and Emergency Response and Office of Superfund Remediation and Technology Innovation.

TABLE C.13 Suggested conversion factors

SUGGESTED CONVERSION FACTORS

ITEM OR SERVICE	UNIT	PARAMETERS USED, EXTRACTED, EMITTED, OR GENERATED						
		ENERGY USED (MMBtu)	CO_2e EMITTED (lbs)	NOx EMITTED (lbs)	SOx EMITTED (lbs)	PM EMITTED (lbs)	HAPs EMITTED (lbs)
Fuel Combustion								
Biodiesel use	gal	0.127	22.3	0.20	0	0.00099	NP
Diesel use	gal	0.139	22.5	0.17	0.0054	0.0034	0.0000052
Gasoline use	gal	0.124	19.6	0.11	0.0045	0.00054	0.000039
Landfill gas use	ccf CH_4	0.103	13.1	0.01	0.0000063	0.00076	0.0000084
Natural gas use	ccf	0.103	13.1	0.01	0.0000063	0.00076	0.0000084
Construction Materials								
Cement	Dry-lb	0.002	0.9	0.0018	0.00105	0.0000032	0.000029
Concrete	lb	0.00041	0.171	0.00035	0.00021	0.00001	0.00001
Gravel/sand/clay	lb	0.000028	0.0034	0.000017	0.000015	0.0000020	2.1E-10
HDPE	lb	0.031	1.9	0.0032	0.0041	0.00064	0.0000034
Photovoltaic system (installed)	Watt	0.034	4.5	0.015	0.032	0.00063	0.0000029
PVC	lb	0.022	2.6	0.0048	0.0076	0.0012	0.0047
Stainless steel	lb	0.012	3.4	0.0075	0.012	0.0044	0.00014
Steel	lb	0.0044	1.1	0.0014	0.0017	0.00056	0.000067
Other refined construction materials	lb	0.014	1.98	0.0037	0.0053	0.0014	0.00014
Other unrefined construction materials	lb	0.000028	0.00335	0.000017	0.000015	0.000002	2.1E-10
Treatment Materials and Chemicals								
Cheese whey	lb	0.0025	0.031	0.000062	0.000033	0.000002	NP
Emulsified vegetable oil	lb	0.0077	3.44	0.0066	0.0019	0.000033	NP
Molasses	lb	0.0044	0.48	0.0011	0.00024	0.000041	NP
Treatment materials and chemicals	lb	0.015	1.7	0.003	0.0065	0.0061	0.000016
Virgin GAC (coal based)	lb	0.015	8.5	0.014	0.034	0.00078	0.0012

(Continued)

TABLE C.13 (*Continued*)

| | | SUGGESTED CONVERSION FACTORS | | | | | | |
| | | PARAMETERS USED, EXTRACTED, EMITTED, OR GENERATED | | | | | | |
ITEM OR SERVICE	UNIT	ENERGY USED (MMBtu)	CO_2e EMITTED (lbs)	NOx EMITTED (lbs)	SOx EMITTED (lbs)	PM EMITTED (lbs)	HAPs EMITTED (lbs)
Fuel Processing							
Biodiesel produced	gal	0.029	−16.8	0.018	0.033	0.00082	NP
Diesel produced	gal	0.019	2.7	0.0064	0.013	0.00034	0.00012
Gasoline produced	gal	0.021	4.4	0.008	0.019	0.00052	0.00016
Natural gas produced	ccf	0.0052	2.2	0.0037	0.0046	0.000072	0.0000061
Public Water	gal × 1,000	0.0092	5	0.0097	0.0059	0.016	0.0000150
Offsite Services							
Offsite wastewater treatment	gal × 1000	0.015	4.4	0.016	0.015	NP	NP
Offsite solid waste disposal	ton	0.16	25	0.14	0.075	0.4	0.0014
Offsite hazardous waste disposal	ton	0.18	27.5	0.154	0.0825	0.44	0.0015
Offsite laboratory analysis	$	0.0065	1	0.0048	0.0036	0.0004	0.00013
Electricity Generation							
Resource Extraction for Electricity							
Coal extraction and processing	MWh	3.1	180	0.77	0.15	0.018	NP
Natural gas extraction and processing	MWh	1.6	270	0.18	13	0.0071	NP
Nuclear fuel extraction and processing	MWh	0.16	250	0.15	0.5	0.0015	NP
Oil extraction and processing	MWh	2.3	270	1.7	0.069	0.042	NP
Electricity Transmission	10% of electricity generation footprint for each parameter						

NP = Not Provided

Source: Based on U.S. EPA (2012) Methodology for Understanding and Reducing a Project's Environmental Footprint. U.S. Environmental Protection Agency, Office of Solid Waste and Emergency Response and Office of Superfund Remediation and Technology Innovation.

TABLE C.14 Estimating carbon stored in planted trees

$$C_s = e^{-2.2 + 2.4 \ln (bhd)} \times \left(1 + e^{\left(-1.63 + \frac{0.74}{b} hd\right)}\right) \times 0.46 \times 1.69$$

C_s = carbon dioxide stored

bhd = breast-height diameter of tree in (cm), assume 1 cm/yr growth rate for up to 40 years and a slower rate thereafter

Source: Based on Jenkins, Jennifer C., Chojnacky, David C., Heath, Linda S., Birdsey, Richard A., National Scale Biomass Estimators for United States tree species, Forest Science. 49: 12–35, 2003

TABLE C.15 Emission Factors and uncertainty estimates

EMISSIONS CATEGORY	FACTOR	UNITS	ESTIMATED ERROR (+/−)	SOURCE
Gasoline (direct)	8,874	gCO_2e/gal	1%	(1)
Gasoline (indirect)	2,307	gCO_2e/gal	15%	(2)
Diesel (direct)	10,153	gCO_2e/gal	1%	(1)
Diesel (indirect)	2,335	gCO_2e/gal	20%	(2)
Vehicle manufacturing	56	gCO_2e/mile	10%	(3)*
Average flight	223	gCO_2/passenger-mile	10%	(4)
Short flights (<400 mi)	254	gCO_2/passenger-mile	10%	(4)
Medium flights (400−1,500)	204	gCO_2/passenger-mile	10%	(4)
Long flights (1,500−3,000)	181	gCO_2/passenger-mile	10%	(4)
Extended flights (>3,000)	172	gCO_2/passenger-mile	10%	(4)
Air travel indirect effects	1.00	× direct emissions	30%	(5)*
Public transportation	179	gCO_2/passenger-mile	10%	(4)
Miles on bus	107	gCO_2/passenger-mile	10%	(4)
Miles on commuter rail (light & heavy)	163	gCO_2/passenger-mile	10%	(4)
Miles on transit rail (subway, tram)	163	gCO_2/passenger-mile	10%	(4)
Miles on Amtrak	185	gCO_2/passenger-mile	10%	(4)
Housing construction	930	gCO_2/sq.ft.	20%	(3)
Electricity usage ($)	11,789	gCO_2/$	19%	(7,8,10)
Electricity indirect factor	0.08		15%	(11,12)
Electricity usage (U.S. average shown)	835	gCO_2/kWh	5%	(7)
Natural gas usage (U.S. average shown)	4,317	gCO_2/$	5%	(7,8)
Therms natural gas (U.S. average shown)	5,470	gCO_2/therm	1%	(1)
Cubic feet natural gas (U.S. average shown)	54.7	gCO_2/cu.ft.	1%	(1)
Natural gas indirect factor	0.14		15%	(11)

(*Continued*)

TABLE C.15 (*Continued*)

EMISSIONS CATEGORY	FACTOR	UNITS	ESTIMATED ERROR (+/−)	SOURCE
Fuel oil and other fuels	682	$CO_2e/\$(2005)$	15%	(3)
Water (California average)	444	$gCO_2e/person$	15%	(9)
Water, sewage, wastes ($)	4,121	$CO_2e/\$(2005)$	15%	(3)*
Waste (California average)		$gCO_2e/person$	15%	(99)
Food	2.92	$gCO_2e/calorie$	15%	(3)*
Meat, fish, & eggs	5.53	$gCO_2e/calorie$	15%	(3)*
Beef, pork, lamb, veal	6.09	$gCO_2e/calorie$	15%	(3)*
Processed meat & other	2.24	$gCO_2e/calorie$	15%	(3)*
Fish & seafood	5.71	$gCO_2e/calorie$	15%	(3)*
Eggs and poultry	4.27	$gCO_2e/calorie$	15%	(3)*
Cereals & bakery products	1.45	$gCO_2e/calorie$	15%	(3)*
Dairy	4.00	$gCO_2e/calorie$	15%	(3)*
Fruits & vegetables	3.35	$gCO_2e/calorie$	15%	(3)*
Other (snacks, beverages, alcohol, oils, etc.)	2.24	$gCO_2e/calorie$	15%	(3)*
Goods (sum of below)	565	$CO_2e/\$(2005)$	15%	(3)*
Clothing	750	$CO_2e/\$(2005)$	15%	(3)*
Furnishings, appliances, other household	614	$CO_2e/\$(2005)$	15%	(3)*
Other goods	971	$CO_2e/\$(2005)$	15%	(3)*
Medical	696	$CO_2e/\$(2005)$	15%	(3)*
Entertainment	1,279	$CO_2e/\$(2005)$	15%	(3)*
Reading	2,100	$CO_2e/\$(2005)$	15%	(3)*
Personal care & cleaning	954	$CO_2e/\$(2005)$	15%	(3)*
Auto parts	558	$CO_2e/\$(2005)$	15%	(3)*
Services (sum of below)	507	$CO_2e/\$(2005)$	15%	(3)*
Vehicle services	433	$CO_2e/\$(2005)$	15%	(3)*
Household maintenance and repair	134	$CO_2e/\$(2005)$	15%	(3)*
Education	1,065	$CO_2e/\$(2005)$	15%	(3)*
Health care	1,151	$CO_2e/\$(2005)$	15%	(3)*
Personal business and finances	197	$CO_2e/\$(2005)$	15%	(3)*
Entertainment & recreation	711	$CO_2e/\$(2005)$	15%	(3)*

(*Continued*)

TABLE C.15 (*Continued*)

EMISSIONS CATEGORY	FACTOR	UNITS	ESTIMATED ERROR (+/−)	SOURCE
Information and communication	291	CO_2e/\$(2005)	15%	(3)*
Organizations and charity	122	CO_2e/\$(2005)	15%	(3)*
Miscellaneous services	720	CO_2e/\$(2005)	15%	(3)*
Water emissions per gallon	27.2	gCO_2e/gal	15%	(9)

*emission factor has been modified (beyond unit conversion), as described elsewhere in this report

(1) EIA(a), Voluntary Reporting of Greenhouse Gases Program

(2) GREET, 2.8a

(3) EIO-LCA, CEDA, authors' calculations

(4) WRI/WBCSK, Greenhouse Gas Protocol

(5) Air indirect effects assumed 0.9 plus 0.1 from airports

(6) Housing construction: Assume 90 tCO_2/50 yrs=1.8tCO_2/sqft

(7) eGRID

(8) EIA(b)

(9) California Air Resources Board

(10) Uncertainty parameter from Weber et al., 2010. See citation 72 for full reference

(11) Jaramillo et al., 2007. See citation 37

(12) Pacca, S., Horvath, A. 2002. See citation 36

Source: Jones, C. M. and Kammen, D.M. Quantifying Carbon Footprint Reduction Opportunities for U.S. Households and Communities: Supporting Materials. Renewable and Appropriate Energy Laboratory (RAEL), University of California, Berkley, CA.

Carbon Sources and Equivalence

TABLE D.1 Physical properties of CO_2

PROPERTY	VALUE
Molecular weight	44.01
Critical temperature	31.3°C
Critical pressure	73.9 bar
Critical density	467 kg/m^{-3}
Triple point temperature	−56.5°C
Triple point pressure	5.18 bar
Boiling (sublimation) point (1.013 bar)	−78.5°C
Gas Phase	
Gas density (1.013 bar at boiling point)	2.814 kg/m^{-3}
Gas density (@ STP)*	1.976 kg/m^{-3}
Specific volume (@ STP)*	0.506 m^3/kg^{-1}
Cp (@ STP)*	0.0364 kJ (mol^{-1} K^{-1})
Cv (@ STP)*	0.0278 kJ (mol^{-1} K^{-1})
Cp/Cv (@ STP)*	1.308
Viscosity (@ STP)	13.72 μN·s m^{-2} (or μPa·s)
Thermal conductivity (@ STP)*	14.65 mW (mK^{-1})
Solubility in water (@ STP)*	1.716 vol vol^{-1}
Enthalpy (@ STP)*	21.34 kJ mol^{-1}
Entropy (@ STP)*	117.2 J mol K^{-1}
Entropy of formation	213.8 J mol K^{-1}
Liquid Phase	
Vapor pressure (at 20°C)	58.8 bar
Liquid density (at −20°C and 19.7 bar)	1032 kg/m^{-3}
Viscosity (@ STP)*	99 μN·s m^{-2} (or μPa·s)
Solid Phase	
Density of carbon dioxide snow at freezing point	1562 kg/m^{-3}
Latent heat of vaporization (1.013 bar at sublimation point)	571.1 kJ/kg^{-1}

*Standard Temperature and Pressure, which is 0°C and 1.013 bar.

Source: Based on Air Liquid gas table; Kirk-Othmer (1985); NIST (2003).

TABLE D.2 Approximate equivalents and other definitions

TO CONVERT	INTO THE FOLLOWING UNITS	MULTIPLY BY
1 t C	t CO_2	3.667
1 t CO_2	m^3 CO_2 (at 1.013 bar and 15°C)	534
1 t crude oil	Bbl	7.33
1 t crude oil	m^3	1.165

TABLE D.3 Characterization of coals by rank (according to ASTM D388-92A)

CLASS GROUP	FIXED CARBON LIMITS (dmmf basis)[a] % Equal to or greater than	Less than	VOLATILE MATTER LIMITS (dmmf basis)[a] % Greater than	Equal to or less than	GROSS CALORIFIC VALUE LIMITS (mmmf basis)[b] MJ kg⁻¹ Equal to or greater than	Less than	Agglomerating Character
Anthracite							Non-agglomerating
Meta-anthracite	98	–	–	2	–	–	
Anthracite	92	98	2	8	–	–	
Semi-anthracite	86	92	8	14	–	–	
Bituminous coal							Commonly agglomerating
Low volatile	78	86	14	22	–	–	
Medium volatile	69	78	22	31	–	–	
High volatile A	–	69	31	–	32.6[d]	–	
High volatile B	–	–	–	–	30.2[d]	32.6	
High volatile C	–	–	–	–	26.7	30.2	
					24.4	26.7	Agglomerating
Sub-bituminous coal							Non-agglomerating
A	–	–	–	–	24.4	26.7	
B	–	–	–	–	22.1	24.4	
C	–	–	–	–	19.3	22.1	
Lignite							
A	–	–	–	–	14.7	19.3	
B	–	–	–	–	–	14.7	

[a] Indicates dry-mineral-matter-free basis (dmmf).

[b] mmmf indicates moist mineral-matter-free basis; moist refers to coal containing its natural inherent moisture but not including visible water on the surface of the coal.

[c] If agglomerating, classified in the low volatile group of the bituminous class.

[d] Coals having 69% or more fixed carbon (dmmf) are classified according to fixed carbon, regardless of gross calorific value.

Source: Based on ASTMD388-92A.

TABLE D.4 Typical ultimate analysis of petroleum-based heating fuels

COMPOSITION %	NO. 1 FUEL OIL (41.5°API)[a]	NO. 2 FUEL OIL (33°API)[a]	NO. 4 FUEL OIL (23.3°API)[a]	LOW SULFUR, NO. 6 FUEL OIL (33°API)[a]	HIGH SULFUR, NO. 6 FUEL OIL (15.5°API)[a]	PETROLEUM COKE[b]
Carbon	86.4	87.3	86.47	87.26	84.67	98.5
Hydrogen	13.6	12.6	11.65	10.49	11.02	3.08
Oxygen	0.01	0.04	0.27	0.64	0.38	1.11
Nitrogen	0.003	0.006	0.24	0.28	0.18	1.71
Sulfur	0.09	0.22	1.35	0.84	3.97	4.00
Ash	<0.01	<0.01	0.02	0.04	0.02	0.50
C/H Ratio	6.35	6.93	7.42	8.31	7.62	29.05

[a] Degree API = $(141.5/s) - 131.5$; where s is the specific density at 15°C

[b] Reference; Kaantee et al. (2003)

TABLE D.5 Typical natural gas composition

COMPONENT	PIPELINE COMPOSITION USED IN ANALYSIS Mol % (dry)	TYPICAL RANGE OF WELLHEAD COMPONENTS (Mol %) LOW VALUE	HIGH VALUE
Carbon dioxide, CO_2	0.5	0	10
Nitrogen, N_2	1.1	0	15
Methane, CH_4	94.4	75	99
Ethane, C_2H_6	3.1	1	15
Propane	0.5	1	10
Isobutane	0.1	0	1
N-butane	0.1	0	2
Pentanes	0.2	0	1
Hydrogen sulphide	0.0004	0	30
Helium	0.0	0	5
Heat of combustion (LHV)	48.252 MJ·kg^{-1}	–	–
Heat of combustion (HHV)	53.463 MJ·kg^{-1}	–	–

TABLE D.6 Chemical analysis and properties of some biomass fuels

	PEAT	WOOD (SAW DUST)	CROP RESIDUES (SUGAR CAN BAGASSE)	MUNICIPAL SOLID WASTE	ENERGY CROPS (EUCALYPTUS)
Proximate Analysis					
Moisture	70–90	7.3	–	16–38	–
Ash	–	2.6	11.3	11–20	0.52
Volatile matter	45–75	76.2	–	67–78	–
Fixed carbon	–	13.9	14.9	6–12	16.9
Ultimate Analysis					
C	45–60	46.9	44.8	–	48.3
H	3.5–6.8	5.2	5.4	–	5.9
O	20–45	37.8	39.5	–	45.1
N	0.75–3	0.1	0.4	–	0.2
S	–	0.04	0.01	–	0.01
Heating Value MJ kg^{-1} (HHV)	**17–22**	**18.1**	**17.3**	**15.9–17.5**	**19.3**

Source: Based on Sami et al., Hower 2003.

TABLE D.7 Direct emissions of non-greenhouse gases from two examples of coal and natural gas plants based on best available control technology, burning specific fuels

EMISSIONS	COAL (SUPERCRITICAL PC WITH BEST AVAILABLE EMISSION CONTROLS)	NATURAL GAS (NGCC WITH SCR)
NOx, g GJ^{-1}	4–5	5
SOx, g GJ^{-1}	4.5–5	0.7
Particulates, g GJ^{-1}	2.4–2.8	2
Mercury, mg GJ^{-1}	0.3–0.5	N/A

Source: Based on Cameron, 2002.

TABLE D.8 Direct CO_2 emission factors for some examples of carbonaceous fuels

CARBONACEOUS FUEL	HEAT CONTENT (HHV) (MJ kG^{-1})	EMISSION FACTOR (GCO_2 MJ^{-1})
Coal		
Anthracite	26.2	96.8
Bituminous	27.8	87.3
Sub-bituminous	19.9	90.3
Lignite	14.9	91.6
Biofuel		
Wood (dry)	20.0	78.4
Natural Gas	(kJ m^{-3})	
	37.3	50
Petroleum Fuel	(MJ m^{-3})	
Distillate Fuel Oil (#1, 2, & 4)	38,650	68.6
Residual Fuel Oil (#5, & 6)	41,716	73.9
Kerosene	37,622	67.8
LPG (average for fuel use)	25,220	59.1
Motor Gasoline	–	69.3

IPCC (2005) IPCC Special Report on carbon Dioxide Capture and Storage. Prepared by Working Group III of the Intergovernmental Panel on Climate Change [Metz, B., Davidson, O., de Coninck, H.C., Loas, M., and Meyer, L.A. (eds.)] Cambridge University Press, Cambridge, United Kingdom and New York, NY, USA. 442pp.

Source: Based on NIES, 2003.

Water Footprint of Products

TABLE E.1 Global average water footprint of primary crops and derived crop products, 1996–2005

FAOSTAT CROP CODE	PRODUCT DESCRIPTION	GLOBAL AVERAGE WATER FOOTPRINT ($m^3 \, ton^{-1}$)			
		GREEN	BLUE	GRAY	TOTAL
15	Wheat	1,277	342	207	1,827
	Wheat flour	1,292	347	210	1,849
	Wheat bread	1,124	301	183	1,608
	Dry pasta	1,292	347	210	1,849
	Wheat pellets	1,423	382	231	2,036
	Wheat, starch	1,004	269	163	1,436
	Wheat gluten	2,929	785	476	4,189
27	Rice, paddy	1,146	341	187	1,673
	Rice, husked (brown)	1,488	433	242	2,172
	Rice, broken	1,710	509	278	2,497
	Rice flour	1,800	535	293	2,628
	Rice groats and meal	1,527	454	249	2,230
44	Barley	1,213	79	131	1,423
	Barley, rolled or flaked grains	1,685	110	182	1,977
	Malt, not roasted	1,662	108	180	1,950
	Malt, roasted	2,078	135	225	2,437
	Beer made from malt	254	16	27	298
56	Maize (corn)	947	81	194	1,222
	Maize (corn) flour	971	83	199	1,253
	Maize (corn) groats and meal	837	72	171	1,081
	Maize (corn), hulled, pearled, sliced, or kibbled	1,018	87	209	1,314
	Maize (corn) starch	1,295	111	265	1,671
	Maize (corn) oil	1,996	171	409	2,575
71	Rye	1,419	25	99	1,544
	Rye flour	1,774	32	124	1,930
75	Oats	1,479	181	128	1,788
	Oats, groats and meal	2,098	257	182	2,536
	Oats, rolled or flaked grains	1,998	245	173	2,416
79	Millet	4,306	57	115	4,478
83	Sorghum	2,857	103	87	3,048
89	Buckwheat	2,769	144	229	3,142

(Continued)

TABLE E.1 (*Continued*)

FAOSTAT CROP CODE	PRODUCT DESCRIPTION	GLOBAL AVERAGE WATER FOOTPRINT (m³ ton⁻¹)			
		GREEN	BLUE	GRAY	TOTAL
116	Potatoes	191	33	63	287
	Tapioca of potatoes	955	165	317	1,436
	Potato flour and meal	955	165	317	1,436
	Potato flakes	694	120	230	1,044
	Potato starch	1,005	173	333	1,512
122	Sweet potatoes	324	5	53	383
125	Manioc (cassava)	550	0	13	564
	Tapioca of cassava	2,750	1	66	2,818
	Flour of cassava	1,833	1	44	1,878
	Dried cassava	1,571	1	38	1,610
	Manioc (cassava) starch	2,200	1	53	2,254
136	Taro (coco yam)	587	3	15	606
137	Yams	341	0	1	343
156	Sugar cane	139	57	13	210
	Raw sugar, cane	1,107	455	104	1,666
	Refined sugar	1,184	487	111	1,782
	Fructose, chemically pure	1,184	487	111	1,782
	Cane molasses	350	144	33	527
157	Sugar beet	82	26	25	132
	Raw sugar, beet	535	167	162	865
176	Beans, dry	3,945	125	983	5,053
181	Broad beans, horse beans, dry	1,317	205	496	2,018
187	Peas, dry	1,453	33	493	1,979
191	Chick peas	2,972	224	981	4,117
195	Cow peas, dry	6,841	10	55	6,906
197	Pigeon peas	4,739	72	683	5,494
201	Lentils	4,324	489	1,060	5,874
217	Cashew nuts	12,853	921	444	14,218
220	Chestnuts	2,432	174	144	2,750
221	Almonds, with shell	4,632	1,908	1,507	8,047
	Almonds, shelled or peeled	9,264	3,816	3,015	16,095
222	Walnuts, with shell	2,805	1,299	814	4,918
	Walnuts, shelled or peeled	5,293	2,451	1,536	9,280
223	Pistachios	3,095	7,602	666	11,363
224	Kola nuts	23,345	26	19	23,391
225	Hazelnuts, with shell	3,813	1,090	354	5,258
	Hazelnuts, shelled or peeled	7,627	2,180	709	10,515

(*Continued*)

TABLE E.1 (*Continued*)

FAOSTAT CROP CODE	PRODUCT DESCRIPTION	GLOBAL AVERAGE WATER FOOTPRINT (m^3 ton^{-1})			
		GREEN	BLUE	GRAY	TOTAL
226	Areca nuts	10,621	139	406	11,165
236	Soya beans	2,037	70	37	2,145
	Soya sauce	582	20	11	613
	Soya paste	543	19	10	572
	Soya curd	2,397	83	44	2,523
	Soy milk	3,574	123	65	3,763
	Soya bean flour and meals	2,397	83	44	2,523
	Soybean oil, refined	3,980	137	73	4,190
	Soybean oilcake	1,690	58	31	1,779
242	Groundnuts in shell	2,469	150	163	2,782
	Groundnuts shelled	3,526	214	234	3,974
	Groundnut oil, refined	6,681	405	442	7,529
	Groundnut oilcake	1,317	80	87	1,484
249	Coconuts	2,669	2	16	2,687
	Copra	2,079	1	12	2,093
	Coconut (husked)	1,247	1	7	1,256
	Coconut (copra) oil, refined	4,461	3	27	4,490
	Coconut/copra oilcake	829	1	5	834
	Coconut (coir) fibre, processed	2,433	2	15	2,449
254	Oil palm	1,057	0	40	1,098
	Palm nuts and kernels	2,762	1	105	2,868
	Palm oil, refined	4,787	1	182	4,971
	Palm kernel/babassu oil, refined	5,202	1	198	5,401
	Palm nut/kernel oilcake	802	0	31	833
260	Olives	2,470	499	45	3,015
	Olive oil, virgin	11,826	2,388	217	14,431
	Olive oil, refined	12,067	2,437	221	14,726
265	Castor oil seeds	8,423	1,175	298	9,896
	Castor oil	21,058	2,938	744	24,740
267	Sunflower seeds	3,017	148	201	3,366
	Sunflower seed oil, refined	6,088	299	405	6,792
	Sunflower seed oilcake	1,215	60	81	1,256
270	Rapeseed	1,703	231	336	2,271
	Rape oil, refined	3,226	438	636	4,301
	Rape seed oilcake	837	114	165	1,115
280	Safflower seeds	6,000	938	283	7,221
289	Sesame seed	8,460	509	403	9,371
	Sesame oil	19,674	1,183	936	21,793
292	Mustard seeds	2,463	1	345	2,809

(*Continued*)

TABLE E.1 (*Continued*)

FAOSTAT CROP CODE	PRODUCT DESCRIPTION	GLOBAL AVERAGE WATER FOOTPRINT (m³ ton⁻¹)			
		GREEN	BLUE	GRAY	TOTAL
296	Poppy seeds	1,723	0	464	2,188
299	Melon seed	5,087	56	41	5,184
328	Seed cotton	2,282	1,306	440	4,029
	Cotton seeds	755	432	146	1,332
	Cotton lint	5,163	2,955	996	9,113
	Cotton linters	1,474	844	284	2,602
	Cotton-seed oil, refined	2,242	1,283	432	3,957
	Cotton-seed oilcake	487	279	94	860
	Cotton, not carded or combed	5,163	2,955	996	9,113
	Cotton yarn waste (including thread waste)	950	544	183	1,677
	Garneted stock of cotton	1,426	816	275	2,517
	Cotton, carded or combed	5,359	3,067	1,934	9,460
	Cotton fabric, finished textile	5,384	3,253	1,344	9,982
333	Linseed	4,730	268	170	5,168
	Linseed oil, refined	8,618	488	310	9,415
	Linseed oilcake	2,816	160	101	3,077
336	Hempseed	3,257	12	417	3,685
358	Cabbages and other brassicas	181	26	73	280
366	Artichokes	478	242	98	818
367	Asparagus	1,524	119	507	2,150
372	Lettuce	133	28	77	237
373	Spinach	118	14	160	292
388	Tomatoes	108	63	43	214
	Tomato juice unfermented & not spirited	135	79	53	267
		539	316	213	1,069
	Tomato juice, concentrated	413	253	171	855
	Tomato paste	270	158	107	534
	Tomato ketchup	360	211	142	713
	Tomato puree	135	79	53	267
	Peeled tomatoes				
	Tomato, dried	2,157	1,265	853	4,276
393	Cauliflowers and broccoli	189	21	75	285
	Brussels sprouts	189	21	75	285
394	Pumpkins, squash and gourds	228	24	84	336
397	Cucumbers and gherkins	206	42	105	353
399	Eggplants (aubergines)	234	33	95	362
401	Chilies and peppers, green	240	42	97	379
402	Onions (incl. shallots), green	167	44	51	272

(*Continued*)

TABLE E.1 (*Continued*)

FAOSTAT CROP CODE	PRODUCT DESCRIPTION	GLOBAL AVERAGE WATER FOOTPRINT (m^3 ton^{-1})			
		GREEN	BLUE	GRAY	TOTAL
403	Onions, dry	192	88	65	345
406	Garlic	337	81	170	589
	Garlic powder	1,297	313	655	2,265
414	Beans, green	320	54	188	561
417	Peas, green	382	63	150	595
423	String beans	301	104	143	547
426	Carrots and turnips	106	28	61	195
430	Okra	474	36	65	576
446	Maize, green	455	157	88	700
461	Carobs	4,557	334	703	5,594
486	Bananas	660	97	33	790
489	Plantains	1,570	27	6	1,602
490	Oranges	401	110	49	560
	Orange juice	729	199	90	1,018
495	Tangerines, mandarins, clement	479	118	152	748
497	Lemons and limes	432	152	58	642
507	Grapefruit	367	85	54	506
515	Apples, fresh	561	133	127	822
	Apples, dried	4,678	1,111	1,058	6,847
	Apple juice unfermented & not spirited	780	185	176	1,141
521	Pears	645	94	183	922
526	Apricots	694	502	92	1,287
530	Sour cherries	1,098	213	99	1,411
531	Cherries	961	531	112	1,604
534	Peaches and nectarines	583	188	422	2,180
536	Plums and sloes	1,570	188	139	910
544	Strawberries	201	109	37	347
547	Raspberries	293	53	67	413
549	Gooseberries	487	8	31	526
550	Currants	457	19	23	499
552	Blueberries	341	334	170	845
554	Cranberries	91	108	77	276
560	Grapes	425	97	87	608
	Grapes, dried	1,700	386	347	2,433
	Grapefruit juice	490	114	71	675
	Grape wines, sparkling	607	138	124	869
567	Watermelons	147	25	63	235

(*Continued*)

TABLE E.1 (*Continued*)

| FAOSTAT CROP CODE | PRODUCT DESCRIPTION | GLOBAL AVERAGE WATER FOOTPRINT (m³ ton⁻¹) | | | |
		GREEN	BLUE	GRAY	TOTAL
569	Figs	1,527	1,595	228	3,350
571	Mangoes, mangosteens, guavas	1,314	362	124	1,800
572	Avocados	849	283	849	1,981
574	Pineapples	215	9	31	255
	Pineapple juice	1,075	45	153	1,273
577	Dates	930	1,250	98	2,277
591	Cashew apple	3,638	34	121	3,793
592	Kiwi fruit	307	168	38	514
600	Papayas	399	40	21	460
656	Coffee, green	15,249	116	532	15,897
	Coffee, roasted	18,153	139	633	18,925
661	Cocoa beans	19,745	4	179	19,928
	Cocoa paste	24,015	5	218	24,238
	Cocoa butter, fat and oil	33,626	7	305	33,938
	Cocoa powder	15,492	3	141	15,636
	Chocolate	16,805	198	193	17,196
667	Green and black tea	7,232	898	726	8,856
677	Hop cones	2,382	269	1,414	4,065
	Hop extract	9,528	1,077	5,654	16,259
687	Pepper of the genus Piper	6,540	467	604	7,611
689	Chilies and peppers, dry	5,869	1,125	371	7,365
692	Vanilla beans	86,329	39,048	1,065	126,505
693	Cinnamon (canella)	14,853	41	632	15,526
698	Cloves	59,834	30	1,314	61,205
702	Nutmeg, mace and cardamoms	30,683	2,623	1,014	34,319
711	Anise, badian, fennel, coriander	5,369	1,865	1,046	8,280
	Coriander seeds	5,369	1,865	1,046	8,280
720	Ginger	1,525	40	92	1,657
748	Peppermint	206	63	19	288
773	Flax fiber and tow	2,637	443	401	3,481
	Flax fiber, otherwise processed but not spun	2,866	481	436	3,783
	Flax tow and waste	581	98	88	767
777	Hemp fiber and tow	1,824	–	624	2,447
	True hemp fiber processed but not spun	2,026	–	693	2,719
780	Jute and other textile bast fibers	2,356	33	217	2,605

(*Continued*)

TABLE E.1 (*Continued*)

FAOSTAT CROP CODE	PRODUCT DESCRIPTION	GLOBAL AVERAGE WATER FOOTPRINT (m³ ton⁻¹)			
		GREEN	BLUE	GRAY	TOTAL
788	Ramie	3,712	201	595	4,507
789	Sisal	6,112	708	222	7,041
	Sisal textile fiber, processed but not spun	6,791	787	246	7,824
800	Agave fibers	6,434	9	106	6,549
809	Manila fiber (Abaca)	19,376	246	766	20,388
	Abaca fiber, processed but not spun	21,529	273	851	22,654
826	Tobacco, unmanufactured	2,021	205	700	2,925
836	Natural rubber	12,964	361	422	13,748

Source: Based on Mekonnen, M.M. and Hoekstra, A.Y. (2011) The green, blue and grey water footprint of crops and derived crop products, Hydrology and Earth System Sciences, 15(5): 1577–1600.

TABLE E.2 Global average water footprint of biofuel for 10 crops providing ethanol and seven crops providing biodiesel, 1996–2005

CROP	WATER FOOTPRINT PER UNIT OF ENERGY				WATER FOOTPRINT PER LITER OF BIOFUEL		
	GREEN	BLUE	GRAY		GREEN	BLUE	GRAY
Crops for ethanol	m³ per GJ ethanol				liters water per liter ethanol		
Barley	119	8	13		2,796	182	302
Cassava	106	0	3		2,477	1	60
Maize	94	8	19		2,212	190	453
Potatoes	62	11	21		1,458	251	483
Rice, paddy	113	34	18		2,640	785	430
Rye	140	2	10		3,271	58	229
Sorghum	281	10	9		6,585	237	201
Sugar beet	31	10	10		736	229	223
Sugar cane	60	25	6		1,400	575	132
Wheat	126	34	20		2,943	789	478
Crops for biodiesel	m³ per GJ biodiesel				liters water per liter biodiesel		
Coconuts	4,720	3	28		156,585	97	935
Groundnuts	177	11	12		5,863	356	388
Oil palm	150	0	6		4,975	1	190
Rapeseed	145	20	29		4,823	655	951
Seed cotton	310	177	60		10,274	5,879	1,981
Soybeans	326	11	6		10,825	374	198
Sunflower	428	21	28		14,200	696	945

Source: Based on Mekonnen, M.M. and Hoekstra, A.Y. (2011) The green, blue and grey water footprint of crops and derived crop products, Hydrology and Earth System Sciences, 15(5): 1577–1600.

TABLE E.3 The water footprint of rain-fed and irrigated agriculture for selected crops (1996–2005)

CROP	FARMING SYSTEM	YIELD (ton ha^{-1})	TOTAL WATER FOOTPRINT RELATED TO CROP PRODUCTION (Gm^3 yr^{-1})				WATER FOOTPRINT PER TON OF CROP (m^3 ton^{-1})			
			GREEN	BLUE	GRAY	TOTAL	GREEN	BLUE	GRAY	TOTAL
Wheat	Rain-fed	2.48	610	0	65	676	1,629	0	175	1,805
	Irrigated	3.31	150	204	58	411	679	926	263	1,868
	Global	2.74	760	204	123	1,087	1,278	342	208	1,828
Maize	Rain-fed	4.07	493	0	85	579	1,082	0	187	1,269
	Irrigated	6.01	104	51	37	192	595	294	212	1,101
	Global	4.47	597	51	122	770	947	81	194	1,222
Rice	Rain-fed	2.69	301	0	30	331	1,912	0	190	2,102
	Irrigated	4.67	378	202	81	661	869	464	185	1,519
	Global	3.90	679	202	111	992	1,146	341	187	1,673
Apples	Rain-fed	8.93	24	0	6	30	717	0	167	883
	Irrigated	15.91	8	8	2	18	343	321	71	734
	Global	10.92	33	8	7	48	561	133	127	822
Soybean	Rain-fed	2.22	328	0	5	333	2,079	0	33	2,112
	Irrigated	2.48	24	12	1	37	1,590	926	85	2,600
	Global	2.24	351	12	6	370	2,037	70	37	2,145
Sugarcane	Rain-fed	58.70	95	0	7	102	164	0	13	176
	Irrigated	71.17	85	74	10	169	120	104	14	238
	Global	64.96	180	74	17	271	139	57	13	210
Coffee	Rain-fed	0.68	106	0	4	110	15,251	0	523	15,774
	Irrigated	0.98	1	1	0	2	8,668	4,974	329	13,971
	Global	0.69	108	1	4	112	15,249	116	532	15,897
Rapeseed	Rain-fed	1.63	62	0	12	74	1,783	0	356	2,138
	Irrigated	1.23	4	9	1	14	1,062	2,150	181	3,394
	Global	1.57	66	9	13	88	1,703	231	336	2,271
Cotton	Rain-fed	1.35	90	0	13	103	3,790	0	532	4,321
	Irrigated	2.16	41	75	13	129	1,221	2,227	376	3,824
	Global	1.73	132	75	25	233	2,282	1,306	440	4,029
All crops	Rain-fed	–	4,701	0	472	5,173	–	–	–	–
	Irrigated	–	1,070	899	261	2,230	–	–	–	–
	Global	–	5,771	899	733	7,404	–	–	–	–

Source: Based on Mekonnen, M.M. and Hoekstra, A.Y. (2011) The green, blue and grey water footprint of crops and derived crop products, Hydrology and Earth System Sciences, 15(5): 1577–1600.

TABLE E.4 Average annual water footprint of one animal, per animal category (1996–2005).

ANIMAL CATEGORY	WATER FOOTPRINT OF LIVE ANIMAL ATE END OF LIFE TIME (m³/ton)	AVERAGE ANIMAL WEIGHT AT END OF LIFE TIME (kg)	AVERAGE WATER FOOTPRINT AT END OF LIFE TIME (m³/ ANIMAL)[1]	AVERAGE LIFE TIME (y)	AVERAGE ANNUAL WATER FOOTPRINT OF ONE ANIMAL (m³/y/ ANIMAL)[2]	ANNUAL WATER FOOTPRINT OF ANIMAL CATEGORY (Gm³/y)	% OF TOTAL WF
Beef cattle	7,477	253	1,889	3.0	630	798	33
Dairy cattle			20,558	10	2,056	469	19
Pigs	3,831	102	390	0.75	520	458	19
Broiler chickens	3,364	1.90	6	0.25	26	255	11
Horses	40,612	4.73	19,189	12	1,599	180	7
Layer chickens			47	1.4	33	167	7
Sheep	4,519	31.3	141	2.1	68	71	3
Goats	3,079	24.6	76	2.3	32	24	1
Total						2.422	100

[1] Calculated by multiplying the water footprint of the live animal at the end of its lifetime in m3/ton and the average animal weight.
[2] Calculated by dividing the average water footprint of the animal at the end of its life time by the average lifetime.

Source: Based on Mekonnen, M.M. and Hoekstra, A.Y. (2012) A global assessment of the water footprint of farm animal products, Ecosystems, 15(3): 401–415.

TABLE E.5 The water footprint of some selected food products from vegetable and animal origin

FOOD ITEM	WATER FOOTPRINT PER TON (m³/ton)				NUTRITIONAL CONTENT			WATER FOOTPRINT PER UNIT OF NUTRITIONAL VALUE		
	GREEN	BLUE	GRAY	TOTAL	CALORIE (kcal/kg)	PROTEIN (g/kg)	FAT (g/kg)	CALORIE (liter/ kcal)	PROTEIN (liter/g protein)	FAT (liter/g fat)
Sugar crops	130	52	15	197	285	0.0	0.0	0.69	0.0	0.0
Vegetables	194	43	85	322	240	12	2.1	1.34	26	154
Starchy roots	327	16	43	387	827	13	1.7	0.47	31	226
Fruits	726	147	89	962	460	5.3	2.8	2.09	180	348
Cereals	1,232	228	184	1,644	3,208	80	15	0.51	21	112
Oil crops	2,023	220	121	2,364	2,908	146	209	0.81	16	11
Pulses	3,180	141	734	4,055	3,412	215	23	1.19	19	180
Nuts	7,016	1,367	680	9,063	2,500	65	193	3.63	139	47
Milk	863	86	72	1,020	560	33	31	1.82	31	33
Eggs	2,592	244	429	3,265	1,425	111	100	2.29	29	33
Chicken meat	3,545	313	467	4,325	1,440	127	100	3.00	34	43
Butter	4,695	465	393	5,553	7,692	0.0	872	0.72	0.0	6.4

(Continued)

TABLE E.5 (*Continued*)

FOOD ITEM	WATER FOOTPRINT PER TON (m³/ton)				NUTRITIONAL CONTENT			WATER FOOTPRINT PER UNIT OF NUTRITIONAL VALUE		
	GREEN	BLUE	GRAY	TOTAL	CALORIE (kcal/kg)	PROTEIN (g/kg)	FAT (g/kg)	CALORIE (liter/ kcal)	PROTEIN (liter/g protein)	FAT (liter/g fat)
Pig meat	4,907	459	622	5,988	2,786	105	259	2.15	57	23
Sheep/goat meat	8,253	457	53	8,763	2,059	139	163	4.25	63	54
Beef	14,414	550	451	15,415	1,513	138	101	10.19	112	153

Source: Based on Mekonnen, M.M. and Hoekstra, A.Y. (2012) A global assessment of the water footprint of farm animal products, Ecosystems, 15(3): 401–415.

References

Mekonnen, M.M., and Hoekstra, A.Y. (2011) "The green, blue and grey water footprint of crops and derived crop products." *Hydrology and Earth System Sciences*, 15(5): 1577–1600.

Mekonnen, M.M., and Hoekstra, A.Y. (2012) "A global assessment of the water footprint of farm animal products." *Ecosystems,* 15(3): 401–415.

Exposure Factors for Risk Assessments

Source: Based on USEPA (2011) Exposure Factors Handbook: 2011 Edition. National Center for Environmental Assessment, Office of Research and Development, U.S. Environmental Protection Agency. Washington, D.C. U.S.A.

TABLE F.1 Summary of exposure factor recommendations

	PER CAPITA INGESTION OF DRINKING WATER				CONSUMERS-ONLY INGESTION OF DRINKING WATER			
	Mean		95th Percentile		Mean		95th Percentile	
	mL/day	mL/kg-day	mL/day	mL/kg-day	mL/day	mL/kg-day	mL/day	mL/kg-day
Children								
Birth to 1 month	184	52	839[a]	232[a]	470[a]	137[a]	858[a]	238[a]
1 to <3 months	227[a]	48	896[a]	205[a]	552	119	1,053[a]	285[a]
3 to <6 months	362[a]	52	1,056	159	556	80	1,171[a]	173[a]
6 to <12 months	360	41	1,055	126	467	53	1,147	129
1 to <2 years	271	23	837	71	308	27	893	75
2 to <3 years	317	23	877	60	356	26	912	62
3 to <6 years	327	18	959	51	382	21	999	52
6 to <11 years	414	14	1,316	43	511	17	1,404	47
11 to <16 years	520	10	1,821	32	637	12	1,976	35
16 to <18 years	573	9	1,783	28	702	10	1,883	30
18 to <21 years	681	9	2,368	35	816	11	2,818	36
Adults								
>21 years	1,043	13	2,958	40	1,227	16	3,092	42
>65 years	1,046	14	2,730	40	1,288	18	2,960	43
Pregnant women	819[a]	13[a]	2,503[a]	43[a]	872[a]	14[a]	2,589[a]	43[a]
Lactating women	1,379[a]	21[a]	3,434[a]	55[a]	1,665[a]	26[a]	3,588[a]	55[a]

[a] Estimates are less statistically reliable based on guidance published in the *Joint Policy on Variance Estimation and Statistical Reporting Standards on NHANES III and CSFII Reports: NHIS/NCHS Analytical Working Group Recommendations* (NCHS, 1993).

	INGESTION OF WATER WHILE SWIMMING			
	Mean		Upper Percentile	
	mL/event[a]	mL/hour	mL/event	mL/hour
Children	37	49	90[b]	120[b]
Adults	16	21	53[c]	71[c]

[a] Participants swam for 45 minutes.

[b] 97th percentile

[c] Based on maximum value.

TABLE F.1 Summary of exposure factor recommendations (*Continued*)

MOUTHING FREQUENCY AND DURATION

	Hand-to-Mouth				Object-to-Mouth			
	Indoor Frequency		Outdoor Frequency		Indoor Frequency		Outdoor Frequency	
	Mean contacts/ hour	95th Percentile contacts/ hour	Mean contacts/ hour	95th Percentile contacts/ hour	Mean contacts/ hour	95th Percentile contacts/ hour	Mean contacts/ hour	95th Percentile contacts/ hour
Birth to 1 month	–	–	–	–	–	–	–	–
1 to <3 months	–	–	–	–	–	–	–	–
3 to <6 months	28	65	–	–	11	32	–	–
6 to <12 months	19	52	15	47	20	38	–	–
1 to <2 years	20	63	14	42	14	34	8.8	21
2 to <3 years	13	37	5	20	9.9	24	8.1	40
3 to <6 years	15	54	9	36	10	39	8.3	30
6 to <11 years	7	21	3	12	1.1	3.2	1.9	9.1
11 to <16 years	–	–	–	–	–	–	–	–
16 to <21 years	–	–	–	–	–	–	–	–

Object-to-Mouth

	Duration	
	Mean minute/hour	95th Percentile minute/hour
Birth to 1 month	–	–
1 to <3 months	–	–
3 to <6 months	11	26
6 to <12 months	9	19
1 to <2 years	7	22
2 to <3 years	10	11
3 to <6 years	–	–
6 to <11 years	–	–
11 to <16 years	–	–
16 to <21 years	–	–

- No data

TABLE F.1 Summary of exposure factor recommendations (*Continued*)

SOIL AND DUST INGESTION

	Soil				Dust		Soil + Dust	
	General Population Central Tendency mg/day	High End			Central Tendency mg/day	General Population Upper Percentile mg/day	General Population Central Tendency mg/day	General Population Upper Percentile mg/day
		General Population Upper Percentile mg/day	Soil-Pica mg/day	Geophagy mg/day				
6 weeks to <1 year	30	–	–	–	30	–	60	–
1 to <6 years	50	–	1,000	50,000	60	–	100	–
3 to <6 years	–	200	–	–	–	100	–	200
6 to <21 years	50	–	1,000	50,000	60	–	100	–
Adult	20	–	–	50,000	30	–	50	–

- No data.

INHALATION

	Long-Term Inhalation Rates	
	Mean m³/day	95th Percentile m³/day
Birth to 1 month	3.6	7.1
1 to <3 months	3.5	5.8
3 to <6 months	4.1	6.1
6 to <12 months	5.4	8.0
1 to <2 years	5.4	9.2
Birth to <1 year	8.0	12.8
2 to <3 years	8.9	13.7
3 to <6 years	10.1	13.8
6 to <11 years	12.0	16.6
11 to <16 years	15.2	21.9
16 to <21 years	16.3	24.6
21 to <31 years	15.7	21.3
31 to <41 years	16.0	21.4
41 to <51 years	16.0	21.2
51 to <61 years	15.7	21.3
61 to <71 years	14.2	18.1
71 to <81 years	12.9	16.6
≥81 years	12.2	15.7

TABLE F.1 Summary of exposure factor recommendations (*Continued*)

Short-Term Inhalation Rates, by Activity Level

	Sleep or Nap		Sedentary/Passive		Light Intensity		Moderate Intensity		High Intensity	
	Mean $m^3/$ minute	95th $m^3/$ minute	Mean $m^3/$ minute	95th $m^3/$ minute	Mean $m^3/$ minute	95th $m^3/$ minute	Mean $m^3/$ minute	95th $m^3/$ minute	Mean $m^3/$ minute	95th $m^3/$ minute
Birth to <1year	3.0E-03	4.6E-03	3.1E-03	4.7E-03	7.6E-03	1.1E-02	1.4E-02	2.2E-02	2.6E-02	4.1E-02
1 to <2 years	4.5E-03	6.4E-03	4.7E-03	6.5E-03	1.2E-02	1.6E-02	2.1E-02	2.9E-02	3.8E-02	5.2E-02
2 to <3 years	4.6E-03	6.4E-03	4.8E-03	6.5E-03	1.2E-02	1.6E-02	2.1E-02	2.9E-02	3.9E-02	5.3E-02
3 to <6 years	4.3E-03	5.8E-03	4.5E-03	5.8E-03	1.1E-02	1.4E-02	2.1E-02	2.7E-02	3.7E-02	4.8E-02
6 to <11 years	4.5E-03	6.3E-03	4.8E-03	6.4E-03	1.1E-02	1.5E-02	2.2E-02	2.9E-02	4.2E-02	5.9E-02
11 to <16 years	5.0E-03	7.4E-03	5.4E-03	7.5E-03	1.3E-02	1.7E-02	2.5E-02	3.4E-02	4.9E-02	7.0E-02
16 to <21 years	4.9E-03	7.1E-03	5.3E-03	7.2E-03	1.2E-02	1.6E-02	2.6E-02	3.7E-02	4.9E-02	7.3E-02
21 to <31 years	4.3E-03	6.5E-03	4.2E-03	6.5E-03	1.2E-02	1.6E-02	2.6E-02	3.8E-02	5.0E-02	7.6E-02
31 to <41 years	4.6E-03	6.6E-03	4.3E-03	6.6E-03	1.2E-02	1.6E-02	2.7E-02	3.7E-02	4.9E-02	7.2E-02
41 to <51 years	5.0E-03	7.1E-03	4.8E-03	7.0E-03	1.3E-02	1.6E-02	2.8E-02	3.9E-02	5.2E-02	7.6E-02
51 to <61 years	5.2E-03	7.5E-03	5.0E-03	7.3E-03	1.3E-02	1.7E-02	2.9E-02	4.0E-02	5.3E-02	7.8E-02
61 to <71 years	5.2E-03	7.2E-03	4.9E-03	7.3E-03	1.2E-02	1.6E-02	2.6E-02	3.4E-02	4.7E-02	6.6E-02
71 to <81 years	5.3E-03	7.2E-03	5.0E-03	7.2E-03	1.2E-02	1.5E-02	2.5E-02	3.2E-02	4.7E-02	6.5E-02
≥81 years	5.2E-03	7.0E-03	4.9E-03	7.0E-03	1.2E-02	1.5E-02	2.5E-02	3.1E-02	4.8E-02	6.8E-02

SURFACE AREA

Total Surface Area

	Mean m^2	95th Percentile m^2
Birth to 1 month	0.29	0.34
1 to <3 months	0.33	0.38
3 to <6 months	0.38	0.44
6 to <12 months	0.45	0.51
1 to <2 years	0.53	0.61
2 to <3 years	0.61	0.70
3 to <6 years	0.76	0.95
6 to <11 years	1.08	1.48
11 to <16 years	1.59	2.06
16 to <21 years	1.84	2.33
<u>Adult Males</u>		
21 to <30 years	2.05	2.52
30 to <40 years	2.10	2.50
40 to <50 years	2.15	2.56
50 to <60 years	2.11	2.55
60 to <70 years	2.08	2.46
70 to <80 years	2.05	2.45
≥80 years	1.92	2.22

TABLE F.1 Summary of exposure factor recommendations (*Continued*)

Adult Females

21 to <30 years	1.81	2.25
30 to <40 years	1.85	2.31
40 to <50 years	1.88	2.36
50 to <60 years	1.89	2.38
60 to <70 years	1.88	2.34
70 to <80 years	1.77	2.13
≥80 years	1.69	1.98

Percent Surface Area of Body Parts

	Head	Trunk	Arms	Hands	Legs	Feet
			Mean Percent of Total Surface Area			
Birth to 1 month	18.2	35.7	13.7	5.3	20.6	6.5
1 to <3 months	18.2	35.7	13.7	5.3	20.6	6.5
3 to <6 months	18.2	35.7	13.7	5.3	20.6	6.5
6 to <12 months	18.2	35.7	13.7	5.3	20.6	6.5
1 to <2 years	16.5	35.5	13.0	5.7	23.1	6.3
2 to <3 years	8.4	41.0	14.4	4.7	25.3	6.3
3 to <6 years	8.0	41.2	14.0	4.9	25.7	6.4
6 to <11 years	6.1	39.6	14.0	4.7	28.8	6.8
11 to <16 years	4.6	39.6	14.3	4.5	30.4	6.6
16 to <21 years	4.1	41.2	14.6	4.5	29.5	6.1
Adult Males ≥21	6.6	40.1	15.2	5.2	33.1	6.7
Adult Females ≥21	6.2	35.4	12.8	4.8	32.3	6.6

Surface Area of Body Parts

	Head		Trunk		Arms		Hands		Legs		Feet	
	Mean m^2	95th m^2	Mean m^2	95th m^2	Mean m^2	95th m^2	Mean m^2	95th m^2	Mean m^2	95th m^2	Mean m^2	95th m^2
Birth to 1 month	0.053	0.062	0.104	0.121	0.040	0.047	0.015	0.018	0.060	0.070	0.019	0.022
1 to <3 months	0.060	0.069	0.118	0.136	0.045	0.052	0.017	0.020	0.068	0.078	0.021	0.025
3 to <6 months	0.069	0.080	0.136	0.157	0.052	0.060	0.020	0.023	0.078	0.091	0.025	0.029
6 to <12 months	0.082	0.093	0.161	0.182	0.062	0.070	0.024	0.027	0.093	0.105	0.029	0.033
1 to <2 years	0.087	0.101	0.188	0.217	0.069	0.079	0.030	0.035	0.122	0.141	0.033	0.038
2 to <3 years	0.051	0.059	0.250	0.287	0.088	0.101	0.028	0.033	0.154	0.177	0.038	0.044
3 to <6 years	0.060	0.076	0.313	0.391	0.106	0.133	0.037	0.046	0.195	0.244	0.049	0.061
6 to <11 years	0.066	0.090	0.428	0.586	0.151	0.207	0.051	0.070	0.311	0.426	0.073	0.100
11 to <16 years	0.073	0.095	0.630	0.816	0.227	0.295	0.072	0.093	0.483	0.626	0.105	0.136
16 to <21 years	0.076	0.096	0.759	0.961	0.269	0.340	0.083	0.105	0.543	0.687	0.112	0.142
Adult Males ≥21	0.136	0.154	0.827	1.10	0.314	0.399	0.107	0.131	0.682	0.847	0.137	0.161
Adult Females ≥21	0.114	0.121	0.654	0.850	0.237	0.266	0.089	0.106	0.598	0.764	0.122	0.146

TABLE F.1 Summary of exposure factor recommendations (*Continued*)

MEAN SOLID ADEHERENCE TO SKIN (mg/cm²)

	Face	Arms	Hands	Legs	Feet
Children					
Residential (indoors)[a]	–	0.0041	0.0011	0.0035	0.010
Daycare (indoors and outdoors)[b]	–	0.024	0.099	0.020	0.071
Outdoor sports[c]	0.012	0.011	0.11	0.031	–
Indoor sports[d]	–	0.0019	0.0063	0.0020	0.0022
Activities with soil[e]	0.054	0.046	0.17	0.051	0.20
Playing in mud[f]	–	11	47	23	15
Playing in sediment[g]	0.040	0.17	0.49	0.70	21
Adults					
Outdoor sports[i]	0.0314	0.0872	0.1336	0.1223	–
Activities with soil[h]	0.0240	0.0379	0.1595	0.0189	0.1393
Construction activities[j]	0.0982	0.1859	0.2763	0.0660	–
Clamming[k]	0.02	0.12	0.88	0.16	0.58

[a] Based on weighted average of geometric mean soil loadings for 2 groups of children (ages 3 to 13 years; $N = 10$) playing indoors.

[b] Based on weighted average of geometric mean soil loadings for 4 groups of daycare children (ages 1 to 6.5 years; $N = 21$) playing both indoors and outdoors.

[c] Based on geometric mean soil loadings of 8 children (ages 13 to 15 years) playing soccer.

[d] Based on geometric mean soil loadings of 6 children (ages ≥8 years) and 1 adult engaging in Tae Kwon Do.

[e] Based on weighted average of geometric mean soil loadings for gardeners and archeologists (ages 16 to 35 years).

[f] Based on weighted average of geometric mean soil loadings of 2 groups of children (age 9 to 14 years; $N = 12$) playing in mud.

[g] Based on geometric mean soil loadings of 9 children (ages 7 to 12 years) playing in tidal flats.

[h] Based on weighted average of geometric mean soil loadings of 3 groups of adults(ages 23 to 33 years) playing rugby and 2 groups of adults (ages 24 to 34) playing soccer.

[i] Based on weighted average of geometric mean soil loadings for 69 gardeners, farmers, groundskeepers, landscapers, and archeologists (ages 16 to 64 years) for faces, arms and hands; 65 gardeners, farmers, groundskeepers, and archeologists (ages 16 to 64 years) for legs; and 36 gardeners, groundskeepers, and archeologists (ages 16 to 62) for feet.

[j] Based on weighted average of geometric mean soil loadings for 27 construction workers, utility workers and equipment operators (ages 21 to 54) for faces, arms, and hands; and based on geometric mean soil loadings for 8 construction workers (ages 21 to 30 years) for legs.

[k] Based on geometric mean soil loadings of 18 adults (ages 33 to 63 years) clamming in tidal flats.

- No data.

	BODY WEIGHT
	Mean Kg
Birth to 1 month	4.8
1 to <3 months	5.9
3 to <6 months	7.4
6 to <12 months	9.2
1 to <2 years	11.4
2 to <3 years	13.8
3 to <6 years	18.6
6 to <11 years	31.8
11 to <16 years	56.8
16 to <21 years	71.6
Adults	80.0

TABLE F.1 Summary of exposure factor recommendations (*Continued*)

FRUIT AND VEGETABLE INTAKE

	Per Capita		Consumers-Only	
	Mean g/kg-day	95th Percentile g/kg-day	Mean g/kg-day	95th Percentile g/kg-day
Total Fruits				
Birth to 1 year	6.2	23.0[a]	10.1	25.8[a]
1 to <2 years	7.8	21.3[a]	8.1	21.4[a]
2 to <3 years	7.8	21.3[a]	8.1	21.4[a]
3 to <6 years	4.6	14.9	4.7	15.1
6 to <11 years	2.3	8.7	2.5	9.2
11 to <16 years	0.9	3.5	1.1	3.8
16 to <21 years	0.9	3.5	1.1	3.8
21 to <50 years	0.9	3.7	1.1	3.8
≥50 years	1.4	4.4	1.5	4.6
Total Vegetables				
Birth to 1 year	5.0	16.2[a]	6.8	18.1[a]
1 to <2 years	6.7	15.6[a]	6.7	15.6[a]
2 to <3 years	6.7	15.6[a]	6.7	15.6[a]
3 to <6 years	5.4	13.4	5.4	13.4
6 to <11 years	3.7	10.4	3.7	10.4
11 to <16 years	2.3	5.5	2.3	5.5
16 to <21 years	2.3	5.5	2.3	5.5
21 to <50 years	2.5	5.9	2.5	5.9
≥50 years	2.6	6.1	2.6	6.1

[a] Estimates are less statistically reliable based on guidance published in the *Joint Policy on Variance Estimation and Statistical Reporting Standards on NHANES III and CSFII Reports: NHIS/NCHS Analytical Working Group Recommendations* (NCHS, 1993).

FISH INTAKE

	Per Capita		Consumers-Only	
	Mean g/kg-day	95th Percentile g/kg-day	Mean g/kg-day	95th Percentile g/kg-day
General Population—Finfish				
All	0.16	1.1	0.73	2.2
Birth to 1 year	0.03	0.0[a]	1.3	2.9[a]
1 to <2 years	0.22	1.2[a]	1.6	4.9[a]
2 to <3 years	0.22	1.2[a]	1.6	4.9[a]
3 to <6 years	0.19	1.4	1.3	3.6[a]
6 to <11 years	0.16	1.1	1.1	2.9[a]
11 to <16 years	0.10	0.7	0.66	1.7
16 to <21 years	0.10	0.7	0.66	1.7
21 to <50 years	0.15	1.0	0.65	2.1
Females 13 to 49 years	0.14	0.9	0.62	1.8
≥50 years	0.20	1.2	0.68	2.0

TABLE F.1 Summary of exposure factor recommendations (*Continued*)

General Population—Shellfish

All	0.06	0.4	0.57	1.9
Birth to 1 year	0.00	0.0[a]	0.42	2.3[a]
1 to <2 years	0.04	0.0[a]	0.94	3.5[a]
2 to <3 years	0.04	0.0[a]	0.94	3.5[a]
3 to <6 years	0.05	0.0	1.0	2.9[a]
6 to <11 years	0.05	0.2	0.72	2.0[a]
11 to <16 years	0.03	0.0	0.61	1.9
16 to <21 years	0.03	0.0	0.61	1.9
21 to <50 years	0.08	0.5	0.63	2.2
Females 13 to 49 years	0.06	0.3	0.53	1.8
≥50 years	0.05	0.4	0.41	1.2

General Population—Total Finfish and Shellfish

All	0.22	1.3	0.78	2.4
Birth to 1 year	0.04	0.0[a]	1.2	2.9[a]
1 to <2 years	0.26	1.6[a]	1.5	5.9[a]
2 to <3 years	0.26	1.6[a]	1.5	5.9[a]
3 to <6 years	0.24	1.6[a]	1.3	3.6[a]
6 to <11 years	0.21	1.4	0.99	2.7[a]
11 to <16 years	0.13	1.0	0.69	1.8
16to <21 years	0.13	1.0	0.69	1.8
21 to <50 years	0.23	1.3	0.76	2.5
Females 13 to 49 years	0.19	1.2	0.68	1.9
≥50 years	0.25	1.4	0.71	2.1

[a] Estimates are less statistically reliable based on guidance published in the *Joint Policy on Variance Estimation and Statistical Reporting Standards on NHANES III and CSFII Reports: NHIS/NCHS Analytical Working Group Recommendations* (NCHS, 1993).

Recreational Population—Marine Fish—Atlantic

	Mean g/day	95th Percentile g/day
3 to <6 years	2.5	8.8
6 to <11 years	2.5	8.6
11 to <16 years	3.4	13
16 to <18 years	2.8	6.6
>18 years	5.6	18

TABLE F.1 Summary of exposure factor recommendations (*Continued*)

Recreational Population—Marine Fish—Gulf

3 to <6 years	3.2	13
6 to <11 years	3.3	12
11 to <16 years	4.4	18
16 to <18 years	3.5	9.5
>18 years	7.2	26

Recreational Population—Marine Fish—Pacific

3 to <6 years	0.9	3.3
6 to <11 years	0.9	3.2
11 to <16 years	1.2	4.8
16 to <18 years	1.0	2.5
>18 years	2.0	6.8

Recreational Population—Freshwater Fish—See Chapter 10

Native American Population—See Chapter 10

Other Populations—See Chapter 10

MEATS, DAIRY PRODUCTS, AND FAT INTAKE

	Per Capita		Consumers-Only	
	Mean g/kg-day	95th Percentile g/kg-day	Mean g/kg-day	95th Percentile g/kg-day
Total Meats				
Birth to 1 year	1.2	5.4[a]	2.7	8.1[a]
1 to <2 years	4.0	10.0[a]	4.1	10.1[a]
2 to <3 years	4.0	10.0[a]	4.1	10.1[a]
3 to <6 years	3.9	8.5	3.9	8.6
6 to <11 years	2.8	6.4	2.8	6.4
11 to <16 years	2.0	4.7	2.0	4.7
16 to <21 years	2.0	4.7	2.0	4.7
21 to <50 years	1.8	4.1	1.8	4.1
≥50 years	1.4	3.1	1.4	3.1
Total Dairy Products				
Birth to 1 year	10.1	43.2[a]	11.7	44.7[a]
1 to <2 years	43.2	94.7[a]	43.2	94.7[a]
2 to <3 years	43.2	94.7[a]	43.2	94.7[a]
3 to <6 years	24.0	51.1	24.0	51.1
6 to <11 years	12.9	31.8	12.9	31.8
11 to <16 years	5.5	16.4	5.5	16.4
16 to <21 years	5.5	16.4	5.5	16.4
21 to <50 years	3.5	10.3	3.5	10.3
≥50 years	3.3	9.6	3.3	9.6

TABLE F.1 Summary of exposure factor recommendations (*Continued*)

Total Fats				
Birth to 1 month	5.2	16	7.8	16
1 to <3 months	4.5	12	6.0	12
3 to <6 months	4.1	8.2	4.4	8.3
6 to <12 months	3.7	7.0	3.7	7.0
1 to <2 years	4.0	7.1	4.0	7.1
2 to <3 years	3.6	6.4	3.6	6.4
3 to <6 years	3.4	5.8	3.4	5.8
6 to <11 years	2.6	4.2	2.6	4.2
11 to <16 years	1.6	3.0	1.6	3.0
16 to <21 years	1.3	2.7	1.3	2.7
21 to <31 years	1.2	2.3	1.2	2.3
31 to <41 years	1.1	2.1	1.1	2.1
41 to <51 years	1.0	1.9	1.0	1.9
51 to <61 years	0.9	1.7	0.9	1.7
61 to <71 years	0.9	1.7	0.9	1.7
71 to <81 years	0.8	1.5	0.8	1.5
≥81 years	0.9	1.5	0.9	1.5

[a] Estimates are less statistically reliable based on guidance published in the *Joint Policy on Variance Estimation and Statistical Reporting Standards on NHANES III and CSFII Reports: NHIS/NCHS Analytical Working Group Recommendations* (NCHS, 1993).

	GRAINS INTAKE			
	Per Capita		Consumers-Only	
	Mean g/kg-day	95th Percentile g/kg-day	Mean g/kg-day	95th Percentile g/kg-day
Birth to 1 year	3.1	9.5[a]	4.1	10.3[a]
1 to <2 years	6.4	12.4[a]	6.4	12.4[a]
2 to <3 years	6.4	12.4[a]	6.4	12.4[a]
3 to <6 years	6.2	11.1	6.2	11.1
6 to <11 years	4.4	8.2	4.4	8.2
11 to <16 years	2.4	5.0	2.4	5.0
16 to <21 years	2.4	5.0	2.4	5.0
21 to <50 years	2.2	4.6	2.2	4.6
≥50 years	1.7	3.5	1.7	3.5

[a] Estimates are less statistically reliable based on guidance published in the *Joint Policy on Variance Estimation and Statistical Reporting Standards on NHANES III and CSFII Reports: NHIS/NCHS Analytical Working Group Recommendations* (NCHS, 1993).

TABLE F.1 Summary of exposure factor recommendations (*Continued*)

HOME-PRODUCED FOOD INTAKE

	Mean g/kg-day	95th Percentile g/kg-day
Consumer-Only Home-Produced Fruits, Unadjusted[a]		
1 to 2 years	8.7	60.6
3 to 5 years	4.1	8.9
6 to 11 years	3.6	15.8
12 to 19 years	1.9	8.3
20 to 39 years	2.0	6.8
40 to 69 years	2.7	13.0
≥70 years	2.3	8.7
Consumer-Only Home-Produced Vegetables, Unadjusted[a]		
1 to 2 years	5.2	19.6
3 to 5 years	2.5	7.7
6 to 11 years	2.0	6.2
12 to 19 years	1.5	6.0
20 to 39 years	1.5	4.9
40 to 69 years	2.1	6.9
≥70 years	2.5	8.2
Consumer-Only Home-Produced Meats, Unadjusted[a]		
1 to 2 years	3.7	10.0
3 to 5 years	3.6	9.1
6 to 11 years	3.7	14.0
12 to 19 years	1.7	4.3
20 to 39 years	1.8	6.2
40 to 69 years	1.7	5.2
≥70 years	1.4	3.5

TABLE F.1 Summary of exposure factor recommendations (*Continued*)

	Consumer-Only Home-Caught Fish, Unadjusted[a]	
1 to 2 years	–	–
3 to 5 years	–	–
6 to 11 years	2.8	7.1
12 to 19 years	1.5	4.7
20 to 39 years	1.9	4.5
40 to 69 years	1.8	4.4
≥70 years	1.2	3.7

Per Capita for Populations that (Garden or (Farm)

	Home-Produced Fruits[b]		Home-Produced Vegetables[b]	
	Mean g/kg-day	95th Percentile g/kg-day	Mean g/kg-day	95th Percentile g/kg-day
1 to <2 years	1.0 (1.4)	4.8 (9.1)	1.3 (2.7)	7.1 (14)
2 to <3 years	1.0 (1.4)	4.8 (9.1)	1.3 (2.7)	7.1 (14)
3 to <6 years	0.78 (1.0)	3.6 (6.8)	1.1 (2.3)	6.1 (12)
6 to <11 years	0.40 (0.52)	1.9 (3.5)	0.80 (1.6)	4.2 (8.1)
11 to <16 years	0.13 (0.17)	0.62 (1.2)	0.56 (1.1)	3.0 (5.7)
16 to <21 years	0.13 (0.17)	0.62 (1.2)	0.56 (1.1)	3.0 (5.7)
21 to <50 years	0.15 (0.20)	0.70 (1.3)	0.56 (1.1)	3.0 (5.7)
50 + years	0.24 (0.31)	1.1 (2.1)	0.60 (1.2)	3.2(6.1)

Per Capita for Populations that Farm or (Raise Animals)

	Home-Produced Meats[b]		Home-Produced Dairy	
	Mean g/kg-day	95th Percentile g/kg-day	Mean g/kg-day	95th Percentile g/kg-day
1 to <2 years	1.4 (1.4)	5.8 (6.0)	11(13)	76 (92)
2 to <3 years	1.4 (1.4)	5.8 (6.0)	11(13)	76 (92)
3 to <6 years	1.4 (1.4)	5.8 (6.0)	6.7 (8.3)	48 (58)
6 to <11 years	1.0 (1.0)	4.1 (4.2)	3.9 (4.8)	28 (34)
11 to <16 years	0.71 (0.73)	3.0 (3.1)	1.6 (2.0)	12 (14)
16 to <21 years	0.71 (0.73)	3.0 (3.1)	1.6 (2.0)	12 (14)
21 to <50 years	0.65 (0.66)	2.7 (2.8)	0.95 (1.2)	6.9 (8.3)
50 + years	0.51 (0.52)	2.1 (2.2)	0.92 (1.1)	6.7 (8.0)

[a] Not adjusted to account for preparation and post cooking losses.

[b] Adjusted for preparation and post cooking losses.

- No data

TABLE F.1 Summary of exposure factor recommendations (*Continued*)

TOTAL PER CAPITA FOOD INTAKE

	Mean g/kg-day	95th Percentile g/kg-day
Birth to 1 year	91	208[a]
1 to <3 years	113	185[a]
3 to <6 years	79	137
6 to <11 years	47	92
11 to <16 years	28	56
16 to <21 years	28	56
21 to <50 years	29	63
≥50 years	29	59

[a] Estimates are less statistically reliable based on guidance published in the *Joint Policy on Variance Estimation and Statistical Reporting Standards on NHANES III and CSFII Reports: NHIS/NCHS Analytical Working Group Recommendations* (NCHS, 1993).

HUMAN MILK AND LIPID INTAKE

	Mean		Upper Percentile	
	mL/day	mL/kg-day	mL/day	mL/kg-day
Human Milk Intake				
Birth to 1 month	510	150	950	220
1 to <3 months	690	140	980	190
3 to <6 months	770	110	1,000	150
6 to <12 months	620	83	1,000	130
Lipid Intake				
Birth to 1 month	20	6	38	8.7
1 to <3 months	27	5.5	40	8.0
3 to <6 months	30	4.2	42	6.1
6 to <12 months	25	3.3	42	5.2

TABLE F.1 Summary of exposure factor recommendations (*Continued*)

ACTIVITY FACTORS

	Time Indoors (total) minutes/day		Time Outdoors (total) minutes/day		Time Indoors (at residence) minutes/day	
	Mean	95th Percentile	Mean	95th Percentile	Mean	95th Percentile
Birth to <1 month	1,440	-	0	-	-	-
1 to <3 months	1,432	-	8	-	-	-
3 to <6 months	1,414	-	26	-	-	-
6 to <12 months	1,301	-	139	-	-	-
Birth to <1 year	-	-	-	-	1,108	1,440
1 to <2 years	1,353	-	36	-	1,065	1,440
2 to <3 years	1,316	-	76	-	979	1,296
3 to <6 years	1,278	-	107	-	957	1,355
6 to <11 years	1,244	-	132	-	893	1,275
11 to <16 years	1,260	-	100	-	889	1,315
16 to <21 years	1,248	-	102	-	833	1,288
18 to <64 years	1,159	-	281	-	948	1,428
>64 years	1,142	-	298	-	1,175	1,440

	Showering minutes/day		Bathing minutes/day		Bathing/Showering minutes/day	
	Mean	95th Percentile	Mean	95th Percentile	Mean	95th Percentile
Birth to <1 year	15	-	19	30	-	-
1 to <2 years	20	-	23	32	-	-
2 to <3 years	22	44	23	45	-	-
3 to <6 years	17	34	24	60	-	-
6 to <11 years	18	41	24	46	-	-
11 to <16 years	18	40	25	43	-	-
16 to <21 years	20	45	33	60	-	-
18 to <64 years	-	-	-	-	17	-
>64 years	-	-	-	-	17	-

TABLE F.1 Summary of exposure factor recommendations (*Continued*)

	Playing on Sand/Gravel minutes/day		Playing on Grass minutes/day		Playing on Dirt minutes/day	
	Mean	95th Percentile	Mean	95th Percentile	Mean	95th Percentile
Birth to <1 year	18	-	52	-	33	
1 to <2 years	43	121	68	121	56	121
2 to <3 years	53	121	62	121	47	121
3 to <6 years	60	121	79	121	63	121
6 to <11 years	67	121	73	121	63	121
11 to <16 years	67	121	75	121	49	120
16 to <21 years	83	-	60	-	30	-
18 to <64 years	0 (median)	121	60 (median)	121	0 (median)	120
>64 years	0 (median)	-	121 (median)	-	0 (median)	-

	Swimming minutes/month	
	Mean	95th Percentile
Birth to <1 year	96	-
1 to <2 years	105	-
2 to <3 years	116	181
3 to <6 years	137	181
6 to <11 years	151	181
11 to <16 years	139	181
16 to <21 years	145	181
18 to <64 years	45 (median)	181
>64 years	40 (median)	181

TABLE F.1 Summary of exposure factor recommendations (*Continued*)

	Occupational Mobility	
	Median Tenure (years) Men	Median Tenure (years) Women
1 All ages, ≥16 years	7.9	5.4
16 to 24 years	2.0	1.9
25 to 29 years	4.6	4.1
30 to 34 years	7.6	6.0
35 to 39 years	10.4	7.0
40 to 44 years	13.8	8.0
45 to 49 years	17.5	10.0
50 to 54 years	20.0	10.8
55 to 59 years	21.9	12.4
60 to 64 years	23.9	14.5
65 to 69 years	26.9	15.6
≥70 years	30.5	18.8

	Population Mobility			
	Residential Occupancy Period (years)		Current Residence Time (years)	
	Mean	95th Percentile	Mean	95th Percentile
All	12	33	13	46

- No data.

LIFE EXPECTANCY	
	Years
Total	78
Males	75
Females	80

TABLE F.1 Summary of exposure factor recommendations (*Continued*)

BUILDING CHARACTERISTICS

	Residential Buildings	
	Mean	10th Percentile
Volume of Residence (m³)	492	154
Air Exchange Rate (air changes/hour)	0.45	0.18

	Non-Residential Buildings	
	Mean (Standard Deviation)	10th Percentile
Volume of Non-residential Buildings (m³)		408
Vacant	4,789	510
Office	5,036	2,039
Laboratory	24,681	1,019
Non-refrigerated warehouse	9,298	476
Food sales	1,889	816
Public order and safety	5,253	680
Outpatient healthcare	3,537	1,133
Refrigerated warehouse	19,716	612
Religious worship	3,443	595
Public assembly	4,839	527
Education	8,694	442
Food service	1,889	17,330
Inpatient healthcare	82,034	1,546
Nursing	15,522	527
Lodging	11,559	1,359
Strip shopping mall	7,891	35,679
Enclosed mall	287,978	510
Retail other than mall	3,310	459
Service	2,213	425
Other	5,236	527
All Buildings	5,575	
Air Exchange Rate (air changes/hour)	1.5(0.87) Range 0.3-4.1	0.60

Benchmarks Used in Conservation Planning

TABLE G.1 Recent estimates of indoor water use with and without conservation

Type of Use	WITHOUT CONSERVATION Amount (gpcd)	Percent of total	WITH CONSERVATION Amount (gpcd)	Percent of total	Savings
Toilets	18.3	28.4%	10.4	23.2%	44%
Clothes washers	14.9	23.1%	10.5	23.4%	30%
Showers	12.2	18.8%	10.0	22.4%	18%
Faucets	10.3	16.0%	10.0	22.5%	2%
Leaks	6.6	10.2%	1.5	3.4%	77%
Baths	1.2	1.9%	1.2	2.7%	0%
Dishwashers	1.1	1.6%	1.1	2.4%	0%
Total indoor water use	64.6	100%	44.7	100%	31%

gpcd = gallons per capita per day

Note: These data are provided for illustrative purposes only and may not be applicable to a given situation. To the extent practical, planners use system – specific assumptions and estimates.

Source: Based on AWWA WaterWiser, "Household End Use of Water Without and With Conservation," 1997 Residential Water Use Summary - Typical Single Family Home (http://www.waterwiser.org/wateruse/tables.html), in USEPA Water Conservation Plan Guidelines, Appendix B, http://www.epa.gov/WaterSense /docs/app_b508.pdf.

TABLE G.2 Benchmarks for estimating residential end uses of water

TYPE OF USES	UNITS	LIKELY RANGE OF AVERAGE VALUES
INDOOR		
Average household size	Pearsons	2.0–3.0
Frequency of toilet flushing	Flushes/person/day	4.0–6.0
Flushing volumes	Gallons/flush	1.6–8.0
Fraction of leaking toilets	Percent	0–30
Showering frequency	Showers/person/day	0–1.0
Duration of average shower	Minutes	5–15
Shower flow rates	Gallons/minute	1.5–5.0
Bathing frequency	Baths/person/day	0–0.2
Volume of water	Gallons/cycle	30–50
Washing machine use	Loads/person/day	0.2–0.5

(Continued)

TABLE G.2 (*Continued*)

TYPE OF USES	UNITS	LIKELY RANGE OF AVERAGE VALUES
Volume of water	Gallons/cycle	45–50
Dishwasher use	Loads/person/day	0.1–0.3
Volume of water	Gallons/cycle	10–15
Kitchen faucet use	Minutes/person/day	0.5–5.0
Faucet flow rates	Gallons/minute	2.0–3.0
Bathroom faucet use	Minutes/person/day	0.5–3.0
Faucet flow rates	Gallons/minute	2.0–3.0
OUTDOOR		
Average lot size[a]	Square feet	5,000–8,000
Average house size[a]	Square feet	1,200–2,500
Landscape area[a]	Square feet	4,000–5,000
Fraction of lot size in turf[a]	Percent	30–50
Water application rates[a]	Feet/year	1–5
Percent of homes with pools	Percent	10–25
Pool evaporation losses	Feet/year	3–7
Frequency of refilling pools	Times per year	1–2
Frequency of car washing	Times/month	1–2

[a] Reflects single-family averages.

Note: These data are provided for illustrative purposes only and may not be current or applicable. To the extent practical, planners should regionally appropriate or system-specific assumptions and estimates.

Source: Based on Duane D. Baumann, John J. Boland, and W. Michael Hanemann, Urban Water Demand Management and Planning. New York: McGraw Hill, 1998. in USEPA Water Conservation Plan Guidelines, Appendix B, http://www.epa.gov/WaterSense/docs/app_b508.pdf.

TABLE G.3 Sample calculation of water savings from showerhead replacement

The following calculations represent the water savings expected as the result of a showerhead retrofit program. The savings rate represents a difference in average winter water use between homes with low-flow showerheads and homes without low-flow showerheads.

- Nonconserving showerhead flow rate = 3.4 gallons/minute
- Low-flow showerhead flow rate = 1.9 gallons/minute
- Estimated showering time = 4.8 minutes/person/day
- Average winter household water use = 200 gallons per household per day
- Average household size = 2.5 persons
- Water use with nonconserving showerhead = (3.4 gal/min) × (4.8 min/person/day) = 16.3 gpcd
- Water use with low-flow showerhead = (3.4 gal/min) × (4.8 min/person/day) = 9.1 gpcd
- Water savings = 16.3 gpcd − 9.1 gpcd = 7.2 gpcd

At an average household size of 2.5 persons, the savings rate would be 18.0 gallons per household per day (2.5 persons × 7.2 GPCD). The formula for calculating the reduction factors representing the fraction of, for example, single-family winter water use is

R = (18.0 GPHD)/(200 GPHD during winter) = 0.09 (or 9 percent)

Note: These data are provided for illustrative purposes only and may not be current or applicable. To the extent practical, planners should regionally appropriate or system-specific assumptions and estimates.

Source: Based on Duane D. Baumann, John J. Boland, and W. Michael Hanemann, Urban Water Demand Management and Planning. New York: McGraw Hill, 1998, in USEPA Water Conservation Plan Guidelines, Appendix B, http://www.epa.gov/WaterSense/docs/app_b508.pdf.

TABLE G.4 Benchmarks for savings from selected conservation measures

CATEGORY	MEASURE	REDUCTION IN END USE	LIFESPAN (YEARS)
Level 1 Measures			
Universal metering	Connection metering	20%	8 to 20
	submetering	20 to 40%	8 to 20
Water accoungting and loss control	System audits and leak Detection	Based on system	n/a
Costing and pricing	10% increase in residential prices	2 to 4%	n/a
	10% increase in nonresidential prices	5 to 8%	n/a
	Increasing-block rate	5%	n/a
Information and education	Public education and behavior changes	2 to 5%	n/a
Level 2 Measures			
End-use audits	General industrial water conservation	10 to 20%	n/a
	Outdoor residential use	5 to 10%	n/a
	Large landscape water audits	10 to 20%	n/a
Retrofits	Toilet tank displacement devices (For toilets using > 3.5 gallons/flush)	2 to 3 gpcd	1.5
	Toilet retrofit	8 to 14 gpcd	1.5
	Showerhead retrofit (aerator)	4 gpcd	1 to 3
	Faucet retrofit (aerator)	5 gpcd	1 to 3
	Fixture leak repair	0.5 gpcd	1
	Governmental buildings (indoors)	5%	n/a
Pressure management	Pressure reduction, system	3 to 6% of total production	n/a
	Pressure-reducing valves, residential	5 to 30%	n/a
Outdoor water-use efficiency	Low water-use plants	15 to 20%	n/a
	Lawn watering guides	10 to 25%	n/a
	Large landscape management	15 to 20%	n/a
	Irrigation timer	10 gpcd	4
Level 3 Measures			
Replacements and promotions	Toilet replacement, residential	16 to 20 gpcd	15 to 25
	Toilet replacement, commercial	16 to 20 gpcd	10 to 20
	Showerhead replacement	8.1 gpcd	2 to 10
	Faucet replacement	6.4 gpcd	10 to 20
	Clothes washers, residential	4 to 12 gpcd	12
	Dishwashers, residential	1 gpcd	12
	Hot water demand units	10 gpcd	n/a
Reuse and recycling	Cooling tower program	Up to 90%	n/a

(Continued)

TABLE G.4 (*Continued*)

CATEGORY	MEASURE	REDUCTION IN END USE	LIFESPAN (YEARS)
Water-use regulation	Landscape requirements for new developments	10 to 20% in sector	n/a
	Graywater reuse, residential	20 to 30 gpcd	n/a
Integrated resource management	Planning and management	Energy, chemical, and wastewater treatment costs	n/a

Source: Based on various sources, in USEPA Water Conservation Plan Guidelines, Appendix B, http://www.epa.gov/WaterSense/docs/app_b508.pdf.

TABLE G.5 Water efficiency standards established by the Energy Policy Act of 1992

Faucets. The maximum water use allowed by any of the following faucets manufactured after January 1, 1994, when measured at a flowing water pressure of 80 pounds per square inch, is as follows:

FAUCET TYPE	MAXIMUM FLOW RATE (GALLONS PER MINUTE OR PER CYCLE)
Lavatory faucets	2.5 gpm
Lavatory replacement aerators	2.5 gpm
Kitchen faucets	2.5 gpm
Kitchen replacement aerators	2.5 gpm
Metering faucets	0.25 gpc

Showerheads. The maximum water use allowed for any showerhead manufactured after January 1, 1994, is 2.5 gallons per minute when measured at a flowing pressure of 80 pounds per square inch.

Water Closets. (1) The maximum water use allowed in gallons per flush for any of the following water closets manufactured after January 1, 1994, is as follows.

WATER CLOSET TYPE	MAXIMUM FLUSH RATE (GALLONS PER FLUSH)
Gravity tank-type toilets	1.6 gpf
Flushometer tank toilets	1.6 gpf
Electromechanical hydraulic toilets	1.6 gpf
Blowout toilets	3.5 gpf

(2) The maximum water use allowed for any gravity tank-type white two-piece toilet which bears an adhesive label conspicuous upon installation of the words "Commercial Use Only" manufactured after January 1, 1994 and before January 1, 1997, is 3.5 gallons per flush.

(3) The maximum water use allowed for flushometer valve toilets, other than blowout toilets, manufactured after January 1, 1997, is 1.6 gallons per flush.

Urinals. The maximum water use allowed for any urinals manufactured after January 1, 1994, is 1.0 gallons per flush.

Note: These standards were developed in 1992. New and emerging technologies can increase the cost effectiveness of conservation measures, affect demand forecasts, and eventually lead to the establishment of new standards.

Source: Based on USEPA Water Conservation Plan Guidelines, Appendix B, http://www.epa.gov/WaterSense/docs/app_b508.pdf.

TABLE G.6 Potential water savings from efficient fixtures

FIXTURE[a]	FIXTURE CAPACITY[b]	WATER USE (gpd)		WATER SAVINGS (GPD)	
		PER CAPITA	2.7-PERSON HOUSEHOLD	PER CAPITA	2.7-PERSON HOUSEHOLD
TOILETS[c]					
Efficient	1.5 gal/flush	6.0	16.2	n/a	n/a
Low-flow	3.5 gal/flush	14.0	37.8	8.0	21.6
Conventional	5.5 gal/flush	22.0	59.4	16.0	43.2
Conventional	7.0 gal/flush	28.0	75.6	22.0	59.4
SHOWERHEADS[d]					
Efficient	2.5 [1.7] gal/min	8.2	22.1	n/a	n/a
Low-flow	3.0 to 5.0 [2.6] gal/min	12.5	33.8	4.3	11.7
Conventional	5.0 to 8.0 [3.4] gal/min	16.3	44.0	8.1	22.0
FAUCETS[e]					
Efficient	2.5 [1.7] gal/min	6.8	18.4	n/a	n/a
Low-flow	3.0 [2.0] gal/min	8.0	21.6	1.2	3.2
Conventional	3.0 to 7.0 [3.3] gal/min	13.2	36.6	6.4	17.2
TOILETS, SHOWERHEADS, AND FAUCETS COMBINED					
Efficient	n/a	21.0	56.7	n/a	n/a
Low-flow	n/a	34.5	93.2	13.4	36.4
Conventional	n/a	54.5	147.2	33.5	90.4

n/a = not applicable

[a] Efficient = post-1994

   Low-flow = post-1980

   Conventional = pre-1980

[b] For showerheads and faucets: maximum rated fixture capacity (measured fixture capacity). Measured fixture capacity equals about two-thirds the maximum.

[c] Assumes four flushes per person per day; does not include losses through leakage.

[d] Assumes 4.8 shower-use-minutes per person per day.

[e] Assumes 4.0 faucet-use-minutes per person per day.

Source: Based on Amy Vickers, "Water Use Efficiency Standards for Plumbing Fixtures: Benefits of National Legislation," American Water Works Association Journal. Vol. 82 (May 1990): 53, in USEPA Water Conservation Plan Guidelines, Appendix B, http://www.epa.gov/WaterSense/docs/app_b508.pdf.

Glossary

A

abiotic resources: Non-living natural resources and raw materials.

Acidification Potential (AP): A common index used to quantify the contribution of chemical species to acid precipitation using SO_2 as benchmark.

activity coefficient: Used to relate the standard chemical activity and the conditional chemical reactivity.

acute health effect: Short-term symptoms are often much more severe in individuals predisposed to adverse respiratory events.

adoption and diffusion of innovations: The processes through which people and organizations begin to use new or different products, or begin to change their own behaviors.

aerobic organisms: Require molecular oxygen (O_2) for respiration.

Air Quality Index: A numerical scale to help communicate air quality trends to the public and to provide warnings to the public on days that a region is not in compliance with the NAAQS.

albedo: Percent of energy reflected by a given area on the earth.

American Society of Heating, Refrigeration, and Air Conditioning Engineers (ASHRAE): Energy conservation standard applies to all commercial buildings except multifamily residences that are three stories high or less.

anaerobic organisms: Organisms that do not require molecular oxygen; they may obtain oxygen from inorganic ions, such as nitrates, sulfates, or proteins.

angle of bulk plane: The angle from the edge of the lot to be developed to the steepest angle of the proposed building.

aphotic zone: Depth below which light does not penetrate into a body of water.

appropriate technology: Engineering design that takes into consideration the key local social, economic, environmental, and technical factors that influence the success or failure of a design solution.

atmosphere: Layer of gases and suspended particulates held in place by gravity that surround the earth.

B

bagasse: The waste fiber from sugarcane.

best management practices: Techniques based on successful engineering practice that consistently achieve results in a cost-effective manner.

beyond code program: A method for acknowledging that a building exhibits a level of energy efficiency that is significantly greater than that required by building energy codes.

biocentric outlook: A center of life pursing its own good in its own way.

biochemical oxygen demand: The amount of oxygen consumed by microorganisms in water in the process of turning a substrate into cell mass, energy and carbon dioxide.

biogeochemical cycle: The biological and chemical reservoirs, agents of change and pathways of flow from one reservoir of a chemical on earth to another reservoir.

biogeochemical repository: Biological and chemical reservoirs on earth.

biomimicry: A design strategy that systematically analyzes biological processes and forms with the intent of using them for engineering solutions.

bioremediation: The process of using microorganisms to break down contaminated waste into nonhazardous substances.

biotic resources: Non-living natural resources and raw materials.

blackbody temperature: The maximum amount of radiation an object or body can emit at a given temperature.

blue water: Water from surface and groundwater sources.

bottoming cycle plant: A CHP facility that uses the waste heat from industrial furnaces and other high-temperature manufacturing processes (metals, glassmaking) is captured to generate electricity that is typically used within the facility itself.

brownfield: An inactive industrial site, usually with some degree of environmental/chemical contamination.

Bruntland definition of sustainability: Development that meets the needs of the present without compromising the ability of future generations to meet their own needs.

building envelope: The region of a parcel of land that may have a structure legally built upon it.

building integrated system: When renewable energy technologies are used to substitute for a building's conventional construction materials.

built environment: All human-made structures, engineered alterations to land forms, and constructions .

C

capacity factor analysis: A tool that assists with the development of sustainable water, sanitation, and household energy solutions by evaluating a community's capacity to manage its own technology.

cap-and-trade: A market-based incentive program that allows polluters to buy-and-sell pollution allowances, but in a manner that caps, or limits, the total amount of pollution that can occur in a region.

carbon flow and repositories: Mass of carbon movement from one repository to another.

carbon intensity: Amount of carbon dioxide emitted compared to that of another compound (often methane).

carbonaceous oxygen demand: The amount of oxygen consumed to utilize a carbon-based substrate.

carbonate hardness: Represents the portion of the diprotic ions that can combine with carbonates to form scaling.

Carnot Limit: The maximum efficient that can physically be achieved by a heat engine.

carrying capacity: The maximum rate of resource consumption and waste discharge that can be sustained indefinitely in a given region without progressively impairing the functional integrity and productivity of the relevant ecosystem.

centralized treatment: Collection of wastewater by sewers and treatment of collected wastewater in one treatment works facility.

chemical activity: A standardized measure of chemical reactivity within a defined system.

chemical reactivity: The chemical's overall tendency to participate in a reaction.

chronic health effects: Long-term effects of exposure which may include chronic inflammation, cancers, or pulmonary emphysema.

cluster development: Subdivisions or building tracts that are designed to concentrate housing and other structures in one area of a development, leaving an open and community-shared green space.

CMFR or CSTR: A reactor in which the contents within the reactor are equally mixed and equal to the concentration of the contents exiting the reactor.

cogeneration: *See* combined heat-and-power generation.

combined heat-and-power generation: The exploitation of usable waste heat to heat and cool buildings or to generate electricity.

command and control regulation: A form of governmental regulation in which the government specifies or mandates the exact ways in which companies must comply with environmental protection laws.

community-based design: Participatory design at the community scale.

condensation: Converts water in the gas phase to liquid water by cooling the water molecules.

Conference on Environment and Development: The first world environmental congress to which heads of state were invited participants held in Rio de Janeiro, Brazil, in 1992.

conservation design: A sophisticated, site-specific approach to environmentally sensitive design that integrates principles of cluster development, low-impact development, and green infrastructure.

consumption: The selection, use, reuse, maintenance, repair, and disposal of goods and services.

corporate social responsibility: The principle that businesses are accountable for more than just their financial profits and losses to shareholders; they are also accountable to society for the environmental and social consequences of their business operations.

cradle: The environment as the source of all materials.

cradle to cradle: A life cycle model that derives from the absence of waste in natural ecological systems and idealizes closed-loop raw material, product, and waste flows.

critical point: Point at which the lowest oxygen concentration is expected in the stream or river.

D

daylighting: The use of skylights and careful placement of windows to reduce the need for artificial light by providing sunlight during the day.

decentralized water treatment: Collection of wastewater and treatment at the source.

deep ecology: Ethical framework in which humans have no greater importance than any other component of our world.

dematerialization: The redesign of products to minimize their materials content.

demographers: People who study trends in population.

density-neutral design: The design of residential developments in a manner that does not affect the total number of housing units that can be built at the development site.

deoxygenation rate: Removal of diatomic oxygen in water due to uptake from microorganisms and higher organisms.

derived demand: The principle that the social demand for a particular energy fuel or energy-producing technology results from the critical need for a service that the energy provides. For example, the demand for coal derives from the social demand for electricity.

design for the environment: The idea that environmental protection should be designed into products and processes rather than managed as an after-the-fact harm.

development: The conversion of open land and natural areas to the built environment, such as residential neighborhoods, airports, industrial parks, hydroelectric dams, shopping centers, parks, and so on.

direct energy: Energy that is consumed in the act of providing an energy service, such as electricity for lighting. The opposite of direct energy is embodied energy.

dispersion model: Mathematical or statistical models used to estimate the transportation and concentration of pollutants.

dissolution: The process of a substance dissolving in solution.

dissolved inorganic carbon: Dissolved carbon dioxide gas, carbonic acid (H_2CO_3), bicarbonate ion (HCO_3^-), and the carbon ion, (CO_3^{2-}).

dissolved oxygen: The amount of diatomic oxygen dissolved in an aqueous solution.

dissolved solids: Consist of salts and minerals that have been dissolved through natural weathering of soils or anthropogenic processes

dose-response curve: A plot of the measured response versus the dose administered.

downcycling: The process of using a recycled material or product in an application that is of lower quality or more limited functionality than its original purpose.

drinking water: Water used for consumption in homes and ingestion.

duty cycle: The amount of time a device operates in order to complete a functional cycle; for example, the number of hours it takes a pump to move a million liters of water.

E

easement: *See* right of way.

Eco-Industrial Park: A community of manufacturing and service businesses that collaborate in the management of environmental and resource issues.

ecological design: The idea that by "looking to nature" for inspiration, we can take advantage of sustainable forms and processes already successfully established in our environment.

ecological footprint: Calculations used to illustrate the relationship between the consumption and supply of natural resources.

ecosystem: A mutually interdependent community of living organisms and their abiotic physical environment.

ecosystem services: The idea that a variety of ecological dynamics support humankind through the generation of tangible and intangible provisioning, regulating, and cultural services.

effective stack height: The sum of the actual stack height and the plume rise.

embodied energy: The total quantity of energy directly and indirectly consumed in the production, distribution and use of a product.

embodied water: The total volume of fresh water consumed directly and indirectly in the production, distribution, and use of a product.

emerging technologies: The "next generation" of energy supply, which require, significant investments in research and development in order to become commercially viable energy systems.

emission factor: Expresses the amount of an air pollutant likely to be released based on a determinant factor in an industrial process.

endorsement labeling: A type of voluntary certification in which an independent, not-for-profit, third party verifies that a product meets certain criteria in its construction, material content, performance, or operation.

end-use sectors: The residential, commercial, industrial, and transportation sectors.

energy balance: Accounting for the energy flow into and out of a control volume.

energy curtailment: Reducing the demand for energy by changing people's behaviors and lifestyle needs for energy, such as turning off the lights in an unoccupied room.

energy efficiency: The technical productivity for an energy consuming device or piece of equipment, usually measured as the physical property of work.

energy intensity: Energy consumed per monetary unit of gross domestic product.

energy ladder: A model that describes how both the quality and quantity of energy used in a household increases as household income rises.

energy poverty: The inability to access or afford commerically provided energy; energy poverty is commonly understood as a lack of access to electricity or clean cooking fuels.

energy recovery: The ability to capture waste heat from industrial processes or the energy embodied in solid waste and use it for energy production.

energy return on investment (EROI): A measure of the usable energy from a source relative to the amount of energy required to produce the resource; viable resources have EROIs greater than 1.

energy security: The ability of a nation to protect itself from the economic, political, and social disruptions due to an interrupted supply of a critical energy resource, of the failure of an important energy infrastructure, or of rapid and steep changes in energy prices.

energy services: The specific functions that energy provides for people, such as lighting, heating, cooling, refrigeration, and industrial production.

environmental ethic: Recognizing that we are, at least at the present time, unable to explain rationally our attitude towards the environment and that these attitudes are deeply felt, not unlike the feeling of spirituality.

environmental impact assessment: An analysis that catalogues the expected environmental impacts a project will have at a site and indicates how those impacts might be mitigated.

environmental justice: An ethical principle that tries to eliminate the spatial discrimination which results in low-income, minority, and tribal communites being unfairly affected by the environmental actions of others.

environmentally sensitive design: A broad range of design strategies and on-the-ground techniques to integrate both development and environmental conservation.

Envision rating system(TM): An assessment tool for evaluating the sustainability of all civil infrastructure.

eolian (wind) erosion: Occurs when wind removes soil particles.

epilimnion: The uppermost layer of a water body characterized by warmer, less dense water.

ethics: Ideas which provide a framework for making difficult choices when we face a problem involving moral conflict.

eukaryotes: Microorganisms that have a nucleus or nuclear envelope and endoplasmic reticulum, or interconnected organelles. Eukaryotes include protozoa, fungi, and green algae.

eutrophication: The nutrient enrichment of aquatic ecosystems, is a natural process that can be accelerated by excessive anthropogenic emissions of nitrogen and phosphorus.

Eutrophication Potential (EP): A common index used to quantify the contribution of chemical spcies to the eutrophication of water bodies using phosphate as benchmark.

evaporation: The process of converting liquid water from surface water sources to gaseous water that resides in the atmosphere.

exponential growth: Rate of change is proportional to the instantaneous value of the variable that is changing.

extended producer responsibility: The principle that a manufacturer is responsile for the environmental consequences of its product over the full range of the product's life cycle.

F

feedstocks: A raw material used as an input in the manufacturing process.

fine particle: Particles are 2.5 micrometers in diameter and smaller.

finite resources: Resources that formed over a period of millions of years and can only be regenerated on a geological timescale.

floor area ratio (FAR): The ratio of the building's footprint (floor area) to the area of the lot on which the building is to be constructed.

fluvian (water) erosion: Occurs when rainwater runoff carries away soil particles.

formal energy sector: Energy fuels or resources that are bought and sold through market exchange.

form-based codes: Land use regulations that use neighborhoods as the primary geographic planning unit rather than larger zoning districts.

fossil water: Groundwater that has been isolated in an aquifer for thousands or millions of years.

functional unit: Basis for comparison of two or more products by identifying a common function and how each product achieves that function over its life.

G

geotextile: Synthetic materials purposefully designed to protect soil and manage water flow.

geothermal power: The generation of electricity from geothermal energy.

Global Warming Potential (GWP): A common index used to quantify the contribution of green house gases to global warming using CO_2 as benchmark.

grave: The environment as the final repository of all materials.

gray water: The wastewater from activities such as bathing, dishwashing, and laundry that can be safely reused/recycled for non-potable water needs.

green building: The practice of reducing the environmental impacts of buildings in their construction, operation, and demolition.

green engineering: A concept similar to sustainable engineering that provides design rules and principles for more sustainable products and processes.

green infrastructure: A landscape pattern that uses a connective network of greenspace to provide essential public services to a community.

green premium: The higher cost of a more sustainable product compared to those that are less sustainable.

green water: Water captured from precipitation that does not become runoff.

greenhouse gas: A gas component of the atmosphere which absorbs and stores energy.

greenwashing: A form of deceptive environmental advertising, marketing, and public relations that leads consumers to believe a product is more sustainable that it truly is.

ground source heat pump: A technology that takes advantage of the natural and stable temperature gradients of the Earth by circulating the heat-exchanging fluids used in heating and cooling systems underground to preheat or pre-chill them.

H

hazardous air pollutant (HAP): The US EPA list of air pollutants (also called air toxics) that have are potentially significant health risks and may be carcinogenic, mutagenic, or teratogenic.

heating degree days: An estimate of the amount of energy required to heat a building; it is calculated by summing the extent to which the average outdoor temperature deviates from a baseline temperature.

heating values: Amount of energy available form a given mass of fuel.

Henry's law constant: The name given to the constant that describes the slope of this

linear relationship between the liquid phase and gas phase concentration.

housing density: The number and type of dwellings allowed, typically per acre of proposed residential space.

Hubbert curve: A concept used to analyse the rates of growth in demand for a fossil fuel and the rate at which new reserves become available; the Hubbert curve is commonly associated with peak oil discussions.

Human Development Index: United Nations Development Program (UNDP) numerical index to compare the state of human development between countries based on three-dimensional indicators: life expectancy, education, and income.

hydrologic cycle: Describes the movement of water from one biogeochemical cycle to another.

hydrology: The science that treats the waters of the Earth; their occurrence, circulation, and distribution; their chemical and physical properties; and their reaction with the environment, including the relations to living things.

hypercapnia: A condition that occurs when too much carbon dioxide builds up in the blood stream causing distress.

hypolimnion: The lower layer of water characterized by cooler, denser water and usually has a lower DO concentration.

hypoxia: Inadequate supply of oxygen to the body.

I

impervious surfaces: Areas in the urban landscape that do not absorb water or allow water to infiltrate into the ground, such as roads and parking areas.

indicator: A measurement or metric based on verifiable data that can be used to communicate important information to decision makers and the public about processes related to sustainable design or development.

industrial ecology: Development of technological systems that mimic the way nature handles materials and energy transformations.

infiltration: Fraction of precipitation that seeps into the ground.

infinitely renewable resources: Resources that derive from the physical workings of

the planet and solar system and are inexhuastible in their supply, such as solar and wind energy.

information asymmetry: Occurs when someone lacks the knowledge, time, or skill to evaluate the long-term consequences of a decision.

inhalable particle: Particles larger than 2.5 microns and smaller than 10 microns in diameter.

inter-generational equity: Equitable distribution of resources between the current generation and future generations.

intergenerational ethics: The obligation to consider the consequences of our actions and decisions for future generations.

International Energy Conservation Code: A model building code developed by the International Code Council that covers both residential and commercial structures.

International Organization for Standardization: An international NGO with the primary mission of developing voluntary international standards governing most aspects of technology, manufacturing, and business management operations.

Intergovernmental Panel on Climate Change: Research group comprised of a group of scientists from many nations on the planet that are intimately involved in better understanding changes to the Earth's climate.

inventory analysis: Determination of the quantitative values of materials and energy inputs and outputs of all stages within the life cycle.

ionic reactions: There is a change in ion–ion interactions and relationships.

ionic strength: The estimate of the overall concentration of dissolved ions in solution.

IPAT Equation: A conceptual formula that shows the relationship between impacts, population, affluence, and technology.

L

land-use and land-cover change: How human activity affects the use of land and changes to vegetation, topography, and landscape characteristics.

lapse rate: The negative temperature gradient of the atmosphere, and it is dependent on the weather, pressure, and humidity.

law of electroneutrality: The sum of all positive ions (cations) in solution must equal the sum of all the negative ions (anions) in solutions, so that the net charge of all natural waters is equal to zero.

life cycle assessment (LCA): The compilation and evaluation of the inputs and outputs and the potential environmental impacts of a product system throughout its life cycle.

life cycle models: Models that analyze the material and energy intensity of a product from the sourcing of raw materials through end-of-life disposal.

lithosphere: The soil crust that lies on the surface of the planet where we live.

Living Building Challenge: A green building certification standard developed by the Living Building Institute, which specifies the actual performance characteristics of the building itself.

M

management system standard: Standards that govern processes and practices rather than products themselves.

market failure: When the general dynamics of supply and demand do not result in the most economically efficient outcome, or result in a socially suboptimal outcome.

market-based incentives: An alternative to command and control regulation that allows private companies some degree of freedom to choose how to comply with governmental environmental or energy efficiency standards.

material safety data sheet: A fact sheet that summarizes key features of the physical and chemical properties of a material, methods of proper use and disposal, and toxicity and risk information.

materials: Substances as well as goods.

Materials Flow Analysis: A method for performing inventory analysis.

measure: A value that is quantifiable against a standard at a point in time.

methane hydrate: A form of natural gas trapped in ocean sediments and Arctic permafrost.

metric: A standardized set of measurements or data related to one or more sustainability indicators.

Millennium Development Goals: United Nations development goals focused on increasing the standard of living for the world's existing population and also for future generations as the world population continues to grow.

mixed used: Neighborhoods that serve more than one purpose, such as a combined residential and commercial space.

morals: The values people adopt to guide the way they ought to treat each other.

N

nasopharyngeal region: The airway including the nose, mouth, and larynx.

National Ambient Air Quality Standards: Goals set to achieve reasonable air quality in all regions of the United States.

natural capital: A theory which views natural resourcs as the Earth's wealth and therefore should not be overconsumed or degraded.

natural resource: A naturally occuring substance or living organism that is exploited for the benefit of human beings.

natural resources management: The field of study concerned with determining how to harvest or extract resources at a rate that is in balance with rate at which the resource can regenrate or reproduce itself.

negative externality: A cost or harm experienced by someone other than the buyers and sellers of the good or service.

Net Zero Program: Structures that produce as much energy as they consume, resulting in no net energy consumption and a net zero carbon footprint.

New Urbanism: A movement within the field of urban design that moves away from traditional land-use planning and single-use zoning by creating walkable communities with open space.

nitrogenous oxygen demand: The amount of oxygen consumed to utilize a nitrogen based substrate.

noncarbonate hardness: The difference between total hardness and carbonate hardness.

nonmarket valuation: Assigning monetary value that is not based on the price of actually buying and selling the good or service in question.

nonrenewable resources: Resources that regenerate either extremely slowly (over hundreds or thousands of years) or not at all.

nutrients: Levels of nitrogen and phosphorus in water.

O

occurrence: The physical presence of a substance in a recognizable form.

oncogenesis: Tumor formation.

opportunity cost: An economic concept that respresents what is gained or lost by choosing one option over another.

oxygen deficit: The difference between the potential saturation oxygen concentration and the actual measured oxygen concentration in a water body.

Ozone Depletion Potential (ODP): A common index used to quantify the contribution of human-made substances to the depletion of the ozone layer using CFC-11 as benchmark.

ozone-depleting substance (ODSs): Compounds which react to convert ozone (O_3) in the atmosphere to diatomic oxygen (O_2).

P

Partial Pressure: The pressure of each part of a gas that it exerts on its surrounding.

participatory design: Design focused on directly involving individuals who use the technology or information as co-designers and on enabling those affected by a design to drive the collaborative design process.

passive heating and cooling: Heating or cooling a building using solar energy and/or natural convection without any external energy inputs.

passive solar design: A form of architecture that takes advantage of solar energy to help heat homes.

pathogen: Disease-causing organisms usually classified as bacteria, protozoa, helminthes, or viruses.

payback period: The amount of time it takes for an investment to pay for itself in terms of profit or savings.

peak oil: The point in time at which the rate of increase in oil production is zero and can no longer keep up with the rate of increase in the social demand for oil.

peak sun hour: The amount of solar energy striking a 1-square-meter area over a period of 1 hour at solar noon in the summer.

performance standards: Specify the minimum energy efficiency characteristics of residential, commerical, and industrial equipment.

Peri-urban areas: Living areas surrounding urban cities that are characterized by very high population densities and lack the infrastructure to distribute energy, water, and sanitation services.

pervious pavers: Paving materials designed to allow rainwater to penetrate through them into the ground.

photic zone: Depth to which light penetrates in a body of water.

Pickens Plan: A fuel-switching proposal created by T. Boone Pickens that suggested the United States invest heavily in wind farms for electric power production so that natural gas may be shifted to the transportation sector.

point-of-use treatment: Treatment of drinking water in the home or at the source of drinking water use.

population density: The number of people living in a square mile or square kilometer.

Potochemical Ozone Creation Potential (POCP): A common index used to quantify the contribution of chemical species to the creation of tropospheric ozone using ethylen as benchmark.

precipitate: The resulting solid form of a substance that forms from oversaturated concentrations in solution.

precipitation: The process whereby substances (liquids and solids) are removed from a fluid (typically air or water) by gravity.

Prevention of Significant Deterioration: Standards to maintain air quality if in compliance with the NAAQS.

process: The transformation, transport and storage of materials.

prokaryotes: Have the simplest cell structure classified by their lack of a nucleus membrane that contains cellular DNA.

R

radiative forcing: The impact a component or condition with the atmosphere has upon the Earth's average surface temperature.

rare earths: Nonrenewable elements that are not necessarily geologically rare but distributed in the Earth's crust in a manner that makes them hard to recover.

raw materials: Naturally occuring resources that are extracted from the environment and processed into food and useful products for humans.

re-aeration rate: How quickly oxygen is reabsorbed and mixed in the water body.

recoverability: The degree to which a natural resource or material substance can be retrieved after its use or embodiment in a physical product.

recyclable: An element or material that can be recovered after use or embodiment in a product and re-processed for use as a material imput.

recycling: The process of separating recoverable materials from waste to process them back into their raw substances for use as feedstocks.

reduction-oxidation process: Occur when the oxidation state of participating atoms change.

remanufacturing: The recovering of the modules or components of a product for refabrication, reassembly, or reuse.

renewable energy credit: Represents 1 megawatt of electricity generated by a clean energy source.

renewable portfolio standards: Requirements that electric power utilities generate a specific amount of electricity from different sources of renewable energy.

renewable resources: Resources that replenish themselves through a natural process, roughly within a human lifetime.

reserve: The total amount of a natural resource that can be physically extracted at economically viable rates; reserve-sare based on highly reliable statistical estimates.

reserve-to-production ratio: The amount of time it would take to deplete current proved reserves if the current yearly amount of consumption was held constant into the future.

reservoir: A system that holds materials for any quantity of time.

residence time: The length of time that a compound remains in a defined system.

resource: Deposits of natural substances that are able to be physically recovered, although it may not necessarily be economically feasible to do so.

right of way: A piece of land adjacent to the building envelope that is preserved for the use of utilities and other types of infrastructure; consists of the street itself and a narrow strip of land.

riprarian buffer zone: Areas along a streambanks that grow a mixed variety of plants and in which structures cannot be built close to the water.

runoff: Collected precipitation that flows overland.

rural-to-urban transect: An approach to urban planning that establishes form-based regulations based on the physical forms of human settlements and the physical environment.

S

sanitation: Effective control and disposal of human waste and wastewater.

saturated dissolved oxygen: Oxygen concentrations are in equilibrium with the oxygen in the atmosphere.

setback: A required distance between a structure and an adjacent feature such as a property line, street, another structure, or a stream.

single-use zoning: Zoning that allows for only one type of land use in a designated area.

smarth growth: A set of land-use development and urban design principles intended to curb sprawl, conserve natural resources, and create more livable cities.

social and environmental accounting: Business accounting techniques that calculate how business activity benefits people and the environment.

social cost: The negative effects of a condition experienced by society as a whole.

social justice: The fair distribution (or sharing) of the advantages and disadvantages (or benefits and burdens) that exist within society.

social movement: When new ways of thinking become institutionalized in a society after being broadly advocated by groups that have political and social power.

solar radiation: Energy emitted from the sun.

solubility: The amount of a substance that can be dissolved into solution by a solvent.

sorbate: The substance that is transferred from one phase to another.

sorbent: The material into or onto which the sorbate is transferred.

source reduction: Taking measures to reduce the amount of natural resources used in the creation of a product such that the resulting scrap and harmful waste products are minimzed.

stability: A classification of atmospheric conditions that influence dispersion of air pollutants.

standard state activity: Reference state of chemical reactivity.

Stefan–Boltzmann constant: Energy radiate per unit area of a blackbody per unit time.

stock: The amount of a resource that is available at a fixed point in time. Also the quantity of materials held in a reservoir.

Stokes Settling law: A method used to estimate particle removal as a function of the runoff flow velocity and expected residence time of water in a sedimentation removal device.

stratosphere: The layer of the atmosphere about 25 to 50 kilometers above the earth.

stratospheric ozone depletion: A decrease in the historical levels of the ozone concentration in the stratosphere due to reactions with halogenated gases.

stream daylighting: Restoring a buried stream in an urban area to natural conditions.

structural insulated panels: An advanced building product used instead of framing lumber and plywood in order to provide higher insulating values.

Superfund: Created through the The U.S. Comprehensive Environmental Response, Compensation, and Liability Act of 1980; provides funding to clean up the most contaminated sites in the United States.

suspended solids: The materials that are floating or suspended in the water.

sustainability: Everything that we need for our survival and well-being depends, either directly or indirectly, on our natural environment. Sustainability creates and maintains the conditions under which humans and nature can exist in productive harmony that permit fulfilling the social, economic, and other requirements of present and future generations. Sustainability is important to making sure that we have and will continue to have the water, materials, and resources to protect human health and our environment.

sustainability index: A numerical-based scale used to compare alternative designs or processes with one another.

Sustainable design: Design of products, processes, or systems that balance our beliefs in the sanctity of human life and promote an enabling environment for people to enjoy long, healthy, and creative lives, while protecting and preserving natural resources for both their intrinsic value and the natural world's value to humankind.

sustainable development: The desire to improve the worldwide standard of living while considering the effects of economic development on natural resources.

sustainable engineering: A discipline of engineering concerned with finding technically viable and cost-effective solutions that also promote human well-being, protect the environment, and preserve the biosphere.

T

take-back program: A program in which a product is returned to the manufacturer at the end of its life for either reprocessing, refurbishing, recycling, or final disposal of the product.

theoretical oxygen demand: The theoretical amount of oxygen consumed to oxidize a substrate to carbon dioxide.

thermocline: The layer in which there is a rapid change in temperature and density over a very small change in depth.

topping cycle plant: A CHP facility that uses waste steam and heat from electric power production to create steam or hot water that is then used for heating and cooling a building.

Toxic Substances Control Act (TOSCA):

tracheobronchial region: Portion of the respiratory system characterized by the branching of the airway bronchi through the lungs, conveying the air to the pulmonary region.

transpiration: Occurs when water is conveyed from living plant tissue, especially leaves, to the atmosphere.

triple bottom line: Represents business outcomes in which "people, planet, and profit" benefit simultaneously.

trophic level: Productivity of a body of water determined by the organic matter (usually represented by chlorophyll) or the nutrient concentration in the water column.

troposphere: The part of the atmosphere in which humans breath the air from the surface of the Earth to approximately 17 kilometers above the surface of the Earth.

U

universal soil loss equation: A method of estimating soil loss through erosion.

urban agriculture: A type of infill land use where small farms and community gardens are interspersed in a city's neighborhoods.

urban growth boundary: A zoning designation that legally limits the geographic spread of growth and new development.

urban infill: Abandoned, derelict, unused, or underutilized space in urban areas that is redeveloped for more productive use.

urban intensification: A form of development that seeks to create higher levels of housing density in combination with mixed-use zoning, the greater availability of green space and natural areas, and enhanced mass transit.

urban sprawl: Low-density housing that spreads along major streets and arterial roadways.

V

vegetated swales: Areas that collect rainwater runoff planted heavily with vegetation; these areas allow rainwater to percolate naturally through the soil as groundwater.

volatile solids: Determined by the weight of any particles remaining on a filter that evaporate after the filter is heated to 550°C.

voluntary certification: A process in which private companies demonstrate the environmental sustainability or energy efficiency of their products through independent, non-governmental certification organizations.

W

waste: A byproduct of a complex system that has no useful purpose and must therefore be disposed of; waste is usually considered a uniquely human concept, as ecological systems do not generate unused or unusable waste.

waste hierarchy: A ranking of the most preferred to least preferred strategies for waste management.

waste management: The collection, transport, processing, disposal, and monitoring of waste materials in a manner that minimizes the impacts of waste on human health and the environment.

waste-to-energy: Waste-to-energy facilities typically burn solid waste to generate steam or electricity.

water footprint: Embodied water with spatial (where the water is extracted) and temporal (when the water is extracted) considerations.

water hardness: The sum of the concentration of the divalent cations (species with a charge of 2+) in water.

water-energy nexus: The interdependence between water resource use and energy production; reflects generally the need for water to generate electricity as well as the energy-intensity of water supply systems, domestic water use, and wastewater treatment.

water-related disease: Diseases which include those due to micro-organisms and chemicals in water people drink; diseases like schistosomiasis which have part of their lifecycle in water; diseases like malaria with water-related vectors; drowning and some injuries; and others such as legionellosis carried by aerosols containing certain micro-organisms.

watershed: The region that collects rainfall.

Z

zero waste: A waste management concept that encourages the elimination of waste through complete recovery and recycling of materials in the waste stream.

zoning ordinances: Rules and regulations that govern the allowed uses for buildings or dwellings in an area.

Index

PRINCIPAL UNITS USED IN MECHANICS

Quantity	International System (SI)			U.S. Customary System (USCS)		
	Unit	Symbol	Formula	Unit	Symbol	Formula
Acceleration (angular)	radian per second squared		rad/s^2	radian per second squared		rad/s^2
Acceleration (linear)	meter per second squared		m/s^2	foot per second squared		ft/s^2
Area	square meter		m^2	square foot		ft^2
Density (mass) (Specific mass)	kilogram per cubic meter		kg/m^3	slug per cubic foot		slug/ft^3
Density (weight) (Specific weight)	newton per cubic meter		N/m^3	pound per cubic foot	pcf	lb/ft^3
Energy; work	joule	J	N·m	foot-pound		ft-lb
Force	newton	N	kg·m/s^2	pound	lb	(base unit)
Force per unit length (Intensity of force)	newton per meter		N/m	pound per foot		lb/ft
Frequency	hertz	Hz	s^{-1}	hertz	Hz	s^{-1}
Length	meter	m	(base unit)	foot	ft	(base unit)
Mass	kilogram	kg	(base unit)	slug		$\text{lb-s}^2\text{/ft}$
Moment of a force; torque	newton meter		N·m	pound-foot		lb-ft
Moment of inertia (area)	meter to fourth power		m^4	inch to fourth power		in.^4
Moment of inertia (mass)	kilogram meter squared		kg·m^2	slug foot squared		slug-ft^2
Power	watt	W	J/s (N·m/s)	foot-pound per second		ft-lb/s
Pressure	pascal	Pa	N/m^2	pound per square foot	psf	lb/ft^2
Section modulus	meter to third power		m^3	inch to third power		in.^3
Stress	pascal	Pa	N/m^2	pound per square inch	psi	lb/in.^2
Time	second	s	(base unit)	second	s	(base unit)
Velocity (angular)	radian per second		rad/s	radian per second		rad/s
Velocity (linear)	meter per second		m/s	foot per second	fps	ft/s
Volume (liquids)	liter	L	$10^{-3}\ \text{m}^3$	gallon	gal.	$231\ \text{in.}^3$
Volume (solids)	cubic meter		m^3	cubic foot	cf	ft^3